HANDBOOK OF GENETICS

Volume 1
Bacteria,
Bacteriophages,
and Fungi

HANDBOOK OF GENETICS

HANDBOOK OF GENETICS

ROBERT C. KING, EDITOR

Professor of Genetics, Department of Biological Sciences
Northwestern University, Evanston, Illinois

Volume 1
Bacteria,
Bacteriophages,
and Fungi

PLENUM PRESS · NEW YORK AND LONDON

Library of Congress Cataloging in Publication Data

King, Robert C
  Bacteria, bacteriophages, and fungi.

  (His Handbook of genetics, v. 1)
  Includes bibliographies.
  1. Microbial genetics. I. Title. [DNLM: 1. Genetics. QH431 K54h]
QH434.K56   589′.2′0415   74-8867
ISBN 0-306-37611-3

First Printing — November 1974
Second Printing — April 1977

© 1974 Plenum Press, New York
A Division of Plenum Publishing Corporation
227 West 17th Street, New York, N.Y. 10011

United Kingdom edition published by Plenum Press, London
A Division of Plenum Publishing Company, Ltd.
4a Lower John Street, London, W1R 3PD, England

Printed in the United States of America

John Raper, Professor of Botany at Harvard University
since 1954, died in Boston, Massachusetts on May 21, 1974
at the age of 62. The results of Professor Raper's
researches on *Schizophyllum* and *Achlya* have exerted a
profound influence upon all workers in Fungal Biology,
and his memory is held in affection and admiration.

Preface

The purpose of this and future volumes of the *Handbook of Genetics* is to bring together a collection of relatively short, authoritative essays or annotated compilations of data on topics of significance to geneticists. Many of the essays will deal with various aspects of the biology of certain species selected because they are favorite subjects for genetic investigation in nature or the laboratory. Often there will be an encyclopedic amount of information available on such a species, with new papers appearing daily. Most of these will be written for specialists in a jargon that is bewildering to a novice, and sometimes even to a veteran geneticist working with evolutionarily distant organisms. For such readers what is needed is a written introduction to the morphology, life cycle, reproductive behavior, and culture methods for the species in question. What are its particular advantages (and disadvantages) for genetic study, and what have we learned from it? Where are the classic papers, the key bibliographies, and how does one get stocks of wild type or mutant strains? The chapters devoted to different species will contain information of this sort.

Only a few hundreds of the millions of species available to biologists have been subjected to detailed genetic study. However, those that have make up a very heterogeneous sample of the living world. The geneticist will often want to see where in the evolutionary scheme of things various "genetically known" species fit and how close they may be to one another in terms of their positions on the evolutionary tree. The essay by Lynn Margulis is placed as the first chapter in this volume to provide a classification scheme appropriate for just such exercises. After reading this chapter the geneticist should have an up-to-date mental picture of the lines of descent of the evolutionarily diverse organisms that inhabit this planet, and also should appreciate the current controversy concerning the systematics of the prokaryotes and primitive eukaryotes.

This volume contains chapters covering various bacteria, bacteriophages, and fungi. Therefore we are hearing from representatives from two of the Kingdoms in the Margulis' Classification (the Monera and the Fungi). In the next volume representatives of the Kingdoms Plantae and Protista will divulge their genetic secrets.

I am particularly grateful for the splendid assistance provided by Patricia DeOca, Pamela Khipple and Lisa Gross during the preparation of this volume.

Robert C. King

Evanston
June, 1974

Contributors

Bruce N. Ames, Department of Biochemistry, University of California, Berkeley, California

Raymond W. Barratt, Dean, School of Science and Director, Fungal Genetics Stock Center, California State University, Humboldt, Arcata, California

Allan Campbell, Department of Biological Sciences, Stanford University, Stanford, California

Enrique Cerdá-Olmedo, Departamento de Genética, Facultad de Ciencias, Universidad de Sevilla, Sevilla, Spain

Keith F. Chater, John Innes Institute, Norwich, England

A. John Clutterbuck, Genetics Department, Glasgow University, Glasgow, Scotland

Roy Curtiss III, Department of Microbiology, The Medical Center, University of Alabama, Birmingham, Alabama

Julian Davies, Department of Biochemistry, College of Agricultural and Life Sciences, University of Wisconsin, Madison, Wisconsin

Bernard Decaris, Laboratoire de Génétique, Université de Paris-Sud, Centre d'Orsay, Orsay, France

Karl Esser, Lehrstuhl für allgemeine Botanik, Ruhr-Universität Bochum, Bochum, Germany

Joseph O. Falkinham III, Department of Microbiology, The Medical Center, University of Alabama, Birmingham, Alabama

Walter Fiers, Laboratory of Molecular Biology, State University of Ghent, Ghent, Belgium

Jacqueline Girard, Laboratoire de Génétique, Université de Paris-Sud, Centre d'Orsay, Orsay, France

Jean Louis Guerdoux, Centre de Génétique Moléculaire, Centre National de la Recherche Scientifique, Gif sur Yvette, France

Herbert Gutz, Institute for Molecular Biology, University of Texas at Dallas, Richardson, Texas

Philip E. Hartman, Biology Department, The Johns Hopkins University, Baltimore, Maryland

Henri Heslot, Institut National Agronomique, Paris-Grignon, France

Robert M. Hoffman, Genetics Unit, Massachusetts General Hospital, Boston, Massachusetts

Robin Holliday, National Institute for Medical Research, Mill Hill, London, England

Bruce W. Holloway, Department of Genetics, Monash University, Clayton, Victoria, Australia

David A. Hopwood, John Innes Institute, Norwich, England

Christopher W. Lawrence, Department of Radiation Biology and Biophysics, University of Rochester School of Medicine and Dentistry, Rochester, New York

Gérard Leblon, Laboratoire de Génétique, Université de Paris-Sud, Centre d'Orsay, Orsay, France

Urs Leupold, Institute für allgemeine Mikrobiologie der Universität Bern, Bern, Switzerland

Mark Levinthal, Department of Biological Sciences, Purdue University, West Lafayette, Indiana

Nicola Loprieno, Istituto di Genetica dell' Università Pisa Laboratorio di Mutagenesi e Differenziamento del Consiglio Nazionale delle Ricerche, Pisa, Italy

K. Brooks Low, School of Medicine, Yale University, New Haven, Connecticut

Francis L. Macrina, Department of Microbiology, The Medical Center, University of Alabama, Birmingham, Alabama

Lynn Margulis, Biological Science Center, Boston University, Boston, Massachusetts

Yoshimi Okada, Department of Biophysics and Biochemistry, Faculty of Science, University of Tokyo, Hongo, Tokyo, Japan

Lindsay S. Olive, Department of Botany, University of North Carolina, Chapel Hill, North Carolina

Elena Ottolenghi-Nightingale, University Affiliated Center for Child Development, Georgetown University Hospital, Washington, D.C.

John R. Raper, Biological Laboratories, Harvard University, Cambridge, Massachusetts

Richard E. Sanders, Department of Biology, Massachusetts Institute of Technology, Cambridge, Massachusetts

Kenneth Sanderson, Department of Biology, University of Calgary, Calgary, Alberta, Canada

Fred Sherman, Department of Radiation Biology and Biophysics, University of Rochester School of Medicine and Dentistry, Rochester, New York

Robert L. Sinsheimer, Division of Biology, California Institute of Technology, Pasadena, California

Ellen G. Strauss, Division of Biology, California Institute of Technology, Pasadena, California

James H. Strauss, Division of Biology, California Institute of Technology, Pasadena, California

Waclaw Szybalski, McArdle Laboratory for Cancer Research, University of Wisconsin, Madison, Wisconsin

Austin L. Taylor, Department of Microbiology, University of Colorado Medical Center, Denver, Colorado

Annamaria Torriani, Department of Biology, Massachusetts Institute of Technology, Cambridge, Massachusetts

Carol Dunham Trotter, Department of Microbiology, University of Colorado Medical Center, Denver, Colorado

Gary A. Wilson, Department of Microbiology, School of Medicine and Dentistry, University of Rochester, Rochester, New York

William B. Wood, Division of Biology, California Institute of Technology, Pasadena, California

Charles Yanofsky, Department of Biological Sciences, Stanford University, Stanford, California

Frank E. Young, Department of Microbiology, School of Medicine and Dentistry, University of Rochester, Rochester, New York

Irving Zabin, Department of Biological Chemistry, School of Medicine and Molecular Biology Institute, University of California, Los Angeles, California

Contents

B.   THE BACTERIOPHAGES

C. THE FUNGI

Chapter 30
    BERNARD DECARIS, JACQUELINE GIRARD, AND GÉRARD
    LEBLON

Chapter 31
    ROBIN HOLLIDAY

Chapter 32
    JOHN R. RAPER AND ROBERT M. HOFFMAN

Chapter 33
    JEAN LOUIS GUERDOUX

1

The Classification and Evolution of Prokaryotes and Eukaryotes

Lynn Margulis

To classify an organism is to assign it to a group. Classification, which is part of the science of systematics, is the essential concern of taxonomy. Taxonomy is the branch of biological science involved with identifying, naming, and classifying organisms into a formal hierarchical system. Systematics is the science that deals with the relationships of organisms as approached through their form and function, their genetic systems, their fossil histories, and the understanding of the evolutionary process from which they arose.

The living world is populated by several million species of organisms, and each species may contain many millions of individuals. An additional 130,000 morphological species have been described from the fossil record. Although each individual may be distinguishable from any other, each has characteristic "traits in common" with the other individuals of the larger group to which it belongs. Organisms may be classified according to color, size, or any other criteria chosen by the person doing the classifying. Classification schemes which are not based on evolutionary relationships are called artificial classifications. Taxonomists seek continual

Lynn Margulis—Biological Science Center, Boston University, Boston, Massachusetts.

improvement of their classifications which, it is hoped, are "natural," i.e., classifications which attempt to group organisms on the basis of evolutionary history. Biologists believe that similarity in development, form, and function is an expression of underlying genetic similarities, and that these in turn imply descent from common ancestors. Unless otherwise stated, the classification schemes in textbooks and in the original literature describing new species are "natural" or "systematic."

In taxonomy, organisms are classified into units, or taxa. Taxa on one formal level are grouped into larger taxa on another level so that the classification as a whole is a hierarchy—a system composed of units that increase in inclusiveness and decrease in number from each level to the next higher one. The levels of taxa, from more to less inclusive, are generally: kingdom, phylum (or division in some botanical classifications), class, order, family, genus, (plural, genera), species. Species may be comprised of diverging populations of organisms that constitute subspecies; organisms belonging to the same species may be divisible into recognizable interfertile groups called varieties. For excellent discussions of the species problem, see Mayr (1970) for animals and Grant (1971) for plants. To allow further subdivisions, the unit "branch" is sometimes used below kingdom to group grades or phyla, the unit "grade" is used below branch to group phyla, the unit phylum is used below branch to group classes, the unit "cohort" is used below class to group orders, and the unit "tribe" may be used below family to group genera. The prefixes super- and sub- may be added to any group name (Blackwelder, 1967). The level of the highest taxon that two organisms belong to should in general reflect the time at which the two populations of organisms diverged from their most recent common ancestors. For example, it is usually the case that two organisms that belong to the same genus (e.g., beans, such as *Phaseolus vulgaris* and *Phaseolus polystachyus,* or frogs, such as *Rana pipiens* and *Rana catesbiana*) diverged more recently from their common ancestor than two organisms that belong only to the same family (e.g., *Phaseolus vulgaris* and *Vicia americana* (purple vetch) or *Rana pipiens* and the toad *Bufo terrestris*). This is not an absolute rule because rates of evolution differ in different groups. Table I shows the formal classification for a blue-green alga (*Oscillatoria*), a protist (*Stentor*), a fungus (mushroom), an animal (man), and a plant (maize).

G. G. Simpson (with coauthor Beck), a leading evolutionist has said:

> A systematic unit of organisms in nature is a population or a group of related populations. Its anatomical and physiological characteristics are simply the total of those characteristics in the individuals making up the population. The pattern of characteristics is neither a real individual nor an idealized abstraction of the character of an individual. A 'systematic unit' is a frequency distribution of the dif-

TABLE 1. *Examples of Classification*

Taxon	Oscillatoria	Stentor	Field mushroom	Man	Maize
Kingdom	Monera	Protoctista	Fungi	Animalia	Plantae
Grade	—	Mitotica	—	Coelomata	Tracheophyta
Phylum	Cyanophyta	Ciliophora	Basidiomycota	Chordata	Angiospermophyta
Class	Hormogoneae	Polyhymenoptera	Homobasidiomycetae	Mammalia	Monocotyledoneae
Order	Nostocales	Heterotrichida	Agaricales	Primates	Cyperales
Family	Nostocaceae	Stentoridae	Agaricaceae	Hominidae	Gramineae
Genus	*Oscillatoria*	*Stentor*	*Agaricus*	*Homo*	*Zea*
Species	*marinus*	*coeruleus*	*campestris*	*sapiens*	*mays*
Common designation	Filamentous blue-green alga	Ciliate protozoan	Edible mushroom	Person	Corn

ferent variants of each character actually present at any given time. Species are populations of individuals of common descent living together in similar environments in a particular region, with similar ecological relationships and tending to have a unified and continuing evolutionary role distinct from that of other species. In biparental species the distinctiveness and continuation of the group are maintained by extensive interbreeding within it, and less or no interbreeding with other species. . . . Systematic units more inclusive than a species are groups of one or more species of common descent." (Simpson and Beck, 1965).

Because of the many features which all organisms on earth share, e.g., the genetic code, the use of adenosine triphosphate (ATP) in energy-yielding reactions, and so forth, we judge that all are ultimately related by common ancestry. By "populations of individuals of common descent" Simpson refers to a time relationship, i.e., to populations of individuals with relatively recent common ancestors.

Phylogenies, often represented in the form of family trees, may be thought of as summaries of available evolutionary information. They graphically relate groups of organisms to each other, plotted against time. A phylogeny of the chordate phylum to which man, *Homo sapiens,* belongs (see Table 6) is presented in Figure 1. Organisms at a branching fork in an evolutionary line represent the most recent common ancestor of organisms on a branch. Phylogenies are eclectic and use systematic information collected from many sources. For example, estimations of the number of mutations that have occurred since two organisms diverged from a common ancestor can be made and partial phylogenies can be drawn on the basis of primary amino acid sequence data in common proteins (such as cytochrome c or hemoglobin). Computers may be employed to make the most consistent tree on the basis of available data (Dayhoff, 1972). These partial phylogenies can be related to those based primarily on studies of skeletal structures of living and fossil organisms.

Classifications, based on phylogenies, although they require continual revision as new evidence accumulates, summarize a large body of information concerning the plants and animals and their evolutionary history. The broad outlines of the history of the multicellular animals and plants during the last 600 million years of the Phanerozoic eon are known [for example, see Banks (1970a,b; 1972a,b) for plants and Romer (1968) for animals]. This information is summarized in Figures 2 and 3. The evolutionary process that resulted in the diversity of organisms both extinct and preserved in the fossil record and extant can, in broad outline, be understood on the basis of the neo-Darwinian synthetic theory of evolution (Lerner, 1968; Simpson, 1954; Mayr, 1970).

However, the systematics of those organisms that are not clearly animals or plants (such as bacteria, algae, protists, and fungi) has been in

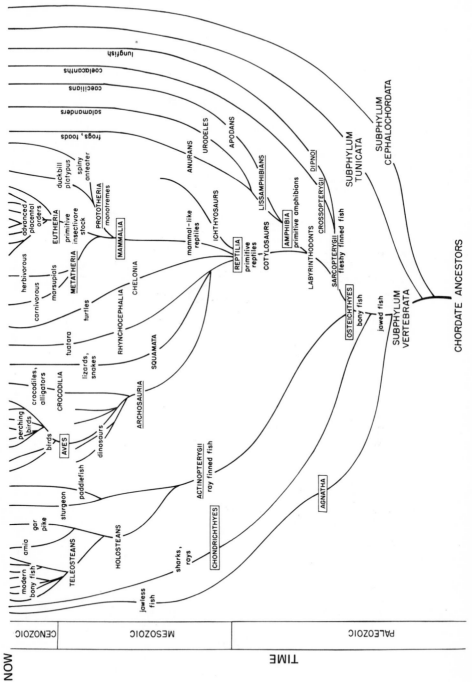

Figure 1. Phylogeny of the chordate phylum. (Living classes of the chordate phylum are boxed, subclasses underlined.)

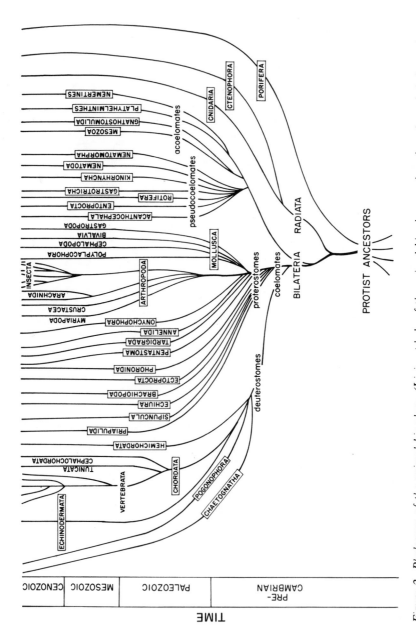

Figure 2. Phylogeny of the animal kingdom. (Living phyla of the animal kingdom are boxed, major subphyla or classes are in capital letters, and grades are in lower case.)

TABLE 2. Various Kingdom Classifications

System 1	System 2	System 3	System 4	System 5	System 6	System 7
Traditional, see Altman and Dittmer, 1972	Dodson, 1971	Curtis, 1968	Stanier et al., 1970	Copeland, 1956	Whittaker, 1969	Whittaker modified by Margulis, 1971
PLANTAE	MONERA (MYCHOTA)	PROTISTA	PROTISTA	MONERA	MONERA	MONERA
Bacteria	Bacteria	Bacteria	Bacteria	Bacteria	Bacteria	Bacteria
Blue-green algae	Blue-green algae	Blue-green algae	Blue-green algae	Blue-green algae	Blue-green algae	Blue-green algae
Green algae	PLANTAE	Protozoa	Protozoa	PROTOCTISTA	PROTISTA	PROTOCTISTA
Chrysophytes	Green algae	Slime molds	Green algae	Protozoa	Protozoa	Protozoa
Brown algae	Chrysophytes	PLANTAE	Chrysophytes	Green algae	Chrysophytes	Green algae
Red algae	Brown algae	Green algae	Brown algae	Chrysophytes	PLANTAE	Chrysophytes
Slime molds	Red algae	Chrysophytes	Red algae	Brown algae	Green algae	Brown algae
True fungi	Slime molds	Brown algae	Slime molds	Red algae	Brown algae	Red algae
Bryophytes	True fungi	Red algae	True fungi	Slime molds	Red algae	Slime molds
Tracheophytes	Bryophytes	True fungi	PLANTAE	True fungi	Bryophytes	FUNGI
ANIMALIA	Tracheophytes	Bryophytes	Bryophytes	PLANTAE	Tracheophytes	True fungi
Protozoa	ANIMALIA	Tracheophytes	Tracheophytes	Bryophytes	FUNGI	PLANTAE
Multicellular animals	Protozoa	ANIMALIA	ANIMALIA	Tracheophytes	Slime molds	Bryophytes
	Multicellular animals	Multicellular animals	Multicellular animals	ANIMALIA	True fungi	Tracheophytes
				Multicellular animals	ANIMALIA	ANIMALIA
					Multicellular animals	Multicellular animals

TABLE 3. Summary of Whittaker's (1969) Five Kingdom System[a]

Kingdom	Examples of organisms	Genetic organization	Approximate time of diversification (millions of years ago), documented first appearance [millions of years ago]	Major traits that environmental selection pressures acted on to produce	Major significant selective factor in the environment
Monera	All prokaryotes: bacteria, blue-green algae, mycelial bacteria, gliding bacteria, etc.	Prokaryote chromoneme merozygotes only; sex unidirectional	Early–middle Precambrian (3000–1000); Fig Tree microfossils [>3000] (Barghoorn, private communication), Bulawayan stromatolites [>3000]	Ultraviolet photoprotection, photosynthesis, motility, and aerobiosis	Solar radiation, increasing atmospheric oxygen concentration, depletion of nutrients
Protoctista	All eukaryotic algae: yellow-green, red and green, brown, and golden-yellow; all protozoa; flagellated fungi; slime molds; and slime net molds	Eukaryote chromosomes; ploidy levels; meiotic sexual systems vary, gametes and zygotes	Late Precambrian, early Paleozoic (1500–500); possibly Bitter Springs microflora [≈900] (Schopf, 1972); Ediacaran animals [750], Glaessner, 1968; see Margulis, 1974a for discussion	Mendelian genetic systems, mitosis and meiosis: obligate recombination each generation; phagocytosis, pinocytosis, intracellular motility	Depletion of nutrients

Animalia	Metazoa: all animals developing from blastulas	Diploids, meiosis precedes gametogenesis	Phanerozoic (600 on); Ediacaran fauna [700] (Glaessner, 1968)	Tissue development for heterotrophic specializations: ingestive nutrition	Transitions from aquatic to terrestrial and aerial environments
Plantae	Metaphyta: all green plants developing from embryos	Alternation between haplo- and diplophases	Phanerozoic (600 on); Rhyniophytes, Downtonian, Wales, Czechoslovakia, New York State [405] (Banks, 1972b)	Tissue development for autotrophic specializations: photosynthetic nutrition	Transitions from aquatic to terrestrial environments
Fungi	Amastigomycota: conjugation fungi, sac fungi (molds), club fungi (mushrooms), yeasts	Haploid and dikaryotic; zygote formation followed by meiosis and haploid spore formation	Phanerozoic (600 on); Rhynie Chert, *Palaeomyces asteroxyli* [400] (Arnold, 1949)	Advanced mycelial development: absorptive nutrition	Transitions from aquatic to terrestrial environments; nature of nutrient source, nature of host

a As modified by Margulis (1971, 1974b)

complete turmoil. It has very recently been recognized that the fossil history of microbial forms dates back over 3 billion years; the evolutionary relationships among these organisms is just now being constructed. Although radiolarians, foraminiferans, blue-green algae, and a few other groups of shelled or sediment-trapping microorganisms are exceptions, in general the microbial fossil record has been so deficient that the construction of phylogenies, and the taxonomy and systematics of such groups have been based primarily on comparisons of living forms. The relatively simple structure of microorganisms provides less information about relationships than the readily visible complex structure of multicellular plants and animals. Many of the smaller organisms have no obvious sexual stages, therefore the definition of "species" must depend solely on morphological and physiological frequency distributions. Development of electron microscopical and biochemical techniques have provided new tools for studying microorganisms. Enough information is now available to permit the realization that more must be known before any current classification of the lower organisms even approaches a phylogenetic one. This confusion is reflected in the different choice of highest taxa: kingdoms, phyla, and classes by different workers. The seven systems of Table 2 reflect the varying schemes in use today. Many people are beginning to follow Whittaker's (1969) five-kingdom system (system 6 in Table 2) for two reasons: It recognizes that the greatest discontinuity in the living world is that between prokaryotes and eukaryotes (Tables 3 and 4), and it acknowledges the vast difference between the fungi and the green plants.

Results of ultrastructural and biochemical studies have entirely supported the concept that the prokaryote–eukaryote dichotomy is unambiguous and far more profound than the traditional animal–plant distinction. Prokaryotic organisms include bacteria and blue-green algae. The organization of the smaller prokaryotic cells is distinctive. In them, DNA is not complexed with RNA and proteins, including a certain class of arginine- and lysine-rich proteins, the histones. Sexual systems are unidirectional in those organisms in which sexual recombination is present. That is, in the sexual process, genes are transferred in one direction, from the donor cell to the partner (the recipient cell). Only genes, and generally only a small fraction of the total genes, are transferred. The new, recombinant, organism consists thus of the recipient cell itself with some genes replaced (those derived from the donor). Eukaryotes contain organelles such as complex chromosomes, mitochondria, chloroplasts, basal bodies, and centrioles, which prokaryotic cells lack. Most familiar organisms (nucleated algae, protozoans, fungi, metazoan animals, and green plants) are

(Text continued on p. 31)

TABLE 4.   *Major Differences Between Prokaryotes and Eukaryotes*

Prokaryotes	Eukaryotes
Mostly small cells (1–10 μ). All are microbes. The most morphologically complex are filamentous or mycelial with fruiting bodies (e.g., actinomycetes, myxobacteria, blue green algae)	Mostly large cells (10–100 μ). Some are microbes, most are large organisms. The most morphologically complex are the vertebrates and the flowering plants
Nucleoid, not membrane-bounded	Membrane-bounded nucleus
Cell division direct, mostly by "binary fission." Chromatin body which contains DNA, but no protein. Does not stain with the Feulgen technique. No centrioles, mitotic spindle, or microtubules	Cell division by various forms of mitosis. Many chromosomes containing DNA, RNA, and proteins; they stain bright red with Feulgen technique. Centrioles present in many; mitotic spindle (or at least some arrangement of microtubules) occurs
Sexual systems absent in most forms; when present, unidirectional transfer of genetic material from donor to host takes place	Sexual systems present in most forms; equal participation of both partners (male and female) in fertilization
Multicellular organisms never develop from diploid zygotes. No tissue differentiation	Meiosis produces haploid forms, diploids develop from zygotes. Multicellular organisms show extensive development of tissues
Includes strict anaerobes (these are killed by O_2), and facultatively anaerobic, microaerophilic, and aerobic forms	All forms are aerobic (these need O_2 to live); exceptions are clearly secondary modifications
Enormous variations in the metabolic patterns of the group as a whole. Mitochondria absent; enzymes for oxidation of organic molecules bound to cell membranes, i.e., not packaged separately	Same metabolic patterns of oxidation within the group (i.e., Embden-Meyerhof glucose metabolism, Krebs cycle oxidations, molecular oxygen combines with hydrogens from foodstuffs, and catalyzed by cytochrome, water is produced). Enzymes for oxidation of 3-carbon organic acids are packaged within membrane-bounded sacs (mitochondria)
Simple bacterial flagella, if flagellated	Complex "9 + 2" flagella or cilia if flagellated or ciliated
If photosynthetic, enzymes for photosynthesis bound to cell membrane (chromatophores), not packaged separately. Anaerobic and aerobic photosynthesis, sulfur deposition, and O_2 elimination	If photosynthetic, enzymes for photosynthesis packaged in membrane-bounded chloroplasts. O_2-eliminating photosynthesis

TABLE 5. *The Five-Kingdom System of Whittaker with a Listing of the Prokaryote Components of Eukaryotic Cells Based on the Cell Symbiosis Theory of Margulis* [a]

Kingdom	Description of major groups in kingdoms	Number of genomes hypothesized
MONERA	In order of evolution, from early to middle precambrian	
	1. Anaerobic heterotrophs Ⓐ (Embden-Meyerhof fermentation = protoeukaryotes)	Monogenomic
	2. Motile anaerobes = spirochetelike protoflagella Ⓕ	Monogenomic
	3. Photoautotrophs (CO_2 fixation, O_2 elimination) = photosynthetic protoplastids Ⓟ	Monogenomic
	4. Aerobic heterotrophs (respiration via the Krebs cycle) = protomitochondria Ⓜ	Monogenomic
PROTOCTISTA	In order of acquisition of organelles	
	Ⓐ + Ⓜ Some amoebae = mitochondria-containing protists, lacking flagellar homologs	Digenomic
	Ⓐ + Ⓜ + Ⓕ = AMF The 9 + 2 amoeboflagellate ancestor, leading to the origin of mitosis and origin of protozoans, slime molds, flagellated fungi, etc.	Trigenomic
	AMF + Ⓟ = AMFP Ancestral flagellate with photosynthetic plastids which evolved into chloroplasts, rhodoplasts, etc. leading to the evolution of nucleated algae: brown and red seaweeds, green algae, euglenids, diatoms, etc.	Tetragenomic
FUNGI	AMF Haploid and dikaryotic true fungi: zygomycetes, ascomycetes, and basidomycetes, from ancestral amoeboflagellate; sacrificed 9 + 2 flagella during the evolution of mitosis and meiosis	Trigenomic
ANIMALIA	AMF Metazoa, derived from the ancestral amoeboflagellates	Trigenomic
PLANTAE	AMFP Green plants; bryophytes to angiosperms, derived from ancestral green algae	Tetragenomic

[a] This table is the key to Figure 4, page 34; this notation is also used in Table 6, pp. 13–31.

TABLE 6.   *Five-Kingdom Classification* [a]

[Footnotes will be found on pp. 30 and 31]

SUPERKINGDOM (CHROMONEMAL ORGANIZATION) [b] PROKARYOTA

KINGDOM MONERA

A,F,P,M, AND OTHERS [c]

Prokaryotic cells, nutrition absorptive, chemosynthetic, photoheterotrophic, or photoautotrophic. Anaerobic, facultative, or aerobic metabolism. Reproduction asexual; chromonemal; recombination unidirectional or viral mediated. Nonmotile or motile by bacterial flagella composed of flagellin proteins or by gliding. Solitary unicellular, filamentous, colonial, or mycelial.

SUBKINGDOM BACTERIA
  GRADE EUBACTERIA; TRUE BACTERIA (INCLUDE A,M,P, AND F)
    PHYLUM 1.   Fermenting bacteria; unable to synthesize porphyrins
      Class 1. Mycoplasmas (*Thermoplasma, Bartonella, Anaplasma*)
      Class 2. Chlamydias (Psittacosis group, bedsonias)
      Class 3. Lactic acid bacteria (*Streptococcus, Diplococcus,* [d] *Leuconostoc*)
      Class 4. Clostridia (*Clostridium*)
    PHYLUM 2.   Spirochaetae; spirochetes ⓔ (*Cristispira, Treponema, Spirochaeta, Leptospira*)
    PHYLUM 3.   Anaerobic sulfate reducers; synthesis of iron hemes limited
      Class 1. Desulfovibrios; spore-forming sulfate reducers (*Desulfovibrio*)
      Class 2. Desulfatomaculum; non-spore-forming sulfate-reducing bacteria
    PHYLUM 4.   Methane bacteria; anaerobic chemotrophs: use CO_2 as electron acceptor for anaerobic respiration reducing it to CH_4 (*Methanococcus, Methanosarcina, Methanobacterium*)
    PHYLUM 5.   Photosynthetic bacteria; synthesis of iron and magnesium chelated hemes, carotenoids
      Class 1. Purple nonsulfur bacteria; photoheterotrophs, organic hydrogen donors (*Rhodopseudomonas, Rhodomicrobium, Rhodospirillum*)
      Class 2. Green sulfur bacteria, H_2S hydrogen donor (*Chlorobium*)
      Class 3. Purple sulfur bacteria, H_2S hydrogen donor (*Chromatium*)
    PHYLUM 6.   Cyanophyta; blue-green algae, blue-green bacteria, oxygen eliminating photosynthesis, photosystem II ⓟ; H_2O hydrogen donor; aerobic photosynthesizers
      Class 1. Coccogoneae; coccoid blue-greens (*sensu lato*)
        1. Order Chroococcales; reproduce by fission (*Chroococcus, Gloeocapsa*)
        2. Order Chamaesiphonales; reproduce by exospores (*Chamaesiphon, Dermocarpa*)
        3. Order Pleurocapsales; reproduce by endospores (*Pleurocapsa*)
      Class 2. Hormogoneae; filamentous blue-greens (*sensu stricto*)
        1. Order Nostocales; false or no branching (*Oscillatoria, Spirulina, Anabaena, Rivularia, Scytonema*)
        2. Order Stigonematales; true branching (*Mastigocladus, Stigonema*)
    PHYLUM 7.   Nitrogen-fixing aerobic bacteria (*Azotobacter, Beijerinckia, Rhizobium*)

Table 6.   Continued

KINGDOM MONERA

PHYLUM 8.   Pseudomonads *(Pseudomonas,[d] Photobacterium, Hydrogenomonas, Halobacterium)*
PHYLUM 9.   Aerobic endospore bacteria, Gram positive *(Bacillus[d])*
PHYLUM 10. Micrococci; Gram-positive aerobes, entire Krebs cycle present *(Micrococcus, Sarcina, Gaffkya)*
PHYLUM 11. Chemoautotrophic bacteria
  Class 1. Sulfur-oxidizing bacteria *(Thiobacillus)*
  Class 2. Ammonia-oxidizing bacteria *(Nitrobacter, Nitrosomonas, Nitrosocystis)*
  Class 3. Iron-oxidizing bacteria *(Ferrobacillus)*
PHYLUM 12. Aerobic Gram-negative heterotrophic bacteria Ⓜ
  Class 1. Enterobacteria; coliforms, facultative anaerobes *(Escherichia coli,[d] Salmonella,[d] Shigella, Serratia)*
  Class 2. Prosthecate bacteria
    1. Order Caulobacters; single polar or subpolar prosthecae, no budding *(Caulobacter, Asticcacaulis)*
    2. Order Hyphomicrobia; budding prosthecate bacteria *(Hyphomicrobium, Ancalomicrobium)*
  Class 3. Sphaerotilus group; form distinctive cell aggregates
    1. Order Sphaerotilus; form tubular sheaths, deposit ferric or manganese oxides *(Sphaerotilus, Leptothrix)*
    2. Order Zoogloea; form flocs, cells in gelatinous slime
  Class 4. Acetic acid bacteria; rectangular sheaths *(Gluconobacter, Acetobacter)*
  Class 5. Moraxella–Neisseria; nonflagellated group *(Neisseria, Moraxella, Acinetobacter)*
  Class 6. Predatory bacteria; reproduce inside host, polar flagellated *(Bdellovibrio)*
  Class 7. Microaerophilic; polar flagellated helical cells *(Spirillum)*; single thick polar flagellum *(Campylobacter)*
  Class 8. Rickettsias; glutamate oxidizers *(Coxiella)*

GRADE GREATER BACTERIA, HETEROTROPHS AND CHEMOAUTOTROPHS WITH DISTINCT MORPHOLOGY AT THE LIGHT–MICROSCOPE LEVEL

PHYLUM 13. Myxobacteria; heterotrophic gliding bacteria
  Class 1. Flexibacteria *(Flexibacter, Saprospira)*
  Class 2. Filamentous gliding bacteria *(Beggiatoa, Leucothrix, Vitreoscilla)*
  Class 3. Fruiting myxobacteria *(Myxococcus, Chondromyces, Podangium)*
PHYLUM 14. Actinomycota; Gram-positive coryniform and mycelial bacteria
  Class 1. Coryniform bacteria *(Arthrobacter, Propionibacterium)*
  Class 2. Proactinomycetes *(Mycobacterium, Actinomyces, Nocardia, Streptomyces[d])*

Table 6.   Continued

SUPERKINGDOM (CHROMOSOMAL ORGANIZATION)[b] EUKARYOTA

KINGDOM PROTOCTISTA[f]

AMF AND AMFP

Eukaryotic cells, nutrition ingestive, absorptive or, if photoautotrophic, in photosynthetic plastids. Premitotic and eumitotic asexual reproduction. In eumitotic forms meiosis and fertilization present but life cycle and ploidy levels vary from group to group. Solitary unicellular, colonial unicellular or multicellular. Complex flagella or cilia composed of microtubules in the 9 + 2 pattern.

GRADE AMITOTICA (AM)

    PHYLUM 1. Caryoblastea; amitotic amoebae *(Pelomyxa palustris)*

GRADE MITOTICA (AMF, AMFP). Microtubular 9 + 2 homolog phyla. Mitotic spindle or equivalent composed of 250-Å microtubules

    PHYLUM 2. Dinoflagellata; mesokaryota, dinoflagellates, AMF, AMFP *(Gymnodinium, Peridinium)*

    PHYLUM 3. Chrysophyta; chrysophytes, golden-yellow algae, AMFP *(Ochromonas, Echinochrysis, Sarcinochrysis)*

    PHYLUM 4. Haptophyta; haptophytes or coccolithophorids, AMFP *(Hymenomonas, Pontosphaera)*

    PHYLUM 5. Euglenophyta; euglenids *(Euglena, Peranema, Astasia)*, AMF, AMFP

    PHYLUM 6. Cryptophyta; cryptomonads, AMF, AMFP *(Cryptomonas, Cyanomonas, Cyathomonas)*

    PHYLUM 7. Zoomastigina; animal flagellates; AMF or AMFP

    Class 1. Opalinida; opalinids *(Opalina)*

    Class 2. Choanoflagellida or craspedomonads; collared flagellates *(Monosiga, Desmarella)*

    Class 3. Bicoecidea; shelled biflagellates, one flagellum attached

    Class 4. Diplomonadida; diplomonads *(Diplomonas, Giardia)*

    Class 5. Kinetoplastida; bodos and trypanosomes *(Bodo, Crithidia, Trypanosoma)*

    Class 6. Oxymonadida; oxymonads *(Oxymonas)*

    Class 7. Trichomonadida; trichomonads *(Trichomonas)*

    Class 8. Hypermastigida; hypermastigotes *(Saccinobaculus, Barbulonympha, Trichonympha)*

    PHYLUM 8. Rhizopodata; rhizopod amoebas; AMF, AMFP *(Difflugia, Rhizochrysis, Chrysarachnion, Arcella)*

    PHYLUM 9. Xanthophyta; yellow-green algae, AMFP *(Tribonema, Botryococcus)*

    PHYLUM 10. Eustigmatophyta; eustigmatophytes, AMFP, *(Pleurochloris, Vischeria)*

    PHYLUM 11. Prasinophyta; AMFP *(Pyramimonas, Platymonas, Prasinocladus)*

    PHYLUM 12. Bacillariophyta; diatoms, AMFP *(Surirella, Navicula, Planktoniella)*

    PHYLUM 13. Phaeophyta; brown algae, AMFP *(Fucus, Dictyota)*

Table 6.   Continued

KINGDOM PROTOCTISTA

PHYLUM 14. Rhodophyta; red seaweeds, AMFP *(Porphyra, Nemalion)*

PHYLUM 15. Gamophyta; conjugating green algae, AMFP *(Spirogyra, Zygnema, desmids)*

PHYLUM 16. Chlorophyta; grass-green algae, AMFP *(Chlamydomonas,[d] Tetraspora, Volvox)*

PHYLUM 17. Siphonophyta; siphonaceous syncitial green algal, AMFP *(Acetabularia, Caulerpa, Codium)*

PHYLUM 18. Charophyta; charophyte green algae, AMFP *(Nitella, Chara)*

PHYLUM 19. Heliozoata; sun animalicules, AMF *(Echinosphaerium, Actinophrys)*

PHYLUM 20. Radiolariata; radiolarians, AMF *(Acantharia, Thalassicola)*

PHYLUM 21. Foraminifera; foraminiferans, AMF *(Globigerina, Nodosaria)*

PHYLUM 22. Ciliophora; ciliates, AMF

    Class 1. Kinetofragminophora (gymnostomes, hypostomes, suctorians) *Stephanopogon, Entodiniomorpha, Tokophyra*

    Class 2. Oligohymenophora (hymenostomes, peritrichs) *Tetrahymena[d], Paramecium[d], Vorticella*

    Class 3. Polyhymenophora (spirotrichs) *Stentor, Stylonychia, Euplotes,[d]* tintinnids

PHYLUM 23. Apicomplexa (Sporozoa, or Telosporidea), parasites, AMF

    Class 1. Sporozoasida

      Subclass 1. Gregarinasina *(Gregarina)*

      Subclass 2. Coccidiasina *(Eimeria, Isospora)* and hemosporidians *(Plasmodium, Haemoproteus)*

    Class 2. Piroplasmasida *(Babesia)*

PHYLUM 24. Cnidosporidia; cnidosporidian parasites, polar capsules

    Class 1. Microsporida *(Nosema)*

    Class 2. Myxosporida *(Myxobolus, Henneguya)*

    Class 3. Actinomyxida *(Sphaeractinomyxon)*

PHYLUM 25. Labyrinthulamycota; slime net amoebas, AMF *(Labyrinthula, Labyrinthorhiza)*

PHYLUM 26. Acrasiomycota; cellular slime molds, AMF

    Class 1. Dictyostelia *(Dictyostelium,[d] Polysphondylium[d])*

    Class 2. Acrasia *(Guttulinopsis, Acrasis)*

PHYLUM 27. Myxomycota; plasmodial slime molds, AMF

    Class 1. Protostelida; protostelids *(Protostelium, Ceratiomyxa)*

    Class 2. Myxogastria; myxomycetous plasmodial slime molds *(Physarum, Echinostelium)*

PHYLUM 28. Plasmodiophoromycota; plasmodiophores, AMF *(Plasmodiophora, Spongospora, Woronina, Polymyxa)*

PHYLUM 29. Hyphochytridiomycota; hyphochytrids, anterior mastigonemate uniflagellum, AMF *(Rhizidiomyces, Anisolpidium).*

PHYLUM 30. Oomycota; oomycetous water molds, AMF *(Saprolegnia, Albugo, Achlya, Pythium, Phytophthora)*

PHYLUM 31. Chytridiomycota; chytrids, posteriorly uniflagellated aquatic fungi, AMF

Table 6.   Continued

KINGDOM PROTOCTISTA

Class 1. Chytridia (*Olpidium, Synchytrium*)
Class 2. Blastocladia (*Allomyces, Blastocladiella*)
Class 3. Monoblepharida (*Monoblepharis, Gonapodya*)
Class 4. Harpochytria (*Harpochytrium*)

KINGDOM FUNGI

AMF 6

Haploid or dikaryotic, mycelial or secondarily unicellular, chitinous walls, absorptive nutrition, lack cells with 9 + 2 homolog motile organelles (i.e., cilia or flagella), body plan branched, coenocytic filament which may be divided by perforate septa. Zygotic meiosis, propagate by haploid spores. (Eumycophytes: Alexopoulos, 1962; amastigote fungi: Copeland, 1956; Whittaker, 1969)

PHYLUM 1. Zygomycota; zygomycetous molds *(Phycomyces,[d] Rhizopus, Mucor)*
PHYLUM 2. Ascomycota; sac fungi or ascomycetes
   Class 1. Hemiascomycetae; yeasts, leaf curl fungi (*Endomyces, Saccharomyces,[d] Schizosaccharomyces[d]*)
   Class 2. Euascomycetes
     Subclass 1. Plectomycetes; scattered asci (black molds, blue molds, *Erysiphe, Aspergillus,[d] Penicillium*)
     Subclass 2. Pyrenomycetes; ascocarp closed (perithecial fungi, *Neurospora,[d] Podospora,[d] Sordaria[d]*)
     Subclass 3. Discomycetes; ascocarp on open apothecium (cup fungi, morels, truffles, *Ascobolus[d]*)
   Class 3. Loculoascomycetae; ascostromatic fungi (*Elsinoe, Ophiobus, Cucurbitana*)
   Class 4. Laboulbeniomycetae; insect parasites (*Laboulbenia, Herpomyces*)
PHYLUM 3. Basidiomycota; club fungi
   Class 1. Heterobasidiomycetae; jelly fungi, rusts, smuts *(Puccinia, Ustilago[d])*
   Class 2. Homobasidiomycetae; mushrooms, shelf fungi, coral fungi, earth stars, stink horns, birds-nests fungi (*Agaricus, Schizophyllum,[d] Coprinus[d]*)
PHYLUM 4. Deuteromycota; fungi imperfecti (*Cryptococcus, Candida, Monilia, Histoplasma, Rhizoctonia*)
PHYLUM 5. Mycophycophyta; lichens, AMF (fungal component) + AMF (blue green algal component) or AMF (fungal component) + AMFP (green algal component)
   Class 1. Lichenized with ascomycetes
     Order 1. Lecanorales; ascomycetous, fruiting bodies apothecia (*Lichina, Lichinella, Collema, Cladonia*)
     Order 2. Caliciales; ascomycetes, fruiting bodies, mazaedia (*Calicium, Coniocybe*)
The following fungal orders contain genera which are frequently lichenized:
     Order 1. Sphaeriales; pyrenomycete ascomycetes (*Pyrenula*)
     Order 2. Myrangiales; loculoascomycetes, asci irregular (*Arthonia, Dermatina*)

Table 6.   Continued

Kingdom Fungi

Order 3. Pleosporales; loculoascomycetes, pseudothecia well-delimited
  (*Anthracothecium, Arthopyrenia*)
Order 4. Hysteriales; loculoascomycetes, pseudothecia, well-delimited,
  resemble apothecia *(Lecanactis, Platygraphopsis, Rocella)*
Class 2. Lichenized with basidiomycetes (Herpothallaceae, Coraceae, Cla-
  variaceae, etc.)
Class 3. Lichenized with fungi imperfecti; fruiting bodies unknown
  (Cystocoleus, Lepraria, Lichenothrix)

Kingdom Animalia

AMF 6

Multicellular animals, develop from diploid blastula, gametic meiosis, gastrula-
tion occurs. Nutrition heterotrophic, ingestive, phagocytosis, extensive cellular
and tissue differentiation.

Subkingdom Parazoa
   Phylum 1. Placozoa; diploblastic, dorsoventral organization, no polarity or
      bilaterality (*Trichoplax*)
   Phylum 2. Porifera (sponges)
   Class 1. Calcarea; calcitic spicules
   Class 2. Demospongiae; spongin network with or without siliceous spic-
      ules
   Class 3. Sclerospongiae; spongin network with or without siliceous spic-
      ules and basal skeleton of aragonite
   Class 4. Hexactinellida; siliceous spicules with 3 axes

Subkingdom Eumetazoa
  Branch Radiata; radially symmetrical organisms
   Phylum 3. Cnidaria (coelenterates)
   Class 1. Hydrozoa (hydroids)
   Class 2. Scyphozoa (true jellyfish)
   Class 3. Anthozoa (corals and sea anemones)
   Phylum 4. Ctenophora (comb jellies)
  Branch Bilateria; bilaterally symmetrical organisms
   Grade Acoelomata
   Phylum 5. Mesozoa (mesozoans)
   Phylum 6. Platyhelminthes (flatworms)
   Class 1. Turbellaria (planarians)
   Class 2. Trematoda (flukes)
   Class 3. Cestoda (tapeworms)
   Phylum 7. Nemertina (nemertine worms)
   Phylum 8. Gnathostomulida (gnathostome worms)
   Grade Pseudocoelomata
   Phylum 9. Acanthocephala (spiny-headed worms)

Table 6. Continued

KINGDOM ANIMALIA

PHYLUM 10. Entoprocta (entoprocts or kamptozoa)
PHYLUM 11. Rotifera (rotifers)
PHYLUM 12. Gastrotricha (gastrotrichs)
PHYLUM 13. Kinorhyncha (kinorhynchs)
PHYLUM 14. Nematoda (round worms, *Ascaris, Caenorhabditis*[d])
PHYLUM 15. Nematomorpha (Gordiaceae) (horsehair worms)
GRADE COELOMATA
 SERIES PROTEROSTOMA
PHYLUM 16. Priapulida (priapulids)
PHYLUM 17. Sipuncula (sipunculids or peanut worms)
PHYLUM 18. Mollusca (mollusks)
  Class 1. Monoplacophora (monoplacophorans)
  Class 2. Aplacophora (solenogasters)
  Class 3. Polyplacophora (chitons)
  Class 4. Scaphopoda (toothshells)
  Class 5. Gastropoda (snails, *Cepaea,*[d] *Arianta,*[d] *Helix,*[d] *Lymnea,*[d] *Partula*[d])
  Class 6. Bivalvia (bivalves, pelecypods, clams, oysters, *Crassostrea,*[d] *Mytilus*[d])
  Class 7. Cephalopoda (squids, octopuses)
PHYLUM 19. Echiura (echiuroids)
PHYLUM 20. Annelida (segmented worms)
  Class 1. Polychaeta (polychaete worms)
  Class 2. Clitellata (leeches, earthworms)
PHYLUM 21. Tardigrada (water bears)
PHYLUM 22. Pentastoma (tongueworms, pentastomes)
PHYLUM 23. Onychophora (*Peripatus*)
PHYLUM 24. Phoronida (phoronids)
PHYLUM 25. Ectoprocta (bryozoa or moss animals)
PHYLUM 26. Brachiopoda (brachiopods or lamp shells)
PHYLUM 27. Arthropoda (jointed-foot animals)
 SUPERCLASS 1. CHELICERATA
  Class 1. Merostomata (horseshoe crabs)
  Class 2. Pycnogonida (sea spiders)
  Class 3. Arachnida (scorpions, harvestmen, ticks, mites, spiders)
 SUPERCLASS 2. CARIDA
  Class 4. Crustacea
   Subclass 1. Cephalocarida (cephalocarids)
   Subclass 2. Malacostraca (lobsters, shrimps, crabs)
   Subclass 3. Branchiopoda (fairy shrimps, water fleas)
   Subclass 4. Maxillipoda
    Order 1. Copepoda (copepods)
    Order 2. Cirripedia (barnacles)
    Order 3. Mystacocarida (mystacocarids)
   Subclass 5. Ostracoda (ostracods)
  Class 5. Myriapoda
    Order 1. Diplopoda (millipedes)
    Order 2. Chilopoda (centipedes)
    Order 3. Pauropoda (pauropods)
    Order 4. Symphyla (symphylans)

Table 6. Continued

KINGDOM ANIMALIA

Class 6. Insecta
- Order 1.   Protura (proturans)
- Order 2.   Collembola (springtails)
- Order 3.   Thysanura (bristletails, diplurans)
- Order 4.   Diplura (campodeids)
- Order 5.   Odonata (dragon flies, damsel flies)
- Order 6.   Ephemeroptera (May flies)
- Order 7.   Blattaria (roaches, *Blattella*[d])
- Order 8.   Isoptera (termites)
- Order 9.   Orthoptera (grasshoppers, crickets, *Melanopus*)
- Order 10. Dermaptera (earwigs)
- Order 11. Plecoptera (stoneflies)
- Order 12. Embioptera (embiids)
- Order 13. Mallophaga (biting lice)
- Order 14. Anoplura (sucking lice)
- Order 15. Psocoptera (bark lice)
- Order 16. Hemiptera (true bugs)
- Order 17. Homoptera (aphids, cicadas, scale insects)
- Order 18. Thysanoptera (thrips)
- Order 19. Mecoptera (scorpionflies)
- Order 20. Neuroptera (lacewings, Dobson flies)
- Order 21. Trichoptera (caddis flies)
- Order 22. Coleoptera (beetles, *Tribolium*[d])
- Order 23. Lepidoptera (moths, butterflies, *Bombyx*,[d] *Ephestia*[d])
- Order 24. Diptera (mosquitos: *Aedes*,[d] *Anopheles*,[d] *Culex*[d] flies, *Chironomus*,[d] *Glyptotendipes*[d], *Sciara*,[d] *Rhynchosciara*[d], *Drosophila*,[d] *Musca*,[d] *Lucilia*[d])
- Order 25. Siphonoptera (fleas)
- Order 26. Hymenoptera (bees, wasps, hornets, *Habrobracon*,[d] *Mormoniella*,[d] *Apis*[d])

SERIES DEUTEROSTOMA
PHYLUM 28. Echinodermata (echinoderms)
- Class 1. Crinoidea (sea lilies)
- Class 2. Asteroidea (starfish)
- Class 3. Ophiuroidea (brittle stars)
- Class 4. Echinoidea (sea urchins)
- Class 5. Holothuroidea (sea cucumbers)

PHYLUM 29. Chaetognatha (arrow worms)
PHYLUM 30. Pogonophora (beard worms)
PHYLUM 31. Hemichordata (acorn worms)
PHYLUM 32. Chordata (notochord-bearing animals)
SUBPHYLUM 1. Tunicata (urochordates, tunicates, ascidians)
SUBPHYLUM 2. Cephalochordata (lancelets, *Amphioxus*, cephalochordates)
SUBPHYLUM 3. Vertebrata (vertebrates)
- Class 1. Agnatha (jawless fish, lampreys, hagfishes)
- Class 2. Chondrichthyes (cartilagenous fish)

Table 6.   Continued

KINGDOM ANIMALIA

    Order 1. Chlamydoselachiformes (frilled sharks)
    Order 2. Hexanchiformes (comb-toothed sharks)
    Order 3. Heterodontiformes (bull-headed sharks)
    Order 4. Squaliformes (sharks)
    Order 5. Rajiformes (rays)
    Order 6. Chamaeriformes (chimaeras)
Class 3. Osteichthyes (bony fish)
    Order 1. Acipenseriformes (sturgeons)
    Order 2. Semionotiformes (gars)
    Order 3. Amiiformes (bowfin)
    Order 4. Elopiformes (tarpons)
    Order 5. Lepisosteiformes (gar pikes)
    Order 6. Clupeiformes (herrings, pikes, mud minnows)
    Order 7. Osteoglossiformes (mooneyes, bony tongued fish)
    Order 8. Salmoniformes (trouts)
    Order 9. Myctophiformes (lantern fish)
    Order 10. Giganturiformes (dep-sea giganturoids)
    Order 11. Ctenothrissiformes
    Order 12. Saccopharyngiformes (gulpers)
    Order 13. Canniformes
    Order 14. Anguilliformes (eels)
    Order 15. Notacanthiformes (spiny eels)
    Order 16. Beloniformes (garfishes)
    Order 17. Cypriniformes (characins, minnows, suckers,
         Oryzias[d] or medaka)
    Order 18. Cyprinidontiformes (*Fundulus*[d] or killifish; *Lebistes*[d] or
         guppy)
    Order 19. Percopsiformes (trout perches)
    Order 20. Gasterosteiformes (pipefishes, sticklebacks, sea-
         horses)
    Order 21. Gadiformes (codfishes)
    Order 22. Lampridiformes (ribbonfishes, oarfishes)
    Order 23. Beryciformes (squirrel fishes)
    Order 24. Zeiformes (dories)
    Order 25. Perciformes (perches, basses, *Tilapia*)
    Order 26. Pegasiformes (seadragons)
    Order 27. Pleuronectiformes (flatfishes, flounders)
    Order 28. Echeneiformes (remoras)
    Order 29. Tetraodontiformes (triggerfishes, puffers)
    Order 30. Atheriniformes (flying fishes, needlefishes, live
         bearers)
    Order 31. Gobiesociformes (clingfishes)
    Order 32. Batrachioidiformes (toadfishes)
    Order 33. Lophiiformes (angler fishes)

Table 6. Continued

KINGDOM ANIMALIA

Order 34. Siluriformes (catfishes)
Order 35. Mormyriformes (elephant snout fishes)
Order 36. Dactylopteriformes (flying gurnards)
Order 37. Scorpaeniformes (scorpion fishes)
Order 38. Cetomimiformes
Order 39. Synbranchiformes (synbranchoid eels, fleshy finned fish (Sarcopterygii)
Order 40. Dipnoi (lungfishes)
Order 41. Crossopterygii (coelacanths; *Latimeria*)
Class 4. Amphibia
Order 1. Apoda (caecilians)
Order 2. Urodela (salamanders, axolotl or *Ambystoma,*[d] *Triturus*[d], *Pleurodeles*[d])
Order 3. Anura (frogs and toads, *Xenopus,*[d] *Rana*[d])
Class 5. Reptilia
Order 1. Chelonia (Testudinata; turtles, tortoises, terrapins)
Order 2. Crocodilia (crocodiles, alligators)
Order 3. Rhynchocephalia (*Sphenodon* or tuatara)
Order 4. Squamata (lizards, snakes)
Class 6. Aves (birds)
Order 1.   Struthioniformes (ostriches)
Order 2.   Casuariiformes (cassowaries and emus)
Order 3.   Aepyornithiformes (elephant birds, recently extinct)
Order 4.   Dinornithiformes (moas, recently extinct)
Order 5.   Apterygiformes (kiwis)
Order 6.   Rheiformes (rheas)
Order 7.   Tinamiformes (tinamous)
Order 8.   Sphenisciformes (penguins)
Order 9.   Gaviiformes (loons)
Order 10. Podicipediformes (grebes)
Order 11. Procellariiformes (albatrosses, fulmars, petrels)
Order 12. Pelecaniformes (pelicans, cormorants, boobies, gannets)
Order 13. Ciconiiformes (herons, storks, ibises, flamingos)
Order 14. Anseriformes (ducks, geese, swans)
Order 15. Falconiformes (vultures, kites, hawks, falcons, eagles)
Order 16. Galliformes (grouse, quail, pheasants, turkeys, *Gallus* [d])
Order 17. Gruiformes (cranes, rails, coots)
Order 18. Charadriiformes (shore birds, gulls, auks)
Order 19. Columbiformes (pigeons, doves)
Order 20. Cuculiformes (cuckoos, roadrunners, anis)
Order 21. Psittaciformes (parrots)
Order 22. Strigiformes (owls)
Order 23. Caprimulgiformes (goatsuckers, oil birds)
Order 24. Apodiformes (swifts, hummingbirds)
Order 25. Coliiformes (colies or mouse birds)

Table 6.   Continued

KINGDOM ANIMALIA

Order 26. Trogoniformes (trogons)
Order 27. Coraciiformes (kingfishers, bee-eaters, hornbills)
Order 28. Piciformes (woodpeckers, jacamars, puffbirds,
            barbets, honey guides, toucans)
Order 29. Passeriformes (perching birds or songbirds)
Class 7. Mammalia
  Subclass 1. Protheria (egg-laying mammals; echidna, platypus)
  Subclass 2. Metatheria (marsupials; opposum, kangaroos)
  Subclass 3. Eutheria (placental mammals)
    Order 1. Insectivora (moles, shrews)
    Order 2. Primates (lemurs, tarsiers, monkeys, apes, people
              or *Homo sapiens* [d])
    Order 3. Chiroptera (bats)
    Order 4. Edentata (sloths)
    Order 5. Rodentia (rodents, *Mus,*[d] *Rattus,*[d] *Peromyscus,*[d]
              Mesocricetus[d], *Cavia*[d], *Chinchilla*[d])
    Order 6. Lagomorpha (hares, rabbits, *Oryctolagus*[d])
    Order 7. Carnivora (carnivores, *Canis*[d], *Felis*[d], *Vulpes*[d], *Mustela*[d])
    Order 8. Pinnipedia (seals)
    Order 9. Cetacea (whales)
    Order 10. Proboscidea (elephants)
    Order 11. Dermoptera (flying lemurs)
    Order 12. Pholidota (scaly anteaters)
    Order 13. Tubulidentata (aardvarks)
    Order 14. Hyracoidea (conies)
    Order 15. Sirenia (manatees)
    Order 16. Perissodactyla (odd-toed hoofed animals, *Equus*[d])
    Order 17. Artiodactyla (even-toed hoofed animals, *Ovis,*[d]
              Bos,[d] *Camelus*)

KINGDOM PLANTAE

Autotrophic green plants, advanced tissue differentiation, diploid phase develops
from embryophyte plants, primarily terrestrial Archegoniates.
GRADE BRYOPHYTA
  PHYLUM 1. Bryophyta
    Class 1. Anthocerotae (hornworts)
    Class 2. Hepaticeae (liverworts)
      Order 1. Sphaerocarpales (*Sphaerocarpos* [d])
      Order 2. Marchantiales (*Marchantia, Riccia*)
      Order 3. Metzgeriales (*Fossumbronia*)
      Order 4. Jungermanniales (*Lepidozia*)
    Class 3. Musci (mosses)
      Order 1. Sphagnales (peat mosses, *Sphagnum*)

Table 6.   Continued

KINGDOM PLANTAE

    Order 2. Andreaeales (*Andreaea*)
    Order 3. Bryales (*Mnium, Polytrichum*)
GRADE TRACHEOPHYTA; vascular plants
  PHYLUM 2. Lycopodophyta (club mosses and quillworts, *Lycopodium, Selaginella, Isoetes*)
  PHYLUM 3. Sphenophyta (Equisetophyta, horsetails, *Equisetum*)
  PHYLUM 4. Filicinophyta (Pteridophytes, ferns)
    Class Filicinae
      Order 1. Ophioglossales (succulent ferns, *Ophioglossum*)
      Order 2. Marattiales (treeferns, *Marattia*)
      Order 3. Filicales (Polypodiales, *Polypodium, Dryopteris,* true ferns)
      Order 4. Marsiliales (water clovers, *Marsilea*)
      Order 5. Salviniales (*Salvinia, Azolla,* water ferns)
  PHYLUM 5. Cycadophyta (cycads, *Zamia, Cycas*)
  PHYLUM 6. Ginkgophyta (ginkgo)
  PHYLUM 7. Coniferophyta (conifers)
    Order 1. Taxales (yews, *Taxus*)
    Order 2. Pinales (pines)
  PHYLUM 8. Gnetophyta
    Order 1. Ephedrales (*Ephedra*)
    Order 2. Welwitschiales (*Welwitschia*)
    Order 3. Gnetales (*Gnetum*)
  PHYLUM 9. Angiospermophyta (Anthophyta, Magnoliophyta, flowering plants) [g]
   Class 1. Monocotyledoneae (Liliatae) (monocots)
   Subclass 1. Alismatidae
    Order 1. Alismatales
      Family Alismataceae (water plantain, arrowhead)
    Order 2. Hydrocharitales
      Family Hydrocharitaceae (elodea, tape grass)
    Order 3. Najadales
      Family Potamogetonaceae (pondweed)
      Family Najadaceae (*Najas, Caulinia,* naiads or water nymphs)
      Family Zosteraceae (Cymodoceacea, Posidoniaceae, eel grass)
    Order 4. Triuriae
   Subclass 2. Commelinidae
    Order 1. Commelinales
      Family Commelinaceae (*Commelina, Tradescantia,* dayflowers and spiderworts)
    Order 2. Eriocaulales
      Family Eriocaulaceae (pipewort)
    Order 3. Restionales
    Order 4. Juncales
      Family Juncaceae (rushes)

Table 6. Continued

KINGDOM PLANTAE

Order 5. Cyperales
    Family Cyperaceae (sedges)
    Family Gramineae (grasses, *Oryza,* [d] *Triticum,* [d] *Zea,* [d] *Hordeum* [d])
Order 6. Typhales
    Family Sparganiaceae (bur-reed)
    Family Typhaceae (cattails)
Order 7. Bromeliales
    Family Bromeliaceae (pineapple, Spanish moss, bromeliads)
Order 8. Zingiberales
    Family Strelitziaceae (bird-of-paradise plants)
    Family Musaceae (banana)
    Family Zingiberaceae (ginger, cardomum)
    Family Cannaceae (cannas)
Subclass 3. Arecidae
  Order 1. Arecales
    Family Palmae (Arecaceae) (palms)
  Order 2. Arales
    Family Araceae (arum, philodendron, jack-in-the-pulpit, skunk cabbage)
    Family Lemnaceae (duckweed)
Subclass 4. Liliidae
  Order 1. Liliales
    Family Pontederiaceae (pickerel weeds)
    Family Liliaceae (lilies, *Colchicum*)
    Family Iridaceae (irises)
    Family Agavaceae (century plants)
    Family Smilacaceae (bull briar)
    Family Dioscoreaceae (yams)
  Order 2. Orchidales
    Family Orchidaceae (Apostasiaceae) (orchids)
Class 2. Dicotyledoneae (Magnoliatae) (dicots)
  Subclass 1. Magnoliidae (L) [h]
  Order 1. Magnoliales
    Family Magnoliaceae (magnolia, tulip tree)
    Family Winteraceae (winter-bark tree)
    Family Annonaceae (pawpaw, custard apple)
    Family Myristicaceae (nutmeg)
    Family Calycanthaceae (spicebush)
    Family Lauraceae (Cassythaceae) (bay, laurels, sassafras, avocados)
  Order 2. Piperales (H) [h]
    Family Piperaceae (black pepper)
  Order 3. Aristolochiales (H)
    Family Aristolochiaceae (wild ginger, Dutchman's pipes)

Table 6.   Continued

KINGDOM PLANTAE

Order 4. Nymphaeales (H)
  Family Nymphaeaceae (yellow and white water lilies, *Nymphaea, Nuphar*)
  Family Cabombaceae (*Cabomba, Brasenia,* water shields)
  Family Nelumbonaceae (*Nelumbo,* lotus)
  Family Barclayaceae (*Barclaya*)
  Family Euryalaceae (*Euryale, Victoria*)
  Family Ceratophyllaceae (hornworts)
Order 5. Ranunculales (H)
  Family Ranunculaceae (buttercups)
  Family Berberidaceae (Nandinaceae, barberries, *Podophyllum*)
  Family Lardizabalaceae (Sargentodoxaceae, Japanese fiveleaf)
  Family Menispermaceae (moonseeds)
Order 6. Papaverales (H)
  Family Papaveraceae (poppies)
  Family Fumariaceae (Hypecoaceae) (bleeding hearts)
Subclass 2. Hamamelidae (L)
  Order 1. Trochodendrales
  Order 2. Hamamelidales
    Family Cercidiphyllaceae (katsura trees)
    Family Platanaceae (sycamores)
    Family Hamamelidaceae (witch hazels)
  Order 3. Eucommiales
  Order 4. Urticales
    Family Ulmaceae (elms)
    Family Moraceae (figs, mulberries)
    Family Cannabaceae (hemp, marijuana)
    Family Urticaceae (stinging nettle, ramie)
  Order 5. Leitneriales (corkwood)
  Order 6. Juglandales
    Family Juglandaceae (walnuts, hickories)
  Order 7. Myricales
    Family Myricaceae (wax myrtle)
  Order 8. Fagales
    Family Fagaceae (beeches, oaks)
    Family Betulaceae (Corylaceae) (birches, alders)
  Order 9. Casuarinales
    Family Casuarinaceae (beefwood or Australian pine)
Subclass 3. Caryophyllidae (H) (or Centrospermae)
  Order 1. Caryophyllales
    Family Phytolaccaceae (pokeweeds)
    Family Nyctaginaceae (four o'clocks, *Mirabilis*[d])
    Family Cactaceae (cacti)
    Family Aizoaceae (ice plants)
    Family Caryophyllaceae (pinks)

Table 6. Continued

KINGDOM PLANTAE

Family Portulacaceae (purslanes)
Family Basellaceae (Madeira vine, Malabar spinach)
Family Chenopodiaceae (beet, spinach, goosefoot)
Family Amaranthaceae (pigweed, tumbleweed)
Order 2. Batales (*Batis*)
Order 3. Polygonales
Family Polygonaceae (buckwheat, rhubarb)
Order 4. Plumbaginales
Family Plumbaginaceae (leadworts)
Subclass 4. Dilleniidae (L)
Order 1. Dilleniales
Family Paeoniaceae (paeonies)
Order 2. Theales
Family Dipterocarpaceae
Family Theaceae (tea)
Family Actinidiaceae (Saurauiaceae) (kiwi berries or New Zealand gooseberries)
Family Guttiferae (Clusiaceae, Hypericaceae) (St. John's worts)
Order 3. Malvales
Family Tiliaceae (basswoods)
Family Sterculiaceae (Byttneriaceae) (cocoa)
Family Bombacaceae (kapok, balsa)
Family Malvaceae (*Gossypium*^{d} or cotton, okra, *Hibiscus*)
Order 4. Lecythidales
Order 5. Sarraceniales (H)
Family Sarraceniaceae (pitcher plant)
Family Nepenthaceae (pitcher-plant vine)
Family Droseraceae (sundews)
Order 6. Violales (Parietales, Bixales, Cistales)
Family Violaceae (violets)
Family Passifloraceae (passion flower)
Family Cistaceae (rock roses)
Family Tamaricaceae (tamarisk)
Family Fouquieriaceae (ocotillo)
Family Caricaceae (papaya)
Family Loasaceae (evening star or blazing star)
Family Begoniaceae (begonias)
Family Cucurbitaceae (*Cucurbita*^{d} or squash, *Cucumis,*^{d} or cucumber, melons)
Order 7. Salicales
Family Salicaceae (willows)
Order 8. Capparales
Family Capparaceae (Koeberliniaceae, Pentadiplandraceae) (capers, spider plants)
Family Cruciferae (Brassicaceae) [mustards, cabbage, radish, *Arabidopsis*^{d} (H)]

Table 6. Continued

KINGDOM PLANTAE

Order 9. Ericales
  Family Clethraceae (sweet pepperbush)
  Family Ericaceae; Vacciniaceae (blueberry, heaths, rhododendrons)
  Family Empetraceae (crowberries)
  Family Pyrolaceae (wintergreens)
  Family Monotropaceae (Indian pipe, pine drops)
Order 10. Diapensiales
  Family Diapensiaceae (*Diapensa, Shortia* or Oconocee bells)
Order 11. Ebenales
  Family Sapotaceae; Sarcospermataceae (star apple, sapodilla, chicle)
  Family Ebenaceae (persimmon, ebony)
  Family Styracaceae (styrax, snowdrop tree)
Order 12. Primulales (H)
  Family Myrsinaceae (Aegicerataceae) (Marlberry or *Ardisia*)
  Family Primulaceae (Coridaceae) (primroses, shooting stars, loose-
    strifes)
Subclass 5. Rosidae (L)
Order 1. Rosales
  Family Hydrangeaceae (Philadelphaceae) (hydrangeas)
  Family Grossulariaceae (currants, gooseberries)
  Family Crassulaceae (stonecrops)
  Family Saxifragaceae (saxifrages)
  Family Rosaceae (roses)
  Family Leguminosae (mimosas or sensitive plant, peas *Pisum*[d], beans
    Phaseolus)
Order 2. Podostemales (H)
  Family Podostemaceae (riverweed)
Order 3. Haloragales
  Family Haloragaceae (water milfoils)
  Family Gunneraceae (*Gunnera*)
Order 4. Myrtales
  Family Lythraceae (*Lythrum,* crepe myrtle)
  Family Thymelaeaceae (leatherwood, mezereum)
  Family Trapaceae (Hydrocaryaceae, water chestnut family)
  Family Myrtaceae (Heteropyxidaceae) (myrtle, *Eucalyptus* or gum
    trees)
  Family Punicaceae (pomegranates)
  Family Onagraceae (evening primroses, *Oenothera*[d]) (H)
  Family Melastomataceae (meadow beauties)
  Family Combretaceae (white mangrove)
Order 5. Proteales
  Family Elaegnaceae (oleaster, silverberries)
  Family Proteaceae (silk oak, macadamia nut)
Order 6. Cornales
  Family Rhizophoraceae (mangrove)

Table 6.   Continued

KINGDOM PLANTAE

    Family Nyssaceae (sour gum or tupelo)
    Family Cornaceae (Masrixiaceae, Torricelliaceae, dogwoods)
    Family Garryaceae (silk-tassel bush)
  Order 7. Santales
    Family Santalaceae (sandalwoods)
    Family Loranthaceae (mistletoe)
  Order 8. Rafflesiales (*Rafflesia, Pilostyles*)
  Order 9. Celastrales
    Family Celastraceae (climbing Bittersweet, strafftree, strawberry
      bush)
  Order 10. Euphorbiales
    Family Buxaceae (boxwoods)
    Family Euphorbiaceae (spurges)
  Order 11. Rhamnales
    Family Rhamnaceae (*Ceanothus,* mountain lilac, buckthorn, jujube)
    Family Vitaceae (Ampelidaceae) (grapes)
  Order 12. Sapindales
    Family Staphyleaceae (bladdernut)
    Family Sapindaceae (soapberries, balloon vines, lychees)
    Family Hippocastanaceae (horse chestnuts, buckeyes)
    Family Aceraceae (maples)
    Family Burseraceae (gumbo limbo, myrrh, frankincense)
    Family Anacardiaceae (poison ivy, mangos, cashews, pistachios)
    Family Simaroubaceae (tree-of-heaven or *Ailanthus*)
    Family Rutaceae (citrus, rue)
    Family Meliaceae (chinaberry tree)
    Family Zygophyllaceae (caltrop or puncture vine, creosote bush)
  Order 13. Geraniales (H)
    Family Oxalidaceae (wood sorrels)
    Family Geraniaceae (geraniums)
    Family Limnanthaceae (meadow foam)
    Family Tropaeolaceae (nasturtiums)
    Family Balsaminaceae (touch-me-nots, balsams)
  Order 14. Linales
    Family Erythroxylaceae (coca)
    Family Linaceae (flax)
  Order 15. Polygalales
    Family Polygalaceae; Diclidantheraceae (milkworts)
  Order 16. Umbellales (H).
    Family Araliaceae (ginsengs)
    Family Umbelliferae (parsley, celery, anise, carrots)
 Subclass 6. Asteridae (H)
  Order 1. Gentianales
    Family Loganiaceae (strychnine, curare, jessamine)
    Family Gentianaceae (gentians)

Table 6. Continued

KINGDOM PLANTAE

Family Apocynaceae (Plocospermaceae) (dogbane, *Vinca*, or periwinkles)
Family Asclepiadaceae (Periplocaceae) (milkweeds)
Order 2. Polemoniales
Family Solanaceae (*Lycopersicon*[d] or tomato, potato, eggplant, red peppers, petunias, *Nicotiana*[d] or tobacco, *Datura*[d])
Family Convolvulaceae (morning glories, sweet potatoes)
Family Cuscutaceae (dodders)
Family Polemoniaceae (Cobaeaceae) (phlox)
Family Hydrophyllaceae (waterleaf, *Phacelia*)
Order 3. Lamiales
Family Boraginaceae (borages, heliotrope, forget-me-nots)
Family Verbenaceae (verbenas)
Family Labiatae (mints)
Order 4. Plantaginales
Family Plantaginaceae (plantain)
Order 5. Scrophulariales (Personales, figworts)
Family Buddlejaceae (butterfly bush)
Family Oleaceae (lilac, olive, privet)
Family Scrophulariaceae (snapdragons, pentstemons, *Antirrhinum*,[d] *Collinsia*[d])
Family Gesneriaceae (African violets)
Family Orobanchaceae (beech drops, broom rape)
Family Bignoniaceae (trumpet vine)
Family Acanthaceae (*Acanthus*, shrimp plant)
Family Pedaliaceae (*Sesamum*, or sesame seed)
Family Lentibulariaceae (bladderworts)
Order 6. Campanulales
Family Campanulaceae (Lobeliaceae) (bellflowers)
Order 7. Rubiales
Family Rubiaceae (coffee, quinine, madder, bedstraw)
Order 8. Dipsacales
Family Caprifoliaceae (honeysuckles, twinberries)
Family Adoxaceae (moschatel)
Family Valerianaceae (valerians)
Family Dipsacaceae (teasel)
Order 9. Asterales
Family Compositae (sunflowers, daisies, *Eupatorium*, *Solidago*, *Haplopappus*[d])

---

[a] Modified from Whittaker, 1969. Living groups only
[b] Whitehouse, 1969; see Table 8 for viruses
[c] Letters (A,M,P,F) refer to genomes identified in Figure 4 and explained in Table 5.
[d] Favorite species for genetic studies.
[e] Whittaker, 1969.
[f] Copeland, 1956.

composed of cells with membrane-bounded nuclei which contain chromosomes complexed with RNA and histone and other proteins, and they follow the patterns of Mendelian inheritance based on sexual reproduction via meiosis and fertilization (i.e., they are eukaryotes). Table 4 summarizes the differences between these two basic cell types.

The classification which will next be presented is system 7 of Table 2, a modification of Whittaker's five-kingdom plan. It is summarized in Table 3 and expanded in. Table 6. Unlike the classical two- or three-kingdom scheme (systems 1–4 in Table 2), systems 5, 6, and 7 are consistent with the cell symbiosis theory (Margulis, 1970). This theory proposes that prokaryotic organisms originated and diversified during the early and middle Precambrian eon (from 3 to 1.5 × 10⁹ years ago) and that the eukaryotes arose as products of a series of specific symbioses during the middle to late Precambrian (from 2 to 0.6 × 10⁹ years ago) in which one kind of cell became an intracellular, self-reproducing inhabitant of another kind of cell. For clarification a brief resume of the cell symbiosis theory follows, yet it must be realized that hypotheses of this sort are currently being tested, and it may be premature to use them as a basis for phylogenies of the lower organisms. For partial phylogenies and a taxonomy based on the nonsymbiotic view, for example, see Klein and Cronquist (1967).

All terrestrial living systems contain certain organic constituents and a minimal amount of DNA-coded hereditary information. As listed in Table 7, the minimal number of biochemical components (enzymes, nucleic acids, etc.) necessary for free-living replication has been estimated to be about 50 (Morowitz, 1967). At least two classes of organelles of eukaryotic cells, chloroplasts and mitochondria (Cohen, 1970), fulfill "minimal self-replicating criteria," i.e., contain most of those items listed in Table 7. I have postulated that these items are present inside chloroplasts and mitochondria precisely because these organelles originated as free-living self-replicating cells (Margulis, 1970). That is, certain free-living prokaryotes are hypothesized to have later become obligate symbionts in the population of complex cellular entities ancestral to all of the organisms in the four eukaryote kingdoms: Protoctista, Plantae, Animalia, and Fungi. These concepts are outlined in Figure 4 and in Table 5.

The earliest Precambrian microorganisms were by inference anaerobic heterotrophs, and probably fermentative bacteria, and it is

---

[g] Only major families included. See Cronquist, (1968); Joly (1966), and Hutchinson (1969) for more complete listings.

[h] (L) are recognized by Hutchinson as groups which are fundamentally woody dicots, Lignoseae; groups labeled (H) are considered basically herbaceous subclasses, Herbaceae see J. Hutchinson, 1959, 1969). All orders within a subclass are woody or herbaceous as noted, unless otherwise indicated.

TABLE 7.   *The Minimum Biochemical Components Found in a Self-Replicating Cellular System*[a]

DNA (of at least 26×10^6 daltons molecular weight)

Messenger RNA colinear with the DNA

Polymerase enzymes: DNA polymerase and DNA-dependent RNA polymerase

Approximately 20 different activating enzymes, one each for each of the 20 amino acids

Approximately 20 different transfer RNA molecules

Ribosome complexes: ribosomal RNA's, ribosomal proteins and protein synthesis factors required for incorporation of amino acids into protein

Lipoprotein membranes, active transport to maintain optimal concentration of K^+, Mg^{++}, Ca^{++}, and Na^+ ions

A fermentable carbohydrate or other source of energy resulting in ATP production

Particular product or activity selected for by the environment to perpetuate replicating system (ability to oxidize volatile amines or abiotically produced amino acids or sugars)

---

[a]Modified from Morowitz, 1967; the calculated minimal size required for a sphere which could contain these components is approximately 0.1 μ in diameter. The smallest known cells are about a factor of ten larger than this.

assumed their source of food was abiotically produced organic matter (Sylvester-Bradley, 1971). These prokaryotes eventually gave rise to an enormous range of bacterial cells, among them spirochetes, which are motile heterotrophs, and photosynthetic cells. Blue-green algae are essentially highly evolved, complex bacterial photosynthesizers (Echlin and Morris, 1965). Mutations of photosynthetic bacteria which permitted the use of water instead of organic hydrogen, molecular hydrogen, or hydrogen sulfide as hydrogen donors in CO_2 reduction during photosynthesis eventually led to cells that eliminated molecular oxygen into the atmosphere. The blue-green algae arose from these, and they intensively speciated and flourished during the Precambrian eon. A highly diversified moneran biota then existed, setting the stage for evolution of eukaryotes.

The production of oxygen forced the development of oxygen tolerance in many of the prokaryotic microbes. In some populations of microoganisms, the evolution of oxygen toleration was followed by the evolution of systems for oxygen utilization. Microaerophilic organisms (those requiring O_2 in quantities less than in the ambient atmosphere) and aerobic bacteria evolved, including a population of bacteria thought to be "protomitochondria." The symbiotic theory states that first such protomitochondria invaded an anaerobic host to form a stable symbiotic com-

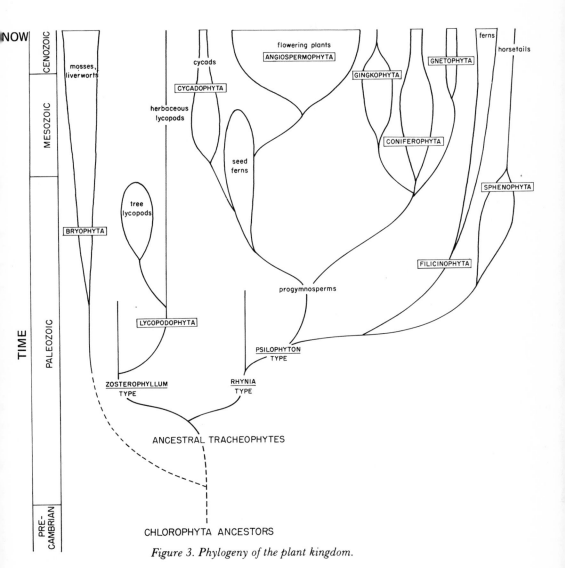

Figure 3. Phylogeny of the plant kingdom.

plex; the primitive amoebae.* Secondly, some of these primitive amoebae became amoeboflagellates when they acquired still another set of symbionts which evolved into the eukaryote flagella.

The common ancestors of all mitotic eukaryotes were thus amoeboflagellate heterotrophs, cells that are thought to be trigenomic, i.e., products of three endocellular symbioses: the protoameboid host (A) plus the protomitochondrial (M) and protoflagellar (F) symbionts (see Figure

* This "invasion" is thought to be analogous to the invasion by *Bdellovibrio* of its bacterial hosts (see Margulis, 1974a; Taylor, 1974 for discussion).

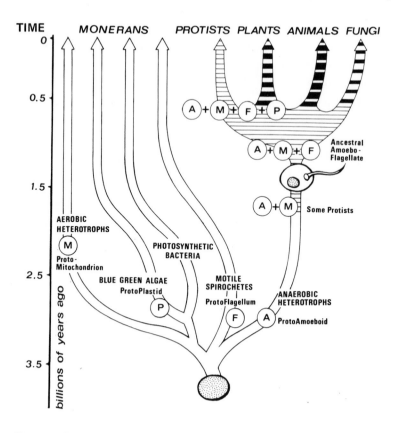

Figure 4. Phylogeny of the five kingdoms based on cell symbiosis theory (see Table 5).

4). Like eukaryotes today, the ancestral eukaryotes were fundamentally aerobic; they depended on many oxygen-mediated metabolic steps including the biosynthesis of polyunsaturated fatty acids and steroids. It was in the ancestral eukaryote population that mitosis and meiosis evolved, giving rise to the higher nucleated algae and plants, the fungi and the metazoans. During the evolution of mitosis and eventually meiosis, some eukaryotes acquired photosynthesis by a further cellular symbiosis, incorporating into their cells prokaryotic algae (such as blue-greens) which, with time, became the plastids $\textcircled{P}$. This scheme implies all nucleate algae and multicellular plants are tetragenomic, i.e., they contain a total of the 4 original genomes abbreviated in Figure 4 by the letters $\textcircled{A}$ (amoeboid host), $\textcircled{M}$ (mitochondria), $\textcircled{F}$ (flagellum), and $\textcircled{P}$ (plastid).

This symbiosis theory is consistent with the fact that over the vast stretches of Precambrian time the earth teemed with microbial, rather than macroscopic life (Schopf, 1972). It utilizes the concept that the metabolism of blue-green algae, during the middle Precambrian eon was the major cause of the transition to an oxygen-rich atmosphere. The blue-green algae differ radically from all other algae and higher plants, except in their photosynthetic metabolism. Not until oxygen was widespread in the terrestrial atmosphere did eukaryotic cells evolve. Not until the elegant mitotic mechanism of eukaryote chromosome segregation evolved in a series of protists could the fungi, plants, and animals diversify (Margulis, 1970). When it is acknowledged that the adaptive radiation leading to mitosis and meiosis resulted in the origin of miscellaneous protist forms such as dinoflagellates, radiolarians, euglenids, cryptomonads, and others, these organisms can be graphed on a consistent phylogenetic diagram (Figure 5). The presumed history and the characteristics of the five kingdoms have been summarized in Table 3. The phylogenies of the five

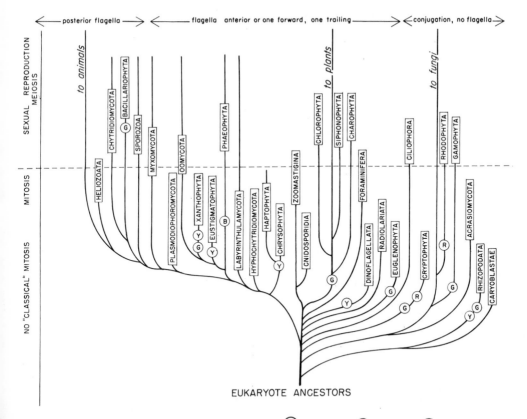

Figure 5. *Phylogeny of the protist kingdom. Plastid color:* Ⓨ *= yellow,* Ⓖ *= green,* Ⓑ *= blue-green or red.*

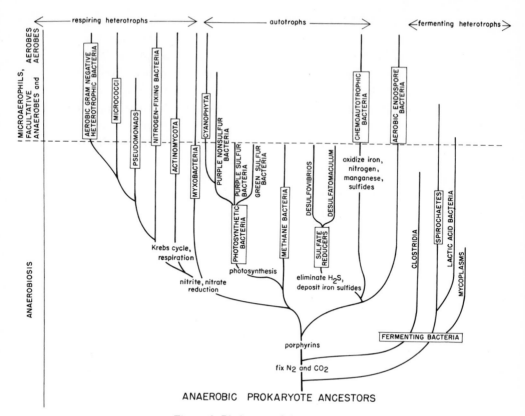

Figure 6. Phylogeny of the monera kingdom.

kingdoms according to these ideas are graphed in Figures 2 (animals), 3 (plants), 5 (protists), 6 (monera), and 7 (fungi).

The classification system of Table 6 is based on the evolutionary views of the symbiotic theory; that is, system 6 (Whittaker, 1969) is accepted with minor modification, system 7 (Margulis, 1971). Whittaker bases his three multicellular eukaryotic kingdoms on concepts of nutrition and ecological role: the plants are photoautotrophic primary producers, the animals are heterotrophic consumers, and the fungi are absorptive saprophytes. However, Whittaker, like Copeland (1956) and Dodson (1971) among others, has wrestled with the problem of "blurred boundaries" between these three kingdoms and the protists from which they presumably arose. Colonial and multicellular organization has evolved convergently in many groups; certainly in no modern classification can all protists be considered single-celled (cf. the volvocine algae or the colonial ciliates). The solution to this dilemma involves the acceptance of Whittaker's basic scheme but the close circumscription of the fungal, animal, and plant kingdoms. Thus the fungal kingdom here excludes all of the flagellated fungi and the amoeboid forms with a variety of life cycles, and it therefore contains only the presumed homologous series of amastigote (flagellaless) haploid or dikaryotic forms. The animal kingdom is limited to those multicellular organisms which develop from a blastula (metazoans), and the plant kingdom comprises only embryophytes, multicellular green plants (land plants). The protists then become that remaining miscellaneous assemblage of phyla in which profound "evolutionary experimentation" occurred (Margulis, 1974a). Concomitant with their diversification, mitosis and Mendelian genetic patterns of inheritance evolved and became stabilized. The name "protist," which implies single-celled forms, should be formally replaced by "protoctist," as Copeland (1956) so cogently recognized.

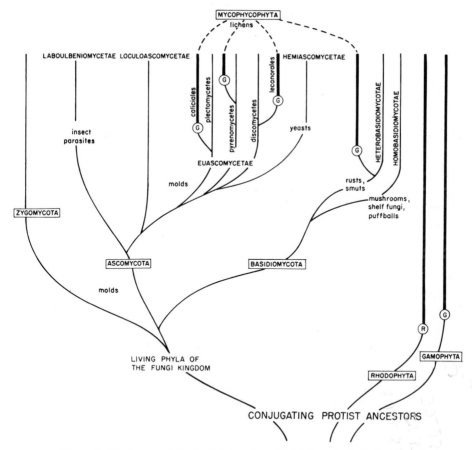

Figure 7. Phylogeny of the fungal kingdom. Thick lines indicate functionally photosynthetic organisms.

TABLE 8. Viruses

Ribovira (RNA viruses)	Ribohelica (nucleocapsid with helical symmetry)	Tobacco mosaic virus, myxovirus (influenza A), pea mosaic virus, ribgrass virus
	Ribocubica (nucleocapsid with cubical symmetry)	Napoviridae (turnip yellow mosaic virus, polio virus, Coxsackie 1, Echovirus, RNA phage)
		Reoviridae (reovirus, neovirus)
		Arboviridae (Arbovirus)
Deoxyvira (DNA viruses)	Deoxyhelica (nucleocapsid with helical symmetry)	Smallpox virus, fowl pox, rabbit myxoma cyanophages
	Deoxycubica (nucleocapsid with cubical symmetry)	(Phage ϕX174, Shope papilloma, polyoma, adenovirus, herpes simplex)
	Deoxybinala (with head and tail)	(Bacteriophages: T phages and λ, 434, ϕ80, and P22)

The viruses, which are not cells, are incapable of autonomous replication. Since they are best thought of as parts of organisms, each virus group would thus be classified in the appropriate taxa of its host. Since this would be unwieldy, a listing of viruses (based on the classification of Lwoff and Tournier, 1966; Whitehouse, 1969) is shown in Table 8.

It should be understood that names of certain groups mentioned here have been improvised, and this has often been done because many of the microorganisms have been considered in the province of both zoologists and botanists. Therefore, traditional names for these taxa have been inconsistently applied, and those in Table 6 have not (yet) been accepted by the international codes of botanical and zoological nomenclature. With the gradual acceptance of five kingdoms, it is anticipated that the contradictions in the classifications of the monerans and protists will eventually be resolved.

Acknowledgment

I am indebted to many systematic biologists whose work has been referred to, especially in Table 6. In some cases the referenced published schemes have been modified based on suggestions and comments by my colleagues. I am particularly grateful to S. Banerjee, E. S. Barghoorn, H. P. Banks, H. C. Bold, H. Booke, B. Cameron, G. Carroll, J. Corliss, J. A. Doyle, S. Duncan, A. Echternacht, S. Golubic, W. Hartman, A. Humes, G. E. Hutchinson, G. Leedale, N. D. Levine, L. S. Olive, G. G. Simpson, J. Staley, T. Varghese, R. Norris, and R. H. Whittaker. I thank NASA (NGR 22-004-025) for support, and the Boston University Graduate School.

Literature Cited

Alexopoulos, C. J. 1962 *Introductory Mycology*, second edition, John Wiley & Sons, New York.

Altman, P. L. and D. S. Dittmer, editors, 1972 *Biology Data Book*, Federation of American Societies for Experimental Biology, Bethesda, Md.

Arnold, C. A., 1949 *Introduction to Paleobotany*, McGraw-Hill Book Co., New York.

Banks, H. P., 1970a Major evolutionary events and the geological record of plants. *Biol. Rev. (Camb.)* **45**:451–454.

Banks, H. P., 1970b *Evolution and Plants of the Past*, Wadsworth Publishing Co., Belmont, Calif.

Banks, H. P., 1972a *Evolutionary History of Plants*, Wadsworth Publishing Co., Belmont, Calif.

Banks, H. P., 1972b The stratigraphic occurrence of early land plants. *Palaeontology* **15**:365–397.

Blackwelder, R. E., 1967 *Taxonomy*, John Wiley & Sons, New York.

*Bold, H. C., 1973 *Morphology of Plants*, fourth edition, Harper and Row, New York.

*Brock, T. D., 1970 *Biology of Microorganisms*. Prentice-Hall, Englewood Cliffs, N.J.

*Campbell, L. L. and J. R. Postgate, 1965 Classification of the spore forming sulfate reducing bacteria. *Bacteriol. Rev.* **29**:359–363.

Cohen, S. S., 1970 Are/were mitochondria and chloroplasts microorganisms? *Am. Sci.* **58**:281–289.

Corliss, J. O., 1974 The changing world of ciliate systematics. *System. Zool.* **23**:91–138.

Copeland, H. F., 1956 *Classification of the Lower Organisms*, Pacific Books, Palo Alto, Calif.

Cronquist, A., 1968 *Evolution and Classification of Flowering Plants*, Houghton Mifflin, Boston.

*Cronquist, A., 1972 *Introductory Botany*, third edition, pp. 365–374, Harper and Row, New York.

Curtis, H., 1968 *Biology*, Worth Publishers, New York.

Dayhoff, M. O., 1972 *Atlas of Protein Sequence and Structure*, National Biomedical Research Organization, Bethesda, Md.

Dodson, E. O., 1971 The kingdoms of organisms. *Syst. Zool.* **20**:265–281.

Echlin, P. and I. Morris, 1965 The relationship between blue-green algae and bacteria. *Biol. Rev.* **40**:143–187.

*Fritsch, F. E., 1935 *The Structure and Reproduction of the Algae*, Vol. 1, Cambridge University Press, Cambridge, England.

Glaessner, M. F., 1968   Biological events and the precambrian time scale. *Can. J. Earth Sci.* **5**:585–590.

Grant, V., 1971   *Plant Speciation,* Columbia University Press, New York.

*Greenwood, P. H., D. E. Rosen, S. H. Weitzman and G. S. Mayer, 1966   Fishes. *Bull. Am. Mus. Nat. Hist.* **131**:345–354.

*Grell, K. G. 1973   *Trichoplax adhaerens* and the origin of metazoa. *Proc. 4th Int. Congr. Protozool., Clermont-Ferrand, France, September, 1973.*

*Hale, M. E., Jr., 1967   *The Biology of Lichens,* Edward Arnold (Publishers) Ltd., London.

*Honigberg, B. M., W. Balamuth, E. C. Bovee, J. O. Corliss, M. Godjics, R. O. Hall, R. R. Kudo, N. D. Levine, A. R. Loeblich, Jr., J. Weiser and D. H. Wenrich, 1964   A revised classification of the phylum Protozoa. *J. Protozool.* **11**:7–20.

*Hutchinson, G. E., 1967   *Treatise on Limnology,* Vol. 2, John Wiley & Sons, New York.

Hutchinson, J., 1959   *The Families of Flowering Plants,* Vol. 1, *Dicotyledons,* second edition, Clarendon Press, Oxford.

Hutchinson, J., 1969   *Evolution and Phylogeny of Flowering Plants. Dicotyledons: Facts and Theory,* Academic Press, New York.

Joly, A. B., 1966   *Botanica: Introdução à Taxonomia Vegetal,* Companhia Editora Nacional, Editôra da Universidade de São Paulo, São Paulo, Brasil.

Lerner, I. M., 1968   *Heredity, Evolution and Society,* Freeman Co., San Francisco, Calif.

Lwoff, A. and M. Tournier, 1966   Classification of viruses. *Ann. Rev. Microbiol.* **20**:45–74.

Klein, R. M. and A. Cronquist, 1967   A consideration of the evolutionary and taxonomic significance of some biochemical, micromorphological and physiological characters in the Thallophyta. *Q. Rev. Biol.* **42**:105–296.

Margulis, L., 1970   *Origin of Eukaryotic Cells,* Yale University Press, New Haven, Conn.

Margulis, L., 1971   Whittaker's five kingdoms: minor modifications based on considerations of the origins of mitosis. *Evolution* **25**:242–245.

Margulis, L., 1974a   On the origin and possible mechanism of colchicine-sensitive mitotic movements. *Biosystems* **6**:16–36.

Margulis, L., 1974b   Five-kingdom classification and the origin and evolution of cells. In *Evolutionary Biology,* Vol. 7, edited by W. Steere, M. Hecht, and T. Dobzhansky, Plenum Press, New York.

Mayr, E., 1970   *Populations, Species and Evolution,* Harvard University Press, Cambridge, Mass.

Morowitz, H. J., 1967   Biological self-replicating systems. *Prog. Theoret. Biol.* **1**:35–58.

*Olive, L. S., 1970   The Mycetozoa: A revised classification. *Bot. Rev.* **36**:59–89.

Romer, A. S., 1968   *The Procession of Life,* World Publishing Co., Cleveland (1972 Anchor Books Edition).

Schopf, J. W., 1972   Precambrian paleobiology. In *Exobiology,* edited by C. Ponnamperuma and R. Buvet, pp. 16–61, North Holland Publishing, Amsterdam.

*Schulthorpe, C. D., 1967   *The Biology of Aquatic Vascular Plants,* Edward Arnold (Publishers) Ltd., London.

*Simpson, G. G., 1953   *Major Features of Evolution,* Columbia University Press, New York.

Simpson, G. G., 1954   *The Meaning of Evolution,* Harper and Row, New York.

*Simpson, G. G., 1961   *Principles of Animal Taxonomy,* Columbia University Press, New York.

Simpson, G. G. and W. S. Beck, 1965   *Life, An Introduction to Biology,* second edition, pp. 492–493, Harcourt, Brace, Ivanovich, New York.

Stanier, R., E. Adelberg and M. Douderoff, 1970   *The Microbial World,* third edition, Prentice-Hall, Englewood Cliffs, N.J.

Sylvester-Bradley, P., 1971   Carbonaceous chondrites and the prebiological origin of food. In *Molecular Evolution,* edited by R. Buvet and C. Ponnamperuma, pp. 499–504. North Holland Publishing Co., Amsterdam.

Taylor, F. J. R., 1974   Implications and extensions of the serial endosymbiosis theory of the origin of eukaryotes. *Taxon* **23:**5–34.

Whitehouse, H. L. K., 1969   *Towards an Understanding of the Mechanism of Heredity,* second edition, St. Martin's Press, New York.

Whittaker, R. H., 1969   New concepts of the kingdoms of organisms. *Science (Wash., D.C.)* **163:**150–160.

* These references were utilized in the development of the classification system presented in the text.

PART A
THE BACTERIA

2

Diplococcus pneumoniae

ELENA OTTOLENGHI-NIGHTINGALE

The bacterium *Diplococcus pneumoniae* was first described less than a century ago. Following its discovery in 1881 by both Pasteur and Sternberg and the demonstration by Fraenkel and Weichselbaum five years later that the *Pneumococcus* was the causative agent of lobar pneumonia in man, interest in this organism had centered on its pathogenic properties and on the defense of the host against them (Breed *et al.*, 1957). Griffith's work on pneumococcal transformations (Griffith, 1928) led to the discovery that the chemical nature of the genetic material for all cellular forms of life is deoxyribonucleate. The classic study of Avery, MacLeod, and McCarty on the nature of the substance inducing transformation of pneumococcal types was published in 1944, and it marked the beginning of modern genetics and of molecular biology.

Biology of *Diplococcus pneumoniae*

Morphology

Diplococcus pneumoniae, or *Pneumococcus,* is a true bacterium (order: Eubacteriales; family: Lactobacillaceae; tribe: Streptococceae). This parasitic organism inhabits the respiratory tract of man and certain other mammals. In fluid media, the bacteria grow singly or in pairs and attain a size of 0.5–1.25 μ. Individual cells are almost spherical. The

ELENA OTTOLENGHI-NIGHTINGALE—University Affiliated Center for Child Development, Georgetown University Hospital, Washington D.C.

distal ends of paired forms are slightly pointed, and therefore they are described as *lancet*-shaped *diplo*cocci. When dried, heat-fixed smears of young pneumococci are stained by the procedure of Gram, the cocci appear purple and are designated Gram-positive. If grown on the surface of solid (agar) media, pneumococci form round, shiny colonies easily visible to the naked eye. When defibrinated blood from various mammals such as rabbit, sheep, or horse is included in the medium, pneumococcal colonies are surrounded by a zone of green discoloration due to partial hemolysis of the erythrocytes. The green discoloration, α-hemolysis, is produced by an acidic protein, pneumolysin, elaborated during growth of pneumococci (Shumway and Klebanoff, 1971).

The outermost envelope or capsule of pneumococci is composed of polysaccharide polymers. More than 75 serologically distinct types of capsular polysaccharide have been described. It is the polysaccharide capsule which, by protecting the *Pneumococcus* from phagocytosis, endows this bacterium with pathogenicity, or the ability to cause disease. Encapsulated bacteria form smooth, glistening colonies on agar media. Unencapsulated bacteria form rough colonies and are not pathogenic.

Pneumococci have a cell wall whose teichoic acid includes an unusual component, choline, which may play a role in the ability of these bacteria to be transformed (Tomasz, 1968). The fine structure of pneumococci has been described (Tomasz *et al.*, 1964) and is similar to that of other Gram-positive cocci.

Life Cycle

Pneumococci replicate by binary fission. Some of the details of cell wall synthesis during division have been elucidated (Briles and Tomasz, 1970). A population of pneumococci in fluid medium is composed of single cells, diplococci, and short chains. For the first half hour after inoculation, the population size remains constant (lag phase). Then, the population increases logarithmically for 6 to 8 hours, following which time the cell number remains about the same for several hours (stationary phase). If incubation continues, the cell count declines (see below).

The different parts of the growth cycle can be determined by following the optical density of the broth culture or by counting the number of viable bacteria at consecutive time intervals by plating on agar media (see Meynell and Meynell, 1965 for details).

The growth of pneumococci under usual conditions is not synchronous. At any point in time, a sample of the culture contains cells at all stages of the division cycle. Synchrony can be imposed by several techniques, e.g., by changing the temperature of incubation (Hotchkiss, 1954) or by using large inocula from stationary cultures (Cutler and

Evans, 1966). Although it is difficult to assess the age of any particular cell, a population or culture is said to be "young" when it is growing most rapidly, and it begins to "age" at the end of the period of logarithmic growth. Older cultures differ from young cultures in various ways, e.g., affinity for dyes or susceptibility to injurious agents (Dubos, 1949). Pneumococci contain an autolytic enzyme that causes the cells to rupture. An old culture lyses completely within a few hours unless it is preserved by storing at 4°C or less in appropriate media (Dubos, 1937). Lyophilized (freeze-dried) bacteria remain viable indefinitely.

Most pneumococci found in nature are surrounded by a polysaccharide capsule. The polysaccharide is water soluble, and it was formerly called the "specific soluble substance." During the normal growth of encapsulated pneumococci in fluid culture, the capsule size of most of the cells gradually increases during the logarithmic phase of growth and decreases as the culture ages. The decrease in capsule size is due, at least in part, to diffusion of capsule substance. Thus, identification of capsule type is best done during the logarithmic phase of growth (Davis *et al.*, 1967).

Pneumococci are parasitic bacteria, and they require complex media such as meat infusions for optimal growth (see Wilson and Miles, 1964 for growth requirements). Limited growth will also occur in semidefined media such as that of Adams and Roe (MacLeod, 1965). Synthetic media have been reported (Rappaport and Guild, 1959; Sicard, 1964), but the generation time in these media is doubled or tripled, and survival is less than normal. The complex media required by pneumococci make genetic studies involving nutritional or metabolic markers very difficult.

The natural habitat of the *Pneumococcus* is the respiratory tract of man and a few other mammals, such as the rat. Experimental infection of mice or rabbits, however, can be readily achieved. Infectivity varies with the pneumococcal type and with the host (White, 1938). The mouse is extremely sensitive to infection by most pneumococci. Invasion of the murine peritoneal cavity by only one type 3 diplococcus can result in a fatal infection. For this reason, mice are often used as selective agents for encapsulated pneumococci in specimens, such as sputum, which contain a mixed bacterial flora.

Genetics

Zygosity

Pneumococci, like other bacteria, are normally haploid; partial heterozygotes can be obtained by transformation (Hotchkiss *et al.*, 1971). The genetic material (DNA) is loosely bound to protein and to RNA and is confined to a nuclear region which is not bounded by a limiting membrane (Tomasz *et al.*, 1964). Replication occurs by binary fission, so that each daughter cell is an exact replica of the parent. The generation time can be as short as 20 minutes, a feature which is very helpful in genetic studies. Genetic variation can occur by spontaneous mutation or by genetic transformation (see below).

Genetic Transformation

Griffith reported in 1928 that mice injected with heat-killed encapsulated pneumococci plus viable rough pneumococci, died from infection by encapsulated bacteria of the same capsular type as that of the dead cells. Many years later, Griffith's observation of transformation of the avirulent rough pneumococci to the virulent encapsulated form was found to be due to DNA released from the dead bacteria (Avery *et al.*, 1944).

As we understand it now, genetic transformation is a process by which a cell acquires a piece of soluble deoxyribonucleate from the environment and, by recombination, substitutes the segment of exogenous DNA for an equivalent endogenous piece. The transformed cell is permanently altered; its progeny carry the genetic information present in the incorporated piece of DNA.

Transformation is a nonreciprocal mechanism of recombination. The transforming DNA can be obtained by chemical extraction of cell lysates (Avery *et al.*, 1944) or by spontaneous release from actively growing cells (Ottolenghi and Hotchkiss, 1962). In either case, some cells act as donors of genetic material, and others as recipients. The recipient cells exchange a small portion of their genome for the transforming DNA, and the extruded portion in usually lost. Thus, unlike recombination in higher forms of life, transformation is a means of genetic transfer, but not of genetic exchange.

Mutants and Transformable Traits

For several years, specific capsular polysaccharides were the main phenotypic traits amenable to direct study in *Pneumococcus*. With the advent of antibiotics and chemotherapeutic agents, and with the emergence, by mutation, of bacteria resistant to these agents (Hotchkiss, 1951), new classes of characters could be studied by transformation. Selection of transformants on a large background of untransformed cells became feasible, and transformation became a quantitative discipline (Hotchkiss, 1957).

Several pneumococcal characters, aside from capsular polysaccharide synthesis and resistance to antibiotics, have been transformed. Among these are enzymatic activities, colonial morphology, and production of type-specific somatic M protein (see Table 1 and Ravin, 1961). In fact, any recognizable trait can be transformed.

Mechanism of Transformation

The steps in transformation of bacterial cells by DNA have been reviewed (e.g., Hotchkiss and Weiss, 1956; Ravin, 1961; Schaeffer, 1964; Braun, 1965; Spizizen *et al.*, 1966; Hayes, 1968; Tomasz, 1969; Hotchkiss and Gabor, 1970). The recipient population of pneumococci must be in a physiologic state called *competence* in order to accept DNA (Tomasz, 1971). Competence is usually maximal midway in the logarithmic phase of growth and involves the production of a small protein, the competence factor (Tomasz, 1969). When competent cells come in contact with DNA molecules of appropriate homology and size (Cato and Guild 1968; Lacks, 1968), the DNA is irreversibly bound to the cell surface and may subsequently be incorporated. In *Pneumococcus,* the incoming DNA is probably converted to a single-stranded intermediate (Lacks, 1962) which pairs with a corresponding segment of the endogenous DNA. Pairing is followed by integration of the transforming DNA into the cell's genome by molecular recombination (see Hotchkiss, 1971 for review). The corresponding endogenous piece of DNA is discarded. The process of integration and expression of the newly transformed trait may take from two to three generations.

The number of transformants to a given trait is a function of DNA concentration (Hotchkiss, 1957; Litt *et al.*, 1958). It has been calculated that the interaction of a single competent bacterium with one molecule of DNA results in one transformant (Hotchkiss, 1957; Goodgal and Herriott, 1957).

Linkage

Linked factors in pneumococcal DNA were first reported by Hotchkiss and Marmur (1954). They found that wild-type cells transformed by DNA from pneumococci which were resistant to streptomycin and which could utilize mannitol picked up both characters 30 times more frequently than expected on the basis of two independent but simultaneous transformation events (for kinetic analysis of linked recombinations see Kent and Hotchkiss, 1964). Hotchkiss and Marmur con-

TABLE 1. Some Transformable Characters and Mutant Traits of Pneumococcus [a]

Capsular polysaccharide synthesis	Unencapsulated pneumococci can be transformed to any capsule type or, occasionally, to binary capsulation	Austrian et al., 1959
	Encapsulated pneumococci can be transformed to a different type or to capsule deficiency	Ravin, 1959
Colonial morphology		Taylor, 1949; Austrian, 1953
Drug or antibiotic resistance	Penicillin, several levels	Hotchkiss, 1951; Shockley and Hotchkiss, 1970
	Ampicillin, several levels	Butler and Smiley, 1970
	Streptomycin, high and low levels	Hotchkiss, 1951; Schaeffer, 1956; Bryan, 1961
	Sulfanilamide, several levels	Hotchkiss and Marmur, 1954; Hotchkiss and Evans, 1958
	Optochin	Lerman and Tolmach, 1959
	Erythromycin, several levels	Green, 1959; Ravin and Iyer, 1961
	Bryamicin	Marmur and Lane, 1960
	Canavanine	Ephrussi-Taylor, 1957
	Amethopterin	Drew, 1957; Sirotnak et al., 1960a

Category	Trait	Reference
	Tetracycline	Percival et al., 1969
	8-Azaguanine	Sirotnak et al., 1960b
	Azauracil	Ghei and Lacks, 1967
Synthesis of specific enzymes	Mannitol dehydrogenase	Marmur and Hotchkiss, 1955
	Salicin fermentation enzyme	Austrian and Colowick, 1953
	Amylomaltase	Lacks and Hotchkiss, 1960
	Lactic acid oxidase	Ephrussi-Taylor, 1954; Udaka et al., 1959
	Endonuclease deficiency[b]	
	Exonuclease deficiency[b]	Lacks, 1970
	Cell wall lytic enzyme deficiency[b]	
Synthesis of specific proteins (not enzymes)	M protein, a type-specific somatic protein	Austrian and MacLeod, 1949
	Bacteriocins: several pneumocins are inhibitory or lethal to related strains of pneumococci but are inactive upon the producing strain	Mindich, 1966
	Sensitivity to ultraviolet irradiation[b]	
Miscellaneous	Integration efficiency of donor markers in transformation[b]	Lacks, 1970

[a] Adapted, updated, and extended from Ravin (1961).
[b] These mutants were obtained by treatment of pneumococci with 1-methyl-3 nitro-1-nitrosoguanidine.

cluded that streptomycin resistance and mannitol utilization were located on the same segment of transforming DNA. A weaker linkage was found between streptomycin resistance and sulfanilamide resistance. Linkage between erythromycin resistance and streptomycin resistance (Ravin and Chen, 1967) and between tetracycline resistance and erythromycin resistance (Butler and Nicholas, 1973) have also been found. Several complex loci have been analyzed in *Pneumococcus*. The first was that involving capsular polysaccharide (Ephrussi-Taylor, 1951; Ravin, 1960*b*). The results of genetic analysis of polysaccharide synthesis elucidated the biochemical basis of capsule formation in *Pneumococcus* (Jackson *et al.,* 1959; Austrian *et al.,* 1959; Mills and Smith, 1962). The genetics of the amylomaltase locus has also been studied (Lacks and Hotchkiss, 1960).

Different levels of resistance to a particular antimicrobial agent such as sulfanilamide (Hotchkiss and Evans, 1958), streptomycin (Rotheim and Ravin, 1961; 1964), erythromycin (Iyer and Ravin, 1962), and penicillin (Shockley and Hotchkiss, 1970) are also determined by linked markers. In *Pneumococcus,* as in some other microorganisms, in several cases genes determining related biochemical events tend to be close to each other on a molecule of DNA. For aminopterin resistance, however, two complex loci, *ami*A and *ami*B, which are not linked have been described (Sicard, 1964; Sicard and Ephrussi-Taylor, 1965).

The interpretation of multiple transformation frequencies, however, depends on factors such as the concentration, mode of preparation and integrity of the transforming DNA,* the identity of recipient cells, and factors influencing integration of DNA (Lacks, 1966; Hayes, 1968; Louarn and Sicard, 1969; Butler, 1970). Thus, the ordering of genes on the pneumococcal "chromosome" is very difficult and disagreement exists even on linkage of one marker pair (erythromycin resistance and streptomycin resistance). Recent experiments indicate that different recipient strains may give different linkage results (Butler and Nicholas, 1973). Butler and Smiley (1973) have attempted to determine the order of replication of several pneumococcal markers by the density-shift method. The pneumococci were synchronized with respect to DNA synthesis, and

---

* Soluble DNA has the advantage that it can be manipulated by physical and chemical means such as heat [to denature and separate DNA strands (Doty, 1959–1960); Marmur and Lane, 1960)], chemicals, or radiation (Ephrussi-Taylor, 1957), all of which would damage intact cells. The biologic effects of such treatment can be assessed by transformation (Roger and Hotchkiss, 1961; Schaeffer, 1964). The dependence of genetic function on molecular structure can thus be studied. On the other hand, the procedures used to extract DNA from cells may cause damage to the DNA which might divide linked markers into separate fragments.

both 5-bromouracil and deuterium were used as density labels. A shift in density of a DNA fraction plus the transforming activity of that particular fraction were used to determine replication order; the assumption was made that the genes are arranged linearly on a single "chromosome" and are replicated sequentially in one direction. The map constructed by the density-shift method does not seem to be the same as that constructed by frequencies of cotransfer of markers in transformation studies. These data support a model of bidirectional replication of the chromosome, so that the order of replication is not the same as the order of genes on the chromosome itself.

To date, pneumococcal recombination can be studied only by transformation; one of the participants in the reaction is chemically purified DNA in solution and the other is a bacterium with very fastidious growth requirements. These factors help to explain the difficulties in studying linkage and mapping in *Diplococcus pneumoniae*.

Interspecific Transformation

Pneumococcus can interact genetically with other bacterial species both *in vitro* (Bracco *et al.*, 1957; Pakula *et al.*, 1958; Krauss and MacLeod, 1963; Yurchak and Austrian, 1966; Ravin and Chen, 1967) and *in vivo* (Ottolenghi-Nightingale, 1969). Transformations occur most readily with closely related species, such as *Streptococcus viridans*. Interspecific transformation efficiency can be correlated to the degree of homology in the base sequences of the DNA of pneumococci and streptococci (Mehta and Hutchison, 1970), although other factors can be involved (Krauss and MacLeod, 1963). A genetic approach to bacterial taxonomy by degree of DNA homology and by transformation is discussed by Ravin (1960a) and Marmur (Marmur *et al.*, 1963).

Biologic Significance of Transformation

Although most transformations have been studied *in vitro*, pneumococcal transformation was first described in the mouse (Griffith, 1928), and later in several other animals (Austrian, 1952). More recently, actively growing pneumococci have been shown to release transforming DNA into their environment. If competent cells are present, they may be transformed by DNA released *in vitro* (Ottolenghi and Hotchkiss, 1962), in mice (Ottolenghi and MacLeod, 1963; Conant and Sawyer, 1967; Ottolenghi-Nightingale, 1969; Auerbach and Ottolenghi-Nightingale, 1971), and possibly in man (Ottolenghi-Nightingale, 1972). Thus, probably pneumococci in their natural environment may achieve genetic variation by transformation as well as by mutation.

Literature Cited

Auerbach-Rubin, F. and E. Ottolenghi-Nightingale, 1971   Effect of ethanol on the clearance of airborne pneumococci and the rate of pneumococcal transformations in the lung. *Infect. Immun.* **3**:688–693.

Austrian, R., 1952   Observations on the transformation of *Pneumococcus in vivo. Bull. Johns Hopkins Hosp.* **91**:189–196.

Austrian, R., 1953   Morphologic variation in *Pneumococcus*. II. Control of pneumococcal morphology through transformation reactions. *J. Exp. Med.* **98**:35–40.

Austrian, R. and M. S. Colowick, 1953   Modification of the fermentative activities of *Pneumococcus* through transformation reactions. *Bull. Johns Hopkins Hosp.* **92**:375–384.

Austrian, R. and C. M. MacLeod, 1949   Acquisition of M-protein by pneumococci through transformation reactions. *J. Exp. Med.* **89**: 451–460.

Austrian, R., H. P. Bernheimer, E. E. B. Smith and G. T. Mills, 1959   Simultaneous production of two capsular polysaccharides by *Pneumococcus*, II. The genetic and biochemical bases of binary capsulation. *J. Exp. Med.* **110**:585–602.

Avery, O. T., C. M. MacLeod and M. McCarty, 1944   Studies on the chemical nature of the substance inducing transformation of pneumococcal types. *J. Exp. Med.* **79**:137–157.

Bracco, R. M., M. R. Krauss, A. S. Roe and C. M. MacLeod, 1957   Transformation reactions between *Pneumococcus* and three strains of *Streptococcus. J. Exp. Med.* **106**:247–259.

Braun, W., 1965   *Bacterial Genetics,* second edition. pp. 239–262. W. B. Saunders Company, Philadelphia, Pa.

Breed, R. S., E. G. D. Murray and N. R. Smith, 1957   *Bergey's Manual of Determinative Bacteriology,* seventh edition, pp. 507–508. Williams and Wilkins Company, Baltimore, Md.

Briles, E. B. and A. Tomasz, 1970   Radioautographic evidence for equatorial wall growth in a Gram-positive bacterium. Segregation of choline-^{3}H-labeled teichoic acid. *J. Cell Biol.* **47**:786–790.

Bryan, B. E., 1961   Genetic modifiers of streptomycin resistance in *Pneumococcus. J. Bacteriol.* **82**:461–470.

Butler, L. O., 1970   The transformation of both recipient strands of pneumococcal DNA. *J. Gen. Microbiol.* **61**:137–140.

Butler, L. O. and G. Nicholas, 1973   Mapping of the *Pneumococcus* chromosome. Linkage between the genes conferring resistances to erythromycin and tetracycline and its implication to the replication of the chromosome. *J. Gen. Microbiol.* **79**:31–44.

Butler, L. O. and M. B. Smiley, 1970   Characterization by transformation of an ampicillin-resistant mutant of *Pneumococcus. J. Gen. Microbiol.* **61**:189–195.

Butler, L. O. and M. B. Smiley, 1973   Mapping of the *Pneumococcus* chromosome: Application of the density-shift method. *J. Gen. Microbiol.* **76**:101:113.

Cato, A., Jr. and W. R. Guild, 1968   Transformation and DNA size, I. Activity of fragments of defined size and a fit to a random double cross-over model. *J. Mol. Biol.* **37**:157–178.

Conant, E. and W. D. Sawyer, 1967   Transformation during mixed pneumococcal infection of mice. *J. Bacteriol.* **93**:1869–1875.

Cutler, R. G. and J. E. Evans, 1966   Synchronization of bacteria by a stationary-phase method. *J. Bacteriol.* **91**:469–476.

Davis, B. D., R. Dulbecco, H. N. Eisen, H. S. Ginsberg and W. B. Wood, Jr., 1967   *Microbiology,* pp. 184–190 and 683–700, Hoeber Medical Division, Harper and Row, New York.

Doty, P., 1959–1960   Inside nucleic acids. *Harvey Lect.* **55**:103–139.

Drew, R. M., 1957   Induction of resistance to A-methopterin in *Diplococcus pneumoniae* by deoxyribonucleic acid. *Nature (Lond.)* **179**:1251–1252.

Dubos, R. J., 1937   Mechanism of the lysis of pneumococci by freezing and thawing, bile, and other agents. *J. Exp. Med.* **66**:101–112.

Dubos, R. J., 1949   *The Bacterial Cell,* pp. 135–143. Harvard University Press, Cambridge, Mass.

Ephrussi-Taylor, H. E., 1951   Transformations allogènes du pneumocoque. *Exp. Cell Res.* **2**:589–607.

Ephrussi-Taylor, H. E., 1954   A new transforming agent determining pattern of metabolism and glucose and lactic acids in *Pneumococcus. Exp. Cell Res.* **6**:94–116.

Ephrussi-Taylor, H. E., 1957   X-ray inactivation studies on solutions of transforming DNA from *Pneumococcus.* In *The Chemical Basis of Heredity,* edited by W. D. McElroy and B. Glass, pp. 299–320, Johns Hopkins Press, Baltimore.

Ephrussi-Taylor, H. E., 1958   The mechanism of desoxyribonucleic acid-induced transformations. In *Recent Progress in Microbiology,* edited by G. Tunevall, pp. 51–68, C. C. Thomas, Springfield, Ill.

Ghei, O. K. and S. Lacks, 1967   Recovery of donor deoxyribonucleic acid marker activity from eclipse in pneumococcal transformation. *J. Bacteriol.* **93**:816–829.

Goodgal, S. H. and R. M. Herriott, 1957   Studies on transformation of *Haemophilus influenzae.* In *The Chemical Basis of Heredity,* edited by W. D. McElroy and B. Glass, pp. 336–340, The Johns Hopkins Press, Baltimore, Md.

Green, D. McD., 1959   A host-specific variation affecting relative frequency of transformation of two markers in *Pneumococcus. Exp. Cell Res.* **18**:466–480.

Griffith, F., 1928   The significance of pneumococcal types. *J. Hyg.* **27**:113–159.

Hayes, W., 1968   *The Genetics of Bacteria and their Viruses,* pp. 574–619, John Wiley & Sons, New York.

Hotchkiss, R. D., 1951   Transfer of penicillin resistance in pneumococci by the deoxyribonucleate derived from resistant cultures. *Cold Spring Harbor Symp. Quant. Biol.* **16**:457–460.

Hotchkiss, R. D., 1954   Cyclical behavior in pneumococcal growth and transformability occasioned by environmental changes. *Proc. Natl. Acad. Sci. USA* **40**:49–55.

Hotchkiss, R. D., 1957   Criteria for quantitative genetic transformation of bacteria. In *The Chemical Basis of Heredity,* edited by W. D. McElroy and B. Glass. pp. 321–335, The Johns Hopkins Press, Baltimore, Md.

Hotchkiss, R. D., 1971   Toward a general theory of genetic recombination in DNA. *Adv. Genet.* **16**:325–348.

Hotchkiss, R. D. and A. H. Evans, 1958   Analysis of the complex sulfonamide resistance locus of *Pneumococcus. Cold Spring Harbor Symp. Quant. Biol.* **23**:85–97.

Hotchkiss, R. D. and M. Gabor, 1970   Bacterial transformation, with special reference to recombination process. *Annu. Rev. Genet.* **4**:193–224.

Hotchkiss, R. D. and J. Marmur, 1954   Double marker transformations as evidence of linked factors in desoxyribonucleate transforming agents. *Proc. Natl. Acad. Sci. USA* **40**:55–60.

Hotchkiss, R. D. and E. Weiss, 1956   Transformed bacteria. *Sci. Am.* **Nov.**:30–35.

Hotchkiss, R. D., M. Abe and D. Lane, 1971   Transfer of heterozygotic properties by DNA in *Pneumococcus*. In *Informative Molecules in Biological Systems,* edited by L. G. H. Ledoux, North-Holland/American-Elsevier Publishing Co., New York.

Iyer, V. N. and A. W. Ravin, 1962   Integration and expression of different lengths of DNA during the transformation of *Pneumococcus* to erythromycin resistance. *Genetics* **47**:1355–1368.

Jackson, S., C. M. MacLeod and M. R. Krauss, 1959   Determination of type in capsulated transformants of *Pneumococcus* by the genome of non-capsulated donor and recipient strains. *J. Exp. Med.* **109**:429–438.

Kent, J. L. and R. D. Hotchkiss, 1964   Kinetic analysis of multiple, linked recombinations in pneumococcal transformation. *J. Mol. Biol.* **9**:308–322.

Krauss, M. R. and C. M. MacLeod, 1963   Intraspecies and interspecies transformation reactions in *Pneumococcus* and *Streptococcus*. *J. Gen. Physiol.* **46**:1141–1150.

Lacks, S., 1962   Molecular fate of DNA in genetic transformation of *Pneumococcus*. *J. Mol. Biol.* **5**:119–131.

Lacks, S., 1966   Integration efficiency and genetic recombination in pneumococcal transformation. *Genetics* **53**:207–235.

Lacks, S., 1968   Transformation and DNA size. II. Theoretical relationship between probability of marker integration and length of donor DNA in pneumococcal transformation. *J. Mol. Biol.* **37**:179–180.

Lacks, S., 1970   Mutants of *Diplococcus pneumoniae* that lack deoxyribonucleases and other activities possibly pertinent to genetic transformation. *J. Bacteriol.* **101**:373–383.

Lacks, S. and R. D. Hotchkiss, 1960   A study of the genetic material determining an enzyme activity in *Pneumococcus*. *Biochim. Biophys. Acta* **39**:508–518.

Lerman, L. S., and L. J. Tolmach, 1959   Genetic transformation, II. The significance of damage to the DNA molecule. *Biochim. Biophys. Acta* **33**:371–387.

Litt, M., J. Marmur, H. E. Ephrussi-Taylor and P. Doty, 1958   The dependence of pneumococcal transformation on the molecular weight of deoxyribose nucleic acid. *Proc. Natl. Acad. Sci. USA* **44**:144–152.

Louarn, J-M. and A. M. Sicard, 1969   Identical transformability of both strands of recipient DNA in *Diplococcus pneumoniae*. *Biochem. Biophys. Res. Commun.* **36**:101–109.

MacLeod, C. M., 1965   The Pneumococci. In *Bacterial and Mycotic Infections of Man,* edited by R. J. Dubos and J. G. Hirsch, fourth edition, pp. 391–411, J. B. Lippincott, Philadelphia, Pa.

Marmur, J. and R. D. Hotchkiss, 1955   Mannitol metabolism, a transferable property of *Pneumococcus*. *J. Biol. Chem.* **214**:383–396.

Marmur, J. and D. Lane, 1960   Strand separation and specific recombination in deoxyribonucleic acids: biological studies. *Proc. Natl. Acad. Sci. USA* **46**:453–461.

Marmur, J., S. Falkow and M. Mandel, 1963   New approaches to bacterial taxonomy. *Annu. Rev. Microbiol.* **17**:329–372.

Mehta, B. M. and D. J. Hutchison, 1970   Nucleic acid homology as a measure of genetic compatibility among streptococci and a strain of *Diplococcus pneumoniae*. *Can. J. Microbiol.* **16**:281–286.

Meynell, G. G. and E. Meynell, 1965 *Theory and Practice in Experimental Bacteriology.* Cambridge University Press, Cambridge.

Mills, G. T. and E. E. B. Smith, 1962 Biosynthetic aspects of capsule formation in the *Pneumococcus. Br. Med. Bull.* **18**:27–30.

Mindich, L., 1966 Bacteriocins of *Diplococcus pneumoniae. J. Bacteriol.* **92**:1090–1098.

Ottolenghi-Nightingale, E., 1969 Spontaneously occurring bacterial transformations in mice. *J. Bacteriol.* **100**:445–452.

Ottolenghi-Nightingale, E., 1972 Competence of pneumococcal isolates and bacterial transformations in man. *Infect. Immun.* **6**:785–792.

Ottolenghi, E. and R. D. Hotchkiss, 1962 Release of genetic transforming agent from pneumococcal cultures during growth and disintegration. *J. Exp. Med.* **116**:491–519.

Ottolenghi, E. and C. M. MacLeod, 1963 Genetic transformation among living pneumococci in the mouse. *Proc. Natl. Acad. Sci. USA* **50**:417–419.

Pakula, R., E. Hulanicka and W. Walczak, 1958 Transformation between streptococci, pneumococci, and staphylococci. *Bull. Acad. Polon. Sci. Ser. Sci. Biol. Classe (II)* **6**:325–328.

Percival, A., E. C. Armstrong and G. C. Turner, 1969 Increased incidence of tetracycline-resistant pneumococci in Liverpool in 1968. *Lancet* **1**:998–1000.

Rappaport, H. P. and W. R. Guild, 1959 Defined medium for growth of two transformable strains of *Diplococcus pneumoniae. J. Bacteriol.* **78**:203–205.

Ravin, A. W., 1959 Reciprocal capsular transformations of pneumococci. *J. Bacteriol.* **77**:296–309.

Ravin, A. W., 1960a The origin of bacterial species. *Bacteriol. Rev.* **24**:201–220.

Ravin, A. W., 1960b Linked mutations borne by deoxyribonucleic acid controlling the synthesis of capsular polysaccharide in *Pneumococcus. Genetics* **45**:1387–1404.

Ravin, A. W., 1961 The genetics of transformation. *Adv. Genet.* **10**:61–163.

Ravin, A. W. and K. C. Chen, 1967 Heterospecific transformation of *Pneumococcus* and *Streptococcus.* III. Reduction of linkage. *Genetics* **57**:851–864.

Ravin, A. W. and V. N. Iyer, 1961 The genetic relationship and phenotypic expression of mutations endowing *Pneumococcus* with resistance to erythromycin. *J. Gen. Microbiol* **66**:1–25.

Roger, M. and R. D. Hotchkiss, 1961 Selective heat inactivation of pneumococcal transforming deoxyribonucleate. *Proc. Natl. Acad. Sci. USA* **47**:653–669.

Rotheim, M. B. and A. W. Ravin, 1961 The mapping of genetic loci affecting streptomycin resistance in *Pneumococcus. Genetics* **46**:1619–1634.

Rotheim, M. B. and A. W. Ravin, 1964 Sites of breakage in the DNA molecule as determined by recombination analysis of streptomycin-resistance mutations in *Pneumococcus. Proc. Natl. Acad. Sci. USA* **52**:30–38.

Schaeffer, P. 1956 Analyse génétique de la résistance à la streptomycine chez le pneumocoque. *Ann. Inst. Pasteur* **91**:323–337.

Schaeffer, P., 1964 Transformation. In *The Bacteria.* Vol. 5: Heredity, pp. 87–153, Academic Press, New York.

Shockley, T. E. and R. D. Hotchkiss, 1970 Stepwise introduction of transformable penicillin resistance in *Pneumococcus. Genetics* **64**:397–408.

Shumway, C. N. and S. J. Klebanoff, 1971 Purification of pneumolysin. *Infect. Immun.* **4**:388–392.

Sicard, A. M., 1964 A new synthetic medium for *Diplococcus pneumoniae* and its use for the study of reciprocal transformation at the *ami-A* locus. *Genetics* **50**:31–44.

Sicard, A. M. and H. E. Ephrussi-Taylor, 1965   Genetic recombination in DNA-induced transformation of *Pneumococcus,* II. Mapping the *ami-A* region. *Genetics* **52**:1207–1227.

Sirotnak, F. M., R. Lunt and D. J. Hutchison, 1960*a*   Genetic studies on amethopterin resistance in *Diplococcus pneumoniae. J. Bacteriol.* **80**:648–653.

Sirotnak, F. M., R. Lunt and D. J. Hutchison, 1960*b*   Transformation of resistance to 8-azaguanine in *Diplococcus pneumoniae. Nature (Lond.)* **187**:800–801.

Spizizen, J., B. E. Reilly and A. H. Evans, 1966   Microbial transformation and transfection. *Annu. Rev. Microbiol.* **20**:371–400.

Taylor, H. E., 1949   Transformations réciproques des formes R et ER chez le pneumocoque. *C. R. Hebd. Seances Acad. Sci. Ser. D. Sci. Nat.* **228**:1258–1259.

Tomasz, A., 1968   Biological consequences of the replacement of choline by ethanolamine in the cell wall of *Pneumococcus:* chain formation, loss of transformability and loss of autolysis. *Proc. Natl. Acad. Sci. USA* **59**:86–93.

Tomasz, A., 1969   Some aspects of the competent state in genetic transformation. *Annu. Rev. Genet.* **3**:217–232.

Tomasz, A., 1971   Cell physiological aspects of DNA uptake during genetic transformation in bacteria. In *Informative Macromolecules in Biological Systems,* pp. 4–18, North-Holland/American-Elsevier Publishing Co., New York.

Tomasz, A., J. D. Jamieson and E. Ottolenghi, 1964   The fine structure of *Diplococcus pneumoniae. J. Cell Biol.* **22**:453–467.

Udaka, S., J. Koukol and R. Vennesland, 1959   Lactic oxidase of *Pneumococcus. J. Bacteriol.* **78**:714–725.

White, B., 1938   *The Biology of the Pneumococcus,* The Commonwealth Fund, New York.

Wilson, G. S. and A. A. Miles, 1964   *Topley and Wilson's Principles of Bacteriology and Immunity,* Vol. 1, pp. 737–738, fifth edition, Williams and Wilkins, Baltimore, Md.

Yurchak, A. M. and R. Austrian, 1966   Serologic and genetic relationships between pneumococci and other respiratory streptococci. *Trans. Assoc. Am. Physicians Phila.* **79**:368–375.

3

Pseudomonas

BRUCE W. HOLLOWAY

Introduction

The genus *Pseudomonas* has interested microbiologists firstly by virtue of its ability to use such a wide range of substrates as energy sources and secondly through its resistance to many deleterious agents, including chemotherapeutic substances. This latter aspect has important epidemiological implications for the major pathogenic species, *Pseudomonas aeruginosa*. Both these characteristics have provided a stimulus for genetic investigations, as well as the need for establishing the generality of bacterial genetic mechanisms in organisms other than enterobacteria.

Pseudomonas is a gram-negative, nonsporing, motile bacillus with a single polar flagellum. The taxonomy of the genus is controversial but some groupings, particularly the fluorescent pseudomonads including *Pseudomonas aeruginosa*, *Pseudomonas putida*, and *Pseudomonas fluorescens*, are well defined (Stanier *et al.*, 1966).

Two species have been the subject of genetic study, *P. aeruginosa* found in human urinary tract infections, wounds, and burns and as a secondary invader, and *P. putida*, a soil saprophyte. Reviews of *Pseudomonas* genetics can be found in Holloway (1969), Holloway *et al.* (1971), Holloway (1974), Holloway and Krishnapillai (1974), and Richmond and Stanisich (1974). Aspects of the biochemistry and regulation of catabolic pathways in *Pseudomonas* are covered by Ornston (1971).

BRUCE W. HOLLOWAY—Department of Genetics, Monash University, Clayton, Victoria, Australia.

59

Experimental Material and Methods

In *P. aeruginosa*, most genetic work has been done with two strains, PAO (ATCC15692) and PAT (ATCC15693) (Holloway, 1969). Since 1969, a variety of plasmid containing strains have been isolated and characterized, (Holloway, 1974). For *P. putida* one strain (A3.12) has been used, but here too a variety of native strains with unusual catabolic features have been examined (Wheelis and Stanier, 1970; Chakrabarty *et al.*, 1968, Rheinwald *et al.*, 1973).

Stocks of genetically marked strains of *Pseudomonas* may be made available as a personal favor by the following scientists: *P. aeruginosa:* Professor B. W. Holloway, Genetics Department, Monash University, Clayton, Victoria, 3168, Australia; Professor J. S. Loutit, Microbiology Department, Otago University, Dunedin, New Zealand; Dr. R. H. Olsen, Department of Microbiology, The University of Michigan Medical School, Ann Arbor, Michigan 48104, U.S.A.; and Professor M. H. Richmond, Microbiology Department, Medical School, Bristol, England. *P. putida*: Dr. M. L. Wheelis, Department of Bacteriology, University of California, Davis, California 95616 U.S.A.; Professor I.C. Gunsalus, Department of Biochemistry, University of Illinois, Urbana, Illinois 61801 U.S.A.; and Dr. A.M. Chakrabarty, Research and Development Center, General Electric Company, P.O. Box 8, Schenectady, New York 12301, U.S.A.

The techniques, media, and equipment employed in *Pseudomonas* genetics are essentially the same as those for *Escherichia coli* (*q.v.*) or *Salmonella typhimurium* (*q.v.*) (Hayes, 1968; Clowes and Hayes, 1968). All species grow well on commercially available complete media, although the dryness of agar plates is an important feature in obtaining reproducible colonial morphology. The optimum temperature for *P. aeruginosa* is 37°C; for most other species 30°C. The chemically defined medium of Vogel and Bonner (1956) is entirely satisfactory as a minimal medium in work with auxotrophs. Stock cultures can be maintained either by lyophilization or by quick freezing bacterial suspensions in a dry ice and alcohol mixture which are then stored at −15°C.

A wide range of mutants is available. Auxotrophs can be isolated following mutagenesis using the penicillin-enrichment technique (Lederberg, 1950) modified by the use of carbenicillin (1000 μg/ml). Drug-resistant mutants, e.g., streptomycin, rifampicin, or nalidixic acid, can be isolated by plating ca. 10^8 cells on nutrient agar containing the antibiotic. Mutants lacking the ability to utilize a particular substrate, say mandelic acid or an amino acid, as the sole carbon source (Kemp and Hegeman,

1968; Wheelis and Stanier, 1970) provide a large range of potential genetic markers.

Many mutants are difficult to obtain at spontaneous mutation frequencies, and a mutagen may be needed. Apart from the usual difficulties encountered with N-methyl-nitro-nitrosoguanidine in introducing "silent mutations," this compound also acts to induce the determinants of aeruginocins or prophage carried by most species of *P. aeruginosa*. In the author's experience the mutagen of choice is ethyl methanesulfonate, used at a concentration of 1/100 (v/v) with stationary-phase broth cultures.

Organization of the *Pseudomonas* Genome

Both chromosomal and plasmid components contribute to the whole genome, and a complete genetic analysis involves the estimation of the role of each of these components. As well as coding for gene products which produce readily recognizable phenotypic characteristics, some plasmids can promote DNA transfer at conjugation. They can be considered under the following headings.

Bacteriophages

Almost all strains of *P. aeruginosa* are lysogenic. By contrast, lysogeny in *P. putida* and *P. fluorescens* is rare, but for these species (as well as for *P. aeruginosa*) virulent or nonlysogenizing phages can be readily

TABLE 1.   *Bacteriophages of Pseudomonas*

P. aeruginosa	
B3	Temperate, noninducible, transducing, shows HCM, double-stranded DNA, mol.wt. ca. 20×10^6 daltons ATCC 15692–B.
F116	Temperate, inducible, transducing, doubled-stranded DNA, mol.wt. 38×10^6 daltons.
G101	Temperate, noninducible, transducing, shows HCM, double-stranded DNA, mol.wt. ca. 20×10^6 daltons
E79	Virulent, nontransducing, does not show HCM, used as male contraselective agent in interrupted mating procedure in strain PAO.
P. putida	
pf16	nonlysogenizing, transducing, double-stranded DNA, mol wt 100×10^6 daltons.

isolated from sewage. Transducing phages of *P. aeruginosa* are common and several are known for *P. putida*, all showing general transduction. Extensive use of transduction for gene mapping has been made, and it is valuable for determining the order of closely linked genes in conjunction with the broader location of genes provided by conjugation. The more widely used phages are listed in Table 1, but a comprehensive description is given by Holloway and Krishnapillai (1974).

Plasmids

These can be defined as genetic elements which can occupy an extrachromosomal situation, may contribute to the bacterial phenotype, may be transmissible to other bacteria, and may at times integrate into the bacterial chromosome.

The identification of plasmids may present difficulties. If it is self-transmissible, has some recognizable contribution to the host genotype, and can be shown to be unlinked to the chromosomal DNA, the situation is clear. However, with other combinations of characteristics, identification may have to rely on more circumstantial evidence. Three main types of *Pseudomonas* plasmids will be considered here:

Sex Factors. These are identified by their ability to promote transfer of bacterial chromosome. These have only been found to date in *P. aeruginosa*. Those studied include:

FP2:   Initially isolated from strain PAT (Holloway and Jennings, 1958), promotes transfer of chromosome in strains PAO and PAT. Carries a determinant for mercury resistance (Loutit, 1971).

FP39: Promotes transfer of chromosome in strain PAO, possibly from a site 10 minutes proximal to that of FP2. Carries a determinant for leucine biosynthesis (Pemberton and Holloway, 1973).

FP5:   Unrelated to either FP2 or FP39. Carries a mercury-resistance determinant, can be cured by acriflavine (unlike FP2 and FP39) (Matsumoto and Tazaki, 1973).

Drug-Resistance Plasmids. These have in common the ability to confer antibiotic resistance on strains carrying them. They have only been isolated initially in *P. aeruginosa* strains. A range of resistance patterns are known, and compatibility studies indicate that a variety of types occur. Some are promiscuous and can be transferred not only to other species of *Pseudomonas* but also to such unrelated genera as *Acinetobacter,*

Rhizobium, and *Neisseria* (Datta *et al.,* 1971; Olsen and Shipley, 1973). Some can promote the transfer of bacterial chromosome in *P. aeruginosa* (Stanisich and Holloway, 1971).

Catabolic Plasmids. It has been found in some species of *Pseudomonas* that the ability of certain strains to catabolize specific substrates is due to possession of particular plasmids. Those characterized (with the species in which they were first identified) include CAM (ability to use D-camphor), *P. putida* (Rheinwald *et al.,* 1973); SAL (salicylate), *P. putida* (Chakrabarty, 1972); NAH (naphthalene), *P. putida* (Dunn and Gunsalus, 1973); and OCT (n-octane), *P. oleovorans* (Chakrabarty *et al.,* 1973). Some of these can be transferred to other species of *Pseudomonas* and can also act as sex factors in promoting the transfer of host chromosome.

Recombination Systems

For the satisfactory genetic analysis of any organism, a system of gene recombination is essential. Two *Pseudomonas* species possess the mechanisms for this.

P. aeruginosa

Transformation. Kahn and Sen (1967) and Lambina and Mikhailova (1964) have reported inter- and intraspecific transformation systems for *Pseudomonas.*

Transduction. Three general transducing phages have been used for genetic analysis—F116 (Holloway *et al.,* 1960; Holloway *et al.,* 1962), and recently a variant, F116L, which transduces larger segments of the bacterial chromosome than F116 (Krishnapillai, 1971), G101 (Holloway and van de Putte, 1968), and B (Loutit, 1958). The molecular weights of F116 and G101 are such that they transduce a sufficiently large fragment of bacterial chromosome to enable cotransduction analysis to be used.

Conjugation. Most mapping studies with *P. aeruginosa* have been carried out using both interrupted matings (Stanisich and Holloway, 1969; Loutit *et al.,* 1968; Pemberton and Holloway, 1972), and plate matings between FP2+ and FP− forms of strain PAO. The techniques used are essentially the same as those used in *E. coli,* and are described in the references quoted.

A chromosome map derived mainly from data obtained by the author and his colleagues is shown in Figure 1. Loutit and his associates have

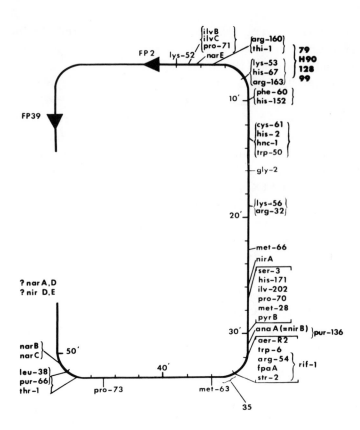

Figure 1. Linkage map of P. aeruginosa strain PAO derived from data of Holloway et al. (1971), van Hartingsveldt and Stouthamer (1973), Krishnapillai and Carey (1972), J. Dodge (private communication), and K. E. Carey (private communication). Marker position is based on times of entry in interrupted matings (±2 min) using FP2 and FP39 donor strains, combined with linkage and transduction data. Origin and direction of chromosome transfer of FP2+ and FP39+ donor strains is indicated. Allele numbers are provided to aid identification of particular loci. Cotransducible loci are given in brackets, square brackets indicate known order, and curved brackets indicate that the precise order remains to be determined. (From Holloway, 1974, with the permission of J. Wiley and Sons.)

Gene symbols: aer-R2, aeruginocinogenic determinant [originally designated pyrR2 (Kageyama, 1970)]; ana, affected in anaerobic growth; arg, arginine biosynthesis; cys, cysteine biosynthesis; fpaA, resistance to p-fluorophenyl-alanine; gly, glycine biosynthesis; his, histidine biosynthesis; ilv, isoleucine and valine biosynthesis; leu, leucine biosynthesis; lys, lysine biosynthesis; met, methionine biosynthesis; nar, dissimilatory nitrate reductase activity; nir, dissimilatory nitrite reductase activity; phe, phenylalanine biosynthesis; pro, proline biosynthesis; pur, purine biosynthesis; 35, 79, 90, 99, 128-prophage locations; pyr, pyrimidine biosynthesis; rif, rifampicin resistance; ser, serine biosynthesis; str, resistance to streptomycin; thi, thiamine biosynthesis; thr, threonine biosynthesis; trp, tryptophan biosynthesis.

presented evidence for two linkage groups (Loutit and Marinus, 1968; Loutit, 1969). A physical examination of the genome of *P. aeruginosa* (Pemberton and Clark, 1973; Pemberton, 1973) provides supporting evidence for only one major chromosomal component in this organism.

P. putida

There is as yet no conjugation system in this species directly comparable to that described for *P. aeruginosa*. Most mapping analysis has been carried out by transduction using the phage pf16 (Chakrabarty *et al.,* 1968). A low-frequency conjugation system has been described using a derivative of pf16 (pf*dm*) as a sex factor (Chakrabarty and Gunsalus, 1969), and systems of conjugation are being developed using catabolic plasmids as sex factors (Rheinwald, *et al.,* 1973; Chakrabarty, 1972).

Host-Controlled Modification (HCM)

Transfer of DNA between bacteria may be limited by enzymatic systems which on the one hand confer specificity on the DNA which replicates within a given strain of bacteria (modification), and on the other destroy the biological activity of DNA not having the homologous specificity (restriction). Thus, phage propagated on one strain of bacteria may plaque at only very low efficiencies (10^{-5}) on another strain because restriction enzymes (endonucleases) recognize a different modification pattern of the phage DNA and act to destroy its biological activity. HCM is known in *P. aeruginosa*, but not in *P. putida*. It is likely to have signifi-

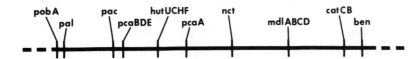

Figure 2. Genetic map of chromosomal genes involved in dissimilatory functions in P. putida as determined by transduction with phage pf16 (Leidigh and Wheelis, 1973, with the permission of Springer-Verlag). The gene symbols and their phenotype or enzyme deficiencies are: ben, benzoate oxidase; catB, muconate lactonizing enzyme; catC, muconolactone isomerase; hutC, constitutive histidine utilization; hutF, formiminoglutamase; hutH, histidase; hutU, urocanase; mdlA, mandelate racemase; mdlB, L-mandelate dehydrogenase; mdlc, benzoyl formate decarboxylase; mdlD, benzaldehyde dehydrogenase; nct, nicotinic acid utilization; pac, phenylalanine utilization; pcaA, protocatechuate, oxygenase; pcaB, carboxymuconate lactonizing enzyme; pcaC, enol lactone hydrolase; pcaE, transferase; pobA, p-hydroxybenzoate hydroxylase.

cance in interstrain and interspecies gene transfer by both conjugation and transduction. It is possible to obtain mutants which lack restriction function and which also may have altered modification function (Holloway *et al.*, 1971).

Gene Arrangement in *Pseudomonas*

In *P. aeruginosa*, and as far as analysis permits in *P. putida*, structural genes controlling biosynthetic functions do not show the clustering of markers affecting sequential biosynthetic steps commonly found in enterbacteria (Holloway *et al.*, 1971). By contrast, the genes affecting catabolic functions are grouped together in both *P. aeruginosa* (Rosenberg and Hegeman, 1969) and *P. putida* (Wheelis and Stanier, 1970; Leidigh and Wheelis, 1973) (see Figure 2). Clearly this arrangement of genes concerned with the catabolism of substrates must have physiological and regulatory significance.

Clarke and her colleagues (Clarke, 1970; Betz *et al.*, 1973) have made an intensive genetic and biochemical study of the genetic basis of amidase formation in *P. aeruginosa*. They have been able to show that a structural gene (*ami*-E) and a regulatory gene (*ami*-R) are contiguous and that this arrangement has regulatory significance. Studies such as this will be aided if a partial diploid structure analogous to say, F-*lac* in *E. coli*, can be developed for *Pseudomonas*.

Conclusion

Genetic studies with *Pseudomonas* have clearly indicated that bacteria are just as variable in their genetic properties as any other group of organisms. The increasing medical importance of these organisms and their potential role in environmental pollution control, clearly signify the need for a greater understanding of their genetic features.

Literature Cited

Betz, J. L., J. E. Brown, P. H. Clarke and M. Day, 1973   Genetic analysis of amidase mutants of *Pseudomonas aeruginosa. J. Gen. Microbiol. (in press).*

Chakrabarty, A. M., 1972   Genetic basis of the biodegradation of salicylate in *Pseudomonas. J. Bacteriol.* **112**:815–823.

Chakrabarty, A. M. and I. C. Gunsalus, 1969   Autonomous replication of a defective transducing phage in *Pseudomonas putida. Virology* **38**:92–104.

Chakrabarty, A. M., C. F. Gunsalus and I. C. Gunsalus, 1968   Transduction and the clustering of genes in fluorescent pseudomonads. *Proc. Natl. Acad. Sci. USA* **60:**168–175.

Chakrabarty, A. M., G. Chou and I. C. Gunsalus, 1973   Genetic regulation and extrachromosomal nature of the octane degradative pathway in *Pseudomonas. Proc. Natl. Acad. Sci. USA* **70:**1137–1140.

Clarke, P. H., 1970   The aliphatic amidases of *Pseudomonas aeruginosa. Adv. Microb. Physiol.* **4:**179–221.

Clowes, R. C. and W. Hayes, 1968   *Experiments in Microbial Genetics,* Blackwell Scientific Publications, Oxford.

Datta, N., R. W. Hedges, E. J. Shaw, R. B. Sykes and M. H. Richmond, 1971   Properties of an R factor from *Pseudomonas aeruginosa. J. Bacteriol.* **108:**1244–1249.

Dunn, N. W. and I. C. Gunsalus, 1973   Transmissible plasmid coding early enzymes of naphthalene oxidation in *Pseudomonas putida. J. Bacteriol.* **114:**974–979.

Hayes, W., 1968   *The Genetics of Bacteria and their Viruses* second edition, Oxford, Blackwell.

Holloway, B. W., 1969   Genetics of *Pseudomonas. Bacteriol Rev.* **33:**419–443.

Holloway, B. W., 1974   Genetic organisation in *Pseudomonas.* In *The Biology of Pseudomonas,* edited by M. H. Richmond and P. H. Clarke, J. Wiley & Sons, London.

Holloway, B. W. and P. A. Jennings, 1958   An infectious fertility factor for *Pseudomonas aeruginosa. Nature (Lond.)* **181:**855–856.

Holloway, B. W. and V. Krishnapillai, 1974   Bacteriophages and Bacteriocins of *Pseudomonas.* In *The Biology of Pseudomonas,* edited by M. H. Richmond and P. H. Clarke, J. Wiley & Sons, New York.

Holloway, B. W. and P. van de Putte, 1968   Lysogeny and bacterial recombination. In *Replication and Recombination of Genetic Material,* edited by W. J. Peacock and R. D. Brock, pp. 175–183 Australian Academy of Science, Canberra.

Holloway, B. W. J. B. Egan and M. Monk, 1960   Lysogeny in *Pseudomonas aeruginosa. Aust. J. Exp. Biol. Med. Sci.* **38:**321–330.

Holloway, B. W., M. Monk, L. M. Hodgins and B. Fargie, 1962   Effects of radiation on transduction in *Pseudomonas aeruginosa. Virology* **18:**89–94.

Holloway, B. W., V. Krishnapillai and V. A. Stanisich, 1971   *Pseudomonas* genetics. *Annu. Rev. Genet.* **5:**425–446.

Kageyama, M., 1970   Genetic mapping of a bacteriocinogenic factor in *Pseudomonas aeruginosa.* II. Mapping of pyocin R2 factor by transduction with phage F116. *J. Gen. Appl. Microbiol.* **16:**531–535.

Kahn, N. V. and S. P. Sen, 1967   Genetic transformation in *Pseudomonas. J. Gen. Microbiol.* **49:**201–209.

Kemp, M. B. and G. D. Hegeman, 1968   Genetic control of the β-ketoadipate pathway in *Pseudomonas aeruginosa. J. Bacteriol.* **96:**1488–1489.

Krishnapillai, V., 1971   A novel transducing phage. Its role in recognition of a possible new host-controlled modification system in *Pseudomonas aeruginosa. Mol. Gen. Genet.* **114:**134–143.

Krishnapillai, V. and K. E. Carey, 1972   Chromosomal location of a prophage in *Pseudomonas aeruginosa* strain PAO. *Genet. Res.* **20:**137–140.

Lambina, V. A. and T. N. Mikhailova, 1964   Quantitative regularities of transformation of streptomycin resistivity in *Pseudomonas fluorescens. Microbiology* **33:**800–806.

Lederberg, J., 1950 Methods of Microbial Genetics. In *Methods in Medical Research*, edited by J. H. Corrie Jr., p. 3, Chicago Year Book Publishers, Chicago, Ill.

Leidigh, B. J. and M. L. Wheelis, 1973 The clustering on the *Pseudomonas putida* chromosome of genes specifying dissimilatory functions. *J. Mol. Evol.* **2**:235–242.

Loutit, J. S., 1958 A transduction-like process within a single strain of *Pseudomonas aeruginosa. J. Gen. Microbiol.* **18**:315–319.

Loutit, J. S., 1969 Investigation of the mating system of *Pseudomonas aeruginosa* strain 1. IV. Mapping of distal markers. *Genet. Res.* **13**:91–98.

Loutit, J. S., 1971 Investigation of the mating system of *Pseudomonas aeruginosa* strain 1. VI. Mercury resistance associated with the sex factor FP. *Genet. Res.* **16**:179–184.

Loutit, J. S. and M. G. Marinus, 1968 Investigation of the mating system of *Pseudomonas aeruginosa* strain 1. II. Mapping of a number of early markers. *Genet. Res.* **12**:37–44.

Loutit, J. S., L. E. Pearce and M. G. Marinus, 1968 Investigation of the mating system of *Pseudomonas aeruginosa* 1. I Kinetic studies. *Genet. Res.* **12**:29–36.

Matsumoto, H. and T. Tazaki, 1973 FP5 factor, an undescribed sex factor of *Pseudomonas aeruginosa. Jap. J. Microbiol.* **17**:409–417.

Olsen, R. H. and P. Shipley, 1973 Host range and properties of the *Pseudomonas aeruginosa* R factor R1822. *J. Bacteriol.* **113**:772–780.

Orston, L. N., 1971 Regulation of catabolic pathways in *Pseudomonas. Bacteriol. Rev.* **35**;87–116.

Pemberton, J. M., 1973 Size of the chromosome of *Pseudomonas aeruginosa* PAO. *Manuscript in preparation.*

Pemberton, J. M. and A. J. Clark, 1973 Detection and characteristation of plasmids in *P. aeruginosa* strain PAO. *J. Bacteriol.* **114**:424–433.

Pemberton, J. M. and B. W. Holloway, 1972 Chromosome mapping in *Pseudomonas aeruginosa. Genet. Res.* **19**:251–260.

Pemberton, J. M. and B. W. Holloway, 1973 A new sex factor for *Pseudomonas aeruginosa. Genet. Res.* **21**:263–272.

Rheinwald, J. G., A. M. Chakrabarty and I. C. Gunsalus, 1973 A transmissible plasmid controlling camphor oxidation in *Pseudomonas putida. Proc. Natl. Acad. Sci. USA* **70**:885–889.

Richmond, M. H. and V. A. Stanisich, 1974 Gene transfer in *Pseudomonas*. In *The Biology of Pseudomonas*, edited by M. H. Richmond and P. H. Clarke, J. Wiley & Sons, London.

Rosenberg, S. L. and G. D. Hegeman, 1969 Clustering of functionally related genes in *Pseudomonas aeruginosa. J. Bacteriol.* **99**:353–355.

Stanier, R. Y., N. J. Palleroni and M. Doudoroff, 1966 The aerobic pseudomonads. *J. Gen. Microbiol.* **43**:159–271.

Stanisich, V. A. and B. W. Holloway, 1969 Conjugation in *Pseudomonas aeruginosa. Genetics* **61**:327–339.

Stanisich, V. A. and B. W. Holloway, 1971 Chromosome transfer in *Pseudomonas aeruginosa* mediated by R factors. *Genet. Res.* **17**:169–172.

van Hartingsveldt, J. and A. H. Stouthamer, 1973 Mapping and characterization of mutants of *Pseudomonas aeruginosa* affected in nitrate respiration in aerobic or anaerobic growth. *J. Gen. Microbiol.* **74**:97–106.

Vogel, H. J. and D. M. Bonner, 1956 Acetyl-ornithinase of *Escherichia coli:* partial purification and some properties. *J. Biol. Chem.* **218**:97–106.

Wheelis, M. L. and R. Y. Stanier, 1970 The genetic control of dissimilatory pathways in *Pseudomonas putida. Genetics* **66**:245–266.

4

Bacillus subtilis

Frank E. Young and Gary A. Wilson

The discovery of DNA-mediated transformation of *Bacillus subtilis* by Spizizen in 1958 stimulated the development of *B. subtilis* as a model procaryotic system for genetic and biochemical experiments. This Gram-positive, spore-forming, soil bacillus grows readily in a chemically defined medium and undergoes genetic exchange by transformation and transduction. The simplicity of the cell wall and the morphogenetic events of sporulation have stimulated the development of this organism as a model system for investigating unicellular differentiation at the molecular level.

This review is aimed at providing a distillate of genetic techniques in *B. subtilis* that have become the established procedures. Particular emphasis will be placed on fundamental and practical methods. For a more comprehensive survey of facets of the genetics of *B. subtilis,* refer to reviews by Schaeffer (1964), Spizizen *et al.* (1966), Armstrong *et al.* (1970), Erickson (1970), and Young and Wilson (1972). Unlike these earlier reviews, we will be less comprehensive and provide practical information to aid the reader in developing the experimental techniques necessary for analysis of the chromosome of *B. subtilis.*

General Methods

The general procedures included in this section summarize techniques and modifications of methods used in our laboratory. The specific

Frank E. Young and Gary A. Wilson—Department of Microbiology, School of Medicine and Dentistry, University of Rochester, Rochester, New York.

methods for genetic analysis will be discussed in their appropriate sections. We have attempted to succinctly describe the methods, but in many cases it will be necessary to consult other publications for additional information.

Growth and Maintenance of *B. subtilis*

Working cultures of *B. subtilis* are maintained on Tryptose Blood Agar Base medium (TBAB, Difco) in Petri dishes stored at room temperature. Sporulation is repressed on this medium. Strains are generally transferred every 3 to 4 weeks and stored in closed boxes to prevent dehydration. Stocks maintained on TBAB slants in screw-capped tubes at 4°C remain useful for over a year For more permanent storage, sporulating strains are grown for 3 to 4 days on AK agar (BBL) supplemented with 4.0 g agar (Difco) and 0.60 g MgSO$_4$ per liter. The culture becomes dark reddish brown as sporulation proceeds on supplemented AK agar, and the slants can be retained in this condition for years. For better preservation, spore suspensions can be treated with lysozyme, washed extensively with distilled water, and then stored in distilled water at 4°C, in 15 percent glycerol at −20°C, or lyophilized. Recently, we have adopted a procedure for storing sporulated cultures in silica gel (F. Sherman and H. Taber, private communication). Chromatographic grade silica gel (Sigma type I, 60–200 mesh) is sterilized by dry heat in small glass vials that are later sealed with rubber-lined screw caps when cooled. Growth from supplemented AK agar is scraped off with wooden applicators, mixed to dryness with the silica gel (while kept cool), and stored at 4°C. Stocks can be recovered even from mutants that sporulate poorly when growth is taken from TBAB medium. The major advantage of this preservation method is that stocks can be recovered many times from the same vial, thus eliminating the need to prepare multiple vials of the same stock.

B. subtilis grows well in Penassay broth (PB, Antibiotic assay medium No. 3, Difco) with a doubling time of 23 minutes at 37°C and 53 minutes at 30°C. In Spizizen's minimal medium, growth is slower (approximately 76 minutes for a doubling of turbidity at 37°C). Sporulation does not occur in either medium and requires the addition of calcium, manganese, or ferric cations and a carbon source such as glutamic acid instead of glucose. Methods have been developed to produce spores in replacement media (Ramaley and Burden, 1970), affording the investigator a more synchronous production of spores without sacrificing the high efficiency of sporulation. Sporulation can also be achieved in potato-extract medium (Spizizen, 1958) or GM2 medium (previously designated SPII medium, Yasbin and Young, 1972).

Recently, stable L forms of *B. subtilis* have been obtained by serially culturing the mutants in Tryptose medium supplemented with 1.2 M NaCl (Young *et al.*, 1970; Gilpin *et al.*, 1973). The stable L forms also grow well in GM1 (previously designated SPI) medium supplemented with 1.2 M NaCl. Although genetic exchange between L forms has not been demonstrated, these forms provide an excellent tool for membrane studies. The ease with which DNA can be extracted from the L forms not only provides a convenient source of DNA, but also produces extremely high molecular weight fragments that have proved to be useful in linking distant markers (Bettinger and Young, 1973).

Isolation of DNA

In addition to the osmotic rupturing of stable L forms, a number of procedures have been developed for isolation of DNA from *B. subtilis*. We have found that a modification combining procedures described by Saito and Miura (1963) and Marmur (1961) yields transforming DNA of a high molecular weight. Cells are grown in Penassay broth until the early stationary phase, harvested by centrifugation, washed with Tris–EDTA buffer [0.5 M tris (hydroxymethyl) aminomethane (Tris) hydrochloride, 0.1 M EDTA, pH 8.0] and concentrated tenfold by resuspending in Tris–EDTA buffer. Cells are incubated for 1 hr at 37°C with lysozyme (1 mg/ml, Calbiochemical Co.). The crude lysate is frozen and thawed several times, incubated with 1 mg/ml pronase (that was heated for 10 min at 60°C to destroy DNase) for 1 hr at 37°C, and finally incubated for 30 min at 37°C with 1 percent Sarkosyl NL-79 (Geigy Co.) and 1 percent sodium dodecyl sulfate (Sigma Co.). The lysate is then mixed with an equal volume of phenol (Mallinckrodt Co.), buffered with Tris (pH 8.1, 0.1 M) for 10 min by gentle rocking, and the phenol layer is discarded. After repeating this procedure twice, the DNA in the aqueous phase is dialyzed against potassium phosphate buffer (pH 7.4, 0.1 M) containing 1.0 M NaCl for 6 hr and then against SSC (0.015 M sodium citrate containing 0.15 M NaCl) twice for a total of 6–8 hr. The DNA can be further purified by precipitation with ethanol in high salt. Teichoic acid can be removed by precipitation of the DNA with isopropanol (Young and Jackson, 1966).

Mechanisms of Genetic Exchange

Transformation

One of the most important prerequisites of genetic experiments is the development of isogenic strains. Unfortunately this precaution is not followed by all workers in this field. In the absence of standardization of strains, one encounters the risk of using mutants with multiple unknown defects. This is particularly important in view of the many closely linked mutations introduced with powerful mutagens such as N-methyl-N-nitro-N´-nitrosoguanidine (Cerdá-Olmedo *et al.*, 1968). The probability of using organisms with multiple defects can be reduced by transferring the newly induced mutation to a standard recipient strain by transformation with 1–10 μg/ml of DNA. With these saturating levels of DNA, 5–10 percent of the transformants will receive more than one fragment of DNA and will be simultaneously transformed for two unlinked genes. The concentration of DNA should not exceed 100 μg/ml since Yoshikawa (1966) observed the induction of mutations with high concentrations of DNA. By this process of congression (Nester *et al.*, 1963), a number of collections of *B. subtilis* 168 have been developed. The largest of these, carrying the BR designation, was constructed by B. E. Reilly. Drs. Hoch, Spizizen, Tevethia, and Young have developed additional collections in the same background. Other collections developed by Nester (designated SB or WB), Sueoka (designated Mu), Dubnau (designated DB), and Anagnostopoulos (designated GSY) are in wide circulation. It would be advisable to utilize only one of these collections whenever possible.

There have been many modifications of the original method for the development of competence as described by Spizizen and co-workers (Anagnostopoulos and Spizizen, 1961; Young and Spizizen, 1961). The most reproducible procedures that yield highly competent populations are those in which the development of competence is correlated with the physiological stage of growth. One of these methods is particularly suited to studies pertaining to the biochemical events accompanying the development of competence because transformability is attained in a single chemically defined medium (Bott and Wilson, 1968). Growth of the cells is monitored to determine the time that the culture leaves the exponential phase of growth. This time is designated as T_0. Three hours after T_0, the cells are maximally competent for the uptake of DNA. Another procedure that yields highly competent cultures in a much shorter time has been used extensively in our laboratory. Cultures are grown overnight without aeration at 20–25°C in growth medium 1 (designated GM1 rather than the previous designation SPI, Yasbin *et al.*, 1973*b*). GM1 medium contains 100 ml Spizizen's minimal salts supplemented with 22 mM glucose, 0.02 percent acid-hydrolyzed casein, *p*H 7.0 (Nutritional Biochemical Co.), 0.1 percent yeast extract (Difco), and 50 μg/ml of the amino acids required to supplement the auxotrophic defects in the recipient strain. The yeast extract may be omitted; however this results in a slight decrease in the frequency of transformation. The overnight culture is transferred aseptically to a nephalometer (250 ml) and incubated at 250 rpm at 37°C. The turbidity of the culture is measured at 15-minute intervals in a Klett–Summerson colorimeter (filter No. 66) after the first hour of incubation. Turbidity is plotted as a function of $\log_2$ until growth departs from a linear relationship (T_0). Ninety minutes after T_0 the cells are diluted 1:10 into growth medium 2 (GM2). The GM2 is similar to GM1 with the exception that $CaCl_2$ and $MgCl_2$ are added to a final concentration of 0.5 and 2.5 mM, respectively. Cells reach maximal competence in GM2 after 60 min and retain a high degree of transformability for 6 hr (Yasbin *et al.*, 1973*b*). Since the GM1 and GM2 media contain casein hydrolysate in addition to yeast extract, care must be taken to avoid batches of casein hydrolysate that contain concentrations of amino acids or heavy metals that inhibit the development of competence (Young and Spizizen, 1961; Wilson and Bott, 1968). Apart from the differences in the methods used to attain competence, no difference in the physiology of the competent cells have been reported among the various procedures.

Although the period of competence is relatively transient, it can be preserved for long periods at −90°C. Cells growing at 37°C in GM2 are mixed with 0.15 volumes of glycerol 15 minutes prior to the period of maximal competence. After the 15-minute incubation period, cells are rapidly cooled in an ice bath and dispensed into tubes that have been previously chilled to −90°C. The cells should be rapidly thawed at 37°C before DNA is added. Although there is some decrease in viability in the population, the frequency of transformation remains unaltered. The cells are usually incubated with DNA for 2.5–30 min at 37°C. Once competence is developed, only divalent ions such as Mg^{++}, Ca^{++}, Sr^{++}, or Ba^{++} are required for the irreversible binding of DNA (Young and Spizizen, 1963; Strauss, 1970). The early events in transformation have been further dissected by studying mutants that are defective in some stage of the transformation process (Polsinelli *et al.*, 1973; Yasbin *et al.*, 1973*a,b*; Joenje *et al.*, 1973), and also by studying the effects on these events of DNA-extraction techniques (Davidoff-Abelson and Dubnau, 1973*a,b*; Bettinger and Young, 1973) EDTA (Morrison, 1971), cyanide (Strauss, 1970), temperature (Strauss, 1970; McCarthy and Nester, 1969; Archer, 1973), cations (Young and Spizizen, 1963; Groves and Young, 1973), and antibodies to single-stranded DNA (Erickson *et al.*, 1969).

The most comprehensive investigations of the early events in the process of transformation have been conducted by Dubnau and his collaborators (Dubnau and Cirigliano, 1972*a,b,c*; Dubnau and Davidoff-Abelson, 1971; Davidoff-Abelson and Dubnau, 1973*a,b*). They have proposed that donor DNA is processed in a stepwise fashion from native

DNA of high molecular weight (approximately 9×10^7) to double-stranded fragments with a molecular weight of about 9×10^6. These fragments are degraded to trichloroacetic acid-soluble 5′-mononucleotides and single-stranded fragments of $2–5 \times 10^6$ molecular weight when isolated carefully in the presence of high concentrations of lysozyme (Davidoff-Abelson and Dubnau, 1973a; Piechowska and Fox, 1971). These single-stranded fragments eventually form a donor–recipient complex either prior to or during recombination. It is important to note that the survival of the transformed clone on minimal glucose agar supplemented with auxotrophic requirements or antibiotics is many steps removed from the initial event.

Transfection

The development of DNA-mediated transformation in *B. subtilis* stimulated the investigation of transfection or infection of the recipient strain by DNA isolated from bacteriophages. To date, a number of bacteriophages (SP50, SP82G, SP01, SP02, $\phi1$, $\phi25$, $\phi29$, $\phi105$, and SPP1 have been extensively utilized in such studies. A number of basic types of experiments were performed in the initial development of the transfection systems (1) genetic analysis of the genome of the bacteriophage, (2) biophysical analysis of the replication of the bacterial and bacteriophage genes following induction of a bacteriophage in lysogenic strains, (3) correction of mutations in the heavy or light strand of the bacteriophage DNA, (4) analysis of the rate and order of entry of bacteriophage DNA, (5) the effect of proteases on transfection, and (6) the relative persistance of double- and single-stranded bacteriophage DNA in nontransformable strains of *B. subtilis*. The reader is referred to the recent reviews by Young and Wilson (1972) and Trautner and Spatz (1973) for detailed discussion of these experimental approaches. Two of these recent developments deserve more attention. First, transfection has been adapted to study gene conversion by the use of heteroduplex molecules of SPP1 DNA (Spatz and Trautner, 1970). The data demonstrate that there is an effective mechanism to convert heterozygous to homozygous molecules prior to replication. Furthermore, both the heavy and the light strands can contribute markers to the progeny. Secondly, transfection can be utilized to dissect the events involved in recombination and repair. For instance, Okubo and Romig (1966) first observed that transfection with DNA isolated from SP01 was diminished in strains carrying the *rec*B mutation, whereas transfection with DNA from SP02 was not reduced. Similar differences were noted by Spatz and Trautner in *rec*4 (Spatz and Trautner,

1971). Yasbin *et al.* (1973*a,b*) have extended these observations and incorporated transfection as one method to classify recombination deficient strains of *B. subtilis*. Although considerable evidence favors the contention that recombination during transfection requires double-stranded fragments, in contrast to use of single-stranded fragments in transformation, further detailed studies with recombination deficient strains will be required.

A promising lead may be the differential effect of lysogeny on transfection and transformation. Yasbin *et al.* (1973*b*) have noted that transformation but not transfection or transduction with PBS1 or SP10 is reduced in lysogenic cultures of *B. subtilis* carrying ϕ105 or SP02. It is not known whether this drastic reduction in transformation is due to abnormalities in recombination enzymes, alterations in the microbial cell surface resulting in a decreased uptake of bacterial DNA, a preferential degradation of bacterial genes by nucleases, or a preferential lysis of competent cells.

The conditions for development of competence for transfection are similar to those required for the development of transformation. Particular care, however, must be taken in the isolation of DNA. One of the most efficient methods for isolation of bacteriophage DNA in our laboratory consists of incubating the pelleted virus for 8 hr at 4°C in potassium phosphate buffer (*p*H 7.4, 0.1 M), extracting the mixture with an equal volume of buffered phenol and then repeating the extraction two more times. The DNA in the aqueous phase is dialyzed first against potassium phosphate buffer (*p*H 7.4, 0.1 M) containing 1.0 M NaCl for 6 hr and then against SSC twice for a total of 6–8 hr. For some viruses, such as ϕ29, it may be necessary to avoid the use of proteolytic enzymes. Hirokawa (1972) demonstrated that a protein which is tightly associated with the viral genome is required for transfection.

Infectious centers can be assayed by plating a sensitive indicator strain with the transfectants in semisolid agar on TBAB medium (Yasbin *et al.*, 1973*b*). Two procedures have been utilized to prevent reinfection of the original transfection mixture. Because glucosylated teichoic acid is required for adsorption of many of the viruses which infect *B. subtilis* (Young, 1967), it is possible to use a highly competent nonglucosylated recipient defective in phosphoglucomutase for transfection experiments with ϕe, ϕ25, ϕ29, and SP01. The indicator in these experiments is the phage-sensitive strain. Alternatively, it is possible to utilize a streptomycin-sensitive recipient in the initial transfection and a streptomycin-resistant indicator strain. In these experiments, streptomycin is added to the agar to kill the streptomycin-sensitive cells. This is of particular importance when the indicator and recipient strains are incompatible due to the production of antibiotics or bacteriocins. In such cases one can be mislead by localized areas of lysis in the bacterial lawn. Although most of the bacteriophage that infect *B. subtilis* form plaques well on TBAB with an overlay of semisolid agar plus glucose (Young, 1967), the bacteriophage SP02 will not. This phage forms plaques well on the modified M medium developed by Yehle and Doi (1967) that contains 1 percent tryptone (Difco), 0.5 percent yeast extract (Difco), 0.17 M NaCl, 0.005 M CaCl$_2$, and 0.0001 M MnCl$_2$. The presence of glucose in the semisolid overlay or in the basal plate media inhibits SP02 infection.

Because transfecting DNA is subject to intracellular inactivation, the frequency of transfection can be greatly increased by superinfection with mutant viruses. A detailed account of the use of superinfection in mapping can be gained from the analysis of the work of Green (1966) and Armentrout and Rutberg (1970, 1971). Essentially the recipient is incubated with transfecting DNA from a mutant bacteriophage for 20 minutes, the reaction is terminated with DNase (50 μg/ml), and the culture is subsequently infected at a multiplicity of infection of 10–20 with a different mutant phage. The recombinant wild-type phages are distinguished from the input superinfecting phages or viruses derived from transfecting DNA by plating on appropriate hosts. The more frequently utilized types of bacteriophage markers include temperature-sensitive mutants and suppressor mutants. The recent development of an isogenic collection of suppressor mutants in *B. subtilis* 168 strain BR151 by Tevethia has provided a constant genetic background for such bacteriophage experiments.

Transduction

Two major types of transduction have been described in bacteria, generalized and specialized. Unfortunately only generalized transduction has been observed in the genus *Bacillus* to date. However, the lysogenic viruses ϕ105 and SP02 may yet prove successful in developing a system for specialized transduction. Two generalized transducing bacteriophages have dominated the genetic studies, SP10 (Thorne, 1962) and PBS1 (Takahashi, 1961). The transducing particle in lysates of both of these viruses contains only bacterial DNA (Okubo *et al.*, 1963; Yamagishi and Takahashi, 1968). Two problems must be surmounted in utilizing these viruses in genetic experiments: first, the development of good methods for the propagation of the viruses, standardization of transducing lysate, and assay of the total viruses in the lysate; and second, the utilization of procedures to ensure the survival of the transduced clones. Most of the methods for transduction with PBS1 are relatively simple if care is taken to select highly motile recipient and donor strains and seed lysates are prepared on *Bacillus pumilus* (Lovett and Young, 1970). Using phase microscopy one can readily assay the effectiveness of the lysate by observing "paralysis" of motile cultures when mixed with seed lysates because PBS1 adsorbs to motile flagella of *B. subtilis* (Joys, 1965). The donor strain is grown for 2–3 hr in PB at 37°C at 250 rpm to produce a final turbidity of 125–150 Klett units (Klett–Summerson colorimeter, filter No. 66), diluted to a final turbidity of 25 Klett units, and mixed with a stock suspension of PBS1 at a multiplicity of infection (moi) of 0.1. After an additional 1 hr of incubation the culture is incubated for 2 hr with chloramphenicol (5 μg/ml), and then incubated for 18 hr at 37°C without aeration. The bacterial debris is removed by centrifugation at 9000 rpm for 20 min, the supernatant liquid incubated with DNase (50 μg/ml) for 15 min at 37°C, and this lysate finally sterilized by filtration through a type-HA filter (0.45 μm, Millipore Corp). Because PBS1 does not readily lyse *B. subtilis* 168, it is difficult to plaque the virus on this host. An estimation of the plaque forming units (pfu) can be obtained by using *B. pumilus* as the indicator strain and determining the pfu after 6 hr of incubation in TBAB medium (Lovett and Young, 1970). In transduction experiments, the recipient strain is grown for 18 hr at 37°C in PB, transferred to 2.5 ml of PB (initial turbidity 25–35 Klett units), incubated at 250 rpm for 5 hr at 37°C, and then infected with PBS1 at a moi of 1 in a final volume of 1.0 ml. After 15 minutes, the culture is diluted with 3.0 ml of Spizizen's minimal medium, centrifuged at 9000 rpm for 5 min, and the pellet suspended in 1.0 ml of minimal medium. Aliquots (0.1–1.0 ml) are plated on minimal medium. To ensure that the lysate contains transducing particles that are capable of transferring 5–6 percent of the genome of *B. subtilis,* we routinely examine the cotransfer of Lys$^+$ and Met$^+$ in strain BR151 carrying *lys*3, *trp*C2, and *met*B10. In standard lysates the cotransfer of these two markers is 20 percent.

SP10 can only be propagated on *B. subtilis* W23; therefore all crosses are between heterologous but closely related strains. For instance *B. subtilis* W23 differs from *B. subtilis* 168 in DNA:DNA hybridization (Lovett and Young, 1969), chromosomal replication (Yoshikawa and Sueoka, 1963a,b), cell wall composition (Glaser *et al.,* 1966), susceptibility to viruses (Glaser *et al.,* 1966) and genetic exchange (Dubnau *et al.,* 1969). Despite these differences between strain W23 and strain 168, fragments of DNA analogous to the size transferred by DNA-mediated transformation can be exchanged. SP10 lysates can be prepared on TBAB plates or in PB. By scraping the soft agar layer off TBAB plates that show confluent lysis and resuspending the agar layer in PB (20–30 ml for 8–10 plates) high titers can be recovered. The agar mixture is centrifuged and the supernate filtered through a millipore filter. In order to obtain lysates in PB, strain W23 is grown in PB with a turbidity of 50 Klett units, and then infected at a moi of 5. Complete lysis occurs in approximately 1 hr at 37°C with aeration. For transduction, the recipient strain is grown in PB to a turbidity of 50 Klett units and then infected at a moi of 5. Plating is essentially the same as for PBS1-mediated transduction.

Mapping Techniques in *Bacillus subtilis*

In the absence of conjugation, mapping has been limited to the exchange of small segments of the chromosome (approximately 1 percent for DNA-mediated transformation and SP10 transduction and 5–8 percent for PBS1 transduction). Extensive linkage of traits can be obtained if the DNA is carefully isolated and protected by large quantities of nontransforming DNA (Kelly, 1967a,b; Bettinger and Young, 1973). Two precautions should be emphasized before discussion of procedures used in mapping. First, low concentrations of DNA must be employed in DNA-mediated transformation to preclude congression (Barat *et al.*, 1965). The most conclusive means of determining linkage between two loci is to examine the number of cells transformed for one or both markers as a function of the DNA concentration. At nonsaturating levels of DNA, the number of transformants for either a single marker or both linked genes will vary linearly with the concentration of DNA, while transformation for both unlinked genes will vary with a power of the DNA concentration (Goodgal, 1961). Secondly, it is important to note that three-factor crosses by PBS1 transduction with closely linked markers will yield anomalous results in a fashion analogous to the study by Crawford and Preiss (1972).

Localization of Mutations

PBS1 transduction is the easiest method for rapid localization of genetic markers. By using approximately six multiply marked auxotrophic strains as donors, one can transduce the unmapped locus to prototrophy on minimal medium supplemented with all of the auxotrophic requirements of the donor strain. The transductants are cloned on minimal medium supplemented with the auxotrophic requirements of the donor strain and subsequently replicated onto minimal medium deficient in these requirements. If clones do not grow on the latter medium, the exact auxotrophic requirement that is linked to the locus under investigation can be readily determined by analysis of the master plate. Fine-structure mapping can then be accomplished by DNA-mediated transformation by the methods outlined below.

Cotransfer Index. The cotransfer index was used by Nester and Lederberg (1961) to determine the degree of linkage between *ind-2* (*trp* C2) and *his-2* (*his* B2). If the cross is made between *AB* (donor DNA) and *ab* (recipient cells), the following classes of transformants can be expected: *AB, Ab,* and *aB,* where *A* and *a* represent wild-type and mutant alleles, respectively. The cotransfer index *r* can be defined:

$$r = \frac{AB}{AB + Ab + aB} = \frac{AB}{A + B - AB}$$

Some investigators approximate *r* from:

$$\frac{AB}{2Ab + AB} \quad \text{or} \quad \frac{AB}{2A - AB}$$

However, this method is generally unreliable because each marker has its own efficiency of integration. These efficiencies can vary three- to fourfold between markers; however, the tenfold difference between high efficiency and low efficiency classes noted in Pneumococcus (Ephrussi-Taylor and Gray, 1966) have not been reported in *B. subtilis*. Map units are generally expressed as $(1 - r)$.

Recombination Index. The recombination index can be used for fine-structure mapping within a gene (or operon) without determining linkage to another gene. Originally developed by Lacks and Hotchkiss (1960), this procedure was later refined and used by Ephrati-Elizur *et al.*, (1961) to order mutations in the histidine locus. The recombination index is calculated by comparing a cross in which wild-type DNA (A_2B) is used to transform an a_2b recipient (where *A* and *B* are unlinked genes) to a cross in which a_1B DNA is used to transform the same recipient. The *B* class of transformants serves as an internal control compensating for differences in the concentration of DNA, the efficiencies of recombination, and the efficiency of the DNA preparation to transform genetic markers. The latter parameter is determined by the average molecular weight of the DNA, number of single-strand nicks, or other forms of physical and chemical damage. The recombination index (RI) is determined from the following relationship:

$$RI = \left(\frac{A \text{ from mutant DNA}}{B \text{ from mutant DNA}}\right)\left(\frac{B \text{ from wild type DNA}}{A \text{ from wild type DNA}}\right)$$

Three-Factor Transformation and Transduction. The most reliable mapping data come from genetic exchange involving 3 linked markers. Because only segments of DNA are exchanged, the least frequent class of recombinants will be that in which a quadruple crossover event was responsible for the formation of a recombinant class. That is, in a cross where *ABC* DNA is used to transform *abc* recipient cells, the *AbC* class will be least frequent if the order is *abc*. Unfortunately some investigators do not construct strains to perform reciprocal analysis. It must be emphasized however that three-factor crosses with closely linked markers may yield anomalous results if PBS1 transduction is the mechanism of exchange.

Marker Frequency Analysis and Synchro-Transfer Analysis. These methods of gross-structure mapping depend on the replication of the genome from a single fixed origin. Synchrony of replication is important in order to obtain unambiguous results. The highly transformable 168 strains, unlike W23, do not have fully replicated, or "arrested," chromosomes in the stationary phase of growth. Therefore, all of the genes in the population of DNA molecules are not in equal amounts. In order to examine crosses with isogenic strains, it is necessary to use spores as the source of DNA. When spores are germinated DNA replication begins in a sequential and synchronous manner. (Borenstein and Ephrati-Elizur, 1969). These characteristics have been used to refine the original mapping techniques that were elegantly developed by Sueoka and Yoshikawa (1963) and Yoshikawa and Sueoka (1963*a,b*). In marker frequency analysis, DNA is isolated from replicating cells or germinating spores after varying periods of replication. The number of transformants for a marker that replicates early is compared to the number of transformants for a marker that replicates later. This ratio is normalized with DNA isolated from spores to correct for differences in the integration frequency for various markers. In synchro-transfer analysis, the cells are allowed to sporulate in medium containing D_2O and other isotopes. The DNA of the spores will have a greater density than that of cells grown in "light" medium. The spores are then germinated in light medium and DNA is isolated after different periods of germination and centrifuged to equilibrium in CsCl. DNA is collected from the gradient and used to transform recipient cells. The map position is determined from the time the marker has shifted to the hybrid density. Both methods have proved to be reliable. A statistical interpretation of replication mapping data is thoroughly explained by O'Sullivan and Sueoka (1967).

Mapping by Enzyme Activity. Kennett and Sueoka (1971) have extended the concept of mapping by gene replication to include mapping by sequential increases in enzyme activity. For example, the relative activity of a variety of enzymes is determined during germination of spores and cell division. Thus this technique provides a functional map that can be compared with the genetic map. Although there is a rough correlation between the order of the genes on the map and the increase in enzyme activity, some enzymes are not expressed during the first round of DNA replication following germination of the spores.

Mapping Studies with Merodiploids. Audit and Anagnostopoulos (1972, 1973*a,b*) have recently developed a method for determination of dominance of mutations by genetic analysis in merodiploids. These workers observed that cells carrying the *trp*E26 locus originally derived from *B. subtilis* 166 readily form heterogenotes over large regions of the chromosome, primarily to the right of the *trp*E26 locus. Merodiploids have been formed by PBS1 transduction as well as by transformation, thus indicating that these heterogenotes probably do not result from some unique mechanism of recombination that is operative only in transformation. Karmazyn *et al.* (1972) utilized this technique to explore the dominance of sporulation mutations in Spo$^+$/Spo$^-$ heterogenotes. They showed that three SpoOa mutants, SpoOa 45, SpoOa 46 and SpoOa 47, (refer to Young and Wilson, 1972 for a description of these mutants), are dominant over the wild-type allele. It is anticipated that this technique will be widely utilized in the future.

Transformation with DNA from Heterospecific Sources. In all of the transformable species of bacteria where adsorption of DNA from other sources has been studied, transformation occurs only when DNA is isolated from a closely related species. Thus organisms which are capable of exchanging DNA are classified as belonging to the same genospecies. Although mutual exchange is generally expected between two members of the genospecies, evidence of exchange may not yet be possible with both organisms. This is especially true when the genospecies is determined by transformation, since only one of the two members might be capable of attaining competence. Transfection may be a special case of genetic alteration by foreign DNA and will be discussed in the next section.

Marmur, Seaman, and Levine demonstrated heterospecific transformation in *B. subtilis* as early as 1963. Subsequently the understanding of the molecular basis for genetic exchange and recombination of foreign DNA has greatly expanded, and a number of models have emerged to explain the phenomenon. Although differences exist in the various systems of transformation, certain features have been shown to be common to all of them. Heterospecific or heterologous DNA is markedly reduced in its ability to transform when compared to homologous or homospecific DNA. Loci are transformed at varying degrees of efficiency and with some markers, the level of exchange is not distinguishable from the level of back mutation or reversion. When efficient exchange is observed in a region of the chromosome, the region is said to be conserved. Regions of genetic conservation have been established between *B. subtilis* and *B. globigii* (Chilton and McCarthy, 1969), between *B. subtilis* 168 and *B. subtilis* W23 (Dubnau *et al.*, 1965), and between *B. subtilis* 168 and *B. subtilis* HSR (Wilson and Young, 1972). Once the fragment of foreign DNA is integrated in the genome of the recipient, this *intergenote* becomes markedly more efficient than the original foreign DNA in its ability to exchange this "foreign" marker when assayed in the original recipient. In those cases where both species are transformable, the intergenote has been found to be impaired in its ability to transform the original donor strain. This phenomenon has been referred to as the "neighborhood effect" (Biswas and Ravin, 1971) and indicates that efficiency of exchange is favored when the recipient has the same nucleotide composition as the neighborhood of DNA surrounding the intergenotic region (Wilson and Young, 1973). In some cases deleterious side effects have been reported as a consequence of integrating foreign DNA (Wilson and Young, 1972).

The quantification of the efficiency of exchange has not always been carefully controlled in many of the experiments reported. Because DNA preparations vary considerably with respect to the length of each single strand in the double-stranded donor fragments, it is not sufficient to compare the number of transformants obtained by intergenotic transformation with those arising from homospecific transformation. The efficiency of exchange is best determined by the use of an "efficiency ratio" (Wilson and Young, 1972). In this method, the exchange of the foreign marker is normalized to an *unlinked* marker that is identical in both the intergenotic and homologous strain. For example, if locus A_1 was transferred from the heterologous strain (e.g., *B. pumilus*) to *B. subtilis* to form an intergenote carrying locus A_1 and the resident reference marker is B, the efficiency of exchange, E, of this DNA when compared to the efficiency of the homologous strain (carrying markers A_2 and B) is calculated from the following equation:

$$E = \left(\frac{A_1 \text{ (intergenotic DNA)}}{A_2 \text{ (homologous DNA)}}\right)\left(\frac{B \text{ (homologous DNA)}}{B \text{ (intergenotic DNA)}}\right)$$

Efficient exchange in *B. subtilis* occurs in only two regions of the map when heterospecific DNA is used as the transforming agent. One of the regions located near the leucine loci has been carefully documented only for *B. subtilis* W23. The other region of conservation lies near and includes the antibiotic-resistance loci which map near *cys*A. Several different species have been shown to be capable of efficient transformation in this region. The earliest studies of heterospecific transformation utilized the streptomycin- and erythromycin-resistance markers. Recently the rifampin-resistance loci have been studied in considerable detail. These studies have extended the genospecies to include the following strains: *B. subtilis* H, *B. subtilis* W23, *B. subtilis* var natto, *B. licheniformis, B. amyloliquefaciens* H, *B. pumilus, B. subtilis* var niger, and *B. tetraultii* (Wilson and Young, 1972; Harford and Mergeay, 1973). Reports of exchange by DNA from much more distantly related organisms such as *Streptomyces coelicolor* (Mergeay *et al.,* 1971) and *Arabidopsis thaliana* (L.) (Ledoux *et al.,* 1971) indicate that transformation is indeed a sensitive tool with which to investigate genetic relatedness.

Organization of Chromosome and the Genetic Map. The genome of *B. subtilis* is arranged in a circle, although vigorous proof of genetic linkage between the origin and terminus has not been presented. However, recent studies of Gyurasits and Wake (1973) and Wake (1973) have provided convincing autoradiographic evidence that the chromosome is circular. Their data also suggest that the molecular weight of the chromosome may be less than the value of 3.0×10^9 daltons previously determined from the DNA content of spores (Dennis and Wake, 1966). The measured length of 900–1100 μm is more consistent with determinations of $1.6–2.8 \times 10^9$ daltons (Bak *et al.,* 1970) for 3 related bacilli and 2.0×10^9 for *B. subtilis* (Klotz and Zimm, 1973).

Most genetic studies support a linear replication of the chromosome beginning at a single origin and ending at the terminus (Yoshikawa and Sueoka, 1963*a,b*; Sueoka and Yoshikawa, 1963). From amino acid starvation experiments in *B. subtilis* 168 and W23, however, Copeland (1969, 1971) has suggested that chromosomal arrest and reinitiation occurs at multiple sites. Recent experiments by Gyurasits and Wake (1973) and Wake (1973) suggest that bidirectional replication occurs over 50 percent of the genome. Bidirectional replication would not be unique to *B. subtilis,* as it has already been reported to occur in *Escherichia coli* (Masters and Broda, 1971) and in *Salmonella typhimurium* (Nishioka and Eisenstark, 1970; Fujisawa and Eisenstark, 1972). These studies have prompted a reexamination of the replication mapping data. If 50 percent of the chromosome is replicated in the bidirection fashion, it is quite possible that the *pur*B region is misplaced. For example, if the origin of replication is considered to be the top of the map, then the *pur*B segment may alternatively reside on the left fork of replication near the origin, thus accounting for its early replication and also the unsuccessful attempts by many investigators to link this region with the rest of the chromosome. One would then have to postulate that a significant portion

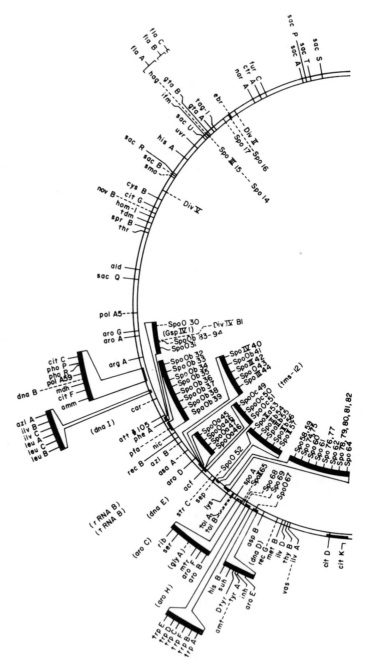

Figure 1. Linkage map of B. subtilis (Young and Wilson, 1975).

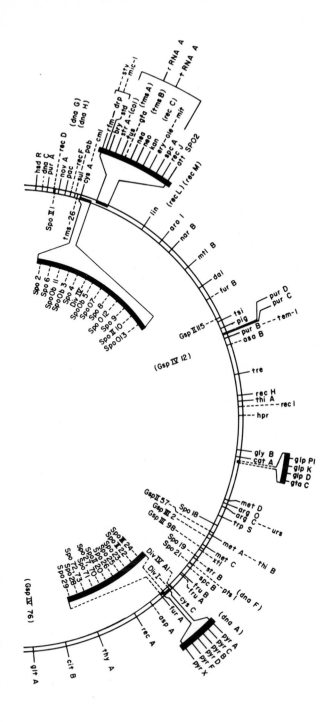

of the map has not been discovered. This is not unreasonable in view of the relatively small numbers of markers that have been identified to date, if one compares the _B. subtilis_ map to the map of _E. coli_ (Taylor and Trotter, 1972). Three newly mapped loci may help to resolve the location of the _pur_B region. One of these markers, _dal,_ that codes for D,L-alanine racemase, has been linked to _pur_B by PBS1 transduction. It appears to be the most distal marker in this region (Dul and Young, 1973). The second marker, _sac_Q maps as the nearest marker to this region in the segment of the chromosome that replicates after _pur_B (Lepesant _et al.,_ 1972). The third marker, _fur_B is cotransduced (PBS1) 10 percent with _pur_B, but its placement to the left or right of _pur_B has not been established. It is not linked to _thr_ or _ery_ (P. Winter and S. A. Zahler, private communication). Therefore, until the genetic location of the _pur_B region is established, we have chosen to leave it in its present position. In addition to the _pur_B region, other markers exist that have not been linked to known portions of the map. These include citric acid-cycle mutants, _cit_A and sporulation-deficient mutants _spo_F (Hoch and Mathews, 1972); a series of _hut_ mutants affecting histidine utilization (Chasin and Magasanik, 1968); suppressor mutations _spr_A (Vapnek and Greer, 1971a,b), Su^+3 and 17 other suppressor mutations of two distinct classes (Georgopoulos, 1969), _sup_-1 (Okubo and Yanagida, 1968), and _su_ (Tevethia, private communication); genes determining biosynthesis of defective phages PBSX, PBSY, and PBSZ (Okamoto _et al.,_ 1968; Garro, _et al.,_ 1970), and PBSH (Haas and Yoshikawa, 1969a,b); and _Nar_B, a gene necessary for nitrate reductase (P. Winter and S. Zahler, private communication).

In addition to these uncertainties in the mapping data, it is difficult to classify genes that regulate sporulation. For instance, many strains have been isolated that are unable to complete the process of sporulation and have been phenotypically designated Spo⁻. However, to date, the only spore-specific deficiency that has been characterized in any asporogenic mutant is the temperature-sensitive serine protease. This mutant is blocked at or before stage 0 of the sporulation process at elevated temperatures (Leighton _et al.,_ 1972). Therefore, the molecular basis for asporogeny is unknown in most mutants. Furthermore, many are blocked in energy producing steps that are secondarily necessary for the completion of sporulation. Therefore, we have introduced a new convention in which the markers which effect differentiation and division are located on the inside of the genetic map. These loci will be designated as phenotypes until the gene product is identified and characterized, and the primary effect of the product is established.

TABLE 1. *Genetic Markers of Bacillus subtilis*

Designation on map	Previous designations and comments	References
afc	*acr₁.* Acriflavin resistance	Ionesco *et al.*, 1970
amm	*amm-35.* Inability to assimilate NH_4^+; will grow on 0.5-percent glutamate	Dubnau *et al.*, 1965
amt	Resistant to 3-amino-tyrosine; cotransformed with *tyrA*	Nasser and Nester, 1967; Polsinelli, 1965
argA	Strict arginine requirement	Barat *et al.*, 1965; Dubnau *et al.*, 1965; Mahler *et al.*, 1963
argO	Requires arginine, ornithine, or citrulline	Mahler *et al.*, 1963; Young *et al.*, 1969
argC	Requires arginine or citrulline	Borenstein and Ephrati-Elizur, 1969; Dubnau *et al.*, 1965; Mahler *et al.*, 1963; Young *et al.*, 1969
aroA	*aro-6.* Requires aromatic amino acids; lacks 3-deoxy-D-arabino heptulosonic acid-7-phosphate synthetase; linked to *aroG* (for position on map, see Dubnau, 1970)	
aroB	*aro-2.* Requires aromatic amino acids; lacks dehydroquinate synthetase	Nasser and Nester, 1967
aroC	*aro-7.* Requires aromatic amino acids; lacks dehydroquinase	
aroD[a]	*aro-4, aro-2.* Requires aromatic amino acids lacks dehydroshikimate reductase; linked to *phe-4* (0.92 map units) by PBS1 transduction	Adams and Oishi, 1972; J. A. Hoch and Mathews, 1972; Staal and Hoch, 1972
aroE	*aro-1.* Requires aromatic amino acids; lacks 3-enolpyruvyl shikimate-5-phosphate synthetase	Nasser and Nester, 1967; Nester *et al.*, 1963
aroF	*aro-5.* Requires aromatic amino acids; lacks chorismate synthetase; position relative to *aroC*, *rib*, and *ser* uncertain	Nasser and Nester, 1967

TABLE 1. Continued

Designation on map	Previous designations and comments	References
*aro*G	CM3. Requires phenylalanine, and tyrosine, lacks chorismic acid mutase; linked to *aro*A, but position uncertain	Nasser and Nester, 1967
*aro*H	CM1,2. All 168 derivatives lack this form of chorismic acid mutase; cotransforms with *trp*	Lorence and Nester, 1967; Nasser and Nester, 1967
*asa*A [a]	*asa.* Arsenate resistance; located between *nic* and *aro*D	Adams and Oishi, 1972; Adams, 1972, 1971
*asa*B [a]	*asa* (W23). Arsenate resistance conferred by transformation of *B. subtilis* 168 by DNA from *B. subtilis* W23; maps near *pur*B; the location of *asa*B to the left of *pur*B is tentative	
*asp*A [a]	Requires either aspartate or glutamate; lacks pyruvate carboxylase; closely linked to *rec*A	J. A. Hoch and Mathews, 1972
*asp*B [a]	Requires either aspartate or glutamate; lacks glutamate-oxalacetate transaminase, single mutant isolated that also decreases cotransduction of *met*B and *trp*C from 55 percent to 10 percent or less; may be the result of an insertion; maps between *trp*C and *met*B	
*att*SP02	Integration site for SP02 bacteriophage; by transduction linked 20 percent to *cys*A14 and 60–65 percent with *ery*; maps to right of *ery*	Inselburg *et al.*, 1969
*att*ø105	Integration site for Ø105 bacteriophage; maps between *ilv*C (*ilv*A1) and *phe*-1 by three-factor transduction	L. Rutberg, 1969
*azl*A	4-Azaleucine resistance, increased levels of biosynthetic enzymes and dihydroxyaciol synthetase; mapped by three-factor transformation	S. A. Zahler and J. Ward, private communication

*azl*B	4-Azaleucine resistance, does not excrete as high levels of leucine as *azl*A, and has normal levels of α-IPM[b] synthetase	S. A. Zahler and J. Ward, private communication
bry	Bryamycin resistance; maps to left of *str*-1 on basis of map distance from *cys*A14 and three-factor transformation	Goldthwaite *et al.*, 1970; Harford and Sueoka, 1970
car	*car*-41. Carbohydrate-negative; originally isolated as maltose⁻; does not respond to trehalose, maltose, or sucrose but grows normally on glucose; very near or distal to *leu*8 by density transfer mapping	Kennett and Sueoka, 1971
*cit*B	Lacks aconitase	B. Rutberg and Hoch, 1970; S. A. Zahler, private communication
*cit*C	*spo*C 15, *spo*C 9, *spo*C 7, *cit*-1. Lacks isocitrate dehydrogenase; maps to the left of *pho* by transformation	J. A. Hoch and Spizizen, 1969; Le Hegarat and Anagnostopoulos, 1969; B. Rutberg and Hoch, 1970
*cit*D	Lacks α-ketoglutarate dehydrogenase; may be a deletion extending into *cit*K region	B. Rutberg and Hoch, 1970
*cit*F	*spo*C 28, *spo*C 24, *spo*C 20. Lacks succinate dehydrogenase	J. A. Hoch and Spizizen, 1969; B. Rutberg and Hoch, 1970
*cit*G	Lacks fumarase	B. Rutberg and Hoch, 1970
*cit*K	*cit*E. Lacks α-ketoglutarate dehydrogenase	B. Rutberg and Hoch, 1970; S. A. Zähler, private communication
col	Colistin resistance; cotransformed with *str*	See Dubnau, 1970
*ctr*A[a]	*nir*-6, *nir*A. Absolute requirement for cytidine in the absence of ammonium; can be satisfied with high levels of ammonium; linked to *sac*A-321B (0.78 map units) by PBS1 transduction	P. Winter and S. A. Zahler, private communication
*cyt*R[a]	Resistance to 5-fluorouracil (40μg/ml); may result in constitutive synthesis of uridylic acid (UMP); not linked to *pyr*A or *fur*A; map location not determined	S. A. Zahler, private communications
*cys*A	*cys*-14. Requires cysteine	Dubnau *et al.*, 1965; Harford and Sueoka, 1970

TABLE 1. Continued

Designation on map	Previous designations and comments	References
cysB	cys-3. Requires cysteine	Dubnau et al., 1965; Harford and Sueoka, 1970
cysC	Requires cysteine	See Dubnau, 1970
dal[a]	Requires D-alanine; defective in D,L alanine racemase; linked to purB (0.91 map units), pig (0.79 map units), and tsi (0.77 map units) by PBS1 transduction	Dul and Young, 1973; Freese et al., 1964
Div IV-A1[a]	Defect in location of division site along cell length resulting in production of minicells; linked to pyr (0.21 map units) and cysC (0.48 map units) by PBS1 transduction	Reeve et al., 1973
Div IV-V1[a]	Defect in location of division site along cell length resulting in production of minicells; linked to pheA (0.45 map units) and leuA (0.91 map units) by PBS1 transduction	
dnaA[a]	ts DNA A. Temperature-sensitive DNA synthesis tsA13 is most likely a mutation in the structural gene of the ribonucleotide reductase; position uncertain with respect to surrounding markers; maps between metA and novB by spore-germination technique of transformation	Bazill and Karamata, 1972; Borenstein and Ephrati-Elizur, 1969; Karamata and Gross, 1970
dnaB	Temperature-sensitive DNA synthesis; maps between argA and leu8 by three-factor transduction	Karamata and Gross, 1970; Mendelson and Gross, 1967
dnaC	Temperature-sensitive DNA synthesis; replicates earlier than purA by spore germination mapping; 30-percent cotransfer with purA	Bazill and Retief, 1969; Borenstein and Ephrati-Elizur, 1969; Karamata and Gross, 1970
dnaD	Temperature-sensitive DNA synthesis; maps to the left of metB and ile-1 by three-factor transduction	Karamata and Gross, 1970
dnaE	Temperature-sensitive DNA synthesis; map position uncertain	

Gene	Description	Reference
*dna*F	Temperature-sensitive DNA synthesis; map position uncertain	Karamata and Gross, 1970
*dna*G	Temperature-sensitive DNA synthesis; maps between *pur*A and *cys*A by three-factor transduction	
*dna*H	Temperature-sensitive DNA synthesis; maps between *pur*A and *cys*A by three-factor transduction	
*dna*I	Temperature-sensitive DNA synthesis; maps between *arg*A and *phe* by three-factor transduction	
drp	*drp*-18. DNA-dependent RNA polymerase; spore morphology altered cotransformation with rifampin resistance (may be same locus)	Korch and Doi, 1971
D-*tyr*	Resistant to D-tyrosine; maps in *tyr*A locus	Champney and Jensen, 1969
ery	*ery*-1. Erythromycin-resistant	Goldthwaite *et al.*, 1970; Harford and Sueoka, 1970; Takahashi, 1965
*fla*A	*fla ts*A. Lacks flagella at 46°C but not at 37°C; order relative to other flagella markers *fla*B and *fla*C uncertain and placement relative to *hag* uncertain; linked to and to the right of *his*A by transduction and loosely grouped around *hag*	Grant and Simon, 1969
*fla*B	*fla ts*B. Lacks flagella at 46°C but not at 37°C; closely linked to *gta*A locus by transduction; segregates from *hag*	
*fla*C	*fla ts*C. Lacks flagella at 46°C but not at 37°C; 16-percent cotransduction to *his*AI	
fru	Unable to grow on fructose; gene order by transformation is F27, F22, F29, F20, F13; F22 and F27 lack fructose-1-phosphate kinase; F13, F20, F29 deficient in fructose transport; mapped between *met*C and *ura*26 by four-factor transduction; 30- to 36-percent cotransduction with *met*C and 32- to 44-percent cotransduction with *ura*26	Gay *et al.*, 1970
*fur*A[a]	Resistant to 5-fluorouracil (1 μg/ml); linked to *pyr*A	Dubnau *et al.*, 1967

TABLE 1. Continued

Designation on map	Previous designations and comments	References
$furB^a$	Resistant to 5-fluorouracil (40 μg/ml); linked to $purB$ (0.90 map units) by PBS1 transduction but order not determined	P. Winter and S. A. Zahler, private communication
$furC^a$	Resistant to 5-fluorouracil (40 μg/ml); linked to $narA$ (0.25 map units), $sacA$-321B (0.55 map units) and $ctrA$ (0.04 map units)	
fus^a	Resistant to fusidic acid; maps between $strA$ and nea	Goldthwaite and Smith, 1972
gfa^a	ts-3. Temperature-sensitive elongation factor G; cotransformable (78–91 percent) with fus; maps between $strA$ and ery; order with respect to nea, neo, and kan not determined	Aharonowitz and Ron, 1972
gl^a	gap. Requires glutamate or aspartate; lacks glutamine-2-ketoglutarate aminotransferase; linked to $citK$ (0.60 map units) by PBS1 transduction	J. A. Hoch and Mathews, 1972
gly	Glycine requirement mapped by unstable linkage transformation between rib and tyr	M. S. Kelly and Pritchard, 1963, 1965; M. S. Kelly, 1967a,b; T. J. Kelly and Smith, 1970
$gtaA$	Glucosylation of teichoic acid; lacks UDP^c glucose:polyglycerol:teichoic acid glucosyl transferase	
$gtaB$	Glucoslyation of teichoic acid; behaves as if lacking UDP:glucose pyrophosphorylase but does not have significant enzymatic defects	Young et al., 1969
$gtaC$	Glucosylation of teichoic acid; lacks phosphoglucomutase; not mapped with respect to hpr	
hag	Flagellar antigen; hag-1 refers to wild-type 168 flagellar antigen, hag-2 to W23, and hag-3 to straight filament; see comment concerning ifm for map placement	Grant and Simon, 1969; Martinez et al., 1968

Gene	Description	References
hisA	his 1,7,8,9,10,12. Requires histidine; probably blocked before imidazoleglycerol phosphate	Borenstein and Ephrati-Elizur, 1969; Dubnau et al., 1965; Ephrati-Elizur et al., 1961; Nasser et al., 1969
hisB	his 2,4,5. Requires histidine, some mutants respond to histidinol; placed on map according to germination mapping technique of Borenstein and Ephrati-Elizur, which may be slightly closer to the origin than indicated by O'Sullivan and Sueoka	Borenstein and Ephrati-Elizur, 1969; Dubnau et al., 1965; O'Sullivan and Sueoka, 1967
hom-1	Requires threonine and methionine, or homoserine plus threonine; lacks asparatakinase and homoserine dehydrogenase; probably a long deletion extending into thr	Anagnostopoulos, private communication; Dubnau, 1970
hpr	Hyperproduction of extracellular protease; maps between spo-18 and argC4	J. A. Hoch and Spizizen, 1969; Prestidge et al., 1971
ifm	Increased flagella and motility; segregates from hag by transduction; maps between uvr-1 and gtaA; tentatively placed between sacU and gtaB because there is 69-percent cotransfer of hisA and sacU, 63-percent cotransfer of hisA and ifm, and 58-percent cotransfer between hisA and hag	Grant and Simon, 1969; Lepesant et al., 1969a
ilvA	ile_1^-, ile_2^-. Isoleucine requirement; lacks threonine deaminase	Anagnostopoulos et al., 1964; Anagnostopoulos and Schneider-Champagne, 1966; Jamet and Anagnostopoulos, 1969
ilvB	Isoleucine and valine requirement; lacks condensing enzyme; order of azlA, ilvB, ilvC, leuB, leuA, leuC, and pheA as determined by Ward and Zahler by three-factor transduction	Anagnostopoulos; J. Ward and S. A. Zahler, private communication; Dubnau, 1970
ilvC	ilvA, ilv1. Isoleucine and valine requirement; lacks reductoisomerase; see ilvB	
ilvD	ilvB, ilva2, ilva4, ilva6. Isoleucine and valine requirement; lacks dehydrase	Anagnostopoulos and Schneider-Champagne, 1966

TABLE 1. *Continued*

Designation on map	Previous designations and comments	References
*inh*491	*inh.* Growth inhibited by histidine; recent biochemical studies indicate a very close relationship to locus coding for prephenate dehydrogenase (*tyr*A)	Nasser and Nester, 1967; Nester *et al.*, 1963
kan	*kan*-2. Kanamycin resistance; by transformation, *kan* maps to right of *neo*-9	Goldthwaite *et al.*, 1970; Harford and Sueoka, 1970
*leu*A	*leu*8. Requires leucine; lacks α-IPM synthetase; fine-structure mapping by three-factor transformation	Dubnau, 1970; J. Ward and S. A. Zahler, private communication; Barat *et al.*, 1965; Borenstein and Ephrati-Elizur, 1969; Dubnau *et al.*, 1965; O'Sullivan and Sueoka, 1967
*leu*B	*leu*6. Requires leucine; lacks IPM isomerase	⎫ J. Ward and S. A. Zahler, private communication
*leu*C	Requires leucine; lacks β-IPM dehydrogenase	⎬
lin	*lin*2, *lin*4. Lincomycin resistance; not linked to *cys*A14 hor *ery*1 by transformation	⎭ Goldthwaite *et al.*, 1970; Harford and Sueoka, 1970
lys	Lysine requirement	Dubnau, 1970; M. S. Kelly and Pritchard, 1965; Anagnostopoulos, private communication
*met*A	Requires methionine, cystathionine, or homocysteine	Borenstein and Ephrati-Elizur, 1969; Dubnau *et al.*, 1965
*met*B	*met*3 *met*4. Requires methionine or homocysteine	Anagnostopoulos and Schneider-Champagne, 1966; Borenstein and Ephrati-Elizur, 1969; Dubnau *et al.*, 1965; O'Sullivan and Sueoka, 1967; Yoshikawa and Sueoka, 1963*a*
*met*C	*met*A. Strict requirement for methionine	Dubnau *et al.*, 1965; J. A. Hoch and Anagnostopoulos, 1970; Young *et al.*, 1969
*met*D	Strict requirement for methionine	Young *et al.*, 1969

Gene	Description	Reference
mic	Micrococcin resistance, altered ribosome, by transduction maps between *cys*A and *str*	Dubnau, 1970; Harford and Sueoka, 1970
mit	Mitomycin resistance; cotransformed with spontaneous Ery^R mutant designated *mac*1	Iyer, 1966
mtr	*mtr*^{r}. Resistance to 5-methyl tryptophan; maps to the left of tryptophan region and linked to it by 48- to 61-percent recombination	S. O. Hoch *et al.*, 1971; Nester *et al.*, 1963
*nar*A[a]	Nitrate reductase; will not grow with nitrate as sole nitrogen source but will use ammonium or nitrite; linked to *fur*C (0.25 map units) by PBS1 transduction	P. Winter and S. A. Zahler, private communication
*nar*B[a]	Does not grow with nitrate as sole nitrogen source; map location not determined	S. A. Zahler, private communication
nea	Neamine resistance; cotransduced with *spc*A, *cys*A, and *ery*; maps to the right of *cys*A and to the left of *lin*-2	Goldthwaite *et al.*, 1970
neo	*neo* 21,26,9. Neomycin resistance; maps between *kan*-2 and *str* by transformation	Goldthwaite *et al.*, 1970; Harford and Sueoka, 1970
nic	*nic*-38, *nia*. Requires nicotinic acid	J. O. Hoch and Anagnostopoulos, 1970; M. S. Kelly, 1967a
*nov*A	*nov*1. Resistant to 5 μg of novobiocin/ml. Maps to the left of *pac* and to the right of *pur*A by transduction (two other novobiocin-resistant mutants, resistant to 3 μg but not 5 μg, do not map between *cys*A14 and *pur*A)	Harford and Sueoka, 1970
*nov*B	*nov*. Resistant to 2 μg of novobiocin/ml; maps at position 0.56 of spore germination map	Borenstein and Ephrati-Elizur, 1969
ole	*ole*-1. Oleandomycin resistance; mapped with *ery*-1 on basis of map distance from *cys*A14 and cross resistance with erythromycin (both act on 50S ribosomal subunit); not ordered with respect to *spc*A or other markers	Goldthwaite *et al.*, 1970; Harford and Sueoka, 1970

TABLE 1. Continued

Designation on map	Previous designations and comments	References
pab	Requirement for paraminobenzoic acid; cotransformed with *sul*	Dubnau, 1970
pac	Pactomycin resistance; three markers mapped to origin proximal side of *cys*A14 by transformation and confirmed by transduction	Harford and Sueoka, 1970
pfa	*p*-Fluorophenylalanine resistance; position relative to *phe*A uncertain	Polsinelli, 1965
*phe*A	*phe*₁⁻. Requires phenylalanine or phenylpyruvic acid; lacks prephenate dehydratase	Barat *et al.*, 1965; Nasser and Nester, 1967; Takahashi, 1965b; J. Ward and S. Zahler, private communication
*pho*P	*pho*-6, *pho*-8, *pho*-11, *pho*-12. Structural gene for alkaline phosphatase	Le Hegarat and Anagnostopoulos 1969; Oishi *et al.*, 1966
*pho*R	Regulatory gene for alkaline phosphatase; position with respect to *pho*P uncertain	Miki *et al.*, 1965; Oishi *et al.*, 1966
pig	*pig*Y1, *pig*18TB, *pig*UT. Spore pigment; contransduces 72–77 percent with *pur*B (60 percent; Siegel and Marmur, 1969); no cotransduction with either *ery* or *thr*	Rogolsky, 1968; Siegel and Marmur, 1969
*pol*A[a]	*pol*A5. Deoxyribonucleic acid polymerase I deficient; linked to *arg*A (0.58 map units), unlinked to *pyr*A by PBS1 transduction; tentatively placed to the left of *rec*A	Laipis and Ganesan, 1972
*pur*A	*ade*-16. Strict adenine requirement	Borenstein and Ephrati-Elizur, 1969; Dubnau *et al.*, 1965; O'Sullivan and Sueoka, 1967
*pur*B	*ade*-6. Requires adenine, guanine, or hypoxanthine; mapped by density transfer and spore germination techniques; no linkage shown to rest of the chromosome; therefore size of the segment is unknown; See text and addendum for discussion of map location	

*pur*C	*ade*-12. Requires adenine or hypoxathine	⎫ Dubnau, 1970
*pur*D	*ade*-11. Requires adenine or hypoxanthine	⎭
*pyr*A	*ura*-1, 26, class I. Pyrimidine requirement; does not lack aspartate transcarbamylase, dihydroorotase, dihydroorotate dehydrogenase, orotidine 5′-phosphate pyrophosphorylase, or orotidine 5′-phosphate carboxylase activity. (*pyr* I–VI classification suggested by R. Kelleher, private communication)	Dubnau, 1970; J. A. Hoch and Anagnostopoulos, 1970; Harford and Sueoka, 1970; O'Sullivan and Sueoka, 1967
*pyr*B	Class II. Pyrimidine requirement; lacks dihydroorotate dehydrogenase activity	
*pyr*C	Class III. Pyrimidine requirement; lacks aspartate transcarbamylase, dihydroorotase, and dihydroorotate dehydrogenase activity	
*pyr*D	Class IV. Pyrimidine requirement; lacks aspartate transcarbamylase and dihydroorotate dehydrogenase activity	⎫ Dubnau, 1970; R. Kelleher, private communication
*pyr*E	Class V. Pyrimidine requirement; lacks dihydroorotase and dihydroorotate dehydrogenase activity	⎬
*pyr*F	Class VI. Pyrimidine requirement; lacks aspartate transcarbamylase, and dihydroorotase activity	⎭
*rec*A	*rec*-1. Radiation sensitive; considerably reduced for transformation and SP10 transduction, PBS1 transduction unaffected	⎫
*rec*B	*mms*, *mc*-1, *rec*B2. Transformation, SP10 and PBS1 transduction reduced; radiation sensitive	⎬ J. H. Hoch and Anagnostopoulos, 1970
*rec*C[a]	*rec*C4. Recombination deficient as assayed by transformation; maps near *mit* and to the right of *str*A; order with respect to *ery* and *mit* not certain	Sinha and Iyer, 1972

TABLE 1. Continued

Designation on map	Previous designations and comments	References
rfm	rfm1, rfm2, rfm3. Resistant to rifampin; altered RNA polymerase was mapped between cysA14 and str by three-factor transformation; indicated to the left of bry	Harford and Sueoka, 1970
rib	Requires riboflavin	M. S. Kelly, 1967b; Anagnostopoulos, private communication
rRNA A	Approximately 80 percent of genes for 5S, 16S and 23S ribosomal RNA located to the left of purB and possibly overlapping str (by density transfer mapping)	Dubnau et al., 1965; Harford and Sueoka, 1970; Smith et al., 1968; 1969
rRNA B	Approximately 20 percent of genes for 5S, 16S and 23S ribosomal RNA; located between argA and metB by density transfer mapping	Smith et al., 1968
sacA[a]	Probable structural gene for sucrase; includes suc locus mapped by Kennett and Sueoka (1971) that replicates as 0.02 on marker density transfer map; not linked to terminal markers	Lepesant et al., 1972
sacB[a]	Probable structural gene for levansucrase; position with respect to sacR is not established	Lepesant et al., 1972
sacP	suc⁻. Included as an individual locus (although this has yet to be proven) on the basis that 33 mutants in sacA map at least 24 percent recombination units away from sacP by transformation; the position of sacT has not been definitely established; cotransformation 86 percent with sacT	Masters and Pardee, 1965; O'Sullivan and Sueoka, 1967; Lepesant et al., 1969a, b
sacQ[a]	Regulatory gene for levansucrase; maps to the origin side of thr and cysB; Linkage by PBS1 transduction to purB or hisA could not be detected	Lepesant et al., 1972

*sac*R[a]	Regulatory gene for levansucrase; maps between *his*A and *cys*B; position with respect to *sac*B not definite	⎫
*sac*S[a]	Regulatory gene that affects synthesis of both sucrase and levansucrase; linked to *pur*A (0.95 map units) and *sac*A (0.32 map units) by PBS1 transduction	⎬ Lepesant *et al.*, 1972
*sac*T[a]	Regulatory gene that preferentially regulates synthesis of sucrase; linked to *sac*A (0.09 map units) by PBS1 transduction	⎭
*sac*U[a]	Regulatory gene for levansucrase	
sep[a]	Serine protease; *ts*-5 is a temperature-sensitive asporogenic mutation in the structural gene for serine protease; linked to aroD and *lys* (0.70 map units) and *str*C2 (0.15 map units) by PBS1 transduction	Leighton *et al.*, 1972; 1973
ser	*ser*-1. Requirement for serine	Dubnau, 1970
smo	*rou*-1. Smooth colony morphology; placed to the left of *sac*R because Lepesant reports average cotransfer of 62 percent between *sac*R and hisA, and Grant shows 60 percent cotransfer between *his*A and *rou*-1	Grant and Simon, 1969; Young *et al.*, 1969; Lepesant *et al.*, 1969*a*
*spc*A	*spc*2D. Spectinomycin resistance	Dubnau, 1970; Goldthwaite *et al.*, 1970; Harford and Sueoka, 1970
*spc*B	Spectinomycin resistance; cotransduced with *pyr*	Harford and Sueoka, 1970
*spr*A[a]	Suppressor mutation that allows isoleucine dependent mutants lacking threonine dehydratase (ilvA) to also grow on homoserine or threonine; map position not known, however, it is not linked to the gene it modifies; causes derepression of homoserine dehydrogenase, homoserine kinase, and a minor threonine dehydratase	Vapnek and Greer, 1971*a*

TABLE 1. Continued

Designation on map	Previous designations and comments	References
sprB[a]	Suppressor mutation that allows threonine dehydratase; sprA mutants to grow on minimal medium; also renders cell sensitive to $10^{-6} M$ methionine, maps between thr and hom by transformation	Vapnek and Greer, 1971a, b
strA[a]	str. Streptomycin resistance; gene determines high level resistance, dependence, sensitivity, and altered 30S ribosomal protein	Dubnau, 1970; Goldthwaite et al., 1970; McDonald, 1969; Takahashi, 1965
strB[a]	Resistant to high levels of streptomycin or dihydrostreptomycin in minimal medium; sensitive during sporulation and germination; probable location is between metC and pyrA	Staal and Hoch, 1972
strC[a]	Resistant to high levels of streptomycin or dihydrostreptomycin in minimal medium; sensitive during sporulation and germination; Maps between aroD and lys; mutants also lack cytochrome a	Staal and Hoch, 1972
suh	Suppressor of hisB2 (confers ability to grow on phenylalanine as well as histidine)	Dubnau, 1970; Nester et al., 1963
sul	Resistance to sulfonilamide; cotransduced with and to the left of cysA14, by four-factor transduction; order is sul, cysA14, ery-1, attSP02	Dubnau, 1970; Nester et al., 1963
tag-1	rod. Lacks teichoic acid (glycerol) at 45°C and forms spheres; maps just to the right of gtaA by three-factor transduction; 50-percent linkage to gtaA12 by transformation	Boylan et al., 1972

tdm[a]	A minor threonine dehydratase not normally detected in wild-type strains; derepressed by suppressor mutation *spr*A; maps between *hom* and *spr*B	Vapnek and Greer, 1971*b*
tem-1	Temperature-sensitive protein and RNA synthesis; cotransduced with *pur*B	Dubnau, 1970
*thi*A	Requires thiamine; linked 10–20 percent to *arg*O by transformation using very high-molecular-weight DNA	Dubnau, 1970; M.S. Kelly, 1967*a*
*thi*B[a]	*thi*-2. Thiamine-requiring mutant; maps close to *met*A	J. A. Hoch and Mathews, 1972
thr	Requires threonine	Borenstein and Ephrati-Elizur, 1969; Dubnau *et al.*, 1965; O'Sullivan and Sueoka, 1967
*thy*A	*thy*Y. Requires thymine if also defective in *thy*B locus; said to specify thymidylate synthetase; maps 0.88 PBS1 transduction units from *gap* and 0.55 units from *cit*B	S. A. Zahler and R. Neubauer, private communication
*thy*B	*thy*-X. Requires thymine if defective in *thy*A also	Anagnostopoulos and Schneider-Champagne, 1966; Wilson *et al.*, 1966
tms	*tms*⁻. Temperature-sensitive growth at 55°C	McDonald, 1969
tre	*tre*-12. Trehalase; maps at position 0.22 by marker density analysis	Kennett and Sueoka, 1971
*t*RNA A	Genes for transfer RNA; maps to the left of *pur*B and possibly overlaps *str* by density transfer	Dubnau, 1970; Oishi *et al.*, 1966; Smith *et al.*, 1968
*t*RNA B	Genes for transfer RNA; maps between *arg*A and *met*B by density transfer	Dubnau, 1970; Smith *et al.*, 1968
*trp*A	Requires tryptophan; lacks tryptophan synthetase A	Anagnostopoulos and Crawford, 1961, 1967; Carlton, 1966; Carlton and Whitt, 1969; Whitt and Carlton, 1968

TABLE 1. Continued

Designation on map	Previous designations and comments	References
trpB	Requires tryptophan; lacks tryptophan synthetase B	Anagnostopoulos and Crawford, 1961, 1967; Carlton, 1966; Carlton and Whitt, 1969; Whitt and Carlton, 1968
trpC	trp-2. Requires tryptophan; lacks indole-3-glycerol phosphate synthetase	Anagnostopoulos and Crawford, 1961, 1967; Carlton, 1966; Carlton and Whitt, 1969; S.O. Hoch et al., 1969; Nester et al., 1963; Whitt and Carlton, 1968
trpD	Requires tryptophan; lacks phosphoribosyl transferase	Anagnostopoulos and Crawford, 1961, 1967; Carlton and Whitt, 1969; S.O. Hoch et al., 1969; Whitt and Carlton, 1968
trpE	Requires tryptophan; lacks anthranilate synthetase	
trpF	Requires tryptophan; lacks N–5′-phosphoribosyl anthranilic acid isomerase	Carlton and Whitt, 1969; S.O. Hoch, et al., 1969; Whitt and Carlton, 1968
trpS	trpS1. Temperature-sensitive mutation in the structural gene coding for tryptophanyl-tRNA synthetase; cotransforms 4–7 percent with argC4, and contransduced 66 percent with metA)	Steinberg and Anagnostopoulos, 1971
tsi	tsi-23. Temperature-sensitive induction of PBS-X	Siegel and Marmur, 1969
tyrA	tyr-1. Requires tyrosine; lacks prephenate dehydrogenase	Dubnau et al., 1965; Nasser and Nester, 1967; Nester et al., 1963
ursᵃ	uraˢ Confers sensitivity to uracil; closely linked by transformation to argC	J. A. Hoch and Mathews, 1972
uvr	uvr-1. Ultraviolet radiation sensitive; transformation, PBS1 transduction, and SP10 transduction unaffected	J. A. Hoch and Anagnostopoulos, 1970

vas	val^{sens}. Sensitive to valine; maps within locus for threonine deaminase	Leibovici and Anagnostopoulos, 1969
xtl	*xtl*-1, *xtl*-2. PBS-X lacking tails; *xtl*-1 refers to strain A044 (K. Bott) and *xtl*-2 to strain BS10 (J. Gross), which upon induction yield tailless PBS-X particles which are kill⁻; maps to right of *met*A and probably to the right of *met*C	Garro *et al.*, 1970

[a] Previous maps of *B. subtilis* have been utilized in the development of the revised map. For additional details consult Dubnau (1970) and Young and Wilson (1972). These markers have been added since the last published map.
[b] IPM, isopropylmalate.
[c] UDP, uridine diphosphate.

Two assemblages of the genetic markers have been recently published. The first by Dubnau (1970) did much to organize the map and alter nomenclature of the markers to conform with the current approved symbols (Demerec *et al.*, 1966).When we began a review on the genetics of sporulation, we found it necessary to also review the map of the markers that were not associated with sporulation as well (Young and Wilson, 1972). Due to limitations of space, it is not warranted to describe the phenotypic markers that influence sporulation and germination that were published earlier by Young and Wilson (1972). Instead, new loci are included in Table 1 and in Figure 1 together with the other markers. Wherever possible we have added mapping data in the Table to aid in the placement of these markers. A few points should be stressed on the method of illustration. Loci that have been ordered with respect to other loci are joined to the map by a solid line. A dashed line indicates that the marker most probably lies where indicated, but its exact location cannot be certain because mapping was not determined with all the closely linked markers. Genes not joined to the map can only be located near the region indicated and could be placed with as much error as 20 percent of the linkage map.

A highly conserved region that bears many of the resistance loci to amino-glucoside antibiotics has recently been examined in detail by Goldthwaite and Smith (1972) and has resulted in some changes in this area. The map also includes nine loci in which DNA synthesis is altered at elevated temperatures. These loci have been redesignated *dna*A–*dna*I instead of *ts* DNAA–*ts* DNAI to conform with suggestions of Demerec *et al.* (1966).

Concluding Remarks

The rapid increase in the size of the known map of *B. subtilis* is an index of the usefulness of this organism as a model for studies of genetic recombination, differentiation, and regulation of the cell surface. The use of isogenic strains is strongly urged in genetic experiments. It is anticipated that comparative genetic studies among the *Bacilli* and experiments involving genetic engineering will become more prominant in the *B. subtilis* genospecies in the next few years as a mechanism for development of unique strains for antibiotic and enzyme production.

Acknowledgments

This review was supported in part by grant VC-27H from the American Cancer Society and Public Health Service grant AI-10141 from the National Institute of Allergy and Infectious Diseases. We wish to thank the many investigators who made preprints available to us and allowed us to report unpublished data.

Addendum

Recent studies by Hara and Yoshikawa (1973) have provided genetic evidence for bidirectional replication of the chromosome. Utilizing mutants with altered patterns of DNA synthesis at elevated temperatures, they have suggested that the origin of bidirectional replication lies to the right or *pur*A, between *pur*A and *ts*8132, and that the replication is asymmetrical. The leftward replication is terminated between *sac*A and *thy*A. Their data show a less extensive region of bidirectional replication than the study of Gyurasits and Wake (1973) who reported 20 percent replication. In a more recent autoradiographic study, Wake (1973) concluded that the chromosome was circular and at least 50 percent of the chromosome was replicated, with both arms showing the same rate of synthesis.

These two conflicting reports may be brought into agreement if it is assumed that there are large "silent" regions of the chromosome between *sac*A and *thy*A. Based on these data and the fact that the *pur*B segment of the chromosome has only been positioned by replication analysis, it is possible that the *pur*B region may be to the left of the origin as we discussed previously in this review. Recent communications by J. C. Copeland indicate that this hypothesis may be valid. He has found a low level of linkage of *pur*B and *met*B and a high level of linkage of *cit*K to *pur*B when *pur*B is used as the selective marker. When *cit*K is the selected marker linkage was not observed. Similar nonreciprocal anomalies were noted earlier in the *his*A region (Young *et al.*, 1969). Therefore it must be emphasized that the location of the *pur*B segment remains to be elucidated.

Literature Cited

Adams, A., 1971   A class of poorly transformable multisite mutations in *Bacillus subtilis*. In *Informative Molecules in Biological Systems*, edited by L. Ledoux, pp. 418–428, North-Holland, Amsterdam.

Adams, A., 1972   Transformation and transduction of a large deletion mutation in *Bacillus subtilis*. *Mol. Gen. Genet.* **118**:311–322.

Adams, A. and M. Oishi, 1972   Genetic properties of arsenate sensitive mutants of *Bacillus subtilis* 168. *Mol. Gen. Genet.* **118**:295–310.

Aharonowitz, Y. and E. Z. Ron, 1972   A temperature sensitive mutant in *Bacillus subtilis* with an altered elongation factor G. *Mol. Gen. Genet.* **119**:131–138.

Anagnostopoulus, C. and I. P. Crawford, 1961   Transformation studies on the linkage of markers in the tryptophan pathway in *Bacillus subtilis. Proc. Natl. Acad. Sci. USA* **47**:378–390.

Anagnostopoulos, C. and I. P. Crawford, 1967   Le groupe des genes régissant la biosynthèse du tryptophane chez *Bacillus subtilis. C. R. Hebd. Seances Acad. Sci. Ser. D Sci. Nat* **265**:93–96.

Anagnostopoulos, C. and A. M. Schneider-Champagne, 1966   Déterminisme génétique de l'exigence en thymine chez certains mutants de *Bacillus subtilis. C.R. Hebd. Seances Acad. Sci. Ser. D Sci. Nat.* **262**:1311–1314.

Anagnostopoulos, C. and J. Spizizen, 1961   Requirements for transformation in *Bacillus subtilis. J. Bacteriol.* **81**:741–746.

Anagnostopoulos, C., M. Barat and A. Schneider, 1964   Étude par transformation, de deux groupes de gènes régissant la biosynthèse de l'isoleucine, de la valine et de la leucine chez *Bacillus subtilis. C. R. Hedb. Seances Acad. Sci. Ser. D Sci. Nat.* **258**:749–752.

Archer, L. J., 1973   Heat sensitivity of competent pre-competent cells of *Bacillus subtilis*. In *Bacterial Transformation,* edited by L. J. Archer, pp. 45–63, Academic Press, New York.

Armentrout, R. W. and L. Rutberg, 1970   Mapping of prophage and mature deoxyribonucleic acid from temperate *Bacillus* bacteriophage ϕ105 by marker rescue. *J. Virol.* **6**:760–767.

Armentrout, R. W., L. Skoog, and L. Rutberg, 1971   Structure and biological activity of deoxyribonucleic acid from *Bacillus* bacteriophage ϕ105: Effects of *Escherichia coli* exonucleases. *J. Virol.* **7**:359–371.

Armstrong, R. L., N. Harford, R. H. Kennett, M. L. St. Pierre and N. Sueoka, 1970   Experimental methods for *Bacillus subtilis*. In *Methods in Enzymology,* Vol. 17a, edited by H. Tabor and C. W. Tabor, pp. 36–59, Academic Press, New York.

Audit, C. and C. Anagnostopoulos, 1972   Production of stable and persistent unstable heterogenotes in a mutant of *Bacillus subtilis*. In *Spores V,* edited by H. O. Halvorson, R. Hanson and L. L. Campbell, pp. 117–125, American Society for Microbiology, Washington, D.C.

Audit, C. and C. Anagnostopoulos, 1973a   Extent of diploidy in stable and unstable merozygotes from *Bacillus subtilis* strains bearing the *trp*E26 marker. In *Bacterial Transformation,* edited by L. J. Archer, pp. 293–305, Academic Press, New York.

Audit, C. and C. Anagnostopoulos, 1973b   Genetic studies relating to the production of transformed clones diploid in the tryptophan region of the *Bacillus subtilis* genome. *J. Bacteriol.* **114**:18–27.

Bak, A. L., C. Christiansen and A. Stenderup, 1970   Bacterial genome sizes determined by DNA renaturation studies. *J. Gen. Microbiol.* **64**:377–380.

Barat, M. C. Anagnostopoulos and A. M. Schneider, 1965   Linkage relationships of genes controlling isoleucine, valine, and leucine biosynthesis in *Bacillus subtilis. J. Bacteriol.* **90**:357–369.

Bazill, G. W. and D. Karmata, 1972   Temperature-sensitive mutants of *B. subtilis* defective in deoxyribonucleotide synthesis. *Mol. Gen. Genet.* **117**:19–29.

Bazill, G. W. and Y. Retief, 1969   Temperature-sensitive DNA synthesis in a mutant of *B. subtilis. J. Gen. Microbiol.* **56**:87–97.

Bettinger, G. E. and F. E. Young, 1973   Transformation of *Bacillus subtilis* using gently typed L-forms; a new mapping technique. *Biochem. Biophys. Res. Commun.* **55**:1105–1111.

Biswas, G. D. and A. W. Ravin, 1971   Heterospecific transformation of Pneumococcus and Streptococcus. IV. Variations in hybrid DNA produced by recombination. *Mol. Gen. Genet.* **110**:1–22.

Borenstein, S. and E. Ephrati-Elizur, 1969   Spontaneous release of DNA in sequential order by *Bacillus subtilis. J. Mol. Biol.* **45**:137–152.

Bott, K. F. and G. A. Wilson, 1968   Metabolic and nutritional factors influencing the development of competence for transfection of *Bacillus subtilis. Bacteriol. Rev.* **32**:370–378.

Boylan, R. J., N. H. Mendelson, D. Brooks and F. E. Young, 1972   Regulation of the bacterial cell wall: Analysis of a mutant of *Bacillus subtilis* defective in biosynthesis of teichoic acid. *J. Bacteriol.* **110**:281–290.

Carlton, B. C., 1966   Fine-structure mapping by transformation in the tryptophan region of *Bacillus subtilis. J. Bacteriol* **91**:1795–1803.

Carlton, B. C. and D. D. Whitt, 1969   The isolation and genetic characterization of mutants of the tryptophan system of *Bacillus subtilis. Genetics* **62**:445–460.

Cerdá-Olmedo, E., P. C. Hanawalt and N. Guerola, 1968   Mutagenesis of the replication point by nitrosoguanidine: map and pattern of replication of the *Escherichia coli* chromosome. *J. Mol. Biol.* **33**:705–719.

Champney, W. S. and R. A. Jensen, 1969   D-Tyrosine as a metabolic inhibitor of *Bacillus subtilis. J. Bacteriol.* **98**:205–214.

Chasin, L. A. and B. Magasanik, 1968   Induction and repression of the histidine-degrading enzymes of *Bacillus subtilis. J. Biol. Chem.* **243**:5165–5178.

Chilton, M.-D. and B. J. McCarthy, 1969   Genetic and base sequence homologies in bacilli. *Genetics* **62**:697–710.

Copeland, J. C., 1969   Regulation of chromosome replication in *Bacillus subtilis:* Effects of amino acid starvation in strain 168. *J. Bacteriol.* **99**:730–736.

Copeland, J. C., 1971   Regulation of chromosome replication in *Bacillus subtilis:* Effects of amino acid starvation in strain W23. *J. Bacteriol.* **105**:595–603.

Crawford, I. P. and J. Preiss, 1972   Distribution of closely linked markers following intragenic recombination in *Escherichia coli. J. Mol. Biol.* **71**:717–733.

Davidoff-Abelson, R. and D. Dubnau, 1973*a*   Conditions affecting the isolation from transformed cells of *Bacillus subtilis* of high-molecular weight single-stranded deoxyribonucleic acid of donor origin. *J. Bacteriol.* **116**:146–153.

Davidoff-Abelson, R. and D. Dubnau, 1973*b*   Kinetic analysis of the products of donor deoxyribonucleate in transformed cells of *Bacillus subtilis. J. Bacteriol.* **116**:154–162.

Demerec, M. E., E. A. Adelberg, A. J. Clark and P. E. Hartman, 1966   A proposal for a uniform nomenclature in bacterial genetics. *Genetics* **54**:61–76.

Dennis, E. S. and R. G. Wake, 1966   Autoradiography of the *Bacillus subtilis* chromsome. *J. Mol. Biol.* **15**:435–439.

Dubnau, D., 1970   Linkage map of *Bacillus subtilis.* In *Handbook of Biochemistry,* second edition, edited by H. A. Sober and R. A. Harte, pp. 39–45, Chemical Rubber Co., Cleveland, Ohio.

Dubnau, D. and C. Cirigliano, 1972*a*   Fate of transforming deoxyribonucleic acid after uptake by competent *Bacillus subtilis:* size and distribution of the integrated donor segments. *J. Bacteriol.* **111**:488–494.

Dubnau, D. and C. Cirigliano, 1972*b*   Fate of transforming DNA following uptake by competent *Bacillus subtilis.* III. Formation and properties of the products isolated from transformed cells which are derived entirely from donor DNA. *J. Mol. Biol.* **64**:9–29.

Dubnau, D. and C. Cirigliano, 1972*c*   Fate of transforming DNA following uptake by competent *Bacillus subtilis*. IV. The endwise attachment and uptake of transforming DNA. *J. Mol. Biol.* **64**:31–46.

Dubnau, D. and R. Davidoff-Abelson, 1971   Fate of transforming DNA following uptake by competent *Bacillus subtilis*. I. Formation and properties of the donor–recipient complex. *J. Mol. Biol.* **56**:209–221.

Dubnau, D., I. Smith, P. Morell and J. Marmur, 1965   Gene conservation in Bacillus species. I. Conserved genetic and nucleic acid base sequence homologies. *Proc. Natl. Acad. Sci. USA* **54**:491–498.

Dubnau, D., C. Goldthwaite, I. Smith and J. Marmur, 1967   Genetic mapping in *Bacillus subtilis*. *J. Mol. Biol.* **27**:163–185.

Dubnau, D., R. Davidoff-Abelson and I. Smith, 1969   Transformation and transduction in *Bacillus subtilis:* Evidence for separate modes of recombinant formation. *J. Mol. Biol.* **45**:155–179.

Dul, M. J. and F. E. Young, 1973   Genetic mapping of a mutant defective in D,L-alanine racemase in *Bacillus subtilis* 168. *J. Bacteriol.* **115**:1212–1214.

Ephrati-Elizur, E., P. R. Srinivasan and S. Zamenhof, 1961   Genetic analysis, by means of transformation of histidine linkage groups in *Bacillus subtilis*. *Proc. Natl. Acad. Sci. USA* **47**:56–63.

Ephrussi-Taylor, H. E. and T. C. Gray, 1966   Genetic studies of recombining DNA in pneumococcal transformation. *J. Gen. Physiol.* **49**:211–231.

Erickson, R. J., 1970   New ideas and data on competence and DNA entry in transformation of *Bacillus subtilis*. *Curr. Top. Microbiol. Immunol.* **53**:149–199.

Erickson, R. J., F. E. Young and W. Braun, 1969   Binding of rabbit gamma globulin by competent *Bacillus subtilis* cultures. *J. Bacteriol.* **99**:125–131.

Freese, E., S. W. Park and M. Cashel, 1964   The developmental significance of alanine dehydrogenase in *Bacillus subtilis*. *Proc. Natl. Acad. Sci. USA* **51**:1164–1172.

Fujisawa, T. and A. Eisenstark, 1972   Evidence for two-way chromosomal replication in *Salmonella typhimurium*. *Abstr. Ann. Meet. Am. Soc. Microbiol*. p. 62.

Garro, A. J., H. Leffert and J. Marmur, 1970   Genetic mapping of a defective bacteriophage on the chromosome of *Bacillus subtilis* 168. *J. Virol.* **6**:340–343.

Gay, P. and G. Rapoport, 1970   Étude des mutants dépourvus de fructose-1-phosphate kinase chez *Bacillus subtilis*. *C. R. Hebd. Seances Acad. Sci. Ser. D Sci. Nat.* **271**:374–377.

Gay, P., A. Carayon and G. Rapoport, 1970   Isolement et localisation génétique de mutants du système métabolique du fructose chez *Bacillus subtilis*. *C. R. Hebd. Seances Acad. Sci. Ser. D Sci. Nat.* **271**:263–266.

Georgopoulos, C. P., 1969   Suppressor system in *Bacillus subtilis* 168. *J. Bacteriol.* **97**:1397–1402.

Gilpin, R. W., F. E. Young, and A. N. Chatterjee, 1973   Characterization of a stable L-form of *Bacillus subtilis* 168. *J. Bacteriol.* **113**:486–499.

Glaser, L., H. Ionesco and P. Schaeffer, 1966   Teichoic acids as components of specific phage receptor in *Bacillus subtilis*. *Biochim. Biophys. Acta* **124**:415–417.

Goldthwaite, C. and I. Smith, 1972   Genetic mapping of aminoglycoside and fusidic acid resistant mutations in *Bacillus subtilis*. *Mol. Gen. Genet.* **114**:181–189.

Goldthwaite, C., D. Dubnau and I. Smith, 1970   Genetic mapping of antibiotic resistance markers in *Bacillus subtilis*. *Proc. Natl. Acad. Sci. USA* **65**:96–103.

Goodgal, S. H., 1961   Studies on transformation of *Hemophilus influenzae*. IV. Linked and unlinked transformations. *J. Gen. Physiol.* **45**:205–228.

Grant, G. F. and M. I. Simon, 1969   Synthesis of bacterial flagella. II. PBS1 transduction of flagella-specific markers in *Bacillus subtilis*. *J. Bacteriol.* **99**:116–124.

Green, D. M., 1966   Physical and genetic characterization of sheared infective SP82 bacteriophage DNA. *J. Mol. Biol.* **22:**15–22.

Groves, D. J. and F. E. Young, 1973   Effect of mercuric chloride on DNA-medicated transformation in *Bacillus subtilis* 168. *Abst. Ann. Meet. Am. Soc. Microbiol.,* p. 58.

Gyurasits, E. B. and R. G. Wake, 1973   Bidirectional chromosome replication in *Bacillus subtilis. J. Mol. Biol.* **73:**55–63.

Haas, M. and H. Yoshikawa, 1969a   Defective bacteriophage PBSH in *Bacillus subtilis.* I. Induction, purification and physical properties of the bacteriophage and its deoxyribonucleic acid. *J. Virol.* **3:**233–247.

Haas, M. and H. Yoshikawa, 1969b   Defective bacteriophage PBSH in *Bacillus subtilis.* II. Intracellular development of the induced prophage. *J. Virol.* **3:**248–260.

Hara, H. and H. Yoshikawa, 1973   Asymmetric bidirectional replication of the *Bacillus subtilis* chromosome. *Nat. New Biol.* **244:**200–203.

Harford, N. and M. Mergeay, 1973   Interspecific transformation of rifampicin resistance in the genus *Bacillus. Mol. Gen. Genet.* **120:**151–155.

Harford, N. and N. Sueoka, 1970   Chromosomal location of antibiotic resistance markers in *Bacillus subtilis. J. Mol. Biol.* **51:**267–286.

Hirokawa, H., 1972   Transfecting deoxyribonucleic acid of Bacillus bacteriophage ϕ29 that is protease sensitive. *Proc. Natl. Acad. Sci. USA* **69:**1555–1559.

Hoch, J. A. and C. Anagnostopoulos, 1970   Chromosomal location and properties of radiation sensitivity mutations in *Bacillus subtilis. J. Bacteriol.* **103:**295–301.

Hoch, J. A. and J. Mathews, 1972   Genetic studies in *Bacillus subtilis.* In *Spores V,* edited by H. O. Halvorson, R. Hanson and L. L. Campbell, pp. 113–116, American Society for Microbiology, Washington, D.C.

Hock, J. A. and J. Spizizen, 1969   Genetic control of some early events in sporulation of *Bacillus subtilis* 168. In Spores IV, edited by L. L. Campbell, pp. 112–120, American Society for Microbiology, Bethesda, Md.

Hock, S. O., C. Anagnostopoulos and I. P. Crawford, 1969   Enzymes of the tryptophan operon of *Bacillus subtilis. Biochem. Biophys. Res. Commun.* **35:**838–844.

Hoch, S. O., C. W. Roth, I. P. Crawford and E. W. Nester, 1971   Control of tryptophan biosynthesis by the methyltryptophan-resistance gene in *Bacillus subtilis. J. Bacteriol.* **105:**38–45.

Inselburg, J. W., T. Eremenko-Volpe, L. Greenwald, W. L. Meadow and J. Marmur, 1969   Physical and genetic mapping of the SP02 prophage on the chromosome of *Bacillus subtilis* 168. *J. Virol.* **3:**627–628.

Ionesco, H., J. Michel, B. Came and P. Schaeffer, 1970   Genetics of sporulation in *Bacillus subtilis.* Marburg. *J. Appl. Bacteriol.* **33:**13–24.

Iyer, V. N., 1966   Mutations determining mitomycin resistance in *Bacillus subtilis. J. Bacteriol.* **92:**1663–1669.

Jamet, C. and C. Anagnostopoulos, 1969   Étude d'une mutation très faiblement transformable au locus de la thréonine desaminase de *Bacillus subtilis. Mol. Gen. Genet.* **105:**225–242.

Joenje, H., W. Admiraal and G. Venema, 1973   Isolation and partial characterization of a temperature-sensitive mutant *Bacillus subtilis* impaired in the development of competence for genetic transformation. *J. Gen. Microbiol.* **78:**67–77.

Joys, T. M., 1965   Correlation between susceptibility to bacteriophage PBS1 and motility in *Bacillus subtilis. J. Bacteriol.* **90:**1575–1577.

Karamata, D. and J. D. Gross, 1970   Isolation and genetic analysis of temperature-sensitive mutants of *B. subtilis* defective in DNA synthesis. *Mol. Gen. Genet.* **108:**277–287.

Karmazyn, C., C. Anagnostopoulos, and P. Schaeffer, 1972   Dominance of *spoOA* mutations in spo⁺/spo merodiploid strains of *Bacillus subtilis*. In *Spores V*, edited by H. O. Halvorson, R. Hanson and L. L. Campbell, pp. 126–132, American Society for Microbiology, Washington, D.C.

Kelly, M. S., 1967a   Physical and mapping properties of distant linkages between genetic markers in transformation of *Bacillus subtilis*. *Mol. Gen. Genet.* **99**:333–349.

Kelly, M. S., 1967b   The causes of instability of linkage in transformation of *Bacillus subtilis*. *Mol. Gen. Genet.* **99**:350–361.

Kelly, M. S., and R. H. Pritchard, 1963   Selection for linked loci in *Bacillus subtilis* by means of transformation. *Heredity* **17**:598.

Kelly, M. S. and R. H. Pritchard, 1965   Unstable linkage between genetic markers in transformation. *J. Bacteriol.* **89**:1314–1321.

Kelly, T. J., Jr. and H. O. Smith, 1970   A restriction enzyme from *Hemophilus influenzae*. II. Base sequence of the recognition site. *J. Mol. Biol.* **51**:393–409.

Kennett, R. H. and N. Sueoka, 1971   Gene expression during outgrowth of *Bacillus subtilis* spores. The relationship between gene order on the chromosome and temporal sequence of enzyme synthesis. *J. Mol. Biol.* **60**:31–44.

Klotz, L. C. and B. H. Zimm, 1972   Size of DNA determined by viscoelastic measurements: Results on bacteriophages, *Bacillus subtilis* and *Escherichia coli*. *J. Mol. Biol.* **72**:779–800.

Korch, C. T. and R. H. Doi, 1971   Electron microscopy of the altered spore morphology of a ribonucleic acid polymerase mutant of *Bacillus subtilis*. *J. Bacteriol.* **105**:1110–1118.

Lacks, S. and R. D. Hotchkiss, 1960   A study of the genetic material determining an enzyme activity in Pneumococcus. *Biochim. Biophys. Acta* **39**:508–517.

Laipis, P. J. and A. T. Ganesan, 1972   A deoxyribonucleic acid polymerase I-deficient mutant of *B. subtilis*. *J. Biol. Chem.* **247**:5867–5871.

Ledoux, L., R. Huart and M. Jacobs, 1971   Fate of exogenous DNA in *Arabidopsis thaliana*. II. Evidence for replication and preliminary results at the biological level. In *Informative Molecules in Biological Systems*, edited by L. Ledoux, pp. 159–175, North-Holland, Amsterdam.

Le Hergarat, J.-C. and C. Anagnostopoulos, 1969   Localisation chromosomique d'un gène gouvernant la synthèse d'une phosphatase alcaline chez *Bacillus subtilis*. *C. R. Hebd. Seances Acad. Sci. Ser. D Sci. Nat.* **269**:2048–2050.

Leibovici, J. and C. Anagnostopoulos, 1969   Propriétés de la thréonine désaminase de la souche sauvage et d'un mutant sensible a la valine de *Bacillus subtilis*. *Bull. Soc. Chim. Biol.* **51**:691–707.

Leighton, T. J., P. K. Freese, R. H. Doi, R. A. J. Warren and R. A. Kelln, 1972   Initiation of sporulation in *Bacillus subtilis*: Requirement for serine protease activity and ribonucleic acid polymerase modification. In *Spores V*, edited by H. O. Halvorson, R. Hanson and L. L. Campbell, pp. 238–246, American Society for Microbiology, Washington, D.C.

Leighton, T. J., R. H. Doi, R. A. J. Warren and R. A. Kelln, 1973   The relationship of serine protease activity to RNA polymerase modification and sporulation in *Bacillus subtilis*. *J. Mol. Biol.* **76**:103–122.

Lepesant, J.-A., F. Kunst, A. Carayon·and R. Dedonder, 1969a   Localisation génétique de mutants du système métabolique de du saccharose chez *Bacillus subtilis*. Localisation par transduction à l'aide du phage PBS1. *C. R. Hebd. Seances Acad. Sci. Ser. D Sci. Nat.* **269**:1712–1715.

Lepesant, J.-A., F. Kunst, A. Carayon and R. Dedonder, 1969*b*   Localisation génétique de mutants du système métabolique du saccharose chez *Bacillus subtilis*. Localisation par transformation. *C. R. Hebd. Seances Acad. Sci. Ser. D Sci. Nat.* **269:**1792–1794.

Lepesant, J.-A., F. Kunst, J. Lepesant-Kejzlarová and R. Dedonder, 1972   Chromosomal location of mutations affecting sucrose metabolism in *Bacillus subtilis* Marburg. *Mol. Gen. Genet.* **118:**135–160.

Lorence, J. H. and E. W. Nester, 1967   Multiple molecular forms of chorismate mutase in *Bacillus subtilis*. *Biochemistry* **6:**1541–1553.

Lovett, P. S. and F. E. Young, 1969   Identification of *Bacillus subtilis* NRRL B-3275 as a strain of *Bacillus pumilus*. *J. Bacteriol.* **100:**658–661.

Lovett, P. S. and F. E. Young, 1970   Genetic analysis in *Bacillus pumilus* by PBS1-mediated transduction. *J. Bacteriol.* **101:**603–608.

McCarthy, C. and E. W. Nester, 1969   Heat-sensitive step in deoxyribonculeic acid-mediated transformation of *Bacillus subtilis*. *J. Bacteriol.* **97:**162–165.

McDonald, W. C., 1969   Linkage of a temperature-sensitive locus to the streptomycin region and its use in recombination studies with streptomycin mutants of *Bacillus subtilis*. *Can. J. Microbiol.* **15:**1287–1291.

Mahler, I., J. Neumann and J. Marmur, 1963   Studies of genetic units controlling arginine biosynthesis in *Bacillus subtilis*. *Biochim. Biophys. Acta* **72:**69–79.

Marmur, J., 1961   A procedure for the isolation of deoxyribonucleic acid from microorganisms. *J. Mol. Biol.* **3:**208–218.

Marmur, J., E. Seaman and J. Levine, 1963   Interspecific transformation in *Bacillus*. *J. Bacteriol.* **85:**461–467.

Martinez, R. J., A. T. Ichiki, N. P. Lundh and S. R. Tronick, 1968   A single amino acid substitution responsible for altered flagellar morphology. *J. Mol. Biol.* **34:**559–564.

Masters, M. and P. Broda, 1971   Evidence for the bidirectional replication of the *Escherichia coli* chromosome. *Nat. New Biol.* **232:**137–140.

Masters, M. and A. B. Pardee, 1965   Sequence of enzyme synthesis and gene replication during the cell cycle of *Bacillus subtilis*. *Proc. Nat. Acad. Sci. U.S.A.* **54:**64–70.

Mendelson, N. H. and J. D. Gross, 1967   Characterization of a temperature-sensitive mutant of *Bacillus subtilis* defective in deoxyribonucleic acid replication. *J. Bacteriol.* **94:**1603–1608.

Mergeay, M., P. Charles, R. Martin, J. Remy and L. Ledoux, 1971   DNA heterogeneity of *B. subtilis* competent cells upon treatment with *Streptomyces coelicolor* DNA. In *Informative Molecules in Biological Systems,* edited by L. Ledoux, pp. 51–63, North-Holland Publishing Co., Amsterdam.

Miki, T., Z. Minami and Y. Ikeda, 1965   The genetics of alkaline phosphatase formation in *Bacillus subtilis*. *Genetics* **52:**1093–1100.

Morrison, D. A. 1971   Early intermediate state of transforming deoxyribonucleic acid during uptake by *Bacillus subtilis*. *J. Bacteriol.* **108:**38–44.

Nasser, D. and E. W. Nester, 1967   Aromatic amino acid biosynthesis: gene-enzyme relationships in *Bacillus subtilis*. *J. Bacteriol.* **94:**1706–1714.

Nasser, D., G. Henderson and E. W. Nester, 1969   Regulated enzymes of aromatic amino acid systhesis: control, isozymic nature, and aggregation in *Bacillus subtilis* and *Bacillus licheniformis*. *J. Bacteriol.* **98:**44–50.

Nester, E. W. and J. Lederberg, 1961   Linkage of genetic units of *Bacillus subtilis* in DNA transformation. *Proc. Natl. Acad. Sci. USA* **47:**52–55.

Nester, E. W., M. Schafer and J. Lederberg, 1963   Gene linkage in DNA transfer: A cluster of genes concerned with aromatic biosynthesis in *Bacillus subtilis*. *Genetics* **48:**529–551.

Nishioka, Y. and A. Eisenstark, 1970   Sequence of genes replicated in *Salmonella typhimurium* as examined by transduction techniques. *J. Bacteriol.* **102**:320–333.

Oishi, M. and Sueoka, 1965   Location of genetic loci of ribosomal RNA on *Bacillus subtilis* chromosome. *Proc. Natl. Acad. Sci. USA* **54**:483–491.

Oishi, M., A. Oishi and N. Sueoka, 1966   Location of genetic loci of soluble RNA on *Bacillus subtilis* chromosome. *Proc. Natl. Acad. Sci. USA* **55**:1095–1103.

Okamoto, K., J. A. Mudd, J. Mangan, W. M. Huang, T. V. Subbaiah and J. Marmur, 1968   Properties of the defective phage of *Bacillus subtilis. J. Mol. Biol.* **34**:413–428.

Okubo, S. and W. R. Romig, 1966   Impaired transformability of *Bacillus subtilis* mutant sensitive to Mitomycin C and ultraviolet radiation. *J. Mol. Biol.* **15**:440–454.

Okubo, S. and T. Yanagida, 1968   Isolation of a suppressor mutant in *Bacillus subtilis. J. Bacteriol.* **95**:1187–1188.

Okubo, S., M. Stodolsky, K. F. Bott, and B. Strauss, 1963   Separation of the transforming and viral deoxyribonucleic acids of a transducing bacteriophage of *Bacillus subtilis. Proc. Natl. Acad. Sci. USA* **50**:679–686.

O'Sullivan, A. and N. Sueoka, 1967   Sequential replication of the *Bacillus subtilis* chromosome. IV. Genetic mapping by density transfer experiment. *J. Mol. Biol.* **27**:349–368.

Piechowska, M. and M. S. Fox, 1971   Fate of transforming deoxyriboncleate in *Bacillus subtilis. J. Bacteriol.* **108**:680–689.

Polsinelli, M., 1965   Linkage relationship between genes for amino acid or nitrogenous base biosynthesis and genes controlling resistance to structurally correlated analogues. *G. Microbiol.* **13**:99–110.

Polsinelli, M., G. Mazza, U. Canosi and A. Falaschi, 1973   Genetical and biochemical characterization of *Bacillus subtilis* mutants altered in transformation. In *Bacterial Transformation*, edited by L. J. Archer, pp. 27–44, Academic Press, New York.

Prestidge, L., V. Gage and J. Spizizen, 1971   Protease activities during the course of sporulation in *Bacillus subtilis. J. Bacteriol.* **107**:815–823.

Ramaley, R. F. and L. Burden, 1970   Replacement sporulation of *Bacillus subtilis* 168 in a chemically defined medium. *J. Bacteriol.* **101**:1–8.

Reeve, J. N., N. H. Mendelson, S. I. Coyne, L. L. Hallock and R. M. Cole, 1973   Minicells of *Bacillus subtilis. J. Bacteriol.* **114**:860–873.

Rogolsky, M., 1968   Genetic mapping of a locus which regulates the production of pigment associated with spores of *Bacillus subtilis. J. Bacteriol.* **95**:2426–2427.

Rutberg, B. and J. A. Hoch, 1970   Citric acid cycle: gene–enzyme relationships in *Bacillus subtilis. J. Bacteriol.* **104**:826–833.

Rutberg, L., 1969   Mapping of a temperate bacteriophage active on *Bacillus subtilis. J. Virol.* **3**:38–44.

Saito, H. and K. Miura, 1963   Preparation of transforming deoxyribonucleic acid by phenol treatment. *Biochim. Biophys. Acta* **72**:619–629.

Schaeffer, P., 1964   Transformation. In *The Bacteria*, Vol. 5, edited by I. C. Gunsalus and R. Y. Stanier, pp. 87–153, Academic Press, New York.

Siegel, E. C. and J. Marmur, 1969   Temperature-sensitive induction of bacteriophage in *Bacillus subtilis* 168. *J. Virol.* **4**:610–618.

Sinha, R. P. and V. N. Iyer, 1972   Isolation and some distinctive properties of a new type of recombination-deficient mutant of *Bacillus subtilis. J. Mol. Biol.* **72**:711–724.

Smith, I., D. Dubnau, P. Morell and J. Mamur, 1968   Chromosomal location of DNA base sequences complementary to transfer RNA and to 5S, 16S, and 23S ribosomal RNA in *Bacillus subtilis. J. Mol. Biol.* **33**:123–140.

Smith, I., C. Goldthwaite and D. Dubnau, 1969   The genetics of ribosomes in *Bacillus subtilis*. *Cold Spring Harbor Symp. Quant. Biol.* **34:**85–89.

Spatz, H. C. and T. A. Trautner, 1970   One way to do experiments on gene conversion? Transfection with heteroduplex SPP1 DNA. *Mol. Gen. Genet.* **109:**84–106.

Spatz, H. C. and T. A. Trautner, 1971   The role of recombination in transfection. *Mol. Gen.Genet.* **113:**174–190.

Spizizen, J., 1958   Transformation of biochemically deficient strains of *Bacillus subtilis* by deoxyribonucleate. *Proc. Natl. Acad. Sci. USA* **44:**1072–1078.

Spizizen, J., B. E. Reilly and A. H. Evans, 1966   Microbial transformation and transfection. *Annu. Rev. Microbiol.* **20:**371–400.

Staal, S. P. and J. A. Hoch, 1972   Conditional dihydrostreptomycin resistance in *Bacillus subtilis*. *J. Bacteriol.* **110:**202–207.

Steinberg, W. and C. Anagnostopoulos, 1971   Biochemical and genetic characterization of a temperature-sensitive, tryptophanyl-transfer ribonucleic acid synthetase mutant of *Bacillus subtilis. J. Bacteriol.* **105:**6–19.

Strauss, N., 1970   Early energy-dependent step in the entry of transforming deoxyribonucleic acid. *J. Bacteriol.* **101:**35–37.

Sueoka, N. and H. Yoshikawa, 1963   Regulation of chromosome replication in *Bacillus subtilis. Cold Spring Harbor Symp. Quant. Biol.* **28:**47–54.

Takahashi, I., 1961   Genetic transduction in *Bacillus subtilis. Biochem. Biophys. Res. Commun.* **5:**171–175.

Takahashi, I., 1965   Localization of spore markers on the chromosome *Bacillus subtilis*. In *Spores III,* edited by L. L. Campbell and H. O. Halvorson, pp. 138–143, American Society for Microbiology, Ann Arbor, Michigan.

Taylor, A. L. and C. D. Trotter, 1972   Linkage map of *Escherichia coli* strain K-12. *Bacteriol. Rev.* **36:**504–524.

Thorne, C. B., 1962   Transduction in *Bacillus subtilis. J. Bacteriol.* **83:**106–111.

Trautner, T. A. and H. C. Spatz, 1973   Transfection in *Bacillus subtilis. Curr. Top. Microbiol. Immunol.* (in press).

Vapnek, D. and S. Greer, 1971*a*   Suppression by derepression in threonine dehydratase-deficient mutants of *Bacillus subtilis. J. Bacteriol.* **106:**615–625.

Vapnek, D. and S. Greer, 1971*b*   Minor threonine dehydratase encoded within the threonine synthetic region of *Bacillus subtilis. J. Bacteriol* **106:**983–993.

Wake, R. G., 1973   Circularity of the *Bacillus subtilis* chromosome and further studies on its bidirectional replication. *J. Mol. Biol.* **77:**569–575.

Whitt, D. D. and B. C. Carlton, 1968   Characterization of mutants with single and multiple defects in the tryptophan biosynthetic pathway in *Bacillus subtilis. J. Bacteriol.* **96:**1273–1280.

Wilson, G. A. and K. F. Bott, 1968   Nutritional factors influencing the development of competence in *Bacillus subtilis* transformation system. *J. Bacteriol.* **95:**1439–1449.

Wilson, G. A. and F. E. Young, 1972   Intergenotic transformation of the *Bacillus subtilis* genospecies. *J. Bacteriol.* **111:**705–716.

Wilson, G. A. and F. E. Young, 1973   Intergenotic and heterospecific transformation: the mechanism of restriction of genetic exchange in *Bacillus subtilis*. In *Bacterial Transformation,* edited by L. J. Archer, pp. 269–292, Academic Press, New York.

Wilson, M. C., J. L. Farmer and F. Rothman, 1966   Thymidylate synthesis and aminopterin resistance in *Bacillus subtilis. J. Bacteriol.* **96:**186–196.

Yamagishi, H. and I. Takahashi, 1968   Transducing particles of PBS1. *Virology* **36:**639–645.

Yasbin, R. E. and F. E. Young, 1972   The influence of temperate bacteriophage ϕ105 on transformation and transfection in *Bacillus subtilis. Biochem. Biophys. Res. Commun.* **47**:365–371.

Yasbin, R. E., M. J. Tevethia, G. A. Wilson and F. E. Young, 1973*a* Analysis of steps in transformation and transfection in transformation defective mutants and lysogenic strains of *Bacillus subtilis.* In *Bacterial Transformation,* edited by L. Archer, pp. 3–26, Academic Press, New York.

Yasbin, R. E., G. A. Wilson and F. E. Young, 1973*b* Transformation and transfection in lysogenic strains of *Bacillus subtilis* 168. *J. Bacteriol.* **113**:540–548.

Yehle, C. O. and R. H. Doi, 1967 Differential expression of bacteriophage genomes in vegetative and sporulating cells of *Bacillus subtilis. J. Virol.* **1**:935–947.

Yoshikawa, H., 1966 Mutations resulting from transformation of *Bacillus subtilis. Genetics* **54**:1201–1214.

Yoshikawa, H. and N. Sueoka, 1963*a* Sequential replication of *Bacillus subtilis* chromsome. I. Comparison of marker frequencies in exponential and stationary growth phases. *Proc. Natl. Acad. Sci. USA* **49**:559–566.

Yoshikawa, H. and N. Sueoka, 1963*b* Sequential replication of *Bacillus subtilis* chromosome. II. Isotopic transfer experiments. *Proc. Natl. Acad. Sci. USA* **49**:806–813.

Young, F. E., 1967 Requirement of glucosylated teichoic acid for adsorption of phage in *Bacillus subtilis* 168. *Proc. Natl. Acad. Sci. USA* **58**:2377–2384.

Young, F. E. and A. P. Jackson, 1966 Extent and significance of contamination of DNA by teichoic acid in *Bacillus subtilis. Biochem. Biophys. Res. Commun.* **23**:490–495.

Young, F. E. and J. Spizizen, 1961 Physiological and genetic factors affecting transformation of *Bacillus subtilis. J. Bacteriol.* **81**:823–829.

Young, F. E. and J. Spizizen, 1963 Incorporation of deoxyribonucleic acid in the *Bacillus subtilis* transformation system. J. Bacteriol. **86**:392–400.

Young, F. E. and G. A. Wilson, 1972 Genetics of *Bacillus subtilis* and other Gram-positive sporulating bacilli. In *Spores V,* edited by H. O. Halvorson, R. Hanson and L. L. Campbell, pp. 77–106, American Society for Microbiology, Washington, D.C.

Young, F. E., and G. A. Wilson, 1975 Chromosomal map of *Bacillus subtilis.* In *Spores VI,* pp. 596–614, American Society for Microbiology, Washington, D.C.

Young, F. E., C. Smith and B. E. Reilly, 1969 Chromosomal location of genes regulating resistance to bacteriophage in *Bacillus subtilis J. Bacteriol.* **98**:1087–1097.

Young, F. E., P. Haywood and M. Pollock, 1970 Isolation of L-forms of *Bacillus subtilis* which grow in liquid medium. *J. Bacteriol.* **102**:867–870.

5

Escherichia coli—An Overview

Roy Curtiss III, Francis L. Macrina, and Joseph O. Falkinham III

Introduction

The pre-eminence of *Escherichia coli* in investigative research has depended upon several characteristics of this organism as well as on decisions and discoveries early in the development of molecular biology. The ubiquitous occurrence of this potential pathogen, its haploid nature (chromosome mol. wt. 2.5×10^9 daltons) allowing the expression of recessive characteristics, its ability to grow in a variety of media, and its rapid growth rate are major factors contributing to the emphasis on *E. coli* as a research organism. Historically, the early studies of Hershey, Delbrück, and Luria (see Cairns *et al.*, 1966) with the T-bacteriophages of *E. coli* (Demerec and Fano, 1945) coincided with the work of Gray and Tatum (1944) demonstrating that ionizing radiation could induce mutations in *E. coli* which resulted in various nutritional requirements. Such auxotrophic mutants of *E. coli* were later employed by Lederberg and Tatum (1946) to demonstrate conjugal gene transmission between strains. The discovery of phage-mediated gene transmission of various genetic markers (generalized

Roy Curtiss III, Francis L. Macrina, and Joseph O. Falkinham III—Department of Microbiology, Institute of Dental Research, Cancer Research and Training Center, University of Alabama in Birmingham, Birmingham, Alabama.

tranduction, see Lennox, 1955) or specific markers (specialized transduction, see Morse *et al.,* 1956) in *E. coli* shortly followed the description of generalized transduction in *Salmonella typhimurium* (Zinder and Lederberg, 1952). All of these events were of importance in the resulting emphasis on the study of *E. coli* as a model biological system. Current areas of interest in *E. coli* research include: the mechanism of genetic recombination and gene transmission (bacterial conjugation), synthesis and regulation of proteins (especially enzymes) and ribonucleic acid (RNA), and the molecular events of deoxyribonucleic acid (DNA) replication and cell division. The techniques of conjugation and transduction are continually used in the genetic manipulation of *E. coli* as well as in chromosomal mapping and complementation studies. With this in mind, it is our goal to present a brief outline of *E. coli* genetics with major emphasis on the conjugation system. A rigorous treatment of the history and methodology of *E. coli* genetics may be found in the work by Hayes (1968). The reader is also directed to the reviews by Curtiss (1969) and Susman (1970) for comprehensive discussions of bacterial conjugation, and to the review by Ozeki and Ikeda (1968) for a complete treatment of generalized and specialized transduction.

Conjugation

In 1953, Hayes (1953a) and Cavalli *et al.* (1953) demonstrated that conjugation in *E. coli* was due to the presence of a conjugal fertility factor that they called F. It was found that strains carrying this factor (F$^+$) were capable of acting as genetic donors in matings with strains containing no F factor (F$^-$) acting as recipients. Matings between donor and recipient strains were found to yield measurable numbers of recombinant progeny inheriting donor chromosomal information (ca. 1 recombinant/10^5 donor cells) whereas matings between two donor strains resulted in very low recombinant frequences (10^{-7}).

The F factor is a member of a class of extrachromosomal elements termed plasmids (Novick, 1969). It may be physically isolated as a covalently closed circular DNA molecule, and is capable of replicating independently of the bacterial chromosome. Furthermore, it is capable of stable integration into the chromosome of *E. coli* and is thus categorized as an episomal plasmid (Jacob and Wollman, 1961; also see Chapter 18 by Campbell). When present in the extrachromosomal state, the F plasmid may occasionally be lost spontaneously during cell division or selectively eliminated by growing F$^+$ cells in the presence of acridine orange (Hirota, 1960). F$^-$ cells may be readily converted to the F$^+$ state by conjugal transmission of the F factor into such cells.

TABLE 1.   *Some Conjugal Fertility Plasmids of E. coli and Their Characteristics*

Plasmid	Capable of chromosomal integration	Molecular weight[a]	Special properties
F	Yes	62×10^6	Classic conjugal fertility factor
F'lac[b]	Yes	74×10^6	F plasmid carrying genes for lactose utilization
F'gal [b]	Yes	96×10^6	F plasmid carrying genes for galactose utilization
F ColM ColB ColV trp	Yes	113×10^6	Carries genes for tryptophan biosynthesis and genes for production of colicins M, B, and V [c]
Col V2	Yes	92×10^6	Carries conjugal fertility genes and gene for colicin V production
R1drd19 [d]	Yes?	63×10^6	Carries conjugal fertility genes and genes which confer resistance to ampicillin, kanamycin, chloramphenicol, streptomycin, sulfonamide, and spectinomycin. Codes for F-like donor pili [e]
R–100–1 [d]	Yes?	70×10^6	Carries conjugal fertility genes and genes which confer resistance to tetracycline, chloramphenicol, sulfonamide, streptomycin, and spectinomycin. Codes for F-like donor pili [e]
R–64–11[d]	No	70×10^6	Carries conjugal fertility genes and genes which confer resistance to tetracycline and streptomycin. Codes for I-like donor pili[e]

[a] Molecular weight expressed in daltons (Clowes, 1972).

[b] These are two representative F factors. The current collection of F factors, however, affords one the ability to choose any region of the *E. coli* chromosomal map (see map of F factors by Low, 1972; and Chapter 7).

[c] Colicins are antibioticlike proteins capable of killing various strains of *E. coli*.

[d] Three of the so-called R factors are presented as representatives of this class of plasmids. There are currently 19 classes of R factors based on their ability to stably coinhabit a cell (incompatibility groupings). Such plasmids are extremely common in nature and pose serious problems in the effective treatment of bacterial infections with antibiotics.

[e] Donor pili are surface appendages coded for by conjugal fertility plasmids and play an important role in conjugal DNA transfer. The I-type and F-type pili are represented by the above R factors. There are undoubtedly numerous other distinct pili types not yet characterized by antigenic classification and phage receptor sites.

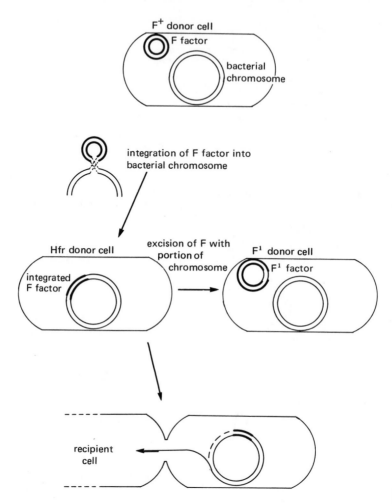

linear transmission of Hfr chromosome to recipient cell

Figure 1. The mating types of E. coli. *The F factor is depicted as two thick circles, each circle representing a complementary strand of the DNA molecule. The two DNA strands of the chromosome are depicted by the thin lines. The molecular weight of the F factor is 62×10^6 daltons. Thus, the length of F is approximately 2.5 percent of the length of the* E. coli *chromosome. F is not drawn to scale in any of the cells depicted in this illustration. Note that only a single strand of the Hfr chromosome is transferred during conjugation and that a portion of the F factor is present on the lead region of the injected chromosome (Glatzer and Curtiss, unpublished).*

In 1951 Lederberg (1951) showed that *E. coli* strains possessing conjugal fertility plasmids were rare (2.5 percent). Today, however, this is no longer the case and almost 30–50 percent of *E. coli* strains isolated from nature possess conjugal fertility plasmids. Table 1 lists a number of different conjugal fertility plasmids and their genetic and physical properties. The presence of conjugal fertility plasmids, such as F, confers several other phenotypic traits on *E. coli* cells. These characteristics are: (1) sensitivity to donor specific bacteriophages (Loeb, 1960) due to the presence of cellular appendages called donor pili (Brinton, 1967; Lawn *et al.*, 1967), (2) decreased ability to accept related plasmids in conjugation (entry exclusion) (Novick, 1969), (3) inability to stably maintain two related plasmids (plasmid incompatibility) (Novick, 1969), and (4) the ability to inhibit the intracellular replication of the so-called female-specific baberiophages (see Curtiss, 1969).

The stable integration of the F plasmid into the circular bacterial chromosome via a reciprocal crossover event results in the formation of an Hfr (high-frequency recombination) donor cell. During conjugation such an Hfr donor strain transfers its chromosome in a linear sequential fashion from the site of F integration (Wollman *et al.*, 1956). This site is termed the point of origin for the particular Hfr strain. The entry of markers with time can be scored following physical separation of donor and recipient cells. Total chromosome transfer takes 90 minutes at 37°C (Taylor and Thoman, 1964). It should be noted that the integrated F factor is transferred at the end of the linearly injected chromosome sequence. Thus, the Hfr trait is not transmitted to recipient cells at a high frequency because of an appreciable rate of spontaneous interruption of mating (Jacob and Wollman, 1961). The foregoing events are depicted in Figure 1.

Hfr donors can frequently revert to the F^+ state (autonomously replicating F plasmid) by excision of F from the bacterial chromosome. In some cases the excision of F involves a crossover that permits the generation of an F plasmid carrying a number of bacterial genes which existed adjacent to the integrated F. These elements are called F-prime (F′) factors and exist as autonomously replicating plasmids (Jacob and Adelberg, 1959) (see Figure 1). They are also capable of integrating into the bacterial chromosome usually at the same but sometimes at a different site. Such F′-factor formation in the Hfr cell may be detected utilizing a modification of the fluctuation test of Luria and Delbrück (1943), as employed by Berg and Curtiss (1967). In this instance, cells carrying F′ plasmids are still haploid in nature (primary F′ donor cell) since the bacterial genes carried by the F′ are deleted from the chromosome. Transfer of such an F′ to an F^- cell with a complete chromosome results in partial diploidy of the F´-carried bacterial genes (secondary F´ donor cell). F´ factors may also be isolated in crosses between Hfr donor strains and F⁻ recipient strains by selecting for recombinants which inherit a terminally transferred marker early in the mating (Jacob and Adelberg, 1959). Alternatively one may select either an early- or late-transferred marker using an F⁻ recipient that is defective in recombinant formation (*recA*) (Low, 1968).

The genetic control of conjugal fertility has been recently reviewed by Willetts (1972), and other aspects of conjugal fertility plasmids have been reviewed by Meynell *et al.* (1968), Novick (1969), Curtiss (1969), and Clowes (1972). Although the following discussion of bacterial conjugation will limit itself to *E. coli*, it should be kept in mind that the F factor is capable of existing in cells of *Salmonella typhimurium* and a conjugation system has also been developed in this organism (Sanderson *et al.*, 1972).

Materials and Methods in *Escherichia coli* Genetics

Foreword.   The laboratory manuals by Clowes and Hayes (1966) and by Miller (1972) elaborate instructions for conducting a great diversity of experiments in the area of *E. coli* genetics. These texts should, therefore, be consulted for a detailed description of experimental protocols. General information regarding history, theory, and practice in *E. coli* genetics may be found in the texts by Jacob and Wollman (1961) and Hayes (1968).

Bacterial Strains.   A very large number of *E. coli* strains are kept at the *E. coli* Genetic Stock Center at Yale University, 310 Cedar Street, New Haven, Connecticut 06510. Pedigrees for many of the major sublines of *E. coli* K12 are given in the review by Bachmann (1972). Strains are also frequently obtained by directly writing to investigators actively working in a given area of *E. coli* genetics. The review article by Taylor and Trotter (1972) may be used to readily identify investigators with their area of research. In addition, Cold Spring Harbor Laboratory, Cold Spring Harbor, New York 11724, provides a strain kit as an accompaniment to Miller's (1972) manual *Experiments in Molecular Genetics*. A list of all the Hfr strains of *E. coli* known to the authors is given in Table 2, and the origins of chromosome transfer for these Hfr donors are depicted in Figure 2. A figure depicting the F´ factors of *E. coli* (Low, 1972) can be found in Chapter 7 of this volume.

Standard nomenclature employed in describing strain genotypes and phenotypes usually follows the proposals put forward by Demerec *et al.* (1966) (also see Taylor and Trotter, 1972; and Chapter 6 in this volume). *E. coli* strains can be routinely stored on Penassay agar slants (Difco

Laboratories, Detroit, Michigan) in the refrigerator for several months, and most investigators keep their stocks frozen for long-term storage as a means of eliminating the necessity of making periodic transfers as well as to minimize genetic changes. We have found that rapid freezing (dry ice–ethanol) of cells from stationary-phase cultures suspended in media containing 1 percent peptone and either 5 or 30 percent glycerol with subsequent storage at −30 to − 40°C provides adequate conditions for long-term culture maintenance. In 30 percent glycerol, the media never solidifies in the deep freeze, which facilitates obtaining samples without having to go through a thaw–refreeze cycle.

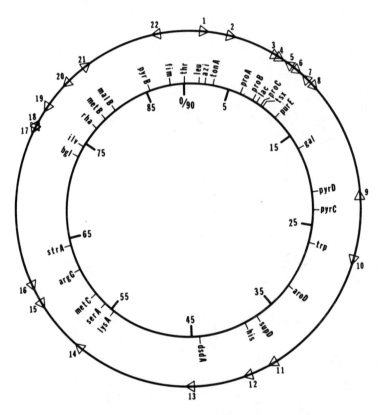

Figure 2. A simplified chromosome map of E. coli K12 showing the origins for chromosome transfer by 22 types of Hfr donors. Map units (0–90) inside the smaller circle are in minutes. Abbreviations for the genetic markers are given by Taylor and Trotter (1972; see Chapter 6). The integrated F factors are depicted as arrowheads on the outer circle. The direction of the arrow corresponds to the direction of chromosome transfer (recall, however, that about two-thirds of the integrated F is transferred distally; see Figure 1). The number designations on the outer circle correspond to the groups of Hfr donor types listed in Table 2.

TABLE 2. Hfr Strains of Escherichia coli[a]

Group[b]	Proximally transferred marker and map position, min[c]		Terminally transferred marker and map position, min[c]		Transfer direction[a]	Hfr designations[e]
1.	leu	1.5	azi	2.0	CC	P1; R3, R5; T4
2.	tonA	3.0	proA	6.5	CC	P7 (P808)
3.	proB	8.3	lac	9.0	CC	P4X6, P804; B1, B2, B3, B4, B5, B6; R2, R4; OR6, OR11, OR55, OR66, OR72, OR73
4.	lacY	9.0	lacI	9.1	CC	OR1
5.	proC	9.6	lac	9.0	C	Hfr 6, Hfr 13; B8, B11, B13; OR21, OR56
6.	tsx	9.8	purE	12.0	CC	OR7; P21
7.	purE	12.0	tsx	9.8	C	OR54
8.	purE	12.0	gal	16.8	CC	Cavalli; P3 (4000); OR5, OR9, OR59, OR69, OR74[f]
9.	pyrC	23.5	pyrD	21.5	C	KL19; KL99
10.	trp	27.0	aroD	32.3	CC	B7, B9, B10; OR20, OR23, OR24, OR81[f]
11.	supD	37.7	his	38.5	CC	Hfr 44
12.	his	38.5	dsdA	44.5	CC	AB311; B12; G1, G3, G8, G9; KL96; OR13, OR14, OR19, OR30, OR40, OR75, OR87[f]
13.	dsdA	44.5	lysA	54.7	CC	KL98; G2; OR57, OR77[f]
14.	lysA	54.7	serA	56.0	CC	KL16; Ra-1; OR12, OR15, OR17, OR37, OR63, OR65, OR78

		[b][c]		[c]	[d]	[e]
15.	argG	61.0	metC	57.7	C	AB312; Hfr 4; G6; KL14; OR18, OR29, OR31, OR32, OR34, OR35, OR36, OR38, OR43, OR48, OR49, OR50, OR51, OR52, OR53, OR58, OR60, OR61, OR62, OR67, OR68, OR70, OR71, OR79, OR82
16.	strA	64.0	argG	61.0	C	G11
17.	ilv	74.7	bgl	73.7	C	R1; P6(P13); G10; KL25; OR8, OR10, OR22, OR39, OR44, OR86
18.	bgl	73.7	ilv	74.7	CC	AB313
19.	rha	77.4	ilv	74.7	C	P5(P72)
20.	metB	78.2	rha	77.4	C	Ra–2
21.	malB	80.6	malB	80.6	CC	Hfr 3; J4(P10)
22.	fim	88.0	pyrB	85.2	C	Hayes; G4, G5, G7; CR34; OR2, OR3, OR4, OR16, OR41, OR42, OR45, OR46, OR47, OR64, OR76, OR80f

[a] The list of Hfr strains is restricted to those arising directly from insertion of F into the chromosome and does not include those arising by transposition of an F plasmid into a nonhomologous region of the chromosome. Of the 137 Hfr strains listed, F integration resulted in gene inactivation in four instances [Hfr 3, (Richter, 1961); Hfr J4(P10), (Schwartz, 1966); Hfr OR1, (Curtiss et al., 1968); Hfr OR77, (Curtiss and Stallions, 1969)].

[b] Number refers to number designations on Figure 2.

[c] Map positions in minutes are as indicated by Taylor and Trotter (1972; Chapter 6).

[d] CC, counter clockwise; C, clockwise (see Figure 2).

[e] The Hfr strains are designated in the following way: Cavalli (Cavalli, 1950); Hayes (Hayes, 1953b); P, Paris collection (Jacob and Wollman, 1961; see also Bachmann, 1972); Hfr with numbers from the Lederberg collection (see Hayes, 1968); R, Reeves collection (see Hayes, 1968); G, Matney et al. (1964); KL, Low collection (Low, 1972; 1973); B, Broda collection (Broda, 1967); AB and T, Taylor and Adelberg (1960); OR, Curtiss collection (see Berg and Curtiss, 1967; Curtiss et al., 1968; 1969; Curtiss and Stallions, 1969; unpublished). Designations in parentheses are synonym designations used in the literature.

[f] The Hfr strains in these groups may not all have exactly the same origins for chromosome transfer.

Media.   *Escherichia coli* will grow well in a diversity of synthetic media including M9 (Adams, 1959), VB (Vogel and Bonner, 1956), Davis (Davis and Mingioli, 1950), and ML (Curtiss, 1965). These minimal media can be supplemented with a diversity of carbon sources, and appropriate concentrations for additions such as amino acids, purines, pyrimidines, and vitamins have been published (Curtiss *et al.*, 1968). Many investigators working on the genetics of *E. coli* use L broth and agar (Lennox, 1955) or Penassay broth and agar (Difco Laboratories, Detroit, Michigan) as complex media.

Mutant Isolation.   Procedures for mutagenizing *E. coli* with a variety of mutagenic agents are given by Miller (1972). A novel method for inducing a diversity of mutations in *E. coli* involves lysogenization with phage Mu1 which results in about 2 percent of the lysogens being mutants (Taylor, 1963). Auxotrophic mutants (requiring some nutritional factor) are generally isolated by penicillin-enrichment techniques (Lederberg, 1950) as modified by Gorini and Kaufman (1960), Lubin (1962), and Curtiss *et al.* (1965). Mutations conferring resistance to bacteriophages, drugs, metabolic analogs, etc., are usually selected by plating high concentrations of bacterial cells on media containing the appropriate agent. Mutants lacking the ability to utilize a given carbon source are often identified by plating cells on indicator media following mutagenesis. Eosin methylene blue (EMB) base agar, MacConkey base agar, and triphenyltetrazolium chloride agar (Lederberg, 1948) supplemented with 1–2 percent of the appropriate carbohydrate are commonly employed. Fermenting and nonfermenting colonies are differentiated by color on such media.

Phage Methods.   Methods for working with the bacteriophages that grow on *E. coli* are treated in a general but thorough way in the Appendix to the text by Adams (1959). Methods for using the generalized transducing phage P1 are given in the papers by Lennox (1955) and Caro and Berg (1971) as well as in the laboratory manual by Miller (1972). The methodology for working with the specialized transducing phages λ and ϕ80 are given by Hershey (1971) and Miller (1972).

Conjugation Methods.   Procedures for growing donor and recipient strains to optimize their respective mating phenotypes have been given by Curtiss *et al.* (1968, 1969). Donors and recipients are mixed to initiate mating and the ratio of donor to recipient cells is usually 1:10 (or greater) to minimize mating events involving one recipient cell and multiple donor cells. Matings are interrupted either by the use of a virulent phage such as T6 (Hayes, 1957; Curtiss *et al.*, 1968) which lyses the wild-type sensitive donor cells but not the T6-resistant mutant recipient cells or by violent agitation which disrupts mating pairs (Wollman and Jacob, 1955; Low and Wood, 1965). In either case, the donor parent is also contraselected by the omission of a donor nutritional requirement and/or by the addition of an agent (e.g., antibiotic) in the selective medium which prevents the growth of the donor but not the recipient. Recombinant frequency for a given genetic marker is expressed as the number of recombinant colonies per number of input donor cells.

Chromosome Mapping. Mapping mutational lesions is usually accomplished by Hfr × F⁻ crosses (Taylor and Thoman, 1964) although satisfactory procedures for utilizing F⁺ donors in both interrupted matings and linkage analysis (as described below) have been developed (Curtiss and Renshaw, 1969). Mapping consists of localizing the unmapped allele to a quadrant of the 90 minute chromosome map (see Chapter 6 of this volume) followed by precise mapping utilizing time-of-entry techniques (Freifelder, 1971).

Mapping relies upon either spontaneous or planned interruption of Hfr chromosome transfer as a function of time following the initiation of mating; i.e., the mixing of donor and recipient cells (Jacob and Wollman, 1961). During conjugal transfer of the chromosome from donor to recipient, it should be kept in mind that interruption of this process may occur at any time due to factors such as breakage of the Hfr chromosome and/or spontaneous separation of mating cell pairs. This constant probability of spontaneous interruption per minute of mating results in decreasing frequencies of inheritance of markers as the distance between the origin of Hfr chromosome transfer and each marker increases. Thus proximally transferred Hfr markers are inherited by F⁻ cells much more often than are distally transferred Hfr markers.

Preliminary mapping can be accomplished by cross-streaking or replica plating Hfr strains with a variety of transmission origins (see Figure 2 and Table 2) against the mutant F⁻ strain on selective medium (Miller, 1972; Low, 1973). Selection of an Hfr demonstrating high-frequency transmission of the mutant allele can be accomplished by this method. Once the allele is located in a specific map quadrant, one employs the Hfr demonstrating high-frequency transmission in a gradient-of-marker-transmission study (Jacob and Wollman, 1961). Such gradient-of-transmission analysis of several genetic markers is performed by plating samples of mating mixtures (usually after 60 minutes of mating) on appropriate selective media. Representative data for such a study are given in part A of Table 3 (also see Figure 2 for Hfr point of origin and location of genetic markers). Note that the recombinant frequency for a given marker decreases as the distance from the point of origin of the Hfr to the marker

TABLE 3.   *Recombinant Frequency Analysis in Hfr* × *F⁻ Cross*[a]

A.   *Gradient of Marker Transmission in 60-Minute Mating*

Selected markers[b]	*Recombinant frequency, recombinants/input Hfr cell*	*Distance from Hfr point of origin, min*	*Position on standard map of E. coli*[c] *min*
Leu⁺ (Str^r)	0.20	5	1.5
Pro⁺ (Str^r)	0.15	10	6.5
Lac⁺ (Str^r)	0.13	13	9.0
Gal⁺ (Str^r)	0.078	21	17
Trp⁺ (Str^r)	0.044	31	27

B.   *Frequency of Unselected Markers among Trp⁺ (Str^r) Recombinants from the Above Cross*

Unselected markers

Selected phenotype	*Frequency of occurrence with selected marker, percent*			
Trp⁺ (Str^r)	Gal⁺	Lac⁺	Pro⁺	Leu⁺
	70	58	55	51

[a]Hfr H prototroph *str^s* × F⁻ *leu⁻ proA⁻ lac⁻ gal⁻ trp⁻ str^r*. See Figure 2 and Table 2 for transmission origin of Hfr chromosome. Abbreviations: *leu⁻*, requires leucine; *proA⁻*, requires proline; *lac⁻*, cannot utilize or ferment lactose; *gal⁻*, cannot utilize or ferment galactose; *trp⁻*, requires tryptophan; *str^r*, resistant to streptomycin.

[b]For example, Leu⁺ Str^r recombinants are selected by plating on minimal agar containing proline, tryptophan, glucose, and streptomycin; Lac⁺ Str^r recombinants by plating on minimal agar containing leucine, proline, tryptophan, lactose, and streptomycin.

[c]Taylor and Trotter (1972).

increases. Such data enable one to obtain the position of an unmapped marker relative to the markers of known location. Using the data in Table 3 as an example, a recombinant frequency of 0.05 recombinants/donor Hfr cell for an unmapped marker would indicate that this marker was located at about minute 25 on the chromosome map of *E. coli*. Alternatively, one may select for a specific recombinant class, e.g., Trp⁺ recombinants as illustrated in part B of Table 3, and analyze such recombinants for the presence of other unselected markers. Crude linkage relationships may then be determined and serve to roughly position unmapped markers.

Precise positioning of a marker is performed by interrupted mating techniques. In order to be accurate, these experiments must be restricted to markers transferred within 30 minutes of the Hfr chromosome origin. In such studies, mating mixtures are sampled at given times, and mating cell pairs are disrupted by agitation or other means (see Conjugation Methods). Samples are then plated on selective media, incubated, and recombinant clones enumerated. The frequency of marker inheritance as a function of time is then computed. Comparison of the time of entry (defined as the time of first appearance of a donor marker in recipients) of the unknown marker to those for known markers transferred by the Hfr gives the position of the new allele on the 90 minute map.

Three- or four-factor linkage analysis is used to establish an unambiguous allele order for markers with similar times of entry. In practice, recombinants inheriting a distally transferred Hfr marker are purified and then the frequencies of single and multiple exchanges within a proximally transferred region containing the markers to be ordered are determined.

Dominance Tests and Complementation. F´ elements are available which include various segments of the entire *E. coli* K12 chromosome (Low, 1972; Chapter 7 of this volume). This collection of F´ plasmids enables one to determine whether the wild-type or mutant allele gene product is dominant. To accomplish this objective it is important to use methods for strain construction that do not predetermine the results. Therefore, a strain harboring the F´ plasmid carrying the wild-type allele is mated with an F⁻ strain possessing a mutation in the gene in question, and the resulting F´ partial diploid strain is isolated either by random testing the recipient exconjugant population for the presence of the F´ (by ability to transfer the F´ and/or sensitivity to F-specific phages) or by selecting for the inheritance of a second wild-type marker on the F´ that is known to be dominant to the corresponding mutant chromosomal allele. The F´ partial diploid strain is then tested for phenotype to determine whether the mutant or wild-type allele is dominant. It is also good practice to repeat these tests with the mutant allele on the F´ and the wild-type allele on the chromosome.

F´ partial diploid strains are also very useful for complementation studies to determine whether independently occurring mutations giving rise to mutants with the same phenotype occur in the same or different genes. Strains are constructed as described above with one mutation on the F´ and the other on the chromosome. Expression of the mutant phenotype implies that the two mutations are in the same gene whereas expression of the wild-type phenotype implies that they are in separate genes. Exceptions to these results sometimes arise. For example, in complementation tests between mutations in genes that specify a protein product (structural genes) and mutations in regulator genes that regulate the activity of the structural genes, the mutant phenotype is often expressed even though the mutations are in separate genes. The wild-type phenotype can also be observed sometimes even though both mutations do occur in the same gene, and this is presumably due to interactions between protein subunits or peptide fragments coded for by the two mutated alleles.

Transduction. The transmission of bacterial genes by a bacterio-phage vector is defined as transduction (see Introduction). The order and relative distances between inter- or intragenic mutational sites is accomplished using the generalized transducing phage P1. An extensive treatment of P1 methods is given by Caro and Berg (1971). The principles of P1-mediated transduction may be summarized as follows. P1 phage is propagated on the strain of *E. coli* to be used as a source of donor genetic information. The resulting phage lysate contains some phage virions (transducing particles) which consist of a P1 protein capsid surrounding bacterial DNA instead of phage DNA (Ikeda and Tomizawa, 1965). Such particles are subsequently capable of injecting bacterial DNA into cells of *E. coli*. This DNA may then recombine with the resident host chromosome. Phage grown on one mutant strain are used to infect a second mutant strain; recombinants are recovered on the appropriate selective medium and analyzed. Relative distances between mutational sites are computed in these two-factor crosses based on the assumption that recombination is less likely to occur between sites which are very close together. An application of two-factor transductional analysis is presented in the classic paper of Yanofsky and Crawford (1959) in which the authors construct a map of mutational sites in the two contiguous tryptophan synthetase genes of *E. coli*.

Transduction can also be utilized to establish linkage relationships between three closely linked markers. Determination of recombinant frequencies with respect to these three markers enables one to establish their linear order on the chromosome. This type of analysis may be extended by examining data in crosses involving four closely linked markers (four-factor analysis). Gross and Englesberg (1959) used both three- and four-factor analysis to establish the order of numerous mutations in the genes governing arabinose utilization. A detailed discussion of the mathematics of three- and four-factor analysis using generalized transduction has been presented by Wu (1966).

Bacterial Genetic Engineering

The above systems of bacterial conjugation and transduction have allowed the manipulation of the topography of the *E. coli* chromosome. Techniques have been developed which enable one to construct strains in which bacterial genes are moved to different positions (transposed) on the *E. coli* chromosome. Cuzin and Jacob (1964) were able to transpose the genes for lactose utilization (*lac* operon) by using a strain carrying a chromosomal deletion of the *lac* genes and harboring an F′ *lac* plasmid unable to replicate at 42°C. In this instance, cells were grown at 42°C in media which contained lactose as a sole source of carbon. Lactose-fermenting cells surviving this treatment were found always to contain F′*lac* which had stably integrated at some (nonhomologous) position on the chromosome of *E. coli.* Other collections of transposition Hfr strains were obtained by Scaife (1966) and Berg and Curtiss (1967) by growing haploid F′ strains in the presence of acridine orange, which prevents F′ replication. Under these conditions, the loss of the F′, which contained genes governing vital cell functions, was lethal and only in those instances in which the genetic material on the F′ had integrated into the chromosome would the cell survive. These techniques have been refined somewhat (Beckwith *et al.,* 1966; Reznikoff *et al.,* 1969) to simultaneously select for a mutation at a specific genetic locus (via insertion of the F′ factor into a gene) while selecting for F′ integration. Reznikoff *et al.* (1969) were able to obtain strains in which the *lac* and *trp* (genes for tryptophan biosynthesis) operons were fused such that the expression of the *lac* genes was now under the control of the *trp* operator. They were also able to isolate specialized ϕ80 transducing phages carrying the *lac* operon, since the ϕ80 prophage integration site is close to the *trp* operon.

An additional innovative means of moving unlinked genes close together was developed by Press *et al.* (1971) who made use of the fact that two related plasmids cannot stably coexist in the same cell due to incompatibility. Press *et al.* (1971), therefore, introduced two F′ plasmids into a cell deficient in genetic recombination so as to minimize integration of the F′ into the chromosome. Stable clones were then isolated in which genetic markers from both F′ plasmids persisted. An examination of these strains revealed the formation of fusion F′ plasmids that consisted of chromosomal markers from both of the original F′ elements. Only one copy of the F factor was present on these so-called fusion F′ plasmids. Using this technique, the ϕ80 attachment site was transposed to the chromosomal region carrying the genes for arginine (*arg*⁺) and methionine (*met*⁺) biosynthesis. This enabled Press *et al.* (1971) to subsequently isolate ϕ80 *arg*⁺ and ϕ80 *met*⁺ specialized transducing phages.

Shimada *et al.* (1972) were able to isolate strains of *E. coli* which carried the integrated prophage λ at various positions on the chromosome. This was achieved by selecting for λ integration in strains which lacked (were deleted in) the normal λ integration site (attachment site near *gal,* see Figure 2). Such a variety of λ integration sites now enables one to construct specialized transducing phage which carry various regions of the *E. coli* chromosome.

In summary, the above techniques in conjunction with the mutant isolation, complementation and mapping procedures previously discussed have provided novel ways of studying the regulation and function of various operons (see Chapters 8, 9, 10, and 11 of this volume) as well as affording a powerful means of enriching for and isolating genes and specific DNA segments via the construction of phage genomes carrying bacterial genes. The latter case has been and will continue to be of great use in the study of transcription and translation of the genetic machinery of *E. coli*.

Conclusion

The short space allotted to discuss the genetics of *E. coli* does not permit one to do justice to the field nor to the organism. This is evidenced by the fact that the genetics of *E. coli* is dealt with in considerable detail in no less than five or six recent books and in some thirty to forty reviews that have appeared in the last several years. The reader is urged to ferret out these more comprehensive treatments of this subject and hopefully the references given herein will lead the way.

Literature Cited

Adams, M. H., 1959 *Bacteriophages*, Interscience Publishers, New York.

Bachmann, B. J., 1972 Pedigrees of some mutant strains of *Escherichia coli* K12. *Bacteriol. Rev.* **36**:525–557.

Beckwith, J. R., E. R. Signer and W. Epstein, 1966 Transposition of the *lac* region of *E. coli. Cold Spring Harbor Symp. Quant. Biol.* **23**:393–401.

Berg, C. M. and R. Curtiss, 1967 Transposition derivatives of an Hfr strain of *Escherichia coli* K12. *Genetics* **56**:503–525.

Brinton, C. C., Jr., 1967 Contributions of pili to the specificity of the bacterial cell surface and a unitary hypothesis of conjugal infectious hereditary. In *The Specificity of Cell Surfaces*, edited by B. D. Davis and H. J. Vogel, pp. 37–70. Prentice-Hall, Englewood Cliffs, N.J.

Broda, P., 1967 The formation of Hfr strains in *Escherichia coli* K12. *Genet. Res.* **9**:35–47.

Cairns, J., G. S. Stent and J. D. Watson, editors, 1966 *Phage and the Origins of Molecular Biology*, Cold Spring Harbor Laboratory, Cold Spring Harbor, N.Y.

Caro, L. G. and C. M. Berg, 1971 P1 transduction. In *Methods in Enzymology*, edited by L. Grossman and K. Moldave, Vol. 21, pp. 443–457. Academic Press, New York.

Cavalli, L. L., 1950 La sessualita nei batteri. *Boll. Ist. Sieroter. Milan* **29**:281–289.

Cavalli, L. L., J. Lederberg and E. M. Lederberg, 1953 An infective factor controlling sex compatibility in *Bacterium coli. J. Gen. Microbiol.* **8**:89–103.

Clowes, R. C., 1972 Molecular structure of bacterial plasmids. *Bacteriol. Rev.* **36**:361–405.

Clowes, R. C. and W. Hayes, 1966 *Experiments in Microbial Genetics*, John Wiley & Sons, New York.

Curtiss, R., 1965 Chromosomal aberrations associated with mutations to bacteriophage resistance in *Escherichia coli. J. Bacteriol.* **89**:28–40.

Curtiss, R., 1969 Bacterial conjugation. *Annu. Rev. Microbiol.* **23**:69–136.

Curtiss, R. and J. Renshaw, 1969 Kinetics of F transfer and recombinant production in F⁺ × F⁻ matings in *Escherichia coli* K12. *Genetics* **63**:39–52.

Curtiss, R. and D. R. Stallions, 1969 Probability of F integration and frequency of Hfr donors in F⁺ populations of *Escherichia coli* K12. *Genetics* **63**:27–38.

Curtiss, R., L. Charamella, C. M. Berg and P. Harris, 1965 Kinetic and genetic analyses of D-cycloserine inhibition and resistance in *Escherichia coli. J. Bacteriol.* **90**:1238–1250.

Curtiss, R., L. J. Charamella, D. R. Stallions and J. A. Mays, 1968 Parental functions during conjugation in *Escherichia coli* K12. *Bacteriol. Rev.* **32**:320–348.

Curtiss, R., L. G. Caro, D. P. Allison and D. R. Stallions, 1969 Early stages of conjugation in *Escherichia coli. J. Bacteriol.* **100**:1091–1104.

Cuzin, F. and F. Jacob, 1964 Délétions chromosomiques et intégration d'un épisome sexuel F-*lac*⁺ chez *Escherichia coli* K12. *C. R. Hebd. Seances Acad. Sci. Ser. D Sci. Nat.* **258**:1350–1352.

Davis, B. D. and E. S. Mingioli, 1950 Mutants of *Escherichia coli* requiring methionine or vitamin B12. *J. Bacteriol.* **60**:17–28.

Demerec, M. and U. Fano, 1945 Bacteriophage-resistant mutants in *Escherichia coli. Genetics* **30**:119–336.

Demerec, M., E. A. Adelberg, A. J. Clark and P. E. Hartman, 1966 A proposal for a uniform nomenclature in bacterial genetics. *Genetics* **54**:61–76.

Freifelder, D., 1971 Genetic mapping by Hfr mating. In *Methods in Enzymology*, edited by L. Grossman and K. Moldave, Vol. 21, pp. 438–443. Academic Press, New York.

Gorini, L. and H. Kaufman, 1960 Selecting mutants by the penicillin method. *Science (Wash., D.C.)* **131**:604–605.

Gray, C. H. and E. L. Tatum, 1944 X-ray induced growth factor requirements in bacteria. *Proc. Natl. Acad. Sci. USA.* **30**:404–410.

Gross, J. and E. Englesberg, 1959 Determination of the order of mutational sites governing L-arabinose utilization in *Escherichia coli* B/r by transduction with phage P1bt. *Virology* **9**:314–331.

Hayes, W., 1953a. Observations on a transmissible agent determining sexual differentiation in *Bacterium coli. J. Gen. Microbiol.* **8**:72–88.

Hayes, W., 1953b The mechanism of genetic recombination in *Escherichia coli. Cold Spring Harb. Symp. Quant. Biol.* **18**:75–93.

Hayes, W., 1957 The kinetics of the mating process in *E. coli. J. Gen. Microbiol.* **16**:97–119.

Hayes, W., 1968 *The Genetics of Bacteria and their Viruses*, John Wiley & Sons, New York.

Hershey, A. D., editor, 1971 *The Bacteriophage Lambda*, Cold Spring Harbor Laboratory, Cold Spring Harbor, N.Y.

Hirota, Y., 1960 The effect of acridine dyes on mating type factors in *Escherichia coli. Proc. Natl. Acad. Sci. USA* **46**:57–64.

Ikeda, H. and J. Tomizawa, 1965 Transducing fragments in generalized transduction by phage P1. I. Molecular origin of the fragments. *J. Mol. Biol.* **14**:85–109.

Jacob, F. and E. A. Adelberg, 1959   Transfert de caractères génétiques par incorporation au factéur sexuel d'*Escherichia coli. C. R. Hebd. Seances Acad. Sci. Ser. D Sci. Nat.* **249**:189–191.

Jacob, F. and E. L. Wollman, 1961   *Sexuality and the Genetics of Bacteria,* Academic Press, New York.

Lawn, A. M., E. Meynell, G. G. Meynell, and N. Datta, 1967   Sex pili and the classification of sex factors in the *Enterobacteriaceae. Nature (Lond)* **216**:343–346.

Lederberg, J., 1948   Detection of fermentative variants with tetrazolium. *J. Bacteriol.* **56**:695.

Lederberg, J., 1950   Isolation and characterization of biochemical mutants of bacteria. In *Methods in Medical Research*, edited by J. M. Comroe, Vol. 3, pp. 5–22, Yearbook Publishers, Chicago, Ill.

Lederberg, J., 1951   Prevalence of *E. coli* strains exhibiting genetic recombination. *Science (Wash. D.C.)* **114**:68–69.

Lederberg, J. and E. L. Tatum, 1946   Gene recombination in *E. coli. Nature (Lond.)* **158**:558.

Lennox, E. S., 1955   Transduction of linked genetic characters of the host by bacteriophage P1. *Virology* **1**:190–206.

Loeb, T., 1960   Isolation of a bacteriophage specific for the F^+ and Hfr mating types of *Escherichia coli* K12. *Science (Wash. D.C.)* **131**:932–933.

Low, K. B., 1968   Formation of merodiploids in matings with a class of rec^- recipient strains of *Escherichia coli* K12. *Proc. Natl Acad. Sci. USA* **60**:160–167.

Low, K. B., 1972   *Escherichia coli* K12 F-prime factors, old and new. *Bacteriol. Rev.* **36**:587–607.

Low, K. B., 1973   Rapid mapping of conditional and auxotrophic mutations in *Escherichia coli* K12. *J. Bacteriol.* **113**:798–812.

Low, K. B. and T. H. Wood, 1965   A quick and efficient method for interruption of bacterial conjugation. *Genet. Res.* **6**:300–303.

Lubin, M., 1962   Enrichment of auxotrophic mutant populations by recycling. *J. Bacteriol.* **83**:696–697.

Luria, S. E. and M. Delbruck, 1943   Mutations of bacteria from virus sensitivity to virus resistance. *Genetics* **28**:491–511.

Matney, T. S., E. P. Goldschmidt, N. S. Erwin, and R. A. Scroggs, 1964   A preliminary map of genomic sites for F-attachment in *Escherichia coli* K12. *Biochem. Biophys. Res. Commun.* **3**:278–281.

Meynell, E., G. G. Meynell and N. Datta, 1968   Phylogenetic relationships of drug-resistance factors and other transmissible bacterial plasmids. *Bacteriol. Rev.* **32**:55–83.

Miller, J. H., 1972   *Experiments in Molecular Genetics*, Cold Spring Harbor Laboratory, Cold Spring Harbor, N.Y.

Morse, M. L., E. M. Lederberg and J. Lederberg, 1956   Transduction in *Escherichia coli* K12. *Genetics* **41**:142–146.

Novick, R. P., 1969   Extrachromosomal inheritance in bacteria. *Bacteriol. Rev.* **33**:210–263.

Ozeki, H. and H. Ikeda, 1968   Transduction mechanisms. *Annu. Rev. Genet.* **2**:245–278.

Press, R., N. Glansdorff, P. Minev, J. deVries, R. Kadner and W. K. Maas, 1971   Isolation of transducing particles of $\phi80$ bacteriophage that carry different regions of the *Escherichia coli* genome. *Proc. Natl. Acad. Sci. USA,* **68**:795–798.

Reznikoff, W. S., J. H. Miller, J. G. Scaife and J. R. Beckwith, 1969   A mechanism for repressor action. *J. Mol. Biol.* **43**:201–213.

Richter, A., 1961   Attachment of wild-type F factor to a specific chromosomal region in a variant strain of *Escherichia coli* K12. The phenomenon of episomic alteration. *Genet. Res.* **2**:333–345.

Sanderson, K. E., H. Ross, L. Ziegler and P. H. Mäkelä, 1972   F⁺, Hfr and F′ strains of *Salmonella typhimurium* and *Salmonella abony*. *Bacteriol. Rev.* **36**:607–637.

Scaife, J., 1966   F-prime factor formation in *E. coli* K12. *Genet. Res.* **8**:189–196.

Schwartz, M., 1966   Location of the maltose A and B loci on the genetic map of *Escherichia coli*. *J. Bacteriol.* **92**:1083–1089.

Shimada, K., R. A. Weisberg and M. E. Gottesman, 1972   Prophage lambda at unusual chromosomal locations. I. Location of the secondary attachment sites and properties of the lysogens. *J. Mol. Biol.* **63**:483–503.

Susman, M., 1970   General bacterial genetics. *Annu. Rev. Genet.* **4**:135–176.

Taylor, A. L., 1963   Bacteriophage-induced mutation in *Escherichia coli*. *Proc. Natl. Acad. Sci. USA.* **50**:1043–1051.

Taylor, A. L., 1963   Bacteriophage-induced mutation in *Escherichia coli*. *Proc. Natl. males of *Escherichia coli*. *Genetics* **45**:1233–1243.

Taylor, A. L. and M. S. Thoman, 1964   The chromosome map of *Escherichia coli* K12. *Genetics* **50**:659–677.

Taylor, A. L. and C. D. Trotter, 1972   Linkage map of *Escherichia coli* strain K12. *Bacteriol. Rev.* **36**:504–524.

Vogel, H. J. and D. M. Bonner, 1956   Acetylornithinase of *Escherichia coli*: partial purification and some properties. *J. Biol. Chem.* **218**:1233–1243.

Willetts, N., 1972   The genetics of transmissible plasmids. *Annu. Rev. Genet.* **6**:257–268.

Wollman, E. L. and F. Jacob, 1955   Sur le mécanisme du transfert de matériel génétique au cours de la recombination chez *Escherichia coli* K12. *C. R. Hebd. Seances Acad. Sci. Ser. D Sci. Nat.* **240**:2499–2451.

Wollman, E. L., F. Jacob and W. Hayes, 1956   Conjugation and genetic recombination in *Escherichia coli*. *Cold Spring Harbor Symp. Quant. Biol.* **21**:141–162.

Wu, T. T., 1966   A model for three-point analysis of random general transduction. *Genetics* **54**:405–410.

Yanofsky, C. and I. Crawford, 1959   The effects of deletions, point mutations, reversions, and suppressor mutations on the two components of the tryptophan synthetase of *Escherichia coli*. *Proc. Natl. Acad. Sci. USA.* **45**:1016–1026.

Zinder, N. D. and J. Lederberg, 1952   Genetic exchange in *Salmonella*. *J. Bacteriol.* **64**:679–699.

6

A Linkage Map and Gene Catalog for *Escherichia coli*

Austin L. Taylor and Carol Dunham Trotter

This contribution contains a diagram which illustrates the position of about 460 genes on the circular linkage map of this bacterium. In Figure 1 the inner circle, which bears the time scale from 0 through 90 minutes is based on the results of interrupted conjugation experiments. The map is graduated in 1-minute intervals beginning arbitrarily with zero at the *thr* locus. Certain parts of the map (e.g., the 9- to 10-minute segment) are displayed on arcs of the outer circle to provide an expanded time scale for crowded regions. The genetic symbols in this figure are defined in Table 1. Markers in parentheses are only approximately mapped at the positions shown. A gene identified by an asterisk has been mapped more precisely than the markers in parentheses, but its orientation relative to adjacent markers is not yet known. The arrows which are placed next to the *phoA* and *argI* genes and next to certain operons show the direction of messenger RNA transcription for these loci. The table provides an alphabetical listing of these genes and also shows the various specific enzyme activities and other phenotypic traits ascribed to the loci. In Table 1 the numbers refer to the time scale shown in Figure 1. Parentheses indicate

Austin L. Taylor and Carol Dunham Trotter—University of Colorado Medical Center, Department of Microbiology, Denver, Colorado.

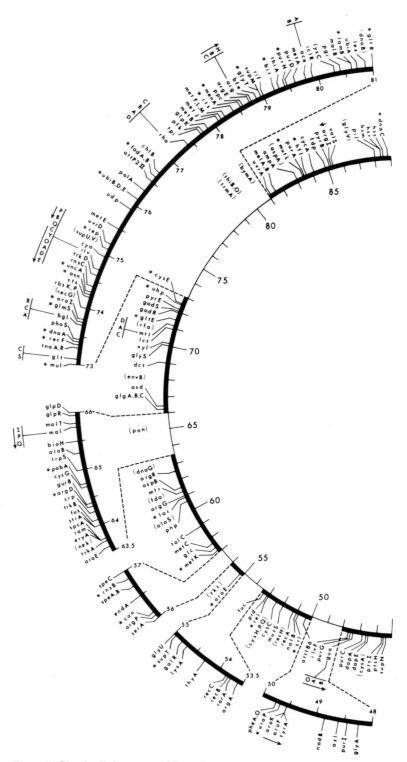

Figure 1. Circular linkage map of E. coli.

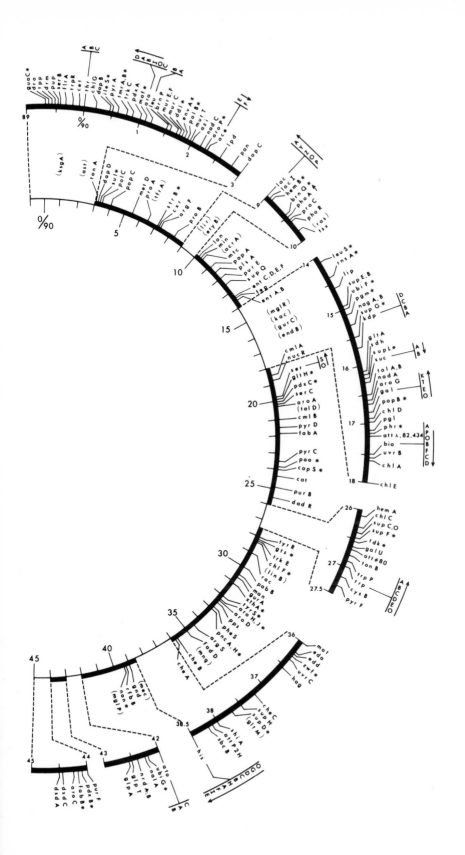

approximate map locations. Citations to the scientific publications containing detailed descriptions of each locus under study are given in Taylor (1970) and Taylor and Trotter (1972). Figure 1 and Table 1 are reproduced with the permission of the American Society for Microbiology.

Literature Cited

Taylor, A. L., 1970   Current linkage map of *Escherichia coli*. Bacteriol. Rev. **34**:155–175.

Taylor, A. L. and C. D. Trotter, 1972   Linkage map of *Escherichia coli* strain K12. Bacteriol. Rev. **36**:504–524.

TABLE 1. List of Genetic Markers of Escherichia coli

Gene symbol	Mnemonic	Map position min	Alternate gene symbols; phenotypic trait affected
aceA	Acetate	80	icl; utilization of acetate: isocitrate lyase
aceB	Acetate	80	mas; utilization of acetate: malate synthetase A
aceE	Acetate	2	aceE1; acetate requirement; pyruvate dehydrogenase (decarboxylase component)
aceF	Acetate	2	aceE2; acetate requirement, pyruvate dehydrogenase (lipoic reductase-transacetylase component)
acrA	Acridine	(10)	Sensitivity to acriflavine, phenethyl alcohol, sodium dodecyl sulfate
alaS	Alanine	(61)	ala-act; alanyl-transfer RNA synthetase
ampA	Ampicillin	83	Resistance or sensitivity to penicillin
araA	Arabinose	1	L-Arabinose isomerase
araB	Arabinose	1	L-Ribulokinase
araC	Arabinose	1	Regulatory gene
araD	Arabinose	1	L-Ribulose 5-phosphate 4-epimerase
araE	Arabinose	55	L-Arabinose permease
araI	Arabinose	1	Initiator locus
araO	Arabinose	1	Operator locus
argA	Arginine	54	argB, Arg1, Arg2; N-acetylglutamate synthetase
argB	Arginine	79	ArgC; α-N-acetyle-L-glutamate-5-phosphotransferase
argC	Arginine	79	argH, Arg2; N-acetylglutamic-γ-semialdehyde dehydrogenase
argD	Arginine	64	argG, Arg1; acetylornithine-δ-transaminase
argE	Arginine	79	argA, Arg4; L-orinthine-N-acetylornithine lyase
argF	Arginine	7	argD, Arg5; ornithine transcarbamylase
argG	Arginine	61	argE, Arg6; argininosuccinic acid synthetase
argH	Arginine	79	argF, Arg7; L-argininosuccinate arginine lyase
argI	Arginine	85	Ornithine transcarbamylase

TABLE 1. Continued

Gene symbol	Map position min	Mnemonic	Alternate gene symbols; phenotypic trait affected
argP	56	Arginine	Arginine permease
argR	63	Arginine	*Rarg*; regulatory gene
argS	35	Arginine	Arginyl-transfer RNA synthetase
aroA	20	Aromatic	3-Enolpyruvylshikimate-5-phosphate synthetase
aroB	65	Aromatic	Dehydroquinate synthetase
aroC	44	Aromatic	Chorismic acid synthetase
aroD	32	Aromatic	Dehydroquinase
aroE	64	Aromatic	Dehydroshikimate reductase
aroF	50	Aromatic	3-Deoxy-D-arabinoheptulosonic acid-7-phosphate (DHAP) synthetase, (tyrosine-repressible isoenzyme)
aroG	17	Aromatic	DHAP synthetase, (phenylalanine-repressible isoenzyme
aroH	32	Aromatic	DHAP synthetase, (tryptophan-repressible isoenzyme)
aroI	74	Aromatic	Function unknown
aroJ	32	Aromatic	Probable operator locus for *aroH*
aroK	50	Aromatic	Operator locus for *aroA*, *tyrA*
aroP	2	Aromatic	General aromatic amino acid transport
asd	66		*dap + hom;* aspartic semialdehyde dehydrogenase
asn	74		Asparagine synthetase
aspA	(83)		Aspartase
aspB	62	Aspartate	*asp;* aspartate requirement
ast	(3)	Astasia	Generalized high mutability
atoA	42	Acetoacetate	Coenzyme A transferase
atoB	42	Acetoacetate	Thiolase II
atoC	42	Acetoacetate	Regulatory gene
attλ	17	Attachment	Integration site for prophage λ

Gene	Category	Position	Description
attP2H	Attachment	38	Phage P2 integration site H
attP2II	Attachment	77	Phage P2 integration site II
attφ80	Attachment	27	Integration site for prophage φ80
att82	Attachment	17	Integration site for prophage 82
att186	Attachment	50	Integration site for prophage 186
att434	Attachment	17	Integration site for prophage 434
azi	Azide	2	*pea, fts*; resistance or sensitivity to sodium azide or phenethyl alcohol; filament formation at 42 C
azl	Azaleucine	49	Regulation of *leu* and *ilv* genes
bfe		79	*cer*; resistance or sensitivity to phage BF23 and colicins E1, E2, E3
bglA	β-glucoside	74	β-*glA*; aryl β-glucosidase
bglB	β-glucoside	74	β-*glB*; β-glucoside permease
bglC	β-glucoside	74	β-*glC*; regulatory gene
bioA	Biotin	18	Group II; 7-oxo-8-aminopelargonic acid (7KAP) → 7, 8-diaminopelargonic acid (DAPA)
bioB	Biotin	18	Conversion of dethiobiotin to biotin
bioC	Biotin	18	Block prior to pimeloyl coenzyme A
bioD	Biotin	18	Dethiobiotin synthetase
bioF	Biotin	18	Pimeloyl coenzyme A → 7KAP
bioH	Biotin	66	*bioB*; block prior to pimeloyl coenzyme A
bioO	Biotin	18	Operator for genes *bioB* through *bioD*
bioP	Biotin	18	Promoter site for genes *bioB* through *bioD*
bir	Biotin retention	79	Biotin uptake, retention, and regulation
brnP	Branched chain	2	Transport of isoleucine, leucine, and valine
brnQ	Branched chain	9	Transport of isoleucine, leucine, and valine
bymA		(81)	Bypass of maltose permease at *malB*
can	Canavanine	56	Canavanine resistance
cap	Capsule		See *crp*
capS	Capsule	24	Regulatory gene for capsular polysaccharide synthesis
cat		24	CR; catabolite repression
cet		90	*ref, reflI*; tolerance to colicin E2

TABLE 1. Continued

Gene symbol	Mnemonic	Map position min	Alternate gene symbols; phenotypic trait affected
cheA	Chemotaxis	36	*motA*; chemotactic motility
cheB	Chemotaxis	36	*motB*; chemotactic motility
cheC	Chemotaxis	37	Chemotactic motility
chlA	Chlorate	18	*narA*; pleiotropic mutations affecting nitrate-chlorate reductase and hydrogen lyase activity
chlB	Chlorate	77	*narB*; pleiotropic mutations affecting nitrate-chlorate reductase and hydrogen lyase activity
chlC	Chlorate	26	*narC*; structural gene for nitrate reductase
chlD	Chlorate	17	*narD*, *narF*; nitrate chlorate reductase
chlE	Chlorate	18	*narE*; nitrate reductase
chlF	Chlorate	28	Structural gene for formate dehydrogenase
chlG	Chlorate	0	Formate nitrate reductase
cmlA	Chloramphenicol	19	Resistance or sensitivity to chloramphenicol
cmlB	Chloramphenicol	21	Resistance or sensitivity to chloramphenicol
crp		64	*cap*; cyclic adenosine monophosphate receptor protein
ctr			See *ptsI*. *ptsH*
cxr		7	Synthesis of methylglyoxal
cya		75	Adenyl cyclase
cycA	Cycloserine	84	First-step resistance to D-cycloserine
cysA	Cysteine	(47)	Requirement
cysB	Cysteine	27	Pleiotropic mutations affecting cysteine biosynthesis
cysC	Cysteine	52	Adenosine 5'-sulfatophosphate kinase
cysE	Cysteine	73	Apparently pleiotropic
cysG	Cysteine	65	Sulfite reductase
cysH	Cysteine	(53)	Adenosine 3'-phosphate 5'-sulfatophosphate reductase
cysP	Cysteine	(53)	Sulfate permease and sulfite reductase

Gene	Category	Position	Description
cysQ	Cysteine	(53)	Sulfite reductase
dadR		26	Regulatory gene for D-amino acid deaminases
dapA	Diaminopimelate	47	Dihydrodipicolinic acid synthetase
dapB	Diaminopimelate	0	Dihydrodipicolinic acid reductase
dapC	Diaminopimelate	3	Tetrahydrodipicolinic acid → N-succinyl-diaminopimelate
dapD	Diaminopimelate	3	Tetrahydrodipicolinic acid → N-succinyl diaminopimelate
dapE	Diaminopimelate	47	dapB; N-succinyl-diaminopimelic acid deacylase
darA			See uvrD
dct		69	Uptake of C$_4$-dicarboxylic acids
ddl		2	D-Alanine: D-alanine ligase
deo	Deoxythymidine		See dra, drm, pup, and tpp
dnaA		73	DNA synthesis: initiation-defective
dnaB		(81)	groP; DNA synthesis
dnaC		89	dnaD; DNA synthesis: initiation-defective
dnaE			See polC
dnaF			See nrdA
dnaG		(63)	DNA synthesis
dra		89	deoC, thyR; deoxyriboaldolase
drm		89	deoB, thyR; deoxyribomutase
dsdA	D-serine	45	D-serine deaminase
dsdC		44	Regulatory gene
eda	D-serine	36	kga, kdg; 2-keto-3-deoxygluconate-6-phosphate aldolase
edd		36	Entner–Doudoroff dehydrase (gluconate-6-phosphate dehydrase)
endA		57	DNA-specific endonuclease I
endB		(16)	DNA-specific endonuclease I
entA	Enterochelin	14	2, 3-Dihydro-2, 3-dihydroxybenzoate dehydrogenase
entB	Enterochelin	14	2, 3-Dihydro-2, 3-dihydroxybenzoate synthetase
entC	Enterochelin	13	Isochorismate synthetase
entD	Enterochelin	13	Unknown step in conversion of 2, 3-dihydroxybenzoate to enterochelin
entE	Enterochelin	13	Unknown step in conversion of 2, 3-dihydroxybenzoate to enterochelin
entF	Enterochelin	13	Unknown step in conversion of 2, 3-dihydroxybenzoate to enterochelin

TABLE 1. Continued

Gene symbol	Mnemonic	Map position min	Alternate gene symbols; phenotypic trait affected
envA	Envelope	2	Anomalous cell division involving chain formation
envB	Envelope	(68)	Anomalous spheroid cell formation
eps	Episome stability		See spc
eryA	Erythromycin	64	50-8 protein of 50S ribosomal subunit
eryB	Erythromycin	(10)	High-level resistance to erythromycin
exbA			See tonB
exr			See lex
fabA		22	β-Hydroxydecanoylthioester dehydrase
fabB		44	Fatty acid biosynthesis
fadA	Fatty acid degradation	77	oldA; thiolase I
fadB	Fatty acid degradation	77	oldB; hydroxyacyl-coenzyme A dehydrogenase
fadD	Fatty acid degradation	35	oldD; acyl-coenzyme A synthetase
fda		(61)	ald; fructose-1, 6-diphosphate aldolase
fdp		85	Fructose diphosphatase
fep		13	Defect of enterochelin-dependent iron transport system
flrA	Fluoroleucine	90	Regulation of leu and ilv genes
ftsA			See azi
fuc	Fucose	53	Utilization of L-fucose
fus	Fusidic acid	64	far; protein chain elongation factor EF G
gadR		72	Regulatory gene for gadS
gadS		72	Glutamic acid decarboxylase
galE	Galactose	17	galD; uridinediphosphogalactose 4-epimerase
galK	Galactose	17	galA; galactokinase
galO	Galactose	17	galC; operator locus
galT	Galactose	17	galB; galactose 1-phosphate uridyl transferase
galR	Galactose	55	Rgal; regulatory gene

galU	Galactose	27	*UDPG*; uridine diphosphoglucose pyrophosphorylase
glc	Glycolate	57	Utilization of glycolate; malate synthetase G
glgA	Glycogen	66	Glycogen synthetase
glgB	Glycogen	66	α-1, 4-Glucan: α-1, 4-glucan 6-glucosyltransferase
glgC	Glycogen	66	Adenosine diphosphate glucose pyrophosphorylase
glmS	Glucosamine	74	L-Glutamine: D-fructose-6-phosphate amino transferase
glpA	Glycerol phosphate	43	L-α-Glycerophosphate dehydrogenase (anaerobic)
glpD	Glycerol phosphate	66	*glyD*; D-α-glycerophosphate dehydrogenase (aerobic)
glpK	Glycerol phosphate	78	Glycerol kinase
glpT	Glycerol phosphate	43	L-α-Glycerophosphate transport system
glpR	Glycerol phosphate	66	Regulatory gene
gltA	Glutamate	16	*glut*; requirement for glutamate; citrate synthase
gltC	Glutamate	73	Operator locus
gltE	Glutamate	72	Glutamyl-transfer RNA synthetase
gltH	Glutamate	20	Requirement
gltM	Glutamate	(38)	Glutamyl-transfer RNA synthetase
gltR	Glutamate	81	Regulatory gene for glutamate permease
gltS	Glutamate	73	Glutamate permease
glyA	Glycine	48	Serine hydroxymethyl transferase
glyS	Glycine	70	*gly-act*; glycyl-transfer RNA synthetase
glyT	Glycine	79	*supA36, su36, sumA*; glycine transfer RNA II
glyU	Glycine	55	*sumB*; glycine transfer RNA I
glyV	Glycine	(86)	Glycine transfer RNA III
gnd		39	Gluconate-6-phosphate dehydrogenase
groN			See *rif*
groP			See *dnaB*
gts		28	Uncharacterized membrane defect
guaA	Guanine	47	*gua_b*; xanthosine-5'-monophosphate aminase
guaB	Guanine	47	*gua_a*; inosine-5'-monophosphate dehydrogenase
guaC	Guanine	89	Guanosine-5'-monophosphate reductase
guaO	Guanine	47	Operator locus

TABLE 1. Continued

Gene symbol	Mnemonic	Map position min	Alternate gene symbols; phenotypic trait affected
gurA	Glucuronide	31	β-glucuronidase
gurB	Glucuronide	64	Utilization of methyl-β-D-glucuronide
gurC	Glucuronide	(16)	Utilization of methyl-β-D-glucuronide
hag	H antigen	37	H; flagellar antigens (flagellin)
hemA	Hemin	26	Synthesis of δ-aminolevulinic acid
hemB	Hemin	9	ncf; synthesis of catalase and cytochromes
hfl		84	High frequency of lysogenization by phage λ
hisA	Histidine	39	Isomerase
hisB	Histidine	39	Imidazole glycerol phosphate dehydrase; histidinol phosphatase
hisC	Histidine	39	Imidazole acetol phosphate transaminase
hisD	Histidine	39	Histidinol dehydrogenase
hisE	Histidine	39	Phosphoribosyl-adenosine triphosphate-pyrophosphohydrolase
hisF	Histidine	39	Cyclase
hisG	Histidine	39	Phosphoribosyl-adenosine triphosphate-pyrophosphorylase
hisH	Histidine	39	Amido transferase
hisI	Histidine	39	Phosphoribosyl-adenosine monophosphate-hydrolase
hisO	Histidine	39	Operator locus
hsm	Host specificity	89	hs, rm, hsp; host modification activity: DNA methylase M
hsr	Host specificity	89	hs, rm, hsp, por; host restriction activity: endonuclease R
hss	Host specificity	89	Specificity determinant for hsm and hsr activities
icl			See aceA
iclR		80	Regulation of the glyoxylate cycle
ileS	Isoleucine	1	Isoleucyl-transfer RNA synthetase
ilvA	Isoleucine-valine	75	ile; threonine deaminase
ilvB	Isoleucine-valine	75	Acetohydroxy acid synthetase I

ilvC	Isoleucine-valine	75	*ilvA*; α-hydroxy-β-keto acid reductoisomerase
ilvD	Isoleucine-valine	75	*ilvB*; dehydrase
ilvE	Isoleucine-valine	75	*ilvC*; transaminase B
ilvF	Isoleucine-valine	48	Possibly acetohydroxy acid synthetase II
ilvO	Isoleucine-valine	75	Operator locus for genes *ilvA, D, E*
ilvP	Isoleucine-valine	75	Operator locus for gene *ilvB*
ilvQ	Isoleucine-valine	75	Induction recognition site for *ilvC*
ilvY	Isoleucine-valine	75	Positive control element for *ilvC* induction
kac	K-accumulation	(16)	Defect in potassium-ion uptake
kdpA–D	K-dependent	16	Requirement for a high concentration of potassium
kga			See *eda*
ksgA	Kasugamycin	(1)	RNA methylase for 16S ribosomal RNA
lacA	Lactose	9	*a, lacAc*; thiogalactoside transacetylase
lacI	Lactose	9	*i*; regulator gene
lacO	Lactose	9	*o*; operator locus
lacP	Lactose	9	*p*; promoter locus
lacY	Lactose	9	*y*; galactoside permease (M protein)
lacZ	Lactose	9	*z*; β-galactosidase
lamB	Lambda	81	*malB*; phage λ receptor site
lar	Large	61	Large cells and radiation resistance
lct	Lactate	71	L-lactate dehydrogenase
leuA	Leucine	2	α-isopropylmalate synthetase
leuB	Leucine	2	β-isopropylmalate dehydrogenase
leuS	Leucine	14	Leucyl-transfer RNA synthetase
lex		81	*exr*; resistance or sensitivity to x rays and ultraviolet light
linB	Lincomycin	(29)	High-level resistance to lincomycin
lip	Lipoic acid	15	Requirement
lir		(9)	Increased sensitivity to lincomycin and/or erythromycin
lon	Long form	10	*capR, dir, muc*; filamentous growth, radiation sensitivity, and regulation of capsular polysaccharide synthesis

TABLE 1. Continued

Gene symbol	Mnemonic	Map position min	Alternate gene symbols; phenotypic trait affected
lpd		3	Lipoyldehydrogenase
lps	Lipopolysaccharide		See *rfa*
lysA	Lysine	55	Diaminopimelic acid decarboxylase
lysC	Lysine	80	*apk*; lysine aspartokinase III
maf		1	Maintenance of autonomous sex factor
malB	Maltose	81	*mal-5*; maltose permease
malI	Maltose	66	Initiator site
malP	Maltose	66	*malA*; maltodextrin phosphorylase
malQ	Maltose	66	*malA*; amylomaltase
malT	Maltose	66	*malA*; positive regulatory gene for the *malPQ* and *malB-lamB* operons
man	Mannose	31	Phosphomannose isomerase
mec		(39)	DNA methylase for cytosine
melA	Melibiose	81	*mel-7*; α-galactosidase
melB	Melibiose	81	*mel-4*; thiomethylgalactoside permease II
menA	Menaquinone	79	Requirement
metA	Methionine	80	*met3*; homoserine O-transsuccinylase
metB	Methionine	78	*met-1*, *met1*; cystathionine synthetase
metC	Methionine	58	Cystathionase
metD	Methionine	6	Utilization of D-methionine
metE	Methionine	76	*met-B12*; N^5-methyltetrahydropteroyl triglutamatehomocysteine methylase
metF	Methionine	78	*met-2*, *met2*; N^5, N^{10}-methyltetrahydrofolate reductase
metJ	Methionine	78	Possible regulatory gene
metK	Methionine	57	S-adenosylmethionine synthetase activity
metL	Methionine	78	Methionine aspartokinase II
metM	Methionine	78	Homoserine dehydrogenase II

Gene	Name	Map	Description
mglP	Methyl-galactoside	(40)	P-MG; methyl-galactoside permease and galactose binding protein
mglR	Methyl-galactoside	(16)	R-MG; regulatory gene
min	Minicell	10	Formation of minute cells containing no DNA
mng	Manganese	(35)	Resistance or sensitivity to manganese
mot	Motility	36	Flagellar paralysis
mtc	Mitomycin C	10	Mb, mbl; sensitivity acridines, methylene blue and mitomycin C
mtlA	Mannitol	71	Mannitol-specific enzyme II of the phosphotransferase (*pts*) system
mtlC	Mannitol	71	Regulatory gene or site
mtlD	Mannitol	71	Mannitol-1-phosphate dehydrogenase
mtr	Methyl tryptophan	61	Resistance to 5-methyltryptophan
mul		73	Mutability of ultraviolet-irradiated phage λ
murC	Murein	2	L-Alanine adding enzyme
murE	Murein	2	meso-Diaminopimelic acid adding enzyme
murF	Murein	2	D-Alanyl-D-alanine adding enzyme
mutL	Mutator	83	Generalized high mutability
mutS	Mutator	52	Generalized high mutability
mutT	Mutator	2	Generalized high mutability; specifically induces AT → CG transversions
nadA	Nicotinamide adenine dinucleotide	17	nicA; nicotinic acid requirement
nadB	Nicotinamide adenine dinucleotide	49	nicB nicotinic acid requirement
nadC	Nicotinamide adenine dinucleotide	2	Quinolinate phosphoribosyl transferase
nagA	N-acetylglucosamine	15	N-acetylglucosamine-6-phosphate deacetylase
nagB	N-acetylglucosamine	15	Glucosamine-6-phosphate deaminase
nalA	Nalidixic acid	42	Resistance or sensitivity to nalidixic acid
nalB	Nalidixic acid	51	Resistance or sensitivity to nalidixic acid
nam			See pncA
nar	Nitrate reductase		See chl
nek		(64)	Resistance to neomycin and kanamycin (30S ribosomal protein)
nic			See nad

TABLE 1. Continued

Gene symbol	Mnemonic	Map position min	Alternate gene symbols; phenotypic trait affected
non	Nonmucoid	39	Block in capsule formation
nrdA		42	*dnaF*; ribonucleoside diphosphate reductase: subunit B1
nrdB		42	Ribonucleoside diphosphate reductase: subunit B2
nucR	Nucleosides	19	*deoR*; regulatory gene for *pup*, *tpp*, and *dra*
old			See *fadA*, *fadB*, *fadD*
pabA	*p*-aminobenzoate	65	Requirement
pabB	*p*-aminobenzoate	30	Requirement
pan	Pantothenic acid	3	Requirement
pdxA	Pyridoxine	1	Requirement
pdxB	Pyridoxine	44	Requirement
pdxC	Pyridoxine	20	Requirement
pfk		78	Structural or regulatory gene for fructose 6-phosphate kinase
pgi		80	Phosphoglucoisomerase
pgl		17	6-Phosphogluconolactonase
pgm		15	Phosphoglucomutase
pheA	Phenylalanine	50	Chorismate mutase P-prephenate dehydratase
pheO	Phenylalanine	50	Operator locus
pheS	Phenylalanine	33	*phe-act*; phenylalanyl transfer RNA synthetase
phoA	Phosphatase	10	*P*; alkaline phosphatase
phoR	Phosphatase	10	*R1 pho*, *R1*; regulatory gene
phoS	Phosphatase	73	*R2 pho*, *R2*; regulatory gene
phr	Photoreactivation	17	Photoreactivation of ultraviolet-damaged DNA (K12-B hybrids)
pil	Pili	88	*fim*; presence or absence of pili (fimbriae)
plsA	Phospholipid	12	Glycerol-3-phosphate acyltransferase
pncA	Pyridine nucleotide cycle	33	*nam*; nicotinamide deamidase

Gene	Category	No.	Description
pmcH	Pyridine nucleotide cycle	33	Hyperproduction of nicotinamide deamidase
pnp		61	Polynucleotide phosphorylase
poa		24	Proline oxidase
polA	Polymerase	76	*resA*; DNA polymerase I
polB	Polymerase	2	DNA polymerase II
polC	Polymerase	4	*dnaE*; DNA polymerase III
pon	P-one	(65)	Resistance or sensitivity to phages P1 and Mu-1
popA	Porphyrin	11	Possibly ferrochelatase
popB	Porphyrin	17	Probably coproporphyrin oxidase
popC	Porphyrin	4	Synthesis of δ-aminolevulinic acid
por	P$_1$ restriction		See *hsr*
ppc		79	*glu, asp*; succinate, aspartate, or glutamate requirement; phosphoenolypyruvate carboxylase
pps	Propanediol	33	Utilization of pyruvate or lactate; phosphopyruvate synthetase
prd	Propanediol	53	1, 2-Propanediol dehydrogenase
proA	Proline	7	*pro$_1$*; block prior to L-glutamate semialdehyde
proB	Proline	8	*pro$_2$*; block prior to L-glutamate semialdehyde
proC	Proline	10	*pro$_3$*; *Pro2*; probably Δ-pyrroline-5-carboxylate reductase
ptsH		46	*ctr, Hpr*; phosphotransferase system: protein cofactor
ptsI		46	*ctr*; phosphotransferase system: enzyme I
pup		90	Purine nucleoside phosphorylase
purA	Purine	84	*ade$_k$, Ad4*; adenylosuccinic acid synthetase
purB	Purine	25	*ade$_h$*, adenylosuccinase
purC	Purine	47	*ade$_g$*; phosphoribosyl-aminoimidazole-succinocarboxamide synthetase
purD	Purine	79	*adt$_{Ha}$*; phosphoribosylglycineamide synthetase
purE	Purine	12	*ade$_3$; ade$_1$, Pur$_2$*; phosphoribosyl-aminoimidazole carboxylase
purF	Purine	44	*purC, ade$_{a,b}$*; phosphoribosyl-pyrophosphate amidotransferase
purG	Purine	47	*adt$_{Hb}$*; phosphoribosylformylglycineamidine synthetase
purH	Purine	79	*ade$_1$*; phosphoribosyl-aminoimidazole-carboxamide formyltransferase
purI	Purine	48	Aminoimidazole ribotide synthetase
pyrA	Pyrimidine	1	*cap, arg + ura*; glutamino-carbamoyl-phosphate synthetase

TABLE 1. Continued

Gene symbol	Mnemonic	Map position min	Alternate gene symbols; phenotypic trait affected
pyrB	Pyrimidine	85	Aspartate transcarbamylase
pyrC	Pyrimidine	24	Dihydroorotase
pyrD	Pyrimidine	21	Dihydroorotic acid dehydrogenase
pyrE	Pyrimidine	72	Orotidylic acid pyrophosphorylase
pyrF	Pyrimidine	27	Orotidylic acid decarboxylase
rac	Recombination activation	34	Suppressor of *recB* and *recC* mutant phenotype in merozygotes
ram	Ribosomal ambiguity	64	P4a protein of 30S ribosomal subunit
ras	Radiation sensitivity	(10)	Sensitivity to ultraviolet and x-ray irradiation
rbsK	Ribose	74	Ribokinase
rbsP	Ribose	74	D-Ribose permease
recA	Recombination	51	Ultraviolet sensitivity and competence for genetic recombination
recB	Recombination	54	Ultraviolet sensitivity, genetic recombination; exonuclease V subunit
recC	Recombination	54	Ultraviolet sensitivity, genetic recombination; exonuclease V subunit
recF	Recombination	73	*uvrF*; ultraviolet sensitivity and competence for genetic recombination
recG	Recombination	(74)	Competence for genetic recombination
recH	Recombination	(52)	Competence for genetic recombination
rel	Relaxed	53	*RC*; regulation of RNA synthesis
rep	Replication	75	Inhibition of lytic replication of temperate phages
rfa	Rough	(71)	*lps*; lipopolysaccharide core defect
rfbB	Rough	39	*som*; thymidine diphosphate-glucose oxidoreductase
rhaA	Rhamnose	77	L-rhamnose isomerase
rhaB	Rhamnose	77	L-rhamnulokinase
rhaC	Rhamnose	77	Regulatory gene
rhaD	Rhamnose	77	L-rhamnulose-1-phosphate aldolase
rif	Rifampicin	79	*sil, stv, groN, ron*; RNA polymerase: β subunit
rne			See *rnsB*

rnsA	Ribonuclease	14	*rns*; ribonuclease I
rnsB	Ribonuclease	57	*me*; ribonuclease II
rnsC	Ribonuclease	74	*SuA*; polarity suppressor; RNA endonuclease A
ron			See *rif*
rorA		54	Resistance to x-ray irradiation
rts		78	*ts-9*; altered electrophoretic mobility of 50S ribosomal subunit
sbcB		38	Exonuclease I; suppressor of *recB*, *recC*
sdh		16	Succinate dehydrogenase
serA	Serine	56	3-Phosphoglyceric acid dehydrogenase
serB	Serine	90	Phosphoserine phosphatase
serC	Serine	20	*pdxF*; 3-phosphoserine-2-oxoglutarate aminotransferase
serO	Serine	20	Operator locus
serS	Serine	20	Seryl-transfer RNA synthetase
shiA	Shikimic acid	38	Shikimate and dehydroshikimate permease
spcA	Spectinomycin	64	*eps*; P4 protein of 30S ribosomal subunit
speA	Spermidine	57	Arginine decarboxylase
speB	Spermidine	57	Agmatine ureohydrolase
speC	Spermidine	57	Ornithine decarboxylase
srl	Sorbitol	51	Utilization of sorbitol
stl	Streptolydigin		See *rif*
strA	Streptomycin	64	P10 protein of 30S ribosomal subunit
strB	Streptomycin	7	Low-level streptomycin resistance
sts		74	Altered ribonuclease II activity
stv	Streptovaricin		See *rif*
SuA			See *rnsC*
sucA	Succinate	16	*suc, lys + met*; succinate requirement; α-ketoglutarate dehydrogenase (decarboxylase component)
sucB	Succinate	16	*suc, lys + met*; succinate requirement, α-ketoglutarate dehydrogenase (dihydrolipoyltransuccinylase component)

TABLE 1. Continued

Gene symbol	Mnemonic	Map position min	Alternate gene symbols; phenotypic trait affected
sul		3	Suppressor of lon mutation
supA36			See glyT
supB	Suppressor	15	su_B; suppressor of ochre mutations (not identical to supL)
supC	Suppressor	26	su_C, su-4; suppressor of ochre mutations (possibly identical to supC)
supD	Suppressor	38	su_1, Su-1; suppressor of amber mutations
supE	Suppressor	15	su_II; suppressor of amber mutations
supF	Suppressor	26	su_III, Su-3; amber suppressor: tyrosine transfer RNA
supG	Suppressor	16	Su-5; suppressor of ochre mutations
supH	Suppressor	38	
supL	Suppressor	16	Suppressor of ochre mutations
supM	Suppressor	79	sup15B; ochre suppressor: tyrosine transfer RNA
supN	Suppressor	45	Suppressor of ochre mutations
supO	Suppressor	26	Suppressor ochre mutations (possibly identical to supC)
supQ	Suppressor	13	
supT	Suppressor	55	
supU	Suppressor	(75)	su7; amber suppressor: glutamine transfer RNA
supV	Suppressor	(75)	su8; suppressor of ochre mutations
tdk		27	Deoxythymidine kinase
tfrA	T-four	(7)	ϕ^r; resistance or sensitivity to phages T4, T3, T7, and λ
thiA	Thiamine	79	thi; synthesis of thiazole
thiB	Thiamine	(78)	Thiamine phosphate pyrophosphorylase
thiO	Thiamine	(78)	Probable operator locus for thiA thiB genes
thrA	Threonine	0	HS, thrD; aspartokinase I-homoserine dehydrogenase I complex
thrB	Threonine	0	Homoserine kinase
thrC	Threonine	0	Threonine synthetase

Gene		Map	Description
thyA	Thymine	54	Thymidylate synthetase
tkt		(55)	Transketolase
tmrA		1	Trimethoprim resistance; dihydrofolate reductase activity
tmrB		1	Trimethoprim resistance; dihydrofolate reductase activity
tnaA		73	*ind*; tryptophanase
tnaR		73	R_{tna}, regulatory gene
tolA	Tolerance	17	*cim*; *tol-2*; tolerance to colicins E2, E3, A, and K
tolB	Tolerance	17	*tol-3*; tolerance to colicins E1, E2, E3, A, and K
tolC	Tolerance	59	*colE1-i*, *tol-8*, *refI*; specific tolerance to colicin E1
tolD	Tolerance	(20)	Tolerance to colicins E2 and E3; ampicillin resistance
tonA	T-one	3	*T1, T5 rec*; resistance or sensitivity to phages T1 and T5
tonB	T-one	27	*T1 rec, exb*; resistance to phages T1, φ80, colicins B, I, V; transport of Fe; enterochelin excretion
tpi		78	Triosephosphate isomerase
tpp		89	*deoA, TP*; thymidine phosphorylase
trkA		64	Transport of potassium
trkB		64	Transport of potassium
trkC		1	Transport of potassium
trkD		75	Transport of potassium
trkE		28	Transport of potassium
trmA		(79)	Methylase for 5-methyluracil in transfer RNA
trpA	Tryptophan	27	*tryp-2*; tryptophan synthetase, A protein
trpB	Tryptophan	27	*tryp-1*; tryptophan synthetase, B protein
trpC	Tryptophan	27	*tryp-3*; *N*-(5-phosphoribosyl) anthranilate isomerase-indolyl-3-glycerol phosphate synthetase
trpD	Tryptophan	27	*tryE*; phosphoribosyl anthranilate transferase
trpE	Tryptophan	27	*tryD, anth, tryp-4*; anthranilate synthetase
trpO	Tryptophan	27	Operator locus
trpP	Tryptophan	27	Tryptophan permease
trpR	Tryptophan	90	*Rtry*; regulatory gene for the *trp* operon and *aroH*
trpS	Tryptophan	65	Tryptophanyl-transfer RNA synthetase

TABLE 1. Continued

Gene symbol	Mnemonic	Map position min	Alternate gene symbols; phenotypic trait affected
tsx	T-six	10	T6 rec; resistance or sensitivity to phage T6 and colicin K
tyrA	Tyrosine	50	Chorismate mutase T-prephenate dehydrogenase
tyrR	Tyrosine	28	Regulation of aroF, aroG, and tyrA genes
tyrS	Tyrosine	32	Tyrosyl-transfer RNA synthetase
ubiA	Ubiquinone	83	4-Hydroxybenzoate → 3-octaprenyl 4-hydroxybenzoate
ubiB	Ubiquinone	76	2-Octaprenylphenol → 2-octaprenyl-6-methoxy-1, 4-benzoquinone
ubiD	Ubiquinone	76	3-Octaprenyl-4-hydroxybenzoate → 2-octaprenylphenol
ubiE	Ubiquinone	76	2-Octaprenyl-6-methoxy-1, 4-benzoquinone → 2-octaprenyl-3-methyl-6-methoxy-1, 4-benzoquinone
ubiF	Ubiquinone	15	2-Octaprenyl-3-methyl-6-methoxy-1, 4-benzoquinone → 2-octaprenyl-3-methyl-5-hydroxy-6-methoxy-1, 4-benzoquinone
ubiG	Ubiquinone	42	2-Octaprenyl-3-methyl-5-hydroxy-6-methoxy-1, 4-benzoquinone → ubiquinone-8
udp		76	Uridine phosphorylase
uhp		72	Uptake of hexose phosphates
uncA	Uncoupling	74	Membrane-bound Mg, calcium adenosine triphosphatase
uraP	Uracil	50	Uracil permease
uvrA	Ultraviolet	81	dar-3; repair of ultraviolet radiation damage to DNA
uvrB	Ultraviolet	18	dar-1, 6; repair of ultraviolet radiation damage to DNA
uvrC	Ultraviolet	36	dar-4, 5; repair of ultraviolet radiation damage to DNA
uvrD	Ultraviolet	75	uvr-502, dar-2, rad; repair of uv radiation damage to DNA
uvrF			See recF
valS	Valine	85	val-act; valyl-transfer RNA synthetase
xthA	Exo-three	31	Exonuclease III
xyl	Xylose	70	Utilization of D-xylose
zuf	Zwischenferment	36	Glucose-6-phosphate dehydrogenase

7

F-Prime Factors of *Escherichia coli*

K. Brooks Low

Basic Properties of F-Prime Factors

An F-prime factor is a circular DNA molecule which consists of most, or all, of the conjugal fertility factor F (see chapter 5 by Curtiss *et al.*) in addition to a portion of the bacterial genome. F-prime factors have been most extensively studied in *Escherichia coli*, and a recent review (Low, 1972) provides detailed references for the properties outlined below.

F-Prime Formation. An F-prime factor may be derived from an Hfr chromosome (see Figure 1). In this case, type IA carries an "early" region from the Hfr, type IB carries a "late" region from the Hfr, and type II carries both early and late regions. An F-prime may also originate from a double Hfr chromosome. In this case, the F-prime carries the region between the two F factors in the parental-double-Hfr chromosome.

F-Prime Strain Types. There are two F-prime strain types, primary and secondary. A primary F-prime strain is derived from the cell in which F-prime was formed by a reciprocal recombination event; the chromosome is deleted for material carried by F-prime. A secondary F-prime strain is derived by conjugal transfer of an F-prime into a normal haploid recipient strain, thereby producing diploidy for the chromosomal region carried by the F-prime.

K. Brooks Low—School of Medicine, Yale University, New Haven, Connecticut.

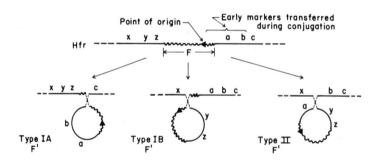

Figure 1. Possible variations in the relation between F and chromosomal DNA during F-prime formation. The top line in the figure represents part of the chromosome of a hypothetical Hfr strain which transfers the genetic markers a, b, and c early, and x, y, and z late in conjugation. Note the convention of using an arrow to denote both the position and orientation of the point of origin. The lower portion of the figure delineates various possible points of cross-over in the formation of the new (circular) F-prime factor.

The Isolation of New F-Prime Strains. Primary F-prime strains can be isolated by replica-plating many colonies of an Hfr strain onto a layer of F⁻ recipient cells on recombinant-selective agar. This allows the detection of rare colonies from which an F-prime factor can be transferred. Secondary F-prime strains can be isolated by a liquid-mating of an Hfr strain with an F⁻ strain. In this case it it possible to select for rare recombinant cells which have inherited an F-prime factor derived from the Hfr strain.

F-Prime Propagation. F-primes are propagated by replication and also by conjugal transfer. Both of these processes are probably analogous to those involving the normal F factor.

Generalizations to Other Systems. F-prime factors have been isolated in *Salmonella* (Sanderson *et al.*, 1972) as well as in *E. coli.* In addition, the potential exists for using other integrated plasmids, such as colicin factors (Kahn, 1969), or R factors (Moody and Runge, 1972, Nishimura *et al.*, 1973) to isolate episomes analogous to F-prime factors.

Complications Involved in the Use of F-Primes. Nonrandom crossover points in the formation of F-primes frustrate efforts to obtain F-primes of certain desired extents. Also, slow cell growth is conferred by certain long F-prime factors.

Active Areas of Research Directly Involving F-Prime Factors. These include the study of the mechanisms of replication and segregation of episomes and chromosomes, mechanisms of conjugal transfer of DNA, and mechanisms of genetic recombination.

Ranges of Sizes. See Figure 2 for typical genetic lengths.

Alterations in F-Prime Factors

Mutations in the F Portion of an F-Prime. These may be induced by mutagenesis or acquired following recombination with a second (mutated) F or F-prime factor. A discussion of the genetics of the F factor is given by Achtman (1973).

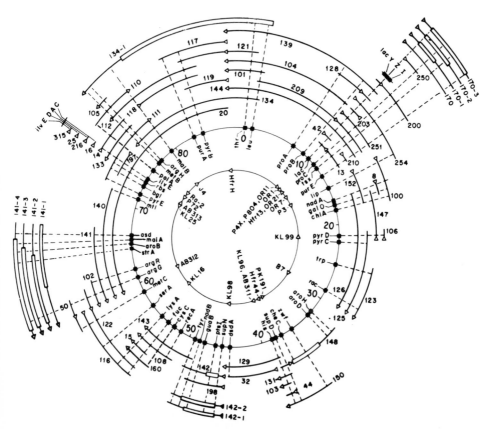

Figure 2. Genetic map of E. coli K12 showing approximate chromosomal regions carried by various F-prime factors. The inner circle shows arrows which represent the points of origin of the Hfr strains from which the F-primes were derived. The dashed lines, which extend radially from the genetic markers on the outer circle, indicate the approximate termini of the F-primes as far as they are known. The order of transfer of markers on any F-prime is simply predicted by considering the F-prime as a closed circle with a point of origin indicated by the arrow as shown in Figure 1. Known deletions are indicated by narrow rectangles, such as in 141-1, 141-2, 141-3, and 141-4 (left side of map) which were all derived from 141 by spontaneous deletion. The group consisting of 16, 216, 25, and 315 were all derived by transductional shortening of 14. References for all of these F-primes are given by Low (1972).

Mutations in the Region of Diploidy. These may be induced by mutagenesis or acquired by recombination with the chromosome or with a homologous region on another F-prime factor.

Deletion of F-Prime Material. This occurs spontaneously at a low rate and is especially noticeable with long F-primes (see Figure 2); the deletion of a sizable portion of the (nonessential) material in the diploid region usually allows the resulting F-prime cells to grow faster than the parental cells.

Transductional Shortening. Transduction of an F-prime factor from one strain to another, using phage P1, sometimes produces shortened derivatives of the original F-prime.

F-Prime Fusion. Two F-primes carrying different chromosomal regions can recombine at low frequency to produce a new (larger) F-prime which contains material from both parental ones, thus producing a new juxtaposition of segments of the chromosome.

Uses of F-Primes

Chromosome Mobilization. Due to frequent recombination between an F-prime and the chromosome in a region of diploidy, the F-prime is transiently contiguous with the chromosome and can cause its transfer to a recipient cell during conjugation. By appropriate choice of F-prime factor, therefore, an F^- strain can be converted to a high-frequency donor for a desired region of the chromsome (see Low, 1973).

Genetic Mapping. There is practically no chromosome mobilization in a primary F-prime strain or in a recombination-deficient ($recA^-$) secondary F-prime strain. These strains therefore can be used to donate well-defined segments of the genome into various mutants to look for the production of wild-type recombinants.

Dominance and Complementation Studies. These work particularly well if the merodiploid strain under study is recombination-deficient in order to prevent the formation of wild-type recombinants.

The Isolation of Haploid-Lethal Mutants.

Fusion of Desired Chromosomal Regions. This can be accomplished by F-prime fusion (see above) or by transposition of F-prime material to a new insertion site in the chromosome.

Isolation of F-prime DNA which Carries Specific Small Regions of the Genome. This use is usually feasible only for F-primes which carry less than five percent of the genome.

Literature Cited

Achtman, M., 1973 Genetics of the F sex factor in *Enterobacteriaceae*. *Curr. Top. Microbiol. Immunol.* **60**:79–123.

Kahn, P. L., 1969 Evolution of a site of specific genetic homology on the chromosome of *Escherichia coli. J. Bacteriol.* **100**: 269–275.

Low, K. B., 1972 *Escherichia coli* K12 F-prime factors, old and new. *Bacteriol. Rev.* **36**:587–607.

Low, K. B., 1973 Rapid mapping of conditional and auxotrophic mutations in *Escherichia coli* K12. *J. Bacteriol.* **113**:798–812.

Moody, E. E. M. and R. Runge, 1972 The integration of autonomous transmissible plasmids into the chromosome of *Escherichia coli* K12. *Genet. Res.* **19**:181–186.

Nishimura, A., Y. Nishimura and L. Caro, 1973 Isolation of Hfr strains from R+ and Col V2+ strains of *Escherichia coli* and derivation of an R′*lac* factor by transduction. *J. Bacteriol.* **116**:1107–1112.

Sanderson, K. E., H. Ross, L. Ziegler and P. H. Mäkelä, 1972 F+, Hfr, and F′ strains of *Salmonella typhimurium* and *Salmonella abony. Bacteriol. Rev.* **36**:608–637.

8

The *Lactose* Operon of *Escherichia coli*

Richard E. Sanders

Lactose Dissimilation

The two enzymes which are specific for the catabolism of the disaccharide lactose in *Escherichia coli* are the lactose permease and beta galactosidase (Watson, 1970). The structural genes for these molecules are regulated as a unit and located adjacent to one another on the *coli* chromosome (see Figure 1). The *Y* gene codes for the permease or M protein which transports lactose into the cell. The *Z* gene carries the information for the synthesis of beta-galactosidase. This enzyme cleaves lactose to glucose and galactose, both of which can directly enter glycolysis and supply the cell with energy and carbon skeletons (Watson, 1970; Zabin and Fowler, 1970; Kennedy, 1970). (The *a* gene codes for thiogalactoside transacetylase. The role of this enzyme in *E. coli* is unknown.)

Regulation of Gene Expression

The *Z* and *Y* genes are transcribed into a single messenger RNA molecule by RNA polymerase which binds to the promoter, P, and polymerizes ribonucleoside triphosphates in a sequence complementary to

---

Richard E. Sanders—Department of Biology, Massachusetts Institute of Technology, Cambridge, Massachusetts.

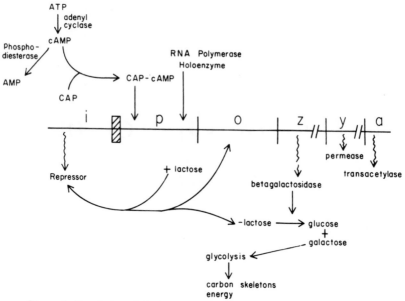

Figure 1. Regulation of the lac operon of E. coli. See text for discussion.

first operator and then Z and Y DNA (Watson, 1970; Reznikoff, 1972; Maizels, 1974).

There are two distinct mechanisms which influence the rate of transcription initiation at the *lac* promoter. Historically, the first device discovered was the repressor–operator interaction. The repressor, coded for by the *i* gene which happens to be adjacent to the region of DNA containing Z and Y, binds tightly to the operator in the absence of lactose. This prevents RNA polymerase from entering Z and Y. When lactose is present, it decreases the repressor's affinity for the operator. Once the repressor diffuses off the operator, RNA polymerase can proceed into Z and Y (Jacob and Monod, 1961; Burstein *et al.*, 1965; Gilbert and Müller-Hill, 1970; Jobe and Bourgeois, 1972; Reznikoff, 1972).

The second mechanism which governs expression of the *lac* operon influences not only this system but also many other operons coding for enzymes involved in catabolic pathways. This general regulatory device is known as the CAP–cAMP system. For efficient transcription of the so-called catabolite-sensitive operons, the RNA polymerase holoenzyme must be accompanied by Catabolite Activator Protein (CAP). The CAP factor is in turn dependent on the presence of cyclic adenosine monophosphate (cAMP) for its transcription-stimulating activity (Zubay *et al.*, 1970; de-Crombrugghe *et al.*, 1971).

Little is known concerning the molecular mechanism(s) by which the availability of cAMP to CAP is regulated, but in general the presence of

TABLE 1. Molecular Control of Transcription in the lac Operon of E. coli

Element	Description	Function
Repressor	Mol wt 160,000, functions as a tetramer, binds only to double helical DNA, apparently synthesized without, regulation from a very inefficient promotor, only 5–10 molecules present per cell	Binds to operator in absence of lactose to prevent transcription
Operator (O)	Approximately 25 nucleotides long, base sequence known	Binding site for repressor
i gene	Structural gene for repressor, transcribed in the same direction as *lac* genes	Genetic information for repressor synthesis
RNA polymerase Subunits:	Alpha mol wt 40,000, beta mol wt 145,000–155,000, beta prime mol wt 160,000, sigma mol wt 85,000	Beta and beta prime implicated in nucleic acid binding, sigma apparently used catalytically at initiation
Core enzyme	1 beta, 1 beta prime, 2 alpha	RNA synthesis from DNA template
Holoenzyme	Core plus one sigma	
Catabolite Activator Protein (CAP)	Mol wt 45,000, stimulates transcription only in the presence of cAMP, mode of action probably by binding to segment of *lac* promotor, synthesis may be regulated	Factor necessary for the transcription of many *E. coli* genes including *lac*
3′:5′ cyclic adenosine monophosphate (cAMP)	Mol wt approximately 360, synthesis from ATP by adenyl cyclase, destruction probably by phosphodiesterase, regulation of levels obscure	Activation of the transcription factor CAP
Promotor (P)	Probably less than 100 base pairs long, appears at least partially distinct from operator, may be divided into CAP and RNA polymerase binding sites	Site where RNA polymerase and its factors initiate transcription

carbon sources which are very efficiently utilized appears to reduce the transcription of the genes coding for enzymes involved in the degradation of poorer carbon sources by decreasing the amount of activated CAP in the cell (Makman and Sutherland, 1965; Ullman and Monod, 1968; Magasanik, 1970).

Molecular Aspects of Gene Regulation

The nucleotide sequence of the operator is known (Gilbert and Maxam, 1973).

CAP, lactose operon DNA, and the *lac* repressor have been purified and studied *in vitro*. The behavior of these isolated components in the cell-free system is qualitatively identical to the results obtained from *in vivo* experiments (Gilbert and Müller-Hill, 1966; Eron and Block, 1971; Adler *et al.*, 1972).

The simplest models for the structure of the promoter envision separate binding sites for CAP and the RNA polymerase holoenzyme within a linear region of double-helical DNA (Beckwith *et al.*, 1966; Smith and Sadler, 1971).

The precise events occurring during initiation of transcription are unknown, but these activities are presumed to include melting of the template and selection of the DNA strand to be transcribed prior to the onset of ribonucleic acid polymerization (Chamberlin, 1970; Burgess, 1971).

Transcription Termination

Very little is known about termination of transcription in *E. coli*. A protein factor *rho* has been isolated from *coli* extracts which appears to terminate transcription on some bacteriophage templates in *in vitro*, but the relationship of this phenomenon to the events occurring at the end of a bacterial operon is obscure (Roberts, 1970; deCrombrugghe *et al.*, 1973).

A barrier to either transcription and translation or simply to translation exists between *i* and *P* in the *lac* region (see crosshatched area in Figure 1) (Miller, 1970). Attempts to isolate mutants in this barrier have produced only deletions fusing *i* to *P* (R. Sanders, unpublished). Unfortunately, this result tells one nothing about the nature of transcription termination.

Current Research

Active areas of research on the *lac* genes include mutant analysis of the steps involved in transcription and translation initiation, the regulation of CAP synthesis, and the control of intracellular levels of cAMP.

Literature Cited

Adler, K., K. Beyreuther, E. Fanning, N. Geisler, B. Gronenborn, A. Klemm, B. Müller-Hill, M. Pfahl and A. Schmitz, 1972 How *lac* repressor binds to DNA. *Nature (Lond.)* **237**:322–327.

Beckwith, J., T. Grodzicker and R. Arditti, 1972 Evidence for two sites in the *lac* promoter. *J. Mol. Biol.* **69**:155–162.

Burgess, R., 1971 RNA polymerase. *Annu. Rev. Biochem.* **40**:711–740.

Burstein, C., M. Cohn, A. Kepes and J. Monod, 1965 Rôle du lactose et de ses produits métaboliques dans l'induction de l'opéron lactose chez *Escherichia coli. Biochim. Biophys. Acta* **95**:634–639.

Chamberlin, M. 1970 Transcription 70: A summary. *Cold Spring Harbor Symp. Quant. Biol.* **35:** 851–873.

deCrombrugghe, B., B. Chen, W. Anderson, P. Nissley, M. Gottesman and I. Pastan, 1971 *Lac* DNA, RNA polymerase and cyclic AMP receptor protein, cyclic AMP, *lac* repressor and inducer are the essential elements for controlled *lac* transcription. *Nat. New Biol.* **231**:139–142.

deCrombrugghe, B., S. Adhya, M. Gottesman and I. Pastan, 1973 Effects of *rho* on transcription of bacterial operons. *Nat. New Biol.* **241**:260–264

Eron, L. and R. Block, 1971 Mechanism of initiation and repression of *in vitro* transcription of the *lac* operon of *Escherichia coli. Proc. Natl. Acad. Sci. USA* **68**:1828–1832.

Gilbert, W. and B. Müller-Hill, 1966 Isolation of the *lac* repressor. *Proc. Natl. Acad. Sci. USA* **56**:1891–1898.

Gilbert, W. and B. Müller-Hill, 1970 The lactose repressor. In *The Lactose Operon,* edited by J. Beckwith and D. Zipser, pp. 93–110, Cold Spring Harbor Laboratory, Cold Spring Harbor, N.Y.

Gilbert, W. and A. Maxam, 1973 The nucleotide sequence of the *lac* operon. *Proc. Natl. Acad. Sci. USA* **70**:3581–3584.

Jacob, F. and J. Monod, 1961 Genetic regulatory mechanisms in the synthesis of proteins. *J. Mol. Biol.* **3**:318–356.

Jobe, A. and S. Bourgeois, 1972 *lac* repressor-operator interaction. VI. The natural inducer of the *lac* operon. *J. Mol. Biol.* **69**:397–408.

Kennedy, E., 1970 The lactose permease system of *E. coli.* In *The Lactose Operon,* edited by J. Beckwith and D. Zipser, pp. 49–93, Cold Spring Harbor Laboratory, Cold Spring Harbor, N. Y.

Magasanik, B. 1970 Glucose effects: inducer exclusion and repression. In *The Lactose Operon,* edited by J. Beckwith and D. Zipser, pp. 189–220, Cold Spring Harbor Laboratory, Cold Spring Harbor, N. Y.

Maizels, N. 1973 The nucleotide sequence of the lactose messenger ribonucleic acid transcribed from the UU5 promoter mutant of *Escherichia coli. Proc. Natl. Acad. Sci. USA* **70**:3585–3590.

Makman, R. and E. Sutherland, 1965 Adenosine $3':5'$ phosphate in *E. coli. J. Biol. Chem.* **240**:1309–1314.

Miller, J., 1970 Transcription starts and stops in the *lac* operon. In *The Lactose Operon,* edited by J. Beckwith and D. Zipser, pp. 173–189, Cold Spring Harbor Laboratory, Cold Spring Harbor, N. Y.

Reznikoff, W., 1972 The operon revisited. *Annu. Rev. Genet.* **6**:133–156.

Roberts, J. 1970   The *rho* factor: termination and antitermination in lambda. *Cold Spring Harbor Symp. Quant. Biol.* **35**:121–126.

Scaife, J. and J. Beckwith, 1966   Mutational alteration of maximal operon expression. *Cold Spring Harbor Symp. Quant. Biol.* **31**:403–408.

Smith, F. and J. Sadler, 1971   The nature of lactose operator constitutive mutations. *J. Mol. Biol.* **59**: 273–305.

Ullman, A. and J. Monod, 1968   Cyclic AMP as an antagonist of catabolite repression in *Escherichia coli. FEBS Lett. (Fed. Eur. Biochem. Soc.)* **2**:57.

Watson, J., 1970   *Molecular Biology of the Gene,* pp. 437–450, W.A. Benjamin, New York.

Zabin, I. and A. Fowler, 1970   Beta galactosidase and thiogalactoside transacetylase. In *The Lactose Operon,* edited by J. Beckwith and D. Zipser, pp. 27–49, Cold Spring Harbor Laboratory, Cold Spring Harbor, N. Y.

Zubay, G., D. Schwartz and J. Beckwith, 1970   Mechanism of activation of catabolite-sensitive genes: a positive control system. *Proc. Natl. Acad. Sci. USA* **66**:104–110.

9

The *lac Z* gene of *Escherichia coli*

Irving Zabin

The Z Gene of the Lactose Operon

The Z gene is a structural gene which specifies the polypeptide monomer of β-galactosidase. This enzyme normally exists as a polymer of four identical chains with a molecular weight for the complete, native protein of 540,000 (cf. Zabin and Fowler, 1970). In Figure 1, the double line represents the gene and the single line represents the corresponding polypeptide chain. Above the gene terminator mutant strains are listed. Polypeptide fragments produced by each of these have been detected by immunological and complementation tests. Molecular weights have been determined in sucrose or urea–sucrose gradients and in several cases by sedimentation equilibrium (Fowler and Zabin, 1966; Brown *et al.*, 1967; Berg *et al.*, 1970; Morrison and Zipser, 1970). The approximate number of residues in the polypeptide chain corresponding to the mutated site is shown below the single line; this was calculated based on an average amino acid residue weight of 115. The amino acid sequences of the first thirteen and the last ten residues have been reported (Zabin and Fowler, 1972).

Irving Zabin—Department of Biological Chemistry, School of Medicine and Molecular Biology Institute, University of California, Los Angeles, California.

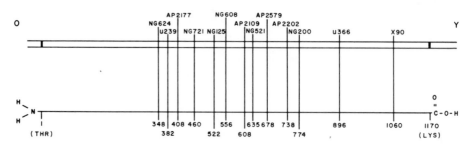

Figure 1. Colinearity of the Z gene of the lactose operon and the corresponding β-galactosidase fragments.

The Z gene has also been subdivided into α, β, and ω segments. The α segment is operator-proximal and is roughly the first quarter of the gene. The ω portion is the operator-distal third, and β is the middle region (Ullmann et al., 1965, 1967, 1968).

Literature Cited

Berg, A. L., A. V. Fowler and I. Zabin, 1970   β-Galactosidase: Isolation of and antibodies to incomplete chains. J. Bacteriol. 101:438–443.

Brown, J. L., D. M. Brown and I. Zabin, 1967   β-Galactosidase: Orientation and the carboxyl-terminal coding site in the gene. Proc. Natl. Acad. Sci. USA 58:1139–1143.

Fowler, A. V. and I. Zabin, 1966   Co-linearity of β-galactosidase with its gene by immunological detection of incomplete polypeptide chains. Science (Wash., D.C.) 154:1027–1029.

Morrison, S. L. and D. Zipser, 1970   Polypeptide products of nonsense mutations. I. Termination fragments from nonsense mutations in the Z gene of the lac operon of Escherichia coli. J. Mol. Biol. 50:359–371.

Ullmann, A., D. Perrin, F. Jacob and J. Monod, 1965   Identification par complémentation in vitro et purification d'un segment peptidique de la β-galactosidase d'Escherichia coli. J. Mol. Biol. 12:918–923.

Ullman, A., F. Jacob and J. Monod, 1967   Characterization by in vitro complementation of a peptide corresponding to an operator-proximal segment of the β-galactosidase structural gene of Escherichia coli. J. Mol. Biol. 24:339–343.

Ullmann, A., F. Jacob and J. Monod, 1968   On the subunit structure of wild-type versus complemented β-galactosidase of Escherichia coli. J. Mol. Biol. 32:1–13.

Zabin, I. and A. V. Fowler, 1970   β-Galactosidase and thiogalactoside transacetylase. In The Lactose Operon, edited by J.R. Beckwith and D. Zipser, pp. 27–47, Cold Spring Harbor Laboratory, Cold Spring Harbor, N.Y.

Zabin, I. and A. V. Fowler, 1972   The amino acid sequence of β-galactosidase. III. The sequences of NH$_2$- and COOH-terminal tryptic peptides. J. Biol. Chem. 247:5432–5435.

10

Tryptophan Synthetase α Chain of *Escherichia coli* and Its Structural Gene

CHARLES YANOFSKY

The tryptophan synthetase of *Escherichia coli* is a tetrameric protein consisting of two pairs of nonidentical polypeptide chains (Yanofsky and Crawford, 1972). These polypeptide chains, designated α and β, are specified by structural genes *trpA* and *trpB*, respectively, which lie adjacent to one another at the operator-distal end of the tryptophan operon. Mutants deficient in the terminal reaction in tryptophan biosynthesis are generally altered in *trpA, trpB,* or both. Extensive genetic studies have been performed with these mutants and fine-structure genetic maps have been prepared for each of the genes (Yanofsky and Crawford, 1972; Yanofsky and Horn, 1972; Yanofsky *et al.*, 1964, 1967). Many *trpA* mutants have also been analyzed biochemically to determine the positions in the tryptophan synthetase α chain affected by mutation. Comparison of the linear correspondence between the genetic map of *trpA* and the positions in the α chain altered by mutation has demonstrated that the map and polypeptide are colinear (Figure 1) (Yanofsky and Horn, 1972; Yanofsky *et al.*, 1964, 1967). Furthermore, relative distance on the genetic map was found to approximate distance between amino acid residues in

CHARLES YANOFSKY—Department of Biological Sciences, Stanford University, Stanford California.

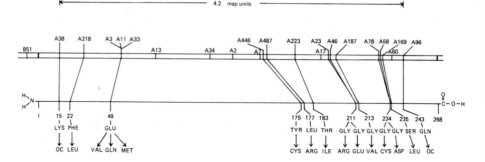

Figure 1. Colinearity of the genetic map of trpA of E. coli and the α polypeptide chain of tryptophan synthetase. The line at the top of the figure represents the map distance (in recombination units) between the mutational sites of mutants A38 and A96. The bar immediately below is a genetic map of trpA, drawn to scale, in which mutant numbers and corresponding map positions are indicated. Shown at the bottom of the figure is the change in the α polypeptide associated with each mutational change, e.g., mutant A218 has leucine instead of phenylalanine at position 22 in the α chain. Mutants A38 and A96 are ochre (oc) chain termination mutants. In order to compare distances on the genetic map and in the polypeptide chain, the polypeptide has also been drawn to scale and the positions of the A38 and A96 changes on the map and in the protein have been aligned. From Yanofsky and Horn (1972), reprinted with the permission of the Journal of Biological Chemistry.

the polypeptide chain. These findings suggest that the nucleotide sequence of a gene corresponds to the linear sequence of amino acids in its specified polypeptide chain.

Literature Cited

Yanofsky, C. and I. P. Crawford, 1972 Tryptophan synthetase. In *The Enzymes*, Vol. VII, edited by P. D. Boyer, pp. 1–31, Academic Press, New York.

Yanofsky, C. and V. Horn, 1972 Tryptophan synthetase α chain positions affected by mutations near the ends of the genetic map of trpA of E. coli. *J. Biol. Chem.* **247**:4494–4498.

Yanofsky, C., B. C. Carlton, J. R. Guest, D. R. Helinski and U. Henning, 1964 On the colinearity of gene structure and protein structure. *Proc. Natl. Acad. Sci. USA* **51**:266–272.

Yanofsky, C., G. R. Drapeau, J. R. Guest and B. C. Carlton, 1967 The complete amino acid sequence of the tryptophan synthetase A protein (α subunit) and its colinear relationship with the genetic map of the A gene. *Proc. Natl. Acad. Sci. USA* **57**:296–298.

11

The Alkaline Phosphatase of *Escherichia coli*

Annamaria Torriani

Introduction

The existence of an alkaline phosphatase (AP) in *Escherichia coli* (E.C. 3.1.3.1.) was first demonstrated in 1958 (Torriani, 1958, 1960; Horiuchi *et al.*, 1959). The active enzyme is a dimer of two identical inactive subunits each of mol wt 43,000 (Plocke *et al.*, 1962; Rothman and Byrne, 1963; Schlesinger and Barrett, 1965; Reynolds and Schlesinger, 1967). The existence *in vitro* of tetramers has been reported (Halford *et al.*, 1972c; Reynolds and Schlesinger, 1969a,b). Each molecule of the active enzyme contains at least 2 Zn^{++} (some authors have found as many as 4 Zn^{++}). It has been established that Zn^{++} binding has a functional and a structural role in the molecule (Applebury and Coleman, 1969; Applebury *et al.*, 1970; Cottam and Ward, 1970; Csopack *et al.*, 1972, 1973; Lazdunski *et al.*, 1969; Petitclerc *et al.*, 1970; Plocke *et al.*, 1962; Reynolds and Schlesinger, 1969a; Simpson and Vallee, 1968). The enzyme has been purified (Garen and Levinthal, 1960; Schlesinger and Olsen, 1970; Simpson *et al.*, 1968) and crystallized (Hanson *et al.*, 1970; Malamy and Horecker, 1964; Knox and Wyckoff, 1973). The "homogeneous" protein, however, has been resolved into four or more different species (isozymes) by starch gel or acrylamide gel electrophoresis and by DEAE chromatography

ANNAMARIA TORRIANI—Department of Biology, Massachusetts Institute of Technology, Cambridge, Massachusetts.

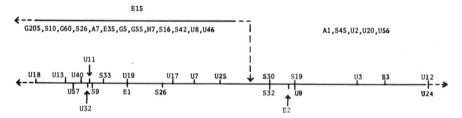

Figure 1. Genetic map of the phoA locus. The mutants were isolated (Garen, 1960; Garen and Levinthal, 1960; Levinthal, 1959; Rothman et al., 1962) from the E. coli K12 strain Hfr K10. The letters preceding the numbers designate the mutagen used (U, ultraviolet light; G, nitrosoguanidine; S, ethylmethanesulfonate; E, x rays; H, hydroxylamine; A, 2-aminopurine). The relative distances correspond approximately to the relative recombination frequencies obtained by two-factor crosses in all possible combinations between the original Hfr K10 mutants and their F⁻ derivatives, constructed by mating (Fan et al., 1966; Garen, 1960). The large deletion E15 has been used to separate 18 nonsense mutants into two classes: those covered by, and those outside the deletion (Suzuki and Garen, 1969). Of the first class, four were used (Natori and Garen, 1970; Suzuki and Garen, 1969) to study short unfinished peptides. They are G205 amber and S10 ochre located to the left of U18, G60 amber located to the right of U18, and S26 amber located close to U17. U24 is an ochre mutant. In the figure, the phoA cistron has been oriented not according to the orientation of the gene linkage, but to the direction of the phoA translation: amino end to the left, carboxyl end to the right.

(Bach *et al.*, 1961; Levinthal *et al.*, 1962; Schlesinger *et al.*, 1963; Signer, 1963, 1965). The enzymatic properties of the separated isozymes have been found to be identical (Lazdunski and Lazdunski, 1967; Piggot *et al.*, 1972; Schlesinger and Andersen, 1968; Signer, 1963).

Some 30 mutants with altered enzyme activity, not ascribable to modifications in the regulation system, have been mapped: they are clustered in a small region at 10 minutes in the *E. coli* genetic map (see Chapter 6, by Taylor and Trotter, in this volume). This site is the location of the phosphatase structural gene *phoA* (originally called P) (Fan *et al.*, 1966; Garen, 1960; Levinthal *et al.*, 1963; Suzuki and Garen, 1969). The two missense mutants farthest apart are U18 and U12 (see Figure 1), and their order relative to the neighboring markers is: *lacI–phoA$_{U12}$–phoA$_{U18}$–proC* (Bracha and Yagil, 1969; Nakata *et al.*, 1971; Yagil *et al.*, 1970). Since it has been found that the U18 end of the *phoA* cistron corresponds to the amino end of the polypeptide (Suzuki and Garen, 1969), translation of *phoA* is anticlockwise relative to the *E. coli* linkage map (Nakata *et al.*, 1971). In the figure are presented also a number of nonsense mutants which were roughly mapped against a deletion (E15) spanning about two thirds of the *phoA* cistron (Gallucci and Garen, 1966; Garen *et al.*, 1965; Gopinathan and Garen, 1970). Four of the nonsense mutants (G205, S10, G60, and S26) were more precisely mapped and used extensively for studying nonsense suppression (Suzuki and Garen, 1969) and phenotypic suppression by 5-fluorouracil (Rosen *et al.*, 1969). Although these studies permitted a partial analysis of codon–amino acid correspondence in the region of the mutations (Andoh and Garen, 1967; Chan and Garen, 1969; 1970; Garen, 1968; Gopinathan and Garen, 1970; Rosen *et al.*, 1969; Suzuki and Garen, 1969; Weigert and Garen, 1965; Weigart *et al.*, 1965, 1967*a,b*), the complete amino acid sequence of the whole protein is still not available, and the complete correlation between base sequence and amino acid sequence is lacking.

Fingerprint determination of the entire enzyme protein was attempted some years ago, but the complexity of the peptides obtained discouraged further analysis (Rothman and Byrne, 1963). Recently, however, the N-terminal end of the protein has been analyzed by studying fragments of about 40 amino acids (mol wt 4000) produced by strains carrying a nonsense mutation close to the amino-terminal side of the structural gene (Natori and Garen, 1970; Suzuki and Garen, 1969). The limited number of peptides obtained by tryptic digestion of these fragments (Natori and Garen, 1970) facilitated the analysis and showed an N-terminal heterogeneity (Natori and Garen, 1970; Suzuki and Garen, 1969). This finding was correlated with the existence of the isozymes, and the correlation was shown correct (Piggot *et al.*, 1972) by analyzing directly the amino terminal of purified monomers obtained from the isolated isozymes constituting the three larger electrophoretic bands. The N-terminal amino acid was found different for each isozyme: aspartate (or asparagine) for isozyme A (the faster moving), probably valine for isozyme B, and threonine for isozyme C. The other minor bands in the electrophoretic pattern could be hybrid dimers between these three types of monomers. Kelley *et al.*, (1973) have recently determined the sequence of two small peptides at the amino and caroboxyl-termini of isolated isozymes. They were found to be (arg)–thr–pro–glu and lys–ala–ala–leu–glu–leu–cys, respectively. Here too a heterogeneity has been observed at the amino terminus: isozyme 3 is devoid of the arginyl residue found in isozyme 1. This heterogeneity cannot be inscribed in the genome, since only one structural gene (*phoA*) is known, and any *phoA* point mutation altering the electrophoretic mobility affects all the isozymes at once (Bach *et al.*, 1961; Levinthal, 1959; Schlesinger and Andersen, 1968; Schlesinger and Barrett, 1965; Schlesinger *et al.*, 1963; Signer, 1963; 1965). The heterogeneity must therefore originate after transcription. Furthermore, it was recently found that ribosomal mutations increasing or restricting translational ambiguity spread or reduce, respectively, the isozymes' heterogeneity (Piggot *et al.* 1972). Therefore events occurring at the translation of the genome and/or post-translational modification could be the origin of isozyme heterogeneity.

The monomers from two different phosphatase negative mutants, if present together, may give active hybrid dimers. In fact, this complementation has been found *in vitro* (Fan *et al.*, 1966; Levinthal *et al.*, 1962; Schlesinger, 1964; Schlesinger and Andersen, 1968; Schlesinger and Barrett, 1965; Schlesinger and Levinthal, 1963, 1965; Schlesinger *et al.*, 1963; Torriani *et al.*, 1965) by co-dimerizing monomers from two different mutants, and *in vivo* (Garen and Garen, 1963) by constructing strains diploid in the phosphatase region from two different phosphatase negative mutants. A correlation between genetic map and ability to complement was sought, but not found (Schlesinger and Levinthal, 1965; Schlesinger *et al.*, 1963; Torriani *et al.*, 1965). This is a predictable result since it is well known that the activity of an enzyme is primarily dependent on the tertiary structure of the peptide chain and only indirectly dependent on its primary amino acid sequence. For a dimeric protein this is even more the case, because activity is not only dependent on the conformation of the active site, but also on the ability to dimerize. A number of *phoA* mutants have been described (Halford *et al.*, 1972a,b; Schlesinger, 1966, 1967) in which the mutation affects the ability to dimerize properly and/or to bind Zn^{++}. Intergeneric hybrids of alkaline phosphatase between monomers from *Serratia marcescens* and from *E. coli* were obtained *in vivo* and *in vitro* in spite of the different DNA composition of the two organisms and the widely different amino acid composition of the two enzymes (Levinthal *et al.*, 1962; Signer, 1963, 1965; Signer *et al.*, 1961). This result supports the concept that complementation involves the tertiary structure of the peptide chain rather than its primary amino acid sequence.

Regulatory Genes

At least three genes control the level of enzyme: *phoR* (or R_1), *phoSa*, and *phoSb* [or R_2a and b (Echols *et al.*, 1961; Garen and Echols, 1962; Garen and Otsuji, 1964; Torriani and Rothman, 1961)]. The regulatory gene *phoR* is closely linked to *phoA* (Bracha and Yagil, 1969; Echols *et al.*, 1961; Yagil *et al.*, 1970), with which it is cotransducible by phage P_1 (Nakata *et al.*, 1971), but it is not contiguous because *proC* is intercalated (Yagil *et al.*, 1970). The *phoS* region (Aono and Otsuji, 1968; Echols *et al.*, 1961) is located at 73 minutes on the *E. coli* map (see Chapter 6, by Taylor and Trotter, in this volume) and has been resolved into two cistrons, *a* and *b* (Garen and Otsuji, 1964). The protein produced by *phoSa* has been chromatographically separated (Garen and Otsuji, 1964).

Phenotypically, the synthesis of the enzyme is repressed by inorganic phosphate (Pi) (Torriani, 1958, 1960); Horiuchi *et al.*, 1959.) but the mechanism of this control is not yet understood. The existence of a class of *phoR* mutants (*phoR^c*) phenotypically phosphatase negative prompted Garen and Echols (1962) to interpret the existing experimental data with the hypothesis that *phoR* controls the synthesis of an internal inducer which, in the presence of Pi, is modified by a reaction directed by *phoSa* and/or *phoSb* generating the repressor:

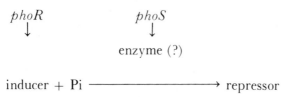

It has since been found that the *phoSa* protein is also repressed by Pi (Garen and Otsuji, 1964), and that the level of phosphatase can be raised even in the presence of Pi by changing the nucleotide pools of the cells (escape synthesis) (Gallant and Spottwood, 1965). Wilkins (1972) suggested that the *phoR* product may not be an inducer, but that it may determine the extent and specificity of the response to alterations in the nucleotide pool. He found that *phoR$^+$* wild type can be induced by adenine nucleotide accumulation, while adenine starvation induces the *phoR^c* mutants. The finding of a *phoR* constitutive dominant mutant (*phoR^d*) suggested that the *phoR* product may be a multimeric protein consisting of 4 subunits (Pratt and Gallant, 1972). Recently, mutants with the *phoR^c* (Garen and Echols, 1962) phenotype have been carefully mapped (Bracha and Yagil, 1973). They appear to complement with *phoR* and therefore are genetically distinct from it, although very closely linked. The new gene has been designated *phoT*. Since the same designation has been previously proposed by Willsky *et al.* (1973) for another *pho* gene in the *phoS* region, the new gene in the *phoR* region has been called *phoB*. As for the *phoS* genetic region, it was found (Jones, 1967, 1969) that at least another gene close to *phoS* in the *xyl* region controls the basal level of the repressed enzyme. Willsky *et al.*, (1973) recently reported that two of the genes— *phoS* (*phoSa*) and *phoT* (*phoSb*)—mapping in this region are involved in the mechanism of transport and accumulation of Pi. Therefore, they probably play an indirect role in the regulation of alkaline phosphatase synthesis.

Literature Cited

Andoh, T. and A. Garen, 1967   Fractionation of a serine transfer RNA containing suppressor activity. *J. Mol. Biol.* **24:**129–132.

Aono, H. and N. Otsuji, 1968   Genetic mapping of regulator gene *pho S* for alkaline phosphatase in *E. Coli. J. Bacteriol.* **95:**1182–1183.

Applebury, M. L. and J. E. Coleman, 1969   *E. coli* alkaline phosphatase. Metal binding protein conformation, and quaternary structure. *J. Biol. Chem.* **244:**308–318.

Applebury, M. L., B. P. Johnson and J. E. Coleman, 1970   Phosphate binding to alkaline phosphatase. Metal ion dependence. *J. Biol. Chem.* **245:**4968–4976.

Bach, M. L., E. R. Signer, C. Levinthal and I. W. Sizer, 1961   The electrophoretic patterns of alkaline phosphatase from various *E. coli* mutants. *Fed. Proc.* **20:**255.

Bracha, M. and E. Yagil, 1969   Genetic mapping of the *pho R* regulator gene of alkaline phosphatase in *E. coli. J. Gen. Microbiol.* **59:**77–81.

Bracha, M. and E. Yagil, 1973   A new type of alkaline phosphatase-negative mutants in *E. coli* K12. *Molec. Gen. Genet.* **122:**53–60.

Chan, T. and A. Garen, 1969   Amino acid substitutions resulting from suppression of nonsense mutations. IV. Leucine insertion by the Su-6+ suppressor gene. *J. Mol. Biol.* **45:**545–548.

Chan, T. and A. Garen, 1970   Amino acid substitution resulting from suppression of nonsense mutations. V. Tryptophan insertion by the Su-9+ gene, a suppressor of UGA nonsense triplet. *J. Mol. Biol.* **49:**231–234.

Cottam, G. L. and R. L. Ward, 1970   Chloride NMR studies on the zinc-binding site in *E. coli* alkaline phosphatase. *Arch. Biochem. Biophys.* **141:**768–770.

Csopack, H., K. E. Falk and H. Szajn, 1972   Effect of EDTA on *Escherichia coli* alkaline phosphatase. *Biochim. Biophys. Acta* **258:**466–472.

Csopack, H. and H. Szajn, 1973   Factors affecting the zinc content of *E. coli* alkaline phosphatase *Arch. Biochem. Biophys.* **157:**374–379.

Echols, H., A. Garen, S. Garen and A. Torriani, 1961   Genetic control of repression of alkaline phosphatase in *E. coli. J. Mol. Biol.* **3:**425–438.

Fan, D. P., M. J. Schlesinger, A. Torriani, K. J. Barrett and C. Levinthal, 1966   Isolation and characterization of complementation products of *E. coli* alkaline phosphatase. *J. Mol. Biol.* **15:**32–48.

Gallant, J. and T. Spottwood, 1965   Escape synthesis of alkaline phosphatase. *Biochim. Biophys. Acta* **103:**109–119.

Gallucci, E. and A. Garen, 1966   Suppressor genes for nonsense mutations. II. The Su-4 and Su-5 suppressor genes of *E. coli. J. Mol. Biol.* **15:**193–200.

Garen, A., 1960   Genetic control of the specificity of the bacterial enzyme, alkaline phosphatase. In *Symposium on Microbial Genetics,* edited by W. Hayes and R. C. Clawes, pp. 239–247, Cambridge University Press, London.

Garen, A., 1968   Sense and nonsense in the genetic code. *Science (Wash., D.C.)* **160:**149–159.

Garen, A. and H. Echols, 1962   Genetic control of induction of alkaline phosphatase synthesis in *E. coli. Proc. Natl. Acad. Sci. USA* **48:**1398–1402.

Garen, A. and S. Garen, 1963   Complementation *in vivo* between structural mutants of alkaline phosphatase from *E. coli. J. Mol. Biol.* **7:**13–22.

Garen, A. and C. Levinthal, 1960   A fine-structure genetic and chemical study of the enzyme alkaline phosphatase of *E. coli.* I. Purification and characterization of alkaline phosphatase. *Biochim. Biophys. Acta* **38:**470–483.

Garen, A. and N. Otsuji, 1964   Isolation of a protein specified by a regulator gene. *J. Mol. Biol.* **8**:841–852.

Garen, A. and O. Siddiqi, 1962   Suppression of mutations in the alkaline phosphatase structural cistron of *E. coli. Proc. Natl. Acad. Sci. USA* **48**:1121–1127.

Garen, A., S. Garen and R. C. Wilhelm, 1965   Suppressor genes for nonsense mutations. I. The Su-1, Su-2, Su-3 genes of *E. coli, J. Mol. Biol.* **4**:167–178.

Gopinathan, K. P. and A. Garen, 1970   A leucyl-transfer RNA specified by the amber suppressor gene Su-6$^+$. *J. Mol. Biol.* **47**:393–401.

Halford, S. E., D. A. Lennette, P. M. Kelley and M. J. Schlesinger, 1972a   A mutationally altered alkaline phosphatase from *E. coli.* I. Formation of an active enzyme *in vitro* and phenotypic suppression *in vivo. J. Biol. Chem.* **247**:2087–2094.

Halford, S. E., D. A. Lennette and M. J. Schlesinger, 1972b   A mutationally altered alkaline phosphatase from *E. coli.* II. Structural and catalytic properties of the activated enzyme. *J. Biol. Chem.* **247**:2095–2101.

Halford, S. E., M. J. Schlesinger and H. Gutfreund, 1972c   *E. coli* alkaline phosphatase. Kinetic studies with the tetrameric enzyme. *Biochem. J.* **126**:1081–1090.

Hanson, A. W., M. L. Applebury, J. E. Coleman and H. W. Wyckoff, 1970   Studies on single crystal of *Escherichia coli* alkaline phosphatase. *J. Biol. Chem.* **245**:4975–4976.

Horiuchi, T., S. Horiuchi and D. Mizuno, 1959   A possible negative feedback phenomenon controlling formation of alkaline phosphatase in *E. coli.* Nature (*Lond.*) **183**:1529–1530.

Jones, T. C., 1967   Genetic control of basal levels of alkaline phosphatase in *E. coli.* Ph.D. Thesis, University of Washington, Seattle, Washington.

Jones, T. C., 1969   Genetic control of basal levels of alkaline phosphatase in *E. coli. Mol. Gen. Genet.* **105**:91–100.

Kelley, P. M. P. A. Neumann, K. Shriefer, F. Cancedda, M. J. Schlesinger and R. A. Bradshaw, 1973   The amino acid sequence of *E. coli* alkaline phosphatase. I. Amino and carboxyl-terminal sequences and variations between two isozymes. *Biochemistry* **12**:3499–3503.

Knox, J. R. and H. W. Wyckoff, 1973   A crystallographic study of alkaline phosphatase at 7.7 Å Resolution *J. Mol. Biol.* **74**:533–545.

Lazdunski, C. and M. Lazdunski, 1967   Les isophosphatases alcalines d'*E. coli.* Séparation, propriétés cinétiques et structurales. *Biochim. Biophys. Acta* **147**:280–288.

Lazdunski, C., C. Petitclerc and M. Lazdunski, 1969   Structure–function relationships for some metalloalkaline phosphatases of *E. coli. Eur. J. Biochem.* **8**:510–517.

Levinthal, C., 1959   Structure and function of genetic elements: Genetic and chemical studies with alkaline phosphatase of *E. coli. Brookhaven Symp. Biol.* **12**:76–85.

Levinthal, C., E. R. Signer and K. Fetherolf, 1962   Reactivation and hybridization of reduced alkaline phosphatase. *Proc. Natl. Acad. Sci. USA* **48**:1231–1237.

Levinthal, C., A. Garen and F. Rothman, 1963   Relationship of gene structure to protein structure: Studies on the alkaline phosphatase of *E. coli. Proc. Fifth Int. Congr. Biochem.* **I**:196–203.

Malamy, M. H. and B. L. Horecker, 1964   Purification and crystallization of alkaline phosphatase of *E. coli. Biochemistry* **3**:1893–1897.

Nakata, A., G. R. Peterson, E. L. Brooks and F. Rothman, 1971   Location and orientation of the *pho A* locus on the *E. coli* K12 linkage map. *J. Bacteriol.* **107**:683–689.

Natori, S. and A. Garen, 1970   Molecular heterogeneity in the amino-terminal region of alkaline phosphatase. *J. Mol. Biol.* **49:**577–588.

Petitclerc, C., C. Lazdunski, D. Chappelet, A. Moulin and M. Lazdunski, 1970   The functional properties of the Zn^{2+} and Co^{2+} alkaline phosphatases of *E. coli*. Labeling of the active site with pyrophosphate, complex formation with arsenate and reinvestigation of the role of the zinc atoms. *Eur. J. Biochem.* **14:**301–308.

Piggot, P. J., M. D. Sklar, and L. Gorini, 1972   Ribosomal alterations controlling alkaline phosphatase isozymes in *E. coli. J. Bacteriol.* **110:**291–299.

Plocke, D. J., C. Levinthal and B. L. Valle, 1962   Alkaline phosphatase of *E. coli:* A zinc metalloenzyme. *Biochemistry* **1:**373–378.

Pratt, C. and J. Gallant, 1972   A dominant constitutive *pho R* mutation in *Escherichia coli. Genetics* **72:**217–226.

Reynolds, J. A. and M. J. Schlesinger, 1967   Conformational states of the subunit of *E. coli* alkaline phosphatase. *Biochemistry* **6:**3552–3559.

Reynolds, J. A. and M. J. Schlesinger, 1969*a*   Alterations in the structure and function of *E. coli* alkaline phosphatase due to Zn^{++} binding. *Biochemistry* **8:**588–593.

Reynolds, J. A. and M. J. Schlesinger, 1969*b*   Formation and properties of a tetrameric form of *E. coli* alkaline phosphatase. *Biochemistry* **8:**4278–4282.

Rosen, B., F. Rothman and M. G. Weigert, 1969   Miscoding caused by 5-fluorouracil. *J. Mol. Biol.* **44:**363–375.

Rothman, F. and R. Byrne, 1963   Fingerprint analysis of alkaline phosphatase of *E. coli* K12. *J. Mol. Biol.* **6:**330–340.

Rothman, F., C. Levinthal and A. Garen, 1962   Genetic control of alkaline phosphatase structure. *J. Gen. Physiol.* **45:**615A–616A.

Schlesinger, M. J., 1964   Subunit Structure of Proteins: Biochemical and Genetic Aspects. *In vitro* complementation and the subunit structure of *E. coli* alkaline phosphatase. *Brookhaven Symp. Biol.* **17:**66–79.

Schlesinger, M. J., 1966   Activation of a mutationally altered form of the *E. coli* alkaline phosphatase by Zn. *J. Biol. Chem.* **241:**3181–3188.

Schlesinger, M. J., 1967   Formation of a defective alkaline phosphatase subunit by mutant of *E. coli. J. Biol. Chem.* **242:**1604–1611.

Schlesinger, M. J. and L. Andersen, 1968   Multiple molecular forms of the alkaline phosphatase of *E. coli. Ann. N. Y. Acad. Sci.* **151:**159–170.

Schlesinger, M. J. and K. Barrett, 1965   The reversible dissociation of the alkaline phosphatase of *E. coli*. I. Formation and reactivation of subunits. *J. Biol. Chem.* **240:**4284–4292.

Schlesinger, M. J. and C. Levinthal, 1963   Hybrid protein formation of *E. coli* alkaline phosphatase leading to *in vitro* complementation. *J. Mol. Biol.* **7:**1–12.

Schlesinger, M. J. and C. Levinthal, 1965   Complementation at the molecular level of enzyme interaction. *Annu. Rev. Microbiol.* **19:**267–284.

Schlesinger, M. J. and R. Olsen, 1970   A new, simple, rapid procedure for purification of *E. coli* alkaline phosphatase. *Anal. Biochem.* **36:**86–90.

Schlesinger, M. J., A. Torriani and C. Levinthal, 1963   *In vitro* formation of enzymatically active hybrid proteins from *E. coli* alkaline phosphatase CRM's. *Cold Spring Harbor Symp. Quant. Biol.* **28:**539–542.

Signer, E. R., 1963   Non-inheritable factors in gene expression. Ph.D. Thesis, Department of Biology, Massachusetts Institute of Technology, Cambridge, Massachusetts.

Signer, E. R., 1965   Gene expression in foreign cytoplasm. *J. Mol. Biol.* **12:**1–8.

Signer, E. R., A. Torriani and C. Levinthal, 1961 Gene expression in intergeneric merozygotes. *Cold Spring Harbor Symp. Quant. Biol.* **26**:31–34.

Simpson, R. T. and B. L. Vallee, 1968 Two differential classes of metal atoms in alkaline phosphatases of *E. coli. Biochemistry* **7**:4343–4350.

Simpson, R. T., B. L. Vallee and G. H. Tait, 1968 Alkaline phosphatase of *E. coli. Composition. Biochem.* **7**:4336–4342.

Suzuki, T. and A. Garen, 1969 Fragments of alkaline phosphatase from nonsense mutants. I. Isolation and characterization of fragments from amber and ochre mutants. *J. Mol. Biol.* **45**:549–566.

Torriani, A., 1958 Effect of inorganic phosphate in the formation of phosphatases of *E. coli. Fed. Proc.* **18**:33.

Torriani, A., 1960 Influence of inorganic phosphates in the formation of phosphatases of *E. coli. Biochim. Biophys. Acta* **38**:460–469.

Torriani, A. and F. Rothman, 1961 Mutants of *E. coli* constitutive for alkaline phosphatase. *J. Bacteriol.* **81**:835–836.

Torriani, A., M. J. Schlesinger and C. Levinthal, 1965 Intracistronic complementation of alkaline phosphatase from *E. coli. Colloq. Int. Cent. Natl. Rech. Sci.* **124**:297–302.

Weigert, M. G. and A. Garen, 1965 Base composition of nonsense codons in *E. coli.* Evidence from amino-acid substitutions at a tryptophan site in alkaline phosphatase. *Nature (Lond.)* **206**:992–994.

Weigert, M. G., E. Lanka and A. Garen, 1965 Amino acid substitutions resulting from suppression of nonsense mutations. II. Glutamine insertion by the Su-2 gene; tyrosine insertion by the Su-3 gene. *J. Mol. Biol.* **14**:522–527.

Weigert, M. G., E. Lanka and A. Garen, 1967*a* Base composition of nonsense codons in *E. coli.* II. The N_2 codon UAA. *J. Mol. Biol.* **23**:391–400.

Weigert, M. G., E. Lanka and A. Garen, 1967*b* Amino acid substitutions resulting from suppression of nonsense mutations. III. Tyrosine insertion by the Su-4 gene. *J. Mol. Biol.* **23**:401–404.

Wilkins, A. S., 1972 Physiological factors in the regulation of alkaline phosphatase synthesis in *E. coli. J. Bacteriol.* **110**:616–623.

Willsky, G. R., R. L. Bennett and M. H. Malamy, 1973 Inorganic phosphate transport in *Escherichia coli:* Involvement of two genes which play a role in alkaline phosphatase regulation. *J. Bacteriol.* **113**:529–539.

Yagil, E., M. Bracha and N. Silberstein, 1970 Further genetic mapping of *phoA–phoR* region for alkaline phosphatase synthesis in *E. coli* K12. *Mol. Gen. Genet.* **109**:18–26.

12

Bacterial Ribosomes

JULIAN DAVIES

Ribosomes

Ribosomes are intracellular ribonucleoprotein particles of about 200 Å in diameter which are essential components of the translation system in all organisms (Tissières and Watson, 1958; Tissières et al., 1959; Watson, 1964; Schlessinger and Apirion, 1969; Nomura, 1970). They are usually isolated from cell-free extracts by sedimentation at 105,000 × g for several hours and then purified from nonribosomal contaminants by high-salt washing (Kurland, 1971). They are characterized by their typical sedimentation properties and their ability to function in protein synthesis *in vitro*. Extensive physical studies of ribosomes have been carried out (Hill et al., 1969). All ribosomes which have been examined are made up of two subunits (small and large). The separation of the ribosome into subunits, which probably has physiological significance (the "ribosome cycle"), is accomplished by dialyzing bacterial ribosomes against low magnesium ion concentrations Staehelin and Maglott, 1971), or eukaryotic ribosomes against solutions containing concentrated potassium chloride (Martin et al., 1971; Staehelin and Falvey, 1971) or urea (Petermann, 1971). The subunits are then separated by sucrose density-gradient centrifugation (McConkey, 1967). The dissociation of ribosomes into subunits is reversible. The gross structure of a prokaryotic (bacterial) ribosome is shown in Figure 1. The structure of a typical eukaryote ribosome is very

JULIAN DAVIES—Department of Biochemistry, College of Agricultural and Life Sciences, University of Wisconsin, Madison, Wisconsin.

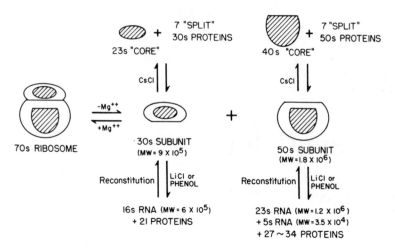

Figure 1. The components of the bacterial (prokaryotic) ribosome. See text for further discussion.

similar; however, ribosomes from prokaryotes and eukaryotes differ in size. The small subunit from eukaryotes is a $40S$ particle which contains one 16–$18S$ RNA molecule (7×10^5 daltons) (Attardi and Amaldi, 1970) and approximately 27 ribosomal proteins (Welfle *et al.*, 1971). The large subunit contains a $5S$ RNA, a 25–$28S$ RNA (1.3–1.7×10^6 daltons) (Attardi and Amaldi, 1970), and approximately 34 ribosomal proteins (Welfle *et al.*, 1971). Estimates of the number of ribosomes per cell vary, and the number changes as a function of growth rate (Maaløe and Kjeldgaard, 1966). In *Escherichia coli* there are approximately 18,000 ribosomes per cell (Goldstein *et al.*, 1964). There is much evidence that protein synthesis takes place on functional complexes of mRNA with two or more ribosomes, called polyribosomes (polysomes); in growing cells the majority of the ribosomes are in the form of polysomes. In eukaryotes, ribosomes are believed to be attached to the endoplasmic reticulum or some other membrane by the large ribosomal subunit (Sabatini *et al.*, 1966; Baglioni *et al.*, 1971). In prokaryotes, there is weak evidence for the attachment of ribosomes to membranes (Schlessinger, 1963; Schaechter, 1963), but it is not known whether this is an absolute requirement for active protein synthesis.

Ribosome Components

RNA. Ribosomal RNA's can be extracted with phenol (Kirby, 1968) or lithium chloride (Kruh, 1967) and purified by sucrose density-gradient centrifugation (McConkey, 1967) or by polyacrylamide gel electrophoresis (Loening, 1970; Dingman and Peacock, 1968). Physical studies have been carried out on ribosomal RNA's and they have been shown to possess substantial helical regions which are apparently retained in this form in the complete ribosome (Cotter *et al.*, 1967; Sarkar *et al.*, 1967). The GC (guanine plus cytosine) contents of ribosomal RNA's are species specific with wide variations in base composition (Attardi and Amaldi, 1970). Ribosomal RNA's also contain a number of methylated bases which vary between species (Attardi and Amaldi, 1970; Nomura, 1970). Partial sequences of ribosomal RNA's have been determined by the "fingerprint" method of Sanger and co-workers (Sanger *et al.*, 1965; Fellner and Sanger, 1968; Fellner, 1969; Fellner *et al.*, 1970). Certain 5*S* RNA's from prokaryotes and eukaryotes have been completely sequenced and show interesting similarities among themselves and with tRNA sequences (Brownlee *et al.*, 1967; Forget and Weissmann, 1967). The role of 5*S* RNA in ribosome function is unknown (Monier *et al.*, 1969), although it is required for the reconstitution of active 50*S* subunits (Nomura and Erdmann, 1970). It is also unclear whether 16*S* and 23*S* RNA have any role other than as skeletons which support the ribosomal proteins in correct conformation.

   Proteins. Ribosomes contain a large number of discrete proteins; the proteins can be removed from the RNA by acetic acid or lithium chloride extraction, and separated into individual components by ion-exchange chromatography or polyacrylamide gel electrophoresis (Kaltschmidt *et al.*, 1967; Hardy *et al.*, 1969; Kurland *et al.*, 1971). Two-dimensional polyacrylamide electrophoresis has been applied to resolve all of the ribosomal proteins (Kaltschmidt and Wittmann, 1970*a,b*). Several research groups have studied the ribosomal proteins of the 30*S* subunit of *E. coli* and correlations of the different numbering systems are given in Table 1 (Wittmann *et al.*, 1971). This table also shows the distribution of the proteins between "core" and "split" fractions, and their function in protein synthesis and RNA binding (Nomura *et al.*, 1969; Nomura, 1970). The complete catalog of 50*S* ribosomal proteins is not yet available. However, there are no proteins common to the 30*S* and 50*S* subunits (Kurland, 1970, 1972; Strnad and Sypherd, 1969; Traut *et al.*, 1969; Kaltschmidt and Wittmann, 1970*b*). Amino acid compositions (Kaltschmidt and Wittmann, 1970*a*) and accurate molecular weights (Kurland *et al.*, 1969; Craven *et al.*, 1969) are known for many of the *E. coli* ribosomal proteins, and there is a strong indication that 30*S* ribosome subunits are heterogeneous. Some of the proteins are found in one copy per particle, while others are fractional (*Kurland et al.*, 1969; Kurland, 1970, 1972; Voynow and Kurland, 1971). Most of the 50*S* ribosomal proteins are found in stoichiometric amounts, with the exception of one protein which is present in *more* than one copy per 50*S* subunit

TABLE 1. Proteins of the 30S Ribosomal Subunit of Escherichia coli

B	Numbering systems[a] U	G	M	Fraction[b]	Assembly of 30S particle[c]	Function in protein synthesis[d]	Binding to 16S RNA	Phenotype associated with altered protein[e]
S1	1	13	P1	S	—	—	—	—
S2	4a	11	P2	S	—	f	—	—
S3	9(+5)	10b	P3	S	a	F	+	str^Rd, ram
S4	10	9	P4a	C	A	F	—	str^Rd, spc^r
S5	3	8a	P4	S	a	f	—	—
S6	2	10a	P3b+P3c	C	—	—	+	—
S7	8	7	P5	C	A	F	+	K-character
S8	2a	8b	P4b	C	A	F	—	—
S9	12	5	P8	S	A	F	—	—
S10	4	6	P6	S	a	F	—	—
S11	11	4c	P7	C	a	F	—	—
S12	15	—	P10	C	—	F	+	str^r, str^d
S13	15b	—	P10a	C	—	f	—	—
S14	12b	—	P11	S	a	F	—	—
S15	14	4b	P10b	C	—	—	+	—
S16	6	4a	P9	C	A	F	—	—
S17	7	3a	—	—	—	—	—	—
S18	12a	2b	P12	C	—	f	·	—
S19	13	2a	P13	C	a	F	+	—
S20	16	1	P14	C	—	f	—	—
S21	15a	0	P15	C	—	f	—	—

[a] B = Berlin (Wittmann), U = Uppsala (Kurland), G = Geneva (Traut), M = Madison (Nomura).
[b] S = split, C = core (see Figure 1).
[c] a, gives 27–28S particles; A, gives 20–25S particles.
[d] f = weak requirement, F = strong requirement.
[e] See Table 2.

(Traut *et al.*, 1969; Mora *et al.*, 1971). The notion of ribosome heterogeneity in bacteria is supported by studies of ribosome composition in bacteria grown under different nutrient conditions (Deusser, 1972). No other bacterial ribosomes have been studied in the same detail as those of *E. coli;* comparative studies of ribosomal proteins of various bacterial species have been carried out and considerable variation has been seen (Waller, 1964; Wittmann, 1970; Otaka *et al.*, 1968; Osawa *et al.*, 1971). Even in closely related genera such as *E. coli* and *Salmonella typhosa*, or related species such as *Salmonella typhosa*, and *Salmonella typhimurium* there are at least four differences in proteins on the 30*S* subunit and a number of different 50*S* ribosomal proteins (Sypherd *et al.*, 1969; Dekio *et al.*, 1970; O'Neil and Sypherd, 1971).

The ribosomal proteins of eucaryotes have not been studied as isolated pure proteins, but proteins from different species have been compared by polyacrylamide gel electrophoresis (Wittmann, 1970); proteins from various organs of the same animal and from taxonomically related species were very similar. Eukaryotic ribosomal proteins can be phosphorylated *in situ;* the function of this modification is unknown (Kabat, 1970; Walton *et al.*, 1971). None of the isolated ribosomal proteins have any demonstrable enzymatic activity.

Cations. Ribosomes require magnesium ions or polyamines (Cohen and Lichtenstein, 1960) (spermidine or putrescine) for maintenance of their structure. However, the exact role of these cations is unclear.

The Function of Ribosomes in Protein Synthesis

All events of protein synthesis take place on ribosomes, with the exception of aminoacyl-tRNA formation and aminoacyl-tRNA-transfer factor complex formation (Lengyel and Söll, 1969; Lipmann, 1969; Nomura, 1970; Schreiber, 1971). These events are:

1. Messenger RNA binding, which involves the ribosome-bound initiation factors
2. N-formylmethionyl-tRNA binding which also requires ribosome-bound initiation factors and guanosine triphosphate
3. Aminoacyl-tRNA binding at a ribosome binding site
4. Peptide-bond formation, which is mediated by the ribosome-bound (50*S*) enzyme, peptidyl transferase
5. Translocation, requiring a ribosome-bound translocation factor (G factor)
6. Polypeptide-chain release, requiring additional ribosome-bound protein factors

This sequence of events is believed to involve only two specific ribosome binding sites (A and P) for aminoacyl-tRNA and peptidyl-tRNA. An alternative model which does not require these sites has been presented (Woese, 1970). It has been found from reconstitution studies that a number of ribosomal proteins may have specific roles in these reactions (see Table 1) (Nomura *et al.*, 1969). The two subunits have separate roles in the protein synthetic process; for example, peptidyl transferase and the translocation factor are associated with the large (50S) subunit, while the initiation steps take place on the small (30S) subunit. The ribosome can be shown to take part in protein synthesis in a cyclical process; the ribosome "cycle" in which the 30S particle is involved in the initiation–recognition process, and the 50S subunit joins subsequently. The 70S particle is found on polyribosomes during the formation of the protein chain, but on completion and release of the protein, the 70S ribosome dissociates into 30S and 50S subunits (Guthrie and Nomura, 1968; Kaempfer and Meselson, 1969; Mangiarotti and Schlessinger, 1967; Nomura *et al.*, 1967; Davis, 1971). There is evidence that the ribosome undergoes subtle conformational changes during the process of protein synthesis (Schreier and Noll, 1971; Chuang and Simpson, 1971). The role of the ribosome in determining the fidelity of the codon–anticodon interaction is well established (Gorini, 1969).

Ribosome Assembly *in Vivo*

In bacterial systems, kinetic studies (Britten and McCarthy, 1962; Osawa, 1965, 1968; Mangiarotti and Schlessinger, 1967) on the flow of radioactive intermediates into ribosomes have indicated that the two ribosome subunits are synthesized by two separate pathways and involve a number of discrete ribonucleoprotein intermediates (Figure 2). The 16S and 23S ribosomal RNA's are probably made as a single transcriptional unit which is cleaved immediately after, or during, its synthesis. There is evidence which suggests that the 30S subunit may have some role in the assembly of the 50S subunit (Guthrie *et al.*, 1969*b*; Nashimoto and Nomura, 1970; Krieder and Brownstein, 1971); this evidence comes from mutants which accumulate ribosomal subunit precursors.

By contrast, in eukaryotic cells, the ribosomes are derived from a common polycistronic precursor RNA (45S RNA) (Figure 3), which becomes a large ribonucleoprotein particle before conversion to intermediates involved in 40S- and 60S-particle formation; this process take place in the nucleolus. The completed ribosomal subunits are subsequently transported out of the nucleus to the cell cytoplasm (Darnell, 1968; Maden, 1968; Warner, 1971).

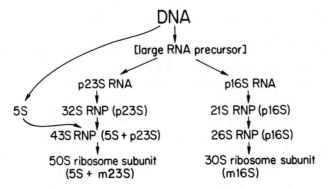

Figure 2. *Ribonucleoprotein intermediates in the formation of bacterial ribosomes. See references in text for further discussion. The numbers in brackets refer to the RNA species found in the particle; the symbol "p" refers to "precursor" and the symbol "m" to "mature" RNA. The mature RNA species differ from the precursor RNA in length (they are shorter) and in their higher content of methylated bases.*

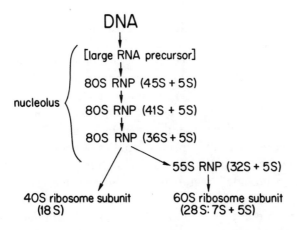

Figure 3. *Known ribonucleoprotein intermediates in the biosynthesis of eukaryote ribosomes. The numbers in brackets indicate the RNA species found in the RNP particles. The early steps occur in the nucleolus; it is not known at what stage the 40S particle (18S RNA) is formed.*

Ribosome Assembly *In Vitro*

Partial reassembly of bacterial ribosome subunits from some component proteins (split) and ribonucleoprotein (core) particles can be easily accomplished (Hosokawa *et al.*, 1966; Staehelin and Meselson, 1966; Maglott and Staehelin, 1971). More recently, the complete assembly (reconstitution) of ribosome subunits *in vitro* (Traub and Nomura, 1968*b*) has given useful clues to intermediates in assembly *in vivo*, and has also provided a way of analyzing the role of different ribosomal proteins in ribosome function in protein synthesis (Traub *et al.*, 1971). With the 30S ribosomal proteins being available as pure single species, it is possible to construct subunits lacking each protein in turn, and assays of these deficient subunits has shown the role of certain of the proteins (see Table 1) (Nomura *et al.*, 1969). Using this reconstitution system it is possible to determine those ribosomal proteins which interact directly with 16S ribosomal RNA (Table 1) (Schaup *et al.*, 1970, 1971; Mizushima and Nomura, 1970; Nomura, 1973) and to construct an assembly map of the 30S subunit which shows the order in which the proteins are believed to add during ribosome assembly in the cell.

Reconstitution of the 50S ribosomal subunits of *Bacillus stearothermophilus* and of *E. coli* (Maruta *et al.*, 1971; Nomura and Erdmann, 1970) have been accomplished, but these systems have not been used for the analysis of 50S-subunit assembly and ribosomal protein function.

Genetics of Ribosomes

The types of mutations used in genetic studies of bacterial ribosomes are shown in Table 2. There are a number of antibiotics which specifically affect protein synthesis (Weisblum and Davies, 1968; Schlessinger and Apirion, 1969; Pestka, 1971) and mutants resistant to these antibiotics have been particularly useful in genetic studies of ribosomes. These mutants have been used in biochemical studies of ribosome genes in *E. coli* and to a lesser extent in *Bacillus subtilis* (Smith *et al.*, 1969). Genetic maps of the "ribosome" region of the chromosomes of *E. coli* and *B. subtilis* are shown in Figure 4. Altered ribosomal proteins have been recognized in some of these mutants (see Table 2). Diploids heterozygous for various antibiotic loci have been studied; in general, antibiotic sensitivity is dominant over resistance (Sparling *et al.*, 1968).

Species differences have been used to study ribosomal proteins, and the differences between *E. coli* K12 and B strains have been shown to be due to a single protein change (the K protein) (see Table 1) (Leboy *et al.*, 1964; Mayuga *et al.*, 1968; Birge *et al.*, 1969; Sypherd, 1969). Inter-

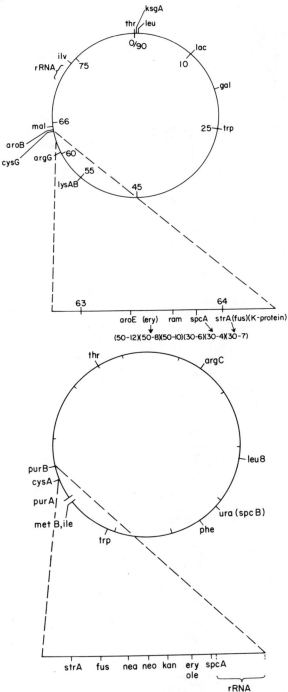

Figure 4. The "ribosome" genes of (a) E. coli and (b) B. subtilis (Davies and Nomura, 1972). Reproduced, with permission from "The Genetics of Bacterial Ribosomes," Annual Review of Genetics, Volume 6, pages 212 and 213, Copyright 1972 by Annual Reviews Inc. All rights reserved.

TABLE 2.   Mutations

	Type	Genetic symbol[a]	Ribosomal subunit alteration
Conditional lethals:	Cold sensitivity	*sad* (subunit assembly defective)	30S or 50S
	Heat sensitivity (a)		30S
	Heat sensitivity (b)		not known
Antibiotic response:	Streptomycin resistance	*str*^{r}	30S
	Streptomycin dependence	*str*^{d}	30S
	Revertants of streptomycin dependence	*str*^{R}*d*	30S
	Spectinomycin resistance	*spc*^{r}	30S
	Neamine resistance	*nea*^{r}	30S?
	Neomycin-kanamycin resistance	*nek*^{r}	30S?
	Kanamycin resistance	*kan*^{r}	30S
	Erythromycin resistance	*ery*^{r}	50S
	Lincomycin resistance	*lin*^{r}	50S?
	Bryamycin (thiostrepton) resistance	*bry*^{r}	50S
	Micrococcin resistance	*mic*^{r}	?
	Lincomycin sensitivity	*lir*	50S
Ribosome function:	Ribosomal ambiguity	*ram*	30S
Ribosomal components:	K protein		30S

[a]For descriptions of the genetic symbols and phenotypes, see references cited.

Affecting Bacterial Ribosomes

Phenotype[a]	Reference
30S and 50S assembly; failure to complete at low temperature	Guthrie et al., 1969b; Tai et al., 1969
Inability to support protein synthesis at high temperature (43°C)	Kang, 1970
High-temperature suppression of nonsense	Phillips et al., 1969; Apirion et al., 1969; Flaks et al., 1966
Resistance to streptomycin	Hashimoto, 1959; Breckenridge and Gorini, 1970; Traub and Nomura, 1968c; Ozaki et al., 1969.
Requirement for streptomycin or organic solvents	Hashimoto, 1959; Momose and Gorini, 1971; Birge and Kurland, 1969
Eliminate requirement for streptomycin in str^d strain, assembly defective	Hashimoto, 1959; Brownstein and Lewandowski, 1967; Apirion et al., 1969; Birge and Kurland, 1970; Kreider and Brownstein, 1971; Deusser et al., 1970; Stöffler et al., 1971
Resistance to spectinomycin, cold sensitivity	Davies et al., 1965; Anderson, 1969; Bollen et al., 1969a,b; Dekio and Takata, 1969; Nashimoto and Nomura, 1970
Resistance to neamine	Goldthwaite et al., 1970; Weisblum and Davies, 1968
Coresistance to neomycin and kanamycin	Apirion and Schlessinger, 1968
Resistance to kanamycin	Goldthwaite et al., 1970; Masukawa et al., 1968
Resistance to erythromycin	Apirion, 1967; Lai and Weisblum, 1971; Tanaka et al., 1968; Takata et al., 1970
Resistance to lincomycin	Apirion, 1967; Chang and Weisblum, 1966
Resistance to bryamycin	Goldthwaite et al., 1970; Smith et al., 1969
Resistance to micrococcin	Goldthwaite et al., 1970
Sensitivity to lincomycin, erythromycin	Krembel and Apirion, 1968
Suppression of nonsense mutants, suppression of str^d	Rosset and Gorini, 1969; Zimmermann et al., 1971; Gorini, 1969; Bjare and Gorini, 1971
Difference in ribosomal protein between K12 and B strains of E. coli	Leboy et al., 1964; Mayuga et al., 1968; Birge et al., 1969; Sypherd, 1969

generic mating has allowed the construction of hybrid *E. coli/Salmonella typhimurium* strains which have hybrid ribosomal proteins; there are at least six differences in ribosomal proteins between the two genera. Genetic studies with the hybrid strains have confirmed the presence of a cluster of ribosomal proteins near the *str* locus in *E. coli* (Dekio *et al.*, 1970; O'Neil and Sypherd, 1971).

To date, no mutants of ribosomal RNA have been found; since there are 6–10 copies of the 16*S* and 23*S* ribosomal RNA genes in *E. coli* (Yankofsky and Spiegelman, 1963), it is not surprising that such mutants have not been detected. Even greater redundancy is found for the ribosomal RNA genes of eukaryotes (Ritossa *et al.*, 1966; Schweizer *et al.*, 1969). The ribosomal RNA genes in bacteria are clustered away from the ribosomal protein genes (minute 64) and have been mapped near minute 74 on the *E. coli* genetic map, using DNA–RNA hybridization techniques (Yu *et al.*, 1970; Birnbaum and Kaplan, 1971). By contrast, similar studies in *B. subtilis* (Smith *et al.*, 1968) have suggested that the ribosomal RNA genes (5*S*, 16*S*, and 23*S*) map as part of a large cluster which also includes the ribosomal protein genes (Goldthwaite *et al.*, 1970; Harford and Sueoka, 1970) (See Figure 4b). The genetics of ribosomes has recently been reviewed (Davies and Nomura, 1972).

The genetics of ribosomes of higher organisms is less well known. A number of antibiotic-resistance characters are inherited cytoplasmically in *Neurospora, Saccharomyces,* and *Chlamydomonas,* and are believed to be mutations in ribosomes of mitochondria or chloroplasts (Wilkie, 1970). In *Saccharomyces cerevisiae* mutation to cycloheximide has been characterized and mapped (Mortimer and Hawthorne, 1966). Cycloheximide resistance is currently the only known mutation affecting a ribosomal protein of the large subunit (60*S*) of the eukaryotic ribosome (Rao and Grollman, 1967; Cooper *et al.*, 1967; Jimenez *et al.*, 1972). Several mutants of *S. cerevisiae* have been obtained which are temperature sensitive for ribosome synthesis (Hartwell *et al.*, 1970).

Literature Cited

Anderson, P., 1969   Sensitivity and resistance to spectinomycin in *Escherichia coli. J. Bacteriol.* **100**:939–947.

Apirion, D., 1967   Three genes that affect *Escherichia coli* ribosomes. *J. Mol. Biol.* **30**:255–275.

Apirion, D. and D. Schlessinger, 1968   Coresistance to neomycin and kanamycin by mutations in an *Escherichia coli* locus that affects ribosomes. *J. Bacteriol.* **96**:768–776.

Apirion, D., S. L. Phillips and D. Schlessinger, 1969 Approaches to the genetics of *Escherichia coli* ribosomes. *Cold Spring Harbor Symp. Quant. Biol.* **34**:117–128.

Attardi, G. and F. Amaldi, 1970 Structure and synthesis of ribosomal RNA. *Annu. Rev. Biochem.* **39**:183–219.

Baglioni, C., I. Bleiberg and M. Zauderer, 1971 Assembly of membrane-bound polyribosomes. *Nature (Lond.)* **232**:8–11.

Birge, E. A. and C. G. Kurland, 1969 Altered ribosomal protein in streptomycin-dependent *Escherichia coli. Science (Wash., D.C.)* **166**:1282–1286.

Birge, E. A. and C. G. Kurland, 1970 Reversion of a streptomycin dependent strain of *Escherichia coli. Mol. Gen. Genet.* **109**:356–369.

Birge, E. A., G. R. Craven, S. J. S. Hardy, C. G. Kurland and P. Voynow, 1969 Structure determinant of a ribosomal protein: K locus. *Science (Wash., D.C.)* **164**:1285–1286.

Birnbaum, L. S. and S. Kaplan, 1971 Localization of a protein of the ribosomal RNA genes in *Escherichia coli. Proc. Natl. Acad. Sci. USA* **68**:925–929.

Bjare, U. and L. Gorini, 1971 Drug dependence reversed by a ribosomal ambiguity mutation, *ram,* in *Escherichia coli. J. Mol. Biol.* **57**:423–435.

Bollen, A., J. Davies, M. Ozaki and S. Mizushima, 1969*a* Ribosomal protein conferring sensitivity to the antibiotic spectinomycin in *Escherichia coli. Science (Wash., D.C.)* **165**:85–86.

Bollen, A., T. Helser, T. Yamada and J. E. Davies, 1969*b* Altered ribosomes in antibiotic-resistant mutants of E. coli. *Cold Spring Harbor Symp. Quant. Biol.* **34**:95–100.

Breckenridge, L. and L. Gorini, 1970 Genetic analysis of streptomycin resistance in *Escherichia coli. Genetics* **65**:9–25.

Britten, R. J. and B. J. McCarthy, 1962 The synthesis of ribosomes in *E. coli.* II. Analysis of the kinetics of tracer incorporation in growing cells. *Biophys. J.* **2**:49–55.

Brownlee, G. G., F. Sanger and B. G. Barrell, 1967 Nucleotide sequence of 5S-ribosomal RNA from *Escherichia coli. Nature (Lond.)* **215**:735–736.

Brownstein, B. L. and L. J. Lewandowski, 1967 A mutation suppressing streptomycin dependence. I. An effect on ribosome function. *J. Mol. Biol.* **25**:99–109.

Chang, F. N. and B. Weisblum, 1966 The specificity of lincomycin binding to ribosomes. *Biochemistry* **8**:836–843.

Chuang, D.-M. and M. V. Simpson, 1971 A translocation-associated ribosomal conformation change detected by hydrogen exchange and sedimentation velocity. *Proc. Natl. Acad. Sci. USA* **68**:1474–1478.

Cohen, S. S. and J. Lichtenstein, 1960 Polyamines and ribosome structure. *J. Biol. Chem.* **235**:2112–2116.

Cooper, D., D. V. Banthorpe and D. Wilkie, 1967 Modified ribosomes conferring resistance to cycloheximide in mutants of *Saccharomyces cerevisiae. J. Mol. Biol.* **26**:347–349.

Cotter, R. I., P. McPhie and W. B. Gratzer, 1967 Internal organization of the ribosome. *Nature (Lond.)* **216**:864–868.

Craven, G. R., P. Voynow, S. J. S. Hardy and C. G. Kurland, 1969 The ribosomal proteins of *Escherichia coli.* II. Chemical and physical characterization of the 30S ribosomal proteins. *Biochemistry* **8**:2906–2915.

Darnell, J. E., Jr., 1968 Ribonucleic acids from animal cells. *Bacteriol. Rev.* **32**:262–290.

Davies, J. E. and M. Nomura, 1972   The genetics of bacterial ribosomes. *Annu. Rev. Genet.* **6**:203–234.

Davies, J., P. Anderson and B. D. Davies, 1965   Inhibition of protein synthesis by spectinomycin. *Science (Wash., D.C.)* **149**:1096–1098.

Davis, B. D., 1971   Role of subunits in the ribosome cycle. *Nature (Lond.)* **231**:153–157.

Dekio, S. and R. Takata, 1969   Genetic studies of the ribosomal proteins is *Escherichia coli.* II. Altered 30*S* ribosomal protein component specific to spectinomycin resistant mutants. *Mol. Gen. Genet.* **105**:219–224.

Dekio, S., R. Takata and S. Osawa, 1970   Genetic studies of ribosomal proteins in *Escherichia coli.* VI. Determination of chromosomal loci for several ribosomal protein components using a hybrid strain between *Escherichia coli* and *Salmonella typhimurium. Mol. Gen. Genet.* **109**:131–141.

Deusser, E., 1972   Heterogeneity of ribosomal populations in *Escherichia coli* cells grown in different media. *Mol. Gen. Genet.* **119**:249–258.

Deusser, E., G. Stöffler, H. G. Wittman and D. Apirion, 1970   Ribosomal proteins. XVI. Altered S4 proteins in *Escherichia coli* revertants from streptomycin dependence to independence. *Mol. Gen. Genet.* **109**:298–302.

Dingman, C. W. and A. C. Peacock, 1968   Analytical studies on nuclear ribonucleic acid using polyacrylamide gel electrophoresis. *Biochemistry* **7**:659–668.

Fellner, P., 1969   Nucleotide sequences from specific areas of the 16*S* and 23*S* ribosomal RNA's of E. coli. *Eur. J. Biochem.* **11**:12–27.

Fellner, P. and F. Sanger, 1968   Sequence analysis of specific areas of the 16*S* and 23*S* ribosomal RNA's *Nature (Lond.)* **219**:236–238.

Fellner, P., C. Ehresmann and J. P. Ebel, 1970   Nucleotide sequences present within the 16*S* ribosomal RNA of *E. coli. Nature (Lond.)* **225**:26–29.

Flaks, J. G., P. S. Leboy, E. A. Birge and C. G. Kurland, 1966   Mutations and genetics concerned with the ribosome. *Cold Spring Harbor Symp. Quant. Biol.* **31**:623–631.

Forget, B. G. and S. M. Weissmann, 1967   Nucleotide suequence of KB cell 5*S* RNA. *Science (Wash., D.C.)* **158**:1695–1699.

Goldstein, A., D. B. Goldstein and L. I. Lowney, 1964   Protein synthesis at 0°C in *Escherichia coli. J. Mol. Biol.* **9**:213–235.

Goldthwaite, C., D. Dubnau and I. Smith, 1970   Genetic mapping of antibiotic resistance markers in *Bacillus subtilis. Proc. Natl. Acad. Sci. USA* **65**:96–103.

Gorini, L. C., 1969   The contrasting role of *str*A and *ram* gene products in ribosome functioning. *Cold Spring Harbor Symp. Quant. Biol.* **32**:101–111.

Gorini, L. C., G. A. Jacoby and L. Breckenridge, 1966   Ribosomal ambiguity. *Cold Spring Harbor Symp. Quant. Biol.* **31**:657–664.

Guthrie, C. and M. Nomura, 1968   Initiation of protein synthesis: A critical test of the 30*S* subunit model. *Nature (Lond.)* **219**:232–235.

Guthrie, C., H. Nashimoto and M. Nomura, 1969*a*   Studies on the assembly of ribosomes *in vivo. Cold Spring Harbor Symp. Quant. Biol.* **34**:69–75.

Guthrie, C., H. Nashimoto and M. Nomura, 1969*b*   Structure and function of *E. coli* ribosomes. VIII. Cold-sensitive mutants defective in ribosome assembly. *Pro. Natl. Acad. Sci. USA* **63**:384–391.

Hardy, S. J. S., C. G. Kurland, P. Voynow and G. Mora, 1969   The ribosomal proteins of *Escherichia coli.* I. Purification of the 30*S* ribosomal proteins. *Biochemistry* **8**: 2897–2905.

Harford, N. and N. Sueoka, 1970   Chromosomal location of antibiotic resistance markers in *Bacillus subtilis. J. Mol. Biol.* **51**:267–286.

Hartwell, L. H., C. S. McLaughlin and J. R. Warner, 1970   Identification of ten genes that control ribosome formation in yeast. *Mol. Gen. Genet.* **109**:42–56.

Hashimoto, K., 1969   Streptomycin resistance in *Escherichia coli* analyzed by transduction. *Genetics* **45**:49–62.

Hill, W. E., G. P. Rossetti and K. E. Van Holde, 1969   Physical studies of ribosomes from *Escherichia coli*. *J. Mol. Biol.* **44**:263–277.

Hosokawa, K., R. K. Fujimura and M. Nomura, 1966   Reconstitution of functionally active ribosomes from inactive subparticles and proteins. *Proc. Natl. Acad. Sci. USA* **55**:198–204.

Jimenez, A., B. Littlewood and J. Davies, 1972   Inhibition of protein synthesis in yeast. In *Proceedings of Symposium on Inhibitors of Protein Synthesis and Membranes, Granada*, Elsevier, Amsterdam 292–306.

Kabat, D., 1970   Phosphorylation of ribosomal proteins in rabbit reticulocytes. Characterization and regulatory aspects. *Biochemistry* **9**:4160–4174.

Kaempfer, R. and M. Meselson, 1969   Studies of ribosomal subunit exchange. *Cold Spring Harbor Symp. Quant. Biol.* **34**:209–222.

Kaltschmidt, E. and H. G. Wittmann, 1970a   Ribosomal proteins. VII. Two-dimensional polyacrylamide gel electrophoresis for fingerprinting of ribosomal proteins. *Anal. Biochem.* **36**:401–412.

Kaltschmidt, E. and H. G. Wittmann, 1970b   Ribosomal proteins. XII. Number of proteins in small and large ribosomal subunits of *Escherichia coli* as determined by two-dimensional gel electrophoresis. *Proc. Natl. Acad. Sci. USA* **67**:1276–1282.

Kaltschmidt, E., M. Dzionara, D. Donner and H. G. Wittmann, 1967   Ribosomal proteins. I. Isolation, amino acid composition, molecular weights, and peptide mapping of proteins from *E. coli* ribosomes. *Mol. Gen. Genet.* **100**:364–373.

Kang, S. -S., 1970   A mutant of *Escherichia coli* with temperature-sensitive streptomycin protein. *Proc. Natl. Acad. Sci. USA* **65**:544–550.

Kirby, K. S., 1968   Isolation of nucleic acids with phenolic solvents *Methods Enzymol.* **12**:87–99.

Kreider, G. and B. L. Brownstein, 1971   A mutation suppressing streptomycin dependence: An altered protein on the 30*S* ribosomal subunit. *J. Mol. Biol.* **61**:135–142.

Krembel, J. and D. Apirion, 1968   Changes in ribosomal proteins associated with mutants in a locus that affects *Escherichia coli* ribosomes. *J. Mol. Biol.* **33**:363–368.

Kruth, J., 1967   Preparation of RNA from rabbit reticulocytes and liver. *Methods Enzymol.* **12**:609–613.

Kurland, C. G., 1970   Ribosome structure and function emergent. *Science (Wash., D.C.)* **169**:1171–1177.

Kurland, C. G., 1971   Purification of ribosomes from *Escherichia coli*. *Methods Enzymol.* **20**:379–381.

Kurland, C. G., 1972   Structure and function of the bacterial ribosome. *Annu. Rev. Biochem.* **41**:377–408.

Kurland, C. G., P. Voynow, S. J. S. Hardy, L. Randall and L. Lutter, 1969   Physical and functional heterogeneity of *E. coli* ribosomes. *Cold Spring Harbor Symp. Quant. Biol.* **34**:17–24.

Kurland, C. G., S. J. S. Hardy and G. Mora, 1971   Purification of ribosomal proteins from *Escherichia coli*. *Methods Enzymol.* **20**:381–391.

Lai, C. J. and B. Weisblum, 1971   Altered methylation of ribosomal RNA in an erythromycin-resistant strain of *Staphylococcus aureus*. *Proc. Natl. Acad. Sci. USA* **68**:856–860.

Leboy, P. S., E. C. Cox and J. G. Flaks, 1964   The chromosomal site specifying a ribosomal protein in *Escherichia coli. Proc. Natl. Acad. Sci. USA* **52**:1367–1374.

Lengyel, P. and D. Söll, 1969   Mechanism of protein biosynthesis. *Bacteriol. Rev.* **33**:264–301.

Lipmann, F., 1969   Polypeptide chain elongation in protein biosynthesis. *Science (Wash., D.C.)* **164**:1024–1031.

Loening, U. E., 1970   The mechanism of synthesis of ribosomal RNA. *Symp. Soc. Gen. Microbiol.* **20**:77–106.

Maaløe, O. and N. O. Kjeldgaard, 1966   *Control of Macromolecular Synthesis*, W. A. Benjamin, New York.

McConkey, E. H., 1967   The fractionation of RNA's by sucrose gradient centrifugation. *Methods Enzymol.* **12**:620–634.

Maden, B. E. H., 1968   Ribosome formation in animal cells. *Nature (Lond.)* **219**:685–689.

Maglott, D. R. and T. Staehelin, 1971   Fractionation of *Escherichia coli* 50*S* ribosomes into various protein-deficient cores and split protein fractions by CsCl density gradient centrifugation and reconstitution of active particles. *Methods Enzymol.* **20**:408–417.

Mangiarotti, G. and D. Schlessinger, 1966   Polyribosome metabolism in *Escherichia coli. J. Mol. Biol.* **20**:123–143.

Mangiarotti, G. and D. Schlessinger, 1967   Polyribosome metabolism in *Escherichia coli.* II. Formation and lifetime of messenger RNA molecules, ribosomal subunit couples and polyribosomes. *J. Mol. Biol.* **29**:395–418.

Martin, T. E., I. G. Wool and J. J. Castles, 1971   Dissociation and reassociation of ribosomes from eukaryotic cells. *Methods Enzymol.* **20**:417–429.

Maruta, H., T. Tauchiya and D. Mizuno, 1971   *In vitro* reassembly of functionally active 50*S* ribosomal particles from ribosomal proteins and RNA's of *Escherichia coli. J. Mol. Biol.* **61**:123–234.

Masukawa, H., N. Tanaka and H. Umezawa, 1968   Localization of kanamycin sensitivity in the 23*S* core of 30*S* ribosomes of *Escherichia coli. J. Antibiot. (Tokyo) Ser. A.* **21**:517–519.

Mayuga, C., D. Meier and T. Wang, 1968   *Escherichia coli:* The K12 ribosomal protein and the streptomycin region of the chromosome. *Biochem. Biophys. Res. Commun.* **33**:203–206.

Mizushima, S. and M. Nomura, 1970   Assembly mapping of 30*S* ribosomal proteins from *Escherichia coli. Nature (Lond.)* **226**:1214–1218.

Momose, H. and L. Gorini, 1971   Genetic analysis of streptomycin dependence in *Escherichia coli. Genetics* **67**:19–38.

Monier, R., J. Feunteun, B. Forget, B. Jordon, M. Reynier and F. Varricchio, 1969   5*S* RNA and the assembly of bacterial ribosomes. *Cold Spring Harbor Symp. Quant. Biol.* **34**:139–148.

Mora, G., D. Donner, P. Thammana, L. Lutter, C. G. Kurland and G. R. Craven, 1971   Purification and characterization of 50*S* ribosomal proteins of *Escherichia coli. Mol. Gen. Genet.* **112**:229–242.

Mortimer, R. K. and D. C. Hawthorne, 1966   Genetic mapping in *Saccharomyces. Genetics* **53**:165–173.

Nashimoto, H. and M. Nomura, 1970   Structure and function of bacterial ribosomes. XI. Dependence of 50*S* ribosomal assembly on simultaneous assembly of 30*S* subunits. *Proc. Natl. Acad. Sci. USA* **67**:1440–1447.

Nashimoto, H., W. Held, E. Kaltschmidt and M. Nomura, 1971   Structure and function of bacterial ribosomes. XII. Accumulation of 21S particles by some cold-sensitive mutants of *Escherichia coli. J. Mol. Biol.* **62:**121–138.

Nomura, M., 1970   Bacterial ribosome. *Bacteriol. Rev.* **34:**228–277.

Nomura, M., 1973   Assembly of bacterial ribosomes. *Science (Wash., D.C.)* **179:**864–873.

Nomura, M. and V. A. Erdmann, 1970   Reconstitution of 50S ribosomal subunits from dissociated molecular components. *Nature (Lond.)* **228:**744–748.

Nomura, M., C. V. Lowry and C. Guthrie, 1967   The initiation of protein synthesis: joining of the 50S ribosomal subunit to the initiation complex. *Proc. Natl. Acad. Sci. USA* **88:**1487–1493.

Nomura, M., S. Mizushima, M. Ozaki, P. Traub and C. V. Lowry, 1969   Structure and function of ribosomes and their molecular components. *Cold Spring Harbor Symp. Quant. Biol.* **34:**49–61.

O'Neil, D. M. and P. S. Sypherd, 1971   Cotransduction of strA and ribosomal protein cistrons in *Escherichia coli–Salmonella typhimurium* hybrids. *J. Bacteriol.* **105:**947–956.

O'Neil, D. M., L. S. Baron and P. S. Sypherd, 1969   Chromosomal location of ribosomal protein cistrons determined by intergeneric bacterial mating. *J. Bacteriol.* **99:**242–247.

Osawa, S., 1965   Biosynthesis of ribosomes in bacterial cells. *Prog. Nucleic Acid Res. Mol. Biol.* **4:**161–188.

Osawa, S., 1968   Ribosome formation and structure. *Annu. Rev. Biochem.* **37:**109–130.

Osawa, S., T. Itoh and E. Otaka, 1971   Differentiation of the ribosomal protein compositions in the genes *Escherichia* and its related bacteria. *J. Bacteriol.* **107:**168–178.

Otaka, E., T. Itoh and S. Osawa, 1968   Ribosomal proteins of bacterial cells; strain- and species-specificity. *J. Mol. Biol.* **33:**93–107.

Ozaki, M., S. Mizushima and M. Nomura, 1969   Identification and functional characterization of the protein controlled by the streptomycin-resistant locus in *E. coli. Nature (Lond.)* **222:**333–339.

Pestka, S., 1971   Inhibitors of ribosome functions. *Annu. Rev. Microbiol.* **25:**487–552.

Petermann, M. L., 1971   The dissociation of rat liver ribosomes to active subunits by urea. *Methods Enzymol.* **20:**429–433.

Phillips, S. L., D. Schlessinger and D. Apirion, 1969   Mutants in *Escherichia coli* ribosomes: a new selection. *Proc. Natl. Acad. Sci. USA* **62:**772–777.

Rao, S. and A. Grollman, 1967   Cycloheximide resistance in yeast: A property of the 60S ribosomal subunit. *Biochem. Biophys. Res. Commun.* **29:**696–704.

Ritossa, F. S., K. C. Atwood, D. L. Lindsley and S. Spiegelman, 1966   On the chromosome distribution of DNA complementary to ribosomal and soluble RNA. *Natl. Cancer Inst. Monogr.* **23:**449–471.

Rosset, R. and L. Gorini, 1969   A ribosomal ambiguity mutation. *J. Mol. Biol.* **39:**95–112.

Sabatini, D. D., Y. Tashiro and G. E. Palade, 1966   On the attachment of ribosomes to microsomal membranes. *J. Mol. Biol.* **19:**503–524.

Sanger, F., G. G. Brownlee and B. G. Barrell, 1965   A two-dimensional fractionation procedure for radioactive nucleotides. *J. Mol. Biol.* **13:**373–398.

Sarkar, P. K., J. T. Yang and P. Doty, 1967   Optical rotatory dispersion of *E. coli* ribosomes and their constituents. *Biopolymers* **5:**1–4.

Schaechter, M., 1963 Bacterial polyribosomes and their participation in protein synthesis *in vivo. J. Mol. Biol.* **7**:561–568.

Schaup, H. W., M. Green and C. G. Kurland, 1970 Molecular interactions of ribosomal components. I. Identification of RNA binding sites for individual 30*S* ribosomal proteins. *Mol. Gen. Genet.* **109**:193–205.

Schaup, H. W., M. Green and C. G. Kurland, 1971 Molecular interactions of ribosomal components. II. Site specific complex formation between 30*S* proteins and ribosomal RNA. *Mol. Gen. Genet.* **112**:1–8.

Schlessinger, D., 1963 Protein synthesis by polyribosomes on protoplast membranes of *B. megaterium. J. Mol. Biol.* **7**,569–582.

Schlessinger, D., and D. Apirion, 1969 *Escherichia coli* ribosomes: recent developments. *Annu. Rev. Microbiol.* **23**:387–426.

Schreiber, G. 1971 Translation of genetic information on the ribosome. *Angew. Chem. Int. Ed. Engl.* **10**:638–651.

Schreier, M. H., and H. Noll, 1971 Conformational changes in ribosomes during protein synthesis. *Proc. Natl. Acad. Sci. USA* **68**:805–809.

Schweizer, E., C. MacKechnie and H. O. Halvorson, 1969 The redundancy of ribosomal and transfer RNA genes in *Saccharomyces cerevisiae. J. Mol. Biol.* **40**:261–277.

Smith, I., D. Dubnau, P. Morell and J. Marmur, 1968 Chromosomal location of DNA base sequences complementary to transfer RNA and to 5*S*, 16*S* and 23*S* ribosomal RNA in *Bacillus subtilis. J. Mol. Biol.* **33**:123–140.

Smith, I., C. Goldthwaite and D. Dubnar, 1969 The genetics of ribosomes in *Bacillus subtilis. Cold Spring Harbor Symp. Quant. Biol.* **34**:85–89.

Sparling, P. F., J. Modolell, T. Takeda and B. D. Davis, 1968 Ribosomes from *Escherichia coli* merodiploids heterozygous for resistance to streptomycin and to spectinomycin. *J. Mol. Biol.* **37**:407–421.

Staehelin, T., and A. K. Falvey, 1971 Isolation of mammalian ribosomal subunits active in polypeptide synthesis. *Methods Enzymol.* **20**:433–446.

Staehelin, T., and D. R. Maglott, 1971 Preparation of *Escherichia coli* ribosomal subunits active in polypeptide synthesis. *Methods Enzymol.* **20**:449–456.

Staehelin, T., and M. Meselson, 1966 *In vitro* recovery of ribosomes and of synthetic activity from synthetically inactive ribosomal subunits. *J. Mol. Biol.* **15**:245–249.

Stöffler, G., E. Deusser, H. G. Wittman and D. Apirion, 1971 Ribosomal proteins: Altered S5 ribosomal protein in an *Escherichia coli* revertant from streptomycin dependence to independence. *Mol. Gen. Genet.* **111**:334–341.

Strnad, B. C., and P. S. Sypherd, 1969 Unique protein moieties for 30*S* and 50*S* ribosomes of *Escherichia coli. J. Bacteriol.* **98**:1080–1086.

Sypherd, P. S., 1969 Amino acid differences in a 30*S* ribosomal protein from two strains of *Escherichia coli. J. Bacteriol.* **99**:379–382.

Sypherd, P. S., D. M. O'Neil and M. M. Taylor, 1969 The chemical and genetic structure of bacterial ribosomes. *Cold Spring Harbor Symp. Quant. Biol.* **34**:77–84.

Tai, P., D. P. Kessler and J. Ingraham, 1969 Cold-sensitive mutations in *Salmonella typhimurium* which affect ribosome synthesis. *J. Bacteriol.* **97**:1298–1304.

Takata, R., S. Osawa, K. Tanaka, H. Teraoka and M. Tamaki, 1970 Genetic studies of the ribosomal proteins in *Escherichia coli:* Mapping of erythromycin resistance mutations which lead to alteration of a 50*S* ribosomal protein component. *Mol. Gen. Genet.* **109**:123–130.

Tanaka, K., H. Teraoka, M. Tamaki, E. Otaka and S. Osawa, 1968 Erythromycin-resistant mutant of *Escherichia coli* with altered ribosomal protein component. *Science (Wash., D.C.)* **162**:576–578.

Tissières, A., and J. D. Watson, 1958 Ribonucleoprotein particles from *Escherichia coli*. *Nature (Lond.)* **182**:778–780.

Tissières, A., J. D. Watson, D. Schlessinger and B. R. Hollingworth, 1959 Ribonucleoprotein particles from *Escherichia coli. J. Mol. Biol.* **1**:221–233.

Traub, P., and M. Nomura, 1968a Structure and function of *E. coli* ribosomes. I. Partial fractionation of the functionally active ribosomal proteins and reconstitution of artificial subribosomal particles. *J. Mol. Biol.* **34**:575–593.

Traub, P., and M. Nomura, 1968b Structure and function of *E. coli* ribosomes. V. Reconstitution of functionally active 30S ribosomal particles from RNA and proteins. *Proc. Natl. Acad. Sci. USA* **59**:777–784.

Traub, P., and M. Nomura, 1968c Streptomycin resistance mutation in *Escherichia coli:* Altered ribosomal protein. *Science (Wash. D.C.)* **160**:198–199.

Traub, P., S. Mizushima, C. V. Lowry and M. Nomura, 1971 Reconstitution of ribosomes from subribosomal components. *Methods Enzymol.* **20**:391–407.

Traut, R. R., H. Delius, C. Ahmad-Zadeh, T. A. Bickle, P. Pearson and A. Tissières, 1969 Ribosomal proteins of *E. coli:* Stoichiometry and implications for ribosome structure. *Cold Spring Harbor Symp. Quant. Biol.* **34**:25–38.

Voynow, P. and C. G. Kurland, 1971 Stoichiometry of the 30S ribosomal proteins of *Escherichia coli. Biochemistry* **10**:517–524.

Waller, J. P., 1964 Fractionation of the ribosomal protein from *Escherichia coli. J. Mol. Biol.* **10**:319–336.

Walton, G. M., G. W. Gill, I. B. Abrass and L. D. Garren, 1971 Phosphorylation of ribosome-associated protein by an adenosine 3′-5′-cyclic monophosphate-dependent protein kinase: Location of the microsomal receptor and protein kinase. *Proc. Natl. Acad. Sci. USA* **68**:880–884.

Warner, J. R., 1971 The assembly of ribosomes in yeast. *J. Biol. Chem.* **246**:447–454.

Watson, J. D., 1964 The synthesis of proteins from ribosomes. *Bull. Soc. Chim. Biol.* **46**:1399–1425.

Weisblum, B., and J. Davies, 1968 Antibiotic inhibitors of the bacterial ribosome. *Bacteriol. Rev.* **32**:493–528.

Welfle, H., J. Stahl and H. Bielka, 1971 Studies on proteins of animal ribosomes: two-dimensional polyacrylamide gel electrophoresis of ribosomal proteins of rat liver. *Biochim. Biophys. Acta* **243**:416–419.

Wilkie, D., 1970 Reproduction of mitochondria and chloroplasts. *Symp. Soc. Gen. Microbiol.* **20**:381–399.

Wittmann, H. G., 1970 A comparison of ribosomes from prokaryotes and eukaryotes. *Symp. Soc. Gen. Microbiol.* **20**:55–76.

Wittman, H. G., G. Stöffler, I. Hindennach, C. G. Kurland, L. Randall-Hazelbauer, E. A. Birge, M. Nomura, E. Kaltschmidt, S. Mizushima, R. R. R. Traut and T. A. Bickle, 1971 Correlation of 30S ribosomal proteins of *Escherichia coli* isolated in different laboratories. *Mol. Gen. Genet.* **111**:327–333.

Woese, C., 1970 Molecular mechanisms of translation, a reciprocating rachet mechanism. *Nature (Lond.)* **226**:817–820.

Yankofsky, S. A., and S. Spiegelman, 1963 Distinct cistrons for the two ribosomal RNA components. *Proc. Natl. Acad. Sci. USA* **49**:538–544.

Yu, M. T., C. W. Vermeulen, and K. C. Atwood, 1970 Location of the genes for 16*S* and 23*S* ribosomal RNA on the genetic map of *Escherichia coli. Proc. Natl. Acad. Sci. USA* **67**:26–31.

Zimmermann, R. A., R. T. Garvin and L. Gorini, 1971 Alteration of a 30*S* ribosomal protein accompanying the *ram* mutation in *Escherichia coli. Proc. Natl. Acad. Sci. USA* **68**:2263–2267.

13

Salmonella

MARK LEVINTHAL AND KENNETH E. SANDERSON

Introduction

Genetic analysis of microorganisms has been valuable in two ways. Information about genetic mechanisms which are fundamental to all organisms, such as the chemical structure of the genetic material and the mechanism of mutation, has been obtained. In addition to their unique and fascinating biological and chemical properties, the microorganisms are important in health and disease, in nutrient cycles in nature, and in industrial processes. The enteric group of Gram-negative bacteria, which includes genera such as *Salmonella, Escherichia,* and *Shigella* has been a favorite subject of study, for the following reasons: Their growth rate is rapid, with a generation time as low as 30 minutes or less. Cell division produces single cells, most of which are uninucleate, at least in stationary phase. The species are facultative aerobes and usually grow on a defined medium and on a wide range of carbon and energy sources. Colonies which can be readily observed and enumerated are formed on agar medium.

Within the enteric group, many studies have been done with *Salmonella* species. *Salmonella typhimurium* has been the most analyzed, primarily due to the extensive genetic studies of M. Demerec and his colleagues (for a review, see Caspari, 1971). Considerable conjugation studies have been done with *Salmonella abony* (Mäkelä, 1963; Mäkelä,

MARK LEVINTHAL—Department of Biological Sciences, Purdue University, West Lafayette, Indiana. KENNETH E. SANDERSON—Department of Biology, University of Calgary, Calgary, Alberta, Canada.

1964). Other species studied are *Salmonella pullorum* (Hoeksma and Shoenhard, 1971), *Salmonella montevideo* (Atkins and Armstrong, 1969, Mäkelä, 1966), *Salmonella minnesota* (Mäkelä *et al.,* 1970), and *Salmonella typhosa* (Johnson and Baron, 1969). Extensive genetic studies have been done with the related organism *Escherichia coli* (summarized in Taylor and Trotter, 1972), and the linkage maps of the two genera are found to be very similar (Sanderson, 1972).

A number of other publications present useful information. Hayes (1968) and Braun (1966) describe the genetics of bacteria and bacterial viruses. Methods of handling bacteria are described in Meynell and Meynell (1970). Clowes and Hayes (1968) present an outline of experiments in microbial genetics, and Burdette (1963) describes methods of basic genetics, including methods of transduction and conjugation. Roth (1972) presents an excellent summary of genetic techniques for the study of bacterial metabolism. Current summaries on relevant fields are given in Annual Reviews of Genetics, Annual Reviews of Microbiology, and Annual Reviews of Biochemistry.

General Methods for Growing, Handling, and Storing

S. typhimurium will grow in a wide variety of minimal and complex media. Recipes for several kinds of media as well as other techniques for *Salmonella* genetics were recently reviewed (Roth, 1972). Vogel and Bonner (1956) described an excellent minimal salts medium. This medium has an advantage over others because it can be prepared as a 50-fold concentrate. The minimal medium of Davis and Mingioli (1950) can be also be used. *S. typhimurium* will grow on a large variety of carbon sources (Gutnick *et al.,* 1969), but 0.2 percent glucose added to either of the above mentioned minimal salts media is customarily used. A special minimal salts medium which lacks citrate (Berkowitz *et al.,* 1968) is used to test the abilities of mutant and wild-type strains to utilize various carbon sources. Several complex indicator media have been used to test the ability of various strains to ferment carbohydrates (Levinthal, 1971; Levinthal and Simoni, 1969; Simoni *et al.,* 1967).

Nutrient broth (Difco), which is the complete media most often used, will support the growth of most auxotrophic mutants, with the exception of cysteine-requiring mutants, which need 15 μg/ml L-cysteine added to the nutrient broth for optimal growth. A complete media for any auxotroph can be devised by adding required growth factors to a minimal medium at a concentration of 15 μg/ml. Both minimal and complete media are solidified by the addition of agar to a concentration of 1.5 percent w/v. Exceptions to this agar concentration include semisoft nutrient agar (0.6 percent w/v) used for overlay in phage titrations and the semisolid nutrient gelatin agar (Enomoto, 1966) used for studying flagella mutants.

The optimal temperature for growth is 37°C, but 30°C and 42°C have been used as the permissive and nonpermissive temperatures, respectively, for the isolation of temperature sensitive mutations. Similarly, 18°C has been used as the nonpermissive temperature for the isolation of cold-sensitive mutations (Abd-el-al and Ingraham, 1969). Cultures in liquid media should be aerated either by shaking on a reciprocal incubator shaker or by bubbling air through cultures warmed in a water bath.

S. typhimurium is pathogenic for humans (Baumberg and Freeman, 1971). Care should therefore be taken not to ingest cultures. Further precautions include spreading a disinfectant on bench tops, adding disinfectant to pipet jars, and autoclaving all materials, especially plastic Petri dishes before discarding.

The strains must be maintained in a viable and genetically unaltered state. Slant cultures and stab cultures require storage on culture medium under conditions in which growth may occur, and mutants frequently accumulate. Frozen and lyophilized storage, which reduces such variability, should be used for all episome-containing strains, and is desirable in all cases. For making stab cultures, small vials (1/4 dram) containing Difco nutrient agar made at half strength are stabbed, corked, dipped in liquid paraffin to seal, and stored at room temperature. For slant cultures, slants containing nutrient agar may be stored at 4°C, but these must be transferred every 3–4 months. Dorset egg slopes (available from Difco), stored at room temperature, are satisfactory for longer periods. For frozen storage, cells from a slant are suspended in sterile saline, centrifuged, resuspended in 15 percent glycerol in aqueous saline solution, and stored at low temperature; the authors store these cells at −76°C, but some find storage at −20°C satisfactory. For lyophilized storage, a dense suspension of cells in 5 percent Difco Bacto Peptone plus 5 percent glucose is spotted onto filter-paper strips in lyophil ampules, evacuated to 30–60 μ Hg pressure, sealed under vacuum, and stored at room temperature.

Isolation of Mutants

Mutant Induction

There are many mutagens available to induce mutations in *Salmonella*. A procedure using the base analogs 2-amino purine and 5-bromodeoxyuridine has been described (Hartman *et al.*, 1965). These mutagens are mild and usually only cause single mutations. Procedures for more powerful mutagens, such as diethylsulfate (Loper *et al.*, 1964) and 1-methyl-3-nitro-1-nitrosoguanidine (Hartman *et al.*, 1971*a*) are available. These compounds should be used with caution because they are dangerous to humans, and also because they cause multiple mutations in the bacterium. Mutations isolated in response to these mutagens should be transferred to another genetic background, if possible.

Nitrous acid can be used to produce deletions (Schwartz and Beckwith, 1969). In order to select for deletion mutants, the mutagenized culture may be subjected to a double selection for mutations in adjacent genes, following nitrous acid treatment. An example of this technique is the simultaneous selection of histidine auxotrophs by penicillin and *rfb* mutants by resistance to phage P22, leading to the production of many *his–rfb* deletions (Nikaido *et al.*, 1967).

Another class of useful mutagens are the acridines which cause frameshifts (Crick *et al.*, 1961). A series of acridine half mustards was synthesized by Hugh J. Creech of the Institute for Cancer Research, and was evaluated by Ames and Whitfield (1966). A procedure for using the most mutagenic of these, ICR-191 and ICR-364, has been developed by Oeschger and Hartman (1970). A new frame-shift-producing mutagen, hycanthone, has been described (Hartman *et al.*, 1971*b*).

Reversion analysis has been used to determine the type of lesion present in a mutant strain (Whitfield *et al.*, 1966). A summary of the results obtained by application of this method to histidine-requiring mutants has recently appeared (Hartman *et al.*, 1971*a*). In addition, special strains of *Salmonella* have been developed which limit repair of genetic lesions. These strains have been incorporated into a procedure for the rapid spot test of potential mutagens (Ames, 1971). This procedure has been used to screen drugs and food additives for possible mutagenicity (see for example Zeiger and Legator, 1971). X rays, fast neutrons, nitrogen mustard, ultraviolet light, and ^{32}P decay have also been used to induce mutations (Loper *et al.*, 1964).

Mutant Selection

After mutagenesis, the mutation-bearing bacterium must be selected out from among nonmutant bacteria. The first step is to allow replication and expression of the newly induced mutation. Certain types of mutations have phenotypes which allow direct selection. That is, conditions can be arranged such that only the mutant strain will grow, or if both mutant and wild type grow, the mutant is easily recognized on a Petri plate containing several hundred colonies. Mutations conferring resistance to drugs, e.g., streptomycin, to an amino acid analog, e.g., 5-methyl tryptophan, or to a bacteriophage, e.g., a clear plaque mutant of P22, P22.*c2*, are good candidates for direct selection. Mutations leading to inability to ferment carbohydrates can be detected using special media (Levinthal and Simoni, 1969). Mutations leading to the inability to incorporate radioactive materials can be screened for by radioautography of colonies (Martin, 1968).

For most auxotrophic mutations, the method of choice is penicillin selection (Davis, 1948). The procedure is based on the observation that cells will grow in the presence of penicillin produce defective cell walls and eventually lyse. Those cells not growing (because their auxotrophic requirement is not met by the medium) survive exposure to penicillin. The technique works well to enrich for auxotrophs provided the cell density is low (about 1×10^8 cells/ml) and the penicillin concentration is correct (200 units/ml for *S. typhimurium* LT2). A variation on the Davis procedure has been described by Gorini and Kaufman (1960). Their procedure involves pregrowth in selective medium which starves auxotrophs and allows prototophs to initiate exponential growth. Then, the culture is challenged with high concentrations of penicillin for short times. Another modification described by Adelberg and Meyers (1953) allows the entire selection to take place in Petri dishes.

Following penicillin selection, the culture is plated on a complete media and replica plated (Lederberg and Lederberg, 1952) to selective media to identify auxotrophs. A general procedure for screening for auxotrophs involves replica plating to a series of amino acid pool plates. The pools are arranged so each contains different amino acids, but two pools contain the same amino acid. An example of such a scheme is given by Roth (1972). Another general screen is to plate the presumptive mutants on media containing suboptimal amounts of nutrients. Under these conditions, auxotrophs will form small colonies (Loper *et al.*, 1964).

Hong and Ames (1971) have used mutagenized phage to isolate temperature-sensitive mutations in a specified region of the chromosome. After treating a transducing lysate of P22 with hydroxylamine, they crossed the phage with suitably marked recipient, and selected prototophs at 30°C. The transductants were tested at 40°C for temperature-sensitive lesions. In this way, it was possible to obtain conditional lethal mutations which are cotransducible with genes of interest.

Methods of Genetic Analysis

General Considerations

When a new genetic mutant has been isolated, perhaps an auxotrophic mutant unable to synthesize an amino acid, the following questions are usually of interest: Are the mutant phenotype(s) due to a single genetic lesion; at what location on the linkage map is the mutation; is this mutation dominant or recessive to normal allele; is it in the same cistron as known mutations? The first question may be answered by reversion or recombination studies, to see if all phenotypic manifestations of the mutant change simultaneously. The genetic linkage is usually tested by conjugation crosses using Hfr strains to detect the general map location, followed by more precise studies by phage mediated transduction. The last two questions require formation of partial diploids, usually accomplished by abortive transduction of the use of F′ factors.

Bacterial Conjugation

The Use of F⁺ and Hfr Strains in Genetic Mapping. Conjugation, involving genetic exchange after cell-to-cell contact, and mediated by a sex factor, was first exploited in *E. coli*. The F factor of *E. coli* K12 was transferred into *S. typhimurium* (Zinder, 1960) and into *S. abony* (Mäkelä, 1963) and the Hfr strains isolated were used in genetic analysis. The mating methods used, which are very similar to those of *E. coli*, have been described for mating following replica plating, for plate mating, for broth mating, and for interrupted mating (Sanderson *et al.*, 1972). In broth mating, a 1 ml amount of the logarithmic-phase broth culture of the donor strain is mixed with 2 ml of a stationary-phase culture of the F⁻strain in 7 ml of Penassay broth, and then incubated at 37°C for 60 minutes. A 0.1 ml amount of the original mating mixture and of a 10² dilution of the mixture are plated in a soft agar overlay onto plates of selective minimal medium, which contain, when the recipient is streptomycin resistant, 600 μg/ml of streptomycin sulfate. The number of donor cells is determined by plating the donor culture, at zero time, onto nutrient agar. In interrupted conjugation, which determines the time of entry of specific genes from the Hfr into the F⁻, separation of the mated pair is accomplisehd by blending for 20 seconds in the micro jar of a Waring blender. The theory of interrupted conjugation is discussed in detail elsewhere (Hayes, 1968); numerous examples of its use in *Salmonella* genetics are found in Sanderson (1972). However, to allow interrupted conjugation analysis, a selectable allele of the gene to be studied must be located near the point of origin of a highly fertile Hfr strain. For example, auxotrophic mutants, which are unable to grow on minimal medium without appropriate supplementation, must be in F⁻ strains, and thus the prototrophic allele from an Hfr strain can be selected in time-of-entry crosses. Alternatively, mutants for resistance to antibiotics must be located in an Hfr strain near the point-of-origin of transfer.

However, frequently the mutant to be studied is not selectable, i.e., recombinants cannot be detected in a cross among the background of donor and recipient cells, or alternatively the mutant may have been isolated in an inappropriate strain. In these cases, interrupted-conjugation studies are not valuable, and classical crossover genetic analysis must be used. Linkage relationships can be determined in crosses to multiply marked strains. For example, suppose that the mutant to be studied, which has the phenotype of amino acid excretion and is hence not selectable in crosses, is in an F$^-$ strain of *S. typhimurium* with no other genetic markers. The strain can be made into a conjugation donor strain by transferring to it an F*lac* episome of *E. coli;* F*lac*$^+$ recombinants can readily be selected in *S. typhimurium* because the species is normally Lac$^-$. The recombinants will behave like an F$^+$ strain, transferring its chromosome at low frequency (10^5–10^6 per donor cell) with no fixed point-of-origin of transfer. This F*lac*$^+$ strain can be crossed to an F$^-$ multiply auxotrophic streptomycin-resistant strain of *S. typhimurium,* prototrophic recombinants selected for genes at various points on the chromosome, and the recombinants analyzed for the donor phenotype of amino acid excretion. If the donor gene for excretion is close to the selected donor gene, 50 percent or more of the recombinants will be donor type, but if it is distant, the frequency will be very low. If joint conjugation of 70 percent or more is detected between the mutant and a specific gene, then the two may be jointly transduced by phage P22, or by phage P1, as described below. For an example of the use of the above system, see Anton (1968).

In *Salmonella,* Hfr strains have been isolated in *S. typhimurium* and *S. abony,* and the points-of-origin of chromosome transfer of these strains are shown in Figure 1. The isolation, genotype, frequency of chromosome transfer, and other properties of these strains have been described (Mäkelä, 1963; Sanderson and Demerec, 1965; Sanderson and Saeed, 1972a; Sanderson *et al.,* 1972). The circularity of the linkage map has been confirmed by many Hfr crosses. Most of the Hfr strains in Figure 1 produce from 1 to 10 recombinants per 100 donor cells for proximal genes, and will yield analyzable time-of-entry data.

The Use of F$'$ Factors. F$'$ factors result from the attachment of a portion of bacterial chromosome to the F factor. Usually some gene activity such as lactose fermentation, in the case of F*lac*$^+$ factors, is associated with this chromosome fragment. Only a part of the chromosome of *S. typhimurium* and of *S. abony* is represented on known F$'$ factors,

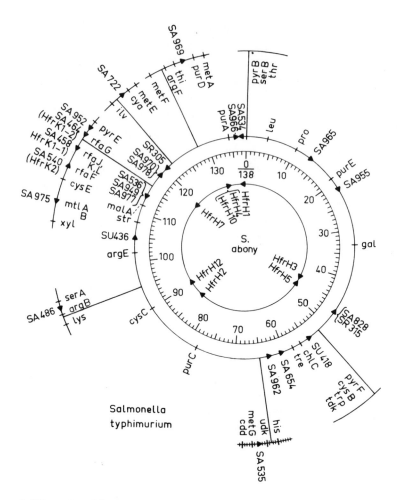

Figure 1. Hfr strains of S. typhimurium (shown outside the circle) and of S. abony (shown inside the circle). The numbers in the circle represent the time-of-entry, in minutes, for Hfr strains. The point-of-origin and orientation-of-transfer of each strain is indicated by the arrows. For S. typhimurium the strain number is usually given, though in some cases, eg., SA464 at min 116, the Hfr description, HfrK1-2 is also given. In those cases in which F is inserted into a known transduction linkage group, the Hfr is displayed on an arc outside the main circle; the genes shown on a crosshatched arc are a P1-mediated transduction linkage group, but all other linkage groups are for P22 phage. Where more than one Hfr strain number is shown in a gene interval (e.g., SA536, SA949, and SA977, displayed within a bracket at min 113), these represent independent isolates for which the point-of-origin may be different. Not all strains with point-of-origin in the rfa genes at min 116 are shown. For more details on the genotypes and fertility of all these strains, see Sanderson et al. (1972) and Sanderson and Saeed (1972a). A detailed linkage map, showing the positions of 323 genes along with descriptions of the phenotypes of each gene, is presented elsewhere (Sanderson, 1972). The Hfr strains of S. typhimurium can be obtained from the Salmonella Genetic Stock Centre, those of S. abony, from Dr. P. H. Mäkelä, Central Public Health Laboratory, Helsinki, Finland.

(Sanderson *et al.*, 1972), but a large number of factors collectively covering the entire *E. coli* chromosome are available, and can be transferred into *Salmonella* (Low, 1972). The F′ factors are useful for dominance studies, for complementation studies, and for genetic mapping studies.

Dominance tests are usually done to determine if the wild-type allele of a specific gene is dominant to the mutant allele. Usually, the wild-type allele is dominant, but in the case of regulatory mutants, suppressor mutants, or mutants for resistance to metabolic analogs, the mutant allele may be dominant. A heterozygous partial diploid is made using an F′ factor which carries the gene being studied, and the phenotype is observed to see which allele is dominant. Such tests have been done with histidine regulatory mutant in *S. typhimurium,* using F′ factors of *E. coli* and of *S. typhimurium* (Fink and Roth, 1968).

Complementation analysis is done to determine if two alleles are located in the same functional unit; such units are called cistrons. A partial diploid cell is produced carrying two mutant genes; if the recombinant has the mutant phenotype, the two mutants do not complement and are in the same cistron. Such partial diploids for complementation tests are usually produced by abortive transduction (see below), but may also be made by F′ factors.

F′ factors are useful in producing chromosome mobilization for linkage analysis, as described above, and in producing genetic mapping data through complementation analysis of unmapped mutations. For example, a group of F*lac* factors carrying greater or lesser parts of the chromosome adjacent to *lac* are available from *E. coli*. If the phenotype of a mutant is returned to wild type by the presence of a specific F′ factor, then the wild-type allele of the mutant gene must be located close to *lac*.

Bacteriophage-Mediated Transduction

Generalized Transduction. In the case of P22 and related phages, Zinder and Lederberg (1952) first showed generalized transduction in *S. typhimurium,* using phage P22. This phage is easy to grow, store, and use in genetic crosses. Complete transduction results in stable recombinants, apparently due to insertion of the donated chromosome fragment into the chromosome of the recipient. Procedures for the use of P22, and an example of fine-structure genetic analysis, are described by Hartman *et al.* (1960).

There are three problems in the use of P22. Firstly, the transductants are always lysogenic, due to superinfection with normal phage particles which are always present in the lysate; further transduction using these lysogenic recombinants as recipients is possible, but the frequency is sharply reduced due to superinfection exclusion (Ebel-Tsipis and Botstein, 1971). Secondly, P22 is a small phage which, based on the following two sets of observations, can only carry about one percent of the *Salmonella* chromosome. (1) Transducing particles and plaque-forming units both carry DNA with molecular weight 2.6–2.7 × 10^6 daltons (Rhoades *et al.*, 1968). Transducing particles are composed of 90 percent bacterial DNA and 10 percent phage DNA (Schmieger, 1972). The molecular weight of *Salmonella* DNA is about 3 × 10^9 daltons (Maaløe and Kjelgaard, 1966). (2) Genes on the same transducing fragment may be separated by no more than a minute of transfer time in Hfr matings, while the total chromosome length in mating time is 138 minutes. Thirdly, mutations leading to an absent or altered O antigen lipopolysaccharide, "rough" mutants, are resistant to P22.

The following means of circumventing these problems have been devised: The isolation by Smith and Levine (1967) of P22 mutants (L4) having a lowered frequency of lysogenization has helped to solve the problem of lysogenic transductants. In order to isolate nonlysogenic transductants from a P22 (L4) cross, it is important to streak out transductant colonies as soon as they become visible on the plate. The single-colony isolates should be restreaked as early as possible, and the resulting single colonies tested for phage sensitivity. To test for phage sensitivity, a clear plaque mutant, P22H5, should be used. A 0.1 ml aliquot of 1 × 10^8 ml P22H5 suspension is spread on an Eosin-methylene blue (EMB) agar plate without sugar. A sterile toothpick is touched to the single-colony isolate, then dipped into a drop of sterile water, mixed, and stabbed into the EMB agar. After 24 hours of incubation at 37°C, sensitive clones will be sectored and pink whereas lysogens will be uniform and purple. Sometimes, even following the above procedure, a strain will remain lysogenic; the following method can be used to obtain a transducing lysate (D. Fankhauser, private communication): The lysogenic strain is grown overnight with aeration in nutrient broth supplemented with 6 percent each of 50X Vogel Bonner E medium and 40 percent glucose. After stationary phase is reached, the cells are separated from the culture fluid by centrifugation. Chloroform is added to the culture fluid and then the culture fluid is titered. Enough phage is liberated by spontaneous lysis to provide a transducing lysate. A recent publication summarizes the current state of P22-mediated transduction methodology (Ely *et al.*, 1974). Another approach to obtain transducing lysates from lysogens is to use heteroimmune phage. MG40 (Grabnar and Hartman, 1968) is partially homoimmune with P22, but can grow on P22 lysogens at multiplicities of infection of ten. Other new phage related to P22 are L phage (Bezdek and Amati, 1967) and PSA68 (Enomoto and Ishiwa, 1972). Unfortunately, neither will grow on P22 lysogens, but P22 will grow on their lysogens. Another transducing phage for *Salmonella,* not yet well characterized, was reported by Kitamura and Mise (1970). An extensive analysis of phage carried by *S. typhimurium* has been reported (Boyd, 1950; Boyd and Bidwell, 1957).

The problems posed by the small size of the P22 transducing fragment and the P22 resistance of rough mutants has been partially solved by the use of phages P1 ES18 (see below).

Hong, Smith, and Ames (1971) have isolated a mutant of P22 (P22*cly*) which lysogenizes with such high efficiency that it will not form plaques on wild-type *Salmonella.* P22 mutants with increased or with decreased frequency of transduction have been isolated (Schmeiger, 1972).

The bacteriophage ES18 differs serologically and morphologically from phage P22 (Kuo and Stocker, 1970). ES18 can attack all classes of smooth as well as of nonsmooth (rough) mutants of *S. typhimurium* LT2 provided they are free of prophage Fels 2. A host range mutant, ES18*h*1, plates on strains lysogenic for Fels 2. ES18 is a general transducing phage, although it is ten- to one hundredfold less efficient than P22 in carrying out complete and abortive transduction, possibly because ES18 is virulent. Transductional analysis of many rough mutants, not normally possible with P22, has been done with ES18 (Kuo and Stocker, 1972). From genetic data, the size of the transducing fragment is similar to that of P22.

The phage KB1 differs from P22 in plague morphology, frequency-of-transductant formation, superinfection-immunity properties, antiserum specificity, and heat sensitivity (Boro and Brenchley, 1971). Transduction of genes in different parts of the linkage map is obtained at higher frequencies per plaque-forming unit than with P22. KB1 belongs to the heat-sensitive B phage group of Boyd and Bidwell (1957), while P22 and related phages are heat-resistant type-A phages.

P1, the generalized transducing phage of *E. coli* K12 will not adsorb to normal, smooth strains of *Salmonella,* but will adsorb to certain classes of rough mutants (Okada and Watanabe, 1968), such as *galE* and *rfaG* mutants in which the LPS is galactose deficient; the techniques for isolation of such mutants have been described (Wilkinson *et al.,* 1972). The P1 transducing fragment is larger than that of P22, so joint tranduction is obtained between genes not jointly transduced by P22.

Abortive Transduction. Abortive transduction appears to result from transcription from the donor transducing fragment with replication of the fragment; this enables the detection of complementation between genes (trans test). In transduction studies using auxotrophic mutants, abortive transduction may result in the formation of very tiny colonies, which are very difficult to detect (Demerec and Ozeki, 1959). Abortive transduction has been used successfully to detect cistrons for many auxotrophic mutants (for example see Loper *et al.*, 1964 for studies with the histidine genes). In studies with nonmotile mutants of the *fla* and *mot* loci, abortive transductants are readily detected as "trails" (due to motile cells) in a semisolid medium (Stocker, 1956a; Yamaguchi *et al.*, 1972).

    Specialized Transduction. The attachment site of P22 lies between the *proA* and *proB* genes (Young and Hartman, 1966). Specialized transduction of *proA* and *proB* by defective transducing particles (P22*dpro*) was observed by Smith-Keary (1966). Patai-Wing (1968), utilizing a *rec⁻* host, detected specialized transducing particles for *proA* and *proB* but not for *proC* which had arisen during lytic growth. Jessop (1972), using a strain carrying the deletion mutation *proAB47ataA*, has isolated several perfect specialized transducing phage (P22-p-*pro*-1).

    G. R. Smith (1971) analyzed the gene order of the *hut* region by producing a specialized λ transducing particle carrying the *Salmonella hut* region (λ *phut*). The technique involved transfer of genes by an Hfr strain of *S. typhimurium* into an *E. coli* recipient. The hybrid was lysogenized with λ, and an induced lysate was screened for plaque formers able to transduce the *hut* region. Voll (1972) was able to produce a high-frequency transducing phage carrying the histidine operon of *Salmonella*, using the phage φ80 from *E. coli*.

The Linkage Map, Nomenclature, and Sources of Strains

    The fourth edition of the linkage map of *S. typhimurium* (Sanderson, 1972) shows the map as a closed circle with 323 genes, and includes descriptions of the phenotypes controlled by each of these genes. The position of a few of these genes is shown in Figure 1. About 250 of these genes have been analyzed by joint transduction studies, usually with phage P22 but occasionally with ES18 or P1, while most of the remainder are mapped only by conjugation. There is a high probability that a newly identified gene, once located by conjugation, can be shown to be jointly transduced with other genes.

    The nomenclature recommended by Demerec *et al.* (1966) is used by *Salmonella* geneticists. Each gene is given a 3-letter designation which may be followed by a capital letter. At the time of isolation, each new mutant is given a gene designation and an allele number. In addition, many investigators maintain their entire set of stocks under a stock designation, followed by a number; for example, some of the Hfr strains of *S. typhimurium* are designated SA464, etc. (Figure 1). To prevent duplication, allele numbers, gene designations, and stock designations may be cleared through the *Salmonella* Genetic Stock Centre, Department of Biology, University of Calgary.

Many of the strains necessary for studies in *Salmonella* genetics can be obtained free of charge from the *Salmonella* Genetic Stock Centre (mentioned above). A detailed listing of Hfr strains and a partial listing of multiply-mutant F⁻ strains has been published (Sanderson *et al.*, 1972). It is also normally possible to get strains by writing to the authors of studies which used these strains. Strains of *E. coli*, especially Hfr and F′ strains, can be obtained from the *Coli* Genetic Stock Centre, Department of Microbiology, Yale University.

Topics of Special Interest in *Salmonella* Genetics

The regulation of gene activity has been of special interest to bacterial geneticists. Several systems in *Salmonella* have been analyzed, some

TABLE 1.   *Summary of Studies of Gene Regulation in Salmonella typhimurium*

Regulatory system	Regulatory gene(s)	Reference
Synthesis of the common precursor of aromatic amino acids	—	Zalkin, 1967
Cysteine biosynthesis	—	Smith, D. A., 1971
Deoxyribose utilization	*deoR*	Robertson *et al.*, 1970; Blank and Hoffee, 1972
Galactose utilization	*galR*	Saier *et al.*, 1973
Histidine biosynthesis	*hisR,S,T,U,* and *W*	Brenner and Ames, 1971
Histidine utilization	*hutC,M,P,Q,R*	Smith and Magasanik, 1971
Isoleucine–valine biosynthesis	*flrB ilvS*	Umbarger, 1971; Alexander and Calvo, 1969
Melibiose utilization	*melC*	Levinthal, 1971
Methionine biosynthesis	*metJ,K*	D. A. Smith, 1971
Proline utilization	*putR*	Newell and Brill, 1972
Purine biosynthesis	—	Gots, 1971
Pyrimidine biosynthesis	*pyrR*	O'Donovan and Gerhardt, 1972
Tryptophan biosynthesis	*trpR*	Margolin, 1971
Tyrosine biosynthesis	*tyrR*	Gollub and Sprinson, 1972

exhaustively while some are still in early stages of development. These systems are summarized in Table 1. In each case, the references are to review articles where possible, and if not, to a central article from which other references may be culled.

Another topic of historical interest to *Salmonella* geneticists is biosynthesis, and biosynthetic control of flagella. These studies have encompassed the interesting subject of phase variation. Reviews on early work (Stocker, 1956*b*), and on recent work (Iino, 1969) have been published.

More recently, *Salmonella* has been used to study the genetics and physiology of transport. The transport of aromatic amino acids (Ames and Roth, 1968), sulfate (Ohta *et al.*, 1971), histidine (Ames and Lever, 1970), methionine (Ayling and Bridgeland, 1970), and iron (Pollack *et al.*, 1970) have been studied. In addition to these systems, a comprehensive study exists of carbohydrate transport mediated by the phosphotransferase system (Simoni *et al.*, 1967).

The genetics of synthesis of the lipopolysaccharide component of the cell wall of *Salmonella* has been studied because this macromolecule carries the endotoxic activity of the cell, is an important phage receptor site, is the somatic antigen used in classification, and is essential for cell viability (Mäkelä and Stocker, 1969; Stocker and Mäkelä, 1971). Transduction studies of mutants with defective lipopolysaccharide, "rough" mutants, have been made possible using phage ES18 (see above), and also by inducing P22 lysogens or by using part-rough mutants as recipients (Kuo and Stocker, 1972; Sanderson and Saeed, 1972*b*).

Acknowledgments

The *Salmonella* Genetic Stock Centre is supported by grants from the National Research Council of Canada and the National Science Foundation of the United States of America. While writing this manuscript, K. E. Sanderson was the recipient of an Alexander von Humboldt fellowship at the Max Planck Institut für Immunbiologie, Freiburg. M. Levinthal would like to thank Dr. H. Koffler and the Department of Biological Sciences of Purdue University for their support and encouragement during the preparation of this manuscript.

Literature Cited

Abd-el-al, A. and J. L. Ingraham, 1969   Cold sensitivity and other phenotypes resulting from mutation in *pyrA* gene. *J. Biol. Chem.* **244**:4039–4045.

Adelberg, E. A. and J. W. Meyers, 1953   Modification of the penicillin technique for the selection of auxotrophic bacteria. *J. Bacteriol.* **65**:348–353.

Alexander, R. R. and J. M. Calvo, 1969   A *Salmonella typhimurium* locus involved in the regulation of isoleucine, valine and leucine biosynthesis. *Genetics* **61**:539–556.

Ames, B. N, 1971   The detection of chemical mutagens with enteric bacteria. In *Chemical Mutagens: Principles and Methods for Their Detection,* Vol. 1, edited by A. Hollaender, pp. 267–282. Plenum Press, New York.

Ames, B. N. and H. J. Whitfield, Jr., 1966   Frameshift mutagenesis in *Salmonella. Cold Spring Harbor Symp. Quant. Biol.* **31**:221–225.

Ames, G. F. and J. Lever, 1970   Components of histidine transport: histidine-binding proteins and *hisP* protein. *Proc. Natl. Acad. Sci. USA* **66**:1096–1103.

Ames, G. F. and J. R. Roth, 1968   Histidine and aromatic permeases of *Salmonella typhimurium. J. Bacteriol.* **96**:1742–1749.

Anton, D. N., 1968   Histidine regulatory mutants in *Salmonella typhimurium.* V. Two new classes of histidine regulatory mutants. *J. Mol. Biol.* **33**:533–546.

Atkins, C. G. and F. B. Armstron, 1969   Electrophoretic study of *Salmonella typhimurium–Salmonella montevideo* hybrids. *Genetics* **63**:775–779.

Ayling, P. D. and E. S. Bridgeland, 1970   Methionine transport systems in *Salmonella typhimurium. Heredity* **25**:687–688.

Baumberg, S. and R. Freeman. 1971   *Salmonella typhimurium* LT-2 is still pathogenic for man. *J. Gen. Microbiol.* **65**:99–100.

Berkowitz, D., J. Hushon, H. J. Whitfield Jr., J. R. Roth and B. N. Ames, 1968   Procedure for identifying nonsense mutations. *J. Bacteriol.* **96**:215–220.

Bezdek, M. and P. Amati, 1967   Properties of P22 and a related *Salmonella typhimurium* phage. *Virology* **31**:272–278.

Blank, J. and P. Hoffee, 1972   Regulatory mutants of the *deo* regulon in *Salmonella typhimurium. Mol. Gen. Genet.* **116**:291–298.

Boro, H. and J. E. Brenchley, 1971   A new generalized transducing phage for *Salmonella typhimurium* LT2 *Virology* **45**:835–836.

Boyd, J. S. K. 1950   The symbiotic bacteriophages of *Salmonella typhimurium. J. Pathol. Bacteriol.* **62**:501–523.

Boyd, J. S. K. and D. E. Bidwell, 1957   The type A phages of *Salmonella typhimurium:* identification by standardized cross-immunity tests. *J. Gen. Microbiol.* **16**:217–233.

Braun, W., 1966   *Bacterial Genetics,* second edition, W. B. Saunders, Philadelphia, Pa.

Brenner, M. and B. N. Ames, 1971   The histidine operon and its regulation. In *Metabolic Regulation,* Vol. V, edited by H. J. Vogel, pp. 350–388, Academic Press, New York.

Burdette, W. L., 1963   *Methodology in Basic Genetics,* Holden-Day, San Francisco, Calif.

Caspari, E. W., editor, 1971   Demerec Memorial Volume. *Adv. Genet.* **16**:V–381.

Clowes, R. and W. Hayes, 1968   *Experiments in Microbial Genetics,* John Wiley and Sons, New York.

Crick, F. H. C., L. Barnett, S. Brenners and R. J. Watts-Tobin, 1961   General nature of the genetic code for proteins. *Nature (Lond.)* **192**:1227–1232.

Davis, B. D., 1948   Isolation of biochemically deficient mutants of bacteria by penicillin. *J. Am. Chem. Soc.* **70**:4267.

Davis, B. D. and E. S. Mingioli, 1950   Mutants of *Escherichia coli* requiring methionine or vitamin B_{12}. *J. Bacteriol.* **60**:17–28.

Demerec, M. and H. Ozeki, 1959   Tests for allelism among auxotrophs of *Salmonella typhimurium. Genetics* **44**:269–278.

Demerec, M., E. A. Adelberg, A. J. Clark and P. E. Hartman, 1966   A proposal for a uniform nomenclature in bacterial genetics. *Genetics* **54**:61–76.

Ebel-Tsipis, J. and D. Botstein, 1971   Super-infection exclusion by P22 prophage in lysogens of *Salmonella typhimurium*. I. Exclusion of generalized transducing particles. *Virology* **45**:629–637.

Ely, B., R. M. Weppelmen, H. C. Massey, Jr. and P. E. Hartman, 1974   Methods in P22 transduction. *Genetics* (in press).

Enomoto, M., 1966   Genetic studies of paralyzed mutants in *Salmonella*. I. Genetic fine structure of the *mot* locus in *Salmonella typhimurium*. *Genetics* **54**:715–726.

Enomoto, M. and H. Ishiwa, 1972   A new transducing phage related to P22 of *Salmonella typhimurium*. *J. Gen. Virol.* **14**:157–164.

Fink, G. R. and J. Roth, 1968   Histidine regulatory mutants in *Salmonella typhimurium*. VI. Dominance studies. *J. Mol. Biol.* **33**:547–557.

Gollub, E. and D. B. Sprinson, 1972   Regulation of tyrosine biosynthesis in *Salmonella*. *Fed. Proc.* **31**:491.

Gorini, L. and H. Kaufman, 1960   Selecting bacterial mutants by the penicillin method. *Science (Wash., D.C.)* **131**:604–605.

Gots, J., 1971   Regulation of purine and pyrimidine metabolism. In *Metabolic Regulation,* edited by H. J. Vogel, pp. 225–256. Academic Press, New York.

Grabnar, M. and P. E. Hartman, 1968   MG40 phage, a transducing phage related to P22. *Virology* **34**:521–530.

Gutnick, D., J. M. Calvo, T. Klopotowski and B. N. Ames, 1969   Compounds which serve as the sole source of carbon or nitrogen for *Salmonella typhimurium* LT-2 *J. Bacteriol.* **100**:215–219.

Hartman, P. E., J. C. Loper and D. Serman, 1960   Fine structure mapping by complete transduction of histidine requiring mutants. *J. Gen. Microbiol.* **22**:323–353.

Hartman, P. E., S. R. Suskind and T. R. F. Wright, 1965   *Principles of Genetics, Laboratory Manual,* p. 43, W. C. Brown, Dubuque, Iowa.

Hartman, P. E., Z. Hartman, R. C. Stahl and B. N. Ames, 1971*a*   Classification and mapping of inspontaneous and induced mutations in the *his* operon of *Salmonella*. *Adv. Genet.* **17**:1–34.

Hartman, P. E., K. Levine, Z. Hartman and H. Berger, 1971*b*   Hycanthone: a frameshift mutagen. *Science (Wash., D. C.)* **172**:1058–1060.

Hayes, W., 1968   *The Genetics of Bacteria and their Viruses,* second edition, John Wiley & Sons, New York.

Hoeksma, W. D. and D. E. Schoenhard, 1971   Characterization of a thermolabile sulfite reductase from *Salmonella pullorum*. *J. Bacteriol.* **108**: 154–158.

Hong, J.-S. and B. N. Ames, 1971   Localized mutagenesis of any specific small region of the bacterial chromosome. *Proc. Natl. Acad. Sci. USA* **68**:3158–3162.

Hong, J.-S., G. R. Smith and B. N. Ames, 1971   Adempsome 3′-5′-cyclic monophosphate concentration in the bacterial host regulates the viral decision between lysogeny and lysis. *Proc. Natl. Acad. Sci. USA* **68**:2258–2262.

Iino, T., 1969   Genetics and chemistry of bacterial flagella. *Bacteriol. Rev.* **33**:454–475.

Jessop, A. P., 1972   A specialized transducing phage of P22 for which the ability to form plaques is associated with transduction of the *proAB* region. *Mol. Gen. Genet.* **114**:214–222.

Johnson, E. M. and L. S. Baron, 1969   Genetic transfer of the Vi antigen from *Salmonella typhosa* to *Escherichia coli. Bacteriology* **99**:358–359.

Kitamura, J. and K. Mise, 1970   A new generalized transducing phage in *Salmonella*. *Jap. J. Med. Sci. Biol.* **23**:99–102.

Kuo, T.-T. and B. A. D. Stocker, 1970   ES18, a general transducing phage for smooth and non-smooth *Salmonella typhimurium. Virology* **42**:621–632.

Kuo, T.-T. and B. A. D. Stocker, 1972   Mapping of *rfa* genes in *Salmonella typhimurium* by P22 and ES18 transduction and by conjugation. *J. Bacteriol.* **112**: 48–63.

Lederberg, J. and E. M. Lederberg, 1952   Replica plating and indirect selection of bacterial mutants. *J. Bacteriol.* **63**:399–406.

Levinthal, M., 1971   Biochemical studies of melibiose metabolism in wild type and *mel* mutant strains of *Salmonella typhimurium. J. Bacteriol.* **105**:1047–1052.

Levinthal, M. and R. Simoni, 1969   Genetic analysis of carbohydrate transport-deficient mutants of *Salmonella typhimurium. J. Bacteriol.* **97**:250–255.

Loper, J., M. Grabnar, R. C. Stahl, Z. Hartman and P. E. Hartman, 1964   Genes and proteins involved in histidine biosynthesis in *Salmonella. Brookhaven Symp. Biol.* **17**:15–52.

Low, K. B.,1972 *Escherichia coli* K-12 F-prime factors, old and new. *Bacteriol. Rev.* **36**:587–607.

Maaløe, O. and N.O. Kjehlgaard, 1966   *Control of Macromolecular Synthesis,* W. A. Benjamin, New York.

Mäkelä, P. H., 1963   Hfr males in *Salmonella abony. Genetics* **48**:423–429.

Mäkelä, P. H., 1964   Genetic homologies between flagellar antigens of *Escherichia coli* and *Salmonella abony.* J. Gen. Microbiol. **35**:503–510.

Mäkelä, P. H., 1966   Genetic determination of the O antigens of *Salmonella* groups B(4, 5,12) and C₁(6,7). *J. Bacteriol.* **91**:1115–1125.

Mäkelä, P. H. and B. A. D. Stocker, 1969   Genetics of polysaccharide biosynthesis. *Ann. Rev. Genet.* **3**:291–322.

Mäkelä, P. H., M. Jahkola and O. Luderitz, 1970   A new gene cluster *rfe* concerned with the biosynthesis of *Salmonella* lipopolysaccharide. *J. Gen. Microbiol.* **60**:91–106.

Margolin, P., 1971   Regulation of tryptophan synthesis. In *Metabolic Pathways,* edited by H. J. Vogel, pp. 389–446, Academic Press, New York.

Martin, R. G., 1968   Polarity in relaxed strains of *Salmonella typhimurium. J. Mol. Biol.* **31**:127–134.

Meynell, G. G. and E. Meynell, 1970   *Theory and Practice in Experimental Bacteriology,* second edition, University Press, Cambridge, England.

Newell, S. L. and W. J. Brill, 1972   Mutants of *Salmonella typhimurium* that are insensitive to catabolite repression of proline degradation. *J. Bacteriol.* **111**:375–382.

Nikaido, H., M. Levinthal, K. Nikaido and K. Nakone, 1967   Extended deletions in the histidine-rough B region of the *Salmonella* chromosome. *Proc. Natl. Acad. Sci. USA* **57**:1825–1832.

O'Donovan, G. A. and J. C. Gerhart, 1972   Isolation and partial characterization of regulatory mutants of the pyrimidine pathway in *Salmonella typhimurium. J. Bacteriol.* **109**:1085–1096.

Oeschger, N. S. and P. E. Hartman, 1970   ICR induced frameshift mutations in the histidine operon of *Salmonella. J. Bacteriol.* **101**:490–504.

Ohta, N., P. R. Glasworthy and A. B. Pardee, 1971   Genetics of sulfate transport by *Salmonella typhimurium. J. Bacteriol.* **105**:1053–1062.

Okada, M. and T. Watanabe, 1968   Transduction with phage P1 in *Salmonella typhimurium. Nature (Lond.)* **218**:185–187.

Patai-Wing, J., 1968   Transduction in P22 in a recombination deficient mutant of *Salmonella typhimurium. Virology* **36**:271–276.

Pollack, J. R., B. N. Ames, and J. B. Neilands, 1970 Iron transport in *Salmonella typhimurium*: mutants blocked in the biosynthesis of enterobactin. *J. Bacteriol.* **104**: 635–639.

Rhoades, M., L. A. MacHattie and C. A. Thomas, Jr., 1968 The P22 bacteriophage DNA molecule. I. The mature form. *J. Mol. Biol.* **37**:21–40.

Robertson, B. C., P. Jargiello, J. Blank and P. A. Hoffee, 1970 Genetic regulation of ribonucleosides and deoxyribonucleoside catabolism in *Salmonella typhimurium*. *J. Bacteriol.* **102**:628–635.

Roth, J. R., 1972 Genetic techniques in studies of bacterial metabolism. *Methods Enzymol.* **174**:3–35.

Saier, M. H., Jr., F. G. Bromberg and S. Roseman, 1973 Characterization of constitutive galactose permease mutants in *Salmonella typhimurium*. *J. Bacteriol.* **113**:512–523.

Sanderson, K. E., 1972 Linkage map of *Salmonella typhimurium, edition IV. Bacteriol. Rev.* **36**:558–586.

Sanderson, K. E. and M. Demerec, 1965 The linkage map of *Salmonella typhimurium, Genetics* **51**:897–913.

Sanderson, K. E. and H. Saeed, 1972*a* Insertion of the F-factor into the cluster of *rfa* (rough A) genes of *Salmonella typhimurium*. *J. Bacteriol.* **112**:64–73.

Sanderson, K. E. and H. Saeed, 1972*b* P22-mediated transduction analysis of the rough A (*rfa*) region of the chromosome of *Salmonella typhimurium*. *J. Bacteriol.* **112**:58–63.

Sanderson, K. E., H. Ross, L. Ziegler and P. H. Mäkelä, 1972 F⁺, F′ and Hfr strains of *Salmonella typhimurium* and *S. abony. Bacteriol Rev.* **36**:608–637.

Schmeiger, H., 1972 The molecular structure of the transducing particles of *Salmonella* phage P22. Density gradient analysis of DNA. *Molec. Gen. Genet.* **109**:323–337.

Schwartz, D. O. and J. R. Beckwith, 1969 Mutagens which cause deletions in *Escherichia coli. Genetics* **61**:371–376.

Simoni, R., M. Levinthal, F. Kundig, W. Kundig, B. Anderson, P. E. Hartman, and S. Roseman, 1967 Genetic evidence for the role of a bacterial phosphotransferase system in sugar transport. *Proc. Natl. Acad. Sci. USA* **58**:1963–1970.

Smith, D. A., 1971 S-amino acid metabolism and its regulation in *Escherichia coli* and *Salmonella typhimurium*. *Adv. Genet.* **16**:142–165.

Smith, G. R., 1971 Specialized transduction of the *Salmonella hut* operons by coliphage λ: deletion analysis of the *hut* operons employing λ *phut. Virology* **45**:208–223.

Smith, G. R. and B. Magasanik, 1971 The two operons of the histidine utilization system in *Salmonella typhimurium*. *J. Biol. Chem.* **246**:3330–3341.

Smith, H. O. and M. Levine, 1967 A phage P22 gene controlling integration of prophage. *Virology* **31**:207–216.

Smith-Keary, P. F., 1966 Restricted transduction by bacteriophage P22 in *Salmonella typhimurium. Genet. Res.* **8**:73–82.

Stocker, B. A. D., 1956*a* Abortive transduction of motility in *Salmonella*, a non-replicated gene transmitted through many generations to a single descendant. *J. Gen. Microbiol.* **15**:575–593.

Stocker, B. A. D., 1956*b* Bacterial flagella: morphology, constitution and inheritance. *Symp. Soc. Gen. Microbiol.* **6**:19–40.

Stocker, B. A. D. and P. H. Mäkelä, 1971 Genetic aspects of biosynthesis and structure of *Salmonella* lipopolysaccharide. In *Microbial Toxins*, Vol. 4, *Bacterial Endotoxins*, edited by G. Weintaum, S. Kadis, and S. J. Ajl, pp. 369–438, Academic Press, New York.

Taylor, A. L. and C. D. Trotter, 1972   Linkage map of *Escherichia coli* strain K12. *Bacteriol. Rev.* **36**:504–524.

Umbarger, H., 1971   The regulation of enzyme levels in the pathways to branched chain amino acids. In *Metabolic Regulation,* Vol. 5, edited by H. J. Vogel, pp. 447–463, Academic Press, New York.

Vogel, H. J. and D. M. Bonner, 1956   Acetyl ornithinase of *Escherichia coli:* partial purification and some properties. *J. Biol. Chem.* **218**:97–106.

Voll, M. J., 1972   Derivation of an F-merogenote and $\phi80$ high frequency transducing phage carrying the histidine operon of *Salmonella. J. Bacteriol.* **109**:741–750.

Whitfield, H. J. Jr., R. G. Martin and B. N. Ames, 1966   Classification of aminotransferase (C gene) mutants in the histidine operon. *J. Mol. Biol.* **21**:335–355.

Wilkinson, R. G., P. Gemski, Jr. and B. A. D. Stocker, 1972   Non-smooth mutants of *Salmonella typhimurium*: differentiation by phage sensitivity and genetic mapping. *J. Gen. Microbiol.* **70**:527–554.

Yamaguchi, S., T. Iino, T. Horiguchi, and K. Ohta, 1972   Genetic analysis of *fla* and *mot* cistrons closely linked to H1 in *Salmonella abortusequi* and its derivatives. *J. Gen. Microbiol.* **70**:59–75.

Young, B. G. and P. E. Hartman, 1966   Sites of P22 and P221 prophage integration in *Salmonella typhimurium. Virology* **28**:265–270.

Zalkin, H., 1967   Control of aromatic amino acid biosynthesis in *Salmonella typhimurium. Biochim. Biophys. Acta* **148**:609–621.

Zeiger, E. and M. S. Legator, 1971   Mutagenicity of N-nitrosomorpholine in the host-mediated assay. *Mutat. Res.* **12**:467–471.

Zinder, N., 1960   Sexuality and mating in *Salmonella. Science (Wash., D.C.)* **131**:813–816.

Zinder, N. and J. Lederberg, 1952   Genetic exchange in *Salmonella. J. Bacteriol.* **64**:679–699.

14

The Histidine Operon of
Salmonella typhimurium

Bruce N. Ames and Philip E. Hartman

The Biosynthesis of Histidine

Salmonella typhimurium uses ten enzymic steps to convert phos-phoribosylpyrophophate and ATP to histidine. The biosynthetic steps are illustrated in Figure 1. Any mutant lacking one of the enzymic activities will grow normally when supplied with exogenous histidine; hence, the pathway has no branch points leading to other metabolites required for growth. About 1500 mutations leading to a requirement for histidine have been described in the cluster of genes specifying the enzymes of histidine biosynthesis (Hartman *et al.*, 1971). These genes, then, are both necessary and sufficient to produce the enzymes to synthesize histidine. The enzymology of histidine biosynthesis has been recently reviewed in detail (Martin *et al.*, 1971).

Regulation of the Pathway

Salmonella has developed two methods for the efficient control of the biosynthetic pathway. These have been reviewed recently (Brenner and Ames, 1971).

Bruce N. Ames—Department of Biochemistry, University of California, Berkeley, California. Philip E. Hartman—Biology Department, The Johns Hopkins University, Baltimore, Maryland.

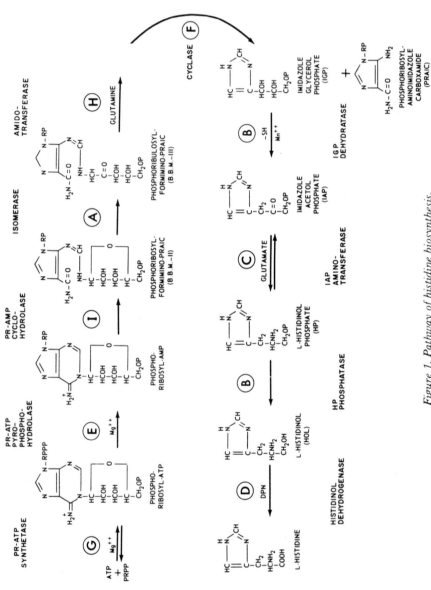

Figure 1. Pathway of histidine biosynthesis.

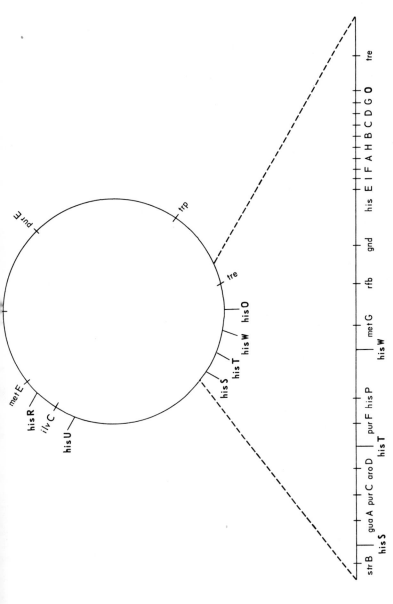

Figure 2. Genetic map of histidine mutations in Salmonella. The genes specifying the histidine biosynthetic enzymes, hisA–hisI, are a cluster near 6 o'clock on the map. Gene hisR is the structural gene for tRNA^{HIS}, hisS is the structural gene for the histidyl-tRNA synthetase, hisO is the operator–promoter region, hisT is the structural gene for an enzyme that converts uridine to pseudouridine in tRNA^{HIS}, and hisU and hisW are concerned with tRNA maturation. The gene hisP is involved with histidine transport.

Feedback Inhibition. Feedback inhibition of the first enzyme of the pathway keeps the pathway adjusted to the availability of external histidine. When sufficient exogenous histidine is present, feedback inhibition completely stops the biosynthesis of histidine.

Repression Control. Repression control regulates the actual amounts of enzymes made. The key pool in this regulation appears to be that of charged tRNA^{His}. Several recent papers deal with this regulation (Chang *et al.*, 1971; Brenner and Ames, 1972; Goldberger and Kovach, 1972; Lewis and Ames, 1972; Singer *et al.*, 1972; Singer and Smith, 1972; and Brenner *et al.*, 1972).

Figure 3. (Pages 227–234) Map of the histidine operon and adjacent gene region in S. typhimurium by P. Hartman, Z. Hartman, R. Stahl, and B. Ames (1971). Reproduced with the permission of Academic Press. Each gene is designated by a capital letter. The map is not drawn to scale.

Each number represents an independently isolated mutation and carries the suffix his. Mutations indicated below the heavy horizontal line are deletions. A wavy line indicates that the extent of the deletion in the wavy region is unknown, and an arrow at the end of the wavy line indicates that the extent of the mutation into the particular gene is unknown. Roman numerals just below the heavy horizontal line designate subregions of each gene as determined by deletion mapping.

Mutational sites are placed above the heavy horizontal line in their most probable map order, determined by deletion mapping and three-point tests. Mutations listed in vertical columns have not been ordered. In a number of instances, more than one column of successively numbered mutations is shown in one gene region. Parentheses indicate that the map order is unknown. Horizontal brackets indicate that the mutation(s) map in a particular region but have not been crossed further. Sets of mutations that fail to recombine with one another are enclosed by vertical brackets. Mutations not mapped in detail but placed only as to gene affected are listed in a cluster, for each gene, at the top of the diagram.

Prefixes indicate complementation patterns (for genes D, B, I and E). Suffix designations are: C = constitutive; F = mutation on E. coli F′; L = lysogenic stock; M = missing (lost); P = polar prototroph; R = rough (P22-resistant; in the case of extended deletions this is generally due to inclusion of the rfb operon in the mutation; and in the case of point mutations to a separate mutation in one of the somatic antigen genes); a = amber (UAG nonsense); c = cold sensitive, feedback hypersensitive (prototrophic at 37°C); f = frameshift; m = missense; n = nonsense (either amber or ochre); o = ochre (UAA nonsense); r = feedback resistant (prototrophic); s = stable (deletions listed below the heavy horizontal line also are stable to reversion on minimal medium); u = UGA nonsense; − = CRM negative; + = CRM positive.

Genes close to the histidine operon are fla (flagella formation), supW (UAG suppressor gene), gnd (gluconate-6-phosphate dehydrogenase), and rfb (somatic antigen synthesis).

Figure 3A

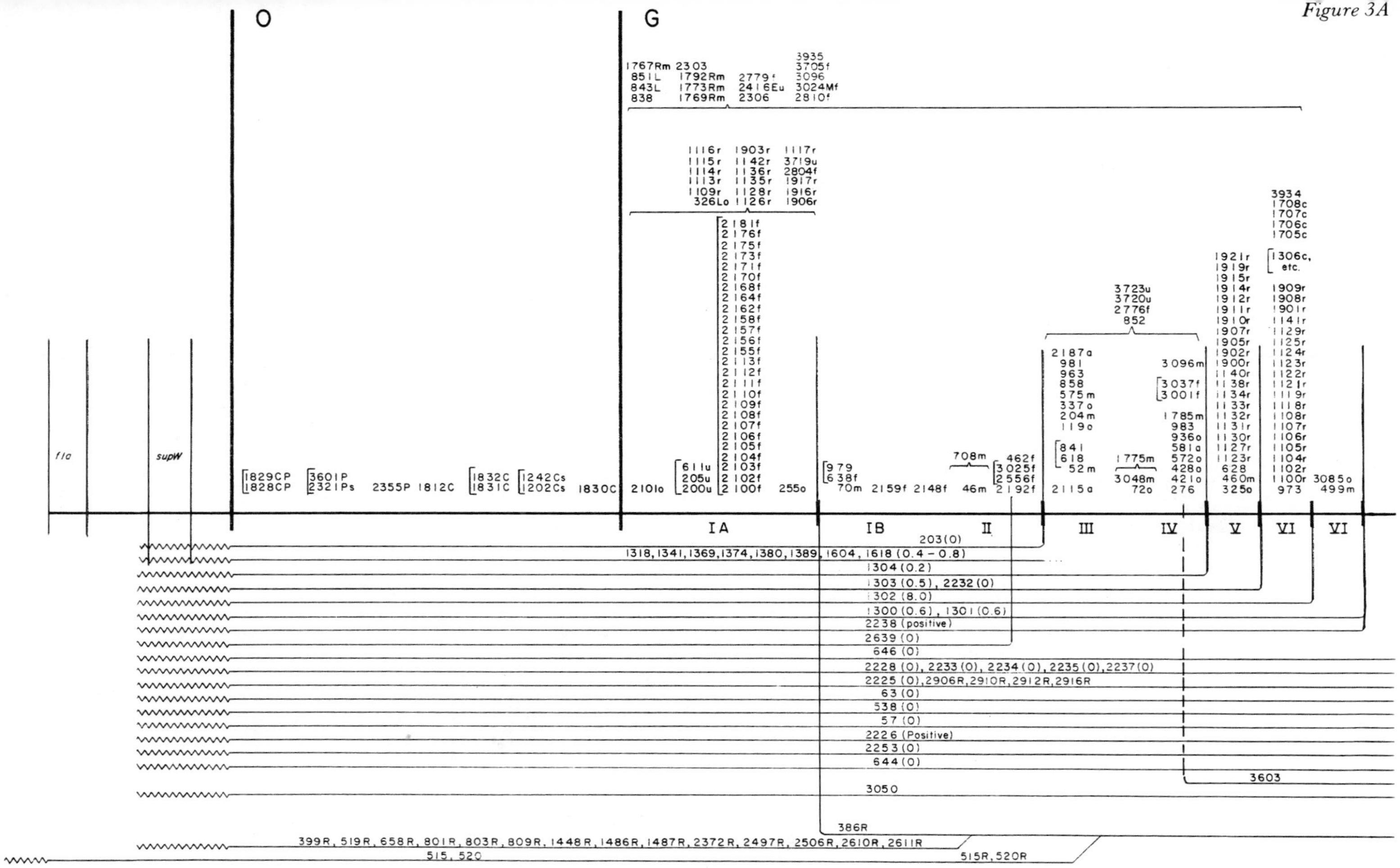

O
G
fla
supW
IA
IB
II
III
IV
V
VI
VI
1829CP
1828CP
3601P
2321Ps
2355P 1812C
1832C
1831C
1242Cs
1202Cs
1830C
2101o
1767Rm 2303
851L 1792Rm 2779f
843L 1773Rm 2416Eu
838 1769Rm 2306
3935
3705f
3096
3024Mf
2810f
1116r 1903r 1117r
1115r 1142r 3719u
1114r 1136r 2804f
1113r 1135r 1917r
1109r 1128r 1916r
326Lo 1126r 1906r
611u
205u
200u
2181f
2176f
2175f
2173f
2171f
2170f
2168f
2164f
2162f
2158f
2157f
2156f
2155f
2113f
2112f
2111f
2110f
2109f
2108f
2107f
2106f
2105f
2104f
2103f
2102f
2100f
255o
979
638f
70m
2159f 2148f
708m
46m
462f
3025f
2556f
2192f
3723u
3720u
2776f
852
2187a
981
963
858
575m
337o
204m
119o
841
618
52m
2115a
1775m
3048m
72o
3096m
3037f
3001f
1785m
983
936o
581a
572o
428o
421o
276
1921r
1919r
1915r
1914r
1912r
1911r
1910r
1907r
1905r
1902r
1900r
1140r
1138r
1134r
1133r
1132r
1131r
1130r
1127r
1123r
628
460m
325o
3934
1708c
1707c
1706c
1705c
1306c, etc.
1909r
1908r
1901r
1141r
1129r
1125r
1124r
1123r
1122r
1121r
1119r
1118r
1108r
1107r
1106r
1105r
1104r
1102r
1100r
973
3085o
499m
203(0)
1318, 1341, 1369, 1374, 1380, 1389, 1604, 1618 (0.4 – 0.8)
1304 (0.2)
1303 (0.5), 2232 (0)
1302 (8.0)
1300 (0.6), 1301 (0.6)
2238 (positive)
2639 (0)
646 (0)
2228 (0), 2233 (0), 2234 (0), 2235 (0), 2237 (0)
2225 (0), 2906R, 2910R, 2912R, 2916R
63 (0)
538 (0)
57 (0)
2226 (Positive)
2253 (0)
644 (0)
3603
3050
386R
399R, 519R, 658R, 801R, 803R, 809R, 1448R, 1486R, 1487R, 2372R, 2497R, 2506R, 2610R, 2611R
515, 520
515R, 520R

Figure 3B

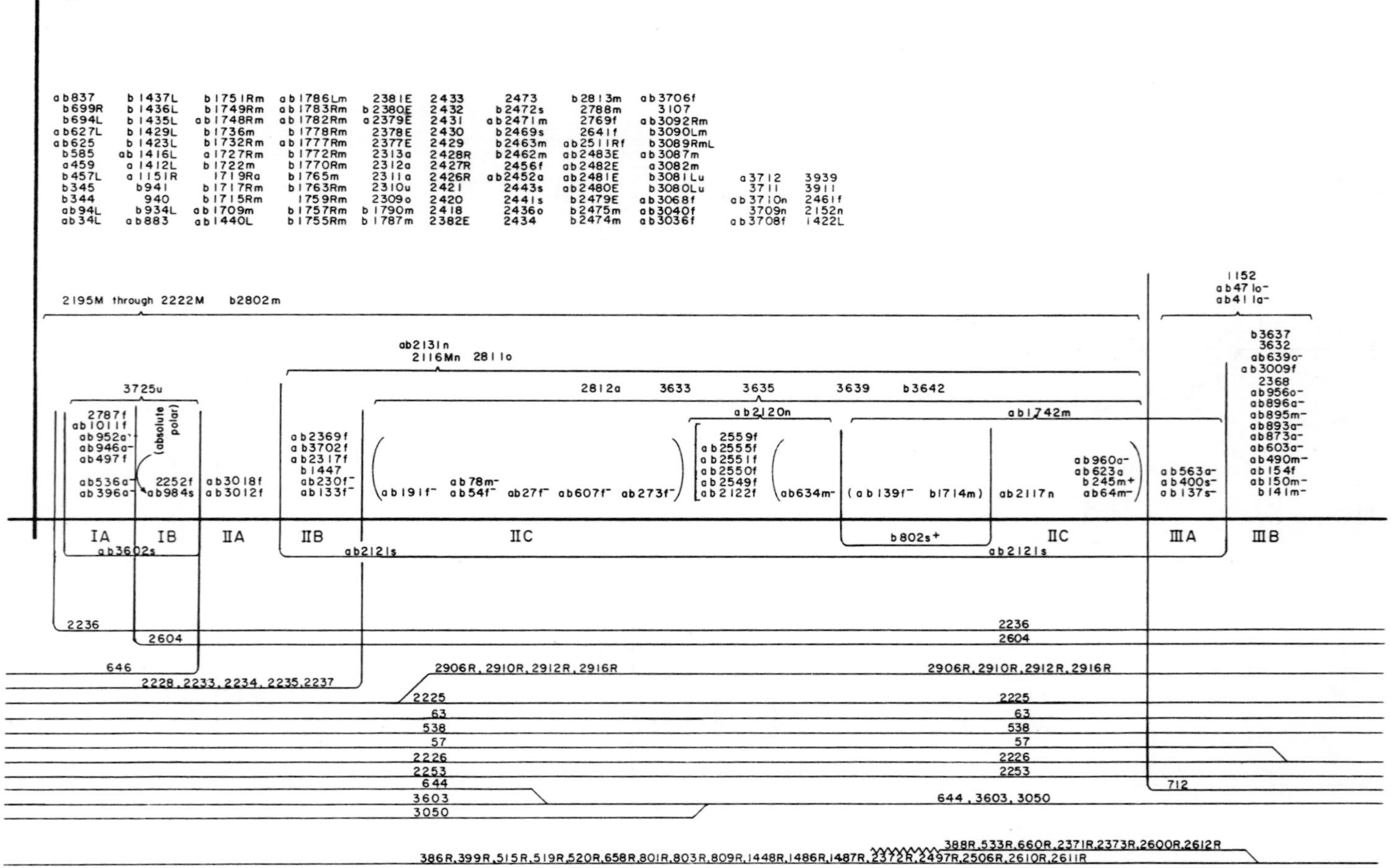

D
ab837 b699R b694L ab627L ab625 b585 a459 b457L b345 b344 ab94L ab34L
b1437L b1436L b1435L b1429L b1423L ab1416L a1412L a1151R b941 940 b934L ab883
b1751Rm b1749Rm ab1748Rm b1736m b1732Rm a1727Rm b1722m 1719Ra b1717Rm b1715Rm ab1709m ab1440L
ab1786Lm ab1783Rm ab1782Rm b1778Rm ab1777Rm b1772Rm b1770Rm b1765m b1763Rm 1759Rm b1757Rm b1755Rm
2381E b2380E a2379E 2378E 2377E 2313a 2312a 2311a 2310u 2309a b1790m b1787m
2433 2432 2431 2430 2429 2428R 2427R 2426R 2421 2420 2418 2382E
2473 b2472s ab2471m b2469s b2463m b2462m 2456f ab2452a 2443s 2441s 2436a 2434
b2813m 2788m 2769f 2641f ab2511Rf ab2483E ab2482E ab2481E ab2480E b2479E b2475m b2474m
ab3706f 3107 ab3092Rm b3090Lm b3089RmL ab3087m a3082m b3081Lu b3080Lu ab3068f ab3040f ab3036f
a3712 3711 ab3710n 3709n ab3708f
3939 3911 2461f 2152n 1422L
2195M through 2222M b2802m
1152 ab471a⁻ ab411a⁻
ab2131n 2116Mn 2811o
3725u
2812a 3633 3635 3639 b3642
ab2120n
ab1742m
2787f ab1011f ab952a⁻ ab946a⁻ ab497f
(absolute polar)
ab536a⁻ ab396a⁻
2252f ab984s
ab3018f ab3012f
ab2369f ab3702f ab2317f b1447 ab230f⁻ ab133f⁻
(ab191f⁻ ab78m⁻ ab54f⁻ ab27f⁻ ab607f⁻ ab273f⁻)
2559f ab2555f ab2551f ab2550f ab2549f ab2122f
(ab634m⁻
(ab139f⁻ b1714m)
ab2117n
ab960a⁻ ab623a b245m+ ab64m⁻
ab563a⁻ ab400s⁻ ab137s⁻
b3637 3632 ab639a⁻ ab3009f 2368 ab956a⁻ ab896a⁻ ab895m⁻ ab893a⁻ ab873a⁻ ab603a⁻ ab490m⁻ ab154f ab150m⁻ b141m⁻
IA IB IIA IIB IIC IIC IIIA IIIB
ab3602s
ab2121s
b802s+
ab2121s
2236
2604
646
2228, 2233, 2234, 2235, 2237
2906R, 2910R, 2912R, 2916R
2225
63
538
57
2226
2253
644
3603
3050
2236
2604
2906R, 2910R, 2912R, 2916R
2225
63
538
57
2226
2253
644, 3603, 3050
712
388R, 533R, 660R, 2371R, 2373R, 2600R, 2612R
386R, 399R, 515R, 519R, 520R, 658R, 801R, 803R, 809R, 1448R, 1486R, 1487R, 2372R, 2497R, 2506R, 2610R, 2611R

C

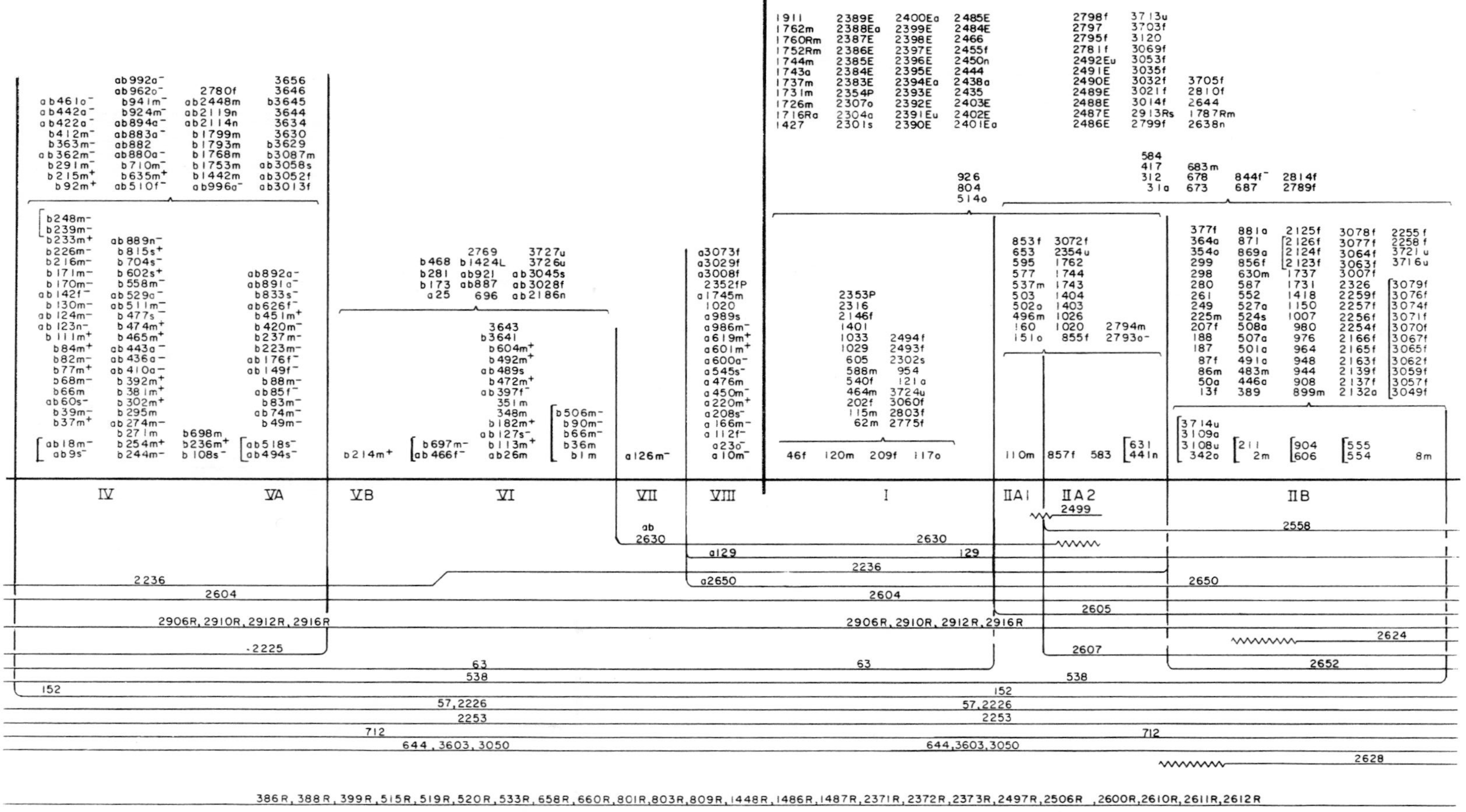
1911
1762m
1760Rm
1752Rm
1744m
1743a
1737m
1731m
1726m
1716Ra
1427
2389E
2388Ea
2387E
2386E
2385E
2384E
2383E
2354P
2307o
2304a
2301s
2400Ea
2399E
2398E
2397E
2396E
2395E
2394Ea
2393E
2392E
2391Eu
2390E
2485E
2484E
2466
2455f
2450n
2444
2438a
2435
2403E
2402E
2401Ea
2798f
2797
2795f
2781f
2492Eu
2491E
2490E
2489E
2488E
2487E
2486E
3713u
3703f
3120
3069f
3053f
3035f
3032f
3021f
3014f
2913Rs
2799f
3705f
2810f
2644
1787Rm
2638n
926
804
514o
584
417
312
31a
683m
678
673
844f⁻
687
2814f
2789f
ab461o⁻
ab442a⁻
ab422a⁻
b412m⁻
b363m⁻
ab362m⁻
b291m⁻
b215m⁺
b92m⁺
ab992a⁻
ab962o⁻
b941m⁻
b924m⁻
ab894a⁻
ab883a⁻
ab882
ab880a⁻
b710m⁻
b635m⁺
ab510f⁻
2780f
ab2448m
ab2119n
ab2114n
b1799m
b1793m
b1768m
b1753m
b1442m
ab996a⁻
3656
3646
b3645
3644
3634
3630
b3629
b3087m
ab3058s
ab3052f
ab3013f
b248m⁻
b239m⁻
b233m⁺
b226m⁻
b216m⁻
b171m⁻
b170m⁻
ab142f⁻
b130m⁻
ab124m⁻
ab123n⁻
b111m⁺
b84m⁺
b82m⁻
b77m⁺
b68m⁻
b66m
ab60s⁻
b39m⁻
b37m⁺
ab18m⁻
ab9s⁻
ab889n⁻
b815s⁺
b704s⁻
b602s⁺
b558m⁻
ab529a⁻
ab511m⁻
b477s⁻
b474m⁺
b465m⁺
ab443a⁻
ab436a⁻
ab410a⁻
b392m⁺
b381m⁺
b302m⁺
b295m
ab274m⁻
b271m
b254m⁺
b244m⁻
b698m
b236m⁺
b108s⁻
ab892a⁻
ab891a⁻
b833s⁻
ab626f⁻
b451m⁺
b420m⁻
b237m⁻
b223m⁻
ab176f⁻
ab149f⁻
b88m⁻
ab85f⁻
b83m⁻
ab74m⁻
b49m⁻
ab518s⁻
ab494s⁻
b214m⁺
b468
b281
b173
a25
2769
b1424L
ab921
ab887
696
3727u
3726u
ab3045s
ab3028f
ab2186n
3643
b3641
b604m⁺
b492m⁺
ab489s
b472m⁺
ab397f⁻
351m
348m
b182m⁺
ab127s⁻
b113m⁺
ab26m
b697m⁻
ab466f⁻
b506m⁻
b90m⁻
b66m⁻
b36m
b1m
a126m⁻
a3073f
a3029f
a3008f
2352fP
a1745m
1020
a989s
a986m⁻
a619m⁺
a601m⁺
a600a⁻
a545s⁻
a476m
a450m⁻
a220m⁺
a208s⁻
a166m⁻
a112f⁻
a23o⁻
a10m⁻
2353P
2316
2146f
1401
1033
1029
605
588m
540f
464m
202f
115m
62m
2494f
2493f
2302s
954
121a
3724u
3060f
2803f
2775f
46f
120m
209f
117o
853f
653
595
577
537m
503
502o
496m
160
151o
3072f
2354u
1762
1744
1743
1404
1403
1026
1020
855f
2794m
2793o⁻
110m
857f
583
631
441n
377f
364a
354o
299
298
280
261
249
225m
207f
188
187
87f
86m
50a
13f
881a
871
869a
856f
630m
587
552
527a
524s
508a
507a
501a
491a
483m
446a
389
2125f
2126f
2124f
2123f
1737
1731
1418
1150
1007
980
976
964
948
944
908
899m
3078f
3077f
3064f
3063f
3007f
2326
2259f
2257f
2256f
2254f
2166f
2165f
2163f
2139f
2137f
2132a
2255f
2258f
3721u
3716u
3079f
3076f
3074f
3071f
3070f
3067f
3065f
3062f
3059f
3057f
3049f
3714u
3109a
3108u
342o
211
2m
904
606
555
554
8m
IV
VA
VB
VI
VII
VIII
I
IIA1
IIA2
IIB
2499
ab 2630
2630
2558
a129
129
2236
2236
a2650
2650
2604
2604
2605
2906R, 2910R, 2912R, 2916R
2906R, 2910R, 2912R, 2916R
2624
-2225
2607
63
63
2652
538
538
152
152
57,2226
57,2226
2253
2253
712
712
644, 3603, 3050
644,3603,3050
2628
386R, 388R, 399R, 515R, 519R, 520R, 533R, 658R, 660R, 801R, 803R, 809R, 1448R, 1486R, 1487R, 2371R, 2372R, 2373R, 2497R, 2506R, 2600R, 2610R, 2611R, 2612R

Figure 3D

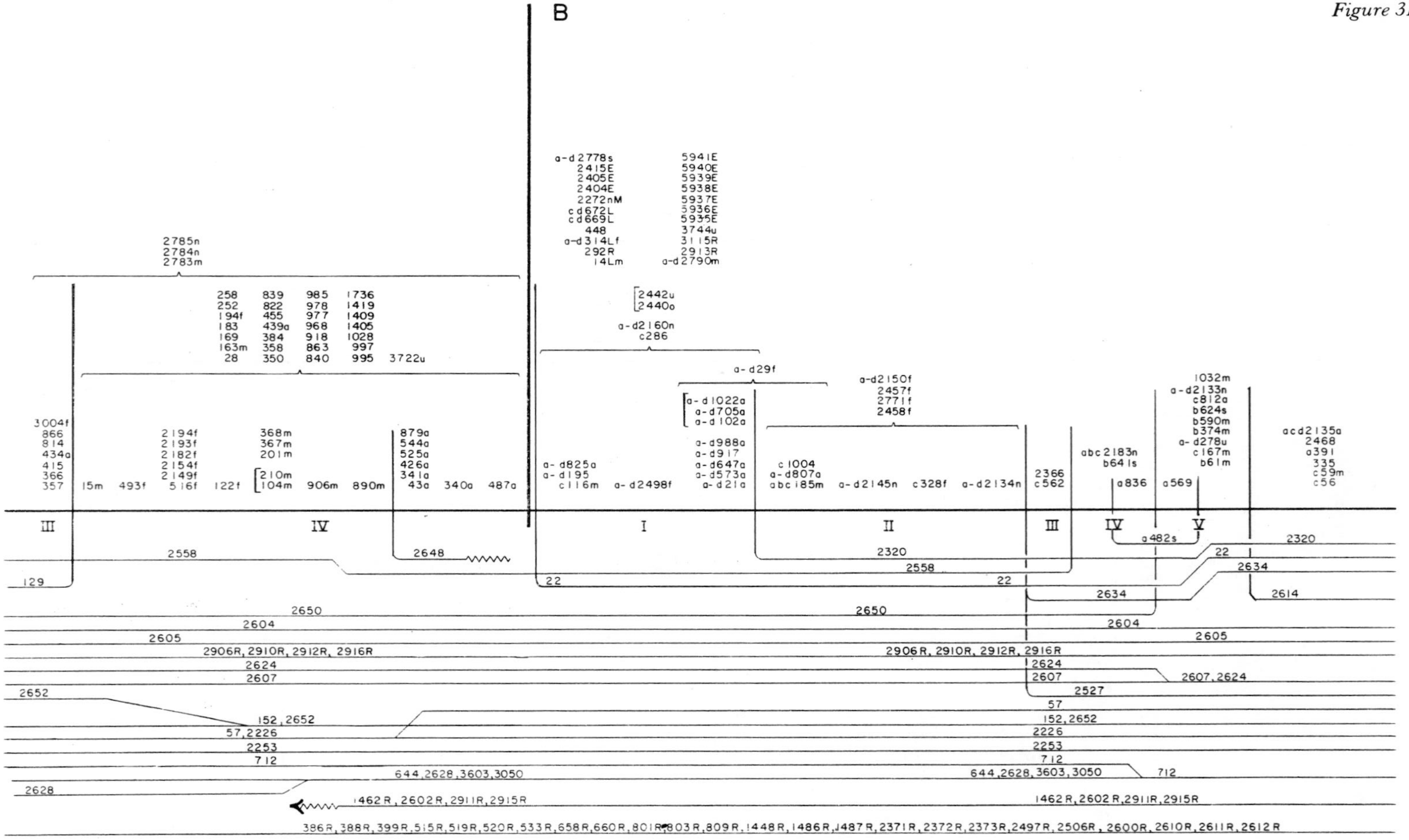
B
2785n
2784n
2783m
258 252 194f 183 169 163m 28
839 822 455 439a 384 358 350
985 978 977 968 918 863 840
1736 1419 1409 1405 1028 997 995
3722u
3004f 866 814 434a 415 366 357
15m 493f
2194f 2193f 2182f 2154f 2149f 516f
122f
368m 367m 201m
210m 104m
906m 890m
879a 544a 525a 426a 341a 43a
340a 487a
a-d2778s 2415E 2405E 2404E 2272nM cd672L cd669L 448 a-d314Lf 292R 14Lm
5941E 5940E 5939E 5938E 5937E 5936E 5935E 3744u 3115R 2913R a-d2790m
2442u
2440o
a-d2160n
c286
a-d29f
a-d1022a
a-d705a
a-d102a
a-d988a a-d917 a-d647a a-d573a a-d21a
a-d825a a-d195 c116m
a-d2498f
a-d2150f 2457f 2771f 2458f
c1004 a-d807a abci85m
a-d2145n c328f a-d2134n
2366 c562
abc2183n b641s
a836 a569
1032m a-d2133n c812a b624s b590m b374m a-d278u c167m b61m
acd2135a 2468 a391 335 c59m c56
III IV I II III IV V
a482s
2558 2648 2320 2558 22 22 22 129
2320 2634 2614 2634
2650 2650
2604 2604
2605 2605
2906R, 2910R, 2912R, 2916R
2906R, 2910R, 2912R, 2916R
2624 2624
2607 2607 2607, 2624
2527
2652 57
152, 2652 152, 2652
57, 2226 2226
2253 2253
712 712 712
644, 2628, 3603, 3050
644, 2628, 3603, 3050
2628
1462R, 2602R, 2911R, 2915R
1462R, 2602R, 2911R, 2915R
386R, 388R, 399R, 515R, 519R, 520R, 533R, 658R, 660R, 801R, 803R, 809R, 1448R, 1486R, 1487R, 2371R, 2372R, 2373R, 2497R, 2506R, 2600R, 2610R, 2611R, 2612R

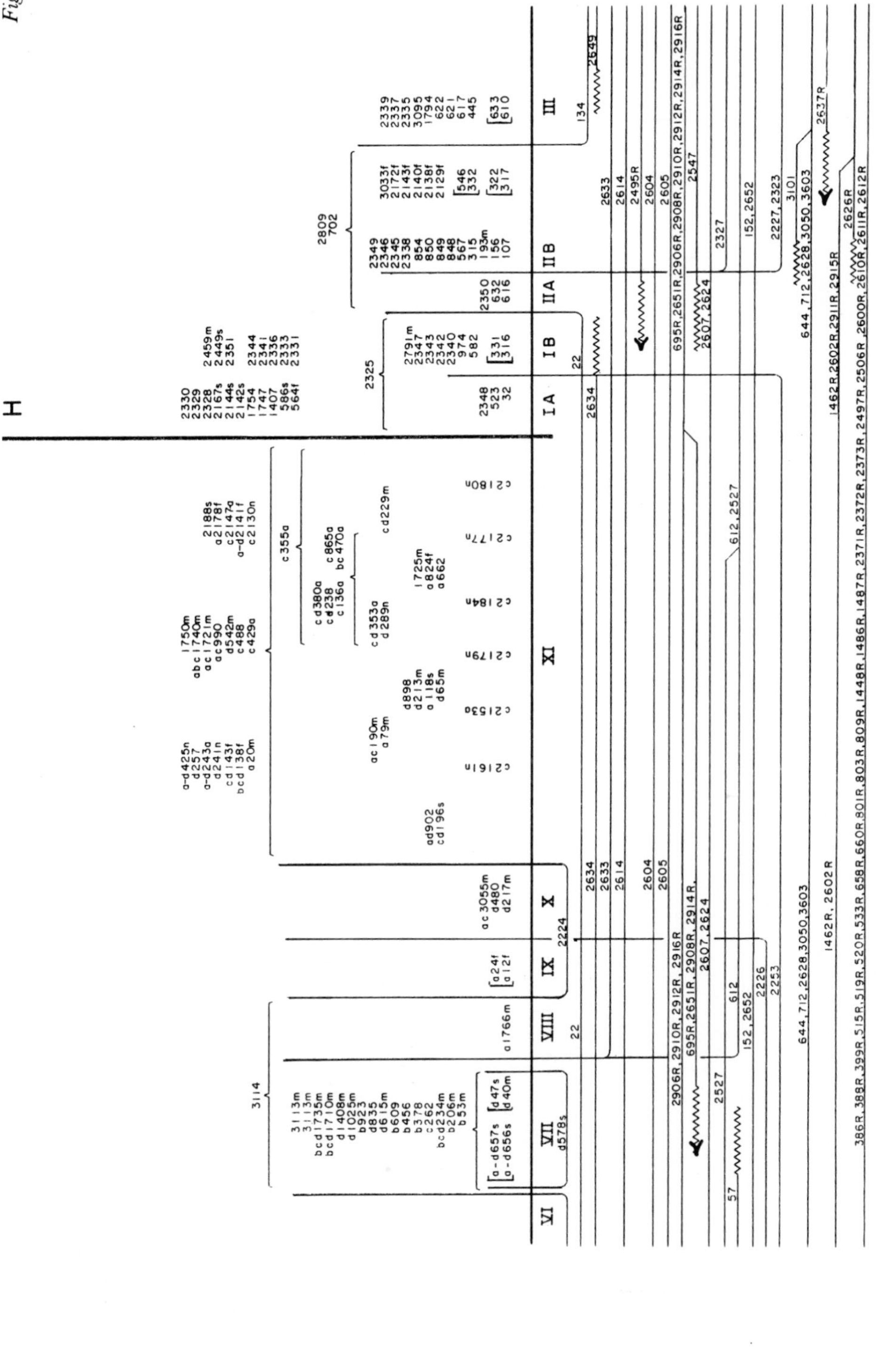

Figure 3E

Figure 3F

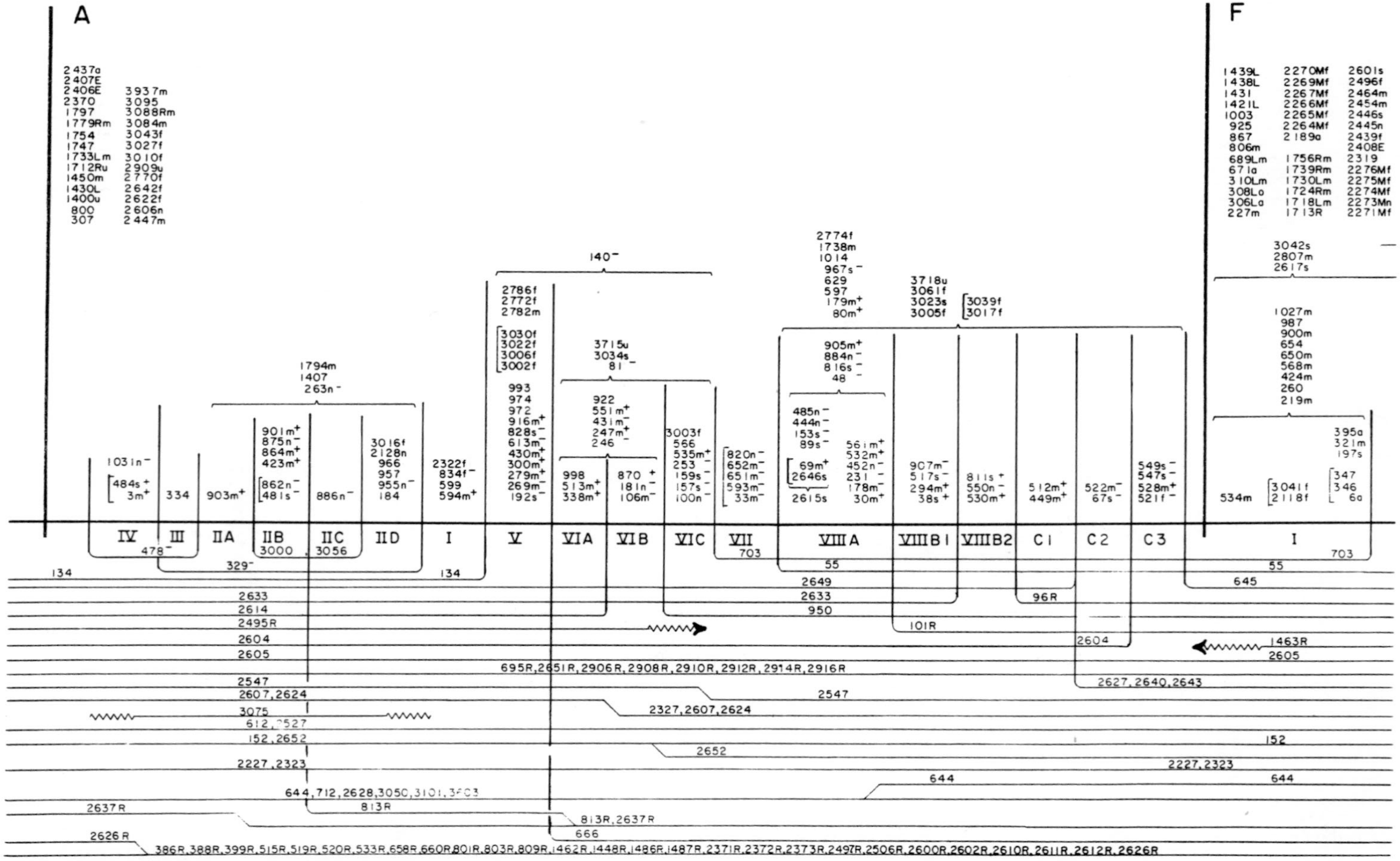
A
2437a 2407E 2406E 2370 1797 1779Rm 1754 1747 1733Lm 1712Ru 1450m 1430L 1400u 800 307
3937m 3095 3088Rm 3084m 3043f 3027f 3010f 2909u 2770f 2642f 2622f 2606n 2447m
103In⁻
484s⁺ 3m⁺
334
903m⁺
1794m 1407 263n⁻
901m⁺ 875n⁻ 864m⁺ 423m⁺
862n⁻ 481s⁻
886n⁻
3016f 2128n 966 957 955n⁻ 184
2322f 834f⁻ 599 594m⁺
140⁻
2786f 2772f 2782m
3030f 3022f 3006f 3002f
993 974 972 916m⁺ 828s⁻ 613m⁻ 430m⁺ 300m⁺ 279m⁺ 269m⁻ 192s⁻
3715u 3034s 81⁻
922 551m⁺ 431m⁻ 247m⁺ 246⁻
998 513m⁺ 338m⁺
870⁺ 181n⁻ 106m⁻
3003f 566 535m⁺ 253⁻ 159s⁻ 157s⁻ 100n⁻
820n⁻ 652m⁻ 651m⁻ 593m⁻ 33m⁻
2774f 1738m 1014 967s⁻ 629 597 179m⁺ 80m⁺
905m⁺ 884n⁻ 816s⁻ 48⁻
485n⁻ 444n⁻ 153s⁻ 89s⁻
69m⁺ 2646s 2615s
561m⁺ 532m⁺ 452n⁻ 231⁻ 178m⁻ 30m⁺
3718u 3061f 3023s 3005f
3039f 3017f
907m⁻ 517s⁻ 294m⁺ 38s⁺
811s⁺ 550n⁻ 530m⁺
512m⁺ 449m⁺
522m⁻ 67s⁻
549s⁻ 547s⁻ 528m⁺ 521f⁻
IV III IIA IIB IIC IID I V VIA VIB VIC VII VIIIA VIIIB1 VIIIB2 C1 C2 C3
478⁻ 3000 3056 329⁻ 703
F
1439L 1438L 1431 1421L 1003 925 867 806m 689Lm 671a 310Lm 308La 306La 227m
2270Mf 2269Mf 2267Mf 2266Mf 2265Mf 2264Mf 2189a 1756Rm 1739Rm 1730Lm 1724Rm 1718Lm 1713R
2601s 2496f 2464m 2454m 2446s 2445n 2439f 2408E 2319 2276Mf 2275Mf 2274Mf 2273Mn 2271Mf
3042s 2807m 2617s
1027m 987 900m 654 650m 568m 424m 260 219m
395a 321m 197s
534m
3041f 2118f
347 346 6a
I
703
134
134
55
55
645
2649
2633
2633
96R
2614
950
2495R
101R
2604
2604
1463R
2605
2605
695R,2651R,2906R,2908R,2910R,2912R,2914R,2916R
2547
2627,2640,2643
2607,2624
2547
3075
2327,2607,2624
152,2652
152
2652
2227,2323
2227,2323
644
644
813R
2637R
813R,2637R
666
2626R
386R,388R,399R,515R,519R,520R,533R,658R,660R,801R,803R,809R,1462R,1448R,1486R,1487R,2371R,2372R,2373R,2497R,2506R,2600R,2602R,2610R,2611R,2612R,2626R

Figure 3G

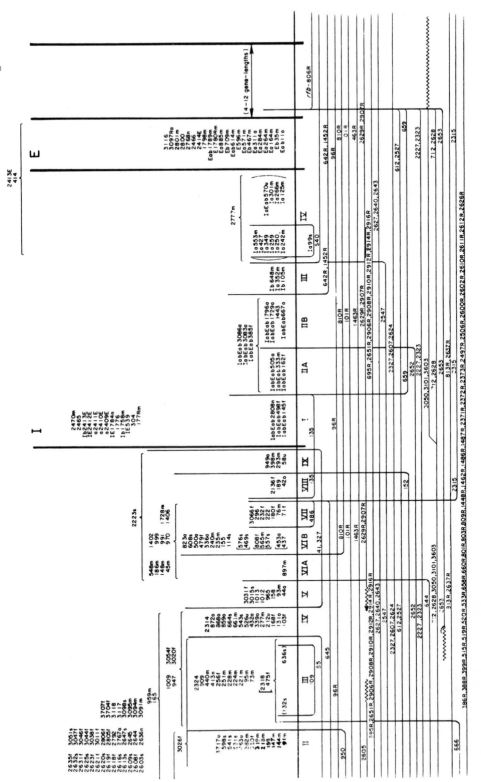

Figure 3H

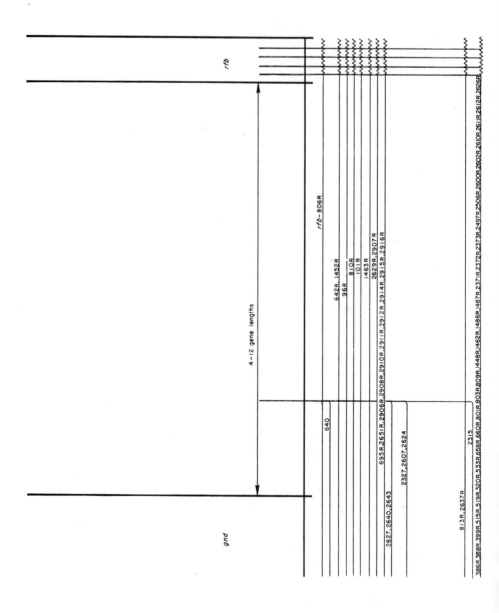

The Histidine Operon Fine-Structure Map

The position of the histidine operon on the *Salmonella* genetic map appears in Figure 2. Over 1000 mutations have been mapped in this operon (see Figure 3), and the genetics of the system have been reviewed recently (Hartman *et al.,* 1971).

Literature Cited

Brenner, M. and B. N. Ames, 1971   The histidine operon and its regulation. In *Metabolic Pathways,* Vol. 5, edited by H. J. Vogel, pp. 349–387, Academic Press, New York.

Brenner, M. and B. N. Ames, 1972   Histidine regulation in *Salmonella typhimurium.* IX. Histidine tRNA of the regulatory mutants. *J. Biol. Chem.* **247**:2302–2307.

Brenner, M., J. A. Lewis, D. S. Straus, F. De Lorenzo and B. N. Ames, 1972   Histidine regulation in *Salmonella typhimurium.* XIV. Interaction of histidyl-tRNA synthetase with histidine tRNA. *J. Biol. Chem.* **247**:4333–4339.

Chang, G. W., J. R. Roth and B. N. Ames, 1971   Histidine regulation in *Salmonella typhimurium.* VIII. Mutations in the hisT gene. *J. Bacteriol.* **108**:410–414.

Goldberger, R. F. and J. S. Kovach, 1972   Regulation of histidine biosynthesis in *Salmonella typhimurium. Curr. Top. Cell. Regul.* **5**:285–308 (edited by B. L. Horecker and E. R. Stadtman).

Hartman, P. E., Z. Hartman, R. C. Stahl and B. N. Ames, 1971   Classification and mapping of spontaneous and induced mutations in the histidine operon of *Salmonella. Adv. Genet.* **16**:1–34.

Lewis, J. A. and B. N. Ames, 1972   Histidine regulation in *Salmonella typhimurium.* XI. The percentage of the tRNAHis charged *in vivo* and its relation to the repression of the histidine operon. *J. Mol. Biol.* **66**:131–142.

Martin, R. G., M. A. Berberich, B. N. Ames, W. W. Davis, R. F. Goldberger and J. D. Yourno, 1971   Enzymes and intermediates of histidine biosynthesis in *Salmonella typhimurium. Methods Enzymol.* **17B**:3–44, (edited by H. Tabor and C. W. Tabor).

Singer, C. E. and G. R. Smith, 1972   Histidine regulation in *Salmonella typhimurium.* XIII. Nucleotide sequence of histidine tRNA. *J. Biol. Chem.* **247**:2989–3000.

Singer, C. E., G. R. Smith, R. Cortese and B. N. Ames, 1972   Mutant tRNA[His] ineffective in repression and lacking two pseudouridine modifications. *Nat. New Biol.* **238**:72–74. **238**:72–74.

15

Streptomyces coelicolor

David A. Hopwood and Keith F. Chater

Introduction

Three features of *Streptomycetes* formed the initial stimulus for making genetic studies of them: their taxonomic distance from the genetically well-studied Gram-negative bacteria, their morphological complexity, unique among prokaryotes and perhaps of general interest in the study of morphogenesis (Chater and Hopwood, 1973), and their role as major industrial producers of antibiotics. The hundred or so genetic markers that have been identified in the most-studied strain, *Streptomyces coelicolor* A3(2) (*Streptomyces violaceoruber* according to Kutzner and Waksman, 1959) have all turned out to be located on a single circular linkage map (see Figure 1) in a curiously symmetrical arrangement (Hopwood, 1967), and a unique fertility system governing various aspects of recombinant formation has become apparent (Hopwood *et al.*, 1969; Vivian and Hopwood, 1970, 1973; Vivian, 1971; Hopwood and Wright, 1973). These novel features make *S. coelicolor* an organism of considerable importance in comparative bacterial genetics.

In all its ultrastructural and molecular features, *S. coelicolor* is a typical Gram-positive prokaryote. It is unusual, however, in its morphological complexity (see Figure 2). In the growth of a colony from a spore on agar medium, germination is followed by the growth across and into the medium of a loosely packed, much-branched, septate mycelium. This initially smooth and colorless colony later becomes covered with

David A. Hopwood and Keith F. Chater—John Innes Institute, Norwich, England.

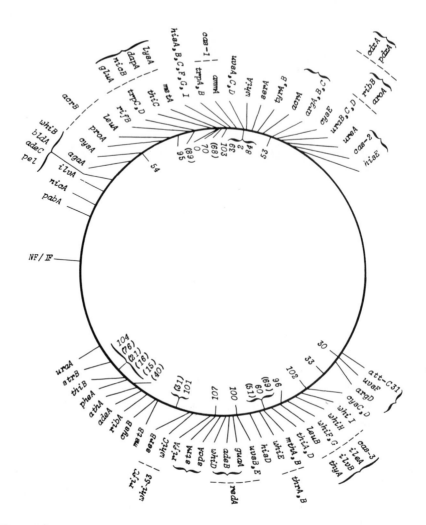

Figure 1. Linkage map of S. coelicolor. The locus designations are as given by Hopwood et al., 1973. Numbers on the inside of the circle represent indispensable temperature-sensitive mutations (Hopwood, 1966a), with parentheses indicating mutations which have now been lost. The orders within groups of bracketed loci are unknown. Certain loci have not been ordered relative to groups of loci covered by dotted lines.

white densely packed aerial growth (Wildermuth, 1970). The white colonies soon turn gray, as the aerial hyphae become spore chains. This metamorphosis involves first the synchronous formation of closely and regularly spaced sporulation septa within the long cells of the aerial hyphae, and then the maturation into spores of the compartments so formed (Wildermuth and Hopwood, 1970). During maturation, thick spore walls are laid down, and each spore changes from a cylindrical to an ellipsoidal shape. Ultimately the spores in a chain are joined to one another only at a very small interface. The mature spores are not resistant to high temperatures as are the very different endospores found in some other groups of bacteria, but they are somewhat dessication-resistant. Mutants have been obtained and partially characterized that are unable to carry out stages in the sporulation process (*white*) (Hopwood *et al.*, 1970; Chater, 1972; Chater and Hopwood, 1973), or in aerial mycelium production (*bald*). It should presumably be possible to obtain other mutants with defects in the processes of spore germination and mycelial branching.

Because of the filamentous growth habit of *S. coelicolor*, direct studies of the mating process are less easy than in unicellular organisms such as *Escherichia coli*. Thus most of what we know of the process has been inferred from the analysis of unselected marker segregation in recombinants. Mating takes place on a solid medium densely seeded with parental spores or hyphal fragments. Recombinants can be detected in the

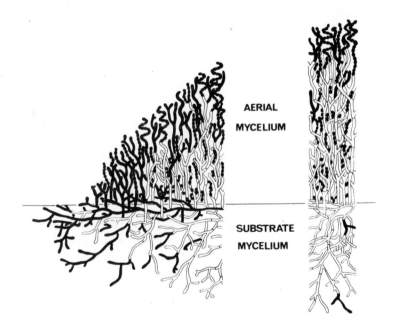

AERIAL MYCELIUM

SUBSTRATE MYCELIUM

Figure 2. Idealized diagram of a vertical section through the center of a Streptomyces coelicolor colony at the climax of sporulation. Black represents intact cells and white represents disintegrating or completely lysed cells (from Wildermuth, 1970).

mycelium of a mating mixture soon after germination of parental spores (Hopwood, 1970b), but because the recombinant chromosomes are contained at this stage in multinucleate hyphae they are not in general suitable for genetic analysis. Normally, mating mixtures are allowed to sporulate before harvesting and plating for recombinants, because the spores are uninucleate and thus segregation occurs in them. It is apparent from genetic analysis of spore progeny that comparatively large fragments of genetic material are usually transferred during mating, on average about one sixth of a whole chromosome, indicating that a kind of conjugation is the means of genetic exchange (Hopwood, 1967).

Certain combinations of strains are much more fertile than others; the proportion of recombinants recovered from a cross (fertility level) ranges from 0.001 percent (or occasionally even lower) to 100 percent. In the last few years three different fertility types have been defined: NF (normal fertility), IF (initial fertility) (Vivian and Hopwood, 1970), and UF (ultra-fertility) (Hopwood et al., 1969). More recently, certain other fertility classes have been added to the first three (Vivian and Hopwood, 1973; Hopwood and Wright, 1973). The behavior of all these fertility types has been interpreted in terms of a plasmid, first postulated and named as SCP1 by Vivian (1971). The plasmid, apart from acting as a sex factor, confers on strains that harbor it the capacity to produce, and be resistant to, a diffusible substance that inhibits UF strains (Vivian, 1971), and these properties have proved to be very useful phenotypic markers for the plasmid. Current hypotheses concerning the nature of the different fertility types are presented below. A fuller account can be found in the review by Hopwood et al. (1973).

In the IF fertility type, SCP1 is autonomous. The IF strains give rise to UF variants, lacking the plasmid, at a high frequency (about 0.1 percent) which is increased up to tenfold by ultraviolet and x irradiation but not by N-methyl-N′-nitro-N-nitrosoguanidine treatment. The plasmid is transferred very efficiently so that when the spore progeny of an IF × UF cross are tested, nearly all those having the genetic markers of the UF parent are found to be IF (Vivian, 1971).

In the NF fertility type the plasmid is integrated, in some configuration which is still obscure, into the chromosome at a position corresponding to 9 o'clock on the linkage map. In an NF × UF cross, as in an IF × UF cross, the plasmid is transferred to virtually 100 percent of the spore progeny. However, since in the NF strain it is attached to the chromosome, it takes with it fragments of chromosome of varying length, but all including the 9 o'clock region. Following crossing over between the resident UF chromosome and the incoming NF fragment, the zygotes give rise to recombinants; all of them inherit the integrated plasmid. Chromosomal markers from the NF parent are inherited by the recombinants with frequencies which fall with the distance of the marker loci from 9 o'clock. A surprising feature of such "ultra-fertile" crosses is that asexual sporulation of the parents does not occur; essentially *all* the spore progeny are recombinants (Hopwood *et al.*, 1969).

The fertility of UF × UF crosses is usually 0.001 percent or less. The fertility is about tenfold higher in IF × UF crosses; some or all of this extra fertility is due to the occurrence, within any IF culture, of clones of various donor classes in which SCP1 has interacted with the chromosome at various points and with varying results. It is possible to isolate many kinds of such donor clones by an indirect selection procedure (Vivian and Hopwood, 1973; Hopwood and Wright, 1973). The simplest kind of donor, so far represented by a single example, is one in which SCP1 is still autonomous but has acquired an insertion of a short chromosomal region carrying the wild-type allele of a chromosomal locus, in this case *cysB*. Thus the strain is analogous with an F′ strain of *Escherichia coli* K12, and has been called SCP1′–*cysB* (Hopwood and Wright, 1973). In a cross with a *cysB* UF strain, virtually 100-percent conversion of the UF parent to phenotypically Cys$^+$ heterozygous progeny occurs. Some of the other classes of donors described by Vivian and Hopwood (1973) may represent SCP1′ strains bearing unmarked chromosomal regions, while still others may represent integrations of the plasmid to give strains resembling NF except for the fact that the fragments of donated chromosome appear to extend on only one side of the site of integration. Much still needs to be learned about these newer classes of donors, and they have not yet found a place in routine genetic studies.

In crosses between strains of the same fertility type (NF × NF; IF × IF; UF × UF), different regions of the chromosome are probably donated with approximately equal frequency (crosses are *symmetrical*), and both parents can act alternatively as donor or recipient in the same cross (crosses are *nonpolarized*) (Hopwood, 1967; Vivian and Hopwood, 1970). Approximate fertility levels in these crosses are 1 percent (NF × NF), 0.01 percent (IF × IF) and 0.001 percent (UF × UF). In contrast, NF × IF crosses, and, as we have seen, NF × UF crosses, are *polarized*, with the IF or UF parent acting as recipient of chromosome fragments from the NF donor; they are also *asymmetrical*, since these fragments are nonrandom segments of the chromosome, always including the 9 o'clock region. The fertility level is in the region of 10 percent in NF × IF crosses, compared with the near 100 percent for NF × UF crosses. In contrast to the situation in NF × UF crosses, where, as we have seen, all progeny are NF, there is a segregation of NF and IF among the progeny of NF × IF crosses; in other words, the NF 9 o'clock region, although always included in the zygotes, is not obligatorily inherited by the recombinant progeny deriving from these zygotes.

The study of the fertility system of *S. coelicolor* is developing rapidly at the present time and a complete picture, particularly in molecular terms, is still lacking. Further information on the possible interactions of SCP1 with the chromosome will undoubtedly lead to increased flexibility of genetic manipulation of the organism.

Further flexibility would be added by the availability of a transduction or transformation system. There have been occasional reports of transformation in *Streptomyces* (Sermonti, 1969; Hopwood, 1972*b*), but no reproducible system has yet been developed. Since *S. coelicolor* produces extracellular DNase (Hopwood and H. M. Wright, unpublished), such studies may require the isolation of DNase-less mutants.

Although actinophages have been isolated and studied for a long time (Welsch, 1969), there is to our knowledge only one published report of transduction in *Streptomyces* (Alikhanian *et al.*, 1960), and this has not led to a workable transducing system. In our laboratory it has proved relatively easy to isolate phages attacking *S. coelicolor* A3(2), both virulent and temperate, and to initiate genetic studies of them (Dowding, 1973; Dowding and Hopwood, 1973), while a recent report by Lomovskaya *et al.* (1971) describes a temperate phage (ϕC31) released from *S. coelicolor* A3(2). Apart from possibly allowing transduction, phage systems will doubtless also prove of value in the recognition of nonsense suppressors and in the study of repair mechanisms, morphogenesis, and restriction/modification (Boyer, 1971).

Technical Considerations

Stock Cultures

Cultures are normally grown on agar slants for about a week, and then stored in a refrigerator. Viability is satisfactory after long periods of storage, even years, provided good sporulation occurred before storage. However, the viability of morphological mutants that produce no spores, and of certain auxotrophs that sporulate badly, is soon lost. The use of simply prepared lyophils (Hopwood and Ferguson, 1970) overcomes the long term storage problem. Recently we have also found it convenient for short- and medium-term storage to suspend spores or mycelial fragments in 20 percent glycerol and store them frozen at −20°C. Repeated freezing and thawing has no detectable effect on viability in these conditions (Chater, unpublished), so that the technique is particularly valuable in the storage of mutagenized spores and the progeny of crosses.

Media

The two agar media for routine use are the "minimal" medium and "complete" medium described by Hopwood (1967). For phage work, Dowding (1973) has used Difco nutrient broth or nutrient agar supplemented with 0.5 percent glucose and 4 mM $Ca(NO_3)_2$.

Liquid cultures are grown in shake flasks inoculated with spore suspensions (for example, 25–50 ml in 250 ml Erlenmeyer flasks, or 500 ml in 2000 ml flasks). Suitable media are liquid minimal medium supplemented with 0.6 percent Bacto Casaminoacids, or nutrient broth containing 34 percent sucrose, 1 percent $MgCl_2$, and 0.5 percent glucose (M. E. Townsend, private communication). Growth may be slower in the latter medium, but lysis of the culture by lysozyme occurs much more readily (Danford and Frea, 1970).

S. coelicolor grows as coherent pellets of mycelium in shaken liquid culture. The pellets are large in most normal media and are formed by the aggregation of numerous germinated spores. In the second liquid medium, clumping is much less marked so that the culture is more uniform and more finely divided.

Preparation of DNA

After growth in the second of the liquid media described above, the mycelium is harvested by centrifugation, washed and suspended in Tris-EDTA buffer (0.2 M tris, 0.1 M EDTA) pH 8. Lysis is achieved by incubating with lysozyme (e.g., 2.5 mg/ml) at 37°C for 15 minutes, followed by the addition of detergent (e.g., 2.5 percent SDS). The rest of the procedure for DNA extraction is essentially as described by Marmur (1961). Recently Wilcockson (1973) has extracted *S. coelicolor* nucleic acids by an alternative method involving treatment with lysozyme, sodium dodecyl sulfate, and perchlorate.

Mutagenesis

Ultraviolet light (UV) (Harold and Hopwood, 1970*a*), N-methyl-N′-nitro-N-nitrosoguanidine (NTG) (Delíc *et al.*, 1970), near ultraviolet light in the presence of 8-methoxypsoralen (NUV + MOP) (Town-send *et al.,* 1971), and ethyl methane sulfonate (EMS) (Engel, 1970) have been used. Tests of forward mutation to auxotrophy have shown that NTG is the most potent mutagen. MOP + NUV is also consider-ably more effective than UV or EMS. A special feature of MOP + NUV, indicated by studies on another microorganism (Scott and Alder-son, 1971), is its apparent lack of cistron specificity; while a peculiarity of NTG is its tendency to induce closely linked multiple mutations (Randazzo *et al.,* 1973). The former phenomenon may be an advantage in the isolation of a *random* selection of mutations; the latter tendency will usually be disadvantageous when the mutants are to be used in analytical studies, though it may be acceptable when NTG offers the best chance of isolating a rare class of mutant.

Mutant Isolation

Apart from the isolation of mutants resistant to antibiotics and phages, and of prototrophic revertants, selective or enrichment methods have not been developed. Instead, replica-plating and visual methods have been used extensively. Replica plating serves to isolate auxotrophs (Hop-wood and Sermonti, 1962), temperature-sensitive mutants, growing poorly or not at all at 38°C though normally at 30°C (Hopwood, 1966*a*), and ultraviolet-sensitive mutants (Harold and Hopwood, 1970*a*); in each case, about 200 colonies per dish can be examined. Morphological mutants, i.e., the *bald* and *white* colony mutants described earlier, are recognized on plates carrying many hundreds or even thousands of colonies (Hopwood *et al.,* 1970). Agar-overlay methods have resulted in the isolation of mutants lacking urease (Hopwood, 1965) and mutants of another strain of *S. coelicolor,* strain K673, lacking an antibiotic active against strain A3(2) and other *Streptomycetes* (Hopwood, 1970*a*).

In the isolation of resistant mutants, an opportunity for phenotypic and segregational delay is provided either by growth of mutagenized spore suspensions in liquid medium for about 6 hours before plating on strep-tomycin (50 μg/ml for high-level resistance, *strA*, or 5 μg/ml for low-level resistance, *strB*), acriflavine (0.0015 percent), spectinomycin (50 μg/ml) or rifampicin (50–200 μg/ml) or by plating mutagenized suspensions directly on cellophane disks placed on medium lacking inhibitor, then transferring the disks after 18 hours of incubation to medium containing inhibitor (Chater, 1974).

Revertants of auxotrophic mutations, some of which have served to identify suppressor loci (V. Najfeld, private communication), can be isolated by direct plating on selective plates.

Figure 1 shows the location of the loci on the circular linkage map. The characteristics of the mutants used to define these loci are listed elsewhere (Hopwood *et al.*, 1973).

Crossing Techniques

By far the most widely used crossing procedure is the preparation of simple mixed cultures of two parent strains on nonselective agar medium (usually the complete medium). When the cross is to be analyzed selectively, all that is necessary is a dry loopful of spores and/or mycelial growth from a recent culture (no more than two months old) of each parent, crudely mixed together on a fresh slant, usually in a 25 × 150 mm test tube. For nonselective analysis, more care is needed to ensure good mixing of the parents and an adequate density of each. Young (no more than two weeks old), well-sporulating cultures are needed from which an inoculum is prepared by scraping a small area of the culture with a loop already loaded with a drop of water. The resulting loopful of spore suspension is transferred to a little liquid (0.2–0.5 ml) at the bottom of a fresh slant and the procedure is repeated with the second parent. The resulting suspension is then mixed with the loop and spread over the entire surface of the slant. Alternatively, nonselective crosses may be made by harvesting a whole slant of each parental culture and preparing a centrifuged pellet of the resulting filtered material. After resuspending the pellets in a small volume of water, they are mixed and drops of the mixture spread over new slants. This procedure is suitable when one or both parents fail to sporulate abundantly.

Crosses are usually harvested and plated after four days incubation at 28–30°C, and, when they are to be analyzed selectively, they may often be harvested successfully after only two days incubation. Full details of harvesting and plating procedures are given in the article by Hopwood (1967).

Identification and Interconversions of Fertility Type

It is often important to know the fertility type of strains and, equally, to be able to change it. The routine identification of fertility type is done by simple tests on plates, with up to 20 tests on each (Hopwood *et al.*,

1969). Strains are inoculated on a master plate and incubated to form patches of growth. These are replica-plated to two complete-medium plates, one spread with an NF, the other with a UF tester strain as a dense spore suspension. During subsequent incubation, mating takes place in each patch, and spore progeny are formed which are then replica-plated to a selective medium. Strains that are UF give dense patches of recombinants with the NF, but virtually none with the UF tester; NF strains give sparse recombinants with the NF but dense patches with the UF tester, and IF strains give intermediate levels with the NF, and very few recombinants with the UF tester.

Interconversion of IF and UF fertility types is straightforward. After ultraviolet irradiation of IF spores to about 0.1 percent survival, about 1 percent of the survivors are UF (Vivian and Hopwood, 1970). These are identified after growth of survivors to give 100–300 colonies per plate, by the standard replica-plating test using an NF strain. Conversion of UF to IF is achieved by crossing the UF strain with a differently marked IF strain, and plating the spore progeny on a medium selective for the UF markers at a low density. The standard replica-plating test against an NF tester will show that nearly all the colonies are now IF (Vivian, 1971).

Interconversions between NF and either IF or UF are less easy and involve the generation of recombinants. Thus, the production of NF strains completely isogenic with IF or UF strains is not yet possible. The simplest transition is UF to NF, since a considerable number of progeny of a UF × NF cross possess NF fertility but the UF parental markers (Hopwood *et al.*, 1969) (the number increases as the distance of the nearest markers to 9 o'clock increases). The conversion of IF to NF can be achieved in a two-stage process via the UF obtained by irradiation of the IF. Apparent conversion of IF to NF by mutagenic treatment in two stages via new kinds of donor strains has also been obtained recently (Vivian and Hopwood, 1973), though it has not been put to any practical use. Since NF strains possess the SCP1 plasmid that UF strains lack, mutation from UF to NF is presumably impossible.

Conversion of NF to IF can only be done by crossing NF with IF, when the NF/IF difference segregates with the 9 o'clock region of the map. Selection has to be such that IF recombinants of the desired genotype can be obtained as a relatively frequent class. Unfortunately, this always means that the generated IF lacks at least one of the NF markers. (If the cross is done nonselectively, IF recombinants having the NF parental markers will usually, though not always, be too rare to detect economically.) The production of a UF from an NF strain has to proceed via the IF because all recombinants of a UF × NF cross are of NF type.

Since all fertility combinations are fertile, at least at a low frequency, it goes without saying that homologous crosses (NF × NF, IF × IF, or UF × UF) can be used to prepare new recombinant strains of a desired fertility type provided that a suitable pair of parent strains of the same fertility type is available; as in the case of the NF to IF or UF changes just described, a counterselected marker of each parent is lost in the process.

Phage Techniques

Methods for the study of phages of *S. coelicolor* have recently been described (Lomovskaya *et al.*, 1971, 1972, 1973; Dowding, 1973; Dowding and Hopwood, 1973). Most of the techniques are variants of well-established bacteriophage methods modified to take account of the mycelial growth habit of the host, and they include the isolation of a temperate phage ϕC31 from strain A3(2), and of virulent and temperate phages from soil; phage mutagenesis and recombination; curing and lysogenization; and genetic mapping of prophage attachment sites.

Genetic Analysis

Genetic Mapping

S. coelicolor is well adapted for coarse genetic mapping. New mutants are crossed with a multiply marked NF strain such as 1258 (see Figure 3), and selection is made among the spore progeny for streptomycin resistance and for, e.g., histidine independence. Other markers are left unselected and, after recombinant colonies have been inoculated in an orderly array on a "master plate," the inheritance of unselected markers is scored by replica plating on suitable diagnostic media. Analysis of the data is done in two stages (Hopwood, 1967, 1972a). Since the frequency of each unselected marker is related to its closeness to the selected marker, a first step is to calculate these frequencies. This gives two possible locations for the new mutation in relation to the standard markers, one in each of the arcs separating the selected markers. In the second step, the segregation of the new mutation with respect to that of a potentially closely linked marker from each arc is tabulated, thus allowing assessment of the extent to which segregation is independent. Independent segregation should only be shown by markers separated by the selection points. This kind of analysis is applicable to UF × UF, IF × IF, NF × NF, and IF × NF crosses, though interpretation of IF × NF data may be

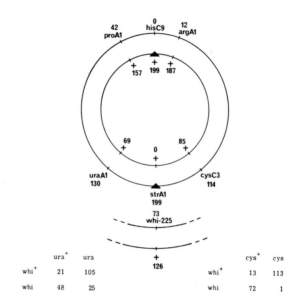

Figure 3. Mapping a new mutation by selective analysis. A prototrophic streptomycin-sensitive strain carrying whi-225 (inner circle) was crossed with strain 1258 (outer circle) and recombinants bearing the markers strA1 and hisC9⁺ (indicated by triangles) were selected. The allele frequencies of nonselected markers in a sample of such recombinants are shown. The frequency of whi-225⁺ gives a map position either between uraA1 and proA1, or between cysC3 and strA1. Analysis of the segregation of whi⁺/whi in relation to that of ura⁺/ura or cys⁺/cys (see tabulations) indicates a location for whi-225⁺ between cysC3 and strA1, since the segregation of whi⁺/whi is much more highly correlated with that of cys⁺/cys than with that of ura⁺/ura (from Chater, 1972).

slightly complicated by the polarization and zygotic asymmetry of such crosses (Hopwood *et al.*, 1970).

An alternative operationally nonselective approach to coarse mapping can be used when the unmapped mutation is carried by either a UF or an NF strain. The strain is crossed with a suitable marked tester NF or UF strain, respectively, and the spore progeny of the mating mixture are spread on nonselective medium at dilutions giving discrete colonies. The inheritance of all markers is scored by replication from master plates in the usual way (Figure 4). Since the frequency of markers originating in the NF parent falls with increasing distance from the 9 o'clock region, the frequency of the unmapped mutation among the progeny immediately gives two possible locations, one being clockwise, the other anticlockwise, of the 9 o'clock region. The choice between these locations is made by considering the degree of independence of segregation of the unmapped mutation with that of potentially closely linked markers. UF × NF crosses may be particularly useful in strain construction since it is not necessary to lose markers by nutritional selection of recombinants.

More precise mapping in *S. coelicolor* is relatively difficult, and it becomes more so as greater resolution is required. This is in part a consequence of the relatively long fragments of genetic material transferred. The very frequent occurrence of heterozygous, partially diploid, plating units (heteroclones, see below) is also a hindrance (Hopwood, 1970*b*). The most reliable approach to sequencing two closely linked mutations is to make a cross in which they lie between two nearby markers which are used in selecting against the parent strains. Of the four possible classes of recombinants, three can be generated by a single crossover between the selected markers, while the fourth requires three crossovers and is therefore rarer (Hopwood, 1967). Where the mutations to be sequenced have a similar phenotype, it is necessary to make such crosses in two different coupling arrangements (Harold and Hopwood, 1970*b*; Chater, 1972). The most laborious aspect of mapping by this technique is the construction of suitably marked strains.

Recent work by Sermonti *et al.* (1971) has suggested that, at least in certain types of cross, the transfer of genetic material proceeds from a fixed origin and may be interrupted by mechanical means. However, there

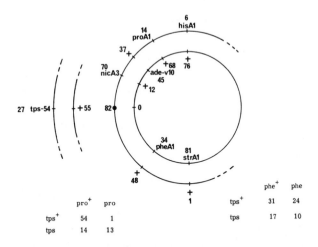

Figure 4. Mapping a new mutation by nonselective analysis. An NF strain carrying tps-54 (outer circle) was crossed with a UF strain (inner circle). The allele frequencies at each locus in a sample of nonselected progeny are shown. The frequency of tps-54 gives a map position either between ade-v10$^+$ and proA1 or between pheA1$^+$ and strA1$^+$. Analysis of the segregation of tps$^+$/tps in relation to that of pro$^+$/pro or phe$^+$/phe (see tabulations) indicates a location for tps-54 between ade-v10$^+$ and proA1, since the segregation of tps$^+$/tps is highly correlated with that of pro$^+$/pro but independent of that of phe$^+$/phe.

are many difficulties inherent in this kind of interrupted mating, which differs from the classical *E. coli* procedure, and it is not likely that it will prove very useful for genetic mapping, though it will undoubtedly be informative about the processes leading up to recombinant formation.

The analysis of the pattern of segregation of heterozygous markers from heteroclones (see below) has also been used for mapping (Hopwood and Sermonti, 1962; Hopwood, 1965) and for calculating map distances over long intervals more accurately than is possible with normal haploid analysis (Hopwood, 1966b).

Complementation and Dominance Tests

As with other prokaryotes, no permanently diploid or heterokaryotic stage exists as a regular part of the life cycle of *S. coelicolor*. This provides an obstacle to the routine performance of functional tests. However, in certain circumstances, unstable partially diploid recombinants (termed heteroclones) can be detected, and it is often possible to utilize them in functional tests. Heteroclones are most frequent when two closely linked nonallelic recessive markers are selected against in the *trans* configuration; this, conveniently, is the very circumstance in which complementation testing is most often needed. Thus, the appearance of heteroclones in given selective conditions indicates complementation between the selected markers. Heteroclones were first observed by Sermonti *et al.* (1960) as small, irregular colonies, each giving rise to genetically heterogeneous populations of spores; most of these contained only one of the selected markers, with the second locus being represented by the recessive counterselected allele. Thus, most heteroclones did not grow when transferred by replica plating to the same medium as that on which they had formed. Moreover, where the parents differed at other nearby loci, both alleles of these loci were represented among the progeny of many individual heteroclones (Hopwood *et al.*, 1963). The simplest interpretation of these observations (Hopwood, 1967) is that in the mating mixture, wherever an odd number of crossovers occurs between a circular recipient chromosome and a linear fragment of donor chromosome, a terminally redundant (that is partially diploid) linear structure is generated (Figure 5, I and II) that may then become incorporated into a spore. If the linear structure contains both selected markers and is heterozygous for at least one of them, the spore will initiate heteroclone formation on the selective medium. Since a single crossover in the diploid region is sufficient to regenerate a closed circular chromosome from the terminally redundant structure (Figure 5, IV and V), heteroclone chromosomes are unstable.

When the selected markers are widely separated, this crossover will often occur between them early in the growth of the heteroclone colony, thus generating a stable recombinant chromosome in a cell which will then outgrow the heteroclone colony to form a colony indistinguishable from a normal haploid recombinant colony. If, on the other hand, the selected markers are very closely linked, the crossover regenerating circularity is likely to occur outside them, and the regenerated circular chromosomes, therefore, do not usually carry the two selected alleles and so do not allow the cells housing them to grow in the prevailing selective conditions. However, such cells are not selected against in the terminal process of colony development, namely sporulation, when no further growth occurs. Thus, the majority of spores on heteroclone colonies, being of this type, are unable to grow when replicated to the same selective medium. This is the major criterion for the recognition of heteroclones in complementation tests.

It is possible directly to test complementation between two *unmapped* auxotrophic mutations, provided that the strains that carry them differ in

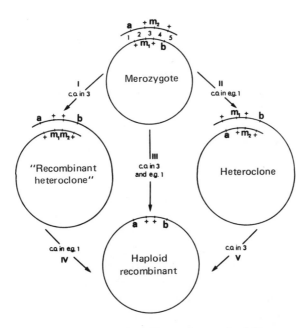

Figure 5. The generation of heteroclone or haploid recombinant genomes by different patterns of crossing-over within a merozygote. In the figure, m_1 and m_2 represent closely linked counter-selected markers, while a and b represent outside, nonselected, markers.

respect of two other selectable markers that are known not to be closely linked. Spores harvested from a mating mixture of the two strains are plated on two different selective media; one selecting only against the standard markers, the other only against the mutations under test. Three results are possible: (1) approximately equal numbers of haploid recombinant colones on both media, indicating that the two new mutations being tested are not closely linked; (2) many more colonies on the medium counterselecting the standard markers than on the other medium, and no heteroclones on the latter, indicating that the new mutations are allelic; and (3) many more haploid colonies on the medium counterselecting the standard markers than on the other medium, but heteroclones present on the latter, usually in numbers not more than an order of magnitude less than the haploids on the first medium, indicating that the two new mutations are closely linked but nonallelic (Hopwood, 1967).

Harold and Hopwood (1970*b*) carried out complementation tests on closely linked, nonselectable, ultraviolet-sensitive (*uvs*) mutants by an indirect heteroclone technique. Selection was made for heterozygosity at two complementing histidine loci close to the *uvs* genes, and the uv sensitivity of such heterozygotes was measured by counting the relative numbers of heteroclones surviving a dose of uv sufficient to kill only 50 percent of spores having wild-type sensitivity, but sufficient to kill more than 99 percent of uv-sensitive spores. Three complementation groups were obtained. More recently, Harold and Hopwood (1972) have successfully adapted this technique to rapid plate complementation tests (up to 6 tests per plate plus controls) by utilizing the high fertility of UF × NF crosses.

Variants of the indirect heteroclone complementation test should also be applicable to dominance and complementation tests with certain resistance and morphological mutations, though such experiments have not been reported.

A complementation test utilizing apparently heterokaryotic growth has been described for certain auxotrophic mutants (Russi *et al.,* 1966). Spore mixtures of two mutants being tested were spread on cellophane lying on medium containing suboptimal quantities of the growth requirement. After a 24-hour incubation, the cellophane was transferred to medium devoid of this requirement. In some cases, tufts of vigorous heterokaryotic growth were obtained after further incubation; in such cases, the mutations were classified as complementing. The absence of these tufts when the cross was heterozygous for streptomycin resistance and made on medium containing streptomycin supported the notion that the tufts were heterokaryotic.

It is to be hoped that, following the recent isolation of a strain harboring the substituted plasmid SCP1′–*cysB* (Hopwood and Wright, 1973), such plasmids will become available for making dominance and complementation tests of the kind so widely done with F-heterogenotes of *E. coli* and *Salmonella typhimurium*.

Conclusion

Although this article is about *S. coelicolor* A3(2), the reader is referred to genetic studies with other actinomycetes of the genera *Streptomyces, Nocardia,* and *Micromonospora* which may be useful because of the applicability of at least some of the techniques worked out for *S. coelicolor* to these other organisms (see Hopwood, 1972*a*, for a review).

In particular, a method of linkage analysis which is merely of historical interest in *S. coelicolor* genetics (Hopwood, 1959), is very useful in the early stages of developing the linkage map of a new actinomycete. This method, which has been called the "four on four" technique (Hopwood, 1972*b*) because it involves the study of four-factor crosses by plating on four different selective media, has been described elsewhere (Hopwood, 1972*a*), and a good example of its use is in a study of the linkage map of *Streptomyces rimosus* (Friend and Hopwood, 1971).

Literature Cited

Alikhanian, S. I., T. S. Iljina and N. D. Lomovskaya, 1960   Transduction in actinomyces. *Nature (Lond.)* **188:**245–246.

Boyer, H. W., 1971   DNA restriction and modification mechanisms in bacteria. *Annu. Rev. Microbiol.* **25:**153–176.

Chater, K. F., 1972   A morphological and genetic mapping study of white colony mutants of *Streptomyces coelicolor. J. Gen. Microbiol.* **72:**9–28.

Chater, K. F., 1974   Rifampicin resistant mutants of *Streptomyces coelicolor* A3(2). *J. Gen. Microbiol.* **80:**277–290.

Chater, K. F. and D. A. Hopwood, 1973   Differentiation in actinomycetes. In *Microbial Differentiation, 23rd Symposium of the Society for General Microbiology,* edited by J. M. Ashworth and J. E. Smith, pp. 143–160, Cambridge University Press, Cambridge, England.

Danford, T. R. and J. I. Frea, 1970   Protoplasts in actinomycetes. *Bacteriol. Proc.*: 37–38.

Delic, V., D. A. Hopwood and E. J. Friend, 1970   Mutagenesis by N-methyl-N′-nitro-N-nitrosoguanidine (NTG) in *Streptomyces coelicolor. Mutat. Res.* **9:**167–182.

Dowding, J. E., 1973   Characterization of a bacteriophage virulent for *Streptomyces coelicolor* A3(2). *J. Gen. Microbiol.* **76:**163–176.

Dowding, J. E. and D. A. Hopwood, 1973   Temperate bacteriophages for *Streptomyces coelicolor* A3(2) isolated from soil. *J. Gen. Microbiol.* **78:**349–359.

Engel, P. E., 1970   Genetic control of tryptophan biosynthesis in *Streptomyces coelicolor*. In *Genetics of Industrial Microorganisms: Actinomycetes and Fungi*, edited by Z. Vaněk, Z. Hoštálek and J. Cudlín, pp. 125–147, Academia, Prague.

Friend, E. J. and D. A. Hopwood, 1971   The linkage map of *Streptomyces rimosus*. *J. Gen. Microbiol.* **68**:187–197.

Harold, R. J. and D. A. Hopwood, 1970a   Ultraviolet-sensitive mutants of *Streptomyces coelicolor*. I. Phenotypic characterisation. *Mutat. Res.* **10**:427–438.

Harold, R. J. and D. A. Hopwood, 1970b   Ultraviolet-sensitive mutants of *Streptomyces coelicolor*. II. Genetics. *Mutat. Res.* **10**:439–448.

Harold, R. J. and D. A. Hopwood, 1972   A rapid method for complementation testing of ultraviolet-sensitive (*uvs*) mutants of *Streptomyces coelicolor*. *Mutat. Res.* **16**:27–34.

Hopwood, D. A., 1959   Linkage and the mechanism of recombination in *Streptomyces coelicolor*. *Ann. N. Y. Acad. Sci.* **81**:887–898.

Hopwood, D. A., 1965   New data on the linkage map of *Streptomyces coelicolor*. *Genet. Res.* **6**:248–262.

Hopwood, D. A., 1966a   Non-random location of temperature-sensitive mutants on the linkage map of *Streptomyces coelicolor*. *Genetics* **54**:1169–1176.

Hopwood, D. A., 1966b   Lack of constant genome ends in *Streptomyces coelicolor*. *Genetics* **54**:1177–1184.

Hopwood, D. A., 1967   Genetic analysis and genome structure in *Streptomyces coelicolor*. *Bacteriol. Rev.* **31**:373–403.

Hopwood, D. A., 1970a   The isolation of mutants. In *Methods in Microbiology*, Vol. 3A, edited by J. R. Norris and D. W. Ribbons, pp. 363–433, Academic Press, London.

Hopwood, D. A., 1970b   Developments in actinomycete genetics. In *Genetics of Industrial Microorganisms: Actinomycetes and Fungi*, edited by Z. Vaněk, Z. Hoštálek and J. Cudlín, pp. 21–46, Academia, Prague.

Hopwood, D. A., 1972a   Genetic analysis in microorganisms. In *Methods in Microbiology*, Vol. 7B, edited by J. R. Norris and D. W. Ribbons, pp. 29–158, Academic Press, London.

Hopwood, D. A., 1972b   Genetics of the actinomycetales. In *Actinomycetales: Characteristics and Practical Importance*, edited by G. Sykes and F. A. Skinner, pp. 131–153, Academic Press, London.

Hopwood, D. A. and H. M. Ferguson, 1970   A rapid method for lyophilizing *Streptomyces* cultures. *J. Appl. Bacteriol.* **32**:434–436.

Hopwood, D. A. and G. Sermonti, 1962   The genetics of *Streptomyces coelicolor*. *Adv. Genet.* **11**:273–342.

Hopwood, D. A. and H. M. Wright, 1973   A plasmid of *Streptomyces coelicolor* carrying a chromosomal locus and its interspecific transfer. *J. Gen. Microbiol.* **79**:331–342.

Hopwood, D. A., G. Sermonti and I. Spada-Sermonti, 1963   Heterozygous clones in *Streptomyces coelicolor*. *J. Gen. Microbiol.* **30**:249–260.

Hopwood, D. A., R. J. Harold, A. Vivian and H. M. Ferguson, 1969   A new kind of fertility variant in *Streptomyces coelicolor*. *Genetics* **62**:461–477.

Hopwood, D. A., H. Wildermuth and H. M. Palmer, 1970   Mutants of *Streptomyces coelicolor* defective in sporulation. *J. Gen. Microbiol.* **61**:397–408.

Hopwood, D. A., K. F. Chater, J. E. Dowding and A. Vivian, 1973   Advances in *Streptomyces coelicolor* genetics. *Bacteriol. Rev.* **37**:371–405.

Kutzner H. J. and S. A. Waksman, 1959   *Streptomyces coelicolor* Müller and *Strep-*

tomyces violaceoruber Waksman and Curtis, two distinctly different organisms. *J. Bacteriol.* **78:**528–538.

Lomovskaya, N. D., L. K. Emeljanova and S. I. Alikhanian, 1971   The genetic location of prophage on the chromosome of *Streptomyces coelicolor. Genetics* **68:**341–347.

Lomovskaya, N. D., N. M. Mkrtumian, N. L. Gostimskaya and V. N. Danilenko, 1972   Characterization of temperate actinophage ϕC31 isolated from *Streptomyces coelicolor* A3(2). *J. Virol.* **9:**258–262.

Lomovskaya, N. D., L. K. Emeljanova, N. M. Mkrtumian and S. I. Alikhanian, 1973   The prophage behavior in crosses between lysogenic and nonlysogenic derivatives of *Streptomyces coelicolor* A3(2). *J. Gen. Microbiol.* **77:**455–463.

Marmur, J., 1961   A procedure for the isolation of deoxyribonucleic acid from microorganisms. *J. Mol. Biol.* **3:**208–218.

Randazzo, R., G. Sermonti, A. Carere and M. Bignomi, 1973   Comutation in *Streptomyces. J. Bacteriol.* **113:**500–501.

Russi, S., A. Carere, B. Fratello and V. Khoudokormoff, 1966   Caratterizzazione biochimica di alcuni mutanti di *Streptomyces coelicolor* richiedenti istidina. *Ann. Ist. Super. Sanità* **2:**506–522.

Scott, B. R. and T. Alderson, 1971   The random (non-specific) forward mutational response of gene loci in *Aspergillus* conidia after photosensitization to near ultraviolet light (365 mm) by 8-methoxypsoralen. *Mutat. Res.* **12:**29–34.

Sermonti, G., 1969   *Genetics of Antibiotic-Producing Microorganisms*, Wiley-Interscience, London.

Sermonti, G., A. Mancinelli and I. Spada-Sermonti, 1960   Heterogenous clones (heteroclones) in *Streptomyces coelicolor* A3(2). *Genetics* **45:**669–672.

Sermonti, G., A. M. Puglia and G. Ficarra, 1971   The time course of recombinant production in *Streptomyces coelicolor. Genet. Res.* **18:**133–145.

Townsend, M. E., H. M. Wright and D. A. Hopwood, 1971   Efficient mutagenesis by near ultraviolet light in the presence of 8-methoxypsoralen in *Streptomyces. J. Appl. Bacteriol.* **34:**799–801.

Vivian, A., 1971   Genetic control of fertility in *Streptomyces coelicolor* A3(2): plasmid involvement in the interconversion of UF and IF strains. *J. Gen. Microbiol.* **69:**353–364.

Vivian, A. and D. A. Hopwood, 1970   Genetic control of fertility in *Streptomyces coelicolor* A3(2): the IF fertility type. *J. Gen. Microbiol.* **64:**101–117.

Vivian, A. and D. A. Hopwood, 1973   Genetic control of fertility in *Streptomyces coelicolor* A3(2): new kinds of donor strains. *J. Gen. Microbiol.* **76:**147–162.

Welsch, M., 1969   Biology of actinophages In *Symposium on Genetics and Breeding of Streptomyces, Dubrovnik*, pp. 43–62, Yugoslav Academy of Sciences and Arts, Zagreb.

Wildermuth, H., 1970   Development and organization of the aerial mycelium in *Streptomyces coelicolor. J. Gen. Microbiol.* **60:**43–50.

Wildermuth, H. and D. A. Hopwood, 1970   Septation during sporulation in *Streptomyces coelicolor. J. Gen. Microbiol.* **60:**51–59.

Wilcockson, J., 1973   The use of sodium perchlorate in deproteinization during the preparation of nucleic acids. *Biochem. J.* **135:**599–561.

PART B
THE
BACTERIOPHAGES

16

Bacterial Viruses of Genetic Interest

ELLEN G. STRAUSS AND JAMES H. STRAUSS

Introduction

The literature on bacterial viruses is very large, and a short treatment such as this cannot hope to be comprehensive with regard to bacteriophage genetics. Many different bacteriophages have been described; for some of these, such as the T-even phage of *Escherichia coli*, extensive information is available, while for others, relatively little is known. An attempt has been made to discuss only those bacteriophages for which the genetic analysis is particularly complete, or phage which are representative of the various types of bacterial viruses. Wherever possible, the references cited are to recent review articles which survey the original literature in detail.

Physical Properties of Bacteriophages

Table I describes the properties of several bacteriophages, chosen to illustrate the various phage systems which have been characterized, and the properties of their nucleic acids. The first of the two references refers

ELLEN G. STRAUSS AND JAMES H. STRAUSS—Division of Biology, California Institute of Technology, Pasadena, California.

TABLE 1. *Properties*

			Nucleic acid		
Phage	*Common hosts*	*Replication*	*Type*	*Mol wt × 10^{-6}*	*Topology[a]*
P1	*E. coli, Shigella*	Temperate	2-DNA	58	Linear, CPS, TR
λ	*E. coli*	Temperate	2-DNA	31	Linear, U, CE
P22	*Salmonella*	Temperate	2-DNA	27	Linear, CPS, TR
P2	*E. coli, Shigella, Serratia*	Temperate	2-DNA	23	Linear, U, CE
P4	*E. coli, Shigella[c]*	Temperate	2-DNA	6.7	Linear, U, CE
Mu-1	*E. coli*	Temperate	2-DNA	28	Linear, U, CE
β	*Cornyebacterium diptheriae*	Temperate	2-DNA	22	Linear, U, TR
T4	*E. coli*	Virulent	2-DNA[d]	130	Linear, CPS, TR
SP82	*B. subtilis*	Virulent	2-DNA[e]	130	Linear, U
T5	*E. coli*	Virulent	2-DNA	75	Linear, U[f]
T7	*E. coli*	Virulent	2-DNA	26	Linear, U, TR
N4	*E. coli* (K12 strain only)	Virulent	2-DNA	44	Linear, U, —
φ29	*B. subtilis*	Virulent	2-DNA	11	Linear, U, CE
PM2	*Marine pseudomonad*	Virulent	2-DNA	6.0	Circular
f1	*E. coli* (F⁺)	Leaks[i]	1-DNA	2.0	Circular
φX174	*E. coli* C	Virulent	1-DNA	1.7	Circular
R17	*E. coli* (F⁺)	Virulent	1-RNA	1.1	Linear, U

[a] U = unique base sequence (nonpermuted); CPS = cyclically permuted base sequences; TR = double-stranded terminal redundancy; CE = cohesive ends (DNA molecule is terminated by single strand regions of complementary base sequence allowing circularization of molecule).

[b] Six spikes 7 × 3 nm surround a central spike 17 × 2 nm, all attached to a base plate 20 nm in diameter.

[c] P4 is defective in maturation, requiring P2 as a helper phage or a host lysogenic for P2.

to studies of the nucleic acid, the second to the morphology of the virion. In some cases, a single reference includes both aspects.

Bacterial viruses vary greatly in complexity, from small icosahedral phages with a genome of only one or two million daltons, such as R17 or φX174, to large complicated structures, such as T4, with a genome of 130 million daltons and a virion composed of a head structure and a contractile tail with various appendages such as a base plate, tail fibers, *of Some Bacteriophages*

Virion		Related phages	References
Morphology	Dimensions, nm		
Icosahedral head	85	—	Campbell, 1969; Walker and Anderson, 1970
contractile tail	220 × 18		
Icosahedral (?) head	54	φ21, φ80, φ82	Davidson and Szybalski, 1971; Kellenberger and Edgar, 1971
noncontractile tail	150 × 17	φ424, φ434	
Icosahedral (?) head	60	—	Rhoades *et al.*, 1968; Yamamoto and Anderson, 1961
tail of 6 short spikes[b]	7 × 3		
Icosahedral head	60	Wφ, P2HyDis	Inman and Bertani, 1968; Bertani and Bertani, 1971
contractile tail	135 × 17		
Icosahedral head	46	—	Inman *et al.*, 1971
contractile tail	135 × 17		
Icosahedral head	54 × 61	—	Daniell *et al.*, 1973; To *et al.*, 1966
contractile tail	100 × 18		
Polygonal head	45 × 50	B, Bh	Singer, 1973b; Mathews *et al.*, 1966
noncontractile tail	250 × 17		
Oblong head	80 × 120	T2, T6	Brenner *et al.*, 1959; MacHattie *et al.*, 1967
contractile tail	95 × 20		
Icosahedral (?) head	100	SP8, SP01,	Davison, 1963; Truffaut *et al.*, 1970
contractile tail	165 × 20	φe, 2c	
Octahedral (?) head	90	BF23	Lang, 1970; Bradley, 1967
noncontractile tail	200		
Octahedral head	63	T3, φII	Studier, 1969; Luftig and Haselkorn, 1968
noncontractile tail	15 × 15		
Hexagonal head	70	—	Sinha *et al.*, 1973; Schito *et al.*, 1966
noncontractile tail	10 × 15		
Oblong head	32 × 42	—	Viñuela *et al.*, 1970; Anderson *et al.*, 1966
noncontractile tail[g]	32 × 60		
Icosahedral[h]	61	—	Espejo *et al.*, 1969; Harrison *et al.*, 1971
Filamentous	870 × 5	fd, M13, HR	Marvin and Hohn, 1969
Icosahedral	25	S13, φR	Sinsheimer, 1968
Icosahedral	25	fr, f2, MS2, M12, Qβ	Kozak and Nathans, 1972

[d] Contains glucosylated 5-hydroxymethylcytosine instead of cytosine.

[e] Contains 5-hydroxymethyluracil instead of thymine.

[f] Contains 3 single-strand breaks in one of the DNA strands (Jacquemin-Sablon and Richardson, 1970).

[g] Also has fibers attached to head.

[h] Lipid-containing phage.

[i] Phage produced by leakage or extrusion; the host cell is not killed.

etc. The protein composition varies in complexity accordingly. The R17 virion contains only two protein species whereas the T4 particle contains approximately 35 different polypeptide chains. PM2 is the only bacteriophage described so far which contains anything other than nucleic acid and protein in the virion; it has a phospholipid bilayer as an integral part of its structure. Bradley (1971) has recently compared the morphology of bacteriophages in a review article.

The genome of all bacterial viruses exists as a single molecule of nucleic acid, either single-stranded or double-stranded, linear or circular. For those bacteriophages containing linear, double-stranded DNA, several possible arrangements exist. The various DNA molecules in a population of virions may contain cyclic permutations of the same gene order; in that case the DNA also contains double-stranded terminal redundancies, i.e., base sequences present at the beginning of the molecule (representing up to 2 percent of the total DNA) are repeated at the end. In the case of other phage species, all the DNA molecules have the same base sequence, resulting in a unique gene order. These DNA's of unique sequence may also contain double-stranded terminal redundancies as discussed above (e.g., T7 DNA), or alternatively, they may contain "cohesive ends," which are single-stranded terminal regions of complementary sequence (e.g., λ DNA). These are called "sticky" or "cohesive" ends because they readily anneal to one another, resulting in the formation of hydrogen-bonded circular molecules.

Bacteriophage Mutants

Many different types of bacteriophage mutations have been used in the study of phage genetics. The first mutations described were plaque morphology mutants. These mutants can occur in only a limited number of functions, but they are easy to select.

Deletion mutants occur only in nonessential functions, making selection for them complicated. One approach is to isolate mutants in functions required in one host, but not in another; a fraction of these are usually deletions. Another method used with T7 is to select for heat-resistant virions; this technique enriches for virus particles which have a portion of their DNA deleted. Mutants containing large deletions have been isolated by repeated selection of virions whose isopycnic density is less than that of wild type. Most deletions large enough to affect the density of the particle have lost whole genes or groups of genes. Since the DNA of the mutated gene is missing totally rather than being altered, by annealing one strand of mutant DNA with the complementary strand from wild-type virus, it is possible to visualize the region of nonhomology in the electron microscope. By means of a series of deletion mutants, one can construct a physical genetic map in the case of DNA's of unique sequence. Such maps have been very useful in comfirming colinearity between the genetic (recombination) map and the physical genome.

Nonessential phage functions have also been looked for by simply assaying various enzymatic activities after virus infection. Once a virus-specific activity is identified, mutants lacking such a function can be obtained.

The introduction of temperature-sensitive mutants and of nonsense (suppressor-sensitive) mutants such as amber, opal, and ochre, has allowed for selection of mutants in essential functions. These mutations are relatively easy to select for and can occur in most bacteriophage cistrons. In the case of very small viruses, all functions are essential and therefore conditional lethal mutations can be obtained in every gene. In the case of the large phages, however, up to half of the viral functions can be dispensed with in the normal bacterial hosts, and other techniques are required to study these nonessential genes. Conditional lethal mutants are in general "point mutations," i.e., they consist of a single altered base pair which results in a single altered amino acid (for temperature-sensitive mutants) or in premature termination of the polypeptide chain (nonsense mutants).

Host range mutants have been isolated for a number of systems. These are mutant viruses (usually of altered surface charge) which can attach to and infect bacterial hosts to which the "wild-type" virus cannot attach. Such mutants have been particularly useful in the study of the initial stages of the infection cycle.

In the case of the temperate bacteriophages, a mutation may affect either the ability to undergo a lytic infection or the ability to lysogenize the host. In the latter case, the mutation is usually in the regulatory proteins or in the operator or promoter regions. The techniques used to isolate and characterize these regulatory mutants are quite different from those used for the mutants described above.

Genetic Analysis of Bacteriophages

Table 2 summarizes genetic information on bacteriophages for which a genetic map has been constructed; the reference in the last column is to a recent publication containing a genetic map of the virus. Column 2 of the table lists the number of genes identified for each phage, usually as determined by complementation tests. Note that complementation assays in general are restricted to functions which are essential under the restrictive condition used. In the case of the smaller phages (e.g., ϕX174, T7 and R17) most or all of the virus-specific proteins have been separated by polyacrylamide gel electrophoresis and the various polypeptides identified with the complementation group coding for them. For large phages, which possess numerous nonessential genes, a comparable analysis has not been possible. However, especially in the case of T4, genetic studies have been

TABLE 2. Genetic Studies of Bacteriophages

Virus	Number of genes known[a]	Approximate number of total genes[b]	Topology of genetic map[c]	References
Temperate Phages				
P1	23	70	Linear	Scott, 1968
λ	33[d]	40	Linear[e]	Campbell, 1971
P22	39[f]	40	Circular[g]	Gough and Levine, 1968; Botstein et al., 1972
P2	20	30	Linear[e]	Bertani and Bertani, 1971; Sunshine et al., 1971
ε[34h]	11	—	Circular	Ikawa et al., 1968
β[i]	15	30	Linear	Singer, 1973a
Mu-1	21	35	Linear[j]	Abelson et al., 1973
Virulent Phages				
T4	70	160	Circular	Epstein et al., 1963; Wood et al., 1968
SP82	26	160	Linear	Kahan, 1966; Green and Laman, 1972
T5	30	100	Segmented[k]	Fattig and Lanni, 1965; Hendrickson and McCorquodale, 1971
T7	25[l]	30	Linear	Studier, 1972; Beier and Hausman, 1973
φ29	13	15–20	Linear	Reilly et al., 1973; Talavera et al., 1972

κ^m	22	—	Circular	Winkler *et al.*, 1970; Wackernagel and Winkler, 1970
N4	27	40	Linear	Schito, 1973
f1	8	8–9	Circular	Lyons and Zinder, 1972
φX174	8	8–9	Circular	Baker and Tessman, 1967; Benbow *et al.*, 1971
R17	3	3	Linear[n]	Jeppesen *et al.*, 1970

[a] May refer either to genes whose functions are known, as in the case of T7, φX174, and R17, or more generally to the number of complementation groups which have been identified.

[b] In the case of T7, φX174, f1, and R17 the number given is the number of phage-specific proteins identified (which saturate the coding capacity of the genome). For the remaining phages an "average gene size" of approximately 8×10^5 daltons of double-stranded DNA has been assumed.

[c] Refers to vegetative phage map for temperate phages.

[d] Number of structural genes coding for protein. In addition numerous mutations are known which affect promoter and operator functions.

[e] The prophage map is also linear and a cyclic permutation of the map for the vegetative phage.

[f] Of the genes identified, 26 are essential; the remaining 13 are nonessential and mostly regulatory.

[g] The prophage map is linear.

[h] ϵ^{34} is a temperate phage of *Salmonella*. It contains 2-DNA, circularly permuted with terminal redundancies. It is a generalized transducing phage, affecting the host surface O antigens and is related to ϵ^{15}.

[i] Corynebacteriophage β is a temperate phage of *Corynebacterium diptheria*. The structural gene for Diptheria toxin was recently shown to be a phage gene, and several mutants of the "*tox*" gene have been described (Uchida *et al.*, 1973).

[j] The genetic maps of the prophage state and of vegetative Mu-1 are linear and identical. However, it appears that Mu-1 is integrated at random into the host chromosome, with either of the two possible orientations of the prophage genes (Boram and Abelson, 1973).

[k] The T5 genetic map consists of 3 small linkage groups of genes and one large group, with maximal recombination between groups. Genetic segments appear to correspond to physical segments of the nicked strand of T5 DNA.

[l] Thirty T7 proteins have been identified, but mutants have only been isolated for 25 of these.

[m] Virulent *Serratia* phage related to Af8 and Chi.

[n] Does not undergo recombination. The gene order has been determined by physical means.

extensive, and very detailed maps of the essential regions of the genome exist. For some viruses, on the other hand, the small proportion of genes characterized reflects the preliminary nature of the investigations of these phages, rather than the nature of their genomes.

Genetic maps of the various phages are obtained by recombinational analysis, (except of the RNA-containing phages). Nonessential functions as well as essential ones can be mapped. The maps obtained are in general colinear with the physical DNA molecule and reflect the peculiarities of the DNA structure. Circular DNA molecules or linear DNA with cyclically permuted base sequences give rise to circular linkage maps. Linear DNA molecules of unique sequence give rise to linear maps. In the case of T5, three single-strand interruptions in one of the DNA strands appears to give rise to a segmented genetic map.

The various phage functions are normally clustered. Genes controlling production of proteins used in the virus head structure are located close to one another on the DNA molecule, genes used in tail structure are also found together, and so forth. Genes required early in infection are normally situated in one part of the genetic map, and "late" genes in another region. This clustering of genes on the map is related, at least in part, to the regulation of gene expression.

Literature Cited

Abelson, J., W. Boram, A. I. Bukhari, M. Faelen, M. Howe, M. Metlay, A. L. Taylor, A. Touissaint, P. van de Putte, G. C. Westmaas and C. A. Wijffelman, 1973   Summary of the genetic mapping of prophage Mu. *Virology* **54**:90–92.

Anderson, D. L., D. D. Hickman and B. E. Reilley, 1966   Structure of bacteriophage $\phi29$ and the length of the $\phi29$ DNA. *J. Bacteriol.* **91**:2081–2089.

Baker, R. and I. Tessman, 1967   The circular genetic map of Phage S13. *Proc. Natl. Acad. Sci. USA* **58**:1438–1445.

Beier, H. and R. Hausman, 1973   Genetic map of bacteriophage T3. *J. Virol.* **12**:417–419.

Benbow, R. M., C. A. Hutchison, III, J. D. Fabricant and R. L. Sinsheimer, 1971   Genetic map of Bacteriophage ϕX174. *J. Virol.* **7**:549–558.

Bertani, L. E. and G. Bertani, 1971   Genetics of P2 and related phages. *Adv. Genet.* **16**:200–239.

Boram, W. and J. Abelson, 1973   Bacteriophage Mu integration: On the orientation of the prophage. *Virology* **54**:102–108.

Botstein, D., R. K. Chan and C. H. Waddell, 1972   Genetics of bacteriophage P22. II. Gene order and gene function. *Virology* **49**:268–282.

Bradley, D. E., 1967   Ultrastructure of bacteriophages and bacteriocins. *Bacteriol. Rev.* **31**:230–314.

Bradley, D. E., 1971   A comparative study of the structure and biological properties of bacteriophages. In *Comparative Virology*, edited by K. Maramorosch and E. Kurstak, pp. 208–255, Academic Press, New York.

Brenner, S., G. Streisinger, R. W. Horne, S. P. Champe, L. Barnett, S. Benzer and M. W. Rees, 1959 Structural components of bacteriophage. *J. Mol. Biol.* **1**:281–292.

Campbell, A., 1969 *Episomes*, pp. 25–34, Harper and Row, New York.

Campbell, A., 1971 Genetic structure. In *The Bacteriophage Lambda*, edited by A. D. Hershey, pp. 13–44, Cold Spring Harbor Laboratory, Cold Spring Harbor, N.Y.

Daniell, E., J. Abelson, J. S. Kim and N. Davidson, 1973 Heteroduplex structures of bacteriophage Mu DNA. *Virology* **51**:237–239.

Davidson, N. and W. Szybalski, 1971 Physical and chemical characteristics of lambda DNA. In *The Bacteriophage Lambda*, edited by A. D. Hershey, pp. 45–82, Cold Spring Harbor Laboratory, Cold Spring Harbor, N.Y.

Davison, P. F., 1963 The structure of bacteriophage SP8. *Virology* **21**:146–151.

Epstein, R. H., A. Bolle, C. M. Steinberg, E. Kellenberger, E. Boy de la Tour, R. Chevalley, R. S. Edgar, M. Susman, G. H. Denhardt and A. Lielausis, 1963 Physiological studies of conditional lethal mutants of bacteriophage T4D. *Cold Spring Harbor Symp. Quant. Biol.* **28**:375–394.

Espejo, R., E. S. Canelo and R. L. Sinsheimer, 1969 DNA of bacteriophage PM2: A closed circular double-stranded molecule. *Proc. Natl. Acad. Sci. USA* **63**:1164–1168.

Fattig, W. D. and F. Lanni, 1965 Mapping of temperature sensitive mutants of bacteriophage T5. *Genetics* **51**:157–166.

Gough, M. and M. Levine, 1968 The circulatory of the P22 linkage map. *Genetics* **58**:161–169.

Green, D. M. and D. Laman, 1972 Organization of gene function in *Bacillus subtilis* bacteriophage SP82G. *J. Virol.* **9**:1033–1046.

Harrison, S. C., D. L. D. Casper, R. D. Camerini-Otero and R. M. Franklin, 1971 Lipid and protein arrangement in bacteriophage PM2. *Nat. New Biol.* **229**:197–201.

Hendrickson, H. E. and D. J. McCorquodale, 1971 Genetic and physiological studies of bacteriophage T5. I. An expanded genetic map of T5. *J. Virol.* **7**:612–618.

Ikawa, S., S. Toyoma and H. Uetake, 1968 Conditional lethal mutants of ϵ^{34}. I. Genetic map of ϵ^{34}. *Virology* **35**:519–528.

Inman, R. B. and G. Bertani, 1968 Heat denaturation of P2 bacteriophage DNA: Compositional heterogeneity. *J. Mol. Biol.* **44**:533–549.

Inman, R. B., M. Schnös, L. D. Simon, E. W. Six and D. H. Walker, Jr., 1971 Structural properties of P4 bacteriophage and P4 DNA. *Virology* **44**:62–72.

Jacquemin-Sablon, A. and C. C. Richardson, 1970 Analysis of the interruptions in bacteriophage T5 DNA. *J. Mol. Biol.* **47**:477–493.

Jeppesen, P. G. N., J. Argetsinger-Steitz, R. F. Gesteland and P. F. Spahr, 1970 Gene order in the bacteriophage R17 RNA: 5′–A protein–coat protein–synthetase–3′. *Nature (Lond.)* **226**:230–237.

Kahan, E., 1966 A genetic study of temperature-sensitive mutants of *Subtilis* phage SP82. *Virology* **30**:650–660.

Kellenberger, E. and R. S. Edgar, 1971 Structure and assembly of phage particles. In *The Bacteriophage Lambda*, edited by A. D. Hershey, pp. 271–296, Cold Spring Harbor Laboratory, Cold Spring Harbor, N.Y.

Kozak, M. and D. Nathans, 1972 Translation of the genome of a ribonucleic acid bacteriophage. *Bacteriol. Rev.* **36**:109–134.

Lang, D., 1970 Molecular weights of coliphages and coliphage DNA III. Contour length and molecular weights of DNA from bacteriophages T4, T5, T7 and from bovine papilloma virus. *J. Mol. Biol.* **54**:557–565.

Luftig, R. and R. Haselkorn, 1968   Comparison of blue-green algae virus LPP-1 and the morphologically related viruses G_{III} and coliphage T7. *Virology* **34**:675–678.

Lyons, L. B. and N. D. Zinder, 1972   The genetic map of the filamentous bacteriophage fl. *Virology* **49**:45–60.

MacHattie, L. A., D. A. Ritchie, C. A. Thomas, Jr. and C. C. Richardson, 1967   Terminal repetition in permuted T2 phage DNA molecules. *J. Mol. Biol.* **23**:355–363.

Marvin, D. A. and B. Hohn, 1969   Filamentous bacterial viruses. *Bacteriol. Rev.* **33**:172–209.

Mathews, M. M., P. A. Miller and A. M. Pappenheimer, Jr., 1966   Morphological observations on some diptherial phages. *Virology* **29**:402–409.

Reilly, B. E., V. M. Zeece and D. L. Anderson, 1973   Genetic Study of suppressor-sensitive mutants of the *Bacillus subtilis* bacteriophage ϕ29. *J. Virology* **11**:756–760.

Rhoades, M., L. A. MacHattie and C. A. Thomas, Jr., 1968   The P22 bacteriophage DNA molecule. I. The mature form. *J. Mol. Biol.* **37**:21–40.

Schito, G. C., 1973   Genetics and physiology of coliphage N4. *Virology* **55**:254–265.

Schito, G. C., G. Rialdi, and A. Pesce, 1966   Biophysical properties of N4 coliphage. *Biochim. Biophys. Acta* **129**:482–490.

Scott, J. R., 1968   Genetic studies on bacteriophage P1. *Virology* **36**:564–574.

Singer, R. A., 1973*a*   Temperature-sensitive mutants of toxinogenic cornyebacteriophage β. I. Genetics. *Virology* **55**:347–356.

Singer, R. A., 1973*b*   Temperature-sensitive mutants of toxinogenic cornyebacteriophage β. II. Properties of mutant phages. *Virology* **55**:357–362.

Sinha, R. K., D. N. Misra and N. N. Das Gupta, 1973   Electron microscopy of DNA from coliphage N4. *Virology* **51**:493–498.

Sinsheimer, R. L., 1968   Bacteriophage ϕX174 and related viruses. *Prog. Nucleic Acid Res. Mol. Biol.* **8**:115–169.

Studier, F. W., 1969   The genetics and physiology of bacteriophage T7. *Virology* **39**:562–574.

Studier, F. W., 1972   Bacteriophage T7. *Science (Wash., D.C.)* **176**:367–376.

Sunshine, M. G., M. Thorn, W. Gibbs and R. Calendar, 1971   P2 phage amber mutants: Characterization by use of a polarity suppressor. *Virology* **46**:691–702.

Talavera, A., F. Jimenez, M. Salas and E. Viñuela, 1972   Mapping of temperature-sensitive mutants of bacteriophage ϕ29. *Mol. Gen. Genet.* **115**:31–35.

To, C. M., A. Eisenstark and H. Toreci, 1966   Structure of the mutator phage Mu-1 of *Escherichia coli. J. Ultrastruct. Res.* **14**:441–448.

Truffaut, N., B. Revet and M.-O. Soulie, 1970   Étude comparative des DNA de phages 2C, SP8, SP82, ϕe, SP01, et SP50. *Eur. J. Biochem.* **15**:391–400.

Uchida, T., A. M. Pappenheimer, Jr. and R. Greany, 1973   Diptheria toxin and related proteins I. Isolation and properties of mutant proteins serologically related to Diptheria toxin. *J. Biol. Chem.* **248**:3838–3844.

Viñuela, E., E. Méndez, A. Talavera, J. Ortîn and M. Salas, 1970   Structural components of bacteriophage ϕ29. In *Macromolecules; Biosynthesis and Function*, edited by S. Ochoa, C. Ascensio, C. F. Heredia, and D. Nachmansohn, pp. 195–202, Academic Press, New York.

Wackernagel, W. and R. Winkler, 1970   Temperature-sensitive mutants of *Serratia* phage kappa. *Virology* **42**:777–779.

Walker, D. H., Jr., and T. F. Anderson, 1970   Morphological variants of coliphage P1. *J. Virol.* **5**:765–782.

Winkler, U., U. Kopp-Scholz and C. Hauz, 1970   Nonsense mutants of *Serratia* phage kappa. I. Their isolation and use for proving circular linkage of the phage genes. *Mol. Gen. Genet.* **106**:239–253.

Wood, W. B., R. S. Edgar, J. King, I. Lielausis and M. Henninger, 1968   Bacteriophage assembly. *Fed. Proc.* **27**:1160–1166.

Yamamoto, N. and T. F. Anderson, 1961   Genomic masking and recombination between serologically unrelated phages P22 and P221. *Virology* **14**:430–439.

17

RNA Bacteriophages

WALTER FIERS

Introduction

RNA phages have attracted much interest because (1) they are one of the smallest autonomous viruses known, (2) they can be prepared in high yield (10,000 plaque forming units/cell), (3) they have been frequently used as a model messenger RNA in studies on the mechanism of *in vitro* translation, and (4) only with this system has net *in vitro* synthesis of fully infective progeny viral RNA been achieved by a mechanism presumably identical to that operative *in vivo*.

Most studies so far have been made with phages of *Escherichia coli* (see below). However, RNA phages specific for other bacterial species have been described, such as the phage φCb5 which infects *Caulobacter crescentus* (Schmidt and Stanier, 1965), and the serologically related phages 7S and PP7 which infect *Pseudomonas aeruginosa* (Bradley, 1967). All RNA phages are physically very similar, and they all infect their host cell by adsorption to special pili or filamentous appendages.

Classification of the RNA Bacteriophages of *E. coli*

Groups

Over 40 RNA coliphages have been isolated in various parts of the world, mostly from sewage or feces. Serologically they belong to four groups (see Table 1). Some types of the same group are virtually identical, while others show up to a 10-fold difference in inactivation rate by type-

WALTER FIERS—Laboratory of Molecular Biology, University of Ghent, Ghent, Belgium.

TABLE 1. Serological Groups of RNA Coliphages[a]

| Group | Serotypes | | Other distinctive properties[b] | | | | | |
| | | | Phage particle | | | RNA | Coat protein[c] | |
		Types	Buoyant density in CsCl, g/cm³	$s_{20,w}$ value	A/U ratio[d]	Amino acids not present	C terminal
I	Main:	f2, MS2, R17, fr, M12	1.46	79S	0.95–0.98	His	— Ile-Tyr
	Others:	f4, μ2, FH5, β, ZR, GR, MY, SN, f can 1, R23, R40					
II	Main:	GA	1.44	76S	0.84–0.86	His, Met, Cys	—(Tyr, Phe)-Ala
	Others:	EI, KJ, SW, SS, SD, SB, MC, SK					
III	Main:	Qβ	1.47[e]	83S	0.78–0.79	His, Met, Trp	— Ala-Tyr
	Others:	CF, HI, NH, NM, SG, VK, ST, SO					
IV[f]	Main:	SP, FI	—	—	—	—	—
Unclassified		ZIK/1, ZJ/1, ZG, ZS/3, ZL/3, α15	—	—	—	—	—

[a]Main references: Scott, 1965; Bishop and Bradley, 1965; Watanabe et al., 1967; Krueger, 1969.
[b]Nishihara et al., 1969; Nishihara and Watanabe, 1969.
[c]The N terminal is Ala for all groups.
[d]Ratio of A and U bases in the viral RNA.
[e]The phage precipitates out in the band.
[f]References: Sakurai et al., 1968; Miyake et al., 1971.

specific antiserum. Some other distinctive properties, in addition to those listed in Table 1, are electrophoretic mobility, adsorption to nitrocellulose filters, uv sensitivity, pH sensitivity, etc.

A theoretically interesting discriminative test is also the assay for template specificity by means of a heterologous purified RNA-dependent RNA polymerase. It seems that the enzymes are strictly group specific, i.e., they only function with RNA derived from a phage of the same group (Haruna *et al.*, 1967; Miyake *et al.*, 1971). However, only the enzyme complex of the group III (Qβ) and IV can readily be obtained.

Phages of the group I (f2, R17, MS2) and of the group III (Qβ) have been most extensively studied.

Host Range

All RNA phages plate on all male *E. coli* strains, either F^+ or Hfr. No restriction or other limitation in host range is known. Coinfection with various DNA phages, however, rapidly excludes the RNA phage. Avoidance of shear (which may break off the pili), presence of divalent ions (usually Mg^{++} or Ca^{++} in a concentration of 0.7 mM or above), and absence of ribonuclease are required for successful infection. Transfer of the *coli* F factor to some *Salmonella, Shigella* and *Proteus* strains confers to these the susceptibility to the RNA coliphages (Zinder, 1965). The viral RNA is infectious to spheroplasts both from male as well as from female strains (Davis *et al.*, 1961).

Some bacterial mutants with restricted resistance have been isolated, e.g., M27, which is still susceptible to Qβ, but not to phages of the f2 group (Valentine *et al.*, 1969).

Structure of the Virion

Physical Properties

The physical properties of phage particles belonging to the same group are virtually identical (see Table 2). Therefore, variations in values reported in the literature are mostly due to experimental limitations. The capsid of the virion is an icosahedron composed of 32 morphological units, 12 pentamers and 20 hexamers (Vasquez *et al.*, 1966).

Virus-Coded Proteins

The molecular weights, some properties, and functions of the virus-coded proteins are listed in Table 3. The amino acid sequence of the f2

TABLE 2.   *Physical Properties of the Virions[a]*

Property	f2, R17, MS2	Qβ
Mol wt, daltons	3.6×10^6 [b,c]	4.2×10^6 [d]
$s_{20, w}$, S	79–80[c]	83–84[d]
Density, g/cm³	1.46[b]	1.47[e]
Partial specific volume $\bar{v}$, ml/g	0.690–0.703[b,d]	0.695[d]
Diameter, Å	260–266[b,f]	—
Inner diameter, Å	210[g]	—
Radius of gyration, R_G	105.2[g]	—
RNA content, percent	31–34[b,h]	—
Specific absorbancy, $A^{1\ \text{mg/ml}}_{260\ \text{m}}$	7.66–8.03[b,c]	8.02[d]
Ratio $A_{260\ m\mu}/A_{280\ m\mu}$	1.79[i]	—
Water content, g H_2O/g	0.9–1.0[f,g]	—
Isoelectric point, pH	3.9[d]	5.3[d]
Refractive increment, ml/g	0.199[d]	0.198[d]

[a] Usually only a single typical reference is given, but often concurrent results have been obtained.
[b] Strauss and Sinsheimer, 1963.
[c] Gesteland and Boedtker, 1964.
[d] Overby *et al.*, 1966a.
[e] Watanabe *et al.*, 1967.
[f] Fischbach *et al.*, 1965.
[g] Zipper *et al.*, 1971.
[h] Enger *et al.*, 1963.
[i] Vasquez *et al.*, 1966.

coat protein and the Qβ coat protein is shown in Figure 1. Although no serological crossreaction occurs, a very weak relationship can nevertheless be detected. From a comparison of the coat proteins of the Group I phages (Table 4), it is clear that they are very closely related, except fr which shows an 18 percent difference.

General Properties of the Viral RNA

Physical Properties and Base Composition

As mentioned in the previous section, differences between reported properties of phages within the same group are most likely due to experimental limitations, and are not of a fundamental nature. Only typical values and references are listed in Table 5.

Biological Properties and Genetic Information Content

The viral RNA is the carrier of all the genetic information, and as such it is infective to spheroplasts (Davis *et al.*, 1961). A duplex form,

TABLE 3. Virus-Coded Proteins

	Mol wt, daltons	No. amino acids	No. molecules/ virion	Function
f2, R17, MS2:				
Coat	13,700	129[a]	180[b]	Main capsid protein
A protein	42,000[c,d]	380 ± 5[e]	1	Adsorption, penetrates, together with the infecting RNA, into the host cell[f,g]
Replicase	63,000[h]	550 ± 50	—	Responsible, together with host proteins, for viral RNA replication
Qβ:				
Coat	14,050	131[i]	180[b]	Main capsid protein
A1 (or IIb)	38,000[j,e]	350 ± 20	3—14[k]	Unknown; read-through product of the coat gene[j,k,l]
A2 (or IIa)	44,000[j,e]	410 ± 20	1	Adsorption
Replicase	67,000 ± 2,000[m,n]	610 ± 20	—	Responsible, together with host proteins, for viral RNA replication

[a] Weber and Konigsberg, 1967.
[b] This number is based on symmetry considerations (triangulation number T = 3); as a few coat molecules may be replaced by the A protein, or as a few coat proteins may be preferentially bound (Sugiyama *et al*, 1967), the real number may be slightly less or slightly more.
[c] Argetsinger Steitz, 1968.
[d] Remaut and Fiers, 1973.
[e] Fiers and colleagues, unpublished.
[f] Kozak and Nathans, 1972.
[g] Krahn *et al*, 1972.
[h] Fedoroff and Zinder, 1971.
[i] Maita and Konigsberg, 1971.
[j] Moore *et al*, 1971.
[k] Weiner and Weber, 1971.
[l] Horiuchi *et al*, 1971.
[m] Kondo *et al*, 1970.
[n] Kamen, 1970.

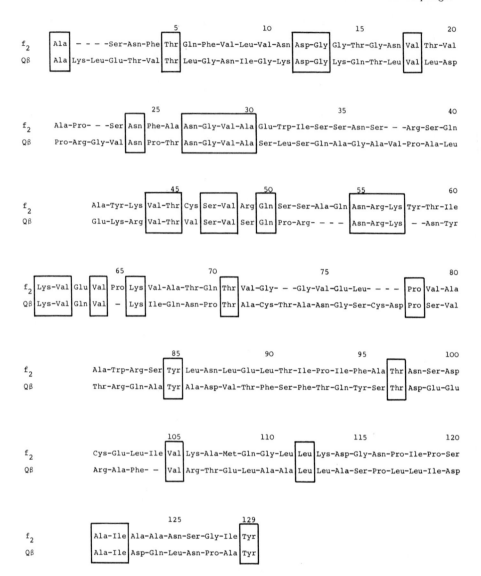

Figure 1. Amino acid sequence of the coat protein of phage f2 and of phage Qβ. Corresponding amino acids are enclosed in boxes; some insertions and deletions have to be invoked in order to obtain optimal alignment (modified from Konigsberg et al., 1970).

consisting of a plus and a minus strand, can be isolated from infected cells (Ammann *et al.*, 1964). The intact minus strand, isolated from such a duplex form, is not infective and does not even bind ribosomes (Schwartz *et al.*, 1969).

Using the plus strand as a messenger and under appropriate conditions, all three virus-coded proteins have been synthesized *in vitro* (Kozak and Nathans, 1972). As no recombination is observed with RNA phages, the order of the three genes had to be determined biochemically by measuring the coding capacity of defined RNA fragments in an *in vitro* system (Jeppesen *et al.*, 1970; Konings *et al.*, 1970). A map of the genes is shown in Figure 2. In the case of the $Q\beta$ phage the map order is likewise A_2 protein–coat protein–replicase, except that the coat gene and the succeeding region also code for the A_1 protein (cf. Table 3).

The regions on the messenger where the ribosomes attach in the formation of an initiation complex for polypeptide synthesis have been

TABLE 4. *Comparison of the Amino Acid Sequence of the Coat Protein of Group I Phages*

Position	Phage			
	$f2^{a,b}$	$R17^{c,d}$	$MS2^e$	fr^f
5	Thr	—	—	Glu
6	Gln	—	—	Glu
11	Asn	—	Asp	—
12	Asp	—	Asn	—
17	Asn	—	Asp	Asp
19	Thr	—	—	Lys
54	Gln	—	—	Asn
60	Ile	—	—	Val
71	Thr	—	—	Val
72	Val	—	—	Gln
86	Leu	—	—	Met
88	Leu	Met	Met	Met
94	Ile	—	—	Val
98	Asn	—	—	Asx
99	Ser	—	—	Asp
102	Glu	—	—	Ala
108	Met	—	—	Leu
111	Leu	—	—	Thr
112	Leu	—	—	Phe
114	Asp	—	—	Thr
116	Asn	—	—	Ile
117	Pro	—	—	Ala
118	Ile	—	—	Pro
119	Pro	—	—	Asn
120	Ser	—	—	Thr

[a] The phage f2 is chosen as prototype, and only differences from this sequence are listed.
[b] Weber and Konigsberg, 1967.
[c] The phage M12 is presumably identical to R17 except for a Lys to Glx change (Enger and Kaesberg, 1965).
[d] Weber, 1967.
[e] Van de Kerckhove and Van Montagu, unpublished; reported in Min Jou *et al.*, 1972.
[f] Wittmann-Liebold and Wittmann, 1967.

TABLE 5. *Physical Properties of the Viral RNA*

Property	f2, R17, MS2	Qβ
Mol wt, daltons	$1.05-1.2 \times 10^6$ [a, b]	1.5×10^6 [c]
No. of nucleotides	3500 ± 200 [d]	4500
$s_{20,w}$, S		
in 0.1 M NaCl	26 [e]	29 [f]
in 0.01 M NaCl	16.5 [b]	
Specific absorbancy,		
$E^{1 \text{ mg/ml}}_{260 \text{ m}\mu}$	22.6–25.1 [g, a]	25.5 [f]
ϵ_p	7400–8600 [h, a]	
Ratio, $A_{260 \text{ m}\mu}/A_{280 \text{ m}\mu}$	2.09 ± 0.1 [i, j, k]	
Hypochromicity, percent	26–27	
Tm, °C		
in 0.1 M NaCl—0.01 M EDTA	59.5 [b]	
in 0.01 M NaCl—0.001 M EDTA	41 [b]	
Helicity, percent		
by spectrophotometry	73–82 [k, e]	
by IR	63 ± 5 [h]	
Buoyant density in		
Cs_2SO_4, g/cm³	1.63 [j]	1.63 [f]
Partial specific volume		
ν, cm³/g	0.495 [g]	
R_G, Å		
in 0.1 M NaCl	183–196 [b]	
in 0.08 M phosphate	210 [g]	
Solvation water,		
g H_2O/g RNA	5.5 [g]	
Base composition [f]		
A	23.3 ± 0.26	22.1 ± 0.05
U	24.8 ± 0.23	29.5 ± 0.15
G	26.4 ± 0.83	23.7 ± 0.23
C	25.5 ± 0.86	24.7 ± 0.35

[a] Strauss and Sinsheimer, 1963.
[b] Gesteland and Boedtker, 1964.
[c] Boedtker, 1971.
[d] Haegeman et al., 1971.
[e] Mitra et al., 1963.
[f] Overby et al., 1966b.

[g] Slegers et al., 1973.
[h] Isenberg et al., 1971.
[i] Fiers et al., 1965.
[j] Billeter et al., 1966.
[k] Boedtker, 1967.

characterized (Table 6). They were isolated by taking advantage of the protection offered by the bound ribosome against nuclease breakdown. In the native form, however, only the coat initiation site is available, and partial denaturation or degradation is required in order to expose the other two sites. No common feature is observed in these regions preceding the initiating AUG, although several contain the sequence U-U-U-G-A and most contain a tract of purine nucleotides. The meaning of this, if any, is not clear.

Some termination signals for protein synthesis as well as the following intercistronic regions are shown in Table 7. The termination signal of the R17 replicase contains UAA, but may be more complex (Capecchi and Klein, 1970). The coat protein of Qβ ends with UGA, but due to leakiness some read-through always occurs, thus producing the A_1 protein (Horiuchi *et al.*, 1971; Weiner and Weber, 1971; Moore *et al.*, 1971). In strains carrying an amber suppressor mutation, some read-through of the MS2 A protein is observed (Remaut and Fiers, 1972).

The complete nucleotide sequence of the coat gene is shown in Figure 3. It is folded into a "flower model," based on the nature of the products obtained by partial degradation and on some theoretical considerations. Forty-nine different codewords are used; these are summarized in Table 8. Also several regions of the replicase (or RNA-polymerase) gene have been sequenced (Contreras *et al.*, 1972); the codewords used here are summarized in Table 9. In total, all 61 sense codons have been found to actually occur in the MS2 genome (Contreras *et al.*, 1973). It is not yet known whether a regulation mechanism exists, based on the choice between synonymous codewords.

At the 5′ end, the MS2 RNA chain starts with an untranslated sequence, 129 nucleotides in length (De Wachter *et al.*, 1971a, b). A secondary-structure model of this region is shown in Figure 4. It is interesting that this region is very resistant to mutational drift, as exactly the same nucleotide sequence is present at the 5′ end of the phage R17 (Adams *et al.*, 1972b). On the other hand, the variation in translated regions is about 4 percent for the comparison f2 and R17 or MS2 and R17 (Robertson and Jeppesen, 1972). Nearly all the changes are transitions (Fiers *et al.*, 1971). The untranslated, 5′ terminal region of Qβ RNA is shown in Figure 5. The population actually consists of chains starting with either 4 or 5 G residues, presumably as a result of slipping during replication (De Wachter and Fiers, 1969).

The phage RNA ends also at the 3′ end with an untranslated region. As the termination signal of the last cistron, the replicase, has not yet been unambiguously identified, the exact length of the untranslated region is unknown. But in the phage MS2 there is evidence that it is only 61 nucleotides in length (Vandenberghe and Fiers, unpublished). These 3′-terminal sequences for the phage MS2 and Qβ are shown in Figure 6 and

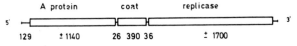

Figure 2. Genetic map of bacteriophage MS2 (R17, f2). The numbers refer to map distances in nucleotide units (Fiers et al., 1973).

TABLE 6. Ribosomal Binding Sites

R17

A protein

C-C-U-A-G-G-A-G-G-U-U-U-G-A-C-C-U-A-U-G-C-G-A-G-C-U-U-U-U-A-G-U-G

fMet – Arg – Ala – Phe – Ser

Reference: Argetsinger Steitz, 1969

Coat protein R17

A-G-A-G-C-C-C-U-C-A-A-C-C-G-G-G-G-U-U-U-G-A-A-G-C-A-U-G-G-C-U-U-C-U-A-A-C-U-U-U

Reference: Argetsinger Steitz, 1969; Adams *et al.*, 1972a

f2

A-G-A-G-C-C-C-U-C-A-A-C-C-G-G-A-G-U-U-U-G-A-A-G-C-A-U-G-G-C-U-U-C-U-A-A-C-U-U-U-A-C-U-U-C-A-G

fMet – Ala – Ser – Asn – Phe – Thr – Glu

Reference: Gupta *et al.*, 1970

Replicase

A-A-A-C-A-U-G-A-G-G-A-U-U-A-C-C-C-A-U-G-U-C-G-A-A-G-A-C-A-A-C-A-A-A-G

fMet – Ser – Lys – Thr – Thr – Lys

Reference: Argetsinger Steitz, 1969

Qβ

A₂ protein

A-A-G-A-G-G-A-C-A-U-A-U-G-C-C-U-A-A-A-U-U-A-C-C-G-G-C

fMet – Pro – Lys – Leu – Pro

Reference: Staples *et al.*, 1971

Coat protein

A-A-A-C-U-U-U-G-G-G-U-C-A-A-U-U-U-G-A-U-C-U-A-U-G-G-C-A-A-A-A-U-U-A-G-A-G-A-C-U-G-U-U

fMet – Ala – Lys – Leu – Glu – Thr – Val

Reference: Argetsinger Steitz, 1972

Replicase

U-A-A-C-U-A-A-G-G-A-U-G-A-A-A-U-G-C-A-U-G-U-C-U-A-A-G-A-C-A-G-C

fMet – Ser – Lys – Thr – Ala

Reference: Staples and Hindley, 1971; Argetsinger Steitz, 1972

TABLE 7. Termination Signals and Intercistronic Regions

		Sequence	Reference
A protein coat	MS2	[U–A–G] A–G–C–C–U–C–A–A–C–C–G–G–A–G–U–U–G–A–A–G–C [A–U–G]	Contreras et al., 1973
Coat replicase	R17	[U–A–A–U–A–G] A–U–G–C–C–G–G–C–C–A–U–U–C–A–A–A–C–A–U–G–A–G–G–A–U–U–A–C–C–C [A–U–G]	Nichols, 1970
	MS2	[U–A–A–U–A–G] A–C–C–G–C–C–G–G–C–C–A–U–U–C–A–A–A–C–A–U–G–A–G–G–A–U–U–A–C–C–C [A–U–G]	Min Jou et al., 1972

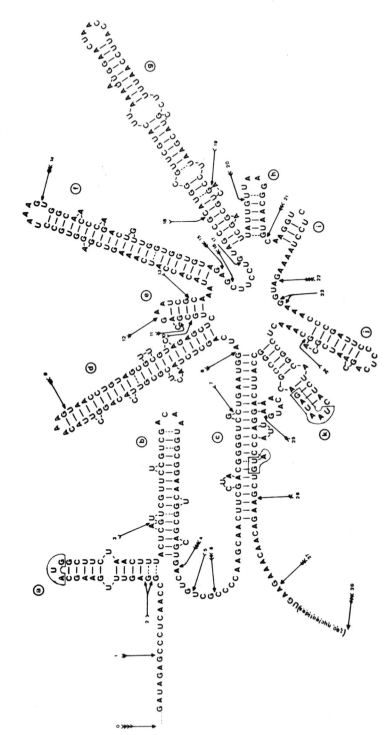

Figure 3. Nucleotide sequence of the MS2 coat-protein gene. The sequence is arranged in the "flower model." The arrows point to links easily split by T₁ ribonuclease. The initiating AUG, the termination signal UAA–UAG, and the initiating AUG of the following replicase gene have been enclosed. Reproduced from Min Jou, Haegeman, Ysebaert, and Fiers, Nature 237:82–88, 1972, by permission of Macmillan Journals, Ltd.

Figure 7, respectively. The sequence of the 3′ end of R17 is the same as for MS2, as far as it has been determined (Cory *et al.*, 1972). It is interesting that all bacteriophage RNA's end with the sequence -C-C-A, just like tRNA's. The minus strand, however, starts with a pppGp . . . both in MS2 (Vandenberghe *et al.*, 1969) as well as in Qβ (Banerjee *et al.*, 1969). This observation means that the 3′-terminal A residue is added by another mechanism than conventional base-pairing. This terminal A is not required for infectivity and *a fortiori* not for replication (Kamen, 1969; Rensing and August, 1969; Weber and Weissmann, 1970).

The meaning of the long, untranslated regions at both ends of the phage RNA's is unknown. Presumably they play an important role in replication, e.g., as recognition sites for specific interaction with the RNA-polymerase enzyme complex.

TABLE 8. Codons Found in the MS2 Coat Protein Gene

	U	C	A	G	
U	Phe 1	Ser 3	Tyr 4	Cys 1	U
	Phe 3	Ser 2	Tyr	Cys 1	C
	Leu 1	Ser 2	Ochre O	Opal O	A
	Leu	Ser 2	Amber O	Trp 2	G
C	Leu 2	Pro 2	His	Arg 3	U
	Leu 2	Pro 1	His	Arg 1	C
	Leu 2	Pro 2	Gln 1	Arg	A
	Leu	Pro 1	Gln 5	Arg	G
A	Ile 4	Thr 4	Asn 4	Ser 4	U
	Ile 4	Thr 4	Asn 6	Ser	C
	Ile	Thr	Lys 5	Arg	A
	Met ⊙ + 2	Thr 1	Lys 1	Arg	G
G	Val 4	Ala 5	Asp 1	Gly 3	U
	Val 4	Ala 2	Asp 3	Gly 3	C
	Val 3	Ala 6	Glu 2	Gly 2	A
	Val 3	Ala 1	Glu 3	Gly 1	G

First letter of code word is indicated at left, second letter on top and third letter at right; e.g. the code word U-U-U is used 4 times for phenylalanine and U-U-C 6 times.

⊙ Codon used for initiation.

O Indicates codons not used in the coat gene.

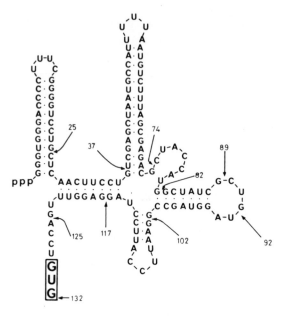

Figure 4. Nucleotide sequence of the untranslated 5'-terminal fragment of MS 2 RNA. The arrows point to positions easily split by T_1 ribonuclease (from De Wachter et al., 1971a; .Volckaert and Fiers, 1973).

RNA Replication

In vitro replication with net synthesis of infectious viral RNA has been achieved with Qβ RNA as template (Spiegelman *et al.*, 1967). The enzyme complex, purified from Qβ-infected cells, consists of 4 polypeptide chains, α, β, γ, δ, with molecular weights of 74,000; 69,000; 47,000; and 33,000 daltons, respectively (Kondo *et al.*, 1970; Kamen, 1970). The α-subunit is identical to the interfering factor *i*, which plays a regulatory

pppG-G-G$_{1-2}$-A-C-C-C-C-C-C-U-U-U-A-G-G-G-G-G-U-C-A-C$\left[\left(\text{A-C}\right)\left(\text{A-C}\right)\right.$

$\left.\left(\text{C-U-C}\right)\right]$A-G-C-A-G-U-A-C-U-U-C-A-C-U-G-A-G-U-A-U-A-A-G-A-G-G-A-C-A-U

$\boxed{\text{A-U-G} \quad \text{A}_2}$ protein

Figure 5. Nucleotide sequence of the untranslated 5'-terminal fragment of Qβ RNA (Billeter et al., 1969; Staples et al., 1971).

TABLE 9.   Codons Found in the MS2 RNA-Polymerase Gene[a]

	U	C	A	G	
U	Phe 4, 6 Leu 4, ②	Ser 6, 5, 5, 5	Tyr ④, 4 Ochre Amber	Cys 5, 2 Opal Trp 3	U C A G
C	Leu 3, 9, 6, ⑤	Pro 3, 1, _, 4	His[b] ②, ② Gln 4, 6	Arg 4, 7, ②, ④	U C A G
A	Ile 4, 7, ④ Met ◉ + 5	Thr 2, 4, ⑥, 1	Asn 6, 5 Lys 3, 10	Ser ③, 1 Arg ⑤, ③	U C A G
G	Val 5, 5, 3, 2	Ala 10, 3, 4, 3	Asp 12, 6 Glu 4, 4	Gly 7, 4, 1, 1	U C A G

[a] 46 percent of the genome.

[b] Histidine does not occur in the coat protein.

First letter of code word is indicated at left, second letter on top and third letter at right; e.g. the code word U-U-U is used 4 times for phenylalanine and U-U-C 6 times.

◉ Codon used for initiation.

○ Indicates codons not used in the coat gene.

◌ Indicates codons for an amino acid (his) absent in the coat gene.

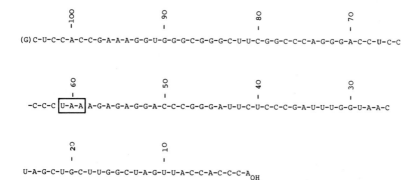

```
        -100              -90               -80               -70
          |                |                 |                 |
(G) C-U-C-C-A-C-C-G-A-A-A-G-G-U-G-G-G-C-G-G-G-C-U-U-C-G-G-C-C-C-A-G-G-G-A-C-C-U-C-C


         -60              -50               -40               -30
          |                |                 |                 |
-C-C-C U-A-A A-G-A-G-A-G-G-A-C-C-C-G-G-G-A-U-U-C-U-C-C-C-G-A-U-U-U-G-G-U-A-A-C


         -20              -10
          |                |
U-A-G-C-U-G-C-U-U-G-G-C-U-A-G-U-U-A-C-C-A-C-C-C-A OH
```

Figure 6. Nucleotide sequence of the 3'-terminal end of MS2 RNA (Contreras et al., 1971). A possible termination codon of the replicase is indicated by a box.

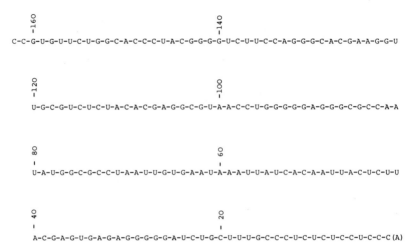

Figure 7. Nucleotide sequence of the 3' -terminal end of Qβ RNA (Goodman et al., 1970; Weissmann et al., 1973).

role in protein synthesis (Groner *et al.*, 1972). The β subunit is the only virus-coded polypeptide (Table 3), while γ and δ correspond to the elongation factors Tu and Ts, respectively (Blumenthal *et al.*, 1972). When Qβ plus strands are used as template, there is an absolute requirement for one or two additional host factors, the function of which in the uninfected cell is unknown (Stavis and August, 1970). No factors are needed when Qβ minus strands or some other polynucleotides like poly C are used as template.

Replication of the phage-f2 RNA apparently involves the same host cell components (Fedoroff and Zinder, 1971).

Genetics

Mutants

Plaque-morphology, temperature-sensitive, suppressor-sensitive, and antiserum-resistant mutants have been obtained after treatment with mutagens like nitrous acid, hydroxylamine, or growth in the presence of fluorouracil (Pfeifer *et al.*, 1964; Van Assche *et al.*, 1972; and references listed in Table 10). Most single-site mutants show a reversion frequency of 10^{-3} or larger. No deletion mutants or frame-shift mutants are known.

The amber mutants have been most extensively studied and were most useful in the characterization of the viral functions. Their main properties are listed in Table 10. Three complementation groups have been found, corresponding to the three physically identified gene products (Figure 2).

TABLE 10. Properties of Amber Mutants

Mutation in cistron		Typical mutants[a]	Growth on[b]				Physiological effect in an su⁻ cell
			su_I^+	su_{II}^+	su_{III}^+	$su_{(UAA)}^+$[c]	
A protein	f2:	sus 1	+	+	+	−	Cell lyses with production of a normal burst of noninfectious, A-protein-defective particles; these are sensitive to ribonuclease, but if degradation is avoided, the phage can be isolated in an intact form, which is infectious in a spheroplast system.
	R17:	A3, A9, A30–A68	+	+	+	+	
	MS2:	class 1: 303, 900 (12 mutants)	+	+	+	−	MS2: Class 1: no nonsense polypeptide detected. Class 2: synthesis of a nonsense polypeptide 88 percent of the normal length.
		class 2: 100, 302, 902 (7 mutants)	+	+	+	+	
	Qβ:	2, 203	+	+	+	−	Qβ: A₁-protein mutants do not lyse host cells.
Coat protein polar: position 6	f2:	sus 3	+	−	−	−	Little effect on cellular growth rate; no phage antigen; no or poor complementation with replicase mutants (polar effect on replicase); decreased level of replicase activity.
	R17:	B2	+	±	−	−	
		B23–B29	+	−	−	−	
	MS2:	908	+	−	−	−	
nonpolar: position 50	R17:	B11, B22	+	±	−	−	Reduced cellular growth rate, but no lysis; normal complementation; enhanced level of virus-induced replicase activity and virus-specific double-stranded RNA.
	MS2:	623	+	−	±	−	
		111	+	±	±	−	
		904	+	−	−	−	
position 54	R17:	B17 to 21	+	±	−	−	
	f2:	sus 11	+	−	−	−	
position 70	MS2:	Mu 9	+	±	−	−	
		601	+	−	−	−	
unidentified	Qβ:	11, 12	+	−	−	−	
Replicase	f2:	sus 10	+	+	+	−	No viral RNA synthesis; no effect on growth of host cell; no phage antigen.
	R17:	C13	±	±	+		
		C8, C14, C15, C16	+	+	+	−	
	MS2:	607	+	+	+		
	Qβ:	12, 01	+	+	+		

[a] References: f2 mutants: Zinder and Cooper, 1964; Lodish and Zinder, 1966. R17 mutants: Gussin, 1966; Tooze and Weber, 1967. MS2 mutants: Viñuela et al., 1968. Over 100 mutants were isolated and characterized by M. Van Montagu (private communication) and partially described in Van Montagu et al., 1967; Fiers et al., 1969; Van de Kerckhove et al., 1971; Vandamme et al., 1972.

Qβ mutants: Horiuchi and Matsuhashi, 1970; Horiuchi et al., 1971.

[b] Very often efficiency of plating (e.o.p.) and not actual growth was tested; ± means either very small plaques or decreased e.o.p.

[c] Ochre suppressor strains suppress both UAA and UAG, but at very low efficiency (1–5 percent).

No ochre mutants are known. Several UGA mutants have been isolated mutated either in the A-protein cistron or the replicase cistron (Van Montagu, 1968; Model *et al.,* 1969). Another class of mutants are the reversed ambers, also called azure, which grow on su$^-$ cells but not on su$^+$ hosts (Horiuchi and Zinder, 1967; Van Montagu private communication).

Regulatory Functions

The coat protein is synthesized in largest amounts, both *in vivo* and *in vitro* (Kozak and Nathans, 1972). With native RNA, only the coat initiation site is open for interaction with the ribosomes (see above). In addition, however, the coat cistron may have an intrinsically higher translation frequency. At least in the case of Qβ, the RNA replication complex can bind to a region which includes the coat initiation site. In this way, the viral RNA molecules which start to function as templates in replication, are stripped free from ribosomes, thus avoiding collision between RNA-polymerase complexes and translating ribosomes (Kolakofsky and Weissmann, 1971).

The A protein is very poorly translated *in vitro*. By using an *in vitro* system derived from *Bacillus stearothermophilus* instead of from *E. coli,* however, only initiation at the A-protein cistron was observed (Lodish, 1969). It is quite possible that *in vivo* A-protein synthesis only takes place on chains *statu nascendi* (Kolakofsky and Weissmann, 1971; Remaut and Fiers, 1974). This would explain the turning off of A-protein synthesis later in infection (Viñuela *et al.,* 1967).

The coat cistron has to be at least partially translated in order to allow translation of the replicase gene. This polar effect is observed by means of position-6 amber mutants (Table 10) both *in vivo* (Roberts and Gussin, 1967) and *in vitro* (Engelhardt *et al.,* 1967). As position-50 mutants are not polar, the regulatory region has to be located between these positions. This fits with the "flower model" of the coat gene (Figure 3), which predicts that translation of the coat in the region between amino acids 24 and 32 releases the opposite region, which contains the ribosome-attachment site for the replicase cistron.

Later in infection, the synthesis of the replicase polypeptide is turned off. This is due to a repressorlike activity of the coat gene, and premature termination of the latter leads to considerable overproduction of the replicase (Lodish and Zinder, 1966). This repressor activity of the coat protein can also be reproduced *in vitro* (Kozak and Nathans, 1972). The specific binding of this coat-protein molecule is to a region preceding and including the replicase initiation site (Bernardi and Spahr, 1972).

Literature Cited

Adams, J. M., S. Cory, and P. F. Spahr, 1972a Nucleotide sequences of fragments of R17 bacteriophage RNA from the region immediately preceding the coat-protein cistron. *Eur. J. Biochem.* **29**:469–479.

Adams, J. M., P. F. Spahr and S. Cory, 1972b Nucleotide sequence from the 5′ end to the first cistron of R17 bacteriophage ribonucleic acid. *Biochemistry* **11**:976–988.

Ammann, J., H. Delius and P. H. Hofschneider, 1964 Isolation and properties of an intact phage-specific replicative form of RNA phage M12. *J. Mol. Biol.* **10**:557–561.

Argetsinger Steitz, J., 1968 Isolation of the A protein from bacteriophage R17. *J. Mol. Biol.* **33**:937–945.

Argetsinger Steitz, J., 1969 Polypeptide chain initiation: nucleotide sequences of the three ribosomal binding sites in bacteriophage R17 RNA. *Nature (Lond.)* **224**:957–964.

Argetsinger Steitz, J., 1972 Oligonucleotide sequence of replicase initiation site in Qβ RNA. *Nat. New Biol.* **236**:71–75.

Banerjee, A. K., C. H. Kuo and J. T. August, 1969 Replication of RNA viruses. VIII. Direction of chain growth in the Qβ RNA polymerase reaction. *J. Mol. Biol.* **40**:445–455.

Bernardi, A. and P. F. Spahr, 1972 Nucleotide sequence at the binding site for coat protein on RNA of bacteriophage R17. *Proc. Natl. Acad. Sci. USA* **69**:3033–3037.

Billeter, M. A., C. Weissmann and R. C. Warner, 1966 Replication of viral ribonucleic acid. IX. Properties of double-stranded RNA from *Escherichia coli* infected with bacteriophage MS2. *J. Mol. Biol.* **17**:145–173.

Billeter, M. A., J. E. Dahlberg, H. M. Goodman, J. Hindley and C. Weissmann, 1969 Sequence of the first 175 nucleotides from the 5′ terminus of Qβ RNA synthesized *in vitro*. *Nature (Lond.)* **224**:1083–1086.

Bishop, D. H. L. and D. E. Bradley, 1965 Determination of base ratios of six ribonucleic acid bacteriophages specific to *Escherichia coli*. *Biochem. J.* **95**:82–93.

Blumenthal, T., T. A. Landers and K. Weber, 1972 Qβ replicase contains the protein biosynthesis elongation factors *Tu* and *Ts*. *Proc. Natl. Acad. Sci. USA* **69**:1313–1317.

Boedtker, H., 1967 The reaction of ribonucleic acid with formaldehyde. I. Optical absorbance studies. *Biochemistry* **6**:2718–2727.

Boedtker, H., 1971 Conformation independent molecular weight determinations of RNA by gel electrophoresis. *Biochim. Biophys. Acta* **240**:448–453.

Bradley, D. E., 1967 Ultrastructure of bacteriophages and bacteriocins. *Bacteriol. Rev.* **31**:230–314.

Capecchi, M. R. and H. A. Klein, 1970 Release factors mediating termination of complete proteins. *Nature (Lond.)* **226**:1029–1033.

Contreras, R., A. Vandenberghe, W. Min Jou, R. De Wachter and W. Fiers, 1971 Studies on the bacteriophage MS2 nucleotide sequence of a 3′-terminal fragment (n = 104). *FEBS (Fed. Eur. Biochem. Soc.) Lett.* **18**:141–144.

Contreras, R., A. Vandenberghe, G. Volckaert, W. Min Jou and W. Fiers, 1972 Studies on the bacteriophage MS2. Some nucleotide sequences from the RNA-polymerase gene. *FEBS (Fed. Eur. Biochem. Soc.) Lett.* **24**:339–342.

Contreras, R., M. Ysebaert, W. Min Jou and W. Fiers, 1973 Nucleotide sequence of the end of the A-protein gene and the intercistronic region. *Nat. New Biol.* **241**:99–101.

Cory, S., J. M. Adams, P. F. Spahr and U. Rensing, 1972 Sequence of 51 nucleotides at the 3′end of R17 bacteriophage RNA. *J. Mol. Biol.* **63**:41–56.

Davis, J. E., J. H. Strauss and R. L. Sinsheimer, 1961   Bacteriophage MS2: Another RNA phage. *Science (Wash., D.C.)* **134**:1427.

DeWachter, R. and W. Fiers, 1969   Sequences at the 5′-terminus of bacteriophage Qβ RNA. *Nature (Lond.)* **221**:233–235.

DeWachter, R., J. Merregaert, A. Vandenberghe, R. Contreras and W. Fiers, 1971*a*   Studies on the bacteriophage MS2. The untranslated 5′-terminal nucleotide sequence preceding the first cistron. *Eur. J. Biochem.* **22**:400–414.

DeWachter, R., A. Vandenberghe, J. Merregaert, R. Contreras and W. Fiers, 1971*b*   The leader sequence from the 5′-terminus to the A-protein initiation codon in MS2-virus RNA. *Proc. Natl. Acad. Sci. USA* **68**:585–589.

Englehardt, D. L., R. E. Webster and N. D. Zinder, 1967   Amber mutants and polarity *in vitro. J. Mol. Biol.* **29**:45–58.

Enger, M. D. and P. Kaesberg, 1965   Comparative studies of the coat proteins of R17 and M12 bacteriophages. *J. Mol. Biol.* **13**:260–268.

Enger, M. D., E. A. Stubbs, S. Mitra and P. Kaesberg, 1963   Biophysical characteristics of the RNA-containing bacterial virus R17. *Biochemistry* **49**:857–860.

Fedoroff, N. V. and N. D. Zinder, 1971   Structure of the poly(G) polymerase component of the bacteriophage f2 replicase. *Proc. Natl. Acad. Sci. USA* **68**:1838–1843.

Fiers, W., 1974   Chemical structure and biological activity of bacteriophage MS2 RNA. in manuscript.

Fiers, W., L. Lepoutre and L. Vandendriessche, 1965   Studies on the bacteriophage MS2. I. Distribution of the purine sequences in the viral RNA and in yeast RNA. *J. Mol. Biol.* **13**:432–450.

Fiers, W., M. Van Montagu, R. DeWachter, G. Haegeman, W. Min Jou, E. Messens, E. Remaut, A. Vandenberghe and B. Van Styvendaele, 1969   Studies on the primary structure and the replication mechanism of bacteriophage RNA. *Cold Spring Harbor Symp. Quant. Biol.* **34**:697–706.

Fiers, W., R. Contreras, R. DeWachter, G. Haegeman, J. Merregaert, W. Min Jou and A. Vandenberghe, 1971   Recent progress in the sequence determination of bacteriophage MS2 RNA. *Biochimie (Paris)* **53**:495–506.

Fiers, W., R. Contreras, R. DeWachter, G. Haegeman, J. Merregaert, W. Min Jou, A. Vandenberghe, G. Volckaert and M. Ysebaert, 1973   In *Viral Replication and Cancer*, edited by J. L. Melnick, S. Ochoa and J. Oro, pp. 34–50, Editorial Labor, Barcelona.

Fischbach, F. A., P. M. Harrison and J. W. Anderegg, 1965   An x-ray scattering study of the bacterial virus R17. *J. Mol. Biol.* **13**:638–645.

Gesteland, R. F. and H. Boedtker, 1964   Some physical properties of bacteriophage R17 and its ribonucleic acid. *J. Mol. Biol.* **8**:496–507.

Goodman, H. M., M. A. Billeter, J. Hindley and C. Weissmann, 1970   The nucleotide sequence at the 5′-terminus of the Qβ RNA minus strand. *Proc. Natl. Acad. Sci. USA* **67**:921–928.

Groner, Y., R. Scheps, R. Kamen, D. Kolakofsky and M. Revel, 1972   Host subunit of Qβ replicase is translation control factor i. *Nat. New Biol.* **239**:19–20.

Gupta, S. L., J. Chen, L. Schaefer, P. Lengyel and S. M. Weissmann, 1970   Nucleotide sequence of a ribosome attachment site of bacteriophage f2 RNA. *Biochem. Biophys. Res. Commun.* **39**:883–888.

Gussin, G. N., 1966   Three complementation groups in bacteriophage R17. *J. Mol. Biol.* **21**:435–453.

Haegeman, G., W. Min Jou and W. Fiers, 1971   Studies on the bacteriophage MS2. IX.

The heptanucleotide sequence present in the pancreatic ribonuclease digest of the viral RNA. *J. Mol. Biol.* **57**:597–613.

Haruna, I., T. Nishihara and I. Watanabe, 1967   Template activity of various phage RNA for replicases of Qβ and VK phages. *Proc. Jap. Acad.* **43**:375–377.

Hindley, J. and D. H. Staples, 1969   Sequence of ribosome binding site in bacteriophage Qβ-RNA. *Nature (Lond.)* **224**:964–967.

Horiuchi, K. and S. Matsuhashi, 1970   Three cistrons in bacteriophage Qβ. *Virology* **42**:49–60.

Horiuchi, K. and N. D. Zinder, 1967   Azure mutants: A type of host-dependent mutant of the bacteriophage f2. *Science (Wash., D.C.)* **156**:1618–1623.

Horiuchi, K., R. E. Webster and S. Matsuhashi, 1971   Gene products of bacteriophage Qβ. *Virology* **45**:429–439.

Isenberg, H., R. I. Cotter and W. B. Gratzer, 1971   Secondary structure and interaction of RNA and protein in a bacteriophage. *Biochim. Biophys. Acta* **232**:184–191.

Jeppesen, P. G. N., J. Argetsinger-Steitz, R. F. Gesteland and P. F. Spahr, 1970   Gene order in the bacteriophage R17 RNA: 5′-A protein-coat protein-synthetase-3′. *Nature (London)* **226**:230–237.

Kamen, R., 1969   Infectivity of Bacteriophage R17 RNA after sequential removal of 3′-terminal nucleotides. *Nature (London)* **221**:321–325.

Kamen, R., 1970   Characterization of the subunits of Qβ replicase. *Nature (London)* **228**:527–533.

Kolakofsky, D. and C. Weissmann, 1971   Possible mechanism for transition of viral RNA from polysome to replication complex. *Nat. New Biol.* **231**:42–46.

Kondo, M., R. Gallerani and C. Weissmann, 1970   Subunit structure of Qβ replicase. *Nature (Lond.)* **228**:525–527.

Konigsberg, W., T. Maita, J. Katze and K. Weber, 1970   Amino-acid sequence of the Qβ coat protein. *Nature (Lond.)* **227**:271–273.

Konings, R. N. H., R. Ward, B. Francke and P. H. Hofschneider, 1970   Gene order of RNA bacteriophage M12. *Nature (Lond.)* **226**:604–607.

Kozak, M. and D. Nathans, 1972   Translation of the genome of a ribonucleic acid bacteriophage. *Bacteriol. Rev.* **36**:109–134.

Krahn, P. M., O'Callaghan, R. J. and W. Paranchych, 1972   Stages in phage R17 infection. VI. Injection of A protein and RNA into the host cell. *Virology* **47**:628–637.

Krueger, R. G., 1969   Serological relatedness of the ribonucleic acid-containing coliphages. *J. Virol.* **4**:567–573.

Lodish, H. F., 1969   Species specificity of polypeptide chain initiation. *Nature (Lond.)* **224**:867–870.

Lodish, H. F. and N. D. Zinder, 1966   Mutants of the bacteriophage f2. VIII. Control mechanisms for phage-specific syntheses. *J. Mol. Biol.* **19**:333–348.

Maita, T. and W. Konigsberg, 1971   The amino acid sequence of the Qβ coat protein. *J. Biol. Chem.* **246**:5003–5024.

Min Jou, W., G. Haegeman, M. Ysebaert and W. Fiers, 1972   Nucleotide sequence of the gene coding for the bacteriophage MS2 coat protein. *Nature (Lond.)* **237**:82–88.

Mitra, S., M. D. Enger and P. Kaesberg, 1963   Physical and chemical properties of RNA from the bacterial virus R17. *Biochemistry* **50**:68–75.

Miyake, T., I. Haruna, T. Shiba, Y. H. Itoh, K. Yamane and I. Watanabe, 1971   Grouping of RNA phages based on the template specificity of their RNA replicases. *Proc. Natl. Acad. Sci. USA* **68**:2022–2024.

Model, P., R. E. Webster, and N. D. Zinder, 1969   The UGA codon *in vitro*: Chain termination and suppression. *J. Mol. Biol.* **43**:177–190.

Moore, C. H., F. Farron, D. Bohnert and C. Weissmann, 1971   Localization of Qβ maturation cistron ribosome binding site. *Nat. New Biol.* **234**:202–206.

Nichols, J. L., 1970   Nucleotide sequence from the polypeptide chain termination region of the coat protein cistron in bacteriophage R17 RNA. *Nature (Lond.)* **225**:147–151.

Nishihara, T. and I. Watanabe, 1969   Discrete buoyant density distribution among RNA phages. *Virology* **39**:360–362.

Nishihara, T., I. Haruna, I. Watanabe, Y. Nozu and Y. Okada, 1969   Comparison of coat proteins from three groups of RNA phages. *Virology* **37**:153–155.

Overby, L. R., G. H. Barlow, R. H. Doi, M. Jacob and S. Spiegelman, 1966a   Comparison of two serologically distinct ribonucleic acid bacteriophages. I. Properties of the viral particles. *J. Bacteriol.* **91**:442–448.

Overby, L. R., G. H. Barlow, R. H. Doi, M. Jacob and S. Spiegelman, 1966b   Comparison of two serologically distinct ribonucleic acid bacteriophages. II. Properties of the nucleic acids and coat proteins. *J. Bacteriol.* **92**:739–745.

Pfeifer, D., J. E. Davis and R. L. Sinsheimer, 1964   The replication of bacteriophage MS2. III. Asymmetric complementation between temperature-sensitive mutants. *J. Mol. Biol.* **10**:412–422.

Remaut, E. and W. Fiers, 1972   Studies on the bacteriophage MS2. XVI. The termination signal of the A-protein cistron. *J. Mol. Biol.* **71**:243–261.

Remaut, E. and W. Fiers, 1974, unpublished.

Rensing, U. and J. T. August, 1969   The 3′-terminus and the replication of phage RNA. *Nature (Lond.)* **224**:853–856.

Roberts, J. W. and G. N. Gussin, 1967   Polarity in an amber mutant of bacteriophage R17. *J. Mol. Biol.* **30**:565–570.

Robertson, H. D. and P. G. N. Jeppesen, 1972   Extent of variation in three related bacteriophage RNA molecules. *J. Mol. Biol.* **68**:417–428.

Sakurai, T., T. Miyake, T. Shiba and I. Watanabe, 1968   Isolation of a possible fourth group of RNA phage. *Jap. J. Microbiol.* **12**:544–546.

Schmidt, J. M. and R. Y. Stanier, 1965   Isolation and characterization of bacteriophages active against stalked bacteria. *J. Gen. Microbiol.* **39**:95–107.

Schwartz, J. H., W. J. Iglewski and R. M. Franklin, 1969   The lack of messenger activity of ribonucleic acid complementary to the viral ribonucleic acid of bacteriophage R17. *J. Biol. Chem.* **244**:736–743.

Scott, D. W., 1965   Serological cross reactions among the RNA-containing coliphages. *Virology* **26**:85–88.

Slegers, H., J. Clauwaert and W. Fiers, 1973   Studies on the bacteriophage MS2. XXIV. Hydrodynamic properties of the native and acid MS2 RNA structures. *Biopolymers,* **12**:2033–2044.

Spiegelman, S., I. Haruna, N. R. Pace, D. R. Mills, D. H. L. Bishop, J. R. Claybrook and R. Peterson, 1967   Studies in the replication of viral RNA. *J. Cell. Physiol.* **70**:35–64.

Staples, D. H. and J. Hindley, 1971   Ribosome binding site of Qβ polymerase cistron. *Nat. New Biol.* **234**:211–212.

Staples, D. H., J. Hindley, M. A. Billeter and C. Weissmann, 1971   Localization of Qβ maturation cistron ribosome binding site. *Nat. New Biol.* **234**:202–204.

Stavis, R. L. and J. T. August, 1970   The biochemistry of RNA bacteriophage replication. *Annu. Rev. Biochem.* **39**:527–560.

Strauss, J. H., Jr. and R. L. Sinsheimer, 1963   Purification and properties of bacteriophage MS2 and of its ribonucleic acid. *J. Mol. Biol.* **7**:43–54.

Sugiyama, T., R. R. Hebert and K. A. Hartman, 1967   Ribonucleoprotein complexes formed between bacteriophage MS2 RNA and MS2 protein *in vitro. J. Mol. Biol.* **25**:455–463.

Tooze, J. and K. Weber, 1967   Isolation and characterization of amber mutants of bacteriophage R17. *J. Mol. Biol.* **28**:311–330.

Valentine, R. C., R. Ward and M. Strand, 1969   The replication cycle of RNA bacteriophages. In *Advances in Virus Research,* Vol. 15, edited by K. M. Smith and M. A. Lauffer, pp. 1–59.

Van Assche, W., J. Van de Kerckhove, J. Gielen and M. Van Montagu, 1972   Antiserum-resistant mutants of the RNA bacteriophage MS2. *Arch. Int. Physiol. Biochim.* **80**:410–411.

Vandamme, E., E. Remaut, M. Van Montagu and W. Fiers, 1972   Studies on the bacteriophage MS2. XVII. Suppressor-sensitive mutants of the A protein cistron. *Mol. Gen. Genet.* **117**:219–228.

Van de Kerckhove, J., J. Gielen, A. Lenaerts, W. Van Assche and M. Van Montagu, 1971   Difference between the nitrous acid-induced and the hydroxylamine-induced amber mutants in the RNA bacteriophage MS2. *Arch. Int. Physiol. Biochim.* **79**:636–637.

Vandenberghe, A., B. Van Styvendaele and W. Fiers, 1969   Studies on the bacteriophage MS2. VI. The nucleoside 5′-triphosphate end group of the replicative intermediate and the replicative form. *Eur. J. Biochem.* **7**:174–185.

Van Montagu, M., 1968   Studies with amber and UGA mutants of the RNA phage MS2. *Arch. Int. Physiol. Biochim.* **76**:393–394.

Van Montagu, M., C. Leurs, P. Brachet and R. Thomas, 1967   A set of amber mutants of bacteriophages λ and MS2 suitable for the identification of suppressors. *Mutat. Res.* **4**:698–700.

Vasquez, C., N. Granboulan and R. M. Franklin, 1966   Structure of the ribonucleic acid bacteriophage R17. *J. Bacteriol.* **92**:1779–1786.

Viñuela, E., I. D. Algranati and S. Ochoa, 1967   Synthesis of virus-specific proteins in *Escherichia coli* infected with the RNA bacteriophage MS2. *Eur. J. Biochem.* **1**:3–11.

Viñuela, E., I. D. Algranati, G. Felix, D. Garwes, C. Weissmann and S. Ochoa, 1968   Virus-specific proteins in *Escherichia coli* infected with some amber mutants of phage MS2. *Biochim. Biophys. Acta* **155**:558–565.

Volckaert, G. and W. Fiers, 1973   Studies on the bacteriophage MS2. *G-U-G* as the initiation codon of the A-protein cistron. *FEBS (Fed. Sur. Biochem. Soc.)* **35**:91–96.

Watanabe, I., T. Nishihara, H. Kaneko, T. Sakurai and S. Osawa, 1967   Group characteristics of RNA phages. *Proc. Jap. Acad.* **43**:210–213.

Weber, K., 1967   Amino acid sequence studies on the tryptic peptides of the coat protein of the bacteriophage R17. *Biochemistry* **6**:3144–3154.

Weber, K. and W. Konigsberg, 1967   Amino acid sequence of the f2 coat protein. *J. Biol. Chem.* **242**:3563–3578.

Weber, H. and C. Weissmann, 1970   The 3′ termini of Qβ plus and minus strands. *J. Mol. Biol.* **51**:215–224.

Weiner, A. M. and K. Weber, 1971   Natural read-through at the UGA termination signal of Qβ coat protein cistron. *Nat. New Biol.* **234**:206–209.

Weissmann, C., M. A. Billeter, H. M. Goodman, J. Hindley and H. Weber, 1973   Structure and function of phage RNA. *Annu. Rev. Biochem.* **42**:303–328.

Wittmann-Liebold, B. and H. G. Wittmann, 1967   Coat proteins of strains of two RNA viruses: Comparison of their amino acid sequences. *Mol. Gen. Genet.* **100:**358–363.

Zinder, N. D., 1965   RNA phages. *Ann. Rev. Microbiol.* **19:**455–472.

Zinder, N. D. and S. Cooper, 1964   Host-dependent mutants of the bacteriophage f2. I. Isolation and preliminary classification. *Virology* **23:**152–158.

Zipper, P., O. Kratky, R. Herrmann and T. Hohn, 1971   An x-ray small angle study of the bacteriophages fr and R17. *Eur. J. Biochem.* **18:**1–9.

18

Episomes

ALLAN CAMPBELL

General Properties

Bacterial episomes (Jacob and Wollman, 1958) are genetic elements able to replicate either separately from or as an integral part of the main chromosome of the bacterium. The process by which one episome inserts itself into the chromosome is depicted in Figure 1 (step 3). A small circular DNA molecule (the episome) undergoes a reciprocal crossover with the chromosome. No material is discarded in the process. All genetic information of both chromosome and episome is preserved in the new, composite structure.

Following insertion, the chromosome bearing the inserted episome must be able to replicate as such. The replication pathway specified by the episome must not interfere with chromosome replication or the cell-division cycle. Thus both ability to insert and ability to be replicated as part of a compound structure are necessary attributes of all episomes.

The only elements whose episomal nature is well documented (Hayes, 1968; Campbell, 1969; Wolstenholme and O'Connor, 1969) are some of the temperate bacteriophages and some of the agents that cause DNA transfer between bacteria. Conjugal transfer requires direct contact between recipient cells and donor cells (those that harbor the transfer agent). Table 1 summarizes for a few bacterial episomes those properties most critical to their episomal existence. This chapter will be concerned with the experimental basis and biological significance of the facts summarized in Table 1. The DNA of one animal tumor virus (SV40) is ap-

ALLAN CAMPBELL—Department of Biological Sciences, Stanford University, Stanford, California.

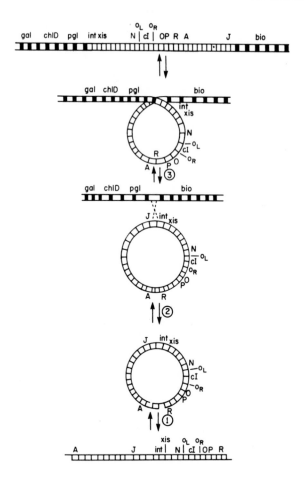

Figure 1. Insertion of bacteriophage λ *into the chromosome of Escherichia coli. At the bottom of the figure is the chromosome as it exists in the virus particle—a linear, double-stranded DNA molecule with short single-stranded ends of complementary base sequence. Phage genes int, xis, N, cI, O, P, and operators O_L and O_R are described in the text. Genes A, J, and R function in the productive cycle of viral infection. When the phage chromosome enters the cell, the single-stranded ends pair, forming a circular molecule with nicks (step 1). The enzyme polynucleotide ligase converts this to a covalently closed circle (step 2). This molecule then undergoes reciprocal breakage and joining with the bacterial chromosome, at a specific site on each partner (step 3). Bacterial genes are: gal, three genes determining enzymes of galactose catabolism; chlD, gene whose product renders the cell sensitive to chlorate under anaerobiosis; pgl, gene determining 6-phos-phogluconolactonase; bio, five genes determining enzymes of biotin biosynthesis. Step 3 inserts the episome into the chromosome. The top line is a linear representation of the structure immediately below it. Downward arrows show the steps in virus production starting from the chromosome bearing the inserted episome.*

TABLE 1. Some Episomes and Their Properties

Episome	Bacteriophage λ	Bacteriophage P2	Bacteriophage P22	Bacteriophage Mu-1	Conjugal transfer agent F	Tumor virus SV-40
Host of origin	Escherichia coli	Shigella dysenteriae	Salmonella typhimurium	Escherichia coli	Escherichia coli	Monkey kidney
Mol wt of DNA	30×10^{6a}	22×10^{6b}	26×10^{6c}	28×10^{6d}	61×10^{6e}	3×10^{6f}
DNA morphology: In virus particles	Linear, unique sequence, single-stranded ends	Linear, unique sequence, single-stranded ends	Linear, cyclically permuted, terminally redundant	Linear	—	Closed circular
Autonomous	Closed circular	Closed circular	Closed Circular	Closed circular	Closed circular	Closed circular
Inserted	Linear, cyclic permutation of order in virus particle	Linear, cyclic permutation of order in virus particle	Linear, unique permutation of order in virus particle	Linear	Linear	Unknown
Result of autonomous replication	Phage production, cell death	Phage production, cell death	Phage production, cell death	Phage production, cell death	F⁺ strain, high-frequency donor of F	Virus production, cell death

TABLE 1. Continued

Episome	Bacteriophage λ	Bacteriophage P2	Bacteriophage P22	Bacteriophage Mu-1	Conjugal transfer agent F	Tumor virus SV-40
Result of insertion	Stably lysogenic strain immune to further infection	Stably lysogenic strain immune to further infection	Stably lysogenic strain immune to further infection	Stably lysogenic strain immune to further infection	Hfr strain, high-frequency donor of bacterial genes	Transformed cell, harboring latent virus
Genes required for autonomous replication:						
Episomal	O, P^g	A, B^b	12, 18^i	Unknown	At least one gene^j	Unknown
Host	$groP$, $(dnaB)^k$	rep^l, $dnaB$, $dnaE$	Unknown	Unknown	At least two genes $(seg)^m$	Unknown
Control of autonomous replication after insertion	Repression by cI product of transcription of O, P, activator gene N, and initiation site	Repression of A and B by C gene product	Repression of 12 and 18 by $c2$ gene product	Unknown	Unknown	Unknown
Genes required for insertion	int^n	int^o	int (also called L)^p	Unknown	Unknown	Unknown

Genes required for excision	*int, xis*[a]	*int*[ρ], *cox*[r]	*int, xis*	Unknown	Unknown	Unknown
Insertion sites:						
On episome	Unique	Unique	Unique	Probably unique	Probably unique	Unknown
On chromosome	Strong preference for one site	Strong preference for one site	Strong preference for one site	Many possible sites	Many possible sites	Unknown
Control of excision after insertion	Repression by *cI* product of transcription of *int, xis*, activator gene *N*, and left prophage terminus	Splitting of *int* operon by insertion	Repression of *int* by *mnt* gene product	Unknown	Unknown	Unknown

[a] Davidson and Szybalski, 1971.
[b] Mandel, 1967.
[c] Rhoades et al., 1968.
[d] Torti et al., 1970.
[e] Sharp, P. A., M. Hsu, E. Ohtsubo, and N. Davidson, private communication.
[f] Dulbecco, 1969.
[g] Packman and Sly, 1968.
[h] Lindahl, 1970.
[i] Levine and Schott, 1971.

[j] Cuzin and Jacob, 1967.
[k] Georgopolous and Herskowitz, 1971.
[l] Calendar et al., 1970.
[m] Cuzin and Jacob, 1967; D. Korn, private communication.
[n] Zissler, 1967.
[o] Bertani, 1970.
[p] Gough, 1970.
[q] Guarneros and Echols, 1970.
[r] Lindahl and Sunshine, 1972.

parently covalently joined to host DNA in virus-transformed cells. Although its status as an episome is still questionable, SV40 is included in Table 1.

Insertion of phage DNA results in stably lysogenic cells, where the latent potentiality to produce phage is transmitted from parent to progeny. Insertion of the transfer agent F produces cell lines (Hfr) that transfer on conjugation the entire bacterial chromosome (starting at the site of F insertion) rather than just the F agent itself.

All the episomes in Table 1 are molecules of double-stranded DNA. The DNA of the free phage particles is linear, but is converted to closed circular form after infection. One mechanism of conversion is shown in the lower part of Figure 1. The DNA molecule of the infecting λ phage particle contains about 47,000 base pairs, plus twelve unpaired bases at each end. The terminal single-stranded segments of the molecule have complementary base sequences. Spontaneous pairing (step 1) generates a circular molecule with two single-stranded nicks. Sealing of these nicks (step 2) by the host enzyme polynucleotide ligase produces a closed circular molecule. The same mechanism is used by bacteriophage P2. Ring closure in phage P22, on the other hand, requires genetic recombination between the two double-stranded ends of the same molecule. Development of this phage is arrested when all recombination pathways of host and virus are blocked by mutation (Botstein and Matz, 1970). The conjugal transfer factor F does not pass through an extracellular phase in its life cycle. F DNA is presumably linear and probably single-stranded during intercellular transfer.

Autonomous Replication

Once generated, the closed circular molecules of episomes can replicate autonomously. Circular molecules, and replication intermediates derived therefrom, can be extracted from infected cells and obtained physically separate from the bacterial chromosome. Autonomous replication of the phages listed in Table 1, if unarrested by repression, terminates in destruction of the infected cell and liberation of progeny—virus particles. On the other hand, the transfer agent F can persist indefinitely in the autonomous state, replicating coordinately with its host.

Autonomous replication requires synthesis of some episome-specified proteins. Mutational inactivation of the genes determining these proteins prevents replication. Such genes have been detected by use of conditionally lethal mutants. For example, some phage mutants forming plaques only at 30°C but not at 43°C have been found defective in DNA replication at high temperature. The known mutations affecting F replication allow maintenance of F at 30°C, but cause F$^+$ cells to produce F$^-$ progeny at 43°C. Some host mutations affecting F replication at high temperature are closely linked to the gene for spectinomycin resistance, whereas others are not and therefore must lie in some other gene(s) (D. Korn, private communication).

Autonomous replication of phage λ is curtailed by mutations inactivating any of three genes, *N, O,* or *P.* The *N* product is not included in Table 1 because it influences replication only indirectly, through enhancing transcription of genes *O* and *P* (Packman and Sly, 1968). Gene *A* of phage P2 is required for expression of many phage genes (Lindahl, 1970) and might likewise affect replication indirectly. For none of the genes listed does available evidence demonstrate that the gene products participate directly in DNA synthesis.

Some of the genes required for autonomous replication probably specify proteins that recognize a specific base sequence on the episome at which replication is initiated. Most or all replication of λ DNA is initiated at a site close to gene *O* and can proceed from there in either direction (Stevens *et al.,* 1971). In phage P2, demonstrable replication is found only in one direction, from a specific site (Schnös and Inman, 1971).

DNA synthesis in *E. coli* likewise starts at one or a few specific sites and proceeds around the circular molecule in both directions. Of those *E. coli* genes in which mutation can render bacterial DNA synthesis temperature sensitive, some determine products required only to initiate synthesis. Mutants of this type will complete at high temperature a round of synthesis begun at low temperature. The products of other genes are needed for continuation of synthesis. Replication ceases very soon after shifting mutants of this type to high temperature. Some of these latter genes are needed also for replication of phage λ. Phage λ will not replicate in these bacterial mutants at high temperature (Fangman and Feiss, 1969). Episomal replication can thus require both specific initiation factors (coded in this case by the episome) and other less specific enzymes (which may be supplied by the host).

Table 1 lists certain host genes whose products are required for autonomous replication of particular episomes. Only those genes are included in which mutations interfere with episomal replication without killing the host as well. The *groP* mutants of *E. coli* do not support growth of wild-type λ, but allow replication by λ mutants bearing specific alterations in gene *P.* The *groP* mutations may be nonlethal alterations of the *dnaB* gene, whose product is required for *E. coli* replication (Georgopolous and Herskowitz, 1971).

Regulation of Autonomous Replication Pathway in Cells Carrying Inserted Episomes

Instead of replicating autonomously, the closed circular molecule of an episome may become inserted into the bacterial chromosome, generating (in the case of phage infection) the chromosome of a lysogenic bacterium that harbors the phage genome in its latent, prophage form. The prophage is then replicated passively, as part of the host chromosome. Neither the replication genes nor the initiation site of the episome is needed for such passive replication. Mutational inactivation of these genes or sites does not interfere with maintenance of the prophage. It is now part of the host chromosome, which has its own initiation site(s).

The replication pathway used in the autonomous state is not only unneeded, but generally unexpressed in the inserted element. Insertion inself has no known regulatory effect on replication. However, insertion of phage DNA is demonstrated only in cells that survive infection. Survival requires repression of viral genes whose products would otherwise kill the cell, and repression shuts off the replication pathway as well. In bacteriophage λ, the repressor determined by the cI gene suppresses replication in three distinct ways: (1) Transcription of genes O and P is prevented by binding of repressor to the operator site O_R (Figure 1). (2) Transcription of gene N is shut off by repressor bound to site O_L (Ptashne and Hopkins, 1968). (3) In preventing rightward transcription from O_R, repressor also eliminates transcription of the replication initiation site near gene O. Transcription at or near the site itself, as well as of the O and P genes, is required for initiation. Replication is not initiated on a phage whose transcription is blocked by repression, even when O and P products are supplied by another, repressor-insensitive phage particle in the same cell (Dove *et al.*, 1971).

Once repression has been established following infection, the repressor itself stimulates further transcription of the cI gene, thus ensuring the stability of the lysogenic state (Heinemann and Spiegelman, 1970). Phages P2 and P22 resemble λ in that autonomous replication is repressed in lysogenic bacteria.

The F factor presents special problems whose solutions are not yet known. Autonomous F maintains itself at a level of about one molecule F DNA per bacterial chromosome, apparently segregating regularly rather than randomly at cell division. If F is introduced into a cell that already harbors another genetically marked F, either the new or the old.F is lost and the other retained. The inability of two F's to persist in the same cell line might result from competition for a membrane site needed for replication and/or segregation, or from production by F of a diffusible inhibitor of F replication. Inserted F, like autonomous F, prevents replication of a second F in the same cell. Replication of the bacterial chromosome is not demonstrably affected by an inserted F, suggesting that the F-specific replication pathway is inoperative following insertion. In bacterial mutants defective in replication initiation, an inserted F can sometimes compensate for the replication defect of its host, perhaps by allowing the host to use the F replication pathway (Nishimura *et al.*, 1971). Similar results have been found with bacteriophage P2 (Lindahl *et al.*, 1971).

Insertion

Like replication, insertion can require both specific gene products and specific recognition sites. Insertion always splits phage λ DNA at the same point, between genes *J* and *int* (Figure 1). Deletion or mutation of this site blocks insertion. Insertion also requires the *int* gene product. Phage mutants whose *int* gene is nonfunctional cannot insert. The DNA of *int* mutants can be inserted, if the *int* gene product is supplied by another phage infecting the same cell. Mutants altered in their insertion site cannot be complemented in this manner.

Phage λ almost always inserts at the same site on the *E. coli* chromosome (between *pgl* and *bio*, Figure 1) and in the same orientation (with the *int* end of the prophage closer to *pgl*). Infection of cells from which this preferred site has been deleted yields rare lysogenic survivors wherein λ has inserted elsewhere (Shimada *et al.*, 1972). Most such insertions are catalyzed by *int* gene product and split the phage chromosome at its usual insertion site. Sometimes, λ inserts within a bacterial gene, thereby inactivating the function of that gene. Excision of the phage restores the integrity of the gene into which it was inserted. Thus no material is discarded in the insertional act.

Phages P2 and P22 resemble λ in preferring one site but occasionally inserting elsewhere. Phage Mu-1 and transfer agent F, on the other hand, insert at many different bacterial sites. Gene inactivation caused by insertion of these elements is generally irreversible (Bukhari and Zipser, 1972; Beckwith *et al.*, 1966), and insertion may entail deletion of some bacterial DNA.

Excision

Once lysogeny is established, phage can be produced only where repression is lifted. Derepression can be spontaneous, or can be induced by various treatments. Phage development following derepression requires excision of the prophage from the lysogenic chromosome. Excision reverses step 3 of Figure 1, and regenerates a closed circular molecule that then replicates autonomously to produce progeny virus particles. In phages λ and P22 (and probably in P2 as well) excision requires, besides *int* product, the product of a second gene (*xis*). The *xis* product is not required for insertion, only for excision (Guarneros and Echols, 1970). This difference in catalytic requirements for insertion and excision implies that the phage and bacterial sites at which breakage and joining take place have different base sequences, a conclusion corroborated by other evidence (Gottesman and Weisberg, 1971). Both *int* and *xis* are site-specific genes. Phage 21, which can recombine with λ, inserts at a different bacterial site, and cannot supply an *int* or *xis* product that will complement λ mutants deficient in these functions (Kaiser and Masuda, 1970).

Regulation of Insertion and Excision

Since insertion and excision require specific gene products, the inserted state can be stabilized by turning off these genes in cells where insertion has occurred. Insertion of λ has no demonstrated regulatory effect; but as with the replication genes, cell survival requires repression, which affects the *int* and *xis* genes both directly and indirectly. Excision also resembles replication in that transcription of one of the sites participating in the exchange itself stimulates the rate of excision (R. W. Davies, W. F. Dove, H. Inokuchi, J. F. Lehman and R. L. Roehrdanz, private communication). The stability of lysogeny for phage P22 requires two repressors, determined by genes *c2* and *mnt*. The *int* gene is probably under direct control of the *mnt* product (Gough, 1970).

Since the *int* gene of λ is repressed in lysogenic cells, infecting a lysogenic bacterium with another λ particle seldom results in insertion of the superinfecting phage. By contrast, in lysogens of phage P2 the *int* gene of a superinfecting phage is expressed. On the other hand, derepression of a P2 lysogen does not cause appreciable excision. The *int* gene of P2 prophage is inactivated by insertion itself. Probably insertion separates the gene from its normal promoter (Bertani, 1970).

Abnormal Excision

Excision of phage λ, catalyzed by the *int* and *xis* products, regenerates the phage chromosome with complete precision. Rarely, breaking and joining of DNA at heterologous sites within prophage and bacterial DNA excises abnormal molecules wherein a segment of viral

DNA has been substituted by host DNA that was contiguous to the prophage. Excision of λ molecules containing the *gal* genes of *E. coli* (Figure 2) does not require *int, xis,* or any known viral or host genes. These abnormal phages are detectable because they confer ability to metabolize galactose on *gal⁻* cells lysogenized by them. Derivatives (F′) of the F agent containing bacterial genes arise in a similar manner. Because

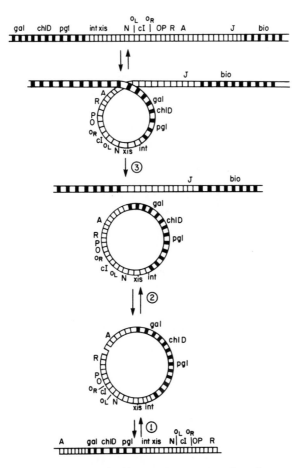

Figure 2. Generation of λ particles carrying the gal genes of E. coli. Step 3 is a rare type of excision in which phage and bacterial chromosomes break and join at heterologous points. The process is diagramed as reciprocal, although it is not known whether both products can result from a single event. Steps 2 and 1 are identical to those of Figure 1, except that the viral chromosome has the substitution generated in step 3.

such fusion episomes retain the property of independent replication, transducing phages and F´ particles are convenient sources of large numbers of identical molecules containing specific segments of viral or host DNA.

Literature Cited

Beckwith, J. R., E. R. Signer and W. Epstein, 1966   Transposition of the *lac* region of *E. coli. Cold Spring Harbor Symp. Quant. Biol.* **21**:393–402.

Bertani, L. E., 1970   Split-operon control of a prophage gene. *Proc. Natl. Acad. Sci. USA* **65**:331–336.

Botstein, D. and M. J. Matz, 1970   A recombination function essential to the growth of bacteriophage P22. *J. Mol. Biol.* **59**:417–440.

Bukhari, A. I. and D. Zipser, 1972   Random insertion of Mu-1 DNA within a single gene. *Nat. New Biol.* **236**:240–243.

Calendar, R., B. Lindqvist, G. Sironi and A. J. Clark, 1970   Characterization of REP⁻ mutants and their interaction with P2 phage. *Virology* **40**:72–83.

Campbell, A., 1969   *Episomes*, Harper and Row, New York.

Cuzin, F. and F. Jacob, 1967   Mutations de l'episome F d'*Escherichia coli* K12. II. Mutants a replication thermosensible. *Ann. Inst. Pasteur (Paris)* **112**:397–418.

Davidson, N. and W. Szybalski, 1971   Physical and chemical characteristics of lambda DNA. In *The Bacteriophage Lambda*, edited by A. D. Hershey, pp. 45–82, Cold Spring Harbor Laboratory, Cold Spring Harbor, N.Y.

Dove, W. F., H. Inokuchi and W. F. Stevens, 1971   Replication control in phage lambda. In *The Bacteriophage Lambda*, edited by A. D. Hershey, pp. 747–771, Cold Spring Harbor Laboratory, Cold Spring Harbor, N.Y.

Dulbecco, R., 1969   Cell transformation by viruses. *Science (Wash. D.C.)* **166**:962–968.

Fangman, W. L. and M. Feiss, 1969   Fate of λ DNA in a bacterial host defective in DNA synthesis. *J. Mol. Biol.* **44**:108–116.

Georgopolous, C. P. and I. Herskowitz, 1971   *Escherichia coli* mutants blocked in lambda DNA synthesis. In *The Bacteriophage Lambda*, edited by A. D. Hershey, pp. 553–564, Cold Spring Harbor Laboratory, Cold Spring Harbor, N.Y.

Gottesman, M. E. and R. A. Weisberg, 1971   Prophage insertion and excision. In *The Bacteriophage Lambda*, edited by A. D. Hershey, pp. 113–138, Cold Spring Harbor Laboratory, Cold Spring Harbor, N.Y.

Gough, M., 1970   Requirement for a functional *int* product in temperature inductions of prophage P22 *ts mnt. J. Virol.* **6**:320–325.

Guarneros, G. and H. Echols, 1970   New mutants of bacteriophage λ with a specific defect in excision from the host chromosome. *J. Mol. Biol.* **47**:565–574.

Hayes, W., 1968.   *The Genetics of Bacteria and Their Viruses*, John Wiley & Sons, New York.

Heinemann, S. F. and W. G. Spiegelman, 1970   Control of transcription of the repressor gene in bacteriophage lambda. *Proc. Natl. Acad. Sci. USA* **67**:1122–1129.

Jacob, F. and E. L. Wollman, 1958   Les épisomes, eléments génétique ajoutés. *C. R. Hebd. Seances Acad. Sci. Ser. D Sci. Nat.* **247**:154–156.

Kaiser, A. D. and T. Masuda, 1970   Evidence for a prophage excision gene in λ. *J. Mol. Biol.* **47**:557–564.

Levine, M. and C. Schott, 1971 Mutations of phage P22 affecting phage DNA synthesis and lysogenization. *J. Mol. Biol.* **62**:53–64.

Lindahl, G., 1970 Bacteriophage P2: Replication of the chromosome requires a protein which acts only on the genome that coded for it. *Virology* **42**:522–533.

Lindahl, G. and M. G. Sunshine, 1972 Excision defective mutants of bacteriophage P2. *Virology* **49**:180–189.

Lindahl, G., Y. Hirota and F. Jacob, 1971 On the process of cellular division in *Escherichia coli*: Replication of the bacterial chromosome under control of prophage P2. *Proc. Natl. Acad. Sci. USA* **68**:2407–2411.

Mandel, M., 1967 Infectivity of phage *P2* DNA in presence of helper phage. *Mol. Gen. Genet.* **99**:88–96.

Nishimura, Y., L. Caro, C. M. Berg and Y. Hirota, 1971 Chrosome replication in *Escherichia coli*. IV. Control of chromosome replication and cell division by an integrated episome. *J. Mol. Biol.* **55**:441–456.

Packman, S. and W. S. Sly, 1968 Constitutive λ DNA replication by λc₁₇, a regulatory mutant related to virulence. *Virology* **34**:778–789.

Ptashne, M. and N. Hopkins, 1968 The operators controlled by the λ phage repressor. *Proc. Natl. Acad. Sci. USA* **60**:1282–1287.

Rhoades, M., L. A. MacHattie and C. A. Thomas, Jr., 1968 The P22 bacteriophage DNA molecule. I. The mature form. *J. Mol. Biol.* **37**:21–40.

Schnös, M. and R. Inman, 1971 Starting point and direction of replication in P2 DNA. *J. Mol. Biol.* **55**:31–38.

Shimada, K., R. A. Weisberg and M. E. Gottesman, 1972 Prophage lambda at unusual chromosomal locations. I. Location of the secondary attachment sites and properties of the lysogens. *J. Mol. Biol.* **63**:483–503.

Stevens, W. F., S. Adhya and W. Szybalski, 1971 Origin and bidirectional orientation of DNA replication in coliphage lambda. In *The Bacteriophage Lambda,* edited by A. D. Hershey, pp. 515–534, Cold Spring Harbor Laboratory, Cold Spring Harbor, N.Y.

Torti, F., C. Barksdale and J. Abelson, 1970 Mu-1 bacteriophage DNA. *Virology* **41**:567–568.

Wolstenholme, G. E. W. and M. O'Connor, 1969 *Bacterial Episomes and Plasmids,* J. & A. Churchill, London.

Zissler, J., 1967 Integration-negative (*int*) mutants of phage λ. *Virology* **31**:189.

19

Bacteriophage Lambda

WACLAW SZYBALSKI

Mature particles of bacteriophage lambda are composed of about equal amounts of protein and DNA. Each phage contains one double-stranded DNA molecule encapsulated in an icosahedral head, which is about 50 nm (0.05 μ) in diameter; a flexible tubular tail, which is about 150 nm (0.15 μ) long and terminates in a fiber, projects from the head (Kellenberger and Edgar, 1971). Lambda is an obligatory parasite of *Escherichia coli*. Growth begins when a phage particle attaches to the host by the tip of its tail and injects its DNA molecule. Lambda is a temperate phage and, thus, can multiply in *E. coli* by one of three ways: (1) In productive growth, the injected DNA molecule directs the synthesis of numerous gene products; these promote replication of the phage DNA, synthesis of the phage heads and tails, packaging of the DNA into mature phage particles, and eventual lysis of the cell. At 37°C it takes about 40 minutes for a lytic cycle, and about 100 infective progeny phage are produced. (2) Lambda DNA can also persist in the host cell as a prophage and replicate passively as an integral part of the bacterial genome. In this case the injected DNA must first direct the synthesis of the gene products that promote its linear insertion into the genome of the host. Next, it turns on the synthesis of the repressor, which blocks transcription of those genes responsible for autonomous lambda DNA replication and most other phage functions. Any fragment of lambda DNA can exist in the form of a

(text continued on page 318)

WACLAW SZYBALSKI—McArdle Laboratory for Cancer Research, University of Wisconsin, Madison, Wisconsin.

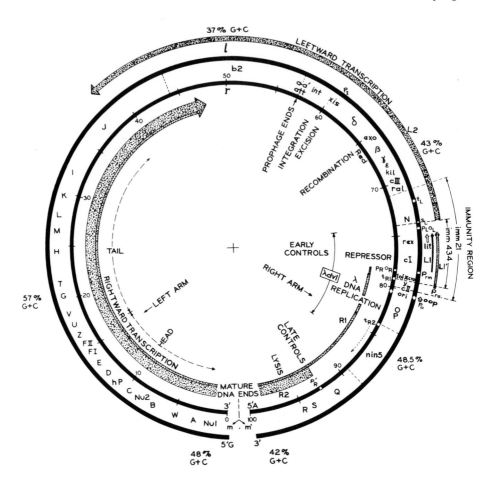

Figure 1. A genetic and molecular map of bacteriophage lambda (λ) of E. coli. (From Szybalski, 1970, 1972; Szybalski and Herskowitz, 1971; and Szybalski et al. 1969, 1970, with modifications.) The base composition of each segment of λ DNA is indicated by %G + C (Davidson and Szybalski, 1971; Skalka et al., 1968). Cytosine-rich clusters are observed on the l strand (gl1 at 60 ± 3 %λ units) and on the r strand (gr1 at 84 ± 3 %λ units, gr2 at 91 ± 2 %λ units, and gr3–gr9 at 0–40 %λ units) (Champoux and Hogness, 1972; Szybalski et al., 1969). The shaded arrows define the orientation of transcription for the indicated operons or scriptons L1, L1′, L2, R1, R2, lit, and oop (Hayes and Szybalski, 1973a,b; Szybalski, 1972; Szybalski et al., 1970). Here λdv1 refers to the fragment of λ DNA that replicates as an autonomous circular plasmid (Davidson and Szybalski, 1971; Matsubara and Kaiser, 1968). The complementary strands of DNA are referred to by l and r. Strand l is transcribed leftward (counterclockwise) and has a 5′-G at its left cohesive (m) terminus. Strand r is transcribed rightward (clockwise) and has a 5′-A at its right cohesive (m′) terminus (Szybalski et al., 1970; Yarmolinsky, 1971). Strand r displays a higher density in the CsCl–poly(U,G) gradient and a lower density in the alkaline CsCl gradient than strand l (Davidson and Szybalski, 1971). The percent λ length (in %λ units; Davidson and Szybalski, 1971) is indicated on the r strand.

TABLE 1.   The Key to the Map of Escherichia coli Bacteriophage Lambda (λ)

Symbol (reading clockwise)	Description[a] (references)
$m \cdot m'$ (included in the *cos* site)	The cohesive termini* (mature ends) of λ DNA extracted from mature phage particles. They are single-stranded, 12 nucleotides long, and complementary. The 5′-terminus of the *l* strand contains a guanine (G) residue (left end, *m*), and that of the *r* strand contains an adenine (A) residue (right end, *m′*). The ends are thought to be generated by the endonucleolytic termination (Ter) function (Campbell, 1971; Davidson and Szybalski, 1971; Wang and Kaiser, 1973; Yarmolinsky, 1971), which recognizes the *co*hesive end recognition *s*ite(s), designated *cos* (Emmons, 1974; Feiss and Campbell, 1974).
*A, W, B, C, hP, D, E, F*I, *F*II	Genes* *A, B, C, hP, D, E,* and *F*I are all required for the *in vivo* formation of the functional phage head and the cohesive $m \cdot m'$ ends of λ DNA by cutting at the *cos* site (Ter function) (Boklage *et al.*, 1973; Campbell, 1971; Kellenberger and Edgar, 1971; Murialdo and Siminovitch, 1971, 1972*a,b*; Wang and Kaiser, 1973; Yarmolinsky, 1971). Gene *A* codes for a protein which has Ter activity *in vitro* (Wang and Kaiser, 1973), but which is not a part of the completed head. Genes *D* and *E* code for two major capsid proteins (Casjens *et al.*, 1970; Murialdo and Siminovitch, 1972*a*), both of which are on the outside of the head (Hendrix and Casjens, private communication). Genes *B* and *C* specify minor capsid proteins which are cleaved during head assembly, and the cleaved *C* protein is fused to an equimolar amount of a cleaved fragment of the *E* protein (Hendrix and Casjens, 1974; Murialdo and Siminovitch, 1972*b*). Gene *F*II (often referred to as *F*) codes for a capsid protein that joins the head to the tail (Boklage *et al.*, 1973; Casjens, 1971; Casjens *et al.*, 1972). Gene *W* codes for a function that is needed for this *F*II-promoted reaction; no *W* protein was found in the finished capsid (Casjens *et al.*, 1972). Genes *Nu*1 (=*m*, or between *m* and *A*), *Nu*2 (between *B* and *C*) and *Nu*3(=*hP*) were identified by C. R. Fuerst and H. Bingham (see Murialdo and Siminovitch, 1972*b*).
Z, U, V, G, T, H, M, L, K, I, J	Genes* required for the formation of the phage tail. Gene *V* codes for a major tail protein (Murialdo and

TABLE 1.   Continued

Symbol (reading clockwise)	Description[a] (references)
	Siminovitch, 1971). Gene *U* controls the length of the tail (Murialdo and Siminovitch, 1972*b*). The *H* protein is cleaved to give tail component *h*3 (R. W. Hendrix, private communication). Gene *J* determines the host specificity (*h* host range mutations) and a tail antigen (Campbell, 1971; Kellenberger and Edgar, 1971).
b2	The b2 region codes for several proteins not essential for vegetative growth and for the Sif function (site-specific *inhibitor function*; Lehman, 1974), and it contains the left component *a* of the *att* site, which is required for DNA insertion and excision. Several proteins coded by region b2 have been identified (Hendrix, 1971; Murialdo and Siminovitch, 1972*a*), including an endonuclease (Rhoades and Meselson, 1973). This region (13 percent of the total λ length) is deleted in mutant λ*b2* (and λ*l*) (*b* = buoyancy). Many other deletions in the *J–N* region have been produced (Davidson and Szybalski, 1971).
att a·a′ (also *att*P·P′	The site of interaction (attachment) between phage and bacterial DNA that is required for the linear insertion of the λ genome during lysogenization. The prophage ends are denoted by the symbols *a·a′* (or P·P′) and the corresponding sites in the bacterial DNA by *b·b′* (or B·B′), (Gottesman and Weisberg, 1971; Signer, 1968).
int	Gene that codes for an enzyme that controls prophage integration (*int* alone) and excision (*int* + *xis*) at the *att* site (Gottesman and Weisberg, 1971; Signer, 1968).
xis	Gene that codes for an enzyme that, in conjunction with the *int* product, controls prophage excision. The symbol *xis* is pronounced "ex-ise" (Gottesman and Weisberg, 1971; Signer, 1968).
Int function	A combination of the products of genes *int* and *xis* that mediates site-specific recombination between the phage and/or bacterial *att* sites (Gottesman and Weisberg, 1971; Signer, 1968). Another function, produced either by "gene" *hen*, located next to gene *xis*, or by a special allele of gene *xis*, appears to be involved in integrative recombination (Echols *et al.*, 1974). Gene *int* is partially under control of an auxiliary promoter p_1 (Shimada and Campbell, 1974).

TABLE 1. Continued

Symbol (reading clockwise)	Description[a] (references)
δ or *del*	Gene whose product interferes with λ growth in P2 lysogens (component of Spi[+] phenotype) (Signer, 1971; Sironi *et al.*, 1971; Tavernier *et al.*, 1974; Zissler *et al.*, 1971).
exo or *red*α (also *red*X)	Gene that codes for the λ exonuclease, which promotes general recombination and growth in *recA*[−] and Feb[−] (e.g., *polA*[−]) hosts, but interferes with growth in P2 lysogens (Signer, 1968; Signer, 1971; Zissler *et al.*, 1971).
β or *red*β (also *bet* or *red*B)	Gene that codes for a protein that, in conjunction with the *exo* product, promotes general recombination; it resembles the *exo* function in its effect on λ growth in various hosts (Signer, 1968; Signer, 1971; Zissler *et al.*, 1971).
Red function	A combination of the products of genes *red*α and *red*β that promotes general recombination; weakly stimulated by the *gam* product (Signer, 1971; Zissler *et al.*, 1971); involved in λ DNA replication (Enquist and Skalka, 1973). Lack of the *red*α, *red*β, *gam*, or *eps* products prevents λ growth in *polA*[−], *lig*[−], or other Feb[−] hosts (Feb[−] phenotype; Signer, 1971; Tavernier *et al.*, 1974).
γ or *gam*	Gene whose product inhibits the host-coded RecB[−]C[−] nuclease (Sakaki *et al.*, 1973; Signer, 1971; Sironi *et al.*, 1971; Tavernier *et al.*, 1974; Ungar *et al.*, 1972; Zissler *et al.*, 1971). Function ε or *eps* is located between genes *gam* and *kil*, since the amber *eps* mutations are frequently polar for *gam* but not for *kil* (Zissler *et al.*, 1971).
Fec[+] phenotype	Ability to grow on *recA*[−] hosts. The products of genes *red*α, *red*β and *gam* (or *eps*), permit λ growth in *recA*[−]*recB*[+]*C*[+] hosts, either by inhibiting the *recB*[+]*C*[+] system (*gam*) or by providing the Red function (*red*α, *red*β) (Enquist and Skalka, 1973; Signer, 1971; Sironi *et al.*, 1971; Tavernier *et al.*, 1974; Ungar *et al.*, 1972; Zissler *et al.*, 1971).
Spi[+] phenotype	Inability of λ to grow on P2 lysogens due to inhibition by the products of genes *del*, *red*α, *red*β, and *gam* or *eps* (Signer, 1971; Sironi *et al.*, 1971; Tavernier *et al.*, 1974; Ungar *et al.*, 1972; Zissler *et al.*, 1971). Phage λ*del*[−]*red*[−]*gam*[−] or λ*red*[−]*eps*[−] can grow in P2 lysogens.
kil	Gene whose product kills the host when λ*N*[+] (deleted for the *gam-int* and *tof-J* segments) lysogens are induced (Greer, 1974).

TABLE 1. Continued

Symbol (reading clockwise)	Description^a (references)
cIII	Gene whose product appears to act together with the cII product in promoting, upon phage infection, early expression of genes cI and *rex* (Echols, 1974; Eisen and Ptashne, 1971; Reichardt and Kaiser, 1971) (see scripton $L1'$).
ral	Gene whose product *a*lleviates the *r*estriction of non-modified phage λ in K12 and B hosts, and phage P2 in B host (Zabeau *et al.*, 1974 a-c). The major Ea10 protein (Hendrix, 1971) is not the *ral* product; Ea10 is coded by a region between genes cIII and *ral* (Zabeau *et al.*, 1974b).
t_L	Site of the *rho*-dependent termination of the leftward p_L-N-t_L transcription and of the antiterminating action of the N product (Echols, 1971; Roberts, 1969; Szybalski *et al.*, 1969, 1970; Szybalski, 1974).
N	Gene* whose product acts as a positive regulator by counteracting the *rho*-dependent termination of transcription at the t termination sites (Couturier *et al.*, 1973; Echols, 1971; Roberts, 1969; Szybalski, 1972, 1974; Szybalski *et al.*, 1969, 1970; Thomas 1970, 1971).
s_L	Startpoint* for the p_L-promoted RNA synthesis, i.e., the base pair on the DNA that codes for the 5'-pppA terminus of the p_L mRNA (Blattner *et al.*, 1972; Szybalski, 1972, 1974).
p_L	Promoter* for the *L2* scripton, i.e., an entry site for the RNA polymerase (Blattner *et al.*, 1972; Echols, 1971; Szybalski, 1972, 1974; Szybalski *et al.*, 1969, 1970; Thomas, 1971). Mutations *sex* or *t27*, which define p_L, are within the *imm*434 region, whereas the s_L startpoint is to the left of this region (Blattner *et al.*, 1972; Szybalski, 1972, 1974).
o_L	Operator for the *L2* scripton, i.e., the binding site for the λ repressor (product of gene cI) (Echols, 1972; Eisen and Ptashne, 1971; Ptashne, 1971; Szybalski, 1972, 1974). Site o_L overlaps with p_L and might overlap or coincide with a site where the products of genes N and *tof (cro)* act (Eisen and Ptashne, 1971; Sly *et al.*, 1971; Szybalski, 1972, 1974; Szybalski *et al.*, 1970).
rex	Gene whose product inhibits the growth of unrelated *E. coli* phages, including T4rII (Ptashne, 1971), T5*lr* (Jacquemin-Sablon and Lanni, 1973), and T1

TABLE 1. *Continued*

Symbol (reading clockwise)	Description[a] (references)
	(Christensen and Geiman, 1973). After induction, only the distal part of gene *rex* is transcribed (*lit* RNA) (Hayes and Szybalski, 1973*b*).
*c*I	Gene that codes for the λ repressor, which binds to the o_L and o_R operators and blocks transcription initiation (Ptashne, 1971). In general, mutations *c*I *ts*A and *c*I *ts*B render the repressor irreversibly or reversibly thermosensitive, respectively. Mutation *ind* protects the repressor from inactivation by the products of host DNA synthesis inhibition. (Ptashne, 1971). Mutation *c*P9 is dominant negative (Oppenheim and Solomon, 1972).
p_{rm} (also p_c, p_M, or *prm*)	Promoter for the *L1* scripton, which includes genes *c*I and *rex* and is responsible for the maintenance of repression in λ lysogens (Echols, 1972; Reichardt and Kaiser, 1971; Szybalski, 1972, 1974; Szybalski *et al.*, 1970; Yen and Gussin, 1973). The hypothetical s_c startpoint, where synthesis of the *L1* mRNA starts, and the o_c operator site, where the *tof* product acts, are associated with the p_{rm} promoter (Szybalski, 1972, 1974; Szybalski, *et al.*, 1970). The *L*1 and *lit* RNAs terminate at the t_i site (Hayes and Szybalski, 1973).
p_R	Promoter* for the *R1* scripton (Echols, 1971; Szybalski, 1972, 1974; Szybalski, *et al.*, 1969, 1970). The hypothetical s_R startpoint, which codes for the 5'-pppA terminus of the *R1* mRNA, is probably located downstream from the p_R promoter, in analogy with the p_L-s_L promoter region (Blattner *et al.*, 1972; Szybalski, 1972, 1974). Probably there is also another weak rightward promoter, p_P, located between gene *cro* and the t_{R1} site that permits weak constitutive expression of gene *P* (Szybalski *et al.*, 1970; Thomas, 1970).
o_R	Operator for the *R1* scripton, i.e., the binding site for the λ repressor. Mutations $v1$, v_s and $v3$ inactivate the o_R operator (Eisen and Ptashne, 1971; Oppenheim and Solomon, 1972; Ptashne, 1971; Sly *et al.*, 1971; Ordal and Kaiser, 1973; Szybalski, 1972, 1974).
tof or *cro*	Gene* in the *x* region that codes for the "second λ repressor," which depresses transcription of the *L1*, *L2*, and *R1* operons, as if acting at the o_c, o_L, and o_R

TABLE 1.   *Continued*

Symbol (reading clockwise)	Description[a] (references)
	sites (Eisen and Ptashne, 1971; Reichardt and Kaiser, 1971; Szybalski, 1972, 1974; Szybalski *et al.*, 1970); also denoted *fed (Fec d*epression) and Ai or *ai* (anti-immunity). Phage λ*c*I$^+$*cro*$^-$ does not form plaques, and phages λ*c*I$^-$*cro*$^-$ and λ*c*II$^-$*cro*$^-$ plate with low efficiencies (Eisen and Ptashne, 1971).
x region	The region* between gene *c*I and the right boundary of the *imm*434 nonhomology segment. It includes elements p_{rm}, p_R, o_R and gene *tof* or *cro* (Eisen and Ptashne, 1971; Szybalski, 1972; Szybalski *et al.*, 1969, 1970). Mutations designated *x*$^-$ usually refer to p_R-promoter mutations.
imm	The "immunity region," which contains the genes that code for the repressor (*c*I), the "antirepressor" (*cro*, *ai*, or *tof*), the sites of action (operators o_L, o_c, and o_R) for these regulatory products, and promoters p_L, p_{rm}, and p_R. This region is usually defined as the segment that is replaced by phage 434 DNA in the hybrid phage λ*imm*434. The "extended immunity region" also contains the elements of immunity establishment, including genes *c*III and *c*II and the p_{re} promoter, and its right boundary roughly corresponds to that of the nonhomology in λ*imm*21 (21*hy*1) (Davidson and Szybalski, 1971; Echols, 1972; Eisen and Ptashne, 1971; Szybalski *et al.*, 1970; Thomas, 1971).
t_{R1}	The site of the *rho*-dependent termination of the rightward p_R-promoted (and p_P-promoted) transcription and of the antiterminating action of the *N* product (Couturier *et al.*, 1973; Echols, 1971; Roberts, 1969; Szybalski, 1972, 1974; Szybalski *et al.*, 1969, 1970; Thomas, 1971).
p_{re} (also p'_c, p_E, or *pre*)	The hypothetical promoter for the *L1′* scripton which is activated by the *c*II and *c*III products. *L1′* includes the leftward-transcribed region *x*-*c*I-*rex* and is responsible for establishment of repression early after infection (Echols, 1972; Eisen and Ptashne, 1971; Reichardt and Kaiser, 1971; Spiegelman *et al.*, 1972).
y region	The region between the right boundary of the *imm*434 substitution and gene *c*II. It includes sites p_{re}, t_{R1}, and the sites of the "clear" and other mutations, e.g., *c*17, which acts as a new constitutive rightward promoter, and *c*42, which is p_{re}^- and often referred

TABLE 1. *Continued*

Symbol (reading clockwise)	Description[a] (references)
	to as a cY^- type of mutation (Echols, 1972; Eisen and Ptashne, 1971; Reichardt and Kaiser, 1971; Szybalski, 1972; Szybalski *et al.*, 1970).
*c*II	Gene whose product appears to act at the p_{re} site. Together, the *c*II and *c*III products stimulate early expression of the *L1'* scripton, which includes genes *c*I and *rex* (Echols, 1972; Eisen and Ptashne, 1971; Reichardt and Kaiser, 1971).
ori	Origin* of λ DNA replication, i.e., the site where replication starts and proceeds in both directions (Dove *et al.*, 1971; Kaiser, 1971; Schnös and Inman, 1970; Stevens *et al.*, 1971; Hayes and Szybalski, 1973*a,b*).
p_0	Promoter for the leftward-transcribed 4S RNA (*oop* or minor leftward RNA), which contains 5'-pppGUU and 3'-U_6A-OH termini, and possibly is a primer for the leftward λ DNA replication (Blattner and Dahlberg, 1972; Dahlberg and Blattner, 1973; Hayes and Szybalski, 1973*a,b*).
O,P	Genes* required for the initiation of λ DNA replication (Dove *et al.*, 1971). Their products may be nucleases (Kaiser, 1971).
t_{R2}	The site(s) of the *rho*-dependent termination similar to t_{R1}; removed by the *nin* deletions (Couturier *et al.*, 1973; Fiandt *et al.*, 1971; Roberts, 1969; Szybalski, 1972, 1974; Szybalski *et al.*, 1966, 1970) or bypassed by the *byp* mutations (Echols, 1971), which map just to the right of the *nin5* deletion (N. Sternberg, private communication).
Q	Gene* whose product acts as a positive regulator by permitting "late" transcription of the *R2* scripton (genes *S-R-A-J*) promoted at p'_R (Dahlberg and Blattner, 1973; Echols, 1971; Szybalski, 1974; Szybalski *et al.*, 1970; Thomas, 1971).
p'_R or p_Q	Promoter* controlling the "late" *R2* scripton, the transcription of which depends on the *Q* product (Echols, 1971; Szybalski, 1972, 1974; Szybalski *et al.*, 1969, 1970; Thomas, 1971). The 5'pppA-initiated and U_6A-OH-terminated RNA that is synthesized in the *Q-S* region is probably promoted by p'_R, but it is strongly impeded at the 15th and terminated at the 198th nucleotide in the absence of the *Q* product (Blattner and Dahlberg, 1972; Dahlberg and Blattner, 1973; Lebowitz *et al.*, 1971; Szybalski, 1972, 1974).

TABLE 1.   *Continued*

Symbol (reading clockwise)	Description[a] (references)
S	Gene* whose product is required for cell (membrane?) lysis (Campbell, 1971).
R	Gene* that codes for the λ endolysin (endopeptidase), which cleaves the amino-carboxyl crosslink between the diaminopimelic acid and D-alanine residues in the murein component of the bacterial cell wall (Campbell, 1971; Taylor, 1971).

[a]Those genes and recognition sites which are marked by asterisks are usually essential for the formation of plaques on standard hosts.

defective prophage. (3) Under special conditions, the lambda genome or its fragment containing the replication genes (expression of which is autoregulated by the *tof* product; Szybalski, 1974) can persist in the "carrier" host as a nonintegrated plasmid (see Chapter 18 by Campbell).

The DNA extracted from mature phage particles of the λpapa strain has a molecular weight of 30.8×10^6 daltons (Davidson and Szybalski, 1971). In Figure 1 the DNA strands are represented by heavy lines drawn as open circles. The linear form of the gene arrangement in mature phage is

$$m\text{-}A\text{-}F\text{-}Z\text{-}J\text{-}b2\text{-}a{\cdot}a'\text{ -}int\text{-}N\text{-}imm\text{-}R\text{-}m'$$

Genes *A–J* form the left arm and genes *int–R* the right arm of the mature phage DNA molecule. Integration of the phage genome into the chromosome of *E. coli* is the result of fusion of the phage DNA termini $(m \cdot m')$ at the *cos* site and opening at the prophage ends $(a{\cdot}a')$ at the *att* site, simultaneous with linear insertion between *E. coli* genes *gal-chlD-pgl-phr* and *bio-uvrB-chlA* (Taylor and Trotter, 1974; gene *pgl* is also designated *blu*). The gene arrangement of the inserted prophage is

$$-gal\text{-}chlD\text{-}pgl\text{-}phr\text{-}b{\cdot}a'\text{ -}int\text{-}N\text{-}imm\text{-}R\text{-}m{\cdot}m'\text{ -}$$
$$-A\text{-}F\text{-}Z\text{-}J\text{-}b2\text{-}a{\cdot}b'\text{ -}bio\text{-}uvrB\text{-}chlA\text{-}$$

Symbols $b \cdot b'$ indicate the left and right elements of the integration site on the bacterial genome, and symbols $b \cdot a'$ and $a \cdot b'$ the left and right junctions between the bacterial and prophage genomes. The genes are positioned according to electron microscopic measurements (Davidson and Szybalski, 1971) and the genetic and molecular maps (Campbell, 1971). The key to the gene symbols and recognition sites is given in

Table 1. The, genes code for structural, catalytic, controlling, and other proteins, and the recognition sites on the DNA interact with various phage and host control factors.

Literature Cited

Blattner, F. R. and J. E. Dahlberg, 1972   RNA synthesis startpoints in bacteriophage λ: Are the promoter and operator transcribed? *Nat. New Biol.* **237**:227–232.

Blattner, F. R., J. E. Dahlberg, J. K. Boettiger, M. Fiandt and W. Szybalski, 1972   Distance from a promoter mutation to an RNA synthesis startpoint on bacteriophage λ DNA. *Nat. New Biol.* **237**:232–236.

Boklage, C. E., E. C. Wong and V. Bode, 1973   The lambda *F* mutants belong to two cistrons. *Genetics* **75**:221–230.

Campbell, A., 1971   Genetic structure. In *Bacteriophage Lambda*, edited by A. D. Hershey, pp. 13–44, Cold Spring Harbor Laboratory, Cold Spring Harbor, N.Y.

Casjens, S., 1971   The morphogenesis of the phage lambda head: the step controlled by gene *F*. In *Bacteriophage Lambda*, edited by A. D. Hershey, pp. 725–732, Cold Spring Harbor Laboratory, Cold Spring Harbor, N.Y.

Casjens, S., T. Hohn and A. D. Kaiser, 1970   Morphological proteins of phage lambda: Identification of the major head protein as the product of gene *E*. *Virology* **42**:496–507.

Casjens, S., T. Hohn and A. D. Kaiser, 1972   Head assembly steps controlled by genes *F* and *W* in bacteriophage λ. *J. Mol. Biol.* **64**:551–563.

Champoux, J. J. and D. S. Hogness, 1972   The topography of lambda DNA: Polyriboguanylic acid binding sites and base composition. *J. Mol. Biol.* **71**:383–405.

Christensen, J. R. and J. M. Geiman, 1973   A new effect of the *rex* gene of phage λ: premature lysis after infection by phage T1. *Virology* **56**:285–290.

Couturier, M., C. Dambly and R. Thomas, 1973   Control of development in temperate bacteriophages. V. Sequential activation of the viral functions. *Mol. Gen. Genet.* **120**:231–252.

Dahlberg, J. E. and F. R. Blattner, 1973   *In vitro* transcription products of lambda DNA: nucleotide sequences and regulatory sites. In *Virus Research* edited by C. F. Fox and W. S. Robinson, pp. 533–543. Academic Press, New York.

Davidson, N. and W. Szybalski, 1971   Physical and chemical characteristics of lambda DNA. In *Bacteriophage Lambda*, edited by A. D. Hershey, pp. 45–82, Cold Spring Harbor Laboratory, Cold Spring Harbor, N.Y.

Dove, W. F., H. Inokuchi and W. F. Stevens, 1971   Replication control in phage lambda. In *Bacteriophage Lambda*, edited by A. D. Hershey, pp. 747–772, Cold Spring Harbor Laboratory, Cold Spring Harbor, N.Y.

Echols, H. 1971   Regulation of lytic development. In *Bacteriophage Lambda*, edited by A. D. Hershey, pp. 247–270, Cold Spring Harbor Laboratory, Cold Spring Harbor, N.Y.

Echols, H. 1972   Developmental pathways for the temperate phage: lysis vs. lysogeny. *Annu. Rev. Genet.* **6**:157–190.

Echols, H., S. Chung and L. Green, 1974   Site specific recombination: genes and regulation. In *Mechanisms in Recombination* edited by R. F. Grell, Plenum Press, New York.

Eisen, H. and M. Ptashne, 1971   Regulation of repressor synthesis. In *Bacteriophage Lambda*, edited by A. D. Hershey, pp. 239–246, Cold Spring Harbor Laboratory, Cold Spring Harbor, N.Y.

Emmons, S. W., 1974   Bacteriophage lambda derivatives carrying two copies of the cohesive end site. *J. Mol. Biol.* **83**:511–525.

Enquist, L. W. and A. Skalka, 1973   Replication of bacteriophage λ DNA dependent on the function of host and viral genes. *J. Mol. Biol.* **75**:185–212.

Feiss, M. and A. Campbell, 1974   Duplication of the bacteriophage lambda cohesive end site: genetic studies. *J. Mol. Biol.* **83**:527–540.

Fiandt, M., Z. Hradecna, H. A. Lozeron and W. Szybalski, 1971   Electron micrographic mapping of deletions, insertions, inversions, and homologies in the DNA's of coliphages lambda and phi 80. In *Bacteriophage Lambda*, edited by A. D. Hershey, pp. 329–354, Cold Spring Harbor Laboratory, Cold Spring Harbor, N.Y.

Gottesman, M. E. and R. A. Weisberg, 1971   Prophage insertion and excision. In *Bacteriophage Lambda*, edited by A. D. Hershey, pp. 113–138, Cold Spring Harbor Laboratory, Cold Spring Harbor, N.Y.

Greer, H. A., 1974   The *kil* gene of bacteriophage lambda. Ph.D. Thesis, M.I.T., Cambridge, Mass.

Hayes, S., and W. Szybalski, 1973a   Synthesis of RNA primer for lambda DNA replication is controlled by phage and host. In *Molecular Cytogenetics* edited by B. A. Hamkalo and J. Papaconstantinou, pp. 277–284 Plenum Press, New York.

Hayes, S. and W. Szybalski, 1973b   Control of short leftward transcripts from the immunity and *ori* regions in induced coliphage lambda. *Mol. Gen. Genet.* **126**:275–290.

Hendrix, R. W., 1971   Identification of proteins coded in phage lambda. In *Bacteriophage Lambda*, edited by A. D. Hershey, pp. 355–370, Cold Spring Harbor Laboratory, Cold Spring Harbor, N.Y.

Hendrix, R. W. and S. Casjens, 1974   Protein fusion: A novel reaction in λ bacteriophage λ head assembly. *Proc. Natl. Ac. Sci. USA.* **71**:1451–1455.

Hershey, A. D., editor, 1971   *The Bacteriophage Lambda.* Cold Spring Harbor Laboratory, Cold Spring Harbor, N.Y.

Jacquemin-Sablon, A. and Y. T. Lanni, 1973   Lambda-repressed mutants of phage T5. I. Isolation and genetical characterization. *Virology* **56**:230–237.

Kaiser, D., 1971   Lambda DNA replication. In *Bacteriophage Lambda*, edited by A. D. Hershey, pp. 195–210, Cold Spring Harbor Laboratory, Cold Spring Harbor, N.Y.

Kellenberger, E. and R. S. Edgar, 1971   Structure and assembly of phage particles. In *Bacteriophage Lambda*, edited by A. D. Hershey, pp. 271–296, Cold Spring Harbor Laboratory, Cold Spring Harbor, N.Y.

Lebowitz, P., S. M. Weissman and C. M. Radding, 1971   Nucleotide sequence of a ribonucleic acid transcribed *in vitro* from λ phage deoxyribonucleic acid. *J. Biol. Chem.* **246**:5120–5139.

Lehman, J. F., 1974   λ Site-specific recombination: Local transcription and an inhibitor specified by the b2 region. *Molec. Gen. Genet.* in press.

Matsubara, K. and A. D. Kaiser, 1968   λdv: an autonomously replicating DNA fragment. *Cold Spring Harbor Symp. Quant. Biol.* **33**:769–775.

Murialdo, H. and L. Siminovitch, 1971   The morphogenesis of bacteriophage lambda. III. Identification of genes specifying morphogenetic proteins. In *Bacteriophage Lambda*, edited by A. D. Hershey, pp. 711–724, Cold Spring Harbor Laboratory, Cold Spring Harbor, N.Y.

Murialdo, H. and L. Siminovitch, 1972a   The morphogenesis of bacteriophage lambda.

IV. Identification of gene products and control of the expression of the morphogenetic information. *Virology* **48:**785–823.

Murialdo, H. and L. Siminovitch, 1972*b* The morphogenesis of phage lambda. V. Form-determining function of the genes required for the assembly of the head. *Virology* **48:**824–835.

Oppenheim, A. B. and D. Salomon, 1972 Studies on partially virulent mutants of lambda bacteriophage. II. The mechanism of overcoming repression. *Mol. Gen. Genet.* **115:**101–114.

Ordal, G. W. and A. D. Kaiser, 1973 Mutations in the right operator of bacteriophage lambda: evidence for operator-promoter interpenetration. *J. Mol. Biol.* **19:**709–722.

Ptashne, M., 1971 Repressor and its actions. In *Bacteriophage Lambda*, edited by A. D. Hershey, pp. 221–238, Cold Spring Harbor Laboratory, Cold Spring Harbor, N.Y.

Reichardt, L. and A. D. Kaiser, 1971 Control of λ repressor synthesis. *Proc. Natl. Acad. Sci. USA* **68:**2185–2194.

Rhoades, M. and M. Meselson, 1973 An endonuclease induced by bacteriophage λ. *J. Biol. Chem.* **248:**521–527.

Roberts, J. W., 1969 Termination factor for RNA synthesis. *Nature (Lond.)* **224:**1168–1174.

Sakaki, Y., A. E. Karu, S. Linn and H. Echols, 1973 Purification and properties of the γ-protein specified by bacteriophage λ: an inhibitor of the host *recBC* recombination enzyme. *Proc. Natl. Acad. Sci. USA* **10:**2215–2219.

Schnös, M. and R. B. Inman, 1970 Position of branch points in replicating λ DNA. *J. Mol. Biol.* **51:**61–73.

Shimada, K., and A. Campbell, 1974 Int-constitutive mutants of bacteriophage lambda. *Proc. Natl. Acad. Sci., U.S.A.* **71:**237–241.

Signer, E., 1968 Lysogeny: the integration problem. *Annu. Rev. Microbiol.* **22:**451–488.

Signer, E., 1971 General recombination. In *Bacteriophage Lambda*, edited by A. D. Hershey, pp. 139–174. Cold Spring Harbor Laboratory, Cold Spring Harbor, N. Y.

Sironi, G., H. Bialy, H. A. Lozeron and R. Calendar, 1971 Bacteriophage P2: interaction with phage λ and with recombination-deficient bacteria. *Virology* **46:**387–396.

Skalka, A. M., E. Burgi and A. D. Hershey, 1968 Segmental distribution of nucleotides in the DNA of bacteriophage lambda. *J. Mol. Biol.* **34:**1–16.

Sly, W. S., K. Rabideau and A. Kolber, 1971 The mechanisms of lambda virulence. II. Regulatory mutations in classical virulence. In *Bacteriophage Lambda*, edited by A. D. Hershey, pp. 575–588, Cold Spring Harbor Laboratory, Cold Spring Harbor, N.Y.

Spiegelman, W. G., L. F. Reichardt, M. Yaniv, S. F. Heinemann, A. D. Kaiser and H. Eisen, 1972 Bidirectional transcription and the regulation of phage λ repressor synthesis. *Proc. Nat. Acad. Sci. USA* **69:**3156–3160.

Stevens, W. S., S. Adhya and W. Szybalski, 1971 Origin and bidirectional orientation of DNA replication in coliphage lambda. In *Bacteriophage Lambda*, edited by A. D. Hershey, pp. 515–34, Cold Spring Harbor Laboratory, Cold Spring Harbor, N.Y.

Szybalski, W., 1970 Genetic and molecular map of *Escherichia coli* bacteriophage lambda (λ). In *Handbook of Biochemistry*, second edition, edited by H. A. Sober, pp. I · 35–I · 38, Chemical Rubber Co., Cleveland, Ohio. (Consult this compilation for earlier references.)

Szybalski, W., 1972 Transcription and replication in *E. coli* bacteriophage lambda. In

Uptake of Informative Molecules by Living Cells, edited by L. Ledoux, pp. 59–82 North-Holland, Amsterdam.

Szybalski, W., 1974   Initiation and regulation of transcription in coliphage lambda. In *Control of Transcription* edited by B. B. Biswas, R. K. Mondal, A. Stevens, and W. E. Cohn, pp. 201–212, Plenum Press, New York.

Szybalski, W. and I. Herskowitz, 1971   Lambda genetic elements. In *Bacteriophage Lambda,* edited by A. D. Hershey, p. 778, Cold Spring Harbor Laboratory, Cold Spring Harbor, N.Y.

Szybalski, W., K. Bøvre, M. Fiandt, A. Guha, Z. Hradecna, S. Kumar, H. A. Lozeron, V. M. Maher, H. J. J. Nijkamp, W. C. Summers and K. Taylor, 1969   Transcriptional controls in developing bacteriophages. *J. Cell Physiol.* **74:**(Suppl. 1) 33–70.

Szybalski, W., K. Bøvre, M. Fiandt, S. Hayes, Z. Hradecna, S. Kumar, H. A. Lozeron, H. J. J. Nijkamp and W. F. Stevens, 1970   Transcriptional units and their controls in *Escherichia coli* phage λ: operons and scriptons. *Cold Spring Harbor Symp. Quant. Biol.* **35:**341–353.

Tavernier, P., K. Barta and J. Zissler, 1974   New recombination genes of bacteriophage lambda. unpublished.

Taylor, A., 1971   Endopeptidase activity of phage λ-endolysin. *Nat. New Biol.* **234:**144–145.

Taylor, A. L. and C. D. Trotter, 1974   A linkage map and gene catalog for *Escherichia coli.* In *Handbook of Genetics,* Vol. 1, edited by R. C. King, Chapter 6, pp. 135–156, Plenum Press, New York.

Thomas, R., 1970   Control of development in temperate bacteriophages. III. Which prophage genes are and which are not *trans*-activable in the presence of immunity? *J. Mol. Biol.* **49:**393–404.

Thomas, R., 1971   Control circuits. In *Bacteriophage Lambda,* edited by A. D. Hershey, pp. 211–220, Cold Spring Harbor Laboratory, Cold Spring Harbor, N.Y.

Ungar, R. C., H. Echols and A. J. Clark, 1972   Interaction of the recombination pathways of bacteriophage and host *Escherichia coli*: effects on λ recombination. *J. Mol. Biol.* **70:**531–537.

Wang, J. C. and A. D. Kaiser, 1973   Evidence that the cohesive ends of mature λ DNA are generated by the gene A product. *Nat. New Biol.* **241:**16–17.

Yarmolinsky, M. B., 1971   Making and joining DNA ends. In *Bacteriophage Lambda,* edited by A. D. Hershey, pp. 97–112, Cold Spring Harbor Laboratory, Cold Spring Harbor, N.Y.

Yen, K-M., and G. N. Gussin, 1973   Genetic characterization of a *prm⁻* mutant of bacteriophage λ. *Virology* **56:**300–312.

Zabeau, M., J. Heip, M. Van Montagu and J. Schell, 1974*a*   The alleviation of DNA restriction by phage λ functions. I. General description of the rescue phenomenon in *E. coli* K12 and *E. coli* B. in manuscript.

Zabeau, M., M. Van Montagu and J. Schell, 1974*b*   The alleviation of DNA restriction by phage λ functions. II. The involvement of the λ genes *red, gam* and *ral* in the rescue phenomenon. unpublished.

Zabeau, M., S. Friedman, M. Van Montagu and J. Schell, 1974*c*   The alleviation of DNA restriction by phage λ functions. III. Mapping of the λ *ral* gene. in manuscript.

Zissler, J., E. Signer and F. Schaefer, 1971   The role of recombination in growth of bacteriophage lambda. II. Inhibition of growth by prophage P2. In *Bacteriophage Lambda,* edited by A. D. Hershey, pp. 469–475, Cold Spring Harbor Laboratory, Cold Spring Harbor, N.Y.

20

Bacteriophage ϕX174

Robert L. Sinsheimer

Bacteriophage ϕX174 is a small icosahedral virus, of particle weight 6.2×10^6 daltons, which grows on certain strains of *Escherichia coli*. The genetic material is a single-stranded, circular DNA of molecular weight 1.7×10^6 daltons (about 5500 nucleotides). During its replication, the single-stranded viral DNA (SS DNA) is converted to a double-stranded "replicative form" DNA (RF DNA) from which the progeny single strands are produced.

The virion itself contains four principal proteins, one of which (capsid) is known to comprise the 20 icosahedral faces, while two others are known to be present in the twelve spikes or projections found at the twelve vertices of fivefold symmetry.

The genetic map for bacteriophage ϕX174 in a *wt* host cell is shown in Figure 1. The order of cistrons was established by three-factor crosses as described in Benbow *et al.*, 1971. Cistron boundaries have been drawn using the cistron-product molecular weights given by Benbow *et al.*, 1972. One map unit (refer to the divisions on the inner circumference of the genetic map) represents 1×10^{-4} *wt* recombinants per total progeny phage. The physical location (in nucleotides) was estimated for *am* 42(D), *am* 10(D), *am* 3(E), *am* 27(E), *am* 6(J), *op* 6(F), *am* H57(F), *am* 88(F), *am* 87(F), *am* 89(F), *am* 9(G), *am* 32(G), and *am* N1(H) on the basis of polypeptide fragments in SDS-polyacrylamide gels. Distances in nucleotides are indicated by the divisions on the outer circumference of the genetic map. Cistron A is arbitrarily selected as the origin. The number of nucleotides in the complete genome is assumed to be 5500 (Sinsheimer,

ROBERT L. SINSHEIMER—Division of Biology, California Institute of Technology, Pasadena, California.

TABLE 1.   The Bacteriophage φX174 Genome

Cistron	Probable function	Mol wt	Location[a]	Major mutants[b]	Location[a]
A	RF DNA replication	65,000	0–1470	am86	110
				am50	220
				ts128	475
				am3	745
				am8	1090
				am30	1280
				am35	1340
				am18	1380
B	SS DNA replication (viral assembly?)	25,000	1470–2035	ts9	1485
				am16	1515
				am14	1765
				ts116	1940
				och5	2010
C	SS DNA replication	7,500	2035–2205	och6	2055
D	SS DNA replication	14,500	2205–2530	amH81	2305
				am42	2305
				am10	2340
E	Host cell lysis	17,000	2530–2915	am27	2635
				am3	2700
J	Structural component of virion	8,500	2915–3105	am6	2925
F	Capsid of virion	49,000	3105–4210	op9	3115
				op6	3130
				tsh6	3505
				amH57	3995
				am88	4030
				am87	4055
				ts41D	4100
				am89	4160
G	Spike of virion	21,000	4210–4685	am9	4235
				ts8	4300
				ts79	4540
				am32	4610
H	Spike of virion	35,000	4685–5500	am90	5135
				am23	5210
				amN1	5255
				am80	5310
				ts4	5340

[a] Approximate location, in nucleotides from an arbitrary origin.
[b] am = amber mutation (UAG), op = opal mutation (UGA), och = ochre mutation (UAA), ts = temperature-sensitive mutant, and tsh = temperature-sensitive and host-range mutant.

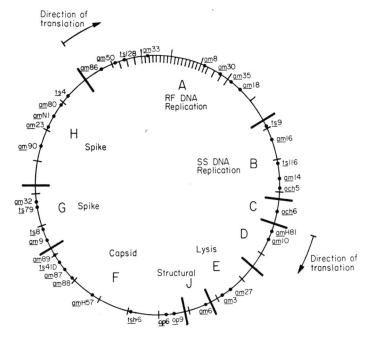

*Figure 1. The genetic map of bacteriophage φX174; shown are the cistron
boundaries and major mutations within each cistron.*

1968). The total length of the bacteriophage φX174 map is 24.4 × 10⁻⁴ *wt*
recombinants per total progeny phage.

The values presented in Table 1 are calculated from the data given in
Benbow *et al.*, 1971; Benbow *et al.*, 1972; and Sinsheimer, 1968. The
numbers presented in fourth and sixth columns do not represent actual
nucleotide position assignments, since they were calculated on the basis of
φX174 protein molecular weights assuming a 5500 nucleotide genome
with no nonessential gene regions. Instead, these numbers are aids for calculating marker separations; nevertheless, the nucleotide distances should
be accurate to 5 percent in most cases, particularly in the region from cistron D through cistron H.

Literature Cited

Benbow, R. M., C. A. Hutchison, III, J. D. Fabricant and R. L. Sinsheimer, 1971   The
genetic map of bacteriophage φX174. *J. Virol.* **7**:549–558.
Benbow, R. M., R. F. Mayol, J. C. Picchi and R. L. Sinsheimer, 1972   Direction of
translation and size of bacteriophage φX174 cistrons. *J. Virol.* **10**:99–114.
Sinsheimer, R. L., 1968   Bacteriophage φX174 and related viruses. *Prog. Nucleic Acid
Res. Mol. Biol.* **8**:115–169.

21

Bacteriophage T4

WILLIAM B. WOOD

The total map length of the bacteriophage T4 genome has been determined by electron microscopy of T4 DNA molecules to be 1.66×10^5 nucleotide pairs (Kim and Davidson, 1974). In Figure 1, each small division on the inner circle indicates 10^3 nucleotide pairs. The zero point has been placed arbitrarily at a genetically and physically well-defined locus, the divide between the rIIA and B cistrons.

Gene positions are indicated by lines or stippled bars on the circular map. The positions shown by lines on the inside of the circle for some genes were determined by a physical mapping procedure independent of recombination frequencies (Mosig, 1968; Edgar and Mosig, 1970; G. Mosig, private communication). These positions have been corrected on the map circle to conform to (1) the lengths of the intervals 38–rIIB, rIIB–e, and e–tRNA, determined by electron microscopy of heteroduplex DNA molecules made using T4 rIIB, e, and tRNA deletion mutants (Kim and Davidson, 1974; Wilson et al., 1972), and (2) the constraints of known minimum gene lengths (see below). Positions of the remaining genes relative to physically mapped loci have been estimated from recombination frequencies (Stahl et al., 1964). Cotranscribed genes are indicated by arrows showing transcription direction (Stahl et al., 1970); O'Farrell et al., 1973; Vanderslice and Yegian, 1974; Hercules and Sauerbier, 1973). Additional information on transcription directions has been reviewed by Szybalski (1969).

WILLIAM B. WOOD—Division of Biology, California Institute of Technology, Pasadena, California.

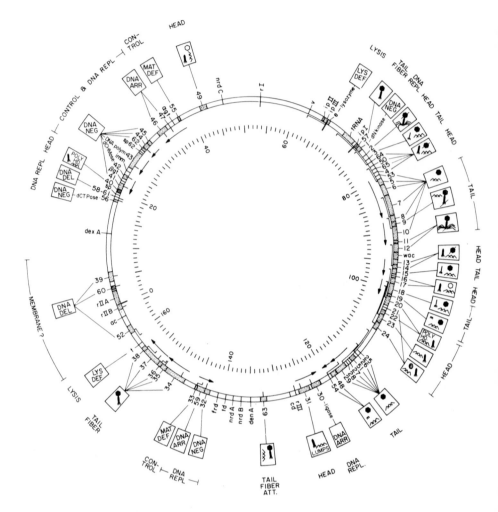

Figure 1. The genetic map of bacteriophage T4.

The minimum lengths of stippled genes are based on gene-product molecular weights estimated by discontinuous polyacrylamide gel electrophoresis in the presence of SDS (Laemmli, 1970; Eiserling and Dickson, 1972; O'Farrell et al., 1973; Vanderslice and Yegian, 1974). Minimum lengths of the two crosshatched genes are estimated from intragenic recombination frequencies (H. Bernstein, private communication).

Functions of essential genes (t, e, and all numbered genes) can be inferred from the consequences of the corresponding gene defects, indicated in the outer circle by the boxed abbreviations or diagrams of structural components that accumulate as a result of mutation (Epstein *et al.*, 1963; Edgar and Wood, 1966; Edgar and Mosig, 1970). Slashes indicate functionally defective structures that appear complete in electron micrographs. Open hexagons indicate heads that appear empty of DNA; solid hexagons represent filled heads. The stippled spheroid (genes 21 and 24) represents the head-related *tau* particle (see Laemmli *et al.*, 1970 for further details on phenotypes of mutants defective in head formation). Gene-11 and 12 mutants produce normal numbers of noninfectious, nonkilling particles with tail fibers. Gene-2 mutants produce a lower number of killing but defective particles with tail fibers, in addition to the unattached components shown (King, 1968). Mutants originally assigned to the two complementation groups 58 and 61 have been shown to belong to the same complementation group, and hence define one gene, designated 58–61 (Yegian *et al.*, 1971). Mutations originally reported to define gene 66, between genes 23 and 24, have been shown to map within gene 23 (Doermann *et al.*, 1973).

Functions or designations of nonessential genes (all others) are as follows: *ac*, acriflavin resistance (Silver, 1965); *αgt, βgt,* DNA-glucosyl transferases (Revel and Luria, 1970); *cd*, dCMP deaminase (Hall and Tessman, 1966); *den*A, endonuclease II (Hercules *et al.*, 1971; Ray *et al.*,1972); *dex*A, exonuclease A (Warner *et al.*, 1972); *frd*, dihydrofolate reductase (Hall, 1967); *imm*, superinfection immunity (Vallée and Cornett, 1972); *ip*I–III, internal proteins of the phage head (Showe and Black, 1973); *nrd*A–C, nucleoside diphosphate reductase system (Yeh and Tessman, 1972; Tessman and Greenberg, 1972); *r*I–III, rapid lysis (Edgar *et al.*, 1962; Edgar *et al.*, 1964); *sp*, "spackle," rapid lysis (Emrich, 1968); *sp*62 (a mutation defining a gene to be designated *reg*A), possible control of early gene expression (Wiberg *et al.*, 1973); *td*, thymidylate synthetase (Simon and Tessman, 1963); *tRNA*, 8 transfer RNA's (Wilson *et al.*, 1972; McClain *et al.*, 1972); *v*, uv sensitivity (Yasuda and Sekiguchi, 1970; L. W. Black, private communication); *wac*, whisker antigen control (Dewey *et al.*, 1974; Follansbee *et al.*, 1974). Another nonessential gene of unknown function, *das* (not shown), has been tentatively defined by mutations mapping between genes 33 and 34 that suppress gene-46 and gene-47 defects (Hercules and Wiberg, 1971). The phenotypically similar *suα* mutations (Krylov and Plotnikova, 1971) may also represent alterations in the *das* function, since they have now been shown to map in the 33–34 region (V. N. Krylov, private communication).

For additional references on gene functions, see Mathews (1971) and Eiserling and Dickson, (1972).

Literature Cited

Dewey, M. J., J. S. Wiberg and F. R. Frankel, 1974   Genetic control of whisker antigen of bacteriophage T4. *J. Mol. Biol.* in press.

Doermann, A. H., F. A. Eiserling and L. Boehner, 1973, Capsid length in bacteriophage T4 and its genetic control. In *Virus Research, Proceedings of the Second ICN-UCLA Symposium on Molecular Biology,* edited by C. F. Fox and W. S. Robinson, pp. 243–258, Academic Press, New York.

Edgar, R. S. and G. Mosig, 1970   In *Handbook of Biochemistry,* second edition, edited by. H. A. Sober, pp. I-32–34, Chemical Rubber Co., Cleveland, Ohio.

Edgar, R. S. and W. B. Wood, 1966   Morphogenesis of bacteriophage T4 in extracts of mutant-infected cells. *Proc. Natl. Acad. Sci. USA* **55**:498–505.

Edgar, R. S., R. P. Feynman, S. Klein, I. Lielausis and C. M. Steinberg, 1962   Mapping experiments with *r* mutants of bacteriophage T4D. *Genetics* **47**:179–186.

Edgar, R. S., G. H. Denhardt and R. H. Epstein, 1964   A comparative genetic study of conditional lethal mutations of bacteriophage T4D. *Genetics* **49**:635–648.

Eiserling, F. A. and R. C. Dickson, 1972   Assembly of viruses. *Annu. Rev. Biochem.* **41**:467–502.

Emrich, J., 1968   Lysis of T4-infected bacteria in the absence of lysozyme. *Virology* **35**:158–165.

Epstein, R. H., A. Bolle, C. M. Steinberg, E. Kellenberger, E. Boy de la Tour, R. Chevalley, R. S. Edgar, M. Susman, G. H. Denhardt and A. Lielausis, 1963   Physiological studies of conditional lethal mutants of bacteriophage T4D. *Cold Spring Harbor Symp. Quant. Biol.* **28**:375–392.

Follansbee, S. E., R. W. Vanderslice, L. D. Chavez and C. D. Yegian, 1974   A new set of adsorption mutants of bacteriophage T4D: Identification of a new gene. *Virology* **58**:180–199.

Hall, D. H., 1967   Mutants of bacteriophage T4 unable to induce dihydrofolate reductase activity. *Proc. Natl. Acad. Sci. USA* **58**:584–591.

Hall, D. H. and I. Tessman, 1966   T4 mutants unable to induce deoxycytidylate deaminase activity. *Virology* **29**:339–345.

Hercules, K. and W. Sauerbier, 1973   Transcription units in bacteriophage T4. *J. Virol.* **12**:872–881.

Hercules, K. and J. S. Wiberg, 1971   Specific suppression of mutations in genes 46 and 47 by *das,* a new class of mutations in bacteriophage T4D. *J. Virol.* **8**:603–612.

Hercules, K., J. L. Munro, S. Mendelsohn and J. S. Wiberg, 1971   Mutants in a nonessential gene of bacteriophage T4 which are defective in the degradation of *E. coli* DNA *J. Virol.* **7**:95–105.

Kim, J.-S. and N. Davidson, 1974   Electron microscope heteroduplex study of sequence relations of T2, T4, and T6 bacteriophage DNA's. *Virology* **57**:93–111.

King, J., 1968   Assembly of the tail of bacteriophage T4. *J. Mol. Biol.* **32**:231–262.

Krylov, V. N. and T. G. Plotnikova, 1971   A suppressor in the genome of phage T4 inhibiting phenotypic expression of mutations in genes 46 and 47. *Genetics* **67**:319–326.

Laemmli, U. K., 1970   Cleavage of structural proteins during the assembly of the head of bacteriophage T4. *Nature (Lond.)* **227**:680–685.

Laemmli, U. K., E. Molbert, M. Showe and E. Kellenberger, 1970   Form-determining function of the genes required for the assembly of the head of bacteriophage T4. *J. Mol. Biol.* **49**:99–113.

Mathews, C. K., 1971   *Bacteriophage Biochemistry*, ACS Monograph 166, chapters 5, 6, and 7, Van Nostrand, New York.

McClain, W. H., C. Guthrie and B. G. Barrell, 1972   Eight transfer RNAs induced by infection of *Escherichia coli* with bacteriophage T4. *Proc. Natl. Acad. Sci. USA* **69**:3703-3707.

Mosig, G., 1968   A map of distances along the DNA molecule of bacteriophage T4. *Genetics* **59**:137-151.

O'Farrell, P. C., L. M. Gold and W.-M. Huang, 1973   The identification of pre-replicative bacteriophage T4 proteins. *J. Biol. Chem.* **248**:5499-5501.

Ray, P., N. K. Sinha, H. R. Warner and D. P. Snustad, 1972   Genetic location of a mutant of bacteriophage T4 deficient in the ability to induce endonuclease II. *J. Virol.* **9**:184-186.

Revel, H. R. and S. E. Luria, 1970   DNA glucosylation in T-even phages: Genetic determination and role in phage-host interaction. *Annu. Rev. Genet.* **4**:177-192.

Showe, M. K., and L. W. Black, 1973   Assembly core of bacteriophage T4: An intermediate in head formation. *Nat. New Biol.* **242**:70-75.

Silver, S., 1965   Acriflavin resistance: A bacteriophage mutation affecting the uptake of dye by the infected bacterial cells. *Proc. Natl. Acad. Sci. USA* **53**:24-30.

Simon, E. H. and I. Tessman, 1963   Thymidine-requiring mutants of phage T4. *Proc. Natl. Acad. Sci. USA* **50**:526-532.

Stahl, F. W., R. S. Edgar and C. M. Steinberg, 1964   The linkage map of bacteriophage T4. *Genetics* **50**:539-552.

Stahl, F. W., J. M. Crasemann, C. D. Yegian, M. M. Stahl and A. Nakata, 1970   Cotranscribed cistrons in bacteriophage T4. *Genetics* **64**:157-170.

Szybalski, W. 1969   Initiation and patterns of transcription during phage development. *Proc. Can. Cancer Res. Conf.* **8**:183-215.

Tessman, I. and D. B. Greenberg, 1972   Ribonucleotide reductase genes of phage T4: Map location of the thioredoxin gene *nrd*C. *Virology* **49**:337-338.

Vallée, M. and J. B. Cornett, 1972   A new gene of bacteriophage T4 determining immunity against superinfecting ghosts and phage in T4-infected E. coli. *Virology* **48**:777-784.

Vanderslice, R. W. and C. D. Yegian, 1974   The identification of late bacteriophage T4 proteins on sodium dodecyl sulfate acrylamide gels. *Virology,* in press.

Warner, H. R., D. P. Snustad, J. F. Koerner and J. D. Childs, 1972   Identification and genetic characterization of mutants of bacteriophage T4 defective in the ability to induce exonuclease A. *J. Virol.* **9**:399-407.

Wiberg, J. S., S. Mendelsohn, V. Warner, K. Hercules, C. Aldrich and J. Munro, 1973   *Sp*62, a viable mutant of bacteriophage T4D defective in regulation of phage enzyme synthesis. *J. Virol.* **12**:775-792.

Wilson, J. H., J.-S. Kim and J. N. Abelson, 1972   Bacteriophage T4 transfer RNA. III. Clustering of the genes for T4 transfer RNA's. *J. Mol. Biol.* **71**:547-556.

Yasuda, S. and M. Sekiguchi, 1970   Mechanism of repair of DNA in bacteriophage. II. Inability of ultraviolet-sensitive strains of bacteriophage in inducing an enzyme activity to excise pyrimidine dimers. *J. Mol. Biol.* **47**:243-245.

Yegian, C. D., M. Mueller, G. Selzer, V. Russo and F. W. Stahl, 1971   Properties of the DNA-delay mutants of bacteriophage T4. *Virology* **46**:900-919.

Yeh, Y.-C. and I. Tessman, 1972   Control of pyrimidine biosynthesis by phage T4. II. *In vitro* complementation between ribonucleotide reductase mutants. *Virology* **47**:767-772.

22

The Lysozyme Cistron of T4 Phage

Yoshimi Okada

The genetic message is translated sequentially, three bases at time, starting from a defined AUG codon at the begining a cistron (Crick *et al.*, 1961; Terzaghi *et al.*, 1966). Deletions or additions of bases cause a shift in the reading frame of the genetic message and, therefore, such changes are called frame-shift mutations.

Proflavine induces frame-shift mutations, exclusively. Frame-shift mutations may be induced by proflavine to revert to pseudowild types by a second frame-shift mutation at a nearby site, thus restoring the correct reading frame. Such reversions also occur spontaneously. In the pseudowild protein, therefore, a changed sequence of amino acids between the sites of the two frame-shift mutations is to be expected. This hypothesis has been directly confirmed by studies on the lysozyme protein synthesized by bacteria infected with pseudowild strains of phage T4 carrying certain pairs of frame-shift mutations in the lysozyme gene (Terzaghi *et al.*, 1966; Streisinger *et al.*, 1966).

Most phages having proflavine-induced mutations in the lysozyme gene form no plaques on the usual media, but they do form plaques on special plates supplemented with egg-white lysozyme. Pseudowild double-mutant strains can be isolated from the progeny of the pairwise crosses of mutants. In some cases, spontaneous pseudowild revertants are isolated from mutant strains.

Yoshimi Okada—Department of Biophysics and Biochemistry, Faculty of Science, University of Tokyo, Hongo, Tokyo, Japan.

Map of Mutants   J16   JD12   J320

Amino Acid Sequence of T4 Lysozyme   10
                                                  Met-Asn-Ile-Phe-Glu-Met-Leu-Arg-Ile-Asp-Glu-Gly-Leu-

Change of Amino Acid Sequence in Pseudowild Revertants   ① Glu-Tyr J16JD12

         JR14 JR13 JD5   J201   JD10   J42 J17   J44
                                                     JD11
                   20   30   40
Arg-Leu-Lys-Ile-Tyr-Lys-Asp-Thr-Glu-Gly-Tyr-Tyr-Thr-Ile-Gly-Ile-Gly-His-Leu-Leu-Thr-Lys-Ser-Pro-Ser-Leu-Asn-Ala-Ala-Lys-Ser-

                    ② Arg-Leu-Leu-His JD5J201   ⑤ Val-His-His-Leu-Met J42J44
                   ③ Glu-Thr-Gln JR14JD5   ⑥ Ser-Val-His-His-Leu-Met J17J44
                   ④ Lys-Thr-Gln JR13JD5   ⑦ Gln-Lys-Cys JD10J17J42
                                                            ⑧ Gln JD10JD11

                                                       J382   M103
              50   60   70
Glu-Leu-Asp-Lys-Ala-Ile-Gly-Arg-Asn-Cys-Asn-Gly-Val-Ile-Thr-Lys-Asp-Glu-Ala-Glu-Ala-Glu-Lys-Leu-Phe-Asn-Gln-Asp-Val-

  J28   JD8 JD2
  JD6
                   80   90   100
Asp-Ala-Ala-Val-Arg-Gly-Ile-Leu-Arg-Asn-Ala-Lys-Leu-Lys-Pro-Val-Tyr-Asp-Ser-Leu-Asp-Ala-Val-Arg-Arg-Cys-Ala-Leu-Ile-Asn-Met-

   ⑨ Gly-Cys-Cys-Cys J28JD2
  ⑩ Gly-Cys-Cys-Cys JD6J2
   ⑪ Val-Asp J28JD8

     L3   L1 L2 J335   M41
                   110   120   130
Val-Phe-Gln-Met-Glu-Glu-Thr-Gly-Val-Ala-Gly-Phe-Thr-Asn-Ser-Leu-Arg-Met-Leu-Gln-Gln-Lys-Arg-Trp-Asp-Glu-Ala-Ala-Val-Asn-Leu-

            M91 J25 JD4 L5
 J304   J37 JD7 JD1   HS92
            J200 JD3
                   140   150   160
Ala-Lys-Ser-Arg-Trp-Tyr-Asn-Gln-Thr-Pro-Asn-Arg-Ala-Lys-Arg-Val-Ile-Thr-Thr-Phe-Arg-Thr-Gly-Thr-Trp-Asp-Ala-Tyr-Lys-Asn-Leu

        ⑫ Met-Val-Tyr J37JD3
          ⑬ Lys-Ile JD7JD4
           ⑭ Ile-Ile J25JD1
          ⑮ Cys-Ile-Ile J200JD4

Figure 1. Amino acid sequence of the wild-type lysozyme and the changed amino acid sequence in pseudowild–frame-shift-type lysozyme together with a map of the position of lysozyme mutants.

Mutations of lysozyme (e) gene with the prefix J are induced by proflavin; mutations with the prefix JD and JR are spontaneous and proflavine-induced mutations isolated from pseudowild strain, respectively.

Mutations with the prefix M and L are nonsense mutations isolated by G. Streisinger and by H. Knesser, respectively.

The lysozyme protein isolated from bacteria infected with wild-type phage and with each of the pseudowild strains is purified from the lysate by fractionation on IRC-50 and Sephadex G-75 columns (Tsugita *et al.,* 1968). Lysozyme activity is assayed by measuring the decrease in optical density of a suspension of dried *Escherichia coli* B cells.

The amino acid sequences of the lysozymes of wild-type strain and of the pseudowild strains were compared. In each case the mutant lysozyme was found to differ from the wild-type strain in a sequence of several amino acids (Inouye *et al.,* 1970a, 1971; Imada *et al.,* 1970; Lorena *et al.,* 1968; Okada *et al.,* 1966, 1968, 1969, 1972; Ocada *et al.,* 1970; Streisinger *et al.,* 1966; Terzaghi *et al.,* 1966; Tsugita *et al.,* 1969).

Using the genetic code, it is possible in each case to assign a sequence of bases that could code for the wild-type sequence of amino acids and that, with the proper addition and/or deletion of bases, could code for the changed sequence of amino acid in the mutant strains.

In Figure 1, the complete amino acid sequence of the wild-type lysozyme (Inouye *et al.,* 1970b) and the changed amino acid sequence in pseudowild lysozyme are summarized, together with a map of the relative position of lysozyme mutants (Owen, 1971). In Figure 2, the sequences of amino acids observed and the base sequences in the mRNA that are compatible with them are indicated for each mutant strain.

Many codons which are in fact utilized *in vivo* have been identified from these results: Leu(UUA, *CUU*), Ser(UCA, *AGU*), Tyr(*UAU*), Cys(UGU, UGC), Trp(*UGG*), Pro(*UCA*), His(CAU, CAC), Gln(*CAA*), Ile(AUA), Met(*AUG*), Thr(ACA), Asn(*AAU*), Lys(AAA, AAG), Arg(AGG), Val(*GUU,* GUC, GUA, GUG), Ala(*GCU*), Asp(GAU, GAC), Glu(*GAA*), and Gly(GGC, GGA). The codons which have been italicized have been identified in the wild-type strain.

It is a corollary of the triplet nature of the genetic code that frame-shift mutations should fall into two classes (Crick *et al.,* 1961). Members of one class should result in a shift of the reading frame one base to the right (one base deletions, two base additions, and five-base additions), and members of the other class should result in a shift one base to the left (one-base additions, four-base additions and two-base deletions).

Frame-shift mutations in phage T4 could occur either at the ends of DNA molecules, or else internally in heterozygous regions. For example, the possible origin of mutation eJD11 (through the addition of a T) is illustrated as follows (Streisinger *et al.,* 1966):

$$\underline{1} \quad \begin{array}{l} - - - - - - \text{A C T A A A A A G T C C A T C A} \\ - - - - - - \text{T G A T T T T T C A G G T A G T} \end{array}$$

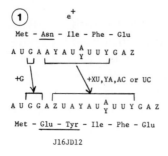

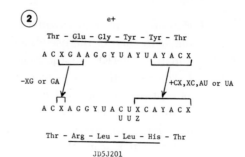

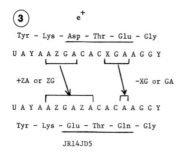

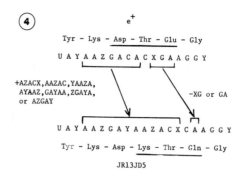

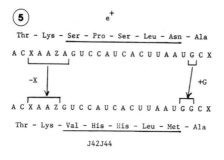

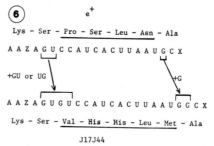

Figure 2. The possible nature of the mutational events. The sequences of the amino acids in the changed region of the wild-type lysozyme (e⁺) and in each of the mutant lysozyme are presented together with sequences of bases that could code for the wild-type sequences of the amino acids and that, with the proper addition and/or deletion of bases, could code for the changed sequences of amino acids in the mutant strains.

The portions of altered amino acid sequences are underlined. All possible choices of base additions and deletions are indicated in each case. X = U, C, A, and G; Y = U or C; Z = A or G.

① *Okada et al., 1968.* ② *Inouye et al., 1971.* ③ *Ocada et al., 1970.* ④ *Ocada et al., 1970.* ⑤ *Terzaghi et al., 1966.* ⑥ *Okada et al., 1966.* ⑦ *Okada et al., 1969.* ⑧ *Okada et al., 1968.* ⑨ *Imada et al., 1970.* ⑩ *Imada et al., 1970.* ⑪ *Imada et al., 1970.* ⑫ *Tsugita et al., 1969.* ⑬ *Lorena et al., 1968.* ⑭ *Tsugita et al., 1969.* ⑮ *Tsugita et al., 1969.*

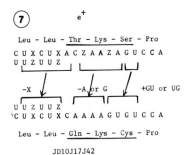

(7)
e+

Leu – Leu – <u>Thr</u> – <u>Lys</u> – <u>Ser</u> – Pro
C U X C U X A C Z A A Z A G U C C A
U U Z U U Z

  –X   –A or G   +GU or UG

U U Z U U Z
C U X C U X C A A A A G U G U C C A

Leu – Leu – <u>Gln</u> – <u>Lys</u> – <u>Cys</u> – Pro

JD10J17J42

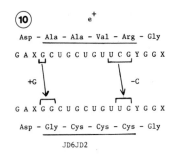

(8)
e+

Leu – Leu – <u>Thr</u> – Lys – Ser – Pro
C U X C U X A C Z A A Z A G U C C A

  –X   +A or G

U U Z U U Z
C U X C U X C A Z A A Z A G U C C A

Leu – Leu – <u>Gln</u> – Lys – Ser – Pro

JD10JD11

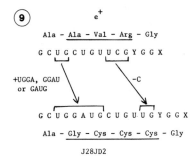

(9)
e+

Ala – <u>Ala</u> – Val – <u>Arg</u> – Gly
G C U G C U G U U C G Y G G X

+UGGA, GGAU   –C
or GAUG

G C U G G A U G C U G U U G Y G G X

Ala – <u>Gly</u> – <u>Cys</u> – <u>Cys</u> – <u>Cys</u> – Gly

J28JD2

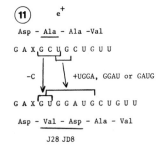

(10)
e+

Asp – <u>Ala</u> – Ala – Val – <u>Arg</u> – Gly
G A X G C U G C U G U U C G Y G G X

+G   –C

G A X G G C U G C U G U U G Y G G X

Asp – <u>Gly</u> – <u>Cys</u> – <u>Cys</u> – <u>Cys</u> – Gly

JD6JD2

(11)
e+

Asp – <u>Ala</u> – Ala –Val
G A X G C U G C U G U U

–C   +UGGA, GGAU or GAUG

G A X G U G G A U G C U G U U

Asp – <u>Val</u> – <u>Asp</u> – Ala – Val

J28 JD8

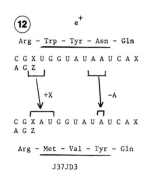

(12)
e+

Arg – <u>Trp</u> – Tyr – Asn – Gln
C G X U G G U A U A A U C A X
A G Z

  +X   –A

C G X A U G G U A U A U C A X
A G Z

Arg – <u>Met</u> – Val – Tyr – Gln

J37JD3

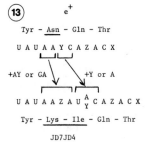

(13)
e+

Tyr – <u>Asn</u> – Gln – Thr
U A U A A Y C A Z A C X

+AY or GA   +Y or A

U A U A A Z A U ^A_Y C A Z A C X

Tyr – <u>Lys</u> – <u>Ile</u> – Gln – Thr

JD7JD4

(14)
e+

Trp – Tyr – <u>Asn</u> – Gln
U G G U A U A A Y C A Z

+UY,YA or AY   +Y or A

U G G U A Y A U ^Y_A A U ^Y_A C A Z

Trp – Tyr – <u>Ile</u> – <u>Ile</u> – Gln

J25JD1

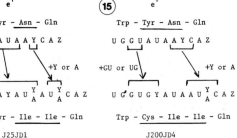

(15)
e+

Trp – <u>Tyr</u> – Asn – Gln
U G G U A U A A Y C A Z

+GU or UG   +Y or A

U G̈ G U G Y A U A A U ^Y_A C A Z

Trp – <u>Cys</u> – <u>Ile</u> – <u>Ile</u> – Gln

J200JD4

2
```
- - - - - - A C T A A A A A G T C C A T C A
- - - - - - T G   T T T
          A T
```

3
```
- - - - - - A C T A A A A A G T C C A T C A
- - - - - - T G   T T T T T C A G G T A G T
          A T
```

Line 1 shows the normal end of a molecule, line 2 shows an end in which one chain has been digested by an exonuclease followed by mispairing, and line 3 shows the appearance of the molecule after resynthesis of the digested chain resulting in the addition of a T.

On the basis of this model, it may be expected that the frequency of mutation would be higher in longer stretches of identical bases, since a longer stretch would increase the chances of mispairing. In fact, a spontaneous mutational "hot spot," eJD11, has been found to contain sequence of six identical bases (Okada *et al.*, 1972).

Proflavin and similar acridines may be mutagenic, not because they stretch the DNA molecule, but rather because they stabilize it (Streisinger *et al.*, 1966).

Literature Cited

Crick, F. H. C., L. Barnett, S. Brenner, and R. J. Watts-Tobin, 1961   General nature of the genetic code for proteins. *Nature (Lond.)* **192**:1227–1232.

Imada, M, M. Inouye, M. Eda and A. Tsugita, 1970   Frame-shift mutation in the lysozyme gene of bacteriophage T4: Demonstration of the insertion of four bases and the preferential occurrence of base addition in acridine mutagenesis. *J. Mol. Biol.* **54**:199–217.

Inouye, M., E. Akaboshi, M. Kuroda and A. Tsugita, 1970*a*   Replacement of all tryptophan residues in T4 bacteriophage lysozyme by tyrosine residues. *J. Mol. Biol.* **50**:71–81.

Inouye, M., M. Imada and A. Tsugita, 1970*b*   The amino acid sequence of T4 phage lysozyme. IV. Dilute acid hydrolysis and the order of tryptic peptides. *J. Biol. Chem.* **245**:3479–3484.

Inouye, M., E. Akaboshi, A. Tsugita, G. Streisinger and Y. Okada, 1971   A frame-shift mutation resulting in the deletion of two base pairs in the lysozyme gene of bacteriophage T4. *J. Mol. Biol.* **30**:39–47.

Lorena, M. J., M. Inouye and A. Tsugita, 1968   Studies on the lysozyme from the bacteriophage T4 eJD7eJD4, carrying two frame-shift mutations. *Mol. Gen. Genet.* **102**:69–78.

Ocada, Y., S. Amagase and A. Tsugita, 1970   Frame-shift mutation in the lysozyme gene of bacteriophage T4: Demonstration of the insertion of five bases, and a summary of *in vivo* codons and lysozyme activities. *J. Mol. Biol.* **54**:219–246.

Okada, Y., E. Terzaghi, G. Streisinger, J. Emrich, M. Inouye and A. Tsugita, 1966   A frame-shift mutation involving the addition of two base pairs in the lysozyme gene of phage T4. *Proc. Natl. Acad. Sci. USA* **56**:1692–1698.

Okada, Y., G. Streisinger, J. Emrich, A. Tsugita and M. Inouye, 1968   Frame-shift mutations near the beginning of the lysozyme gene of bacteriophage T4. *Science (Wash., D.C.)* **162**:807–808.

Okada, Y., G. Streisinger, J. Emrich, A. Tsugita and M. Inouye, 1969   The lysozyme of a triple frame-shift mutant strain of bacteriophage T4. *J. Mol. Biol.* **40**:299–304.

Okada, Y., G. Streisinger, J. Emrich, J. Newton, A. Tsugita and M. Inouye 1972   The molecular basis of a mutational hot spot in the lysozyme gene of bacteriophage T4. *Nature (Lond.)* **236**:388–341.

Owen, E. J., 1971   Some genetic studies of bacteriophage T4. Ph.D. Thesis, University of Oregon, Eugene, Oregon.

Streisinger, G., Y. Okada, J. Emrich, J. Newton, A. Tsugita, E. Terzaghi and M. Inouye, 1966   Frame-shift mutations and the genetic code. *Cold Spring Harbor Symp. Quant. Biol.* **31**:77–84.

Terzaghi, E., Y. Okada, G. Streisinger, J. Emrich, M. Inouye and A. Tsugita, 1966   Change of a sequence of amino acids in phage T4 lysozyme by acridine-induced mutations. *Proc. Natl. Acad. Sci. USA* **56**:500–507.

Tsugita, A., M. Inouye, E. Terzaghi and G. Streisinger, 1968   Purification of bacteriophage T4 lysozyme. *J. Biol. Chem.* **243**:391.

Tsugita, A., M. Inouye, T. Imagawa, T. Nakanishi, Y. Okada, J. Emrich and G. Streisinger, 1969   Frame-shift mutations resulting in the changes of the same amino acid residue (140) in T4 bacteriophage lysozyme and *in vivo* codons for Trp, Tyr, Met, Val and Ile. *J. Mol. Biol.* **41**:349–364.

PART C
THE FUNGI

23

Phycomyces

ENRIQUE CERDÁ-OLMEDO

Why *Phycomyces*?

Since its discovery 150 years ago biologists have been attracted to *Phycomyces* for several reasons. The most persistent fascination has been due to its remarkable sensory responses. Its large aerial hyphae, the sporangiophores, react to four kinds of environmental variables: light intensity, gravity, mechanical pressure, and the concentration of a gas produced by the fungus. The responses consist of localized changes in the growth rate, which is usually about 3 mm/hr. Asymmetrical stimuli lead to asymmetrical growth responses around the sporangiophore and result in tropisms. The sensory physiology of the sporangiophore is the central subject of an extensive review written by Bergman and eleven others (1969), but the reader should refer to it for other aspects of the biology of this organism as well. Light has additional effects on the mycelium, modifying the pattern of appearance of new sporangiophores (Bergman, 1972; Thornton, 1973) and activating the synthesis of β-carotene (Garton et al., 1951; Chichester et al., 1954). The mycelial growth pattern reflects the existence of chemotropisms (Schmidt, 1925). Other sensory reactions take place in relation to sexual processes. The recent development of an automatic tracking machine (Foster, 1974) has permitted a more detailed study of the response to light (Foster and Lipson, 1973); it makes feasible the exhaustive characterization of vast numbers of behavioral mutants.

After the discovery that *Phycomyces* and other Mucorales possess

ENRIQUE CERDÁ-OLMEDO — Departamento de Genética, Facultad de Ciencias, Universidad de Sevilla, Sevilla, Spain.

two mating types (Blakeslee, 1904), a great effort was devoted to the physiology of their sexual reactions. This led to the discovery of sexual hormones (Burgeff, 1924; Plempel, 1963). Trisporic acids (Caglioti *et al.,* 1966; Cainelli *et al.,* 1967), which sexually stimulate single mycelia of *Mucor mucedo* and activate β-carotene synthesis in *Blakeslea trispora,* are produced by mated cultures of *Phycomyces blakesleeanus* (R. P. Sutter, private communication). For reviews of the sexual behavior of the Mucorales, see Van den Ende and Stegwee (1971) and Gooday (1973).

The observation that *Phycomyces* requires thiamine for growth (Burgeff, 1934; Schopfer, 1934) was the basis for a biological assay of this vitamin (Schopfer and Jung, 1937) which was widely used some thirty years ago. Many biochemical and physiological aspects of *Phycomyces* have been investigated, particularly its synthesis of β-carotene [reviewed in the books edited by Goodwin (1965) and Isler (1971)], the activation of its spores (recent work by Van Assche *et al.,* 1972; Furch, 1972; Keyhani *et al.,* 1972; Delvaux, 1973; and references therein), and the effects of low doses of x radiation (Forssberg, 1969; and references therein).

The large size of the sporangiophores, their astonishing ability to regenerate (Gruen and Ootaki, 1972), and their ability to accept injections of foreign objects, make possible daring cytological experiments (Zalokar, 1969; Villet, 1972).

A number of interesting subcellular structures have been discovered (Tu *et al.,* 1971; Ootaki and Wolken, 1973; Malhotra and Tewari, 1973), mostly by electron microscopy.

The study of the genetics of *Phycomyces* was begun by Burgeff in 1911, hoping to answer some of the same questions that made *Drosophila* famous. He discovered heterokaryosis (Burgeff, 1914) and other genetic phenomena, but the difficulty in finding appropriate spontaneous markers, the sluggishness of the sexual cycle, and the irregularities of the results led him to write that "ein güngstigeres Material für solche Untersuchungen gefunden werden kann" (Burgeff, 1928) and give up. However, the research in sensory physiology stimulated new interest in the genetics of *Phycomyces,* as we shall shortly see.

Systematics

Phycomyces nitens was first described by Agardh (1817) and correctly named by Kunze (1823). The species was split by Burgeff (1925) into *Phycomyces nitens* and *Phycomyces blakesleeanus* on the basis of an apparent fertility barrier between the two. Benjamin and Hesseltine (1959) reviewed in detail these two species and others attributed to the genus at different times. *P. blakesleeanus* has smaller spores and larger zygospores than *P. nitens,* but they are very similar in most other respects. *P. blakesleeanus* is the species used in modern research.

The genus *Phycomyces* belongs to the family Mucoraceae, with such common bread molds as *Mucor* and *Rhizopus,* and to the order Mucorales and class Zygomycetes. It is thus one of the less-advanced Fungi which used to be grouped together as Phycomycetes.

Although this fungus has been found in various environments in widely separated parts of the world, practically nothing is known about its way of life in nature. The standard strain for studies of genetics and sensory physiology came in 1940 from L. H. Leonian of West Virginia University, and perhaps before that from A. F. Blakeslee, but its natural source is unknown. It may be obtained from the Northern Marketing and Nutrition Research Division of the United States Department of Agriculture, Peoria, Illinois, and is listed by code number NRRL1555.

Life Cycle

Figure 1 represents the principal stages in the life cycle of *Phycomyces.* The asexual cycle begins with the germination of a *spore* into a branched mycelium. The spores are usually activated by heating at 48°C for 15 minutes (Rudolph, 1960) just before placing them in one of many suitable solid and liquid media. A synthetic medium containing glucose, asparagine, and thiamine (Ødegård, 1952), sometimes supplemented with yeast extract, is commonly used in genetic studies. The sporangiophores grow out of the medium and into the air. Sporangiophores are large, unbranched hyphae which form an apical *sporangium.* The mature sporangium contains some 10^5 spores in a sticky matrix. The sporangiophores are able to grow at temperatures from near freezing to 28°C (Castle, 1928; Petzuch and Delbrück, 1970); at the usual temperature of 23°C the cycle from spore to spore takes three or four days.

All strains of *Phycomyces* belong to one of two mating types, called (+) and (−). The mating type is not recognizable by any morphological feature, but when mycelia of different mating types come together under suitable conditions, such as when grown on potato dextrose agar at 17°C, a complex "mating dance" is executed by specialized hyphae, the *zygophores,* each of which forms a separate cell, the *gametangium,* at the tip. Two gametangia of different mating type fuse and develop into a *zygospore,* a large, black sphere, surrounded by thorns, which remains dormant for a long time. Left on nutrientfree agar, the zygospore eventually germinates and produces a sporangiophore whose sporangium

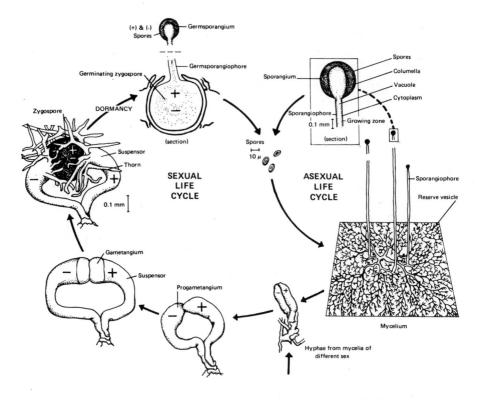

Figure 1. The life cycle of P. blakesleeanus(from Bergman et al., 1969). The germination of the spores gives rise to branched, coenocytic, medium-invading mycelia, which develop aerial, usually unbranched sporangiophores. The spores are formed in apical sporangia, thus completing the asexual cycle. The contact of mycelia of different mating type under suitable conditions originates a succession of specialized mating structures, culminating in the appearance of conspicuous zygospores. After a dormancy period, the zygospore produces a germsporangium, whose germspores enter again the asexual cycle with the production of mycelia.

contains up to several thousand spores. These structures closely resemble their namesakes of the asexual cycle; when confusion might arise, they receive the prefix *germ-*, such as in germsporangiophore, germspores, etc.

The duration of the zygospore dormancy varies between different strains and can be influenced by extraneous substances (Kozlova *et al.*, 1971). E. W. Goodell (unpublished) tested many strains and found that UBC21 and UBC24 (from R. J. Bandoni, Botany Department, University of British Columbia, Vancouver, B.C., Canada) showed short dormancies when crossed with other strains. The zygospores of crosses UBC21 × NRRL1555 have dormancies of less than three months. The development, by repeated backcrosses of UBC21 to NRRL1555, of a (+)

strain nearly isogenic with NRRL1555 and giving short dormancies when crossed with it, is an interesting project. NRRL1555 and the new strain would then become the standard strains for *Phycomyces* research and avoid the unknown variations in genetic background between different natural isolates.

Mutants

Mutant collections are kept by M. Delbrück at the Division of Biology, California Institute of Technology, Pasadena, California; by E. Cerdá-Olmedo at the Departamento de Genética, Facultad de Ciencias, Universidad de Sevilla, Sevilla, Spain; by the American Type Culture Collection, Rockville, Maryland; and by other researchers. The genetic nomenclature developed by Demerec *et al.* (1966) is used in *Phycomyces.* Details of its application and information on mutant strains can be obtained from the scientists listed above.

Morphological mutants are frequently found. Burgeff (1912, 1914, 1928) used several kinds which had spontaneously appeared in his cultures, such as *piloboloides (pil), nanus (nan)*, a colonial type called *arbusculus (arb)*, and others. Still used are some colonial types (*col*), a dwarf colonial suitable for replica plating (*dwf*), and several developmental mutants. Some of the advantages of the colonial types can be obtained by growing the wild type in a medium at *p*H 3.3 where it presents a colonial morphology and allows counting of viable spores and other manipulations.

Convenient markers for genetic studies are provided by hereditary alterations of carotenogenesis, resistance to inhibitory drugs, or requirement of special nutritional supplements. Mutations that block the synthesis of β-carotene and result in the loss of the yellow color of the wild type are known at three loci. Mutants *carA* are white and lack forty-carbon compounds. Mutants *carB* are white and accumulate phytoene. Mutants *carR* are red and accumulate lycopene. Other *car* mutants produce more β-carotone than the wild type and have a brighter yellow color (Meissner and Delbrück, 1968; Ootaki *et al.*, 1973). There are mutants resisting crystal violet (*xtv*), nystatin (*nys*), amphotericine B (*amp*), and canavine (*can*). Those isolated after plating large numbers of untreated spores on drug-containing medium are usually dominant. Auxotrophs are commonly given the same three-letter codes used in bacterial genetics. Auxotrophs can be isolated by direct tests after strong mutagenic treatments. Several counterselection procedures have been tried, but none of them has yet been particularly successful.

Mutants characterized by an altered phototropism are labeled *mad*; they are isolated because of the inability of their sporangiophores to turn downward when grown in a glass-bottom box, where all light comes from below (M. Heisenberg, unpublished). They present several different phenotypes with pleiotropic blocks in other sensory reactions and allow an integrated view of the organism's sensory physiology (Bergman *et al.*, 1973). Surprisingly, the relatively complex behavior of this organism is determined by few genes. Only five genes, *madA* to *madE*, are represented among the phototropically defective mutants used so far in complementation studies (Ootaki, private communication).

To obtain most of these mutants, the spores are treated with the mutagen N-methyl-N′-nitro-N-nitrosoguanidine in a buffer. The peculiarities of this agent (Guerola *et al.*, 1971) have led to a reconsideration of the procedures for mutant isolation, as can be seen in the discussion by Bergman *et al.* (1973).

Nuclear Behavior in the Asexual Cycle

The vegetative nuclei of *Phycomyces* are about 2 μm in diameter. They contain several small chromatin granules and a relatively large nucleolus. The chromosome number has been been reliably determined. The DNA content amounts to approximately 4×10^7 nucleotide pairs per nucleus, with unique copies making up about 70 percent of it (R. Dusenbery, private communication). The DNA is about 40 percent [G + C] (Storck and Alexopoulos, 1970) and probably lacks histones (Leighton *et al.*, 1971). The nucleic acids cannot be labeled with radioactive thymidine or uridine; the different species of RNA have been studied after being labeled with [^{32}P] phosphate (Gamow and Prescott, 1972).

The nuclei divide by elongation and constriction, without apparent loss of the nuclear envelope (Robinow, 1957). With the electron microscope, a set of microtubules has been found lying across the nucleus and anchored at the nuclear envelope. These tubules may direct nuclear division through their own elongation (Franke and Reau, 1973). Nuclear fusion and the phenomena of the parasexual genetic cycle have not been detected, but their existence as rare events cannot be excluded.

Hyphae at all stages of the asexual cycle are multinucleate. The mycelia and sporangiophores contain many thousands or millions of nuclei. The spores of the standard strain NRRL1555 grown in glucose-asparagine-yeast agar present the distribution of nuclei per spore given in Table 1. Different distributions are found in other strains (Johannes, 1950). The multinucleate nature of most of the spores demands that the mutagenic treatments be coupled with the death of most nuclei so that enough functionally uninucleate spores are produced for the detection of recessive mutations (Cerdá-Olmedo and Reau, 1970) or that the few uninucleate spores be isolated and tested. Reau (1972) has devised a procedure for the isolation of uninucleate spores through sedimentation. She could not find culture conditions that would render most the spores uninucleate; even drastic changes in media produced only small displacements in the distribution of nuclei per spore.

The hyphae of *Phycomyces*, unlike those of other Fungi, do not anastomose spontaneously upon contact, and thus heterokaryons cannot be formed by the contiguous growth of two different strains. Heterokaryons originate by mutation of a nucleus in a multinucleate cell, by the germination of some germspores, or by surgical manipulations (Burgeff, 1914; Weide, 1939; Ootaki, 1973). Ootaki's procedure is particularly simple and efficient. The tips of two young sporangiophores of different genetic constitution are cut off, and the sporangiophores are allowed to regenerate with their lesions in contact in a humid atmosphere. Small sporangiophores and sporangia are formed which often contain heterokaryotic spores.

The asexual spores are formed by the division of a large mass of cytoplasm into multinucleate portions which develop strong cell walls (Swingle, 1903). A heterokaryon containing two types of nuclei can thus form three kinds of spores: heterokaryons, homokaryons of one type, and homokaryons of the other type. Assuming no nuclear divisions while the spores are formed and random packaging of the two kinds of nuclei into spores with a certain distribution of nuclei per spore, Heisenberg and Cerdá-Olmedo (1968) calculated the proportions of each type of spore to

TABLE 1. *Distribution of Nuclei per Spore in Strain NRRL1555 of Phycomyces blakesleeanus Grown in Glucose-Asparagine-Yeast Agar* [a]

Number of nuclei	Proportion of spores
1	0.003
2	0.090
3	0.420
4	0.410
5	0.074
6	0.003
≥7	<0.001

[a] From Heisenberg and Cerdá-Olmedo (1968).

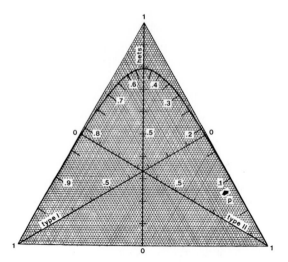

Figure 2. Theoretical relationship between the genetic constitution of a heterokaryon and the genetic constitution of its spores. The nuclear proportion in the heterokaryon containing p nuclei of type I and 1 − p nuclei of type II is read along the curve. The proportions of the three possible kinds of spores (heterokaryons, homokaryons type I, and homokaryons type II) are given by the distances from the p point of the curve to the sides of the triangle, conveniently read along the medians.

be expected from a mycelium with a given nuclear proportion (Figure 2). They devised an experimental assay for the spores and found that the experimentally determined values coincided with points in the theoretical curves, allowing them to read directly the value of the original nuclear proportion. They also found that the nuclear proportions are the same in different parts of the mycelium and even in mycelia derived from it by transplantation of a piece to a new medium. This constancy must be due to the strong cytoplasmic streams observed inside the hyphae and the lack of septa or crosswalls in the relevant parts of the mycelium. Old mycelia tend to dry up and to maintain only small viable propagula where the constancy of nuclear proportions no longer holds.

Heterokaryons containing nuclei of different mating types generally behave as do homokaryons, but they do produce more carotenoids and less sporangiophores, and they form small aerial hyphae called *pseudophores*, which are believed to represent frustrated attempts at conjugation (Burgeff, 1924).

The exceptional stability of the heterokaryons and the relative ease of the calculation of nuclear proportions allow the study of *quantitative complementation,* i.e., the quantitative study of a biochemical variable in heterokaryons containing a known mixture of nuclei presenting genetic differences for that variable. The values of the biochemical variable are thus obtained as functions of the nuclear proportion. The same functions can be obtained theoretically if a certain organization and activity of the biochemical elements involved is assumed. De la Guardia *et al.* (1971; and unpublished) were thus able to conclude that the synthesis of β-carotene in *Phycomyces* results from the successive action of four copies of phytoene dehydrogenase and two copies of lycopene cyclase, organized into an enzyme aggregate. Similar quantitative studies are possible for behavioral traits, such as phototropic thresholds (Villet, 1970). While heterokaryons are not as easy to make routinely in *Phycomyces* as in other Fungi, systematic studies of complementation in a set of mutants can be carried out, as proven by Ootaki *et al.* (1973) for carotenogenesis.

The detailed knowledge of the genetic composition of the spore crop produced by a heterokaryon is the basis of a procedure for the determination of the causes of cell death. Cerdá-Olmedo and Reau (1970) were thus able to estimate the relative incidence of mitotic, recessive, dominant, and cytoplasmic lethal lesions after treatments with x and ultraviolet

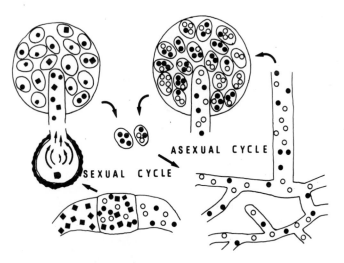

Figure 3. Nuclear behavior in the life cycle of P. blakesleeanus Squares represent nuclei of one of the mating types, say (+). Circles represent nuclei of the other mating type, (−). Three shades of nuclei may be taken as three marking alleles. For details and discussion see text.

radiation, nitrosoguanidine, and heat; the procedure should be useful for other agents as well.

The right part of Figure 3 summarizes the usual behavior of the nuclei in the asexual cycle of *Phycomyces*.

Nuclear Behavior in the Sexual Cycle

The sexual cycle of *Phycomyces* presents many difficulties to the cytogeneticists: the events of greatest interest may take place inside the thick-walled zygospore over a long period of time; the preparations are difficult to fix and stain; the nuclei are small and their behavior difficult to ascertain from fixed preparations. Many researchers have devoted considerable effort to this subject (Moreau, 1913; Burgeff, 1915; Keene, 1919; Baird, 1924; Ling-Young, 1930–1931; Cutter, 1942; Sjöwall, 1945, 1946), but their reports are often contradictory. I described their findings in Bergman *et al.* (1969), and, therefore, will not discuss them here. A new study, using modern techniques in light and electron microscopy, is clearly needed.

Many thousands of nuclei of each mating type are believed to enter the zygospore, and up to tens of thousands of viable germspores may be found in the resulting germsporangium. Both the input and the output nuclei are believed to be haploid. The genetic problem is to discover the relationship between them.

To this end, crosses were carried out by Burgeff (1915, 1928), using the spontaneous markers he found in his cultures, and by myself, intermittently from 1968 to the present, using the new markers isolated after mutagenesis. P. Reau and A. Ferrús have contributed to the work at different times. I have carried out the following crosses:

1. $carA5$ (−) * $carR21$ (−) × (+)
2. $carB10$ (−) * $carR21$ (−) × $mad\text{-}54$ (+)
3. $carA57\ mad\text{-}209$ (−) * $carR21\ aux$ (−) × $xtv\text{-}51$ (+)

Here the asterisk separates the components of a heterokaryon and the multiplication sign separates the two mates of a cross. The following questions can receive tentative answers:

How Many Parental Nuclei Are There?   Table 2 indicates the colors found in the progeny of cross 3, in which nuclei carrying red ($carR21$), white ($carA57$), and yellow (car^+) markers participated. Only one germsporangium, out of 32, contained all three colors. Two other germsporangia containing the three colors were found in cross 2. It is concluded that, however many nuclei enter the zygospore, only a pair of nuclei, rarely more, become parents of the progeny. The other nuclei are presumably used as food and energy reserves.

Which Recombination Process Takes Place? When the germspores of each germsporangium were classified as to their genotypes, there was a total lack of discernible regularities. There was little symmetry in the kinds of genotypes found in one germsporangium, so that sometimes all the genotypes carried the same allele of one locus, but different alleles of other loci. Most of the germsporangia contained more than four different genotypes, and thus excluded the occurrence of meiosis in a single diploid nucleus because this would give rise to a maximum of four different haploid products. The number of germspores with each genotype was also variable, some genotypes making up most of the progeny and others being scarcely represented. These irregularities in the progeny of each zygospore contrast with the regular tetrads of other Fungi and remind one of the variety of genotypes found in each single burst after multiple infection with phage.

The irregular progeny found in each germsporangium can be explained in meiotic terms if we assume that: one or a few diploid nuclei are produced by fusion of parental haploid nuclei, all the other nuclei present in the zygospore are destroyed, the diploid survivor(s) divide mitotically and then undergo meiosis, and the haploid meiotic products divide again mitotically at rates which are very different from one another to produce the progeny nuclei.

As an alternative to this complicated explanation, one may propose that there is no meiosis at all, but that the diploid survivor(s) suffer repeated mitotic divisions in which frequent mitotic recombination and

TABLE 2. *Colors of the Mycelia Produced by the Germspores from 32 Germsporangia of the Cross carA57 mad-209 (−) * carR21 aux (−) × xtv-51 (+)* [a]

Colors	Number of germsporangia
Sterile	11
All white	3
All red	0
All yellow	7
White and red	0
White and yellow	8
Red and yellow	2
White, red, and yellow	1

[a] The (−) heterokaryon contained white- and red-marked nuclei, with a majority of the former; the (+) homokaryon contained yellow-marked nuclei.

haploidization would occur, leading to the haploid progeny nuclei. This parasexual alternative fits well the present results, but we have no experimental confirmation yet. If parasexuality is proven in the future, it would establish *Phycomyces* as a premeiotic organism and would point out a likely root for the evolution of meiosis.

The summation of the results from several germsporangia smoothes out the irregularities found in each of them. Adding just seven germsporangia from cross 3, each allele at each locus is found at frequencies close to 50 percent. Recombination frequencies, required for classical genetic analysis, may then be reliably calculated. To this end, many zygospores are left on a plate and when most of them have germinated, the germspores are harvested and tested.

How Are Germspores Formed?   The primordia of the spores in the asexual cycle are known to be multinucleate; this leads to a high frequency of heterokaryotic spores in the output of heterokaryotic mycelia. On the other hand, the germspores usually give rise to stable homokaryons. There is a low frequency of heterokaryons among them, varying from one germsporangium to the next. This finding indicates that the germspores are derived from primorida which generally contain only one nucleus. These must divide mitotically to produce the multinucleated germspores. Such uninucleated primordia have been observed by P. Reau and myself in a germsporangium of cross 2.

The left part of Figure 3 summarizes present knowledge of nuclear behavior in the sexual cycle of *Phycomyces*.

The genetic analysis of *Phycomyces*, while still fraught with difficulties, is a viable endeavor that should soon produce genetic maps and other powerful tools for biochemical and behavioral studies. This provides a better stimulus for work than Burgeff's suggestion (1928) that "wir können nicht stets die weichen Stellen aus dem Kuchen der Wissenschaft herausessen und unsern Nachfolgern das Kauen der harten überlassen."

Literature Cited

Agardh, C. A., 1817   *Ulva nitens. Synopsis algarum Scandinaviae (Lund)*, p. 46.

Baird, E. A., 1924   The structure and behavior of the nucleus in the life history of *Phycomyces nitens* (Agardh) Kunze and *Rhizopus nigricans* Ehrbg. *Trans. Wis. Acad. Sci. Arts Lett.* **21**:357–380.

Benjamin, C. R. and C. Hesseltine, 1959   Studies on the genus *Phycomyces. Mycologia* **51**:751–771.

Bergman, K., 1972   Blue-light control of sporangiophore initiation in *Phycomyces. Planta (Berl.)* **107**:53–67.

Bergman, K., P. V. Burke, E. Cerdá-Olmedo, C. N. David, M. Delbrück, K. W. Foster, E. W. Goodell, M. Heisenberg, G. Meissner, M. Zalokar, D. S. Dennison and W. Shropshire, 1969   *Phycomyces*. *Bacteriol. Rev.* **33**:99–157.

Bergman, K., A. P. Eslava and E. Cerdá-Olmedo, 1973   Mutants of *Phycomyces* with abnormal phototropism. *Mol. Gen. Genet.* **123**:1–16.

Blakeslee, A. F., 1904   Sexual reproduction in the Mucorineae. *Proc. Am. Acad. Arts Sci.* **40**:205–319.

Burgeff, H., 1912   Über Sexualität, Variabilität und Vererbung bei *Phycomyces nitens*. *Ber. Dtsch. Bot. Ges.* **30**:679–685.

Burgeff, H., 1914   Untersuchungen über Variabilität, Sexualität und Erblichkeit bei *Phycomyces nitens* Kunze I. *Flora (Jena)* **107**:259–316.

Burgeff, H., 1915   Untersuchungen über Variabilität, Sexualität und Erblichkeit bei *Phycomyces nitens* Kunze II. *Flora (Jena)* **108**:353–448.

Burgeff, H., 1924   Untersuchungen über Sexualität und Parasitismus bei Mucorineen. I. *Botan. Abh.* **4**:1–135.

Burgeff, H., 1925   Uber Arten und Artkreuzung in der Gattung *Phycomyces* Kunze. *Flora (Jena)* **118–119**:40–46.

Burgeff, H., 1928   Variabilität, Vererbung und Mutation bei *Phycomyces blakesleeanus* Bgff. *Z. Indukt. Abstammungs-Vererbungsl.* **49**:26–94.

Burgeff, H., 1934   Pflanzliche Avitaminose und ihre Behebung durch Vitaminzufuhr. *Ber. Dtsch. Bot. Ges.* **52**:384–390.

Caglioti, L., G. Cainelli, B. Camerino, R. Mondelli, A. Prieto, A. Quilico, T. Salvatori and A. Selva, 1966   The structure of trisporic-C acid. *Tetrahedron (Suppl.)* **7**:175–187.

Cainelli, G., P. Graselli and A. Selva, 1967   Struttura dell' acido trisporico B. *Chim. Ind. (Milan)* **49**:628–629.

Castle, E. S., 1928   Temperature characteristics for the growth of the sporangiophores of *Phycomyces*. *J. Gen. Physiol.* **11**:407–413.

Cerdá-Olmedo, E. and P. Reau, 1970   Genetic classification of the lethal effects of various agents on heterokaryotic spores of *Phycomyces*. *Mutat. Res.* **9**:369–384.

Chichester, C. O., P. S. Wong and G. Mackinney, 1954   On the biosynthesis of carotenoids. *Plant Physiol.* **29**:238–241.

Cutter, V. M., 1942   Nuclear behavior in the Mucorales. II. The *Rhizopus, Phycomyces* and *Sporodinia* patterns. *Bull. Torrey Bot. Club* **69**:592–616.

De la Guardia, M.D., C. M. G. Aragón, F. J. Murillo and E. Cerdá-Olmedo, 1971   A carotenogenic enzyme aggregate in *Phycomyces:* Evidence from quantitative complementation. *Proc. Natl. Acad. Sci. USA* **68**:2012–2015.

Delvaux, E., 1973   Some aspects of germination induction in *Phycomyces blakesleeanus* by an ammonium-acetate pretreatment. *Arch. Mikrobiol.* **88**:273–284.

Demerec, M., E. A. Adelberg, A. J. Clark and P. E. Hartman, 1966   A proposal for a uniform nomenclature in bacterial genetics. *Genetics* **54**:61–76.

Forssberg, A. 1969   The radiation induced reactions of *Phycomyces blakesleeanus*— Pulsed transfer of cellular particles. *Radiat. Bot.* **9**:391–395.

Foster, K. W. 1974   How to track sporangiophores of the fungus *Phycomyces*. *Rev. Sci. Instrum.* in press.

Foster, K. W. and E. D. Lipson, 1973   The light growth response of *Phycomyces*. *J. Gen Physiol.* **62**:590–617.

Franke, W. W. and P. Reau, 1973   The mitotic apparatus of a Zygomycete, *Phycomyces blakesleeanus*. *Arch. Mikrobiol.* **90**:121–129.

Furch, B. 1972   Zur Wärmeaktivierung der Sporen von *Phycomyces blakesleeanus*. Das Auftreten von Gärungen unter aeroben Bedingungen. *Protoplasma* **75**:371–379.

Gamow, E. and D. M. Prescott, 1972   Characterization of RNA synthesized by *Phycomyces blakesleeanus*. *Biochim. Biophys. Acta* **259**:223–227.

Garton, G. A., T. W. Goodwin and W. Lijinsky, 1951   Studies in carotenogenesis. I. General considerations governing β-carotene synthesis by the fungus *Phycomyces blakesleeanus* Burgeff. *Biochem. J.* **48**:154–163.

Gooday, G. W., 1973   Differentiation in the Mucorales. *Symp. Soc. Gen. Microbiol.*, **23**:269–294.

Goodwin, T. W., editor, 1965   *Chemistry and Biochemistry of Plant Pigments*, Academic Press, London.

Gruen, H. E. and T. Ootaki, 1972   Regeneration on *Phycomyces* sporangiophores. III. Grafting of sporangiophore segments, weight changes, and protoplasmic streaming in relation to regeneration. *Can. J. Bot.* **50**:139–158.

Guerola, N., J. L. Ingraham and E. Cerdá-Olmedo, 1971   Induction of closely linked multiple mutations by nitrosoguanidine. *Nat. New Biol.* **230**:122–125.

Heisenberg, M. and E. Cerdá-Olmedo, 1968   Segregation of heterokaryons in the asexual cycle of *Phycomyces*. *Mol. Gen. Genet.* **102**:187–195.

Isler, O., editor, 1971   *Carotenoids*, Birkhäuser, Basel and Stuttgart.

Johannes, H., 1950   Ein sekundäres Geschlechtsmerkmal des isogamen *Phycomyces Blakesleeanus* Burgeff. *Biol. Zentralbl.* **69**:463–468.

Keene, M. L., 1919   Studies of zygospore formation in *Phycomyces nitens* Kunze. *Trans. Wis. Acad. Sci. Arts Lett.* **19**:1195–1220.

Keyhani, J., E. Keyhani and S. H. Goodgal, 1972   Studies on the cytochrome content of *Phycomyces* spores during germination. *Eur. J. Biochem.* **27**:527–534.

Kozlova, A. N., G. V. Komarova, G. I. El-Registan and N. A. Krasilnikov 1971   Effect of metabolites of actinomycetes on the sexual functions of *Phycomyces blakesleeanus* (in Russian). *Dokl. Akad. Nauk SSSR, Ser. Biol.* **200**:230–232.

Kunze, G., 1823   *Phycomyces*. In *Mykologische Hefte*, zweites Heft, edited by G. Kunze and J. C. Schmidt, pp. 113–116, Voss, Leipzig.

Leighton, T. J., B. C. Dill, J. J. Stock and C. Phillips, 1971   Absence of histones from the chromosomal proteins of fungi. *Proc. Natl. Acad. Sci. USA* **68**:677–680.

Ling-Young, 1930–1931   Etude biologique des phénomènes de la sexualité chez les Mucorinées. *Rev. Gén. Bot.* **42**:145–158, 205–218, 283–296, 348–365, 409–428, 491–504, 535–552, 618–639, 681–704, 722–752; **43**:30–43.

Malhotra, S. K. and J. P. Tewari, 1973   Molecular alterations in the plasma membrane of sporangiospores of *Phycomyces* related to germination. *Proc. R. Soc. Lond. B* **184**:207–216.

Meissner, G. and M. Delbrück, 1968   Carotenes and retinal in *Phycomyces* mutants. *Plant Physiol.* **43**:1279–1283.

Moreau, F., 1913   Recherches sur la reproduction des mucorinées et de quelques autres thallophytes. *Botaniste* **13**:1–136.

Ødegård, K., 1952   On the physiology of *Phycomyces blakesleeanus* Burgeff. I. Mineral requirements on a glucose-asparagine medium. *Physiol. Plant.* **5**:583–609.

Ootaki, T., 1973   A new method for heterokaryon formation in *Phycomyces*. *Mol. Gen. Genet.* **121**:49–56.

Ootaki, T. and J. J. Wolken 1973   Octahedral crystals in *Phycomyces*. II. *J. Cell Biol.* **57**:278–288.

Ootaki, T., A. C. Lighty, M. Delbrück and W. J. Hsu, 1973   Complementation between mutants of *Phycomyces* deficient with respect to carotenogenesis. *Mol. Gen. Genet.* **121**:57–70.

Petzuch, M. and M. Delbrück, 1970   Effects of cold periods on the stimulus-response system of *Phycomyces. J. Gen. Physiol.* **56**:297–308.

Plempel, M., 1963   Die chemischen Grundlagen der Sexualreaktion bei Zygomyceten. *Planta (Berl.)* **59**:492–508.

Reau, P., 1972   Uninucleate spores of *Phycomyces. Planta (Berl.)* **108**:153–160.

Robinow, C. F., 1957   The structure and behavior of the nuclei in spores and growing hyphae of Mucorales. II. *Phycomyces blakesleeanus. Can. J. Microbiol.* **3**:791–798.

Rudolph, H., 1960   Weitere Untersuchungen zur Wärmeaktivierung der Sporangiosporen von *Phycomyces blakesleeanus*. II. Mitteilung. *Planta (Berl.)* **55**:424–437.

Schmidt, R., 1925   Untersuchungen über das Myzelwachstum der *Phycomyceten. Jahrb. Wiss. Bot.* **64**:509–586.

Schopfer, W. H., 1934   Les vitamines crystallisées B₁ comme hormones de croissance chez un microorganisme (*Phycomyces*). *Arch. Mikrobiol.* **5**:511–549.

Schopfer, W. H. and A. Jung, 1937   Un test végétal pour l'aneurine. Méthode, critique, et résultats. *C. R. V Cong. Int. Tech. Chim. Ind. Scheveningen* **1**:22–34.

Sjöwall, M., 1945   Studien über Sexualität, Vererbung und Zytologie bei einigen diözischen Mucoraceen. *Akad. Abh.* 1–97 (Lund. Gleerupska Universitet-Bokhandeln).

Sjöwall, M., 1946   Über die zytologischen Verhältnisse in den Keimschläuchen von *Phycomyces blakesleeanus* und *Rhizopus nigricans. Bot. Not.* **3**:331–334.

Storck, R. and C. J. Alexopoulos, 1970   Deoxyribonucleic acid of Fungi. *Bacteriol. Rev.* **34**:126–154.

Swingle, D. B., 1903   Formation of the spores in the sporangia of *Rhizopus nigricans* and of *Phycomyces nitens. US Dep. Agric. Bur. Plant Ind. Bull.* **37**:1–40.

Thornton, R. M., 1973   New photoresponses of *Phycomyces. Plant Physiol.* **51**:570–576.

Tu, J. C., S. K. Malhotra and S. Prasad, 1971   Electron microscopy of stage IV sporangiophore of *Phycomyces. Microbios* **3**:143–151.

Van Assche, J. A., A. R. Carlier and H. I. Dekeersmaeker, 1972   Trehalase activity in dormant and activated spores of *Phycomyces blakesleeanus. Planta (Berl.)* **103**:327–333.

Van den Ende, H. and D. Stegwee, 1971   Physiology of sex in Mucorales. *Bot. Rev.* **37**:22–36.

Villet, R. H., 1970   Genetic curing of blindness in *Phycomyces blakesleeanus:* A quantitative assessment of dominance. *Nature (Lond.)* **225**:453–454.

Villet, R. H., 1972   Microinjection of *Phycomyces*. Selection of a strain for possible biological assay of photoreceptor pigment. *Plant Physiol.* **49**:273–274.

Weide, A., 1939   Beobachtungen an Plasmaexplantaten von *Phycomyces. Arch. Exp. Zellforsch.* **23**:299–337.

Zalokar, M., 1969   Intracellular centrifugal separation of organelles in *Phycomyces. J. Cell Biol.* **41**:494–509.

24

Saccharomyces

FRED SHERMAN AND CHRISTOPHER W. LAWRENCE

Introduction

Fruitful genetic investigations with *Saccharomyces* were initiated after the elucidation of their life cycle by Winge in 1935. By 1950, Winge, Lindegren, and their collaborators had clarified the genetic control of the mating system, and demonstrated Mendelian segregation of numerous markers, along with linkage relationships. However, yeast did not become immediately the favorite microorganism for genetic studies, probably because of the early difficulty in dissection of asci. Also there were erroneous beliefs that yeast is "less controllable genetically than *Neurospora* and permits less detailed genetic analysis" (Catcheside, 1951). Today the virtues of working with yeast are widely recognized, and it has become an ideal eucaryotic microorganism for biochemical and genetic studies. While yeasts have greater genetic complexity than bacteria, they still share many of the technical advantages which permitted rapid progress in the molecular genetics of procaryotes and their viruses. Some of the properties which make yeast particularly suitable for genetic studies include the existence of both stable haploids and diploids, rapid growth, clonability, and the ease of replica plating and mutant isolation. Unlike many other microorganisms, strains of *Saccharomyces* are viable with a great many markers. The difficulties of ascus dissection, which were sometimes exaggerated, have now been reduced with the use of snail juice (see page

FRED SHERMAN AND CHRISTOPHER W. LAWRENCE—Department of Radiation Biology and Biophysics, University of Rochester School of Medicine and Dentistry, Rochester, New York.

368), and most workers can master the technique after a few days of practice. Yeast has been successfully employed in all phases of genetics, such as mutagenesis, recombination, regulation and gene action, and especially mitochondrial genetics and other aspects distinct to eucaryotic systems.

Much of the early literature concerning yeast genetics can be found in a book by Lindegren (1949) and in a review by Winge and Roberts (1958). Recent general reviews by Mortimer and Hawthorne (1966; 1969) and by Hartwell (1970) have appeared as well as reviews of methodology (Fink, 1970), recombination (Fogel and Mortimer, 1971), mutagenesis (Mortimer and Manney, 1971), and regulation (de Robichon-Szulmajster and Surdin-Kerjan, 1971; Calvo and Fink, 1971; Fink, 1971). A comprehensive review of the biochemical genetics of metabolic pathways, including a compilation of gene products in yeast, has been prepared by de Robichon-Szulmajster and Surdin-Kerjan (1971). There are two comprehensive books covering the entire field of fungal genetics, including yeast genetics (Fincham and Day, 1971; Esser and Kuenen, 1967). Detailed coverage of a variety of topics concerned with yeast can be found in the three-volume treatise, *The Yeasts* (Rose and Harrison, 1969; 1970; 1971). The biannual newsletter "Yeast," which is the official publication of the International Commission on Yeasts and Yeast-like Microorganisms of the International Association of Microbiological Societies (Herman J. Phaff, editor, University of California, Davis, California 95616) contains summaries and notes of interest to all yeast workers, including geneticists.

While genetic analyses has been undertaken with a number of taxonomically distinct varieties of yeast, extensive studies have been restricted primarily to the many freely interbreeding species of *Saccharomyces* and to *Schizosaccharomyces pombe* (Leupold, 1970). Although *Saccharomyces cerevisiae* is commonly used to designate many of the laboratory stocks of *Saccharomyces* used throughout the world, it should be pointed out that most of these strains originated from the interbred stocks of Winge, Lindegren, and others who employed fermentation markers from not only *S. cerevisiae* but also from *S. bayanus*, *S. carlsbergensis*, *S. chevalieri*, *S. chodati*, *S. diastaticus*, etc. However, it is still recommended that the interbreeding laboratory stocks of *Saccharomyces* be denoted as *S. cerevisiae*, in order to conveniently distinguish them from the more distantly related species of *Saccharomyces*.

Care should be taken in choosing strains for genetic studies as well as certain biochemical studies. Unfortunately there are no truly wild-type *Saccharomyces* strains that are commonly employed in genetics studies. Also most strains of brewers' yeast and probably many strains of bakers' yeast and wild-type strains of *S. cerevisiae* are not genetically compatible with laboratory stocks. It is usually not appreciated that many "normal" laboratory strains contain mutant characters, a fact not too surprising since they were derived from pedigrees involving mutagenized strains. The haploid strain S288C is often used because it gives rise to well-dispersed cells and because of the availability of many isogenic mutant derivatives. However, S288C contains abnormally low cytochrome *c* and assimilates several carbon sources at reduced rates. One should check the specific characters of interest before initiating a study with any strain. Many strains containing well-characterized auxotrophic and temperature-sensitive markers can be obtained from the Yeast Genetics Stock Center that is maintained by Drs. R. K. Mortimer, S. Fogel, and J. Bassel (Donner Laboratory, University of California, Berkeley, California 95720).

Genetic Nomenclature

A recommendation for the nomenclature used in yeast genetics was published as the November 1969 supplement to the Microbial Genetics Bulletin, and except for the persistence of a few of the older notations, these proposals are usually followed. For the sake of clarity and convenience, a few minor changes appear appropriate; these are included in the following rules of nomenclature which have been expanded and updated:

1. The two mating-type alleles are designated, respectively by **a** and *α*. (In the old system, the italicized *a*, in contrast to the bold face **a**, was sometimes difficult to differentiate from *α*.)

2. Gene symbols are usually designated by three italicized letters which should be consistent with the proposal of Demerec *et al.* (1966) whenever applicable, e.g., *arg*. The genetic locus is identified by a number immediately following the gene symbol, e.g., *arg2*. Alleles are designated by a number separated from the locus number by a hyphen, e.g., *arg2-6*. While locus numbers must agree with the original assignments, allele numbers may be particular to each laboratory.

3. Complementation groups of a gene or a gene cluster can be designated by capital letters following the locus number, e.g., *his4A*, *his4B*, etc.

4. Dominant and recessive genes are denoted by upper and lower case letters, respectively, e.g., *SUP6* and *arg2*.

5. When there is no confusion, wild-type genes are designated simply as +; the + may follow the locus number to designate a specific wild-type gene, e.g., *sup6+* and *ARG2+*.

6. While superscripts should be avoided, it is sometimes expedient to distinguish genes conferring resistance and sensitivity by the superscript R and S, respectively. For example the genes controlling resistance to canavanine sulfate (*can*1) and copper sulfate (*CUP*1) and their sensitive alleles could be denoted, respectively, as *can*^{R}1, *CUP*^{R}1, *CAN*^{S}1, and *cup*^{S}1.

7. Mitochondrial and other non-Mendelian genotypes can be distinguished from the chromosomal genotype by enclosure in square brackets. Whenever applicable, it is advisable to employ the above rules for designating non-Mendelian genes, and to avoid the use of Greek letters; however, it is less confusing to either retain the original symbols ρ^+, ρ^-, ψ^+, and ψ^- or to use their transliteration, respectively, [*rho*+], [*rho*−], [*psi*+], and [*psi*−]. The mitochondrial genes conferring resistance and sensitivity can be denoted by superscripts as indicated in rule 6. Since the identification of distinct genetic loci and the dominance relationships of alleles are less clear, genes can be denoted by capital letters and mutant numbers, e.g., [*ERY*^{S} *CHL*^{R}321], [*ERY*^{R}221 *CHL*^{S}], etc.

Life Cycles

S. cerevisiae appears to be primarily a heterothallic species, and most laboratory stocks are either stable haploids or diploids, both of which apparently can be grown indefinitely without change of ploidy. Laboratory strains can be found, however, which are secondarily homothallic due to the influence of alleles at two loci, and these generally exist as stable diploids. These alleles seem to be not uncommon in stocks of wild or commercial origin (Fowell, 1969b). Life cycles for heterothallic and homothallic strains are shown in Figure 1 and will be discussed in terms of growth, mating, and sporulation.

Growth

In *Saccharomyces* species of any ploidy, cell multiplication is characteristically achieved by budding; a daughter is initiated as an outgrowth from the mother cell followed by nuclear division, cross-wall formation, and finally by cell separation. Under optimal conditions in vigorous strains this process takes a minimum of about 70 minutes (Burns, 1956). Each mother cell usually forms no more than 20–30 buds, and its "age" can be determined by the number of bud scars left on the cell wall (Beran, 1968).

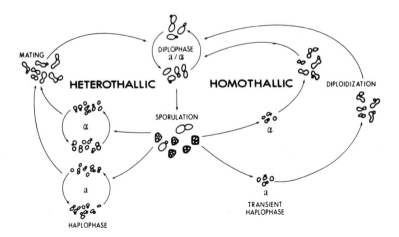

*Figure 1. Life cycles of Saccharomyces species. Two haploids of mating-type **a** and two of mating-type α comprise a four-spored ascus which is formed after sporulation of a heterozygous diploid **a**/α. The stable states of heterothallic strains are diploid and haploid while the stable state of homothallic strains are only diploid since the transient haploid cells rapidly diplodize as a result of the two genes $HO_α$ and HM.*

Mating

Mating is basically under the control of the complex mating-type locus, which is also involved in sporulation and other cell processes. Most haploid strains possess either the **a** or α mating-type allele, and mating can only take place between an **a** and an α cell. Mating proceeds in two stages. In the first, an aggregate forms containing cells of opposite mating type (Sakai and Yanagishima, 1971), and although no cell fusion has yet taken place, these aggregates are sufficiently coherent to withstand gentle suspension in liquid. During the aggregation stage, cells of the **a** mating type accumulate at the start of the cell cycle due to the effect of the "hormone" secreted by α cells which prevents bud initiation and DNA synthesis (Throm and Duntze, 1970). In addition, it has been postulated that the **a** cells may produce a mating factor (Manney, private communication). Mating occurs only between unbudded cells. The second stage involves cell fusion and the formation of a zygote. In most strains nuclear fusion follows shortly after zygote formation, and the first bud produced by the zygote contains a diploid nucleus. In some strains, however, nuclear fusion is delayed, thus giving rise to a transient dikaryon which produces haploid buds (Fowell, 1969b).

Homothallic strains possess certain alleles at the $HO_α$ and HM loci which cause the specific and directed mutation of one mating-type allele to the other (Takano and Oshima 1967, 1970), but do not affect the basic processes of mating. In haploid strains, the HO_α hm genotype causes the mutation of α to **a** with a frequency as high as 0.5 mutations per cell division (Hawthorne, 1963), but has no effect on the **a** allele. The HM allele reverses the effect of HO_α, such that the **a** allele is specifically mutated to α at a similar frequency. The HM has no known effect except to modify the action of HO_α, and neither allele has any influence in diploid cells. The combined effects of HO_α and HM were previously ascribed to a single locus, designated D (Winge and Roberts, 1949). Since the **a** or α cells that arise by this directed mutation also mate readily with cells of the opposite mating type, initially haploid clones of the genotype α HO_α hm or **a** HO_α HM rapidly diploidize, and the haplophase is transient. Haploid cells of these genotypes can, however, be used in controlled matings by using ascospores. Sporulation in homothallic diploids is identical to that in heterothallic strains.

Sporulation

Sporulation is induced by a variety of "starvation" media, such as medium containing potassium acetate as the primary carbon source (see Fowell, 1969a). Usually 10–20 percent of cells undergo a meiotic division and develop into asci, but strains with up to 70 percent asci or more can be found. Usually four haploid ascospores are formed within the wall of the mother cell, which becomes the ascus sac, but in some strains a varying proportion of asci with three or less spores are found, presumably due to inviability of some of the meiotic products. In most tetrads ascospores are arranged tetrahedrally, but some strains give linear tetrads in which spores one and three, or two and four, are the products of the second meiotic divisions. While most isolates of *S. cerevisiae* produce a maximum of four ascospores per ascus, some produce a maximum of eight ascospores and others regularly produce two large spores. In the latter case a single ascospore clone can give rise to a sporulating culture without the intervention of mating (Grewal and Miller, 1972), but the reasons for this are not yet known.

Techniques in Yeast Genetics

Media

The nonsynthetic complete medium, YPD (1 percent Bacto-yeast extract, 2 percent Bacto-peptone, 2 percent dextrose, and agar if solidification is desired) is often used as a general-purpose growth medium and for storage of culture stocks. A commonly employed synthetic medium, that was formulated by Wickerham (1946) and contains vitamins, trace elements, salts, and a nitrogen source, is now available commercially from Difco Laboratories (Bacto-Yeast Nitrogen Base w/o amino acids). Auxotrophic markers can be scored with this synthetic medium after the addition of a carbon source, usually 2 percent dextrose, and appropriate nutrilites. Between 10 and 20 mg/liter is normally sufficient, but it is necessary to add higher concentrations of tyrosine, leucine, isoleucine, lysine (30 mg/liter), phenylalanine (50 mg/liter), glutamic acid, aspartic acid, valine (100 mg/liter), threonine (200 mg/liter), and serine (375 mg/liter). Fermentation markers are scored either by gas formation in Durham tubes or more conveniently by acid formation on plates containing a pH indicator such as bromthymol blue or bromcresol purple. The fermentation tests are carried out in media having 2 percent of the appropriate sugar and yeast extract plus peptone, as above. The *pet* and ρ^- phenotype, i.e., the inability to utilize nonfermentable carbon sources for growth, is usually tested on medium containing 3 percent (v/v) glycerol and yeast extract and peptone as above. Sporulation of most laboratory diploid strains is achieved by first growing on a presporulation medium (0.8 percent Bacto-yeast extract, 0.3 percent Bacto-peptone, and 10 percent dextrose) for 1–2 days and then transferring the cells to sporulation medium (1 percent potassium acetate, 0.1 percent Bacto-yeast extract, and 0.05 percent dextrose) (McClary *et al.*, 1959) for 3–5 days.

Preservation of Stocks

Most yeast strains remain viable on YPD slants for one to two years, if refrigerated. Longer periods of preservation can be achieved by drying yeast in cooled, sterile silica gel (Type I, chromatographic grade, 60–200 mesh, Sigma Chemical Co.) and storing under refrigeration. The viability can be increased by first suspending the yeast in sterile milk before transferring to the silica gel.

Mutagenesis

Except for the base analogs, most of the mutagens employed with bacteria are also suitable with yeast after only slight modifications of the methods of treatments. A compilation of mutagens used in various studies with yeast has been prepared by Mortimer and Manney (1971). Ultraviolet light, nitrous acid, and ethyl methanesulfonate are potent agents that have been successfully employed in both forward and reverse mutagenic studies, and these mutagens are recommended since they are less hazardous than most. Some mutagens are particularly liable to induce ρ^- cytoplasmic mutants (Schwaier et al., 1968) which may interfere with further studies. Flocculent strains should be avoided in mutagenic studies, and low-intensity sonication treatments may be used for dispersing cells, thus allowing a higher recovery of the mutants with certain detection systems. In addition, some workers (Zimmerman et al., 1966) recommend a post-incubation period in complete medium to allow for mutant expression.

Mutant Detection

Reasonable yields of auxotrophic mutants can be isolated simply from mutagenized haploid strains by replica plating the resulting colonies to synthetic medium. Over 2 percent mutants can be recovered after treatments of ethyl methanesulfonate or nitrosoguanidine that result in approximately 50 percent survival. Techniques have been developed that increase the proportion of mutants by killing off the nonmutant population with nystatin (Snow, 1966; Thouvenot and Bourgeois, 1971). Also, some workers have used medium containing the dye Phloxin B (Magdala Red) in order to detect some nutritional mutants on the bases of colony color (Horn and Wilkie, 1966). Other systems that have been used to detect or enrich for forward mutations of more specific loci include the following: (1) ade1 and ade2 mutants can be detected since they give rise to red colonies; (2) additional mutants of one of several genes in the biosynthesis of adenine can be detected since they suppress the red pigment of ade1 and ade2 strains; (3) many types of resistance mutants can be directly selected on media that contain such agents as canavanine, ethionine, 5-fluorouracil, nystatin, etc.; (4) certain uracil auxotrophs can be selected on the basis of resistance to ureidosuccinate (Bach and Lacroute, 1972); (5) cytochrome c mutants can be selected by their resistance to chlorolactate (see Sherman and Stewart, 1973); (6) super-suppressors can be selected by the concomitant reversion of two or more auxotrophic markers (see Mortimer and Gilmore, 1968); and (7) mutants that are sensitive to antibiotics can be selectively rescued with the use of a temperature-sensitive mutant (Littlewood, 1972).

Formation and Isolation of Hybrids

In the heterothallic strains, the isolation of diploids after hybridization is usually achieved by one of three methods, the choice primarily depending on the presence of complementary markers. The most generally applicable method is micromanipulation of zygotes in mass matings. Approximately equal numbers of growing haploid cells of each mating type are thoroughly mixed and grown on the surface of complete medium at 30°C for 4–5 hours. A suspension of cells from this mating mixture is then streaked onto an agar slab or plate, and several zygotes are isolated from the cells by micromanipulation; each zygote is allowed to grow into a separate colony. Zygotes can be recognized by their characteristic dumbbell shape and by the position of the first bud, which in most strains forms at the point of fusion of the two parental cells.

The most convenient method of isolating diploid clones, however, is the prototrophic-selection technique of Pomper and Burkholder (1949), in which mating is carried out between haploids carrying different recessive auxotrophic markers. Cell suspensions made from overnight mass matings are streaked onto minimal medium, on which the diploid but neither haploid can form colonies. It is often necessary to subclone diploids isolated in this way. If only one parental haploid carries an auxotrophic marker, diploids can be isolated in an analogous manner by converting the prototrophic haploid to a neutral cytoplasmic mutant (ρ^-) and streaking a mating suspension on minimal medium containing a nonfermentable carbon source, on which ρ^- mutants cannot grow. Such ρ^- mutants can readily be produced by growth in the presence of acriflavine or ethidium bromide (see below). Where both parents are prototrophic, diploids can often be isolated by streaking a 4- to 5-hour mating mixture on minimal medium and noting the position of the largest colonies as soon as they become visible. After further incubaton, cells from these clones can be transferred and their diploid status verified by sporulation and lack of mating.

Controlled crosses with homothallic strains are usually carried out using ascospores, which are released from the ascus sac by digestion with snail juice or mushroom extract (see page 368). Pairs of spores can then be isolated by micromanipulation, the spores in each pair being placed close together on growth medium to facilitate subsequent mating. This method has the disadvantage that on the average only half the pairs will contain spores of opposite mating type and self-diploidization may intervene in place of or before mating. Genetic markers should be used to exclude this possibility. Alternatively, complete asci of different genotypes can be mixed, a dense suspension plated, and diploids isolated by prototrophic selection.

Finally, it should be pointed out that while hybridization generally involves haploid strains, which normally are the only cells to show a mating reaction, mating can also be carried out between diploids, or cells of higher ploidy, which are homozygous for the mating-type locus. These homozygotes arise by mitotic crossing-over between the centromere and the mating-type locus or by mutation. Such homozygotes can be used to construct a series of derivative strains of varying ploidy. It is usually more convenient, however, to isolate the zygote clones which arise from the mating of such homozygotes using prototrophic selection. Rare matings between α haploids can be also isolated in this way (e.g., Pomper *et al.*, 1954; Mortimer, 1958; Laskowski, 1960).

Meiotic Analysis

Since ascospores can be isolated from individual asci, yeast offers opportunities for both tetrad and random spore analysis. For tetrad analysis, sporulated cultures are treated with snail juice (Johnston and Mortimer, 1959). This is commercially available as "Glusulase" (Endo Laboratories, Garden City, N.Y.) and "suc d'*Helix pomatia*" (L'Industrie Biologique Francaise, Génévilliers, France). Mushroom extract (Bevan and Costello, 1964) may also be used to digest the ascus sac. A suspension of the digested culture is gently streaked along the edge of a thin agar slab mounted on a glass slide or along one side of a Petri plate. If care is taken to avoid overdigestion or shaking of the suspension, the four spores in each tetrad remain associated. Agar slabs are inverted over a dissection chamber, tetrads are picked up on the end of a microneedle, viewed under a magnification of about 150–300×, and individual spores are placed at regular intervals across the slab. Twenty or so tetrads can be dissected on a single 6 × 2.5-cm slab, which is then placed on complete medium, spore side up, to allow each spore to form a colony (Figure 2). Dissection on the surface of a plate is usually achieved with the aid of a microloop, and the spores germinate *in situ*. The two commercially available micromanipulators that are most commonly used for isolating yeast ascospores are the Sensaur-deFonbrune Micromanipulator (Curtin Scientific Co., St. Louis, Mo.) and the Singer Micromanipulator MK-III (Singer Instrument Co., Berkshire, England*). Also the custom-made models that were specifically designed for this purpose by Dr. R. K. Mortimer (University of California, Berkeley) and Dr. F. Sherman (1973) are often used.

For random spore analysis, it is necessary to separate unsporulated diploid cells from asci or spores, and this can be carried out by either physical or genetic methods. One type of physical separation involves the

---

* The Singer Micromanipulator is distributed in the U.S.A. by Melpro Co., Brooklyn, N.Y.; and identical model (Mark III) is sold by the Eric Sobotka Co., Inc. Farmingdale, N.Y.

treatment of the asci with snail juice, sonication to disperse the ascospores, and finally separating of the vegetative cells from the ascospores by a liquid-paraffin phase (Emeis and Gutz, 1958; Siddiqi, 1971). Random ascospores have also been prepared from sporulated spheroplasts which were passed through a French pressure cell (Douglas and Hawthorne, 1972). Genetic separation can be carried out by sporulating diploids heterozygous for a recessive resistance marker, such as *can*[R]. Sporulated cultures are treated with snail juice, briefly sonicated to disrupt the tetrads, and are plated on a canavanine-containing medium on which the diploids and half of the haploids cannot grow (Sherman and Roman, 1963).

In both tetrad and random spore analysis, the phenotypes of spore clones can be easily scored by transferring them to complete medium to

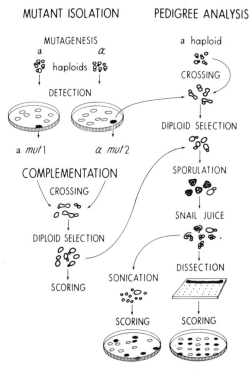

*Figure 2. Schematic representation of some of the common techniques used in yeast genetics. Two mutants of opposite mating types, **a** mut1 and α mut2, may be directly crossed for a complementation test. Pedigree analysis can be performed on a heterozygous diploid **a** × α mut2 or a doubly heterozygous diploid **a** mut1 × α mut2. The diploids are sporulated, treated with snail juice to rupture the ascus sacs, and the clusters of four spores are separated on an agar slab with a microneedle. Alternatively, the spores may be separated by sonication and plated on a medium which is selective against diploids. A master plate can be made from the spore clones to enable scoring by replica plating.*

form a master plate, which is then replicated onto plates containing test media. Replication with velveteen is widely used, but the use of cell suspensions and a "point" replicator is often useful where more subtle discriminations are required. The techniques are schematically illustrated in Figure 2.

Mitotic Analysis

Spontaneous crossing over in vegetative diploid cells, leading to homozygosis for recessive markers distal to the point of crossing-over, is a rare event and normally occurs at a frequency of less than one in 10^5 cells. The frequency of mitotic crossing over can, however, be much increased by x rays, ultraviolet light, or by a variety of chemical recombinogens (see Mortimer and Manney, 1971), up to levels where sectored colonies can be detected by replica plating. Also such events can be detected by visual examination of colonies, using recessive markers that affect morphology, or by selective techniques, using recessive resistance markers. Mutants at various adenine loci are convenient signal markers. Homozygous *ade2* colonies accumulate a red pigment which can easily be detected among normal white colonies, and recessive *ade2* mutants can therefore be used to signal homozygosis for chromosome XV. In addition, since many other *ade* mutations block this pigment formation, these *ade* loci also can be used as signal markers in strains homozygous for *ade2* (Roman, 1971). Selective systems can be devised using suppressible canavanine-resistant mutations and suppressors (Mortimer and Hawthorne, 1973).

Chromosome Mapping

Mapping studies by Lindegren and co-workers, and in particular by Mortimer and Hawthorne (see Mortimer and Hawthorne, 1973), have located approximately 150 genes on 17 chromosomes and 3 fragments (see Figure 3). Four methods can be used to map unlocated mutants (Mortimer and Hawthorne, 1973). Tetrad analysis is used to establish a single-gene basis for the mutant phenotype, and also to test for centromere linkage. Genes which exhibit centromere linkage can then be located by the analysis of crosses with tester strains carrying known centromere markers. Sixteen of the seventeen chromosomes have centromere markers, and some 56 centromere-linked genes have been identified.

Genes that are not linked to a centromere can be located by (1) random spore analysis using an assortment of markers dispersed over the genome, (2) mitotic crossing over using adenine signal markers or selective systems, and (3) trisomic analysis. Mortimer and Hawthorne (1973) found that the most efficient method of using trisomic analysis was to cross multiple disomes, generated by sporulating triploids, to stocks carrying the unlocated genes. A comparison between the segregation, either normal 2:2 or aberrant, of the gene in question and that of various markers serves to indicate on which chromosome the locus is situated or, usually of greater importance, those chromosomes upon which it does *not* reside (Mortimer and Hawthorne, 1973).

The linkage map of *S. cerevisiae* is unusual not only with respect to the high number of chromosomes but also to their exceptional genetic length, which in total exceeds 2600 centimorgans. Genetic lengths of individual chromosomes appear to correlate well with physical lengths (Petes and Fangman, 1972; Blamire *et al.*, 1972) except for chromosome I, which contains 70 percent of all ribosomal RNA cistrons (Finkelstein *et al.*, 1972). With 7×10^8 daltons of DNA, the average chromosome is little more than five times the size of that of bacteriophage T4, or only about a quarter the size of the *Escherichia coli* chromosome. These data also suggest that each chromosome contains a single DNA duplex.

Fine-Structure Analysis

Fine-structure analysis in *S. cerevisiae* can be carried out with vegetative or sporulated diploids using selective techniques, or in unselected samples of asci. Aberrant segregation due to gene conversion occurs in about 0.5–10 percent or more tetrads, depending on the particular allele, with an average frequency of about 1.5 percent. With such a high frequency, it has been possible to investigate gene conversion in unselected tetrads (Fogel *et al.*, 1971), and notable advances have been made in understanding this phenomenon.

Fine-structure mapping is generally carried out using vegetative diploids rather than tetrads, however, because these provide more uniform cell samples and because the disadvantage of a much lower spontaneous frequency of gene conversion can be offset by selective techniques and the use of recombinogens which enhance this frequency tens or hundreds of times. The most common method is x-ray mapping, in which diploids heteroallelic for auxotrophic markers are exposed to x rays and are then plated on minimal media to select the prototrophic recombinants. X irradiation is the best recombinogenic agent, since the yield of prototrophs is linearly related to dose and the slope of the dose–response curve usually exhibits additivity between mutant sites (Manney and Mortimer, 1964). In the case of methyl methanesulphonate, the yield of prototrophs is pro-

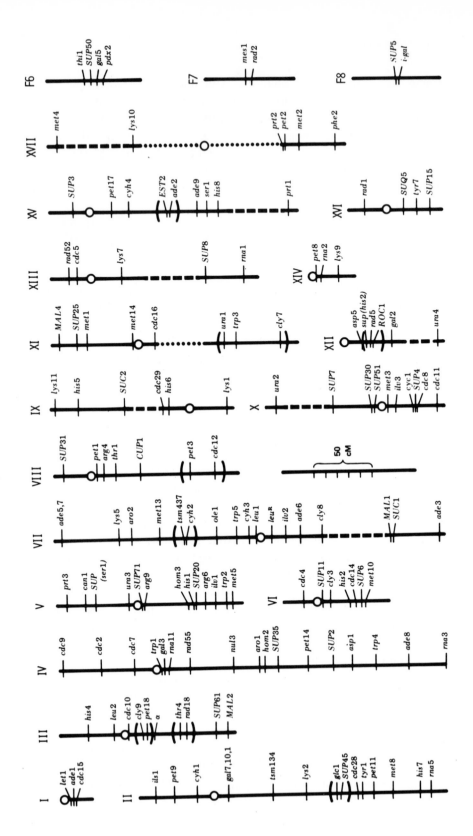

Figure 3. The genetic map of Saccharomyces. Linkages established by tetrad analysis, mitotic segregation, and trisomic analysis are denoted, respectively, by solid lines, dashed lines, and dotted lines. The order is unknown for the genes indicated within brackets. The numbers following the gene symbols designate the particular loci, except after tsm, where they designate mutant numbers. The gene symbols refer to the following phenotypes: nutritional requirements for adenine (ade), arginine (arg), aromatic amino acids phenylalanine, tryptophan, tyrosine, and p-aminobenzoate (aro), histidine (his), homoserine or threonine plus methionine (hom), isoleucine or isoleucine plus valine (ilv), leucine (leu), lysine (lys), methionine (met), oleate (ole), phenylalanine (phe), pyridoxine (pdx), serine (ser), thiamine (thi), threonine (thr), tryptophan (trp), tyrosine or tyrosine plus phenylalanine (tyr), and uracil (ura); defects of asparagine and aspartate metabolism (asp); fermentation of maltose (MAL), and sucrose (SUC); nonfermentation of galactose (gal) and its regulation (i-gal); conditional mutants affecting cell division (cdc), isoleucyl-tRNA synthetase (ils), methionyl-tRNA synthetase (mes), cell lysis (cly), protein synthesis (prt), RNA synthesis (rna) and unknown functions (tsm); lack of utilization of nonfermentable carbon sources (pet); deficiency of cytochrome c (cyc); electrophoretic mobility of esterases (EST); glycogen accumulation (glc); resistance to canavanine (can), copper salts (CUP), cycloheximide (cyh), and benzalkonium chloride, "Roccal" (ROC); sensitivity to radiation (rad) (rad1–rad49 are primarily or exclusively uv sensitive and rad50 upwards are primarily or exclusively X-ray sensitive); mating type (α) and nonmaters (nul); supersuppressors and specific suppressors (SUP or sup); recessive lethals (let). All data are from Mortimer and Hawthorne (1973) and Hartwell et al., (1973) except for the location of cyc1 and the ordering of it and adjacent genes (Sherman et al., 1974), and the location of EST2 (Strobel and Wöhrmann, 1972). The gene products of some of these markers are listed in Table 1.

portional to the square of the time of treatment (Snow and Korch, 1970); ultraviolet light gives a curvilinear dose–response curve.

X-ray mapping with two-point crosses has been used successfully to give consistent maps on a number of occasions (Fink, 1966; Esposito, 1968; Parker and Sherman, 1969; Jones, 1973). However, this method suffers from certain limitations, and it cannot be assumed that map distances reflect physical distances or even that the indicated order of the mutant sites is necessarily correct. Work with the structural gene for iso-1-cytochrome c, in which the position of a mutant site can be established independently by determining the primary structure of the protein, has shown that some mutants give consistently abnormal recombination frequencies. Used qualitatively, however, these results show that it is usually possible to establish the nonidentity of allelic mutants, even when they involve alterations in adjacent codons. These results also show that one x-ray mapping unit (one prototroph/10^8 survivors/rad) is approximately equal to 60 base pairs, but it is not known whether this value applies to other loci (see Sherman and Stewart, 1971).

The best method for ordering mutant sites is deletion mapping, in which mutant strains are crossed in turn with a set of tester strains containing deletions. Prototrophic recombinants can be obtained only when the mutant site lies outside the deleted region, and the deletions themselves are also ordered by the presence or absence of recombination. Unfortunately, suitable deletions are rare in yeast. Fink (private communication), however, has been able to obtain numerous deletions by exploiting the polar nature of certain nonsense mutations in the *his*4 complex. The order of mutant sites obtained with their aid differs in a number of instances from that obtained by x-ray mapping, which emphasizes the need for caution when interpreting x-ray mapping results.

Non-Mendelian Genetics

A mutant is defined as non-Mendelian or extrachromosomal when it proves to give rise to exceptional Mendelian segregations. Unfortunately, the pattern of segregation of many novel chromosomal conditions may superficially resemble a non-Mendelian inheritance. Nevertheless, after such phenomena as polygenic control, polyploidy, or aneuploidy have been excluded by appropriate crosses, one can be reasonably sure that a trait is non-Mendelian when its hybrid crosses exhibit primarily 0:4 or 4:0 segregation. In addition, a mutant would be suspected of being non-Mendelian if hybrid crosses give rise to diploids with mixed phenotypes, or if the trait can be induced or lost at a high frequency with certain specific mutagens. However, it should be pointed out that the property of giving rise to mixtures of zygote clones, referred to as "suppressiveness" or "mixedness," is associated only with *some* non-Mendelian mutants. Also, similar heterogeneity of hybrids may occur with chromosomal mutants in which case it is termed "penetrance."

The heterokaryon test is an equally decisive method for establishing non-Mendelian inheritance, and in addition it indicates whether the determinant resides outside of the nucleus, i.e., whether it is "cytoplasmic" or "nonnuclear." Certain strains of yeast can exist as a transient heterokaryon after zygote formation, and haploid cells from newly formed zygotes can have an interchange of the nuclear and cytoplasmic markers in comparison to the parental strains. While this heterokaryon test is difficult to perform and cannot be used with all strains, results of this kind have established the cytoplasmic inheritance of the ρ^- determinant (Wright and Lederberg, 1957) and of certain drug-resistant markers (Wilkie and Thomas, 1973).

One can test the hypothesis that a given non-Mendelian determinant is contained within the mitochondrial genome by using ρ^- mutants induced with high concentrations of ethidium bromide, since these mutants have no detectable mitochondrial DNA (see Linnane *et al.*, 1972; Nagley and Linnane, 1972). Thus the concomitant loss of drug-resistance markers with the loss of mitochondrial DNA is consistent with their mitochondrial locations (Saunders, *et al.*, 1971). In contrast, the non-Mendelian determinants controlling ψ^+ (Young and Cox, 1972), *killer* (Fink and Styles, 1972), and the uptake of ureiodosuccinic acid (Lacroute, 1971) are retained under these conditions. It should be pointed out that positive results indicating a loss should be interpreted with caution, since it is possible that ethidium bromide and similar agents could eliminate more than one type of non-Mendelian determinant.

Mapping Mitochondrial Mutants

Mapping and ordering of mitochondrial markers are being pursued by various investigators, especially by Slonimski and his co-workers (see Coen *et al.*, 1970; Bolotin *et al.*, 1971). As will be discussed below, the three methods used by the French group to order these drug-resistant markers in specified strains involve (1) determining the ratio of recombinant types in the diploids of hybrid crosses; (2) determining the rate of loss of the marker in diploid strains after uv irradiation of one of the parental strains; and (3) mapping deletions by observing the concomitant loss of two or more markers in certain ρ^- mutants.

Zygotes obtained by crossing cells which carry drug-resistant markers in their mitochondrial genome usually give rise to diploids with a variety of genotypes, some parental and some recombinant. Often the recombinant types occur with unequal frequencies. These rates can be measured by allowing the zygotes to form 10^3–10^4 cells and then plating sample populations on media selective for the diploids. The resulting diploid colonies are then tested for the mitochondrial markers. It has been proposed from the results of such experiments that the preferential transmission of mitochondrial markers involves the hypothetical sex factors ω^+ and ω^- which are at a fixed site in the mitochondrial genome. The underlying hypothesis is that mitochondrial genomes which arise from recombination of heterosexual $\omega^+ \times \omega^-$ crosses all contain the ω^+ determinant; in contrast, the parental sex markers are retained in non-recombinant diploids from both heterosexual and homosexual ($\omega^+ \times \omega^+$ and $\omega^- \times \omega^-$) crosses as well as in recombinant diploids from homosexual crosses. Thus in heterosexual crosses, the markers closer to ω^+ would have a greater probability of being transmitted to the diploids. Using this argument, the order of markers can be assigned in respect to ω^+ site by determining the ratio of the recombinant types. For example, [ω^+ CHL^R ERY^R] would be the major recombinant type from the cross [ω^+ CHL^R ERY^S] $\times$ [ω^- CHL^S ERY^R], if the CHL^R marker is closer to ω^+. Also, an excessive frequency of the parental type [ω^+ CHL^R ERY^S] would be expected since they should include a class that underwent recombination which remained undetected. Those mitochondrial markers that are far from the ω site would be expected to be difficult to map by this method. It remains to be seen whether or not the model involving the ω sex factors can explain the results from other laboratories (see Rank and Bech-Hansen, 1972; Kleese *et al.*, 1972; Wilkie and Thomas, 1973).

Mapping also has been undertaken by observing the rates of loss of transmission of the various mitochondrial markers as a function of uv dose. The ω^+ haploid is uv irradiated, crossed to an ω^- haploid, and the diploids are examined for the mitochondrial markers of the ω^+ strain. It is believed that the transmission is diminished to a smaller extent by uv, the closer the markers are to the ω site.

Independent ρ^- mutants contain different apparent deletions of the mitochondrial genome, and some of these still contain one or more of the drug-resistant markers. If the apparent deletions cover contiguous regions, one also should be able to order the sites. While full reports have not been published, it is claimed that there is a consistency in the order of the mitochondrial markers by all three methods (Slonimski, private communication).

TABLE 1. Selected List of Mutant Genes

Mutant genes	Linkage groups [a]	Phenotypes, requirements and properties	Gene products [b] or activities lost in mutants
ade1	I	Adenine; red	Phosphoribosyl-aminoimidazolesuccinocarboxamide synthetase
ade2	XV	Adenine; red	Phosphoribosyl-aminoimidazole carboxylase
ade3	VII	Adenine + histidine	One isoenzyme of methylenetetrahydrofolate dehydrogenase
ade4	—	Adenine	Phosphoribosyl-pyrophosphate amidotransferase
ade5	VII	Adenine	Phosphoribosyl-glycineamide synthase
ade6	VII	Adenine	Phosphoribosyl-formylglycineamide synthetase?
ade7	VII	Adenine	Phosphoribosyl-aminoimidazole synthase?
ade12	—	Adenine	Adenylosuccinate synthetase
ade13	—	Adenine	Adenylosuccinate lyase
arg1	—	Arginine	Argininosuccinate synthetase
arg3	—	Arginine	Ornithine carbamoyltransferase
arg4	VIII	Arginine	Argininosuccinate lyase
arg5	—	Arginine	N-Acetyl-γ-glutamylphosphate reductase
arg7	—	Arginine	N-2-Acetylornithine deacetylase
aro1A	IV	Phenyl.+ tryp.+ tyros.+ p-aminobenzoate	3-Enolpyruvylshikimate-5-phosphate synthase
aro1B	IV	Phenyl.+ tryp.+ tyros.+ p-aminobenzoate	Shikimate kinase
aro1C	IV	Phenyl.+ tryp.+ tyros.+ p-aminobenzoate	5-Dehydroquinate synthase
aro1D	IV	Phenyl.+ tryp.+ tyros.+ p-aminobenzoate	Shikimate dehydrogenase
aro2	VII	Phenyl.+ tryp.+ tyros.+ p-aminobenzoate	Chorismate synthase
aro3	—	Growth inhibited by tyrosine	7-Phospho-2-keto-3-deoxyheptonate aldolase
aro4	—	Growth inhibited by phenylalanine	7-Phospho-2-keto-3-deoxyheptonate aldolase
asp1	IV	Asparagine in asp5 strains	Asparaginase
asp5	XII	Aspartate + methionine	Aspartate aminotransferase

TABLE 1. Continued

Mutant genes	Linkage groups [a]	Phenotypes, requirements and properties	Gene products [b] or activities lost in mutants
can1	V	Resistance to canavanine	Arginine permease
cpa1	—	Growth inhibited by arginine in the absence of uracil	Carbamoylphosphate synthase
cpa2	—	Growth inhibited by arginine in the absence of uracil	Carbamoylphosphate synthase
cyc1	X	Low cytochrome c	Iso-1-cytochrome c
fas2A	—	Myristic acid	"Dehydratase step" in fatty acid synthetase
fas2B	—	Myristic acid	"Second reductase step" in fatty acid synthetase
fas3A	—	Myristic acid	"First reductase step" in fatty acid synthetase
fas3B	—	Myristic acid	"Condensation step" in fatty acid synthetase
fas3C	—	Myristic acid	"Condensation step" in fatty acid synthetase
gal1	II	Galactose nonfermenter	Galactokinase
gal2	XII	Galactose nonfermenter	Galactose permease
gal4	—	Galactose nonfermenter	Regulator of gal1,7,10
gal5	—	Galactose nonfermenter	Phosphoglucomutase
gal7	II	Galactose nonfermenter	Galactose-1-phosphate uridylyltransferase
gal10	II	Galactose nonfermenter	UDP glucose epimerase
i-gal	F8	Constitutive mutant	Repressor controlling enzymes of gal 1,7,10
glu1	—	Glutamate	Aconitate hydratase
glu4	—	Glutamate	Aconitate hydratase
his1	V	Histidine	ATP-phosphoribosyl transferase
his2	VI	Histidine	L-Histidinolphosphatase
his3		(same as his8)	
his4A	III	Histidine	Phosphoribosyl-AMP cyclohydrolase

his4B	III	Histidine	Phosphoribosyl-AMP pyrophosphorylase
his4C	III	Histidine	Histidinol dehydrogenase
his5	IX	Histidine	L-Histidinolphosphate aminotransferase
his6	IX	Histidine	"BBM-II isomerase"
his7	II	Histidine	"Enzyme causing BBM III → AICAR + IGP"
his8	XV	Histidine	Imidazoleglycerolphosphate dehydratase
hom2	IV	Threonine + methionine	Aspartate semialdehyde dehydrogenase
hom3	V	Threonine + methionine	Aspartate kinase
hom6	—	Threonine + methionine	Homoserine dehydrogenase
ils1	II	Conditional lethal	Isoleucyl-tRNA synthetase
ilv1	V	Isoleucine	Threonine dehydratase
ilv2	—	Isoleucine + valine	Acetolactate synthase
ilv3	X	Isoleucine + valine	Dihydroxyacid dehydratase
ilv4	—	Isoleucine + valine	Hydroxyacid reductoisomerase
ilv5	—	Isoleucine + valine	Hydroxyacid reductoisomerase
leu1	VII	Leucine	Isopropylmalate isomerase
leu2	III	Leucine	Isopropylmalate dehydrogenase
lys1	IX	Lysine	Saccharopine dehydrogenase
lys2	II	Lysine	2-Aminoadipate-semialdehyde dehydrogenase
lys4	—	Lysine	Homoaconitate hydratase
lys5	VII	Lysine	2-Aminoadipate-semialdehyde dehydrogenase
lys7	XIII	Lysine	Homocitrate dehydrase
lys9	XIV	Lysine	α-Aminoadipic-δ-semialdehyde glutamic reductase
lys11	IX	Lysine	Homoisocitrate dehydrogenase
lys12	—	Lysine	Homoisocitrate dehydrogenase
lys13	II?	Lysine	α-Aminoadipic-δ-semialdehyde glutamic reductase
lys14	VII?	Lysine	α-Aminoadipic-δ-semialdehyde glutamic reductase
mes1	F7	Methionine; conditional lethal	Methionyl-tRNA synthetase
met2	XVII	Methionine	Homoserine acetyltransferase

TABLE 1. Continued

Mutant genes	Linkage groups[a]	Phenotypes, requirements and properties	Gene products[b] or activities lost in mutants
met8	II	Methionine	Homocysteine synthetase
met10	VI	Methionine	Sulfate adenylyltransferase?
ole1	VII	Oleate	Δ⁹ Fatty acid desaturase
phe2	XVII	Phenylalanine	Prephenate synthase + phenylalanine transferase
rib5	—	Riboflavin	Riboflavin synthetase
ser1	XV	Serine	L-glutamate:phosphohydroxypyruvate transaminose
ser2	—	Serine	Phosphoserine phosphatase
SUP2	IV	Super-suppressor	Tyrosyl-tRNA?
SUP3	XV	Super-suppressor	Tyrosyl-tRNA?
SUP4	X	Super-suppressor	Tyrosyl-tRNA?
SUP5	F8	Super-suppressor	Tyrosyl-tRNA?
SUP6	VI	Super-suppressor	Tyrosyl-tRNA?
SUP7	X	Super-suppressor	Tyrosyl-tRNA?
SUP8	XIII	Super-suppressor	Tyrosyl-tRNA?
SUP11	VI	Super-suppressor	Tyrosyl-tRNA?
thr1	VIII	Threonine	Homoserine kinase
thr4	III	Threonine	Threonine synthase
trp1	IV	Tryptophan	N-(5'-Phosphoribosyl) anthranilate isomerase
trp2	V	Tryptophan	"Anthranilate synthetase"
trp3	XI	Tryptophan	Indole-3-glycerol-phosphate synthase
trp4	IV	Tryptophan	Anthranilate phosphoribosyltransferase
trp5	VII	Tryptophan	Tryptophan synthase
tyr1	II	Tyrosine	Prephenate dehydrogenase + tyrosine aminotransferase?

tyr7	XVI	Tyrosine + phenylalanine	
*ura*1	XI	Uracil	Dehydroorotate dehydrogenase
*ura*2A	X	Uracil	Aspartate carbamoyltransferase
*ura*2C	X	Uracil	Carbamoylphosphate synthase
*ura*3	V	Uracil	Orotidine-5'-phosphate decarboxylase
*ura*4	XII	Uracil	Dihydroorotase

[a] The linkage group assignments are from Mortimer and Hawthorne (1973) (see Fig. 3).

[b] Original references for the associated gene products can be found in de Robichon-Szulmajester and Surdin-Kerjan (1971) and von Borstel *et al.*, (1970), except for *ade*4, *ade*5, and *ade*7 (Gross and Woods, 1971), *fas* mutants (Kühn *et al.*, 1972), some *lys* mutants (Bhattacharjee and Sinha, 1972), *rib*5 (Baur *et al.*, 1972), *ser* mutants (Ulane and Ogur, 1972), and *ura*2A and *ura*2C (see Denis-Duphil and Lacroute, 1972).

Types of Yeast Mutants

A selected compilation of gene products and associated enzymic deficiencies is given in Table 1 and the known linkage map, containing 146 markers, is presented in Figure 3. While most of these types of mutants can be found in other microorganisms, some are best studied in yeast, and others, like the ρ^- mutants, may be unique to yeast.

Biosynthetic Pathways

Mutational blocks in the biosynthetic pathway of amino acids, purines, and pyrimidines of yeast have been tabulated and critically reviewed by deRobichon-Szulmajster and Surdin-Kerjan (1971). Mutants affecting every step of many of the pathways have been uncovered, including the biosynthesis of arginine, histidine, threonine, tryptophan, and uracil. While the functionally related genes controlling biosynthetic pathways are usually dispersed throughout the genome, clusters of genes occur at the *aro*1, *fas*2, *fas*3, *his*4, and *ura*2 loci, and possibly at *ade*5,7 locus. Several of the genes, especially *trp*5 (see Manney, 1970), *his*4 (see Fink, 1971), and *ura*2 (see Denis-Duphil and Lacroute, 1971) have been the subject of intensive investigations from both genetic and biochemical points of view.

Fermentation

Wild-type genes that control the fermentation of the sugars galactose (*gal*), sucrose (*SUC*), maltose (*MAL*), melibiose (*MEL*), and α-methylglucoside (*MGL*) were uncovered in the early studies of Winge, Lindegren, Hawthorne, and others, who analyzed pedigrees from interbred species of *Saccharomyces*. With some sugars like maltose and sucrose, fermentation takes place if any one of the several dominant genes are present. The fermentation of α-methylglucoside requires complementary gene pairs like *MGL*4 *MAL*1, or *MGL*1 *MGL*2, or other combinations (see ten Berge, 1972; Khan and Haynes, 1972). The fermentation of galactose is diminished by the wild-type recessive *gal*3 gene or is blocked by mutations at any one of six other loci controlling either the steps in its fermentation pathway, its permease, or its regulation.

Regulation

An array of regulatory mutants that affect activity or rate of synthesis of enzymes in various pathways has been obtained in yeast. There are several cases where the feedback inhibitions of enzymes were altered by selecting mutants either resistant to analogs or sensitive to normal metabolites (see de Robichon-Szulmajster and Surdin-Kerjan, 1971). Interesting examples of gene regulation include the regulatory mutants affecting the synthesis of carbamoylphosphate synthetase (*cpa*1) which meet the requirements of the classical-type operator mutants and repressor mutants (Thuriaux *et al.*, 1972). The three linked genes *gal*1, *gal*7, and *gal*10, which determine the galactose-pathway enzymes, appear to be under positive control by the product of the unlinked *gal*4 gene, which in turn is under negative control of a repressor (*i-gal*) acting on an operator linked to *gal*4 (see Douglas and Hawthorne, 1972). The numerous regulatory mutants controlling the enzymes in the biosynthetic and degradative pathways of arginine has been reviewed by Wiame (1971).

Resistance

Mutants which are resistant to toxic agents can arise by diverse mechanisms, some of which have been biochemically elucidated. Mutations of the genes controlling specific permeases, or in some instances, the general amino acid permease (*gap* and *aap*) (see Grenson and Hennaut, 1971), can result in the loss of active uptake of such analogs as canavanine, ethionine, thiosine, and fluorouracil, along with the normal metabolites. Resistance to trifluoroleucine or fluorouracil can occur by altered feedback inhibition of mutant enzymes, while resistance to low concentrations of canavanine in the presence of ornithine is caused by constitutive synthesis of some enzymes in the arginine pathway. Mutants resistant to the fungicide nystatin are associated with altered sterol compositions (Bard, 1972). Certain semidominant mutants resistant to the antibiotic cycloheximide may be due to altered ribosomes. Mitochondrial mutants resistant to antibiotics and other toxic agents are described below.

Macromolecular Synthesis

Hartwell, McLaughlin, and co-workers have isolated and characterized large numbers of mutants that are unable to grow on enriched medium at 36°C but have normal or near normal growth at 23°C (Hartwell 1967, 1971; Hartwell and McLaughlin 1968, 1969; Hartwell *et al.*, 1970*a,b*, 1973; Culotti and Hartwell, 1971; Warner and Udem, 1972). Examinations at the restricted temperatures indicated that the defects in some of these conditional mutants were due to thermolabile syntheses or thermolabile products which are involved in protein synthesis (*prt*), RNA synthesis (*rna*), the cell-division cycle (*cdc*), and cell lysis (*cly*). Two of the protein-synthesis mutants were shown to specifically affect, respectively, isoleucyl-tRNA synthetase (*ils*1) and methionyl-tRNA synthetase (*mes*1).

Sporulation

Mutations (*spo*) that cause deficiency in the ability to complete meiosis and to form asci at an elevated temperature (34°C) have been isolated and investigated by Esposito and co-workers (see Esposito *et al.*, 1972). Several of these conditional mutants are being used to examine the process of sporulation, meiosis, and recombination.

Conjugation

Mutations of at least five genes (*ste* or *nul*) prevent conjugation of haploid strains of opposite mating type (see MacKay, 1972). Only one of the loci, *ste*1, maps at the mating-type locus. More than half of the *ste* mutants no longer produce the diffusible α factor (see above) necessary for mating.

Iso-1-cytochrome *c*

Special mention should be made of the *cyc*1 (or *cy*1) mutants because of the accessibility of determining the primary structures of the gene product, iso-1-cytochrome *c*. Mutationally altered iso-1-cytochromes *c* have been used to identify punctuation codons and to investigate mutagenic specificity, structure–function relationships, and the action of suppressors (see Sherman and Stewart, 1971; 1973).

Super-suppressors

A large number of distinguishable suppressors have been shown to act on certain alleles of numerous loci in yeast. Some of them (Table 1) insert residues of tyrosine at the site of ochre (UAA) (Gilmore *et al.*, 1971) and amber(UAG) (Sherman *et al.*, 1973) mutations. It is believed that the suppression may be due to altered forms of tRNA, as is the case for the nonsense suppressors in *E. coli*. It is of interest that the efficiency or mode of action of certain super-suppressors are modified by a non-Mendelian determinant ψ (Cox, 1971; Young and Cox, 1972).

Recombination

Apart from those which may have been isolated as mutants defective in sporulation, recombination-deficient mutants have been obtained by mutagenizing disomic strains which carry heteroallelic markers on the duplicated chromosome. Rodarte-Ramon and Mortimer (1972) used a disome of chromosome VIII and examined x-ray- or uv-induced heteroallelic reversion. They found mutants which fell into four complementation groups, designated *rec*1, *rec*2, *rec*3, and *rec*4.

Roth and Fogel (1971) used disomes for chromosome III that were heteroallelic for a test locus and also mating type. Such disomes undergo abortive sporulation during which spontaneous heteroallelic reversion takes place; it was therefore possible to isolate mutants defective in this process. However, to date, no mutants have been described which are deficient in recombination or gene conversion throughout the whole genome.

Radiation Sensitivity

More than 25 loci concerned with sensitivity to ultraviolet light or ionizing radiation have been described (see Game and Cox, 1971). They have been designated *rad* if the mutants were isolated on the basis of the lethal effects of radiation, or *rev* where radiation-induced mutation was involved. Locus numbers *rad*1–*rad*49 are reserved for loci primarily or exclusively concerned with uv sensitivity, and *rad*50 upward for those involved in sensitivity to ionizing radiations. Mutants at several loci are sensitive to both types of radiation, some are cross sensitive to methyl methanesulfonate or nitrous acid (Zimmermann, 1968), and a few are unable to sporulate or do so at reduced frequencies. Strains carrying *rad*1 or *rad*2 mutations are unable to excise pyrimidine dimers from their DNA (Unrau et al., 1971; Resnick and Setlow, 1972). A mutant *phr*1 has also been described which lacks photoreactivating enzyme activity (Resnick, 1969). The *rev*1, *rev*2, and *rev*3 mutants show a reduced frequency of uv-induced reversion of *arg*4–17, and are also slightly sensitive to the lethal effects of uv and x rays (Lemontt, 1971).

Pet Mutants

The symbol *pet* designates the multitude of chromosomal mutants that are unable to effectively utilize nonfermentable carbon sources such as glycerol, ethanol, lactate, and acetate. These include mutants which lack one or more of the cytochromes and lack respiration, as well as mutants which retain normal or near-normal levels of all the cytochromes and respiration. Thus, defects in electron transport and in numerous other components can lead to the inability to use nonfermentable substrates for growth. Common types are those that lack cytochromes $a \cdot a_3$ alone and those that lack cytochromes $a \cdot a_3$, b, and c_1 (Sherman and Slonimski, 1964; Lachowicz *et al.*, 1969), a defect similar to the ρ^- mutation (see below). Genetic blocks in the biosynthetic pathway of porphyrins result in mutants deficient in all hemoproteins. Some of the mutants, e.g., *pet3*, may control retention or synthesis of mitochondrial DNA, since they are never observed in the ρ^+ state. A compilation of many of the *pet* genes as well as other genes that affect mitochondrial function can be found in the article by Beck *et al.* (1970). Methods for the preparation and detection of *pet* and related mutants have been summarized by Sherman (1967).

Mitochondrial Mutants

The cytoplasmically-inherited ρ^- mutants all lack cytochromes $a \cdot a_3$, b, and c_1, and contain the physiological and enzymic defects that are the results of these cytochrome deficiencies. The common denominator of the multiple deficiencies can now be identified as the absence of mitochondrial protein synthesis. It is generally accepted that the ρ^- mutation is a direct result of the loss or gross alteration of mitochondrial DNA, which codes for several components of the translational machinery (see reviews by Linnane *et al.*, 1972; Sager, 1972).

Apparently simple mutations of mitochondrial DNA can result in resistance to agents that selectively inhibit the mitochondrial system. All of these resistance mutants have been selected on media containing the inhibitors and nonfermentable carbon sources, such as glycerol, to ensure functioning of the mitochondria. Mitochondrial mutants have been obtained which exhibit resistance to antibiotics that specifically inhibit protein synthesis of the mitochondria but not that of the cytoplasmic system (eythromycin, chloramphenicol, spiramycin, paromycin, etc.) and which exhibit resistance to inhibitors of the energy-transferring chain of oxidative phosphorylation (oligomycin and triethyl tin) (see Linnane *et al.*, 1972; Lancashire and Griffiths, 1971).

Killer

Many laboratory stocks of yeast are killer strains which secrete a toxic substance into the medium that kills cells of sensitive strains. Genetic analysis of the killer condition (Somers and Bevan, 1969; Bevan and

Somers, 1969) shows that it depends on a self-replicating cytoplasmic factor called K which requires the presence of a particular allele (M) of a nuclear gene for its maintenance. The killer factor fails to replicate in strains carrying an alternative recessive allele m, and is eventually diluted out of the cell population. Treatment of killer strains with cyclohexamide also leads to the loss of the killer factor (Fink and Styles, 1972). Apart from killer and sensitive strains, some strains are neutral (n), neither killing or being killed, and these seem to contain an inactive k factor. Since sensitive strains appear to lack a killer factor, either active or inactive, they are designated (o).

The killer factor has not been positively identified, but it may consist of double-stranded RNA (Berry and Bevan, 1972). The killer toxin appears to be protein in nature, since its activity can be destroyed by proteolytic enzymes, and contains several different components, some of which have a molecular weight in excess of 2×10^6 (Bussey, 1972). The action of the toxin is also little understood, but it may adsorb to cell walls and act in a manner formally analogous to bacterial colicins.

Acknowledgments

We are indebted to Dr. R. K. Mortimer for an advanced copy of his paper (Mortimer and Hawthorne, 1973). The writing of this manuscript was supported in part by the U.S. Atomic Energy Commission at the University of Rochester Energy Project, Rochester, New York and has been designated USAEC Report No. UR-3490-237.

Literature Cited

Bach, M. L. and F. Lacroute, 1972   Direct selective techniques for the isolation of pyrimidine auxotrophs in yeast. *Mol. Gen. Genet.* **115:**126–130.

Bard, M., 1972   Biochemical and genetic aspects of nystatin resistance in *Saccharomyces cerevisiae. J. Bacteriol.* **111:**649–657.

Baur, R., A. Bacher and F. Lingens, 1972   The structural gene for riboflavin synthetase in *Saccharomyces cerevisiae. FEBS (Fed. Eur. Biochem. Soc.) Lett.* **23:**215–216.

Beck, J. C., J. H. Parker, W. X. Balcavage, and J. R. Mattoon, 1970   Mendelian genes affecting development and function of yeast mitochondria. In *Autonomy and Biogenesis of Mitochondria and Chloroplasts,* edited by N. K. Boardman, A. W. Linnane, and R. M. Smilie, pp. 194–204, North-Holland, Amsterdam.

Beran, K., 1968   Budding of yeast cells, their scars and ageing. *Adv. Microbial. Physiol.* **2:**143–171.

Berry, E. A. and E. A. Bevan, 1972   A new species of double-stranded RNA from yeast. *Nature (Lond.)* **239:**279–280.

Bevan, E. A. and W. P. Costello, 1964   The preparation and use of an enzyme which breaks open yeast asci. *Microbiol. Genet. Bull.* **21:**5.

Bevan, E. A. and J. M. Somers, 1969   Somatic segregation of the killer (*k*) and neutral (*n*) cytoplasmic genetic determinants in yeast. *Genet. Res.* **14:**71–77.

Bhattacharjee, J. K. and A. K. Sinha, 1972   Relationship among the genes, enzymes, and intermediates of the biosynthetic pathway of lysine in *Saccharomyces*. *Mol. Gen. Genet.* **115:**26–30.

Blamire, J., D. R. Cryer, D. B. Finkelstein and J. Marmur, 1972   Sedimentation properties of yeast nuclear and mitochondrial DNA. *J. Mol. Biol.* **67:**11–24.

Bolotin, M., D. Coen, J. Deutsch, B. Dujon, P. Netter, E. Petrochilo and P. P. Slonimski, 1971   La Recombinaison des mitochondries chez *Saccharomyces cerevisiae*. *Bull. Inst. Pasteur* **69:**215–239.

Burns, V. W., 1956   Temporal studies of cell division. I. The influence of ploidy· and temperature on cell division in *S. cerevisiae*. *J. Cell. Comp. Physiol.* **47:**357–376.

Bussey, H., 1972   Effects of killer factor on sensitive cells. *Nat. New Biol.* **235:**73–75.

Calvo, J. M. and G. R. Fink, 1971   Regulation of biosynthetic pathways in bacteria and fungi. *Annu. Rev. Biochem.* **40:**943–968.

Catcheside, D. G., 1951   The Genetics of Micro-organisms, Pitman and Sons., London.

Coen, D., J. Deutsch, P. Netter, E. Petrochilo and P. P. Slonimski, 1970   Mitochondrial genetics. I. Methodology and phenomenology. *Symp. Soc. Exp. Biol.* **24:**449–496.

Cox, B. S., 1971   A recessive lethal super-suppressor mutation in yeast and other ψ phenomena. *Heredity* **26:**211–213.

Culotti, J. and L. H. Hartwell, 1971   Genetic control of the cell division cycle in yeast. III. Seven genes controlling nuclear division. *Exp. Cell Res.* **67:**389–401.

Demerec, M., E. A. Adelberg, A. J. Clark and P. E. Hartman, 1966   A proposal for a uniform nomenclature in bacterial genetics. *Genetics* **54:**61–76.

Denis-Duphil, M., and F. Lacroute, 1971   Fine structure of the *ura2* locus in *Saccharomyces cerevisiae*. I. *In vivo* complementation studies. *Mol. Gen. Genet.* **112:**354–364.

de Robichon-Szulmajster, H., and Y. Surdin-Kerjan, 1971   Nucleic acid and protein synthesis in yeast; regulation of synthesis and activity. In *The Yeasts*, Vol. 2, edited by A. H. Rose and J. S. Harrison, pp. 335–418, Academic Press, New York.

Douglas, H. C. and D. C. Hawthorne, 1972   Uninducible mutants in the *gal i* locus of *Saccharomyces cerevisiae*. *J. Bacteriol.* **109:**1139–1143.

Emeis, C. C. and H. Gutz, 1958   Eine einfache Technik zur Massenisolation von Hefesporen. *Z. Naturforsch. Teil B* **13b:**647–650.

Esposito, M. S., 1968   X-ray and meiotic fine structure mapping of the adenine-8 locus in *Saccharomyces cerevisiae*. *Genetics* **58:**507–527.

Esposito, R. E., N. Frink, P. Bernstein and M. S. Esposito, 1972   The genetic control of sporulation in *Saccharomyces*. II. Dominance and complementation of mutants of meiosis and spore formation.*Mol. Gen. Genet.* **114:**241–248.

Esser, K. and R. Kuenen, 1967   *Genetics of Fungi*, Springer-Verlag, New York.

Fincham, J. R. S. and P. R. Day, 1971   *Fungal Genetics*, F. A. Davis Co., Philadelphia, Pa.

Fink, G. R., 1966   A cluster of genes controlling three enzymes in histidine biosynthesis in *Saccharomyces cerevisiae*. *Genetics* **53:**445–459.

Fink, G. R., 1970   The biochemical genetics of yeast. *Methods Enzymol.* **17A:**59–78.

Fink, G. R., 1971   Gene clusters and the regulation of biosynthetic pathways in fungi. In *Metabolic Regulation*, edited by H. J. Vogel, pp. 199–223, Academic Press, New York.

Fink, G. R. and C. A. Styles, 1972   Curing of a killer factor in *Saccharomyces cerevisiae*. *Proc. Natl. Acad. Sci. USA* **69:**2846–2849.

Finkelstein, D. B., J. Blamire and J. Marmur, 1972   Location of ribosomal RNA cistrons in yeast. *Nat. New Biol.* **240:**279–281.

Fogel, S. and R. K. Mortimer, 1971   Recombination in yeast. *Annu. Rev. Genet.* **5:**219–236.

Fogel, S., D. D. Hurst and R. K. Mortimer, 1971   Gene conversion in unselected tetrads from multipoint crosses. In *The Second Stadler Symposium,* edited by G. Kimber and G. P. Rédei, pp. 89–110, University of Missouri, Agricultural Experiment Station, Columbia, Mo.

Fowell, R. R., 1969a   Sporulation and hybridization of yeasts. In *The Yeasts,* Vol. 1, edited by A. H. Rose and J. S. Harrison, pp. 303–383, Academic Press, New York.

Fowell, R. R., 1969b   Life cycles in yeast. In *The Yeasts,* Vol. 1, edited by A. H. Rose and J. S. Harrison, pp. 461–471, Academic Press, New York.

Game, J. C. and B. S. Cox, 1971   Allelism tests of mutants affecting sensitivity to radiation in yeast and a proposed nomenclature. *Mutat. Res.* **12:**328–331.

Gilmore, R. A., J. W. Stewart and F. Sherman, 1971   Amino acid replacements resulting from super-suppression of nonsense mutants of iso-1-cytochrome *c* from yeast. *J. Mol. Biol.* **61:**157–173.

Grenson, M. and C. Hennaut, 1971   Mutation affecting activity of several distinct amino acid transport systems in *Saccharomyces cerevisiae*. *J. Bacteriol.* **105:**477–482.

Grewal, N. S. and J. J. Miller, 1972   Formation of asci with two diploid spores by diploid cells of Saccharomyces. *Can. J. Microbiol.* **18:**1897–1906.

Gross, T. S. and R. A. Woods, 1971   Identification of mutants defective in the first and second steps of *de novo* purine synthesis in *Saccharomyces cerevisiae*. *Biochim. Biophys. Acta* **247:**13–21.

Hartwell, L. H., 1967   Macromolecular synthesis in temperature-sensitive mutants of yeast. *J. Bacteriol.* **93:**1662–1670.

Hartwell, L. H., 1970   Biochemical genetics of yeast, *Annu. Rev. Genet.* **4:**373–396.

Hartwell, L. H., 1971   Genetic control of the cell division cycle in yeast. II. Genes controlling DNA replication and its initiation. *J. Mol. Biol.* **59:**183–194.

Hartwell, L. H. and C. S. McLaughlin, 1968   Temperature-sensitive mutants of yeast exhibiting a rapid inhibition of protein synthesis. *J. Bacteriol.* **96:**1664–1671.

Hartwell, L. H. and C. S. McLaughlin, 1969   A mutant of yeast apparently defective in the initiation of protein synthesis. *Proc. Natl. Acad. Sci. USA* **62:**468–474.

Hartwell, L. H., J. Culotti and B. Reid, 1970a   Genetic control of the cell-division cycle in yeast. I. Detection of mutants. *Proc. Natl. Acad. Sci. USA* **66:**353–359.

Hartwell, L. H., C. S. McLaughlin and J. R. Warner, 1970b   Identification of ten genes that control ribosome formation in yeast. *Mol. Gen. Genet.* **109:**42–56.

Hartwell, L. H., R. K. Mortimer, J. Culotti and M. Culotti 1973   Genetic control of the cell division cycle in yeast: V. Genetic analysis of *cdc* mutants. *Genetics* **74:**267–286.

Hawthorne, D. C., 1963   Directed mutation of the mating type alleles as an explanation of homothallism in yeast. *Proc. 11th Int. Congr. Genet.* **1:**34–35.

Horn, P. and D. Wilkie, 1966   Use of Magdala Red for the selection of auxotrophic mutants of *Saccharomyces cerevisiae*. *J. Bacteriol.* **91:**1388.

Johnston, J. R. and R. K. Mortimer, 1959   Use of snail digestive juice in isolation of yeast spore tetrads. *J. Bacteriol.* **78:**292.

Jones, G. E., 1973   A fine-structure map of the yeast L-asparaginase gene. *Mol. Gen. Genet.* **121:**9–14.

Khan, N. A. and R. H. Haynes, 1972 Genetic redundancy in yeast: non-identical products in a polymeric gene system. *Mol. Gen. Genet.* **118**:279–285.

Kleese, R. A., R. C. Grotbeck and J. R. Snyder, 1972 Recombination among three mitochondrial genes in yeast (*Saccharomyces cerevisiae*). *J. Bacteriol.* **112**:1023–1025.

Kühn, L., H. Castorph and E. Schweizer, 1972 Gene linkage and gene–enzyme relations in the fatty-acid-synthetase of *Saccharomyces cerevisiae*. *Eur. J. Biochem.* **24**:492–497.

Lackowicz, T. M., Z. Kotylak, J. Kolodynski and Z. Sniegocka, 1969 New types of respiratory deficient mutants in *Saccharomyces cerevisiae*. II. Physiology and genetics of a series of segregational mutants induced by ultraviolet irradiation or nitrous acid treatment. *Arch. Immunol. Ther. Exp.* **17**:72–85.

Lacroute, F., 1971 Non-Mendelian mutation allowing ureidosuccinic acid uptake in yeast. *J. Bacteriol.* **106**:519–522.

Lancashire, W. E. and D. E. Griffiths, 1971 Biocide resistance in yeast: isolation and general properties of trialkyl tin resistant mutants. *FEBS Lett.* **17**:209–214.

Laskowski, W., 1960 Inaktwierungsversuche mit homozygoten Hefestämmen verschiedenen ploidiegrades. I. Aufbau homozygoter Stämme und Dosiseffektkurven für ionisierende Strahlen, UV und organische Peroxyde. *Z. Naturforsch. Teil B* **15b**:495–506.

Lemontt, J. F., 1971 Mutants of yeast defective in mutation induced by ultraviolet light. *Genetics* **68**:21–33.

Leupold, U., 1970 Genetical methods for *Schizosaccharomyces pombe*. *Methods Cell. Physiol.* **4**:169–177.

Lindegren, C. C., 1969 *The Yeast Cell, Its Genetics and Cytology*, Educational Publishers, St. Louis, Mo.

Linnane, A. W., J. M. Haslam, H. B. Lukins, and P. Nagley, 1972 The biogenesis of mitochondria in microorganisms. *Annu. Rev. Microbiol.* **26**:163–198.

Littlewood, B. S., 1972 A method for obtaining antibiotic-sensitive mutants in *Saccharomyces cerevisiae*. *Genetics* **71**:305–308.

MacKay, V. L., 1972 Genetic and functional analysis of mutations affecting sexual conjugation and related process in *Saccharomyces cerevisiae*. Ph.D. Thesis, Case Western Reserve University, Cleveland, Ohio.

McClary, D. O., W. L. Nulty and G. R. Miller, 1959 Effect of potassium versus sodium in the sporulation of *Saccharomyces*. *J. Bacteriol.* **78**:362–368.

Manney, T. R., 1970 Physiological advantage of the mechanism of the tryptophan synthetase reaction. *J. Bacteriol.* **102**:483–488.

Manney, T. R., and R. K. Mortimer, 1964 Allelic mapping in yeast by X-ray induced mitotic reversion. *Science* (*Wash., D.C.*) **143**:581–583.

Mortimer, R. K., 1958 Radiobiological and genetic studies on a polyploid series (haploid to hexaploid) of *Saccharomyces cerevisiae*. *Radiat. Res.* **9**:312–326.

Mortimer, R. K. and R. A. Gilmore, 1968 Suppressors and suppressible mutations in yeast. *Adv. Biol. Med. Phys.* **12**:319–331.

Mortimer, R. K. and D. C. Hawthorne, 1966 Yeast genetics. *Annu. Rev. Microbiol.* **20**:151–168.

Mortimer, R. K. and D. C. Hawthorne, 1969 Yeast genetics. In *The Yeasts*, Vol. 1, edited by A. H. Rose and J. S. Harrison, pp. 385–460, Academic Press, New York.

Mortimer, R. K. and D. C. Hawthorne, 1973 Genetic mapping in *Saccharomyces*. IV.

Mapping of temperature-sensitive genes and use of disomic strains in localizing genes and fragments. *Genetics* **74**:33–54.

Mortimer, R. K. and T. R. Manney, 1971   Mutation induction in yeast. In *Chemical Mutagens*, Vol. 1, edited by A. Hollaender, pp. 289–310, Plenum Press, New York.

Nagley, P. and A. W. Linnane, 1972   Biogenesis of mitochondria. XXI. Studies on the nature of the mitochondrial genome in yeast: The degenerative effects of ethidium bromide on mitochondrial genetic information in a respiratory competent strain. *J. Mol. Biol.* **66**:181–193.

Parker, J. H. and F. Sherman, 1969   Fine-structure mapping and mutational studies of gene controlling yeast cytochrome *c*. *Genetics* **62**:9–22.

Petes, T. D. and W. L. Fangman, 1972   Sedimentation properties of yeast chromosomal DNA. *Proc. Natl. Acad. Sci. USA* **69**:1188–1191.

Pomper, S. and P. R. Burkholder, 1949   Studies on the biochemical genetics of yeast. *Proc. Natl. Acad. Sci. USA* **35**:456–464.

Pomper, S., K. M. Daniels and D. W. McKee, 1954   Genetic analysis of polyploid yeast. *Genetics* **39**:343–355.

Rank, G. H. and N. T. Bech-Hansen, 1972   Somatic segregation, recombination, asymmetrical distribution and complementation tests of cytoplasmically-inherited antibiotic-resistance mitochondrial markers in *S. cerevisiae*. *Genetics* **72**:1–15.

Resnick, M. A., 1969   A photoreactivationless mutant of *Saccharomyces cerevisiae*. *Photochem. Photobiol.* **9**:307–312.

Resnick, M. A. and J. K. Setlow, 1972   Repair of pyrimidine dimer damage induced in yeast by ultraviolet light. *J. Bacteriol.* **109**:979–986.

Rodarte-Ramon, U. S. and R. K. Mortimer, 1972   Radiation-induced recombination in *Saccharomyces:* isolation and genetic study of recombination deficient mutants. *Radiat. Res.* **49**:133–147.

Roman, H., 1971   Induced recombination in mitotic diploid cells of *Saccharomyces*. In *Genetic Lectures*, Vol. 2, edited by R. Bogart, pp. 43–59, Oregon State University Press, Corvalis, Oregon.

Rose, A. H. and J. S. Harrison, editors, 1969   *The Yeasts*, Vol. 1: *Biology of the Yeasts*, Academic Press, New York.

Rose, A. H. and J. S. Harrison, editors, 1970   *The Yeasts*, Vol. 3: *Yeast Technology*, Academic Press, New York.

Rose, A. H. and J. S. Harrison, editors, 1971   *The Yeasts*, Vol. 2: *Physiology and Biochemistry of Yeasts*. Academic Press, New York.

Roth, R. and S. Fogel, 1971   A system selective for yeast mutants deficient in meiotic recombination. *Mol. Gen. Genet.* **112**:295–305.

Sager, R., 1972   *Cytoplasmic Genes and Organelles*. Academic Press, New York.

Sakai, K. and N. Yanagishima, 1971   Mating reaction in *Saccharomyces cerevisiae*. I. Cell agglutination related to mating. *Arch. Mikrobiol.* **75**:260–265.

Saunders, G. W., E. B. Gingold, M. K. Trembath, H. B. Lukins, and A. W. Linnane, 1971   Mitochondrial genetics in yeast: Segregation of a cytoplasmic determinant in crosses and its loss or retention in the petite. In *Autonomy and Biogenesis of Mitochondria and Chloroplasts*, edited by N. K. Boardman and A. W. Linnane, pp. 185–193, North-Holland, Amsterdam.

Schwaier, R., N. Nashed and F. K. Zimmermann, 1968   Mutagen specificity in the induction of karyotic versus cytoplasmic respiratory deficient mutants in yeast by nitrous acid and alkylating nitrosamides. *Mol. Gen. Genet.* **102**:290–300.

Sherman, F., 1967   The preparation of cytochrome-deficient mutants of yeast. *Methods Enzymol.* **10**:610–616.

Sherman, F., 1973   Micromanipulator for yeast studies. *Appl. Microbiol.* **28**:829.

Sherman, F. and H. Roman, 1963   Evidence for two types of allelic recombination in yeast. *Genetics* **48**:255–261.

Sherman, F. and P. P. Slonimski, 1964   Respiration-deficient mutants of yeast. II. Biochemistry. *Biochim. Biophys. Acta* **90**:1–15.

Sherman, F. and J. W. Stewart, 1971   Genetics and biosynthesis of cytochrome *c*. *Annu. Rev. Genet.* **5**:257–296.

Sherman, F. and J. W. Stewart, 1973   Mutations at the end of the iso-1-cytochrome *c* gene of yeast. In *The Biochemistry of Gene Expression in Higher Organisms,* edited by J. K. Pollak and J. W. Lee, pp. 56–86, Australian and New Zealand Book Co., Sydney.

Sherman, F., S. W. Liebman, J. W. Stewart and M. Jackson, 1973   Tyrosine substitutions resulting from suppression of amber mutants of iso-1-cytochrome *c* in yeast. *J. Mol. Biol.* **78**:157–168.

Sherman, F., M. Jackson, C. W. Lawrence and R. A. Gilmore, 1974   Mapping and orientation of the gene determining iso-1-cytochrome *c* in yeast. (in manuscript).

Siddiqi, B. A., 1971   Random-spore analysis in *Saccharomyces cerevisiae. Hereditas* **69**:67–76.

Snow, R., 1966   An enrichment method for auxotrophic yeast mutants using the antibiotic "nystatin." *Nature (Lond.)* **211**:206–207.

Snow, R. and C. T. Korch, 1970   Alkylation induced gene conversion in yeast: Use in fine structure mapping. *Mol. Gen. Genet.* **107**:201–208.

Somers, J. M. and E. A. Bevan, 1969   The inheritance of the killer character in yeast. *Genet. Res.* **13**:71–83.

Strobel, R. and K. Wöhrmann, 1972   Untersuchungen zur Genetik eines Esterasepolymorphismus bei *Saccharomyces cerevisiae. Genetica (The Hague)* **43**:274–281.

Takano, I. and Y. Oshima, 1967   An allele specific and a complementary determinant controlling homothallism in *Saccharomyces oviformis. Genetics* **57**:875–885.

Takano, I. and Y. Oshima, 1970   Mutational nature of an allele specific conversion of the mating type by the homothallic gene HOα *Saccharomyces. Genetics* **65**:421–427.

ten Berge, A. M., 1972   Genes for the fermentation of maltose and α-methyl-glucoside in *Saccharomyces carlsbergensis. Mol. Gen. Genet.* **115**:80–88.

Thouvenot, D. R. and C. M. Bourgeois, 1971   Optimisation de la sélection de mutants de *Saccharomyces cerevisiae* par la nystatine. *Ann. Inst. Pasteur* **120**:617–625.

Throm, E., and W. Duntze, 1970   Mating-type-dependent inhibition of deoxyribonucleic acid synthesis in *Saccharomyces cerevisiae. J. Bacteriol.* **104**:1388–1390.

Thuriaux, P., F. Ramos, A. Piérard, M. Grenson and J. M. Wiame, 1972   Regulation of the carbamoylphosphate synthetase belonging to the arginine biosynthetic pathway of *Saccharomyces cerevisiae. J. Mol. Biol.* **67**:277–287.

Ulane, R. and M. Ogur, 1972   Genetic and physiological control of serine and glycine biosynthesis in *Saccharomyces. J. Bacteriol.* **109**:34–43.

Unrau, P., R. Wheatcroft and B. S. Cox, 1971   The excision of pyrimidine dimers from DNA of ultraviolet irradiated yeast. *Mol. Gen. Genet.* **113**:359–362.

von Borstel, R. C., R. K. Mortimer and W. E. Cohn, 1970   Genetic markers and associated gene products in *Saccharomyces.* In *Handbook of Biochemistry,* second edition, edited by H. A. Sober, pp. I-83–87, The Chemical Rubber Co., Cleveland, Ohio.

Warner, J. R. and S. A. Udem, 1972 Temperature-sensitive mutations affecting ribosome synthesis in *Saccharomyces cerevisiae. J. Mol. Biol.* **65**:243–257.

Wiame, J. M., 1971 The regulation of arginine metabolism in *Saccharomyces cerevisiae:* exclusion mechanisms. *Curr. Top. Cell. Regul.* **4**:1–38.

Wickerham, L. J., 1946 A critical evaluation of the nitrogen assimilation tests commonly used in the classification of yeasts. *J. Bacteriol.* **52**:293–301.

Wilkie, D. and D. Y. Thomas, 1973 Mitochondrial genetical analysis by zygote cell lineages in *Saccharomyces cerevisiae. Genetics* **73**:367–377.

Winge, Ö. and C. Roberts, 1949 A gene for diploidization in yeast. *C. R. Trav. Lab. Carlsberg Ser. Physiol.* **24**:341–346.

Winge, Ö. and C. Roberts, 1958 Yeast genetics. In *Chemistry and Biology of Yeasts,* edited by A. H. Cook, pp. 123–156, Academic Press, New York.

Wright, R. E. and J. Lederberg, 1957 Extranuclear transmission in yeast heterokaryons. *Proc. Natl. Acad. Sci. USA* **43**:919–923.

Young, C. S. H. and B. S. Cox, 1972 Extrachromosomal elements in a super-suppression system of yeast. II. Relations with other extrachromosomal elements. *Heredity* **28**:189–199.

Zimmermann, F. K., 1968 Sensitivity to methylmethanesulfonate and nitrous acid of ultraviolet light-sensitive mutants in *Saccharomyces cerevisiae. Mol. Gen. Genet.* **102**:247–256.

Zimmermann, F. K., R. Schwaier and U. von Laer, 1966 The effect of residual growth on the frequency of reverse mutations induced with nitrous acid and 1-nitroso-imidazolidone-2 in yeast. *Mutat. Res.* **3**:171–173.

25

Schizosaccharomyces pombe

HERBERT GUTZ, HENRI HESLOT, URS LEUPOLD,
AND NICOLA LOPRIENO

Life Cycle and Mating-Type System

The fission yeast *Schizosaccharomyces pombe* Lindner has a haplontic life cycle (Leupold, 1950, 1958). The vegetative cells are haploid. When cells of compatible mating types are grown together, a strong sexual agglutination occurs at the end of vegetative growth, followed by a pairwise copulation of cells. In the resulting zygotes, karyogamy is followed immediately by meiosis, and four haploid ascospores are formed within the cell wall of the original zygote (zygotic asci). Thus, the diplophase is normally confined to the zygote. However, diploid strains can also be selected. The life cycle of *S. pombe* is shown in Figure 1. The genetic methods used with *S. pombe* as well as the major results obtained with this yeast have previously been reviewed by Leupold (1970). The gene loci which are known in *S. pombe* are listed in Table 1.

The *S. pombe* strains used in genetic research were isolated by Leupold (1950) from a culture of *S. pombe* Lindner str. *liquefaciens* (Osterwalder) which he had obtained from the Yeast Division of the "Centraalbureau voor Schimmelcultures" in Delft. From this culture,

HERBERT GUTZ—Institute for Molecular Biology, The University of Texas at Dallas, Richardson, Texas. HENRI HESLOT—Institut National Agronomique, Paris-Grignon, France. URS LEUPOLD—Institut für allgemeine Mikrobiologie der Universtät Bern, Bern, Switzerland. NICOLA LOPRIENO—Istituto di Genetica dell' Università Pisa e Laboratorio di Mutagenesi e Differenziamento del Consiglio Nazionale delle Richerche, Pisa, Italy.

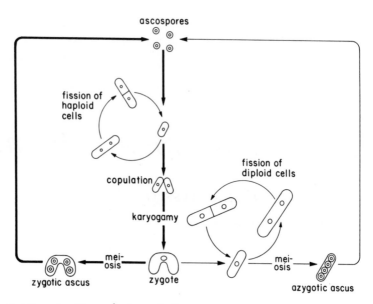

Figure 1. Life cycle of S. pombe (homothallic strain). The left part of the figure (heavy lines) shows the normal haplontic life cycle. The right part (thin lines) demonstrates the events which take place when zygotes develop to diploid cells. In the case of heterothallic strains, copulation occurs only when cultures of opposite mating types are mixed. The mating-type genes segregate in a 2:2 ratio in each ascus.

Leupold was able to isolate heterothallic strains representing two opposite mating types, h^+ and h^-, and homothallic strains. The latter strains are self-fertile as well as cross-fertile with heterothallic + and − strains. Leupold (1950) obtained two homothallic mating types, h^{90} and h^{40}, which were characterized by the formation of 90 percent and 40 percent ascospores, respectively. However, the h^{40} strain has since been lost.

The presence of spores in *S. pombe* colonies can easily be made visible by treating the plates briefly with iodine vapors (Leupold, 1955a). After this treatment, sporulating colonies turn black because the ascospores contain an amyloselike substance which is absent in vegetative cells. However, an exception is a similar iodine reaction exhibited by *vir*1 mutants (Gutz and J. H. Meade, unpublished). Heterothallic *vir*1 colonies, although consisting only of vegetative cells, turn brown after exposure to iodine vapors.

In 1958, Leupold found that in his *S. pombe* strains, two closely linked mating-type genes are present which cooperate in the determination of the mating reaction. He was able to distinguish four mating types: one homothallic, two heterothallic +, and one heterothallic −. The two + mating types were called h^{+N} and h^{+R}. Recently, Gutz and Doe

TABLE 1.   *Gene Loci Known in S. pombe*

Mating type	*mat*1		*gua*2
	*mat*2		*aza*1
Auxiliary genes	*mam*1		*dap*1
	*mam*2		*pur*1
	*map*1		*dea*2
	*map*2		*ins*1
			*ins*2
Copulation	*fus*1		
		Pyrimidines	*ura*1
Meiosis	*meI*1		*ura*2
	*meI*2		*ura*3
	*meI*3		*ura*4
	*meI*4		*ura*5
	*meII*1		
		Arginine	*arg*1
Sporulation	*spo*1		*arg*2
	*spo*2		*arg*3
	*spo*3		*arg*4
	*spo*4		*arg*5
	*spo*5		*arg*6
	*spo*6		
	*spo*7	Glutamic acid	*glu*1
	*spo*8		
	*spo*9	Histidine	*his*1
	*spo*10		*his*2
	*spo*11		*his*3
	*spo*12		*his*4
	*spo*13		*his*5
	*spo*14		*his*6
	*spo*15		*his*7
	*spo*16		*his*8
	*spo*17		*his*9
	*spo*18		
		Leucine	*leu*1
Vegetative iodine reaction	*vir*1		*leu*2
			*leu*3
Purines	*ade*1		
	*ade*2	Lysine	*lys*1
	*ade*3		*lys*2
	*ade*4		*lys*3
	*ade*5		*lys*4
	*ade*6		*lys*5
	*ade*7		
	*ade*8	Methionine	*met*1
	*ade*9		*met*2
	*ade*10		*met*3
	*gua*1		*met*4

TABLE 1. *Continued*

Proline	*pro*1		*rad*9
			*rad*10
Aromatic amino acids	*aro*1		*rad*11
	*aro*2		*rad*12
	*trp*1		*rad*13
	*trp*2		*rad*14
	*trp*3		*rad*15
	*trp*4		*rad*16
			*rad*17
p-Aminobenzoic acid	*pab*1		*rad*18
			*rad*19
Thiamine	*thi*1		*rad*20
			*rad*21
Alcohol dehydrogenase	*adh*1		*rad*22
Hexokinase	*hex*1	Mutators	*mut*1
			*mut*2
Mannosephosphate iso-merase	*man*1		*mut*3
		Suppressors	*sup*1
Acid phosphatase	*pac*1		*sup*2
			*sup*3
DNA, RNA, protein syn-thesis	*tsl*1		*sup*4
	*tsl*2		*sup*5
	*tsl*3		*sup*6
	*tsl*4		*sup*7
	*tsl*5		*sup*8
	*tsl*6		*sup*9
	*tsl*7		*sup*10
	*tsl*8	Antisuppressors	*sin*1
	*tsl*9		*sin*2
	*tsl*10		*sin*3
	*tsl*11		*sin*4
	*tsl*12		*sin*5
			*sin*6
Cycloheximide resistance	*cyh*1		*sin*7
			*sin*8
Radiation sensitivity	*rad*1		*sin*9
	*rad*2		*sin*10
	*rad*3		*sin*11
	*rad*4		*sin*12
	*rad*5		*sin*13
	*rad*6		*sin*14
	*rad*7		
	*rad*8		

(1973) described a second heterothallic − mating type, h^{-U}. The different strains and their genotypes are shown in Table 2. In Leupold's (1958) two-gene scheme, it is assumed that the mating-type genes *mat*1 and *mat*2 have a similar function. For *mat*1 alleles of opposite heterothallic activities (+ or −) and for *mat*2, an active + allele and an inactive allele are postulated. Two haploid strains are compatible with each other if the first has at least one active + and the second at least one active − gene. A homothallic strain results in the case of $mat1^-$ $mat2^+$, i.e., if one mating-type gene with a − and one with a + activity are present in the same haploid cell. For further discussion of Leupold's scheme and the possible existence of additional mating types, see Gutz and Doe (1973). In diploid strains, the mating-type genes determine the capability to copulate and/or to undergo meiosis (see below).

As indicated in Table 2, four of the five mating types listed show mutations to other mating activities (Leupold, 1950, 1958; Bresch *et al.*, 1968; Gutz and Doe, 1973). The genetic basis of most of these mutations is not yet known. It is only safe to say that mutations from $mat1^+$ to $mat1^-$ occur relatively frequently (Leupold, 1958; Gutz and Doe, 1973). In h^{90} cultures, heterothallic strains originate not only from mutations in the mating-type genes, but also from mutations in auxiliary genes of mating-type expression (Table 3) (Egel, 1969, 1973*b*). The mutations

TABLE 2.   Mating Types of S. pombe[a]

Mating reaction	Mating-type symbol	Genotype[b]	Spontaneous mutations to[c]
Homothallic	h^{90}	$mat1^-$ $mat2^+$	Heterothallism + Heterothallism −
Heterothallic +	h^{+N}	$mat1^+$ $mat2^+$	Homothallism Heterothallism −
	h^{+R}	$mat1^+$ $mat2°$	Heterothallism −[d]
Heterothallic −	h^{-S}	$mat1^-$ $mat2°$	None
	h^{-U}	$mat1°$ $mat2^-$ (?)	Homothallism Heterothallism +

[a] Mating types h^{90}, h^{+N}, h^{+R}, h^{-S} and their genotypes as postulated by Leupold (1958). The symbol h^{-S} was suggested by Gutz and Doe (1973) for the stable − mating type instead of Leupold's symbol h^-. The unstable − mating type, h^{-U}, was found by Gutz and Doe (1973); the shown genotype is hypothetical.

[b] The genes *mat*1 and *mat*2 are 1.1 map units apart and cooperate in the determination of the mating reaction. The superscripts, + and −, indicate alleles of opposite heterothallic function, whereas ° indicates an inactive allele. Originally, Leupold (1958) used the gene symbols *h*1 and *h*2; Gutz and Doe (1973) suggested that these be replaced by *mat*1 and *mat*2.

[c] Compiled from Leupold (1958), Bresch *et al.* (1968), and Gutz and Doe (1973). With regard to mutations in h^{90} strains, see also text.

[d] These − mutants are stable like h^{-S}.

TABLE 3. Genes Regulating the Sexual Cycle of S. pombe

Gene	Properties [a]	References
Mating type		
mat1	Polymeric genes; complimentary	Leupold, 1958; see
mat2	heterothallic activities are necessary for copulation and meiosis	also footnote b of Table 1
Auxiliary genes		
mam1	No agglutination with h^+ cells, fertile with h^- strains	Egel, 1969, 1973b
map1	No agglutination with h^- cells, fertile with h^+ strains, no meiosis in map1/map1 diploids	Egel, 1969, 1973b
mam2 [b]	Agglutination, no copulation with h^+ cells, fertile with h^- strains	Egel, 1969, 1973b
map2 [b]	Agglutination, no copulation with h^- cells, fertile with h^+ strains	Egel, 1969, 1973b
Fusion		
fus1	After start of copulation there is no dissolving of cell walls	Bresch et al., 1968; Egel, 1969, 1973a,b
Meiosis		
meI1 [c]	Cell fusion and karyogamy; no	Bresch et al., 1968;
meI2	meiosis, zygotes develop to dip-	Egel, 1969, 1973a
meI3	loid colonies wen transferred to a fresh medium	
meI4	No meiosis, zygotes do not develop to diploid colonies	Bresch et al., 1968; Egel, 1969, 1973a
meII1	No second meiotic division	Bresch et al., 1968
Sporulation		
spo1–spo18	Zygotes contain four nuclei; no ascospores are formed	Bresch et al., 1968

[a] For mat1 and mat2, the functions of the active genes are listed. For the other genes, the properties of h^{90} strains are described which have mutations in these loci.

[b] Egel (1969) had called these genes mam6 and map6, but later on he changed the designations to mam2 and map2, respectively (Egel, 1973b).

[c] The meI1 mutants map in the mating-type region.

from h^{+N} to h^{90} are of practical importance, since old h^{+N} stock cultures always have accumulated h^{90} mutants.

In general, h^{+N} and h^{-S} strains are used in genetic experiments with S. pombe; most authors use only the symbols h^+ and h^- when such strains are employed. The h^{+N} and h^{-S} wild-type cultures of Leupold carry the strain numbers L975 and L972, respectively.

The physiology of sexual agglutination, copulation, and sporulation has been studied by Egel (1969, 1971). Bresch et al. (1968) selected 300 nonsporulating mutants from a h^{90} strain. Among these mutants, Bresch et al. (1968) and Egel (1969, 1973a,b) have identified 28 separate genes which, in addition to the mating-type genes, are involved in the different steps of the sexual cycle of *S. pombe,* from the onset of agglutination to the formation of spores (Table 3). The *meI*1 mutations are closely linked to the mating-type genes (Bresch *et al.,* 1968). One particular *meI*1 mutant, B102, appears to have a mutation in *mat2* which inactivates the function of this gene with respect to meiosis but not with respect to copulation (Egel, 1969, 1973a).

S. *pombe* strains always contain a few diploid cells which presumably originate by endomitosis (Leupold, 1955a). In crosses with haploid heterothallic strains of opposite mating types, these diploid cells will copulate with haploid cells of the partner strain so as to form rare triploid zygotes. The latter yield the various types of tetrad segregations characteristic of triploids (Leupold, 1956; Ditlevsen and Hartelius, 1956).

Diploid S. *pombe* cultures can be easily selected by crossing haploid strains with complementary growth requirements and subsequently plating the young zygotes on minimal medium. Part of the zygotes will develop to diploid colonies which will sporulate at the end of vegetative growth. The diploid cells are transformed directly to asci (i.e., without copulation as an initiating event); these asci are called azygotic (Leupold, 1970). On the basis of their morphology, azygotic asci can easily be differentiated from zygotic asci. Nonsporulating, and therefore stable, diploid strains can be obtained by replating vegetative cells from h^{+N}/h^{-S} colonies. Due to mitotic crossing over between the centromere and the mating-type region, some homzygous h^+/h^+ and h^-/h^- cells will arise. These cells, in contrast to the original h^+/h^- cells, have lost the capacity to self-sporulate; they will form nonsporulating (iodine-negative) sectors (Leupold, 1970). The procedures for isolating diploid strains have been described in detail by Flores da Cunha (1970).

Angehrn and Gutz (1968, and unpublished) have found that h^{90}/h^{-S} cultures and other diploid strains being *mat1*⁻/*mat1*⁻, show frequent mitotic crossings over (approximately 1 in 45 cell divisions) which are strictly localized between *his*7 and *mat2* (cf. the chromosome map, Figure 2). This specific event is not exhibited by *mat1*⁺/*mat1*⁺ and *mat1*⁺/*mat1*⁻ strains. The *mat2* genotype has no influence on the frequency of mitotic crossover in the mating-type region. Another peculiarity of diploid cultures is the spontaneous occurrence of recessive lethal mutations which appear to map in *mat2*. These mutations originate approximately once in 360 cell divisions (Gutz and Angehrn, 1968, and unpublished).

In diploid S. *pombe* strains, the capability to undergo meiosis (as evidenced by the formation of azygotic asci) and/or to copulate is determined by the two mating-type genes. If a strain has at least one active + and one active − gene, it is able to undergo meiosis (this rule is equivalent to that stated above for the compatibility of haploid strains). Diploid strains having only active + or active − mating-type genes cannot self-sporulate, but they are able to copulate and sporulate with strains of compatible mating types (Leupold, 1956, 1958). The ability to undergo meiosis does not necessarily exclude the ability to copulate. For example, h^{90}/h^{90} cells are ambivalent in this respect: they are able to form azygotic asci or, alternatively, to copulate with other cells (Gutz, 1967a,b).

When diploid cells copulate with each other, their nuclei may or may not fuse. In the first case (karyogamy), a tetraploid meiosis takes place in the zygotes and four diploid spores are formed. In the second case (no karyogamy), the separate diploid nuclei undergo individual meioses (twin meiosis) in the common cytoplasm, and asci with eight haploid spores are formed. The relative frequencies of these alternative events depend on the physiological conditions under which the diploid strains are crossed (Gutz, 1967a,b; cf. Leupold, 1956).

Several papers which are related to the physiology and cytology of S. pombe are also of interest to the geneticist. A summary of these topics, with the focus on the vegetative cell cycle, has been published by Mitchison (1970) (see also Mitchison and Creanor, 1971a,b). During exponential growth, S. pombe cells have no detectable G1 phase and only a short S phase. The DNA content of stationary-phase cells varies with the phosphate content of the medium. At low phosphate concentrations, cells stop growing in G1, whereas in high phosphate concentrations growth ceases in G2. The DNA content per haploid spore is 1.46×10^{-14} g (Bostock, 1970). Colcemid inhibits several stages of the cell cycle; mutants which are resistant to this drug have been isolated (Lederberg and Stetten, 1970; Stetten and Lederberg, 1973). Barras (1972) has found a β-glucan endohydrolase which is closely associated with the cell walls of growing S. pombe cells. Shahin (1971) and Foury and Goffeau (1973) have published methods for preparation of protoplasts.

McCully and Robinow (1971) have made a detailed cytological study of mitosis in S. pombe. Duffus (1969, 1971) has isolated nuclei and nucleohistones from this yeast. A method for rapid preparation of DNA molecules for electron microscopy has been described by Egel-Mitani and Egel (1972).

Chromosome Map

The genetic markers which are known in S. pombe are listed in Tables 1–7. In this yeast, mapping can be carried out either by meiotic or mitotic analysis. Quantitative meiotic recombination frequencies can be obtained by tetrad analyses or by plating of free ascospores (cf. Leupold, 1970). Two methods are used in mitotic mapping: haploidization of diploid cells by parafluorophenylalanine, and mitotic crossing over (Gutz, 1966; Flores da Cunha, 1970; Kohli, 1973). The principles of mitotic mapping have been discussed by Pontecorvo (1958).

The first detailed linkage group in *S. pombe,* the one carrying the mating-type genes, was established by Leupold (1958). Flores da Cunha (1970), employing mitotic mapping, was able to assign 30 genetic markers

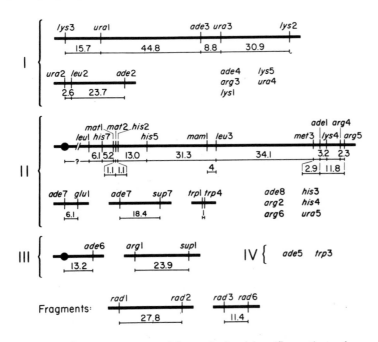

Figure 2. Chromosome map of S. pombe involving 47 genetic markers (redrawn from Kohli, 1973). Forty-three markers have been assigned to four different chromsomes, I–IV, by haploidization and/or mitotic crossover. Meiotic map distances are shown for those markers for which they have been determined. Centromeres are indicated by circles. With respect to chromsome II, it is known from mitotic crossover experiments that ade7, glu1, his3, arg2, arg6, ura5, and sup7 are located between the centromere and the mating-type loci (mat1, mat2). ade8, his4, trp1, and trp4 map distal to mat1 and mat2. For mam1 it is not known whether its position is proximal or distal to leu3. On chromosome III, arg1 is located distal to ade6.

The map is based mainly on data of Leupold (1958, unpublished), Flores da Cunha (1970), and Kohli (1973), but also includes results of Ahmad, Barben, Dietrich, Egel, and Schüpbach (for complete references, see Kohli, 1973).

to four separate chromosomes (for two inconsistencies in Flores da Cunha's paper, see the next paragraph). Other linkage relationships were determined by several authors (see Kohli, 1973). Recently, a substantial extension of the chromosome map was made by Kohli (1973). The present state of mapping in *S. pombe* is shown in Figure 2, 43 loci have been allocated to four different chromosomes, chromosome II carrying more than half of the mapped genes.

Two errors in Flores da Cunha's (1970) publication should be mentioned. Flores da Cunha worked with 32 genes which he allocated to six different chromosomes. However, Kohli (1973) found some discrepancies with respect to the "*arg2*" and "*arg3*" strains used by Flores da Cunha. Kohli could not confirm chromosome IV on which, according to Flores da Cunha, "*arg3*" was the only marker. Nor could Flores da Cunha's chromosome V be confirmed, since the only marker representing this chromosome, *arg6*, turned out to be allelic with *glu1* (P. Thuriaux, unpublished), which is located on chromosome II (Flores da Cunha, 1970; Kohli, 1973). Like other mutants mapping at the *glu1* (glutamate) locus (Megnet, unpublished), the mutant which was originally classified as an arginine mutant of constitution *arg6* (Leupold, unpublished) responds to a number of other amino acids as well. In order to have consecutive chromosome numbers in Figure 2, the following redesignations have been made: Flores da Cunha's chromosome VI is now IV, and his chromosomes IV and V have been deleted. Leupold's *arg7* has been renamed *arg6*.

In stained cells of *S. pombe*, McCully and Robinow (1971) have observed 5–7 rodlets which they regard as chromosomes. This observation does not conflict with the genetically determined number of four chromosomes in the haploid complement. The latter number is, of course, only a minimal estimate; mapping of additional genes may lead to the detection of new chromosomes. At present, only approximately $\frac{1}{3}$ of the known *S. pombe* genes have been mapped.

Fine-Structure and Complementation Maps

S. pombe is a suitable eukaryotic organism for experiments on genetic fine structure and intragenic recombination. It is possible to plate large numbers of free ascospores on minimal agar (10^6–10^7 spores per plate); thus, rare intragenic meiotic recombinants can easily be detected in crosses between nonidentical alleles. The total number of spores plated on minimal agar are determined by plating diluted spore samples on enriched medium (Leupold, 1970; Gutz, 1971*b*). Fine-structure analyses can also be performed by means of intragenic mitotic recombination (e.g., Leupold, 1958; Adondi and Heslot, 1970).

The methods employed for meiotic fine-structure mapping will at the same time give information on the occurrence of interallelic complementation between heteroallelic mutations. Since in the haploid parental cultures a few diploid cells are always present, rare triploid zygotes are formed; these will yield, with a frequency of 1 in 10^3–10^4 ascospores, diploid and aneuploid spores which are heterozygous for the parental alleles. If the two alleles are capable of interallelic complementation, the heterozygous ascospores will form colonies on minimal medium. Colonies resulting from interallelic complementation rather than from recombination are characterized by their variable colony size (normal-sized colonies are formed only in cases of strong complementation) and by their deeper staining on minimal medium containing vital stains like magdala red (this is due to the higher percentage of dead cells present in diploid and aneupolid colonies in comparison to those in haploid colonies). If, as it is generally the case, heterothallic strains have been crossed, the most useful criterion for the detection of interallelic complementation is the presence of azygotic asci in part of the colonies on minimal medium. Azygotic asci will be formed in those colonies which, in addition to carrying both heteroalleles under examination, are also heterozygous for the mating types. The sporulating colonies can be detected by exposing the plates to iodine vapors. However, it should be noted that in cases of weak complementation (i.e., if only small colonies are formed), sporulation may be absent (Leupold, 1970; Gutz, 1963; Megnet and Giles, 1964). For a general discussion of interallelic complementation and of the principles employed in the drawing of complementation maps, see Leupold (1961), Fincham (1966), Fincham and Day (1971), and Gillie (1966, 1968).

In *S. pombe*, extensive fine-structure analyses have been made for the genes *ade*7, *ade*6, and *ade*1. Mutants of *ade*7 and *ade*6 form a red pigment if grown on media with limiting amounts of adenine. Using *ade*7 and *ade*6 mutants induced by ultraviolet light, nitrous acid, and x rays, Leupold (1958, 1961), Gutz (1961, 1963), and Leupold and Gutz (1965) have demonstrated the occurrence of mutagen specific hot spots in both genes. Fifty two different mutation sites have been identified in the *ade*7 locus, and 120 in the *ade*6 locus. Loprieno (1967) has mapped additional *ade*7 mutants induced by alkylating agents or hydroxylamine.

Part of the *ade*6 and *ade*1 mutants show interallelic complementation if crossed with strains carrying appropriate nonidentical alleles; among the *ade*7 mutants this phenomenon was not observed (Leupold, 1961; Gutz, 1963; Friis *et al.*, 1971). The complementation map of the *ade*6 locus has a complicated geometrical shape (Leupold and Gutz, 1965; Gutz, 1963). Frame-shift mutations are expected either to be noncomplementing or, if they are complementing, to exhibit only polarized complementation patterns. Munz and Leupold (1970) studied *ade7* and *ade6* mutants induced by ICR-170; this mutagen is supposed to induce frameshift mutations. Part of Munz and Leupold's *ade6* mutants, besides having other qualities characteristic of frame-shift mutations, behaved with regard to interallelic complementation as stated above. Loprieno (1969), working with temperature-sensitive *ade6* mutants, reported some cases of negative interallelic complementation (cf. Fincham, 1966; Fincham and Day, 1971). Further topics related to the fine structure of *ade7* and *ade6* are discussed in the sections Biochemical Genetics and Suppression.

Both the recombination and complementation maps of the *ade1* locus show a characteristic bipartite structure (Clarke, 1965c; Ramirez *et al.*, 1974) which is paralleled by two different activities of the enzyme coded by *ade1*. The features of this gene are described in more detail in the sections Biochemical Genetics and Suppression.

Fine-structure analyses have also been performed for the genes *ade2* (Treichler, 1964), *ade8* (Leupold, 1961; Angehrn, 1964; Megnet and Giles, 1964), and *ade9* (Adondi and Heslot, 1970), as well as for *ade4*, *ade10*, *his2*, and *pac1*. The latter four genes are discussed in the section Biochemical Genetics, where the pertinent references are also given. Among the above genes, interallelic complementation has been found in *ade2*, *ade8*, *ade10*, *his2*, and *pac1*.

In connection with fine-structure mapping, several topics related to the mechanism of genetic recombination should be mentioned. For a general discussion of genetic recombination, see Holliday (1968), Fincham and Day (1971), and Gutz (1971a,b). In all cases in whch intragenic three-factor crosses have been analyzed, strong negative interference has been observed (*ade2*: Treichler, 1964; *ade6*: Gutz, 1963; Angehrn, 1964; *ade8*; Angehrn, 1964). Most meiotic fine-structure maps show some degree of map expansion, i.e., pairs of relatively distant sites give more recombination than one would expect from adding the intervals delimitated by intervening sites. The distinct map expansion observed in the *ade6* locus has been discussed by Holliday (1968).

Angehrn (1964) studied the intragenic recombination in *ade6* and *ade8* by meiotic tetrad analyses as well as by mitotic half-tetrad analyses; Treichler (1964) did the same for *ade2*. In all three loci, the meiotic and mitotic recombination was mostly nonreciprocal, i.e., the formation of prototrophic spores or prototrophic diploid cells was mainly due to gene conversion at one of the two heterozygous sites. The nonreciprocal meiotic recombination in *ade6* and *ade2* was partially polarized. Nonreciprocal mitotic recombination has also been observed by Leupold (1958) in the *ade7* locus and by Adondi and Heslot (1970) in the *ade9* locus.

Gutz (1971*b,c*) studied an *ade6* mutant, M26, which shows a specific marker effect with respect to meiotic intragenic recombination and gene conversion in this locus. For example, M26, if crossed with the wild-type or nonidentical alleles, yields more conversion tetrads than any other *ade6* mutant tested. At the M26 site, conversions from mutant to wild type are 12 times more frequent than in the opposite direction. The marker effect of this mutant can be best explained by assuming that single DNA strand breaks occur preferentially at the M26 site which are starting points for gene-conversion events. In *S. pombe,* other interesting marker effects with regard to intragenic recombination have been observed in experiments with suppressor genes; pertinent references are given in the section Suppression.

Irradiation with x rays or ultraviolet light increases the frequency of mitotic intragenic recombination; the increase is dependent on the radiation dose. The slopes of recombinant induction curves vary for different allele pairs; the slopes increase with increasing map distance (for references, see Adondi and Heslot, 1970; Fincham and Day, 1971). In *S. pombe,* Adondi and Heslot (1970) have applied x rays and ultraviolet light for mitotic mapping of four sites within the *ade9* gene. As has been already mentioned, the recombinants originated by nonreciprocal events (mitotic gene conversion).

Addition of fluorouracil or fluorodeoxyuridine to the crossing medium was found to increase intergenic as well as intragenic meiotic recombination; the genetic regions studied were *mat1–mat2* and the *ade2* locus (Megnet, 1966*a*). Caffeine reduces recombination between *his7* and *his2* (Loprieno and Schüpbach, 1971).

Abbreviations

AICAR	5-Aminoimidazole-4-carboxyamide nucleotide
AIR	5-Aminoimidazole ribonucleotide
CAIR	5-Aminoimidazole-4-carboxylic acid ribonucleotide
DAHP	3-Deoxy-D-arabino heptulosonate-7-phosphate
FAICAR	5-Formamidoimidazole-4-carboxamide ribonucleotide
FGAM	N-Formylglycinamidine ribonucleotide
FGAR	N-Formylglycinamide ribonucleotide
THFA	Tetrahydrofolate

GAR	Glycinamide ribonucleotide
PRA	5-Phosphoribosyl-1-amine
PRPP	5-Phosphoribosyl-1-pyrophosphate
SAICAR	5-Aminoimidazole-4-N-succinocarboxamide ribonucleotide
s-AMP	Adenylosuccinate
AMP	Adenosine 5′ monophosphate
GMP	Guanosine 5′ monophosphate
IMP	Inosine 5′ monophosphate
XMP	Xanthosine 5′ monophosphate
A	Adenine
G	Guanine
Hx	Hypoxanthine
X	Xanthine

Mutation and Repair

Studies on spontanous and induced mutation have been carried out in a number of gene loci of *S. pombe* with two essentially distinct scoring systems. A widely used system is based on the selection of prototrophic cells arising from auxotrophic cell populations grown on a medium lacking the required nutritional factor. With this method, similar to equivalent mutagenic systems used in other organisms, prototrophic reversions occuring at several *ade* and other loci have been studied (*ade*1: Clarke, 1962, 1963, 1965*a,b*; Segal *et al.*, 1973; Friis *et al.*, 1971; *ade*6: Loprieno, 1973; Munz and Leupold, 1970; *ade*7: Clarke, 1968; Michel *et al.*, 1966; *arg*1: Heslot, 1962; *his*2: Loprieno and Schüpbach, 1971; *met*4: Clarke, 1968; Guglielminetti *et al.*, 1966; Loprieno, 1964; Loprieno and Clarke, 1965; *leu*3: Heslot, 1962; *ura*1: Heslot, 1962; *his*7: Loprieno and Schüpbach, 1971). The second mutational system is based on scoring a change in the phenotype of the colonies arising from a mutant cell. There exist two experimental possibilities: (1) When mutations are induced in the wild-type strain at any site in the locus *ade*6, which controls the conversion of AIR* to CAIR, or in the locus *ade*7, which controls the conversion of CAIR into SAICAR, the phenotype of the colonies changes from white (wild type) to red (*ade*6⁻ or *ade*7⁻) due to the accumulation of a red pigment which is a polymer derived from AIR (see section Biochemical Genetics). This phenotypic change is easily scored on media such as yeast-extract agar which contains a limiting amount of adenine. (2) Strains carrying a mutation in the *ade*6 or *ade*7 locus, and which therefore form red

---

* Abbreviations are defined in the previous section.

colonies, change their phenotype to white if additional mutations occur in one of the five genes which control early reactions preceding the pigment block, *ade*1, *ade*3, *ade*4, *ade*5, or *ade*9. In such a case, mutants are scored as white colonies among a majority of red colonies.

In mutation studies with *S. pombe,* two enrichment methods have been used. One is the method of inositolless-death. Cells are plated, after mutagenic treatment, on a medium lacking inositol (one of the four growth factors required by *S. pombe;* see Appendix, sections Media and Methods) for 5–10 days, and then inositol is added to the plates. During the incubation period in the absence of inositol, the growing nonmutant cells die (Megnet, 1964*a*). The other method is an enrichment method using 2-deoxyglucose (Megnet, 1965*b,c*).

In some of the adenine loci of *S. pombe,* spontaneous mutations have been investigated both during mitosis and meiosis. Friis *et al.* (1971), on analyzing spontaneous mutations at five adenine loci (*ade*1, *ade*3, *ade*4, *ade*5, *ade*9) which control early enzymatic reactions in the adenine pathway, have established that the total meiotic forward mutation rate for all five loci is of the order of 14.6×10^{-6} mutations per meiosis; the corresponding mitotic mutation rate is only 2.3×10^{-6} mutations per mitosis.

This study has confirmed the relative mutagenic effect of meiosis already described in *Saccharomyces cerevisiae* by Magni and Von Borstel (1962). Nevertheless, it seems that in *S. cerevisiae* the rate of spontaneous mutations during mitosis is 30–40 times lower compared to that of *S. pombe.* In the *ade*6 locus, Loprieno (1973) has shown that the spontaneous forward mutation rate is 4.1×10^{-7} mutations per mitosis, a value comparable to the spontaneous mutation rate observed during mitosis in *Neurospora* (Drake, 1969).

With respect to the molecular nature, at the DNA level, of spontaneous genetic alterations occurring in *S. pombe,* studies have been carried out on the pattern of spontaneous mutations occurring at the *ade*6 and *ade*1 loci. In both cases, the information is based on a series of physiological and genetic analyses of spontaneous mutations (leakiness, temperature sensitivity and osmotic remediability of the growth-factor requirement, interallelic complementation, sensitivity to nonsene suppressors, revertibility), and it has been concluded that during meiosis in both the *ade*6 (Loprieno *et al.,* 1969*a*) and the *ade*1 locus (Friis *et al.,* 1971) spontaneous mutations are mainly represented by base-pair substitutions, whereas the frequency of frame-shift mutations is increased during meiosis (Friis *et al.,* 1971). The presence of a mutator background (see below) does not change the proportions of base-pair substitutions occurring at the *ade*6 locus during mitosis (Loprieno, 1973). These data are consistent with similar conclusions reached in *S. cerevisiae* (Magni and Von Borstel, 1962).

The occurrence in yeasts of frame-shift mutations during the meiotic cycle (possibly associated with a high frequency of genetic exchanges) makes it possible to compare these studies with those carried out in bacteriophage (Strigini, 1965), where it has been shown that reversion of proflavine-revertible spontaneous mutants is associated with recombination of outside markers. The higher mutagenic effect of meiosis in yeasts is not correlated with the production of frame-shift mutants only, since base-pair substitutions are also produced, as shown by the analysis carried out in *ade*1 by Friis *et al.* (1971). Among the mutants produced in *Bacillus subtilis* by transformation (which is also mutagenic and associated with recombination), some 30 percent were leaky or temperature-sensitive mutants (and thus known to be due to base-pair substitutions) (Yoshikawa, 1966).

Mutations induced in *S. pombe* by chemical compounds or physical agents have been studied both in reversion and forward-mutation systems, mainly with the aim of elucidating the chemical nature of induced mutations. The majority of these studies have used the analysis of reversions as a mutation system, but contrary to the bacteriophage experiments, they have not produced consistent results to fit all the hypotheses regarding the chemical mechanisms responsible for induced mutations.

As pointed out by Clarke (1965*a*), in *S. pombe* as in other organisms what seems to be a specific response to a given mutagen by a particular mutant allele in reversion tests, may be due to a variety of effects such as back mutation, suppressor mutation, cellular effects of the mutagen, etc.. Using a variety of auxotrophic alleles (*arg*1, *leu*3, *ura*1: Heslot, 1962; *ade*1, *met*4: Clarke, 1963, 1965*a*; Loprieno and Clarke, 1965; Guglielminetti *et al.*, 1966), a large number of chemical mutagens have been tested in *S. pombe*. Other investigations have used forward-mutation systems, namely *ade*6 and *ade*7, i.e., the white to red systm (Abbondandolo and Loprieno, 1967; Loprieno, 1966; Loprieno *et al.*, 1969*b*; Munz and Leupold, 1970), or *ade*1, i.e., the red to white system (Heslot, 1960; Segal *et al.*, 1973). The conclusions from these studies are based mainly on a series of tests of a physiological (*p*H sensitivity, temperature sensitivity, leakiness, osmotic remediability), genetic (complementation, suppressor sensitivity), or mutagenic nature (specific response of auxotrophic mutants to various mutagens). They may be summarized as follows: Hydroxylamine, nitrous acid, ethyl methanesulfonate, N-methyl-N´-nitro-N-nitrosoguanidine, and methyl methanesulfonate produce base-pair substitutions in the majority of the cases. ICR-170 (2-methoxy-6-

chloro-9-[3-(ethyl-2-chloroethyl)aminopropylamino]acridine dihydrochloride) produces 10–30 percent mutants of the frame-shift type (*ade6/ ade7*: Munz and Leupold, 1970; *ade1*: Segal *et al.*, 1973).

It has long been known that after mutagenic treatment of microorganisms producing single cells, colonies which are mosaics for the induced mutation are formed along with colonies which are entirely mutated. Since most treatments affect only one strand of DNA, the origin of complete mutations needs an explanation (Nasim and Clarke 1965; Nasim and Auerbach, 1967; Auerbach, 1967*a*).

The current hypotheses have been reviewed by Nasim and Auerbach (1967), who also tested them experimentally using fission yeast. Their results were compatible with two of the tested hypotheses ("dual mechanism" and "repair") and ruled out the "master-strand hypothesis." The last mentioned hypothesis, which had received experimental support from experiments in *S. pombe* with hydroxylamine and nitrous acid (Guglielminetti *et al.*, 1967; Guglielminetti, 1968) was definitely discarded as the only mechanism producing completes because of two types of evidence. The first type comes from the observation that uv-induced mutations are found in some cases in all descendants of cells which show no lethal sectoring in micromanipulation experiments (Häfner, 1967*a,b*). This was also observed in nitrous acid-treated G_1 cells (Abbondandolo and Bonatti, 1970). The second line of evidence comes from the finding that the proportion of mosaics induced by either nitrous acid or uv (Abbondandolo and Simi, 1971) in G_1 cells of *S. pombe* increases with dose, in contradiction to the prediction of the lethal-hit hypothesis. The work done with synchronized cells also implies that a master-strand model of DNA replication in *S. pombe* is not acceptable.

For this line of research, *S. pombe* has been successfully utilized for clarifying some aspects of the problem of genetic mosaicism. When all results are considered, the repair hypothesis is left as the most likely explanation for the origin of complete mutations (Abbondandolo and Simi, 1971; Auerbach and Kilbey, 1971).

It has also been observed in several cases that after mutagenic treatment of microorganisms and *Drosophila*, mutations may arise many cell generations later (Auerbach, 1967*b*). This phenomenon has been called "replicating instability" and has also been found in *S. pombe* (Nasim, 1967; James *et al.*, 1972). It is due to a genetic instability produced by the mutagenic treatment of the cells. The experimental evidence obtained in *S. pombe* indicates that the mutants isolated from the progeny of a mosaic colony produced by nitrous acid or ethyl methanesulfonate treatment, all contain identical genetic alterations that

TABLE 4.　Genetic Control of Radiation Sensitivity in S. pombe

Locus and allele designation [a]	Relative uv sensitivity [b]	Relative γ-ray sensitivity
Wild type	−	−
rad1–1	+++	++
rad2–44	++	−
rad3–136	+++	++++
rad4–138	++	++
rad5–158	++	−
rad6–165	++	++
rad7–185	+	−
rad8–190	++	++
rad9–192	+++	++++
rad10–198	+++	++
rad11–404	+++	++
rad12–502	+++	−
rad13–A	++	−
rad14–G	++	−
rad15–P	+++	+
rad16–U	+++	+
rad17–W	++	+++
rad18–X	+	++
rad19–M9	+++	++
rad20–M25	++	+
rad21–45	−	++
rad22–67	−	+

[a] The loci rad-1–rad22 have all been shown to be nonallelic in allelism tests (A. Nasim and B. P. Smith, unpublished). The alleles indicated in the table have been isolated by M. Schüpbach (rad1–rad12), F. Fabre (rad3), A. Nasim and B. P. Smith (rad14–rad18, rad21, and rad22), and R. Megnet (rad19, rad20). For additional alleles, see Schüpbach (1971) and A. Nasim and B. P. Smith (unpublished). The locus designations rad1–rad10 and rad13 replace the previous locus designations uvs1–uvs10 (Schüpbach, 1971) and uvsA (Fabre, 1971). It is proposed that future rad loci giving radiation-sensitive phenotypes upon uvs are numbered consecutively in the order of the time of their discovery, irrespective of their phenotype with regard to the radiations to which they are sensitive.

[b] −, normal sensitivity corresponding to that of the wild type; +, ++, +++, ++++, increasing degrees of sensitivity as compared to that of the wild type. These are semiquantitative classifications based on quantitative determinations of the survival as a function of dose (A. Nasim and B. P. Smith, unpublished; for earlier references, see text).

have occured at the same genetic site in the ade6 (Loprieno et al., 1968) or ade1 locus (Nasim and Grant, 1973).

　　Radiation-sensitive mutants have been isolated in S. pombe and their analysis has shown that the ability of repairing the damage produced by radiation or chemicals is under genetic control (Häefner and Howrey, 1967; Schüpbach, 1971; Fabre, 1971; Nasim and Smith, unpublished; see Table 4). It is possible that such genetic mechanisms also control the repair of spontaneously occurring DNA lesions or that they are part of the normal DNA replication processes (Nasim and Saunders, 1968;

Schüpbach, 1971). Fabre (private communication) has shown that in one strain (*rad*13) of *S. pombe* the excision of the TT or CT dimers produced by uv is blocked. Analysis of these strains has demonstrated the presence in the wild-type strain of different and independently acting mechanisms which influence the lethality of the mutagenic lesions occurring spontaneously or induced by radiation and chemicals (Nasim, 1968; Fabre, 1970, 1971, 1972*a*; Schüpbach 1971). It has further been shown that the repair pathways which control lethality also control mechanisms involving recombination or prerecombinational lesions which are necessary for reciprocal chromosomal exchanges (Fabre, 1972*b*). Finally, radiation-sensitive mutants show a reduced genetic instability following uv treatment (Dubinin *et al.*, 1972). Extensive analyses have been made of the spontaneous mutability of one of these radiation-sensitive strains (*rad*10) by Loprieno (1973). The conclusions of this study indicate that missense mutations occur in the sensitive strain at a rate 100 times higher than in the wild type both in a forward-mutation and in a reverse-mutation system.

TABLE 5.   *Temperature-Sensitive Mutants in S. pombe*

Locus and allele designation[a]	Relative growth rate (37°C)[b]	Relative amount of macromolecular synthesis (37°C)		
		DNA	RNA	Protein
Wild type	+ + + +	+ + + +	+ + + +	+ + + +
*tsl*1–347	+ + +	+ +	+ + +	+ + +
*tsl*2–327	+ +	+	+ +	+ + + +
*tsl*3–357	+ +	+	+	+ + + +
*tsl*4–74	+	+ + +	+ + +	+ + + +
*tsl*5–307	+	+ + +	+ + +	+ + + +
*tsl*6–591	+ +	+ +	+ + + +	+ + + +
*tsl*7–543	+	+	+ +	+ +
*tsl*8–437	+ +	+ +	+ +	+ +
*tsl*9–454	+	+	+	+
*tsl*10–477	+ + +	+ + +	+ + +	+ + +
*tsl*11–479	+	+	+	+
*tsl*12–28	+ +	+ +	–	+ +

[a] The loci *tsl*1–*tsl*12 have all been shown to be nonallelic in allelism tests. At the locus *tsl*12, two homoalleles have been obtained which could not be resolved by intragenic recombination.

[b] + + + +, normal growth rate or normal amount of macromolecular synthesis at 37°C corresponding to that of the wild type; + + +, + +, +, –, decreasing degrees of growth rate or macromolecular synthesis at 37°C as compared to that of the wild type. These are semiquantitative classifications based on quantitative determinations of the optical density, cell number, and cell viability, and of the increase of the total amount of DNA, RNA, and protein as a function of time, following a temperature shift from 25°C to 37°C (Bonatti *et al.*, 1972).

Enhancement of spontaneous mutability has also been observed in mutants of *S. pombe* isolated as temperature-sensitive mutants in which macromolecular syntheses are blocked (Bonatti *et al.,* 1972; see Table 5). A direct selection for mutants exhibiting increased rates of spontaneous mutation to ethionine resistance has permitted the identification of three mutator loci, *mut*1–*mut*3. In the presence of mutant alleles at any one of these loci, rates of spontaneous reversion from adenine dependence (strain 49C8-1: *ade*7) to adenine independence (*ade*7$^+$) are increased by factors ranging between 10× and 20× (P. Munz unpublished).

Biochemical Genetics

Purine Metabolism

Structural Genes of the *De Novo* Pathway.   In *S. pombe,* the *de novo* pathway appears to be identical with that described for other organisms (Heslot, 1972). Twelve loci have been identified, i.e., *ade*1–*ade*10, *gua*1, and *gua*2. Mutants of constitution *ade*1, *ade*3, *ade*4, *ade*5, *ade*6, and *ade*7 have a requirement for adenine (or hypoxanthine). Mutants of constitution *ade*2 and *ade*8 grow only on adenine. Mutants of constitution *ade*9 and *ade*10 require adenine (or hypoxanthine) and histidine. Mutants of constitution *gua*1 and *gua*2 require guanine. As shown in Table 6, the enzymatic deficiencies corresponding to these twelve genes have been identified by several workers.

The *ade*1 locus comprises two adjacent regions which code for GAR* synthetase and AIR synthetase, respectively (Heslot *et al.,* 1966; Fluri *et al.,* 1974). The *ade*2 mutants are blocked between IMP and s-AMP, as shown both by cross-feeding techniques (Wanner, 1955) and by direct enzymatic measurement of s-AMP synthetase (Nagy *et al.,* 1973). The *ade*3 mutants lack FGAR-transamidase (Fluri, unpublished). Heslot *et al.* (1966, 1969) and Nagy (1970) have shown that the *ade*4 mutants lack PRPP amidotransferase. Whitehead and Nagy (unpublished) showed that *ade*5 mutants were devoid of GAR transformylase. This has been confirmed by Wyss (1971) using a different method.

Fisher (1969) established that *ade*6 is blocked between AIR and CAIR, and *ade*7 is blocked between CAIR and SAICAR. The *ade*7 mutants were shown to be defective in SAICAR synthetase activity. On a growth medium containing limiting amounts of adenine, mutants of constitution *ade*6 and *ade*7 form red colonies. This pigment is a polymer

---

* The abbreviations are defined on pp. 407–408.

TABLE 6. Examples of Gene-Enzyme Relationships of S. pombe [a]

Gene	Phenotype	Gene product [b]	Analytical procedures [c]	References
Purine metabolism				
*ade*1	Adenine requiring	{ GAR synthetase (EC.6.3.1.3.) [b]	A	Heslot *et al.*, 1966;
		{ AIR synthetase (EC.6.3.3.1)	A	Fluri *et al.*, 1974
*ade*2	Adenine requiring	s-AMP synthetase (EC.6.3.4.4)	A	Wanner, 1955; Nagy *et al.*, 1973
*ade*3	Adenine requiring	FGAR transamidase (EC.6.3.5.3)	A	Fluri, unpublished
*ade*4	Adenine requiring	PRPP amidotransferase (EC.2.4.2.14)	A	Heslot *et al.*, 1966; Nagy, 1970
*ade*5	Adenine requiring	GAR tetrahydrofolate 5, 10 formyl-transferase (EC.2.1.2.2.)	A	Whitehead, unpublished; Wyss, 1971
*ade*6	Adenine requiring	AIR carboxylase (EC.4.1.4.21)	B	Fisher, 1969
*ade*7	Adenine requiring	SAICAR synthetase (EC.6.3.2.6)	A	Fisher, 1969
*ade*8	Adenine requiring	SAICAR lyase (EC.4.3.2.2)	A	Megnet and Giles, 1964
*ade*9	(Adenine + histidine) requiring	{ N^{10} formyl THFA synthetase (EC.6.3.4.3)	A	Nagy, *et al.*, 1969
		{ N^5,N^{10} methylene THFA dehydrogenase (EC.1.5.1.5)	A	
*ade*10	(Adenine + histidine) requiring	{ AICAR transformylase (EC.2.1.2.3)	A	Whitehead *et al.*, 1966; Richter, unpublished
		{ IMP cyclohydrolase (EC.3.5.4.10)	A	
*gua*1	Guanine requiring	IMP dehydrogenase (EC.1.2.1.14)	A	Leupold, 1961; Pourquié, 1969
*gua*2	Guanine requiring	GMP synthetase (EC.6.3.5.2)	B	Leupold, 1961; Pourquié, 1969
*aza*1	Resistant to 8-azaguanine	PRPP amidotransferase (EC.2.4.2.14)	A	Heslot *et al.*, 1966; Nagy, 1970
*dap*1	Resistant to 2,6-diaminopurine	Adenine phosphoribosyltransferase (EC.2.4.2.7)	A	De Groodt *et al.*, 1971
*pur*1	Resistant to 8-aza-thioxanthine	Hx-G-X-phosphoribosyltransferase (EC.2.4.2.8)	A	De Groodt *et al.*, 1971

TABLE 6. Continued

Gene	Phenotype	Gene product [b]	Analytical procedures [c]	References
dea2	Wild	Adenine aminohydrolase (EC.3.5.4.2)	A	Abbondandolo et al., 1971
ins1	Wild	Nucleoside phosphorylase (EC.2.4.2.1)	A	Pourquié, 1972
ins2	Wild	Guanine aminohydrolase (EC.3.5.4.3)	A	Pourquié, 1972
Pyrimidine metabolism				
ura1	Uracil requiring	Blocked before N-carbamyl aspartic acid	B	Megnet, 1959
ura2	Uracil requiring	Dihydroorotase (EC.3.5.2.3)	B	Magnet, 1959
ura3	Uracil requiring	Dihydroorotate dehydrogenase (EC.1.3.3.1)	B	Megnet, 1959
ura4	Uracil requiring	Blocked after orotic acid	B	Magnet, 1959
Aromatic amino acid metabolism				
aro1	Inhibited by phenylalanine	3-Deoxy-d-arabino heptulosonate-7-phosphate synthetase (tyr) (EC.4.1.2.15)	A	Schweingruber and Wyssling, 1974a,b
aro2	Inhibited by tyrosine	3-Deoxy-d-arabino heptulosonate-7-phosphate synthetase (phe) (EC.4.1.2.15)	A	Schweingruber and Wyssling, 1974a,b
trp1	Tryptophan requiring	N-(5-phosphoribosyl)-anthranilate isomerase / Indoleglycerol phosphate synthetase (EC.4.1.1.48) / Anthranilate synthetase in cooperation with trp3	A / A / A	Schweingruber and Dietrich, 1973

trp2	Tryptophan requiring	Tryptophan synthetase (EC.4.2.1.20)	A	Maurer, 1968; Hanni, 1969; Schweingruber and Dietrich, 1973
trp3	Tryptophan requiring	Anthranilate synthetase	B	Schweingruber and Dietrich, 1973
trp4	Tryptophan requiring	Anthranilate-5-phosphoribosyl-1-pyrophosphate phosphoribosyl transferase (EC.2.4.2.18)	A	Schweingruber and Dietrich, 1973
Histidine metabolism				
his1	Histidine requiring	Histidinol phosphate phosphatase (EC.3.1.3.15)	B	R. B. Drysdale, unpublished
his2	Histidine requiring	Histidinol dehydrogenase (EC.1.1.1.23)	B	R. B. Drysdale, unpublished; Althaus, 1968
his3	Histidine requiring	Imidazole acetol phosphate transaminase (EC.2.6.1.9)	B	R. B. Drysdale, unpublished
his5	Histidine requiring	Imidazole glycerolphosphate dehydratase (EC.4.2.1.19)	B	R. B. Drysdale, unpublished
his7	Histidine requiring	Blocked before imidazole glycerol phosphate	A	Nagy, unpublished

[a] For other examples of gene-enzyme relationships, see the section Biochemical Genetics.

[b] Numbers in parentheses are those assigned by the 1964 Recommendation of the International Union of Biochemistry on the Nomenclature and Classification of Enzymes.

[c] A, Identified by enzymatic assays in vitro; B, inferred from an analysis of the utilization and/or accumulation of intermediates.

derived from AIR. Under the conditions mentioned, these two mutants usually grow somewhat slower than the white ones. They readily give rise to white derivatives which grow faster and, upon genetical analysis, turn out to be double mutants of the type *ade*7 *ade*1 (or *ade*3, *ade*4, *ade*5, *ade*9). The second block is an early one, and it suppresses the formation of AIR.

Megnet and Giles (1964) showed that *ade*8 mutants lack SAICAR (and s-AMP) lyase. Regarding *ade*9 mutants, Nagy *et al.* (1969) established that their purine requirement is not a direct result of a block in the *de novo* pathway. These mutants lack two enzymes involved in the interconversion of folic acid coenzymes, namely N^{10} formyl THFA synthetase and N^5 N^{10} methylene THFA dehydrogenase. This deficiency results in the lack of two coenzymes causing two apparent blocks, at the level of AICAR $\rightarrow$ FAICAR and GAR $\rightarrow$ FGAR transformations. Whitehead *et al.* (1966) and R. Richter (unpublished) established that the *ade*10 locus controls AICAR transformylase and IMP cyclohydrolase.

Fine-Structure, Complementation, and Enzymatic Studies. Fine-structure mapping has been carried out for *ade*1, *ade*2, *ade*4, *ade*6, *ade*7, *ade*8, and *ade*9. Interallelic complementation was found in *ade*1, *ade*2, *ade*6, and *ade*8. (For references, see the section Fine-Structure and Complementation Maps).

An interesting situation has been found by Friis *et al.* (1971), Ramirez *et al.* (1974), and Fluri *et al.* (1974) for mutants of constitution *ade*1. They map in either of two distinct regions, designated *ade*1A and *ade*1B, which are at opposite ends of the gene. There are three complementation groups. Mutants in the first group are located in both the A and B regions and do not complement. Mutants in the second group map in the A region and complement with those of the third group, which are located in region B. In some of the pairwise combinations between mutants of region B, a weak complementation is observed.

It has been shown that complementing *ade*1B mutants lack AIR synthetase but possess variable amounts of GAR synthetase. Complementing *ade*1A mutants lack or have low levels of GAR synthetase but possess considerable amounts of AIR synthetase, between 80 and 100 percent of that present in wild type. All of the noncomplementing *ade*1 mutants tested lacked both activities and some of them have been proved, by use of appropriate suppressors (Barben, 1966; Hawthorne and Leupold, 1974), to be nonsense mutants. AIR synthetase and GAR synthetase went together throughout a hundredfold purification. In the last stages, however, the AIR synthetase activity was lost, whereas GAR synthetase activity persisted. The molecular weight of the enzyme was found to be about 200,000.

Matzinger (1973) has done the fine-structure mapping of the *ade*4 gene, including both mutants devoid of enzymatic activity and feedback-deficient mutants resistant to 8-azaguanine, as described by Heslot *et al.* (1966). It was found that the latter are not located in a distinct segment of the genetic map, but distributed over several segments and mixed with the sites of the inactive mutants.

Strauss (1973) found that at increased CO_2 concentrations many of the complementing *ade*6 mutants grow much better and may even reach the growth level of the wild type. These mutants are localized in a definite segment of the fine-structure map of the *ade*6 locus. Two possible interpretations of these facts are: (1) AIR carboxylation could involve two successive steps, i.e., activation of CO_2, then carboxylation proper, with the reparable mutants being affected in the first of these reactions. (2) The *ade*6 gene product possesses two active sites, one concerned with carboxylation, and the other with the synthesis of SAICAR, in conjunction with the *ade*7 gene product.

Regarding *ade*10, R. Richter (unpublished) found two classes of mutants. Those of Class I lack AICAR transformylase, but possess IMP cyclohydrolase activity. Those of Class II lack both AICAR transformylase and IMP cyclohydrolase. Mutants of Class I and II map in adjacent distinct genetic regions, and mutants of one class complement with those of the other. Interallelic complementation has been found to occur between some mutants of Class I. It is not yet known if the *ade*10 locus codes for a single bifunctional protein.

Regulation of the *De Novo* Pathway. There are at least three points of feedback control, namely, at the level of PRPP amidotransferase (EC.2.4.2.14)*, of s-AMP synthetase (EC.6.3.4.4), and of IMP dehydrogenase (EC.1.2.1.14). These three enzymes have been partially purified from the wild-type strain and subjected to kinetic studies. With PRPP amidotransferase, Nagy (1970) has found that the two substrates, PRPP and glutamine, show cooperative saturation kinetics. IMP and GMP are powerful inhibitors of the enzymes, and show both homotropic and heterotropic effects. AMP inhibition is ten times less effective.

The growth of *S. pombe* is strongly inhibited by 8-azaguanine, but resistant mutants are easily obtained. Heslot *et al.* (1966) showed that a small proportion of these mutants form red colonies, although they have no purine requirement. Most of them belong to an *aza*1 locus, which has been shown to be allelic with *ade*4. These *aza*1 mutants excrete both inosine and hypoxanthine, and their phenotype has been attribted to an alteration of their PRPP amidotransferase resulting in a lowering of its feedback control by the end products of the pathway.

* See Table 6, footnote *b*.

PRPP amidotransferase has been purified from *aza*1 mutants and studied kinetically by Nagy (1970). In the mutant enzyme, the homotrophic effect of glutamine is not observed, the sensitivity towards IMP and GMP is reduced 10 times, and the binding of these nucleotides follows first-order kinetics. In the wild-type strain, two forms of PRPP amidotransferase have been identified by Nagy *et al.* (1972), i.e., a labile form (L) and a stable form (S), whose relative proportions vary according to the growth cycle. The function of these two forms is still unclear.

s-AMP synthetase has been purified from the wild-type strain by Nagy *et al.* (1973). The enzyme is inhibited by AMP and GMP. Whereas GMP seems to act by a direct competition for the GTP-binding site, AMP appears as an allosteric effector, showing, at nonsaturating substrate concentration, a homotropic as well as heterotropic effect upon the GTP and aspartate binding. The mechanism of substrate binding appears to be fully random.

IMP dehydrogenase has been partially purified from the wild-type strain by Pourquié (1969, 1972), who also showed that the corresponding structural gene was *gua*1. The substrate NAD^+ shows homotropic effects, but there is no heterotropic effect between NAD^+ and the other substrate, IMP. The latter shows neither homotropic nor heterotropic effects. GMP, the end product of the pathway, exerts an heterotropic effect on the kinetics of fixation of NAD^+ and shows an homotropic effect for its own fixation.

The Salvage Pathway. Apart from the *de novo* pathway, *S. pombe* is able to transform, in a single enzymatic step, free purine bases into the corresponding nucleotides. These reactions are catalyzed by phosphoribosyltransferases.

De Groodt *et al.* (1969, 1971) have induced mutants resistant to purine analogs. An 8-azathioxanthine-resistant mutant, *pur*1, lacks hypoxanthine phosphoribosyltransferase, xanthine phosphoribosyltransferase, and guanine phosphoribosyltransferase (EC.2.4.2.8), and appears to carry a single mutation. Two 2,6-diaminopurine-resistant mutants, *dap*1-retained these activities, but lacked adenine phosphoribosyltransferase (EC.2.4.2.7). These two enzymes have been partially purified and subject to heat inactivation. The overall results strongly suggest that there are two phosphoribosyltransferase enzymes for purine bases in *S. pombe*, one active with adenine, and the other with hypoxanthine, xanthine, and guanine.

Interconversions of Purine Derivatives. From the behavior of double mutants combining a block in the *de novo* pathway with either *pur*1 or *dap*1, Pourquié and Heslot (1971) concluded that AMP, GMP, and IMP cannot enter the cells as such, but are degraded outside the cells into adenosine, guanosine, and inosine.

Abbondandolo *et al.* (1971) have established that the enzyme adenase (adenine aminohydrolase, EC.3.5.4.2) is active in cellfree extracts of *S. pombe* and is able to catalyze the transformation adenine → hypoxanthine. They also showed that several purine analogs (such as 6-chloropurine), bearing various substitutents in position 6, support the growth of some of the mutants of the *de novo* purine nucleotides pathway.

These 6-substituted purines are transformed *in vitro* into hypoxanthine, as shown by Dorée *et al.* (1972) and Guern *et al.* (1972), and it seems that this reaction is catalyzed by an enzymatic system distinct from that of adenase (EC.3.5.4.2) which is coded for by gene *dea2*.

Other enzymatic activities have also been detected, e.g., adenosine kinase by Dorée *et al.* (1972) and Quesnay (1973), nucleoside phosphorylase (coded for by *ins*1), and guanine aminohydrolase (coded for by *ins*2) by Pourquié (1972).

Uptake of Purines. Pourquié (1970, 1972) has shown that when wild-type cells which have been precultivated in unsupplemented minimal medium, are incubated with labeled guanine, they accumulate radioactivity. As this process is linear during the first minutes, one can define an initial speed of penetration. This uptake has the following characteristics: (1) an energy source is necessary; (2) a saturation kinetics is observed when the external guanine concentration is increased, and (3) one observes a competitive inhibition of uptake by structural analogs of gunanine. The saturation kinetics is hyperbolic—guanine, adenine, hypoxanthine, and 2,6-diaminopurine being competitive inhibitors, whereas xanthine exerts no inhibition.

Similar methods have been used by Cummins and Mitchison (1967) to study adenine uptake. The kinetics was found to be hyperbolic. Several purine bases served as competitive inhibitors, the most efficient being 2,6-diaminopurine and hypoxanthine.

Pourquié (1972) also studied xanthine uptake, but found that the process was slow when compared to guanine under identical conditions. By lowering the ammonium sulfate content of the growth medium, the speed of entry can be considerably increased. The saturation kinetics is hyperbolic and the process is inhibited by a number of purine compounds, but only closely related xanthine analogs (2-thioxanthine and uric acid) act as competitive inhibitors.

Purine Catabolism. Fournier-Méric (1972) has established that the wild type of *S. pombe* can utilize, as sole nitrogen source, all the compounds involved in the classical purine-catabolism pathway, i.e., hypoxanthine, xanthine, uric acid, allantoin, allantoic acid, and urea. A number of mutants have been induced which are unable to utilize one or several of these compounds. The observed pattern leaves little doubt regarding the existence of this pathway. However, it has been found that *S. pombe* is fully insensitive to allopurinol, a strong inhibitor of xanthine oxidase and, up to now, this enzymatic activity could not be detected in the extracts. However, urate oxidase is easily detectable and one mutant lacks it.

Pyrimidine Metabolism

Five gene loci (*ura1–ura5*) are involved in the biosynthesis of UMP. Megnet (1959) has studied both the accumulation of intermediates and the capacity of mutants of constitution *ura1*, *ura2*, *ura3*, and *ura4* to grow on these intermediates and has reached the following conclusions (see Table 4): *ura1* is blocked before N-carbamyl aspartic acid, *ura2* is blocked between N-carbamyl aspartic acid and L-dihydroorotic acid, *ura3* is blocked between L-dihydroorotic acid and orotic acid, and *ura4* is blocked after orotic acid. The exact location of the *ura5* block has not been established.

Uracil inhibits the synthesis of N-carbamyl aspartic acid, presumably at the level of aspartate transcarbamylase by feedback, repression, or both.

Aromatic Amino Acid Metabolism

Schweingruber and Wyssling (1974a,b) have studied the enzyme 3-deoxy-D-arabino heptulosonate-7-phosphate (DAHP) synthetase (EC.4.1.2.15), which is the first enzyme of the aromatic amino acid pathway, leading to tyrosine, phenylalanine and tryptophan.

Using genetic and enzymatic methods, they have shown the existence of two isozymes, DAHP synthetase (tyr) and DAHP synthetase (phe), which are feedback-inhibited by tyrosine and phenylalanine, respectively. Their structural genes *aro1* and *aro2* are unlinked.

DAHP synthetase (tyr), the enzyme active in *aro1* mutants, shows a molecular weight of 75,000. In dialyzed extracts of the wild type and of *aro2* mutant strains, two forms of the enzyme are present. Their molecular weights are 45,000 and 125,000, respectively. The high-molecular-weight form is an aggregate in which probably DAHP synthetase (phe), but not DAHP synthetase (tyr) participates. This aggregate is not inhibited by phenylalanine.

Schweingruber and Dietrich (1973) have studied 32 auxotrophic mutants requiring tryptophan. Four different loci have been found: *trp1*, *trp*2, *trp*3, and *trp*4. The loci *trp*1 and *trp*4 are closely linked, while *trp*2 and *trp*3 are neither linked with one another nor with *trp*1 and *trp*4.

From a series of investigations concerning product accumulation, capacity to grow on intermediate metabolites, and enzymatic assays (Maurer, 1968; Hänni, 1969; Schweingruber and Dietrich, 1973), the following situation has been found: *trp*1 codes for both N-(5-phosphoribosyl)-anthranilate isomerase and indoleglycerol phosphate synthetase, *trp*2 codes for tryptophan synthetase, and *trp*4 codes for anthranilate-5-phosphoribosyl-1-pyrophosphate phosphoribosyl transferase. Anthranilate synthetase (whose activity could not be measured *in vitro*) appears to be coded for by *trp*1 and *trp*4. From preliminary complementation studies, it seems that the *trp*1 region represents two genes, one coding for N-(5-phosphoribosyl)-anthranilate isomerase and the other for indoleglycerol phosphate synthetase; the latter forming an active anthranilate synthetase by aggregation with the gene product of *trp*3.

Arginine and Methionine Metabolism

Ali (1967) tried to characterize mutants corresponding to the seven arginine loci found by Leupold. As described in the section Chromosome Map, the number of known gene loci yielding mutants with a *specific* arginine requirement has meanwhile been reduced to six loci, *arg*1–*arg*6. Ali mentioned that *arg*1 was ornithine-requiring. He thought that *arg*3 and *arg*4 lacked arginosuccinate synthetase and arginosuccinase, respectively. As no enzymatic tests have been performed, these conclusions are far from convincing.

Ali (1970) also studied the four distinct methionine loci described by Leupold, but no clear conclusion could be drawn regarding the biochemical pathway and the precise location of the blocks.

Histidine Metabolism

Nine distinct loci have been identified in this pathway by Leupold (1957, 1958, unpublished) and Althaus (1968). R. B. Drysdale (unpublished) worked out the correspondence between some of these loci and enzymes by analyzing the intermediates accumulated by their mutants. Althaus (1968) made a detailed study of the capacity of 29 *his*2 mutants to grow on and accumulate histidinol.

A fine-structure map was constructed, and complementation investigated. The results showed that there is an analogy between *his*2 mutants of *S. pombe* and *his*3 mutants of *N. crassa* or *his*4 mutants of *S. cerevi-*

siae. In both of the latter cases it has been demonstrated that the locus codes for PR-AMP hydrolase, PR-AMP phosphorylase, and histidinol dehydrogenase. Whitehead (1967) studied the kinetic behavior of PR-ATP synthetase in the wild strain of *S. pombe* and showed that this enzyme was inhibited by the concerted action of histidine and IMP.

Ethyl Alcohol Metabolism

Megnet (1965a, 1967) showed that mutants resistant to allyl alcohol were partially deficient in alcohol dehydrogenase (EC.1.1.1.1). Eleven mutants have been mapped at three different sites of the structural gene of alcohol dehydrogenase, *adh*1.

The toxicity of allyl alcohol is probably related to the production of acrolein and catalyzed by ADH, and resistance occurs through partial inactivation of ADH. The alcohol dehydrogenase of the mutants differs from the wild-type enzyme in an increase in the apparent Michaelis constants for ethanol, NAD, acetaldehyde, and NADH, as well as in the relative oxidation rates of aliphatic alcohols.

Glucose and Mannose Metabolism

A mutant resistant to 2-deoxyglucose and unable to use glucose as a carbon source was found by Megnet (1965b) to be partially deficient in hexokinase (EC.2.7.1.1).

Jannsen and Megnet (1972) established that the growth of mutants which are partially deficient in D-mannose-6-phosphate ketol-isomerase (EC.5.3.1.8) is inhibited by mannose at concentrations higher than 1.5 g/liter. This effect is the result of the trapping of GMP in the form of GDP-mannose. In media lacking mannose, the mutants produce large, viable cells.

The acid phosphatase, a mannose containing glycoprotein, has a K_m for *p*-nitrophenyl phosphate which is lower in cells grown in the presence of, than in cells lacking, mannose. It seems therefore that the carbohydrate side chain has another function than that of anchoring the enzyme in the cell wall.

Acid Phosphatase

Acid phosphatase has been studied by Dibenedetto (1972) and Bauer *et al.* (1973). The *p*H activity curve shows an optimum value at *p*H = 3 and a shoulder between 3.5 and 4.5. The specificity of the enzyme for the phosphomonoester bond is fairly broad. All the monoesters tested (including 5′ mononucleotides) were hydrolyzed. The acid phosphatase is sensitive to several inhibitors, particularly phosphate, arsenate, fluoride, and sulfate.

The acid phosphatase certainly has an external location in the cell. After sonication, up to 90 percent of the activity is found in the supernatant. Apparently there is only one acid phosphatase. Orthophosphate appears to be a repressor of the enzyme. The specific activity of the latter is 15 times higher in cells grown on a poor (3 mg/liter) as compared to cells grown on a rich (450 mg/liter) phosphate medium. Bauer *et al.* (1973) isolated, after mutagenic treatment, 160 mutants with low acid phosphatase activity.

Respiratory Metabolism

Heslot *et al.* (1970) have shown that *S. pombe* grows well on glucose and glycerol. However, L-lactate, ethanol, malate, and succinate, although oxidized, are poor growth substrates. A continuous increase in cell respiration during growth on repressive concentrations of glucose has been observed, suggesting that respiration was repressed by glucose.

Taeter (1972) spectrophotometrically examined the respiratory chain and found the following absorption bands: 603 nm ($a + a_3$), 561 nm (b), 553 nm (c_1), 547 and 523 (c).

Schaeffernicht (1972) detected the following enzymatic activities in the wild type: NADH-CoQ-oxidoreductase (complex I), succinate-CoQ-oxidoreductase (complex II), succinate-cytochrome c-oxidoreductase (complex III), and cytochrome oxidase (complex IV).

Michel *et al.* (1971) have established that in the presence of erythromycin (0.01 mg/ml) growth of *S. pombe* on glycerol is reduced to 5–15 percent of that of the control, whereas growth in fermentable substrate (5 percent glucose) is left unaffected by concentrations up to 5 mg/ml.

Reduction of growth on glycerol is paralleled by inhibition of the formation of cytochromes ($a + a_3$) and b, and mitochondrial protein synthesis is inhibited to about 50 percent. Consequently, the primary effect of erythromycin is probably the inhibition of mitochondrial protein synthesis.

Heslot *et al.* (1972) found that no viable cytoplasmic RD (respiratory-deficient) mutants could be obtained by acriflavine and ethidium bromide treatments. Microcolonies appear which, after a growth lag of a few days, develop further into normal, respiratory-competent colonies. These results suggest that unstable petites were induced.

Segregational respiratory-deficient mutants resistant to cobalt sulfate inhibition were isolated. Some of these strains are deficient in cytochrome $(a + a_3)$, and respire at low rates. The morphology of their mitochondrial membranes is altered.

Bachofen *et al.* (1972) found that the indicator triphenyltetrazolium chloride (TTC) exhibits a strong growth inhibition in respiratory-competent cells, but shows only a minor effect in RD mutants. Use of this dye thus allows rapid selection of rarely occuring respiratory-deficient mutants, which show Mendelian inheritance. TTC seems to interfere with cytochrome oxidase.

Wolf *et al.* (1971) used nitrosoguanidine and uv to induce respiration-deficient mutants and concentrated them with triphenyl-tetrazolium chloride. The RD mutants showed a decrease or a loss in cytochrome c oxidase activity and in succinate-cytochrome c reductase activity. The nutritional blocks could be localized in complexes II, III, and IV of the respiratory chain. Genetical analysis showed that all mutants were chromosomal.

Schwab *et al.* (1971) showed that ethidium bromide inhibits respiratory competence, as measured by growth on glycerol, by respiration, and by decrease of cytochromes $(a + a_3)$. After transfer in ethidium bromide-free medium, the cells underwent reversion to respiratory competence.

Goffeau *et al.* (1972a,b), Landry and Goffeau (1972), Devroede (1971), and Colson *et al.* (1973) have isolated mutants unable to grow on glycerol. Although some of these mutants still respire, most are RD mutants showing deficiencies in the respiratory cytochromes. Three types of deficiencies are common: (1) absence of $(a + a_3)$, (2) absence of $(a + a_3)$ and b, and (3) absence of $(a + a_3)$, b and c_1. About 50 percent of the RD mutants had also deficiencies in mitochondrial ATPase. All the mutants are pleiotropic and show various cytochrome deficiencies, a fact with suggests the existence of a common mechanism controlling the synthesis both of the oligomycin-sensitive ATPase complex and of respiratory cytochromes.

Catabolic Repression

In *S. pombe*, Megnet and Schlanderer (1971) and Megnet *et al.* (1972) have shown that synthesis of invertase and α-glucosidase is subject to catabolite repression (no synthesis above 1 percent glucose concentration). In a catabolite-resistant mutant the enzyme levels are much higher even at high glucose concentrations. The intracellular concentration of cAMP in both strains was found to be inversely related to the glucose content of the medium. The enzyme concentration could be directly related to the intracellular cAMP content, and enzyme synthesis was also induced by cAMP. The authors have good evidence that, in this system, cAMP acts at the level of transcription.

Foury and Goffeau (1972) established that, compared to glycerol-grown cells, the respiration of wild-type cells grown in the presence of a 10 percent glucose concentration is decreased by a factor of 3. This factor is 24 for the mutant *Cob5* (super-repressed); it is 1 for the mutant *Cob6*, which is therefore totally insensitive to glucose repression. By use of a derepressing medium, full restoration of respiration and respiratory pigments could be achieved for the super-repressed strain *Cob5*.

Suppression

In *S. pombe*, allele-specific suppressors capable of suppressing only certain alleles at a given locus were first described by Barben (1966). Several of these suppressors are clearly of the "super-suppressor" type (Hawthorne and Mortimer 1963), i.e., they are allele-specific but not locus-specific in their action. Detailed studies on the patterns of their allele-specific activity have been carried out in three adenine loci for which extensive fine-structure maps are available, *ade7*, *ade6*, and *ade1*. The properties of the mutants sensitive to these suppressors are those of nonsense rather than of missense mutants in that none of the suppressible mutants was found to be leaky, temperature sensitive, or osmotic remedial (Barben, 1966; Munz and Leupold, 1970; Friis *et al.*, 1971). For osmotic-remedial mutants (i.e., mutants with a defect which can be corrected by providing a high osmotic pressure in the growth medium) in *S. pombe*, see also Megnet (1964*b*, 1966*b*).

In those of the three adenine loci which show interallelic complementation, *ade6* and *ade1*, the nonsense nature of the mutants sensitive to these suppressors is also demonstrated by the polarity of their complementation patterns (Barben, 1966). This is particularly clear in the *ade1* locus, which has been shown to code for a bifunctional enzyme with two activities involved in purine biosynthesis, GAR synthetase, and AIR synthetase (Fluri *et al.*, 1974), as described in the section Biochemical Genetics. Out of 31 suppressible mutants, all those mapping in the *ade1*A region (23 mutants) are of a noncomplementing type, whereas 5 out of 8 suppressible mutants mapping in the *ade1*B region are capable of complementing missense mutations in the *ade1*A region (Friis *et al.*, 1971; Segal

TABLE 7. Classification of Nonsense Suppressors in S. pombe [a, b]

Suppressor-sensitive mutants [c]		1st class	2nd class		3rd class	
		I	II	II'	III	III'
		sup1–oa [e] sup2–oa	sup3–o	sup8–o	sup3–a	sup8–a
ade7–	262 (+1)	+	−	−	+	−
	489	−	−	−	+	−
	695	+	+	+	−	−
	413 (+3)	+	−	−	+	+
	84 (+1)	+	+	+	−	−
	540	−	+	+	−	−
	608	+	−	−	−	−
	451 (+7)	+	−	−	+	+
	419	+	−	−	+	+
	461	+	+	+	−	−
	606	+	−	−	+	+
	572	+	−	−	−	−

		sup1–oa	sup3–o sup9–o	sup8–o sup10–o
ade6–	706	−	+	+
	712	−	−	+
	704	−	+	+
	611	+	−	−
	469	−	+	+
	588*	+	−	−

		sup1–oa	sup3–o sup9–o	sup8–o sup10–o
ade1A–	40	+	+	+
	3 (+2)	+	−	−
	H538	−	+	+
1B–	25*	+	+	+
	H259	+	−	−

et al., 1973; Ramirez *et al.*, 1974). In agreement with their complementation behavior, suppressible *ade*1A and *ade*1B mutants of the non-complementing type lack both activities, whereas suppressible *ade*1B mutants of the complementing type retain GAR synthetase activity, i.e., the activity which is coded for by the *ade*1A region (Fluri *et al.*, 1974). From these observations, it can be concluded that the suppressible *ade*1 mutants are of a chain-terminating type and that translation proceeds from left (*ade*1A) to right (*ade*1B).

The partially overlapping patterns of allele-specific suppression exhibited by three classes of known suppressors which include a total of six suppressor loci (*sup*1–*sup*3 and *sup*8–*sup*10) are those expected of nonsense suppressors capable of suppressing either one or the other or both of the chain-terminating codons UAA (ochre) and UAG (amber) (Table 7). Two additional classes of allele-specific suppressors which include a total of four different suppressor loci (*sup*4–*sup*7) do not overlap in their patterns of allele-specific suppression. They are active on few mutants only and may represent missense suppressors (Barben, 1966; Hawthorne and Leupold, 1974).

Two inefficient suppressors of a first class of nonsense suppressors, *sup*1 and *sup*2, show identical patterns of allele-specific action. They are capable of suppressing two different types of chain-terminating mutations. These are likely to correspond to the ochre and amber mutations demonstrated in *Saccharomyces cerevisiae* (Hawthorne, 1969; Stewart *et al.*, 1972; Stewart and Sherman, 1972), although direct evidence for this is still lacking in *S. pombe*. Suppressors *sup*1 and *sup*2 may correspond to bacterial ochre-amber suppressors which are known to sup-

---

[a] The table is taken from Hawthorne, D. C. and Leupold, U.: Suppressors in Yeast. Current Topics in Microbiology and Immunology *64*, pg. 14 (1974). Berlin/Heidelberg/New York: Springer.

[b] Patterns of allele-specific action of nonsense suppressors among uv-induced nonsense mutants mapping in the *ade*7, *ade*6, and *ade*1 loci of *S. pombe*.

[c] Mutants are ordered according to their relative map location, from left to right (Barben, 1966), in the fine-structure maps of *ade*7 (Leupold, 1961), *ade*6 (Leupold and Gutz, 1965), and *ade*1 (Clarke, 1965c; Ramirez *et al.*, 1974). Mutants distinguishable by suppression but not by recombination are separated by smaller distances between lines. Numbers of additional mutants which are indistinguishable by both suppression and recombination are indicated in parentheses following the collection number of the mutant with which they are homoallelic. The *ade*6 and *ade*1 mutants capable of interallelic complementation, in combination with suppressor-insensitive missense mutants mapping toward their left, are indicated by an asterisk. Additional nonsense mutants of spontaneous or chemically induced origin which are sensitive to one or both of the two nonsense suppressors *sup*1–oa and *sup*3–o, include 2 *ade*7 mutants (Munz and Leupold, 1970), 1 *ade*6 mutant (noncomplementing; Gutz, 1971b), 24 *ade*1 mutants (18 *ade*1A, noncomplementing; 6 *ade*1B, 3 complementing and 3 noncomplementing; Friis *et al.*, 1971; Segal *et al.*, 1973) and 2 *glu*1 mutants (Barben, 1966).

[d] As far as tested, suppressor sensitivity is indicated by "+", insensitivity as "−".

[e] Suppressors *sup*1–oa, *sup*2–oa, and *sup*3–o correspond to suppressors *84f*, *413c* and *84h* of Barben (1966).

press both ochre (UAA) and amber (UAG) but not opal (UGA) codons and which have been shown, in one case analyzed in *E. coli,* to code for a mutant tRNA carrying the mutant anticodon UUA (Altman *et al.,* 1971). In the text that follows, the suppressor-active alleles of these two suppressors will therefore tentatively be designated as *sup*1-oa and *sup*2-oa.

Four efficient suppressors of a second class, *sup*3, *sup*8, *sup*9, and *sup*10, show a more restricted pattern of allele-specific action in that they suppress only a subset of those mutants which are sensitive to the suppressors *sup*1 and *sup*2 of the first class. As judged from their efficiency, which appears to be higher in most mutant combinations than that of *sup*1 and *sup*2 when comparing growth rates of suppressed mutants in minimal medium (Barben, 1966, and unpublished), these suppressors might well correspond to the ochre-specific suppressors of *S. cerevisiae* (Mortimer and Hawthorne, 1969). As a working hypothesis and in order to facilitate discussion, it will be assumed in the following that this is indeed the case. The suppressor-active alleles of these loci, which are likely to code for mutant tRNA's carrying the anticodon IUA (obtained from AUA by deamination of adenosine to inosine; Bock, 1967), will therefore tentatively be designated as *sup*3-o, *sup*8-o, *sup*9-o, and *sup*10-o. Minor differences in their allele-specific patterns suggest that the amino acid transferred by the suppressor tRNA's of *sup*3-o and *sup*9-o may differ from that transferred by the suppressor tRNA's of *sup*8-o and *sup*10-o (Hawthorne and Leupold, 1974).

If the preliminary classification of the four suppressors of the second class as ochre-specific suppressors is correct, then a third class involving two inefficient suppressors is likely to correspond to the amber-specific suppressors of *S. cerevisiae* (Mortimer and Hawthorne, 1969) and to code for a mutant tRNA with the anticodon CUA as shown for an amber-specific suppressor in *E. coli* (Goodman *et al.,* 1968). Both of these suppressors, *sup*3-a and *sup*8-a, have been derived by mutation from efficient suppressors of the second class, *sup*3-o and *sup*8-o. They behave as alleles of the suppressors from which they were derived, but they suppress a different subset of those mutants which are affected by the suppressors of the first class, *sup*1-oa and *sup*2-oa. Again, minor differences distinguishing the allele-specific patterns of *sup*3-a and *sup*8-a suggest that the same nonsense codon is read as a different amino acid by *sup*3 and *sup*8.

According to this tentative classification, nonsense mutants sensitive to the second class of suppressors are assumed to carry ochre mutations (UAA), whereas mutants sensitive to the third class of suppressors are postulated to carry amber mutations (UAG). Results of a preliminary analysis of the interconversion of the two nonsense codons and their suppressors following mutagenesis with uv and, more specifically, with ethyl methanesulfonate, are not conclusive enough to confirm this classification. They do show, however, that the two types of nonsense codons can be interconverted in both directions by a single base-pair substitution (A. Ahmad and Leupold, unpublished).

The assumption that the nonsense suppressors of the second and third class represent mutant alleles of tRNA structural genes is supported by a meiotic recombination analysis of the genetic fine structure of the suppressor loci *sup3* (Hubschmid and Leupold, 1974), *sup8* (Häsler and Leupold, 1974), and *sup9* (Hofer and Leupold, 1974), and also by a corresponding mitotic analysis (based on methyl methanesulfonate-induced recombination; Wyssling and Leupold, 1974 of *sup3* and *sup9* (for a review, see Hawthorne and Leupold, 1974). Fine-structure maps were constructed on the basis of the frequencies of suppressor-active recombinants obtained from pairwise crosses of suppressor-inactive revertant alleles which had been derived from the suppressor-active allele by mutation. A comparison of the genetic length of the meiotic fine-structure maps of *sup8*-o, *sup9*-o, and *sup3*-o with those of the fine-structure maps of protein structural genes mentioned in the section Fine-Structure and Complementation Maps, shows that most of the latter maps are considerably longer. This would indeed be expected if *sup8*, *sup9*, and *sup3* code for tRNA's which are only about 80 nucleotides long. A comparison of the mitotic map length of *sup9*-o and *sup3*-o with that of *ade7* further supports the conclusion that these nonsense suppressors represent structural genes for tRNA's.

For a more detailed discussion of the suppressors obtained in *S. pombe,* and of modifying mutations mapping at several loci (*sin1*–*sin14*) which restrict their expression (F. Hofer, M. Minet and P. Thuriaux, unpublished), the reader is referred to a recent review of the work on suppressors in yeast (Hawthorne and Leupold, 1974).

Appendix

Media

The following media are used for genetic experiments with *S. pombe;* all media are prepared with distilled water:

1. Yeast-extract agar (YEA). 3 percent glucose, 0.5 percent Bacto-yeast extract, and 2 percent agar (Leupold, 1955*b*).

2. Liquid yeast-extract medium (YEL). The same as YEA, but without agar.

3. Malt-extract agar (MEA). 3 percent Bacto-malt extract, 2% agar. (Leupold, 1955*b*).

4. Liquid malt-extract medium (MEL). 3 percent Bacto-malt extract in 0.05 M phosphate buffer, *p*H 5.9. (Egel, 1971).

5. Minimal agar (MMA; modified Wickerham agar). In 1000 ml water: 1.0 g KH_2PO_4; 0.5 g $MgSO_4 \cdot 7H_2O$; 0.1 g NaCl; 0.1 g $CaCl_2 \cdot 2H_2O$; 5.0 g $(NH_4)_2SO_4$; 500 μg H_3BO_3; 40 μg $CuSO_4 \cdot 5H_2O$; 100 μg KI; 200 μg $FeCl_3 \cdot 6H_2O$; 400 μg $MnSO_4 \cdot 1H_2O$; 160 μg $H_2MoO_4 \cdot 2H_2O$; 400 μg $ZnSO_4 \cdot 7H_2O$; 10 g glucose; 10 μg biotin (in 1 ml 50 percent ethanol); 1 mg calcium pantothenate; 10 mg nicotinic acid; 10 mg meso-inositol; and 20 g agar. (Leupold, 1955*b*).

6. Liquid minimal medium (MML). The same as MMA (without agar), but with 1.5 g asparagine added (Angehrn, 1964). Another version of the minimal medium has been described by Mitchison (1970).

7. Synthetic sporulation agar (SPA). 1 percent glucose, 0.1 percent KH_2PO_4, and four vitamins as in MMA, 3 percent agar. (Angehrn, 1964; Leupold, 1970).

8. Liquid synthetic sporulation medium (SPL). The same as SPA, but without agar. Egel (1971) has used a modified MML as sporulation liquid.

In experiments with auxotrophic mutants it is advisable to supplement media 1–4 as well as 7 and 8 with the substances required by the mutants in order to ensure optimal development. In general, 50 μg/ml are added of each substance needed. As to MMA and MML, it depends, of course, on the purpose of the experiment whether or not supplements are added. *ade*7 and *ade*6 mutants accumulate a red pigment on YEA and also on MMA supplemented with 20 μg/ml adenine. If YEA and MMA are supplemented with 75 μg/ml adenine, the above mutants no longer form the red pigment.

Methods

In the following paragraphs, some general information for the handling of *S. pombe* will be given and several laboratory techniques will be described which are specific for this yeast. Leupold (1955*a,b*, 1957) has developed most of the basic methods used with *S. pombe*.

Incubation Temperatures. A temperature of 30°C is optimal for vegetative growth; if copulation and/or sporulation are desired, 25°C is more suitable.

Media for Vegetative Growth. In general, YEA and YEL are used for this purpose. On YEA, copulation and sporulation are not as good as they are on MEA and SPA (the latter media are used for crosses, see below). Gutz (unpublished) found that on YEA prepared with some lots of Bacto-yeast extract (Difco), copulation did not take place at all.

Stock Cultures on Silica Gel. *S. pombe* strains remain viable for at least five years if the cells are dried on silica gel. The following procedure is a modification of a technique originally described by Perkins (1962) for *Neurospora* conidia:

1. Screw-cap tubes (17 × 60 mm) are filled to a height of 25 mm with white, 12–28-mesh silica gel, and are dry sterilized without the caps (they are plugged with cotton stoppers) for 2 hours at 180°C. The caps are autoclaved separately in Petri dishes and dried at approximately 100°C. After dry sterilization, the silica-gel tubes are cooled down in a desiccator containing silica gel with humidity indicator. Before use, the cotton plugs are replaced by the sterile screw caps.
2. The yeast strains to be stored are grown for 3 days at 30°C on YEA slants.
3. The cells of each culture are suspended in 1.0 ml of sterile skim milk (50 g Bacto-skim milk in 1000 ml distilled water, autoclaved for 15 minutes at 121°C).
4. Each cell suspension is slowly transferred with a 1.0-ml pipette to a silica-gel tube, avoiding complete saturation of the silica gel.
5. The tubes are stored for 1 week in a dry and dustfree place at room temperature, with the screw caps loosely screwed on.
6. To control the viability of the dried cells, the cultures may be checked by transferring a small amount of silica gel to a tube with YEL; growth should occur within 3 days at 30°C.
7. The screw caps are tightened and sealed with paraffin and the tubes are kept refrigerated (ca. 3°C).
8. The stored strains can be grown up by transferring small amounts of silica gel to YEL (see 6). After removing of some silica gel, the tubes can be resealed for further storage.

Purification of Strains. Haploid *S. pombe* strains always contain some diploid cells, and spontaneous mating-type mutations can also occur (cf. Table 2). It is therefore advisable to reisolate a haploid clone of the desired mating type before a strain is used for an experiment. This is mandatory if one uses old stock cultures.

Haploid clones can be reisolated by streaking the strains on YEA with magdala red (phloxin type) added at a concentration of 10 ppm. On this medium, diploid colonies become dark red, whereas haploid colonies are pink. As to mating-type mutations, the procedures vary with the mating type used. Of most practical importance in experiments with *S. pombe* are the mutations from h^{+N} to h^{90}. From h^{+N} strains, pure clones can be reisolated by streaking the strains on MEA. After incubation, the plates are treated with iodine vapors (see below); h^{+N} colonies become yellow, whereas h^{90} colonies (or h^{90} sectors in h^{+N} colonies) turn black. In streaks of h^{90} strains, nonsporulating mutants can be detected as iodine-negative colonies or sectors.

Treatment with Iodine Vapors.   The bottom of a culture plate with the colonies to be tested is inverted on a tin can containing a few iodine crystals or platelets. The can is then heated slightly over a gas burner until some iodine vapors are visible. A moderate iodine treatment (2–4 seconds exposure) kills only part of the cells in the colonies. Therefore, new cultures can be grown up from the treated colonies.

Mating-Type Determination.   Homo- and heterothallic strains can easily be distinguished by iodine treatment: colonies of h^{90} cultures are iodine-positive, whereas colonies of heterothallic cultures are iodine-negative. The mating types of the latter strains can be determined by the following method. From the cultures to be tested, small inocula are arranged on YEA master plates (up to 70 strains per plates); each plate is needed in duplicate. When colonies have developed on these plates, they are replicated on MEA plates on which the wild-type strains L975 (h^{+N}) and L972 (h^{-S}), respectively, have been spread. From each pair of YEA master plates, one plate is replicated on MEA + L975, the other on MEA + L972. The test plates are incubated at 25°C for 4 days, and are then treated with iodine vapors. The h^{-} strains will give positive iodine reaction with the L975 lawn and no reaction with the L972 lawn, whereas h^{+} strains will exhibit the reciprocal activity. (Leupold, 1955a).

Crosses.   Crosses are made by mixing haploid strains of compatible mating types on MEA or SPA. The parental strains needed for the crosses are grown up on YEA slants for two days at 30°C.

For routine crosses, the following procedure is convenient. The yeast on the YEA slants are suspended with saline (0.85 percent NaCl), 4–5 ml are used for each slant. If heterothallic cultures are crossed, equal amounts of the suspended parental strains are mixed. If a h^{90} culture is to be crossed with a h^{+} or h^{-} strain, the heterothallic strain is added in threefold excess to the cross mixture; this is done to minimize selfings between the h^{90} cells. From each cross mixture, 0.2 ml are distributed on the agar surface of a MEA slant. The slants are placed on a slanted tray, so that their agar surfaces remain horizontal, and are incubated at 25°C. It should be noted that MEA is a medium which supports vegetative growth of *S. pombe*. On MEA, the mixed strains first grow side by side; copulation and sporulation take place at the end of vegetative growth. After 2–3 days, MEA slants contain many asci.

SPA does not contain any nitrogen source and, therefore, does not support vegetative growth. This medium is used if some degree of synchrony in copulation and spore formation is desired. Since no vegetative growth occurs on SPA, high cell densities are needed for inoculation. This may be achieved by placing a loop of nonsuspended cell material (from YEA slants) of each parental strain at the same spot on a SPA plate. One droplet (ca. 0.05 ml) saline is added, and the two strains are thoroughly mixed on the agar surface. In doing this, the cross mixture should be spread over an area of approximately 150 mm². The SPA plates are incubated at 25°C. This crossing procedure is especially used when diploid strains are to be selected. After 13–15 hours, many zygotes (but not yet asci) are present on SPA. If strains with complementary growth requirements had been crossed, part of the above zygotes will develop to diploid colonies when streaked on MMA. (For a more detailed description of the isolation of diploid strains, see Flores da Cunha, 1970.)

Tetrad Analyses. Asci from 2- or 3-day-old crosses are placed at distinct positions on SPA with the help of a micromanipulator. It is convenient to work on thin layers of SPA; a simple apparatus for producing such agar layers has been described by Haefner (1967c). The layers with the asci are kept for 17–20 hours at room temperature (ca. 22°C). After this time, the walls of about 80 percent of the asci have dissolved spontaneously. The spores of each former ascus are then separated by micromanipulation, and the SPA layers are transferred to YEA plates which are incubated for 3–4 days at 30°C.

Plating of Free Ascospores. MEA slants with *S. pombe* crosses are incubated for 7 days at 25°C and are subsequently stored for additional 7 days in the refrigerator (ca. 3°C). After this time, the walls of the asci have dissolved spontaneously. The mixtures of free ascospores and vegetative cells are now treated for 15–30 minutes with 30 percent ethanol at 18°C; the alcohol selectively kills the vegetative cells (Leupold, 1957). After convenient dilution, the spores are plated on appropriate media.

The time needed for the disruption of asci can be abbreviated by treatment with snail enzyme. Material from crosses on MEA or SPA, of at least 4 days old, is suspended in "Suc d'Helix pomatia" (diluted 1:1000 in distilled water) for 6 hours at 30°C (Leupold, 1970). This ascus-wall digestion is followed by an alcohol treatment as described above. "Suc d'Helix pomatia" can be obtained from *Industrie Biologique Française S.A.*, Gennevilliers, France.

"Criss-Cross" Technique. A qualitative genetic classification of auxotrophic mutants blocked in the same biosynthetic pathway may be achieved by crossing, via replica plating, mutant strains of opposite mating types on MEA. Up to 36 crosses, in a 6 × 6 matrix, can be made per plate. Later on, the crosses are replicated on MMA to test for the formation of prototrophic recombinants. A detailed description of this technique has been given by Leupold (1955*b*) and Heslot (1960).

Nonsporulating Diploid Cultures. The h^+/h^+ or h^-/h^- strains can be haploidized by growing them on media with p- or m-fluoropheny-lalanine (FPA). Flores da Cunha (1970) used YEA with 0.1 percent p-FPA; Adondi and Heslot (1970) used supplemented MMA with 0.04 percent m-FPA. The latter authors found that m-FPA is more effective for haploidization, as well as less toxic, than p-FPA.

A genetic analysis of nonsporulating diploid strains is also feasible by crossing them with other diploid strains of compatible mating types. The cross mixtures are spotted on YEA plates which are then incubated at 30°C. This condition is favorable for the occurrence of "twin meioses" (Gutz, 1967*a,b*). Eight-spored asci are dissected by micromanipulation. In these asci, karyogamy has not taken place; 4 spores carry the markers of one "parent," the remaining 4 spores carry the markers of the other "parent."

Further Experimental Procedures. These have been mentioned in the main text of this article and, for more details, can be looked up in the cited references.

Acknowledgments

The work carried out in the laboratories of the authors was supported by the following institutions: H. Gutz: NSF Grant GB-15148, NIH Grant GM-13234, The University of Texas at Dallas Research Fund. H. Heslot: Institut de la Recherche Agronomique, EURATOM, Commissariat à l'Energie Atomique, Délégation à la Recherche Scientifique et Technique. U. Leupold: Schweizerischer Nationalfonds. N. Loprieno: Consiglio Nazionale delle Ricerche.

Literature Cited

Abbondandolo, A. and S. Bonatti, 1970 The production, by nitrous acid, of complete and mosaic mutations during defined nuclear stages in cells of *Schizosaccharomyces pombe*. *Mutat. Res.* **9:**59–69.

Abbondandolo, A. and N. Loprieno, 1967 Forward mutation studies with N-nitroso-N-methylurethane and N-nitroso-N-ethylurethane in *Schizosaccharomyces pombe*. *Mutat. Res.* **4:**31–36.

Abbondandolo, A. and S. Simi, 1971 Mosaicism and lethal sectoring in G_1 cells of *Schizosaccharomyces pombe*. *Mutat. Res.* **12:**143–150.

Abbondandolo, A., A. Weyer, H. Heslot and M. Lambert, 1971 Study of adenine aminohydrolase in the yeast *Schizosaccharomyces pombe*. *J. Bacteriol.* **108:**959–963.

Adondi, G. and H. Heslot, 1970 Etude des conversions mitotiques au niveau du gène *ad-9* de *Schizosaccharomyces pombe*. *Mutat. Res.* **9:**41–58.

Ali, A. M. M., 1967 Biochemical characterization of arginine mutants in *Schizosaccharomyces pombe* and the possibility of repression. *Can. J. Genet. Cytol.* **9:**462–472.

Ali, A. M. M., 1970 Complementation studies within *met3* and the location of *met4* in *Schizosaccharomyces pombe*. *Genetica (The Hague)* **41:**334–341.

Althaus, M., 1968 Histidin-2-Mutanten von *Schizosaccharomyces pombe*. Diplomarbeit, Universität Bern, Berne.

Altman, S., S. Brenner and J. D. Smith, 1971 Identification of an ochre-suppressing anticodon. *J. Mol. Biol.* **56:**195–197.

Angehrn, P., 1964 Untersuchungen über intragene Rekombinationsmechanismen und allele Komplementierung an Adeninmutanten von *Schizosaccharomyces pombe*. Dissertation, Universität Zürich, Zurich.

Angehrn, P. and H. Gutz, 1968 Influence of the mating type on mitotic crossing over in *Schizosaccharomyces pombe*. *Genetics* **60:**1581.

Auerbach, C., 1967*a* Lethal sectoring and the origin of complete mutants in *Schizosaccharomyces pombe*. *Mutat. Res.* **4:**875–878.

Auerbach, C., 1967*b* Changes in the concept of mutation and the aims of mutation research. In *Heritage from Mendel,* edited by R. A. Brink, pp. 67–80, University of Wisconsin Press, Madison, Wisc.

Auerbach, C. and B. J. Kilbey, 1971 Mutation in eukaryotes. *Annu Rev. Genet.* **5:**163–218.

Bachofen, V., R. J. Schweyen, K. Wolf and F. Kaudewitz, 1972 Quantitative selection of respiratory deficient mutations in yeast by triphenyl tetrazolium chloride. *Z. Naturforsch. Teil B* **27b:**252–256.

Barben, H., 1966 Allelspezifische Suppressormutationen von *Schizosaccharomyces pombe*. *Genetica (The Hague)* **37:**109–148.

Barras, D. R., 1972 A β-glucan endo-hydrolase from *Schizosaccharomyces pombe* and its role in cell wall growth. *Antonie van Leeuwenhoek J. Microbiol. Serol.* **38:**65–80.

Bauer, C., A. Cinci, G. Bronzetti and N. Loprieno, 1974 Use of biochemical parameters for the individuation of genetic variants: study of the relative efficiency of some parameters determined on a number of mutants of *Schizosaccharomyces pombe* with altered acid phosphatase activity. *Biochem. Genet.* **11:**1–15.

Bock, R. M., 1967   Prediction of a topaz supressor ribonucleic acid. *J. Theor. Biol.* **16**:438–439.

Bonatti, S., M. Simili and A. Abbondandolo, 1972   Isolation of temperature-sensitive mutants of *Schizosaccharomyces pombe*. *J. Bacteriol.* **109**:484–491.

Bostock, C. J., 1970   DNA synthesis in the fission yeast *Schizosaccharomyces pombe*. *Exp. Cell Res.* **60**:16–26.

Bresch, C., G. Müller and R. Egel, 1968   Genes involved in meiosis and sporulation of a yeast. *Mol. Gen. Genet.* **102**:301–306.

Clarke, C. H., 1962   A case of mutagen specificity attributable to a plating medium effect. *Z. Verebungs* **93**:435–440.

Clarke, C. H., 1963   Suppression by methionine of reversions to adenine independence in *Schizosaccharomyces pombe*. *J. Gen. Microbiol.* **31**:353–363.

Clarke, C. H., 1965a   Methionine as an antimutagen in *Schizosaccharomyces pombe*. *J. Gen. Microbiol.* **39**:21–31.

Clarke, C. H., 1965b   Mutagen specificity among reversions of ultraviolet-induced adenine-1 mutants of *Schizosaccharomyces pombe*. *Genet. Res.* **6**:433–441.

Clarke, C. H., 1965c   Recombination among UV-induced adenine-1 mutants of *Schizosaccharomyces pombe*. *Experientia (Basel)* **21**:582–583.

Clarke, C. H., 1968   Differential effects of caffeine in mutagen-treated *Schizosaccharomyces pombe*. *Mutat. Res.* **5**:33–40.

Colson, A. M., C. Colson and A. Goffeau, 1973   Systems for membrane alteration: genetic perturbations of mitochondria in a "petite-négative" yeast. In *Biomembranes, Cell Organelles and Membranous Components,* edited by S. Fleischer, L. Packer and R. Estabrook, Academic Press, New York.

Cummins, J. E. and J. M. Mitchison, 1967   Adenine uptake and pool formation in the fission yeast *Schizosaccharomyces pombe*. *Biochim. Biophys. Acta.* **136**:108–120.

De Groodt, A., H. Heslot, L. Poirier, J. Pourquié and M. Nagy, 1969   La spécificité des pyrophosphorylases des nucléotides puriques chez le *Schizosaccharomyces pombe*. *C. R. Hebd. Séances Acad. Sci. Ser. D Sci. Nat.* **269**:1431–1433.

De Groodt, A., E. Whitehead, H. Heslot and L. Poirier, 1971   The substrate specificity of purine phosphoribosyltransferases in *Schizosaccharomyces pombe*. *Biochem. J.* **122**:415–420.

Devroede, M., 1971   Obtention par radiations ionisantes de mutants respiratoires modifiés dans l'ATPase mitochondriale chez la levure *Schizosaccharomyces pombe*. Licence Université Louvain, Leuven.

Dibenedetto, G., 1972   Acid phosphatase in *Schizosaccharomyces pombe*. I. Regulation and preliminary characterization. *Biochim. Biophys. Acta* **286**:363–374.

Ditlevsen, E. and V. Hartelius. 1956   A "gigas" mutant in *Schizosaccharomyces pombe* induced by X rays. *C. R. Trav. Lab. Carlsberg Sér. Physiol.* **26**:41–49.

Dorée, M., J. J. Leguay, J. Guern and H. Heslot, 1972   Métabolisme de la kinétine chez le *Schizosaccharomyces pombe*. *C. R. Hedb. Seances Acad. Sci. Ser. D Sci. Nat.* **275**:59–62.

Drake, J. W., 1969   Spontaneous mutation: Comparative rates of spontaneous mutation. *Nature (Lond.)* **221**:1132.

Dubinin, N. P., O. N. Kurennaya, L. D. Kurlapova and V. A. Taracov, 1972   UV-induced replicating instability in fission yeast *Schizosaccharomyces pombe*. *Mutat. Res.* **16**:249–264.

Duffus, J. H., 1969   The isolation of nuclei from the yeast *Schizosaccharomyces pombe*. *Biochim. Biophys. Acta* **195**:230–233.

Duffus, J. H., 1971   The isolation and properties of nucleohistone from the fission yeast *Schizosaccharomyces pombe*. *Biochim. Biophys. Acta* **228**:627–635.

Egel, R., 1969   Zur Genetik der Paarungstypausprägung und der Meiosis bei *Schizosaccharomyces pombe*. Dissertation, Universität Freiburg im Breisgau.

Egel, R., 1971   Physiological aspects of conjugation in fission yeast. *Planta (Berl.)* **98**:89–96.

Egel, R., 1973*a*   Commitment to meiosis in fission yeast. *Mol. Gen. Genet.* **121**:277–284.

Egel, R., 1973*b*   Genes involved in mating type expression of fission yeast. *Mol. Gen. Genet.* **122**:339–343.

Egel-Mitani, M. and R. Egel, 1972   A rapid visualization of Kleinschmidt-type DNA preparations by phosphotungstic acid. *Z. Naturforsch. Teil B.* **27b**:480.

Fabre, F., 1970   UV-sensitivity of the wild type and different UVS mutants of *Schizosaccharomyces pombe*. Influence of growth stages and DNA content of the cells. *Mutat. Res.* **10**:415–426.

Fabre, F., 1971   A UV-supersensitive mutant in the yeast *Schizosaccharomyces pombe*. Evidence for two repair pathways. *Mol. Gen. Genet.* **110**:134–143.

Fabre, F., 1972*a*   Photoreactivation in the yeast *Schizosaccharomyces pombe*. *Photochem. Photobiol.* **15**:367–373.

Fabre, F., 1972*b*   Relation between repair mechanisms and induced mitotic recombination after UV irradiation, in the yeast *Schizosaccharomyces pombe*. *Mol. Gen. Genet.* **117**:153–166.

Fincham, J. R. S., 1966   *Genetic Complementation*, W. A. Benjamin, New York.

Fincham, J. R. S. and P. R. Day, 1971   *Fungal Genetics,* third edition, Blackwell Scientific Publications, Oxford.

Fisher, C. R., 1969   Enzymology of the pigmented adenine-requiring mutants of *Saccharomyces* and *Schizosaccharomyces*. *Biochem. Biophys. Res. Commun.* **34**:306–310.

Flores da Cunha, M., 1970   Mitotic mapping of *Schizosaccharomyces pombe*. *Genet. Res.* **16**:127–144.

Fluri, R., A. Coddington and U. Flury, 1974   The product of the *ade*1 gene in *Schizosaccharomyces pombe*: a bifunctional enzyme catalysing two distinct steps in purine biosynthesis. in manuscript.

Fournier-Méric, A., 1972   Etude du catabolisme des purines chez le *Schizosaccharomyces pombe*. Diplôme Université de Paris, Paris.

Foury, F. and A. Goffeau, 1972   Glucose superrepressed and derepressed respiratory mutants in a "petite-negative" yeast: *Schizosaccharomyces pombe* 972h⁻. *Biochem. Biophys. Res. Commun.* **48**:153–160.

Foury, F. and A. Goffeau, 1973   Combination of 2-desoxyglucose and snail-gut enzyme treatments for preparing sphaeroplasts of *Schizosaccharomyces pombe*. *J. Gen. Microbiol.* **75**:227–229.

Friis, J., F. Flury and U. Leupold, 1971   Characterization of spontaneous mutations of mitotic and meiotic origin in the *ad-1* locus of *Schizosaccharomyces pombe*. *Mutat. Res.* **11**:373–390.

Gillie, O. J., 1966   The interpretation of complementation data. *Genet. Res.* **8**:9–31.

Gillie, O. J., 1968   Interpretations of some large non-linear complementation maps. *Genetics* **58**:543–555.

Goffeau, A., A. M. Colson, Y. Landry and F. Foury, 1972*a*   Modification of mitochondrial ATPase in chromosomal respiratory-deficient mutants of a "petite negative"

yeast: *Schizosaccharomyces pombe* 972h⁻ *Biochem. Biophys. Res. Commun.*
48:1448–1454.

Goffeau, A., Y. Landry, A. M. Colson and F. Foury, 1972*b*   Dual control of synthesis of
oligomycin-sensitive ATPase in a "petite-negative" yeast. *FEBS (Fed. Europ.
Biochem. Soc.) Proc. Meet.* **8:**610.

Goodman, H. M., J. Abelson, A. Landy, S. Brenner and J. D. Smith, 1968   Amber suppression: A nucleotide change in the anticodon of a tyrosine transfer RNA. *Nature
(Lond.)* **217:**1019–1024.

Guern, J., M. Dorée, J. J. Leguay and H. Heslot, 1972   Sur la dégradation enzymatique
de la kinétine et de quelques adénines N6 substituées chez le *Schizosaccharomyces
pombe. C. R. Hebd. Seances Acad. Sci. Ser. D Sci. Nat.* **275:**377–380.

Guglielminetti, R., 1968   The role of lethal sectoring in the origin of complete mutations
in *Schizosaccharomyces pombe. Mutat. Res.* **5:**225–229.

Guglielminetti, R., S. Bonatti and N. Loprieno, 1966   The mutagenic activity of N-nitroso-N-methylurethane and N-nitroso-N-ethylurethane in *Schizosaccharomyces
pombe. Mutat. Res.* **3:**152–157.

Guglielminetti, R., S. Bonatti, N. Loprieno and A. Abbondandolo, 1967   Analysis of the
mosaicism induced by hydroxylamine and nitrous acid in *Schizosaccharomyces
pombe. Mutat. Res.* **4:**441–447.

Gutz, H., 1961   Distribution of X-ray- and nitrous acid-induced mutations in the genetic
fine structure of the *ad₇* locus of *Schizosaccharomyces pombe. Nature (Lond.)*
191:1125–1126.

Gutz, H., 1963   Untersuchungen zur Feinstruktur der Gene *ad₇* und *ad₆* von *Schizosaccharomyces pombe* Lind. Habilitationsschrift, Technische Universität Berlin,
Berlin.

Gutz, H., 1966   Induction of mitotic segregation with p-fluorophenylalanine in
Schizosaccharomyces pombe. J. Bacteriol. **92:**1567–1568.

Gutz, H., 1967*a*   "Twin Meiosis" and other ambivalences in the life cycle of *Schizosaccharomyces pombe. Science (Wash., D.C.)* **158:**796–798.

Gutz, H., 1967*b*   Zwillingsmeiose—eine neue Beobachtung bei einer Hefe. *Ber. Dtsch.
Bot. Ges.* **80:**555–558.

Gutz, H., 1971*a*   Gene conversion: remarks on the quantitative implications of hybrid
DNA models. *Genet. Res.* **17:**45–52.

Gutz, H., 1971*b*   Site specific induction of gene conversion in *Schizosaccharomyces
pombe. Genetics* **69:**317–337.

Gutz, H., 1971*c*   Marker effect in gene conversion. *Genetics* **68:**s26.

Gutz, H. and P. Angehrn, 1968   Lethal mutations in the mating-type region of
Schizosaccharomyces pombe. Genetics **60:**186.

Gutz, H. and F. J. Doe, 1973   Two different *h*⁻ mating types in *Schizosaccharomyces
pombe. Genetics* **74:**563–569.

Haefner, K., 1967*a*   Concerning the mechanism of ultraviolet mutagenesis. A
micromanipulatory pedigree analysis in *Schizosaccharomyces pombe. Genetics*
57:169–178.

Haefner, K., 1967*b*   A remark to the origin of pure mutant clones observed after UV
treatment of *Schizosaccharomyces pombe. Mutat. Res.* **4:**514–516.

Haefner, K., 1967*c*   A simple apparatus for producing agar layers of uniform thickness
for microbiological micromanipulator work. *Z. allg. Mikrobiol.* **7:**229–231.

Haefner, K. and L. Howrey, 1967   Gene-controlled UV-sensitivity in *Schizosaccharomyces pombe. Mutat. Res.* **4:**219–221.

Hänni, C., 1969   Ein Beitrag zur Charakterisierung der Tryptophan-Synthetase von *Schizosaccharomyces pombe*. Diplomarbeit, Universität Bern, Berne.

Häsler, K. and U. Leupold, 1974   The genetic fine structure of nonsense suppressors in *Schizosaccharomyces pombe*. II. *sup*8. unpublished.

Hawthorne, D. C., 1969   Identification of nonsense codons in yeast. *J. Mol. Biol.* **43**:71–75.

Hawthorne, D. C. and U. Leupold, 1974   Suppressors in yeast. *Curr. Top. Microbiol. Immunol.* **64**:1–47.

Hawthorne, D. C. and Mortimer, 1963   Super-suppressors in yeast. *Genetics* **48**:617–620.

Heslot, H., 1960   *Schizosaccharomyces pombe:* un nouvel organisme pour l'étude de la mutagenèse chimique. *Abh. Dtsch. Akad. Wiss. Berl., Kl. Med., Jahrg. 1960 Nr* 1: 98–105.

Heslot, H., 1962   Etude quantitative de reversions biochemiques induites chez la levure *Schiz. pombe* par des radiations et des substances radiomimetiques. *Abh. Dtsch. Akad. Wiss. Berl., Kl. Med., Jahrg. 1962 Nr.* 1:193–228.

Heslot, H., 1972   Genetic control of the purine nucleotide pathway in *Schizosaccharomyces pombe*. *Proc. 4th Int. Ferm. Symp. (Kyoto) Ferm. Techol. Today* 867–876.

Heslot, H., M. Nagy and E. Whitehead, 1966   Recherches génétiques et biochimiques sur la première enzyme de la biosynthèse des purines chez le *Schizosaccharomyces pombe*. *C. R. Hebd. Seances Acad. Sc. Ser. D Sci. Nat.* **263**:57–58.

Heslot, H., M. Nagy and E. Whitehead, 1969   Genetical and biochemical investigations on the first enzyme of purine biosynthesis in *Schizosaccharomyces pombe*. *Second Symposium Yeasts, Bratislava (1966)*, 269–271.

Heslot, H., A. Goffeau and C. Louis, 1970   Respiratory metabolism of a "petite-negative" yeast *Schizosaccharomyces pombe* 972 h⁻. *J. Bacteriol.* **104**:473–481.

Heslot, H., C. Louis and A. Goffeau, 1972   Segregational respiratory-deficient mutants of a "petite negative" yeast *Schizosaccharomyces pombe* 972 h⁻. *J. Bacteriol.* **104**:482–491.

Hofer, F. and U. Leupold, 1974   The genetic fine structure of nonsense suppressors in *Schizosaccharomyces pombe*. I. *sup*9. unpublished.

Holliday, R., 1968   Genetic recombination in fungi. In *Replication and Recombination of Genetic Material*, edited by W. J. Peacock and R. D. Brock, pp. 157–174, Australian Academy of Science, Canberra.

Hubschmid, F. and U. Leupold, 1974   The genetic fine structure of nonsense suppressors in *Schizosaccharomyces pombe*. III. *sup*3. unpublished.

James, A. P., A. Nasim and R. S. McCullough, 1972   The nature of replicating instability in yeast. *Mutat. Res.* **15**:125–133.

Jannsen, S. and Megnet, R. 1972   Mannose dependent mutants in *Schizosaccharomyces pombe*. *Fourth Int. Ferm. Symp. (Kyoto)* abstract G 16-16.

Kohli, J., 1973   Mitotische und meiotische Chromosomenkartierung bei *Schizosaccharomyces pombe*. Diplomarbeit, Universität Bern, Berne.

Landry, Y. and A. Goffeau, 1972   Isolement et caracterisation du complexe ATPasique mitochondrial sensible à l'oligomycine chez la souche sauvage et les mutants respiratoires d'une levure "petite-négative" *Schizosaccharomyces pombe*. *Arch. Int. Physiol. Biochim.* **80**:604–606.

Lederberg, S. and G. Stetten, 1970   Colcemid sensitivity of fission yeast and the isolation of colcemid-resistant mutants. *Science (Wash., D.C.)* **168**:485–487.

Leupold, U., 1950   Die Vererbung von Homothallie und Heterothallie bei *Schizosaccharomyces pombe*. *C. R. Trav. Lab. Carlsberg Sér. Physiol.* **24**:381–480.

Leupold, U., 1955a   Metodisches zur Genetik von *Schizosaccharomyces pombe*. *Schweiz. Z. allg. Pathol. Bakteriol.* **18**:1141–1146.

Leupold, U., 1955b   Versuche zur genetischen Klassifizierung adeninabhängiger Mutanten von *Schizosaccharomyces pombe*. *Arch. Julius Klaus-Stift. Vererbungsforsch. Sozialanthropol. Rassenhyg.* **30**:506–516.

Leupold, U., 1956   Some data on polyploid inheritance in *Schizosaccharomyces pombe*. *C. R. Trav. Lab. Carlsberg Sér. Physiol.* **26**:221–251.

Leupold, U., 1957   Physiologisch-genetische Studien an adeninabhängigen Mutanten von *Schizosaccharomyces pombe*. Ein Beitrag zum Problem der Pseudoallelie. *Schweiz. Z. allg. Pathol. Bakteriol.* **20**:535–544.

Leupold, U., 1958   Studies on recombination in *Schizosaccharomyces pombe*. *Cold Spring Harbor Symp. Quant. Biol.* **23**:161–170.

Leupold, U., 1961   Intragene Rekombination und allele Komplementierung. *Arch. Julius Klaus-Stift. Vererbungsforsch. Sozialanthropol. Rassenhg.* **36**:89–117.

Leupold, U. 1970   Genetical methods for *Schizosaccharomyces pombe*. In *Methods in Cell Physiology*, Vol. 4, edited by D. M. Prescott, pp. 169–177, Academic Press, New York.

Leupold, U. and H. Gutz, 1965   Genetic fine structure in *Schizosaccharomyces*. *Proc. XI Int. Congr. Genet. (The Hague)* **2**:31–35.

Loprieno, N., 1964   Cysteine protection against reversion to methionine independence induced by N-nitroso-N-methyl-urethane in *Schizosaccharomyces pombe*. *Mutat. Res.* **1**:469–472.

Loprieno, N., 1966   Differential response of *Schizosaccharomyces pombe* to ethyl methanesulfonate and methyl methanesulfonate. *Mutat. Res.* **3**:486–493.

Loprieno, N., 1967   Intragenic mapping of chemically induced *ad-7* mutants of *Schizosaccharomyces pombe*. *J. Bacteriol.* **94**:1162–1165.

Loprieno, N., 1969   Negative interallelic complementation of *ad-6* temperature-dependent mutants of *Schizosaccharomyces pombe*. *Rend. Accad. Naz. Lincei* **47**:211–217.

Loprieno, N., 1973   A "mutator" gene in the yeast *Schizosaccharomyces pombe*. *Genetics* (suppl.) **73**:161–164.

Loprieno, N. and C. H. Clarke, 1965   Investigations on reversions to methionine independence induced by mutagens in *Schizosaccharomyces pombe*. *Mutat. Res.* **2**:312–319.

Loprieno, N. and M. Schüpbach, 1971   On the effect of caffeine on mutation and recombination in *Schizosaccharomyces pombe*. *Mol. Gen. Genet.* **110**:348–354.

Loprieno, N., A. Abbondandolo, S. Bonatti and R. Guglielminetti, 1968   Analysis of the genetic instability induced by nitrous acid in *Schizosaccharomyces pombe*. *Genet. Res.* **12**:45–54.

Loprieno, N., S. Bonatti, A. Abbondandolo and R. Guglielminetti, 1969a   The nature of spontaneous mutations during vegetative growth in *Schizosaccharomyces pombe*. *Mol. Gen. Genet.* **104**:40–50.

Loprieno, N., R. Guglielminetti, S. Bonatti and A. Abbondandolo, 1969b   Evaluation of the genetic alterations induced by chemical mutagens in *Schizosaccharomyces pombe*. *Mutat. Res.* **8**:65–71.

McCully, E. K. and C. F. Robinow, 1971   Mitosis in the fission yeast *Schizosac-*

charomyces pombe: A comparative study with light and electron microscopy. *J. Cell Sci.* **9**:475–507.

Magni, G. E. and R. C. von Borstel, 1962   Different rates of spontaneous mutations during mitosis and meiosis in yeast. *Genetics* **47**:1097–1108.

Matzinger, P., 1973   Feinstruktur-Kartierung von Mutanten mit katalytischen und regulatorischen Defekten in der PRPP-Amidotransferase von *Schizosaccharomyces pombe*. Dissertation, Universität Bern, Berne.

Maurer, R., 1968   Isolierung und partielle Reinigung der Tryptophansynthetase aus der Hefe *Schizosaccharomyces pombe*. Diplomarbeit, Universität Bern, Berne.

Megnet, R., 1959   Untersuchungen über die Biosynthese von Uracil bei *Schizosaccharomyces pombe*. *Arch. Julius Klaus-Stift. Vererbungsforsch. Sozialanthropol. Rassenhyg.* **33**:299–334.

Megnet, R., 1964*a*   A method for the selection of auxotrophic mutants of the yeast *Schizosaccharomyces pombe*. *Experientia (Basel)* **20**:320–321.

Megnet, R., 1964*b*   Mutants of the yeast *Schizosaccharomyces pombe* requiring a high concentration of potassium. *Experientia (Basel)* **20**:638–639.

Megnet, R., 1965*a*   Alkoholdehydrogenasemutanten von *Schizosaccharomyces pombe*. *Pathol. Microbiol.* **28**:50–57.

Megnet, R., 1965*b*   Effect of 2-deoxyglucose on *Schizosaccharomyces pombe*. *J. Bacteriol.* **90**:1032–1035.

Megnet, R., 1965*c*   Screening of auxotrophic mutants of *Schizosaccharomyces pombe* with 2-deoxyglucose. *Mutat. Res.* **2**:328–331.

Megnet, R., 1966*a*   The effect of fluorouracil and fluorodeoxyuridine on the genetic recombination in *Schizosaccharomyces pombe*. *Experientia (Basel)* **22**:151–152.

Megnet, R., 1966*b*   Osmotic-remedial and osmotic-sensitive mutants of *Schizosaccharomyces pombe*. *Experientia (Basel)* **22**:216–218.

Megnet, R., 1967   Mutants partially deficient in alcohol dehydrogenase in *Schizosaccharomyces pombe*. *Arch. Biochem. Biophys.* **121**:194–201.

Megnet, R. and N. H. Giles, 1964   Allelic complementation at the adenylosuccinase locus in *Schizosaccharomyces pombe*. *Genetics* **50**:967–971.

Megnet, R. and G. Schlanderer, 1971   Induction of invertase by cyclic AMP in the yeast *Schizosaccharomyces pombe*. *FEBS (Fed. Eur. Biochem. Soc.) Proc. Meet.* **8**:127.

Megnet, R., G. Schlanderer and H. Dellweg, 1972   Action of cyclic AMP on the synthesis of invertase and α-glucosidase in *Schizosaccharomyces pombe*. *Fourth Int. Ferm. Symp. (Kyoto)* Abstract G6-5.

Michel, E., M. L. Melen and N. Loprieno, 1966   Reversion studies with *ad-7* "hot spot" mutants of *Schizosaccharomyces pombe*. *Rend. Accad. Naz. Lincei* **40**:1123–1128.

Michel, R., R. J. Schweyen, F. Kaudewitz, 1971   Inhibition of mitochondrial protein synthesis *in vivo* by erythromycin in *Schizosaccharomyces pombe* and *Saccharomyces cerevisiae*. *Mol. Gen. Genet.* **111**:235–241.

Mitchison, J. M., 1970   Physiological and cytological methods for *Schizosaccharomyces pombe*. In *Methods in Cell Physiology*, Vol. 4 edited by D. M. Prescott, pp. 131–165, Academic Press, New York.

Mitchison, J. M. and J. Creanor, 1971*a*   Induction synchrony in the fission yeast *Schizosaccharomyces pombe*. *Exp. Cell Res.* **67**:368–374.

Mitchison, J. M. and J. Creanor, 1971*b*   Further measurements of DNA synthesis and enzyme potential during cell cycle of fission yeast *Schizosaccharomyces pombe*. *Exp. Cell Res.* **69**:244–247.

Mortimer, R. K. and D. C. Hawthorne, 1969   Yeast genetics. In *The Yeasts,* Vol. I, edited by A. H. Rose and J. S. Harrison, pp. 385–460, Academic Press, New York.

Munz, P. and U. Leupold, 1970   Characterization of ICR-170-induced mutations in *Schizosaccharomyces pombe. Mutat. Res.* **9:**199–212.

Nagy, M., 1970   Regulation of the biosynthesis of purine nucleotides in *Schizosaccharomyces pombe.* I. Properties of the phosphoribosylpyrophosphate: glutamine aminotransferase of the wild strain and of a mutant desensitized towards feedback modifiers. *Biochim. Biophys. Acta* **198:**471–481.

Nagy, M., H. Heslot and L. Poirier, 1969   Conséquences d'une mutation affectant la biosynthése des coenzymes foliques chez le *Schizosaccharomyces pombe. C. R. Hebd. Seances Acad. Sci. Ser. D Sci. Nat.* **269:**1268–1271.

Nagy, M., H. Heslot and U. Riviére, 1969.

Nagy, M., U. Reichert and A. M. Ribet, 1972   Two forms of PRPP-amidotransferase, the first enzyme of the purine *de novo* pathway in *Schizosaccharomyces pombe. FEBS (Fed. Eur. Biochem. Soc.) Proc. Meet.* **8:**411.

Nagy, M., M. Djembo-Taty and H. Heslot, 1973   Regulation of the biosynthesis of purine nucleotides in *Schizosaccharomyces pombe.* III. Kinetic studies of adenylosuccinate synthetase. *Biochim. Biophys. Acta,* in press.

Nasim, A., 1967   The induction of replicating instability by mutagens in *Schizosaccharomyces pombe. Mutat. Res.* **4:**753–763.

Nasim, A., 1968   Repair-mechanisms and radiation-induced mutations in fission yeast. *Genetics* **59:**327–333.

Nasim, A. and C. Auerbach, 1967   The origin of complete and mosaic mutants from mutagenic treatment of single cells. *Mutat. Res.* **4:**1–14.

Nasim, A. and C. H. Clarke, 1965   Nitrous acid-induced mosaicism in *Schizosaccharomyces pombe. Mutat. Res.* **2:**395–402.

Nasim, A. and C. Grant, 1973   Genetic analysis of replicating instabilities in yeast. *Mutat. Res.* **17:**185–190.

Nasim, A. and A. S. Saunders, 1968   Spontaneous frequencies of lethal-sectoring and mutation in radiation-sensitive strains of *Schizosaccharomyces pombe. Mutat. Res.* **6:**475–478.

Perkins, D. D., 1962   Preservation of *Neurospora* stock cultures with anhydrous silica gel. *Can. J. Microbiol.* **8:**591–594.

Pontecorvo, G., 1958   *Trends in Genetic Analysis,* Columbia University Press, New York.

Pourquié, J., 1969   Regulation of the biosynthesis of purine nucleotides in *Schizosaccharomyces pombe.* II. Kinetic studies of IMP dehydrogenase. *Biochim. Biophys. Acta* **185:**310–315.

Pourquié, J., 1970   Antagonism by adenine in the nutrition of *Schizosaccharomyces pombe* mutants. Inhibition at the level of guanine uptake. *Biochim. Biophys. Acta* **209:**269–277.

Pourquié, J., 1972   Contrôle génétique de la biosynthèse du GMP chez *Schizosaccharomyces pombe.* Thèse Doct. es-Sciences, Université de Paris, Paris.

Pourquié, J. and H. Heslot, 1971   Utilization and interconversions of purine derivatives in the fission yeast *Schizosaccharomyces pombe. Genet. Res.* **18:**33–44.

Quesnay, M., 1973   Etude de l'adénosine-kinase chez le *Schizosaccharomyces pombe.* Diplôme, Université de Paris, Paris.

Ramirez, C., J. Friis and U. Leupold, 1974   The *ade*1 gene of *Schizosaccharomyces* *pombe*: A complex locus controlling two different steps in purine biosynthesis. unpublished.

Schaffernicht, G., 1972 Charakterisierung karyotischer atmungsdefekter Mutanten von *Schizosaccharomyces pombe* durch Cytochromspektren und Enzymtests. Diplomarbeit, Universität Müchen, Munich.

Schwab, R., M. Sebald and F. Kaudewitz, 1971 Influence of ethidium bromide on respiration in *Schizosaccharomyces pombe. Mol. Gen. Genet.* 110:361–366.

Schweingruber, M. E. and R. Dietrich, 1973 Gene-enzyme relationships in the tryptophan pathway of *Schizosaccharomyces pombe. Experientia (Basel)* 29:1152–1154.

Schweingruber, M. E. and H. Wyssling, 1974a Genes and isoenzymes controlling the first step in the aromatic amino acid biosynthesis in *Schizosaccharomyces pombe. Biochim. Biophys. Acta* 350:319–327.

Schweingruber, M. E. and H. Wyssling 1974b Studies concerning the biochemical genetics of activity and feedback inhibition mutants of *Schizosaccharomyces pombe* 3-deoxy-D-arabino-heptulosonate 7-phosphate synthase. *Biochim. Biophys. Acta,* 350:328–335.

Schüpbach, M., 1971 The isolation and genetic classification of UV-sensitive mutants of *Schizosaccharomyces pombe. Mutat. Res.* 11: 361–371.

Segal, E., P. Munz and U. Leupold, 1973 Characterization of chemically induced mutations in the *ad-1* locus of *Schizosaccharomyces pombe. Mutat. Res.* 18:15–24.

Shahin, M. M., 1971 Preparation of protoplasts from stationary phase cells of *Schizosaccharomyces pombe. Can. J. Genet. Cytol.* 13:714–719.

Stetten, G. and S. Lederberg, 1973 Colcemid sensitivity of fission yeast. II. Sensitivity of stages of the cell cycle. *J. Cell Biol.* 56:259–262.

Stewart, J. W. and F. Sherman, 1972 Demonstration of UAG as a nonsense codon in baker's yeast by amino-acid replacements in iso-1-cytochrome *c. J. Mol. Biol.* 68:429–443.

Stewart, J. W., F. Sherman, M. Jackson, F. L. X. Thomas and N. Shipman, 1972 Demonstration of the UAA ochre codon in bakers' yeast by amino-acid replacements in iso-1-cytochrome *c. J. Mol. Biol.* 68:83–96.

Strauss, A., 1973 Genetische und physiologische Untersuchungen an *ad-6* und *ad-7* Mutanten von *Schizosaccharomyces pombe.* Diplomarbeit, Universität Bern, Berne.

Strigini, P., 1965 On the mechanism of spontaneous reversion and genetic recombination in bacteriophage T4. *Genetics* 52:759–776.

Taeter, H., 1972 Analyse spectrophotométrique de la chaine respiratoire de la levure *Schizosaccharomyces pombe. Mém. Univ. Louvain.*

Treichler, H., 1964 Genetische Feinstruktur und intragene Rekombinationsmechanismen im *ad₂*-Locus von *Schizosaccharomyces pombe.* Dissertation, Universität Zürich, Zurich.

Wanner, H., 1955 Die Akkumulation von Hypoxanthin durch adenin-abhängige Mutanten von *Schizosaccharomyces pombe. Arch. Julius Klaus-Stift. Vererbungsforsch, Sozianthropol. Rassenhyg.* 30:516–520.

Whitehead, E., 1967 La cinétique allostérique de l'inhibition de la première enzyme de la biosynthèse de l'histidine chez *Schizosaccharomyces pombe. Bull. Soc. Chim. Biol.* 49:1529–1535.

Whitehead, E., M. Nagy and H. Heslot, 1966 Interactions entre la biosynthèse des purines nucléotides et celle de l'histidine chez le *Schizosaccharomyces pombe. C. R. Hebd. Seances Acad. Sci. Ser. D Sci. Nat.* 263:819–821.

Wolf, K., M. Sebald-Althaus, R. J. Schweyen and F. Kaudewitz, 1971   Respiration deficient mutants of *Schizosaccharomyces pombe* I. *Mol. Gen. Genet.* **110**:101–109.

Wyss, J., 1971   Enzymatische Analyse adenin-auxotropher Mutanten von *Schizosaccharomyces pombe*. Diplomarbeit, Universität Bern, Berne.

Wyssling, H. and U. Leupold, 1974   Genetic fine-structure analysis by means of induced mitotic recombination in *Schizosaccharomyces pombe*. unpublished.

Yoshikawa, H. 1966   Mutations resulting from the transformation of *Bacillus subtilis*. *Genetics* **54**:1201–1214.

26

Aspergillus nidulans

A. John Clutterbuck

Introduction

Aspergillus nidulans is a sexually reproducing member of the Aspergillaceae (Raper and Fennell, 1965). The history of genetic investigations of this mold dates from 1945, when Pontecorvo started to look for an organism suitable for a genetic approach to "certain problems of the spacial organisation of the cell" (Pontecorvo *et al.*, 1953). These problems amounted to an attempt to define the gene and its relation to cell metabolism. The reasons for the choice of *A. nidulans* as well as the story of the early development of genetical studies are fully given in Pontecorvo *et al.* (1953). Briefly, the advantages of this fungus include the fact that it is a haploid eukaryote which rapidly forms colonies on simple media and can therefore be treated as a microorganism. The asexual spores (conidia) are uninucleate and of a striking green color which is modified in a variety of spore-color mutants (see Table 1) to give conspicuous markers which are invaluable in genetic manipulations. The fungus is homothallic, so any strain can be crossed to any other, and, furthermore, stocks such as those built up in Glasgow can be assumed to be relatively isogenic since they are derived from a single wild isolate.

Accounts of the early studies on the nature of the gene are given in a number of reviews by Pontecorvo (1950, 1952*a,b*, 1955, 1958, 1959, 1963) and Pontecorvo and Roper (1956). One important field to develop from this approach was the study of intragenic recombination, which began with Roper's demonstration of recombination between biotin

A. John Clutterbuck—Genetics Department, Glasgow University, Glasgow, Scotland.

alleles (Roper, 1950). Later work on recombination in *A. nidulans* has made particular use of the advantages of mitotic recombination in diploids for studies in which recombination is influenced by environmental or genetic variables (see Table 2).

Perhaps the most important contribution of *Aspergillus* to genetics has been in the elucidation of the parasexual cycle (Pontecorvo, 1954, 1956; Roper, 1961, 1966*b*). In addition to the importance of this process in other fungi, forms of parasexual genetics are now proving to be very valuable tools in approaches to mammalian genetics through tissue cultures, as predicted by Pontecorvo (1962).

For elucidation of the parasexual cycle and in mitotic recombination studies, diploids are a *sine qua non,* and highly colored, uninucleate conidia are a great asset. Examination of many of the other fields in which *A. nidulans* has been employed suggests that these two features, along with the colonial growth form and homothallic sexual system, constitute the main advantages of *A. nidulans* as an experimental organism.

Techniques

The standard reference for *Aspergillus* methods is Pontecorvo *et al.* (1953). Practical procedures for experiments with the parasexual cycle have also been described by Barron and MacNeill (1962) and for heterokaryosis and intragenic recombination by Pritchard (1968).

Strains

The majority of genetic work with *A. nidulans* makes use of the large collection of genetically marked strains built up in Glasgow (Barratt, *et al.,* 1965; Clutterbuck, 1969*a*). A variety of other wild strains has also been collected at Birmingham, England (Grindle, 1963*a,b*).

The Glasgow strains all derive from a single wild isolate (Pontecorvo *et al.,* 1953). Some variation at unanalyzed loci exists, due partly, no doubt, to mutagenic treatments and partly to unconscious selection during subculture. However, map distances, for instance, vary little between crosses as long as the parent strains are free from chromosome aberrations.

Since all genetic techniques with *A. nidulans* depend on the presence of nutritional markers, new mutants are normally obtained in an auxotrophic strain, and for this purpose, as well as for biochemical experiments, the biotin-requiring strain (*biA 1*) is frequently used in place of the wild type.

Strains can be stored as conidia on slants for six months to a year. Ascospores can be stored for longer periods, but strains are best stored as lyophils or on silica gel. A suitable procedure for storage of conidia or ascospores on silica gel is described by Ogata (1962).

Media

Media are basically as described by Pontecorvo *et al.* (1953). Minimal medium (MM) consists of 6 g NaNO$_3$, 0.52 g KCl, 0.52 g MgSO(7H$_2$O), 1.52 g KH$_2$PO$_4$, 10 g glucose, traces of iron and zinc salts, and 1 liter distilled water, pH adjusted before autoclaving to 6.5. Complete medium (CM) differs from one laboratory to another; that used in Glasgow consists of MM with the addition per liter of: 1.5 g hydrolyzed casein, 2 g peptone, 0.5 g yeast extract, 2 μg biotin, 100 μg p-aminobenzoic acid; 500 μg pyridoxine and aneurin, 1000 μg nicotinamide (or nicotinic acid) and riboflavin. Media are solidified with 1.0–1.5 percent agar and are sterilized by autoclaving at 10 lb for 10 minutes. MM may be supplemented as required at the following rates: 100 μg/ml amino acids, 1 μg/ml vitamins, 200 μg/ml adenine. Higher concentrations than these may be required for some auxotrophs, and it should also be noted that the wild type, which will normally grow on MM, may require biotin at high temperatures (Roper, 1966*a*) or under certain nutritional conditions (Strigini and Morpurgo, 1961). Riboflavin is destroyed by light and by high concentrations of thiosulfate, and biotin is occasionally absorbed or destroyed in certain batches of CM. For classification of drug-resistant mutants and for other special media, the relevant literature should be consulted.

The addition of 0.8 percent sodium deoxycholate to the medium (Mackintosh and Pritchard, 1963) reduces the colony size to the extent that up to 1000 discrete, conidiating colonies can be plated on a dish. This is very valuable for such purposes as mutant isolation, although it may reduce the viability of some auxotrophs.

Spore Plating

Conidia are normally collected in a dilute solution of the detergent Tween 80 (0.01–0.1 percent) and shaken well with a benchtop mixer to break up spore chains. Conidia can be counted in a hemocytometer. Concentrations of Tween greater than 0.001 percent in the medium inhibit growth, so conidia may need to be washed by centrifugation before plating. Ascospores can be suspended in water alone. Neither type of spore requires any special treatment for germination, and they both will withstand temperatures up to 45°C, so that they can be suspended in agar at this temperature for top-layer platings.

A. *nidulans* will not grow under totally anaerobic conditions, therefore, cultures may fail if, for instance, the rim of a Petri dish becomes sealed to the lid by condensed moisture. Another problem not infrequently met with when subcultures are made from old conidia is the appearance of "fluffy" colonies carrying an excess of aerial mycelium (see below, under Hyphal Structure and Development). It is best to discard such colonies and attempt to purify the strain by plating or streaking out further spores.

Incubation

A. *nidulans* will grow at temperatures up to 45°C. At 37°C, which is the temperature most commonly selected, the vegetative cycle from conidium to conidium takes two days.

Mutant Isolation

Conidia can be mutated by any of the usual agents. Ultraviolet light has frequently been used at doses giving 5 percent survival, but it has the disadvantage that a high proportion of the resulting mutants carry translocations (Käfer, 1965). A very powerful mutagen frequently used at present is N-methyl-N´-nitro-N-nitrosoguanidine. Martinelli and Clutter-buck (1971) described a convenient procedure in which handling of the mutagen is minimized by employing it at pH 9 under which conditions it is destroyed as it acts.

Many special procedures have been described for the isolation of mutants, e.g., the biotin starvation (Pontecorvo *et al.*, 1953) or nystatin selection (Ditchburn and Macdonald, 1971) for auxotrophs. However, the most generally useful method is the replica-plating technique described by Mackintosh and Pritchard (1963), in which damp velvet is employed along with medium containing sodium deoxycholate. Earlier, a multipin replicator had been used for the same purpose (Roberts, 1959).

Crosses

Since A. *nidulans* is homothallic, any strain can be crossed with any other. Cleistothecia (also referred to as "perithecia") of three types will be formed: selfed cleistothecia of each parent type, and hybrids. Within each cleistothecium, however, all asci will usually be of similar origin. Strains to be crossed normally differ both in auxotrophic markers and in spore color, so that hybrid cleistothecia can readily be recognized. In addition, complementing auxotrophs favor the formation of hybrid cleistothecia when crosses are made on MM, as is the normal practice. Under these conditions, hybrid cleistothecia are usually larger than selfed ones (Baracho, *et al.*, 1970). Thick media and partial anaerobiosis favor cleistothecium formation, crosses are therefore normally initiated by mixing conidia of the two parent strains with a little CM in a restricted area (1–2 cm in diameter) in the center of a thick MM plate. Alternatively, a previously made heterokaryon may serve as the inoculum. Once growth has started, but before conidiation occurs, air exchange to the dish is restricted by sealing it with cellulose tape. Cleistothecia then take 7–10 days at 37°C to mature, when they form blackish-red spheres, up to 0.5 mm in diameter, covered with yellowish, thick-walled Hülle cells. If the cleistothecia are to be analyzed singly (cleistothecium analysis), they are cleaned of adhering conidia and Hülle cells by rolling them on 3-percent water agar and then crushed in a few milliliters of water. The asci within the cleistothecia normally break open spontaneously. It is then convenient to make trial platings on medium containing sodium deoxycholate (0.08 percent), one plate per cleistothecium, in order to determine whether the cleistothecium is hybrid and what dilution is required. "Twinned" cleisothecia which are partly of selfed and partly hybrid origin are not uncommon, so the ratio among the progeny of spore-color markers from the two parents should be checked. Full platings are then made at dilutions giving 20–30 colonies per dish.

The proportion of hybrid cleistothecia in a cross varies from 0 to 100 percent; this phenomenon has been termed "relative heterothallism" by Pontecorvo *et al.* (1953). If only selfed cleistothecia are found in a particular cross, more success may be obtained if the same cross is repeated or if slightly different parents are chosen. Increasing the sugar content of the medium may also help (Pontecorvo *et al.*, 1953; Acha and Villanueva, 1961; Leal and Villanueva, 1962).

An alternative procedure for analyzing crosses is to plate suspensions of ascospores from mixed, uncleaned cleistothecia on a medium which selects against both parents but will allow calculation of the relative frequencies of particular types of recombinants. This method is essential for analysis of intragenic recombination.

Classification of Progeny

Colonies resulting from a cross are inoculated onto a CM "masterplate" on which they are arranged in an array corresponding to the positions of needles on the multipin replicator with which they will be transferred to various test media. The replicator used in Glasgow consists of a perspex sheet carrying 25 pins in a 5 × 5 array, plus one eccentric pin as an orientation marker. Since conidia are readily scattered, the inoculating needle or wire is always stabbed upwards into the medium in an inverted dish. Control colonies of the two parents should always be classified along with the progeny of any cross.

Epistasis complicates the analysis of some crosses; spore-color mutants, for instance, form a series in which *w* is epistatic to *fw* which is epistatic to *y* and *yg*. Other mutants, e.g., *p* and *cha*, modify these colors. Clutterbuck (1972) has described methods by which *y* can be classified in the presence of markers epistatic to it.

Some auxotrophies, particularly amino acid requirements, may interact to reduce viability. This should be borne in mind when choosing markers for a cross. Wild-type colonies excrete biotin; therefore, plates for classification of this requirement should not be incubated for more than one day, to prevent "breast feeding" of requiring colonies by prototrophic ones (Pontecorvo *et al.,* 1953).

The nutritional requirement of individual segregants can be checked either by testing on various media as above or by auxanography (Ponte-corvo, 1949).

Heterokaryons

Heterokaryons of *A. nidulans* are unstable unless forced by balanced deficiencies (usually auxotrophies) in the parents. They can be obtained by inoculating conidia of the parent strains not more than 5 mm apart on solid CM; after one or two days, small pieces of mixed mycelium from the junction of the strains are transferred to solid MM (or other medium which will support the growth of the heterokaryon but not the parents). If selection against the parents is relaxed, the heterokaryon will break down, and this may occur if one component strain escapes by crossfeeding. Once beyond the region where crossfeeding can occur, growth of this component will stall, but its presence will block the growth of the genuine heterokaryon. Roberts (1964) found that heterokaryons could not be grown satisfactorily in liquid culture due to strong selection for diploids, and, in fact, this can be used as a deliberate procedure for obtaining diploids (Clutterbuck, 1969*c*).

When heterokaryons are used for complementation tests between auxotrophic mutants it is advisable first to establish the heterokaryon on MM supplemented with the requirement the two strains have in common. The heterokaryon can then be subcultured onto the test medium lacking this requirement.

Heterokaryon incompatibilities occur among wild strains of *A. nidulans* (Grindle, 1963a,b), and a "barrage" phenomenon has also been described by Patel (1973), but failure to form heterokaryons among Glasgow strains is rare (Da Cunha, 1970).

Diploids

Since the conidia of *A. nidulans* are uninucleate, conidia from a heterokaryon will contain only nuclei from either one or the other component. Roper (1952) used this fact to select for diploid conidia which contain the potentialities of both parents. The technique normally used to obtain diploids is to plate about 10^7 conidia from a heterokaryon (preferably by the top-layer method) on MM. Diploids are estimated to comprise 10^{-7} of such conidia, but they can be swamped by an excessive inoculum. It is customary to use complementing spore colors in addition to auxotrophies to aid in the recognition of the diploids, e.g., a heterokaryon between yellow-spored, proline-requiring and white-spored, biotin-requiring strains will yield a green, prototrophic diploid. An additional criterion for recognition of diploids is conidial diameter (normally measured in chains of three conidia); this is approximately 3 μ for haploids and 4 μ for diploids.

Diploids can also be obtained from heterokaryons in liquid cultures (see above), from the haploid component of a haploid/diploid heterokaryon (Clutterbuck and Roper, 1966), and from ascospores (Pritchard, 1954).

Haploidization

The technique previously used for obtaining haploid segregants from a heterozygous diploid was to select for homozygosity at two unlinked loci simultaneously (Pontecorvo and Käfer, 1958; Forbes, 1959a,b). Since mitotic crossing over is unlikely to occur simultaneously at two different sites, the majority of selected colonies are more likely to be the products of haploidization. Markers frequently used were *suAadE20* (a recessive suppressor of adenine requirement) and *acrA1* (semidominant acriflavine resistance). Lhoas (1961), however, developed observations of Morpurgo (1963b) into a technique by which haploids could be obtained from diploid colonies of *A. niger* without selection. The technique as applied to *A. nidulans* is to stab diploid conidia into a medium consisting of CM plus 7 μg/ml *p*-fluorophenylalanine. The colonies resulting after 4–7 days incubation will contain many haploid sectors which can be purified and classified for the markers they carry. Some amino acid auxotrophs may be inhibited by *p*-fluorophenylalanine, in which case more success may be obtained with other haploidizing agents such as arsenate (Van Arkel, 1963), benlate (Hastie, 1970), or acridine yellow (70 μg/ml) (Apirion and Kinghorn, private communications).

Mapping

A new mutant can most readily be localized to a linkage group by means of the parasexual cycle. Master strains have been developed by McCully and Forbes (1965) which have markers on each of the eight linkage groups, and diploids made with them can be haploidized with p-fluorophenylalanine. Since crossing over does not normally accompany haploidization, linkage groups segregate as units, and a new marker should be found to be totally linked to one of the eight master-strain markers. Linkage between any two of the master-strain markers themselves is evidence of a translocation in the tested strain (Käfer, 1962, 1965). In meiotic crosses translocations may also give rise to disomic colonies which can be used to identify the chromosomes concerned (Käfer and Upshall, 1973). Strong selection may occur against some mutants during haploidization (Lanier, 1967), but this can be alleviated in many cases by adequate supplementation of the medium in the case of an auxotroph or by the choice of a different haploidizing agent. Since linkage is normally complete, only small numbers of haploids are required to locate a mutant and detect any translocations (if mitotic crossing over happens to occur before haploidization, this will usually lead to total loss of one allele or else to unexpected coupling of alleles from different parents, rather than to apparent partial linkage).

After localization of new mutants to a linkage group, detailed meiotic mapping is normally carried out by cleistothecium analysis. Dorn (1972) has devised a computer program for the automatic calculation of linkages from cross data.

Mapping by mitotic crossing over can also be carried out in diploids (Pontecorvo et al., 1954; Pontecorvo and Käfer, 1956, 1958). The technique requires a series of markers in coupling, including, in a distal position, a marker such as a spore color or resistance mutant which can be selected for homozygosity. Homozygous sectors are picked up from colonies obtained by plating diploid conidia on CM or appropriate resistance selective medium. The homozygous diploid colonies so obtained are then purified and tested for the other markers employed. Vegetative crossing over, if followed by a suitable segregation of the strands at mitosis, leads to homozygosity for all markers distal to the crossover point. Analysis of the products of mitotic crossing over can therefore give the

TABLE 1. Gene Loci of Aspergillus nidulans

Gene symbol	Linkage group	Phenotype and enzyme systems affected[a]	References
aax	(VI)	Allantoic acid non-ut, allantoicase	Darlington and Scazzocchio, 1967, 1968; Darlington et al., 1965; Scazzocchio and Darlington, 1967, 1968
aauA	(VII)	Amino acid upt.	Pateman and Kinghorn, 1974
aauB	VII	Amino acid upt.	
aauC	II	Amino acid upt.	
aauD	VIII	Amino acid upt.	
abA (abl)	II	α-aminobutyric acid req.	Forbes, 1959a
abaA (aba)	VIII	"abacus" aconidial	Clutterbuck, 1969c
ac (see fac, mas, icl)			
aco	—	Aconidial	Barbata et al., 1973
acrA (Acr1)	II	Acriflavine res.	Ball and Roper, 1966; Roper and Käfer, 1957
acrB (acr2)	II	Acriflavine res.	Warr and Roper, 1965
actA (Act)	III	Actidione res.	Foley et al., 1965; Sadasivam et al., 1969
adA	I	Adenine req., adenylosuccinase	Foley et al., 1965
adB	VIII	Adenine req., AMP succinate synthetase	Foley et al., 1965; Katz and Rosenberger, 1970c; Pontecorvo et al., 1953
adC (ad1)	II	Adenine req., AIR carboxylase	
adD (ad3)	II	Adenine req., AIR carboxylase	
adE (ad8)	I	Adenine req.	Pontecorvo et al., 1953
adF (ad9)	I	Adenine req.	Calef, 1957; Käfer, 1958
adG (ad14)	I	Adenine req.	
adH (ad23)	II	Adenine req.	Käfer, 1958
adI (ad50)	III	Adenine req.	Dorn and Rivera, 1965
ade (see ad)			
ag-r (see azgA)			

TABLE 1. Continued

Gene symbol	Linkage group	Phenotype and enzyme systems affected[a]	References
alX	III	Allantoin non-ut.; allantoinase	Darlington and Scazzocchio, 1967; Scazzocchio and Darlington, 1967, 1968
alcA	(VII)	Ethanol non-ut., alcohol dehydrogenase	Page and Cove, 1972
aldA	(VIII)	Ethanol non-ut., aldehyde dehydrogenase	Darlington and Scazzocchio, 1967
alpA (alp)	?	Allantoin non-ut., allantoin permease	
am (see nir)			
amdR	(II)	Acrylamide non-ut., acetamidase regl.	Dunsmuir and Hynes, 1973; Hynes, 1970, 1972, 1973a, b; Hynes and Pateman, 1970a, b, c
amdS	III	Amide non-ut., acetamidase	
amdT (cf. areA)	(III)	Acrylamide ut., acetamidase regl.	
ammE (see amrA)			
amrA (ammE)	(II)	Ammonium derepressed	Kinghorn and Pateman, 1973a; Pateman et al., 1973
anA (an1)	I	Aneurine req.	Pontecorvo and Käfer, 1958
anB (an2)	II	Aneurine req.	Forbes, 1959a
aniA	?	Arginase repressed	Bartnik et al. 1973b
anthA–D	?	Anthranilate req.	Noronha, 1970
apA (ap)	(VIII)	Aminopterin res.	Apirion et al., 1963
aplA (allp)	(VI)	Allopurinol res., xanthine dehydrogenase regl.	Darlington and Scazzocchio, 1967, 1968; Darlington et al., 1965; Holl and Scazzocchio, 1970; Scazzocchio and Darlington, 1967; Scazzocchio et al., 1973
apsA	IV	Anucleate primary sterigmata	Clutterbuck, 1969b
apsB	VI	Anucleate primary sterigmata	
areA (amdT, xprD)	III	Ammonium derepressed	Arst and Cove, 1973
argA (arg1)	(VI)	Arginine req., arginosuccinase	Bainbridge et al., 1966; Cybis and Weglenski, 1969; Cybis et al., 1972b; Dorn,

Gene	Chromosome	Phenotype	Reference
argB (arg2)	III	Arginine req.	1967a; Forbes, 1959a; Pontecorvo et al., 1953
argC (arg3)	VIII	Arginine req.	
aroA	VIII	Aromatic metabolite req.	Roberts, 1969; Zaudy, 1969
aroB	(V)	Aromatic metabolite req.	Darlington and Scazzocchio, 1967
aroC	I	Aromatic metabolite req.	
azgA	?	8-Azaguanine res., purine upt.	
bgaA	?	Galactose non-ut., β-galactosidase	Fantes and Roberts, 1973
bgaB	?	Galactose non-ut., β-galactosidase	
bgaC	?	Galactose non-ut., β-galactosidase	
benA (ben1)	VIII	Benlate res.	Hastie and Georgopoulos, 1971
benB (ben2)	II	Benlate res.	
bge (see fw)			
biA (bil)	I	Biotin req.	Roper, 1950
bio (see bi)			
b1A (bl1)	(II)	Blue ascospores	Apirion, 1963
br (see brl)			
brlA (brl, br)	VIII	"Bristle" aconidial	Clutterbuck, 1969c, 1970b, 1973b
bwA (Bw)	(VI)	Brown conidia	Käfer, 1958
chaA (cha)	VIII	Chartreuse conidia	Clutterbuck, 1972; Käfer, 1961
choA (cho)	VII	Choline req.	Arst, 1968, 1971; Käfer, 1958
clA (cl4)	(IV)	Colorless ascospores	Apirion, 1963
clB (cl6)	I	Colorless ascospores	
cnxA, B, C (ni50)	VIII	Nitrate and hypoxanthine non-ut	Arst, et al., 1970; Cove and Pateman, 1963;
cnxE (ni3)	II	Nitrate and hypoxanthine non-ut.	Dorn and Rivera, 1965; Downey, 1973a,b; Holl, 1971 Käfer, 1958; Pateman et al., 1964, 1967; Scazzocchio et al., 1973; Sorger, 1963
cnxF	(VII)	Nitrate and hypoxanthine non-ut.	
cnxG	(VI)	Nitrate and hypoxanthine non-ut.	
cnxH	III	Nitrate and hypoxanthine non-ut.	
coA (co)	VIII	Compact morphology	Forbes, 1959a; Käfer, 1958
creA	I	Carbon derepressed	Arst and Cove, 1973
csuA	(V)	Choline-O-sulfate non-ut.	Arst, 1971; Scott and Spencer, 1968
cys (see sC)			

TABLE 1. Continued

Gene symbol	Linkage group	Phenotype and enzyme systems affected[a]	References
dilA (dil)	(III)	Dilute conidial color	Jansen, unpublished
drkA (drk)	VII	Dark conidial color	Clutterbuck, 1969c
drkB	II	Dark conidial color, sup. of brlA12	Clutterbuck, 1970b
est	—	Esterase variants	Dorn and Rivera, 1965
f (see fac)			
facA	V	Fluoroacetate res.; acetate non-ut., acetyl-coA synthetase	Apirion, 1965, 1966; Romano and Kornberg, 1968, 1969
facB	VIII	Fluoroacetate res., acetate non-ut.	Apirion, 1965, 1966
facC	VIII	Fluoroacetate res., acetate non-ut.	
fanA	(V)	Fluoroacetate res.	
fanB	(VII)	Fluoroacetate res.	
fanC	(VI)	Fluoroacetate res.	Apirion, 1965
fanD	(VIII)	Fluoroacetate res.	
fanE	(VI)	Fluoroacetate res.	
flA (fl)	VII	Fluffy morphology	Azevedo, 1965; Ball and Azevedo, 1964
fluA (flu2)	VIII	Fluffy morphology	Dorn, 1970; Dorn et al., 1967; Purnell and Martin, 1973b
fluB (flu3)	VIII	Fluffy morphology	
fluC (flu7)	I	Fluffy morphology	
fluD (flu11)	VIII	Fluffy morphology	Dorn, 1970
fmdS	(III)	Formamide non-ut., formamidase	Hynes, 1970, 1972; Hynes and Pateman, 1970a, b, c
fp (see fpa)			
fpaA (fpA, pf, pfp, tyrA)	I	p-Fluorophenylalanine res., Prephenate dehydrogenase	Sinha, 1967, 1972; Warr and Roper, 1965
fpaB (fpB)	I	p-Fluorophenylalanine res.	Sinha, 1970, 1972
fpaD (fpD, fpC)	VIII	p-Fluorophenylalanine res.	Sinha, 1969, 1972

Gene	Linkage group	Description	References
fpaE (see *trypA*)			
frA (*fr, suc*)	IV	Fructose non-ut.	Roberts, 1973a
fwA (*fu, bge*)	VIII	Fawn conidia	Clutterbuck, 1965; Wood and Käfer, 1967
galA (*gal1*)	III	Galactose non-ut., galactokinase and galactose–1–P uridyl transferase regl.	
galB (*gal3*)	(II)	Galactose and arabinose non-ut.	Arst and Cove, 1970; Katz and Rosenberger, 1970a; Roberts, 1963a, b, 1970
galC (*gal4*)	VIII, IV?	Galactose non-ut., molybdate res.	
galD (*gal5*)	I	Galactose non-ut. galactose–1–P uridyl transferase	
galE (*gal9*)	III	Galactose non-ut., galactokinase	
galF (*gal2*)	(VIII)	Galactose non-ut.	
galG	VIII	Galactose non-ut., sup of *brlA12*	Clutterbuck, 1970b, unpublished
gamA	(II)	Galactose non-ut., molybdate res.	
gamB	VIII	Galactose non-ut., molybdate res.	Arst and Cove, 1970
gamC	?.	Galactose non-ut., molybdate res.	
gdhA	III	Ammonium sensitive; NADP-glutamate dehydrogenase	Pateman, 1969; Arst and MacDonald, 1973; Kinghorn and Pateman, 1973a, b, c; Pateman, et al., 1973
gdhB	(IV)	Glutamate non-ut., NAD-glutamate dehydrogenase	
gdhC	(III)	Glutamate ut., NAD-glutamate dehydrogenase regl.	
glu	?.	Glutamate req.	Clutterbuck, 1969a
hisA	IV	Histidine req.	
hisB	I	Histidine req., IGP dehydrase	
hisC	VIII	Histidine req., IAP transaminase	
hisD	VIII	Histidine req., histidinol dehydrogenase	
hisE	VIII	Histidine req., PR-ATP pyrophosphorylase	Berlyn, 1967; Pees, 1966
hisF	(VII)	Histidine req.	
hisG	(II)	Histidine req., PRP-ATP pyrophosphorylase	
hisH	(VIII)	Histidine req.	
hisI	VIII	Histidine req., PR-AMP cyclohydrase	

TABLE 1. *Continued*

Gene symbol	Linkage group	Phenotype and enzyme systems affected[a]	References
hisJ (hisEL, his122)	VII	Histidine req.	Darlington and Scazzochio, 1967, 1968; Darlington et al., 1965; Holl and Scazzochio, 1970; Scazzocchio and Darlington, 1967; Scazzochio et al., 1973
hxA	V	Hypoxanthine non-ut., xanthine dehydrogenase I	
hxB	(VII)	Hypoxanthine non-ut., xanthine dehydrogenase I and II	
hxnC	(VIII)	Nicotinate non-ut., xanthine dehydrogenase II	Scazzochio et al., 1973
iclA (icl)	V	Acetate non-ut., isocitrate lyase	Armitt et al., 1970
ileA (ile)	(II)	Isoleucine req.	Pees, 1966
indA (ind)	?	Indole req.	Noronha, 1970
inoA	(II)	Inositol req.	Forbes, unpublished
inoB	IV	Inositol req.	Clutterbuck, unpublished
iodA (Iod)	II	Iodoacetate res.	Warr and Roper, 1965; Georgopoulos and Georgadis, 1969
ivoA	III	Ivory conidiophores	Clutterbuck, 1969c, 1973b; Oliver, 1972;
ivoB	VIII	Ivory condiophores, phenol oxidase	Clutterbuck, 1969c, 1973b
lacA (lac1)	VI	Lactose non-ut.	Roberts, 1963a
lacB (lac3)	II	Lactose non-ut.	
lacC	(VII)	Lactose non-ut., β-galactosidase	
lacE	?. ?.	Lactose non-ut., lactose upt.	
lacF	?	Lactose non-ut., lactose upt.	Gajewski, et al., 1972; Paszewski et al., 1970
lacG	(VI)	Lactose non-ut., β-galactosidase	
lacH	?	Lactose non-ut., β-galactosidase	
lacI	(VI)	Lactose non-ut., β-galactosidase	
leu	?	Leucine req.	Dhillon and Garber, 1970

Gene	Group	Description	References
luA (lu)	I	Leucine req.	Forbes, 1959a; Sinha, 1970
lysA (lys1)	VI	Lysine req.	Aspen and Meister, 1962; Käfer, 1958; Pees, 1966; Pollard et al., 1968; Pontecorvo et al., 1953
lysB (lys5)	V	Lysine req.	
lysC (lys6)	(VII)	Lysine req.	
lysD (lys7)	VII	Lysine req.	
lysE (lys10)	V	Lysine req.	
lysF (lys51)	I	Lysine req.	
m	—	Mycelial growth form	Kwiatowski and Bohdanowicz, 1962; Roper, 1958
malA (mal)	VII	Maltose non-ut.	Roberts, 1963a
masA (mas)	I	Acetate non-ut., malate synthetase	Armitt et al., 1971
mauA (mau)	IV	Monoamine non-ut., monoamine oxidase	Arst and Cove, 1969; Page and Cove, 1972
mauB	II	Monoamine non-ut., monoamine oxidase	Page and Cove, 1972
mauC	(II)	Monoamine non-ut. in pu strains	
meaA (mea8)	IV	Methylammonium res., ammonium derepressed	Arst and Cove, 1969, 1973; Arst and Page, 1973; Pateman et al., 1973
meaB (mea6)	III	Methylammonium res.	
mecA (meth-i)	I	Inhibited by methionine, cystathionine β-synthetase	
mecB	I	Inhibited by methionine, γ-cystathionase	Pieniazek et al., 1973a, b; Paszewski and Grabski, 1973
mecC	?	Inhibited by methionine, methionine adenosyltransferase	
medA (med)	I	"Medusa" morphology	Clutterbuck, 1969c
melA	(VII)	Excess melanin formation	Martinelli and Bainbridge, 1974
melB	VII	Reduced melanin formation	
met (see meth)			
methA	II	Methionine req.	Forbes, 1959a; Gajewski and Litwenska, 1968; Käfer, 1958; Putrament et al., 1970
methB (meth3)	VI	Methionine req.	
methC	(I)	Methionine req.	
methD	III	Methionine req.	
methE	(VII)	Methionine req.	

TABLE 1. Continued

Gene symbol	Linkage group	Phenotype and enzyme systems affected[a]	References
methF	(IV)	Methionine req.	
methG (meth1)	IV	Methionine req.	
methH (meth2)	III	Methionine req.	
methi-i (see mecA)			
mgA (mg1)	?	Malachite green res.	Warr and Roper, 1965
moA (mol)	I	Morphologically abnormal	Bainbridge, 1966, 1970; Bainbridge and Trinci, 1969
moB (mo9)	VIII	Morphologically abnormal	
moC (mo96)	III	Morphologically abnormal	
molA	VI	Molybdate res.	Arst and Cove, 1970; Arst et al., 1970
molB	(II)	Molybdate res.	
ni (see nia, nir, cnx)			
niaD (ni7)	VIII	Nitrate non-ut., nitrate reductase	Cove, 1966, 1967, 1969, 1970; Cove and Coddington, 1965; Cove and Pateman, 1963, 1969; Downey, 1971, 1973a, b; Downey and Cove, 1971; Holl 1971; Pateman et al., 1967
nicA (nic2)_	V	Nicotinic or anthranilic acid req.	
nicB (nic8)	VII	Nicotinic or anthranilic acid or tryptophan req.	Käfer, 1958; Pontecorvo et al., 1953; Roberts, 1968
nicC (nic10)	VI	Nicotinic acid req.	Käfer, 1958
niiA	VIII	Nitrite non-ut., nitrite reductase	Cove, 1966, 1967, 1969, 1970; Cove and Coddington, 1965; Dorn and Rivera, 1965; Pateman and Cove, 1967; Pateman et al., 1964, 1967
niiB (see nir)			
nirA (ni51, niiB, nir, am2)	VIII	Nitrite non-ut., nitrate and nitrite reductases regl.	

npeA	?	Reduced penicillin production	Holt and Macdonald, 1968*a, b*
oli	(VII)	Oligomycin res.	Rowlands and Turner, 1973
(*oliA*)	Cytoplasmic	Oligomycin res.	Rowlands and Turner, 1973
ornA (*orn4*)	IV	Ornithine req.	Cybis and Weglenski, 1969; Cybis *et al.*, 1972*b*; Käfer, 1958; Pontecorvo *et al.*, 1953
ornB (*orn7*)	VIII	Ornithine req.	Cybis and Weglenski, 1969; Cybis *et al.*, 1972*b*; Forbes, 1959*a*
otaA (*ota*)	(VII)	Arginine non-ut., ornithine transaminase	Cybis and Weglenski, 1969; Cybis *et al.*, 1972*a, b*; Piotrowski *et al.*, 1968; Bartnik *et al.*, 1973*a, b*; Stevens and Heaton, 1973
oxpA (*oxp*)	?	Oxallopurine res.	Holl and Scazzochio, 1970
pA (*p*)	V	Pale conidia	Clutterbuck, 1968*a*; Van Arkel, 1962
pab (see *paba*)			
pabaA (*paba1*)	I	*p*-aminobenzoic acid req.	Pontecorvo *et al.*, 1953
pabaB (*paba22*)	IV	*p*-aminobenzoic acid req.	Luig, 1962; Siddiqi, unpublished
pacA	(IV)	Acid phosphatase deficient	
pacB	VIII	Acid phosphatase deficient	
pacC	VI	Acid phosphatase deficient	
palA	III	Alkaline phosphatase deficient	
palB	VIII	Alkaline phosphatase deficient	
palC	IV	Alkaline phosphatase deficient	
palD	VII	Alkaline phosphatase deficient	
palE	VIII	Alkaline phosphatase deficient	
palF	VII	Alkaline phosphatase deficient	
palcA	II	Acid and alkaline phosphatase deficient	Arst and Cove, 1970; Dorn, 1965*a,b*, 1967*b*, 1968; Harsanyi and Dorn, 1972; Purnell and Martin, 1971
palcB	III	Acid and alkaline phosphatase deficient	
palcC	VIII	Acid and alkaline phosphatase deficient	
pantoA (*panto*)	III	Pantothenic acid req.	Käfer, 1958

TABLE 1. Continued

Gene symbol	Linkage group	Phenotype and enzyme systems affected[a]	References
pantoB	(VII)	Pantothenic acid req.	Rever, unpublished
pcnbA (pcnb)	III	Pentachloronitrobenzene res.	Trelfall, 1968, 1972
pdhA (pdh)	?	Acetate req., pyruvate dehydrogenase	Romano and Kornberg, 1968, 1969
pdx (see pyro)			
penA (pen)	?	Increased pencillin production	Holt and Macdonald, 1968a, b
pf and pfp (see fpa)			
phenA (phen2)	III	Phenylalanine req.	Käfer, 1958; Sinha, 1967
phenB	VII	Phenylalanine req.	Sinha, 1967
ppaA	?	Phenylpyruvate req.	Sinha, unpublished
pppA	?	Pentose phosphate pathway defective	Cove, 1970; Hankinson and Cove, 1972a
pppB	?	Pentose phosphate pathway defective	
pr (see xpr)			
prnA	(VII)	Proline non-ut.	Arst and Cove, 1973
prnB	(VII)	Proline non-ut.	
proA (pro1)	I	Proline req.	Cybis et al., 1972b; Forbes, 1959b; Weglen-
proB (pro3)	I	Proline req.	ski, 1966
puaA (pu)	II	Putrescine req.	Sneath, 1955
purA	?	Purine res.	Arst et al., 1970
pycA	III	Glutamate req., pyruvate carboxylase	Skinner and Armitt, 1972
pyrA	(VII)	Pyrimidine req., pyrimidine-specific carbamyl phosphate synthetase	
pyrB	(VIII)	Pyrimidine req., aspartate transcarbamylase	
pyrC	(VIII)	Pyrimidine req., aspartate transcarbamylase	
pyrD	VIII	Pyrimidine req., dehydro-orotase	
pyrE	I	Pyrimidine req., dehydro-orotate dehydrogen-ase	L. M. Palmer and Cove, unpublished

Gene	Group	Description	Reference
pyrF	I	Pyrimidine req., orotidine MP-pyrophosphorylase	
pyrG	(I)	Pyrimidine req., orotidine MP decarboxylase	
pyroA (*pyro4*)	IV	Pyridoxine req.	Käfer, 1958
rA	I	Enhanced phosphatase	Dorn, 1965*a*; Dorn and Rivera, 1966
rB	II	Enhanced phosphatase	
rC	VIII	Enhanced phosphatase	Dorn and Rivera, 1965
rib (see *ribo*)			
riboA (*ribo1*)	I	Riboflavin req.	Cove, unpublished; Dorn, 1967*a*; Forbes, Shanmugasundaram and Roberts, un-
riboB (*ribo2*)	VIII	Riboflavin req.	published; Käfer, 1958; Panicker and
riboC (*ribo3*)	(V)	Riboflavin req.	Shanmugasundaram, 1962; Radha and
riboD (*ribo5*)	V	Riboflavin req. for nitrate ut.	Shanmugasundaram, 1962; Sadique *et*
riboE (*ribo6*)	II	Riboflavin req.	*al.*, 1966*a, b, c, d*
riboF (*ribo8*)	(I)	Riboflavin req.	Devi and Shanmugasundaram, 1969
rinA	VIII	Riboflavin req. for nitrate ut.	
rinB	I	Riboflavin req. for nitrate ut.	Cove, unpublished
rinC	V	Riboflavin req. for nitrate ut.	
sA (*s1*)	III	Sulfate non-ut., PAPS reductase	
sB (*s3*)	VI	Sulfate non-ut., sulfate upt.	Arst, 1968; Arst and Cove, 1970; Dorn
sC (*s0, s12, cys2*)	III	Sulphate non-ut., ATP sulfurylase	and Rivera, 1965; Gravel *et al.*, 1970;
sD (*s50*)	VIII	Sulfate non-ut., APS kinase	Käfer, 1958; Spencer and Moore, 1973
sE	VIII	Sulfate non-ut., PAPS reductase	
sF	(VII)	Sulfate non-ut.	
saA (*sa*)	II	Sage conidia	Kilbey, 1960
sbA (*sb*)	(VI)	Sorbitol non-ut.	Roberts, 1963*a*, 1964
sgpA (*sgp1*)	(VII)	Slow growing, noncleistothecial	
sgpB (*sgp2*)	(V)	Slow growing, noncleistothecial	Houghton, 1970, 1971
sgpC (*sgp3*)	III	Slow growing, noncleistothecial	
sgpD (*sgp4*)	I	Slow growing, noncleistothecial	

TABLE 1. Continued

Gene symbol	Linkage group	Phenotype and enzyme systems affected[a]	References
sgpE (sgp5)	(IV)	Slow growing, noncleistothecial	Käfer, 1958
smA (sm)	III	Small colony	Elorza and Arst, 1971
sorA	(I)	Sorbose res., sorbose upt.	
sorB	III	Sorbose res., phosphoglucomutase	
stuA (stu)	I	Stunted conidiophores	Clutterbuck, 1969c
suAadE20 (su1ad20)	I	Suppressor of adE20	Pritchard, 1955
su-flu	—	Suppressors of fluffy morphology	Dorn, 1970
suApabaB22 (su1paba22)	IV	Suppressor of pabaB22	Luig, 1962
suApalB7	VIII	Suppressor of palB7	Arst and Cove, 1970, Dorn, 1965 a, b
suBpalB7	(VI)	Suppressor of palB7	Dorn, 1967a
suCpalF15	V	Suppressor of palF15	Dorn, 1967a
suDpalA1	(I)	Suppressor of palA1	Arst and Cove, 1970
suXpalC4	?	Suppressor of palC4	
suApro (Su1pro)	III	Suppressor of pro	Cybis et al., 1970, 1972a, b; Forbes, unpublished; Piotrowska et al., 1969; Weglenski; 1966, 1967
suBpro (Su4pro)	III	Suppressor of pro	
suCpro (su6pro)	III	Suppressor of pro	
suDpro (su19pro)	I	Suppressor of pro, arginine regl.	
suEpro (su11pro)	(V)	Suppressor of pro, arginine regl.	
suFpro	?	Suppressor of pro, carbon repression	Bartnik et al., 1973b
suc (see fr)			
sulA (sulI)	I	Sulfanilamide res.	Jansen, unpublished
sul-reg	?	Arylsulfatase derepressed	Siddiqi et al., 1966
sup	—	Suppressors of meth	Ayling, 1969; Gajewski and Litwinska, 1968; Lilly, 1965; Putrament et al., 1970; Paszewski and Grabski, 1973

Gene	Linkage	Phenotype	Reference
TB1 (I; II)	I	Translocation breakpoint	Azevedo and Roper, 1970; Cooke *et al.*, 1970; Nga and Roper, 1968, 1969; Roper and Nga, 1969
TB1 (I; VII)	VII	Translocation breakpoint	Käfer, 1965
TB1 (III; VIII)	III and VIII	Translocation breakpoint	Bainbridge, 1970; Bainbridge and Roper, 1966; Käfer, 1962, 1965
TB2 (III; VIII)	VIII	Translocation breakpoint	Clutterbuck, 1970b
TB1 (III; V)	(III)	Translocation breakpoint	Ball, 1967
tcnb (see pcnb)			
teA (te6)	(III)	Teoquil res.	Warr and Roper, 1965
telA	VII	Mound (tel)-shaped colony	Clutterbuck, unpublished
thiA (thi4)	II	Thiazole req.	Käfer, 1958
ths (see s)			
try (see tryp)			
trypA (fpaE)	II	Tryptophan req., anthranilate synthetase	
trypB	I	Tryptophan req., tryptophan synthetase	
trypC (ind2)	VIII	Tryptophan req., IGP synthetase, PRA isomerase, anthranilate synthetase	Hutter and DeMoss, 1967; Käfer and DeMoss, 1973; Noronha, 1970; Roberts, 1967, 1968; Sinha, 1967, 1972
trypD	(II)	Tryptophan req., PR transferase	
trypE	(VI)	Tryptophan req.	
tsA	II	Temperature sensitive	
tsB	VI	Temperature sensitive	Forbes and Sinha, 1966
tsC	II	Temperature sensitive	
tsD (ts)	VIII	Temperature sensitive	Forbes, unpublished
tsE (ts6)	?	Temperature sensitive	Cohen *et al.*, 1969; Katz and Rosenberger, 1970b, 1971b
tyrA (see fpaA)			
tyrB	(III)	Tyrosine req. (with *fpaA*), phenylalanine hydroxylase	Sinha, 1967
uX	VIII	Urea non-ut.	Darlington *et al.*, 1965

TABLE 1. Continued

Gene symbol	Linkage group	Phenotype and enzyme systems affected[a]	References
uY	(VII)	Urea non-ut.	Darlington et al., 1965
uZ	VIII	Urea non-ut.	
uaX	VI	Uric acid non-ut.	
uaY	VIII	Uric acid non-ut., xanthine dehydrogenase and urate oxidase regl.	Darlington and Scazzocchio, 1967, 1968; Darlington et al., 1965; Scazzocchio and Darlington, 1967; 1968; Scazzocchio et al., 1973
uaZ	VIII	Uric acid non-ut., uric acid upt.	Darlington and Scazzocchio, 1967
uapA (uap)	(I)	Uric acid res., uric acid upt.	Dunn and Pateman, 1972
uruA	(VIII)	Thiourea res., urea upt.	Kameneva and Romanova, 1969; Kameneva and Evseeva, 1973
uvr	?	UV repair defective	Jansen, 1967
uvsA (uvs1)	I	UV sensitive	
uvsB	IV	UV sensitive	Fortuin, 1971 a, b, c, d; Jansen, 1970a,b, 1972
uvsC	VIII	UV sensitive	
uvsD	V	UV sensitive	
uvsE	V	UV sensitive	
uvsF[b]	I	UV sensitive	Shanfield and Käfer, 1969
uvsG	(VIII)	UV sensitive	Käfer and de la Torre, unpublished
uvs (further mutants)		UV sensitive	Lanier et al., 1968; Wohlrab and Tuveson, 1969; Wright and Pateman, 1970
v-1	V	"Deteriorated variants": only these three out of 18 mapped are shown in Figure 1	
v-10	VII		Azevedo and Roper, 1970
v-11	V		
veA (ve)	VIII	Velvet morphology	Clutterbuck, 1965; Käfer and Chen, 1964

Gene	Group	Phenotype	Reference
wA (w)	II	White conidia	Pontecorvo et al., 1953
wetA (wet)	VII	Wet-white conidia	Clutterbuck, 1969c
xprA	?	Extracellular protease deficient	Cohen, 1973b
xprB	?	Extracellular protease deficient	
xprC	?	Extracellular protease deficient	
xprD (cf. *areA*)	III	Ammonium derepressed	Cohen, 1972, 1973b, c; Pateman et al., 1973
yA (y)	I	Yellow conidia, p-diphenol oxidase	Clutterbuck, 1972; Pontecorvo et al., 1953
yel see *y*			
ygA (yg)	II	Yellow-green conidia	Clutterbuck, 1972. Dorn, 1967a

[a] Abbreviations: regl. = regulation, req. = requiring, res. = resistant, sup. = suppressor, upt. = uptake, ut. = utilizing.
[b] In Shanfield and Käfer (1969), *uvsB*.

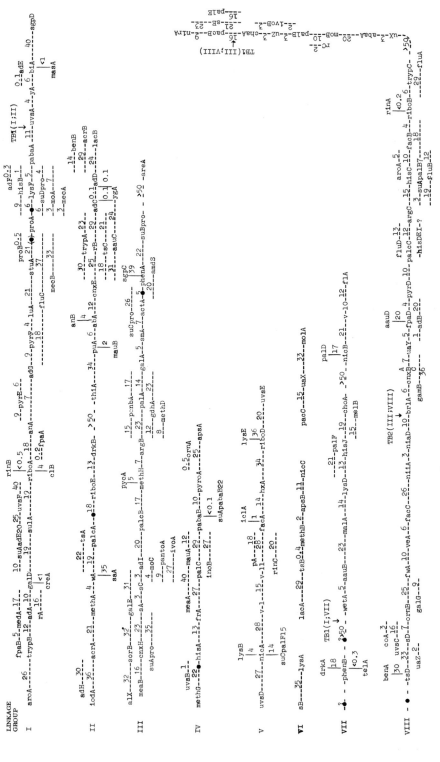

Figure 1. Linkage maps of Aspergillus nidulans.

order of the markers relative to the centromere and distances between them in relative frequencies of mitotic recombination

Genetic Loci

Table 1 contains a list of all locus symbols known to have been used in *A. nidulans*. Most symbols believed to be in current use have been modified by the addition of a locus-specific letter, according to the proposals made by Clutterbuck (1973*a*). The proposals are the outcome of some discussions on gene nomenclature in the *Aspergillus Newsletter* (Sermonti, 1968; Clutterbuck, 1968*b*, 1969*d*; Roper and De Azevedo, 1970). Earlier versions of the symbols are shown in parenthesis. Where the linkage group of a locus is shown in parenthesis, the mutant has only been located by haploidization and not mapped meiotically; therefore, it is not shown on the map (Figure 1). Phenotypes are necessarily described only very briefly; the references should be consulted for details. Mention of an enzyme affected by a locus does not mean that the relationship between the two has been fully determined, and, in fact, this is rarely the case. Where the locus is said to be concerned with regulation, this usually indicates that hyperactive mutants have been mapped at this locus. The references in Table 1 are concerned mainly with either the physiology or map position of the locus; references to the use of loci for mutation or intragenic recombination studies will be found in Tables 2 and 3.

Linkage Map

Figure 1 gives an updated version of the maps published by Käfer (1958), Dorn (1967*a*), and Clutterbuck and Cove (1974). Most of the data from which it is compiled will be found in the references to individual loci given in Table 1, but some have been obtained through private communications. Although map distances are usually self-consistent in *A. nidulans*, this map has been compiled from a variety of sources and should not be taken as standard. The majority of the distances, which are given as percentage of recombinants, have been calculated from crosses involving not less than 200 progeny, but since viability disturbances, etc., could not be taken into account, decimals and standard errors have been deliberately omitted.

Translocations are readily detected in *A. nidulans* (Käfer, 1962, 1965; Käfer and Upshall, 1973) and have been taken into account where necessary. Inversions and intrachromosomal transpositions, however, are less easily detected and none has yet been firmly established. They may, nevertheless account for some of the discrepancies in mapping data. Such discrepancies are known to occur in the following regions: linkage group III, *suBpro–suCpro*, linkage group IV *inoB–apsA;* linkage group VIII, most of the proximal half. Where such disputed map distances exist, this map for the most part incorporates the longest estimate for a particular interval, on the basis that crosses are more likely to be heterozygous than homozygous for an aberration, and this will lead to spurious diminution of map distances.

Locus symbols are in the form proposed by Clutterbuck (1968*b*, 1969*d*, 1973*a*); earlier forms of the symbols can be found in Table 1. Centromeres are shown thus: --●--. The centromere of linkage group I behaves as if it were diffuse, hence two locations are given (Käfer, private communication). The relationship between the three fragments of linkage group VI is unknown, although they are shown by haploidization to form one linkage group. The position relative to the centromere of the left-hand fragment of linkage group VII is also uncertain.

Research Topics

The Parasexual Cycle

A. nidulans was the main fungus involved in the discovery of the parasexual cycle. The cycle consists of the following independent steps (see reviews by Pontecorvo, 1954, 1956; and Roper, 1961, 1966*b*): (1) heterokaryon formation, (2) nuclear fusion to form a heterozygous diploid, (3) mitotic crossing over, and (4) haploidization. Practical details of these steps have already been given above, but some theoretical points remain.

Heterokaryons in *A. nidulans* are quite readily formed but are only maintained under selection. A heterokaryon, in fact, consists of a mixture of homokaryotic and heterokaryotic hyphae with various nuclear ratios (Pontecorvo, 1947; Clutterbuck and Roper, 1966) and it seems likely that the heterokaryon is maintained only by selection among these hyphae for those with optimal nuclear ratios.

Diploids are estimated to form at a frequency of approximately 10^{-7} per mitosis (Roper, 1952). Ultraviolet light or camphor can increase this frequency. Diploid strains behave in a manner similar to haploid ones, but they differ in that they have conidia with twice the volume and DNA content of haploid ones (Heagy and Roper, 1952). Diploid hyphae have the same dimensions as haploid ones, but have half the number of nuclei per unit volume (Clutterbuck, 1969*b*). Diploid strains have also been shown to be more pathogenic than haploid ones when injected into mice (Purnell and Martin 1973*a*). Diploid strains form cleistothecia which are largely sterile (Pontecorvo *et al.* 1953, Elliott 1960*a*) but do contain some diploid products as well as a few recombinant haploid ascospores (Garber *et al.*, 1961). Triploids can also be obtained, but they are less stable than diploids (Elliott, 1956).

Mitotic crossing over at some point in the genome is estimated (Käfer, 1961) to occur at about two percent of diploid mitotic divisions, but this frequency may be increased by various agents (see below, Recombination). Recombinants are normally detected as homozygous products resulting from the segregation of a recombinant chromatid along with one of the parental chromatids; however, recombination between loosely linked markers is reciprocal and both products may be recovered from half-tetrads (Roper and Pritchard, 1955) or from "twin spots" (Wood and Käfer, 1967).

Käfer (1961) has concluded that spontaneous haploidization results from a nondisjunction event which yields one $2n + 1$ and one $2n - 1$ nucleus. The former normally reverts to a diploid (which may be homozygous for markers formerly heterozygous), while the latter loses more chromosomes until the stable haploid is reached. The various disomics and trisomics formed during haploidization have been individually described by Käfer and Upshall (1973). Lhoas (1968) has shown that in *A. niger*, haploidization with *p*-fluorophenylalanine is accompanied by the loss of visible fragments from diploid nuclei. Spontaneous nondisjunction occurs at approximately two percent of diploid mitoses (Käfer, 1961). Agents increasing this rate have already been described (see Techniques: Haploidization, above). Spontaneous nondisjunction is also influenced by mutagenic treatments (e.g., Fratello *et al.*, 1960; Käfer, 1960) or by the presence of translocations in the component strains (Pollard *et al.*, 1968).

Recombination

Table 2 lists recombination studies with *A. nidulans*. Many of the studies make use of the suitability of mitotic recombination for examination of the effects of chemical or other agents since these can be applied at any time to vegetative diploid cultures. In *A. nidulans*, tetrads are unordered and ascospore color is not determined by the genotype of the spore (Apirion, 1963), so tetrad analysis is less rewarding than in some fungi and has only rarely been employed (Strickland, 1958*a,b*). Centromeres are more readily mapped by mitotic recombination (Pontecorvo and Käfer, 1958), and perithecium analysis is generally used for mapping of mutant loci. Reports of interference in *A. nidulans* are conflicting (Käfer, 1958; Strickland 1958*b*; Elliott, 1960*b*; Bandiera *et al.*, 1970;

TABLE 2.   Recombination Studies with Aspergillus nidulans

Locus	Recombina-tion	Conditions	References
adE	Meiotic	—	Pritchard, 1955, 1960*a, b*
adE	Mitotic	—	Roper and Pritchard, 1955
adF	Meiotic	—	Calef, 1957
adF	Mitotic	Replication effects	Putrament, 1967*a*
adF	Mitotic	Adjacent loci	Putrament, 1967*b*
biA	Meiotic	—	Roper, 1950
fpaA	Mitotic	—	Morpurgo and Volterra, 1966, 1968; Bandiera *et al.*, 1970
fpaA	Mitotic	FUDR, FU[a]	Beccari *et al.*, 1967
hisB	Mitotic	Transcription effects	Millington-Ward, 1970
lysF	Meiotic	—	Pees, 1965, 1967
methA	Meiotic	Temperature	Putrament *et al.*, 1971
pabaA	Meiotic	—	Siddiqi, 1962*b;* Siddiqi and Putrament, 1963; Millington-Ward *et al.*, 1971
pabaA	Mitotic	—	Putrament, 1964; Bandiera *et al.*, 1973
pabaA	Mitotic	Ultraviolet light	Jansen, 1964
pabaA	Mitotic	Diepoxybutane	Putrament, 1966
pabaA	Mitotic	Replication effects	Putrament, 1967*a*
pabaA	Mitotic	Adjacent loci	Putrament, 1967*b*
pabaA	Mitotic	*uvs* mutants	Jansen, 1970*b;* Fortuin, 1971*b, d*
pabaA	Mitotic	Transcription effects	Millington-Ward, 1970
proA,B	Meiotic	—	Forbes, 1959*b;* Weglenski, 1966
Inter-genic	Meiotic	Tetrad analysis	Strickland, 1958*a, b*
Inter-genic	Meiotic	Temperature	Elliott, 1960*b*
Inter-genic	Meiotic	*uvs* mutant	Shanfield and Käfer, 1969
Inter-genic	Mitotic	—	Pontecorvo *et al.*, 1953, 1954; Käfer, 1961
Inter-genic	Mitotic	Formaldehyde, nitrogen mustard	Fratello *et al.*, 1960
Inter-genic	Mitotic	Ultraviolet light	Fratello *et al.*, 1960; Kwiatowski and Grad, 1965; Wood and Käfer, 1967, 1969; Käfer, 1969; Shanfield and Käfer, 1971
Inter-genic	Mitotic	Gamma irradiation	Kwiatowski, 1962; Käfer, 1963; Kwiatowski and Grad, 1965
Inter-genic	Mitotic	Bifunctional alkylating agents	Morpurgo, 1963*a*
Inter-genic	Mitotic	Chelating agents	Kwiatowski, 1965

TABLE 2.   *Continued*

Locus	Recombination	Conditions	References
Intergenic	Mitotic	FU, FUDR[a]	Beccari *et al.*, 1967; Bandiera *et al.*, 1970
Intergenic	Mitotic	*uvs* mutants	Shanfield and Käfer, 1969
Intergenic	Mitotic	FUDR, FPA, NTG[a]	Shanfield and Käfer, 1971
Intergenic	Mitotic	Nitrogen mustards	Kovalenko and Tkachenko, 1973

[a]Abbreviations: FPA = *p*-fluorophenylalanine, FU = 5-fluorouracil, FUDR = 5-fluorodeoxyuridine, NTG = N-methyl-N'-nitro-N-nitrosoguanidine.

Dorn, 1972), but they agree that in both meiotic and mitotic recombination, interference, whether positive or negative, is slight. One exception is the case of localized negative interference (Pritchard, 1960a) which is now probably better interpreted in terms of gene conversion. It should be noted that apparent negative interference may result from either chromosome aberrations or twinned cleistothecia.

Intragenic Complementation

The most extensive complementation analysis performed in *A. nidulans* dealt with the adenylosuccinase (*adA*) locus (Foley *et al.*, 1965). A feature of this work was the quantitative description of negatively complementing (i.e., partially dominant) alleles which give less than 50 percent of the normal activity in diploids with the wild-type allele. Such alleles are probably not uncommon and may help to explain quantitative differences between heterokaryons and diploids (Pontecorvo, 1963; Roberts, 1964; Apirion, 1966). *A. nidulans* is a very suitable organism in which to make such comparisons, but quantitative differences are also to be predicted from the very heterogeneous nature of heterokaryotic hyphae (Pontecorvo, 1947; Clutterbuck and Roper, 1966).

Among loci concerned with nitrate reductase, *niaD* mutants, which do not show any intragenic complementation, have been contrasted with *cnx* mutants, in which mutants of groups *B* and *C* complement readily with one another but do not complement with *cnxA* mutants; all of these mutants are closely linked (Hartley, 1970a). A novel experiment in interspecies *in vitro* complementation for components of anthranilate synthetase from *Aspergillus* and *Neurospora* has been performed by Käfer and DeMoss (1973).

TABLE 3. Studies of Mutation in Aspergillus nidulans

Mutation system	References
Adenine reversion	Chang et al., 1968; Wohlrab and Tuveson, 1969
Arginine reversion	Kovalenko, 1964, 1972; Kovalenko et al., 1969a, b, 1970a, b, 1971
Auxotrophs	Katsatkina, 1959
8-Azaguanine resistance	Arlett, 1966b, c,; Morpurgo, 1962; Strigini et al., 1963
p-Fluorophenylalanine resistance	Calvori and Morpurgo, 1966; Morpurgo, 1962; Strigini et al., 1963
Lethals and chromosome aberrations	Azevedo, 1970; Azevedo and Roper, 1967; Käfer and Chen, 1964; Tector and Käfer, 1962
6-Methyl purine resistance	Arlett, 1966b, c
Methionine reversion	Alderson and Clark, 1966; Ball and Roper, 1966; Chang et al., 1968; Duarte, 1972; Klimczuk, 1970; Lilly, 1965; Prasad, 1970; Scott et al., 1973; Sharma, 1970; Siddiqi, 1962a; Wohlrab and Tuveson, 1969
Morphologicals	Hockenhull, 1948; Kuzyurina, 1959b; Martinelli and Clutterbuck, 1971
Nitrate reductase reversion	Hartley, 1969
Proteases	Brotskaya, 1960
Selenate resistance	Jansen, 1972
Spore color	Martinelli and Clutterbuck, 1971; Siddiqi, 1962a
2-Thioxanthine resistance	Alderson and Hartley, 1969; Alderson and Scazzocchio, 1967; Alderson and Scott; 1970, 1971; Martinelli and Clutterbuck, 1971; Scott and Alderson, 1971
Xanthine dehydrogenase	Alderson, 1969; Hartley, 1970b

Mutagenesis

Systems which have been used for assessment of mutation rates in *A. nidulans* have been reviewed by Roper (1971) and are listed in Table 3; the mutagens used are to numerous to list. Worthy of special mention are two multilocus systems: methionine reversion by suppressors whose nature has been investigated by Lilly (1965), Ayling (1969), Putrament et al. (1970), and Paszewski and Grabski (1973), and the 2-thioxanthine resistance system described by Darlington and Scazzocchio (1967) and Alderson and Scazzocchio (1967). The latter method yields xanthine dehydrogenase-deficient mutants whose backmutation can also be studied. Another two-way system involves the production of *p*-fluorophenylalanine-resistant mutants which are partially defective in tyrosine synthesis (Morpurgo, 1962; Calvori and Morpurgo, 1966; Sinha, 1967), and these also can be selected for reversion.

The vegetative diploids of *A. nidulans* provide an opportunity to investigate recessive lethals and chromosome abnormalities (see Table 3) which can then be analyzed via the parasexual cycle.

Many of the agents investigated for effects on mitotic recombination are mutagenic (see Table 2), and a second link between mutation and recombination is provided by the *uvs* mutants (see Table 1) which affect recombination as well as sensitivity to mutagens.

Growth and Nutrition

A variety of growth systems can be employed with *Aspergillus,* and characteristics of many of them are described in the literature. Trinci (1970*b*) has described the kinetics of growth of pellets in shaken liquid cultures and also compared this type of growth with that obtained on agar (Trinci, 1969; Bainbridge and Trinci, 1969) where the growth of the colony depends on a "peripheral growth zone" (Clutterbuck and Roper, 1966; Trinci, 1971).

The homogeneity of shaken batch cultures may be increased sufficiently to allow repeated sampling by the addition of sodium deoxycholate (Dorn and Rivera, 1966), the agent used to produce microcolonies on agar (Mackintosh and Pritchard, 1963). Continuous-flow and chemostat cultures are also convenient for repeated sampling, and have been used to study metabolic states under various nutritional regimes (Carter and Bull, 1969, 1971; Carter *et al.,* 1971; Bainbridge *et al.,* 1971; Rowley and Bull, 1973). Where many replicate cultures are required, thin-layer growth in unshaken Petri dishes is convenient (Cohen, 1973*a*), while surface growth in unshaken flasks is suitable for harvesting accumulated metabolic products (Naguib, 1959). Where conidiating mycelium or cleistothecia are required, material can be harvested from agar plates with or without a liquid top layer (Clutterbuck, 1972; Zonneveld, 1972*a*).

A great many studies have been made to show the effects of nutritional conditions on growth in batch cultures: Hockenhull (1950), Agnihotri (1961, 1962*a,b*, 1963*a,b*, 1964). Mehrotra and Agnihotri (1961, 1962), Agnihotri and Mehrotra (1961), and Rao and Modi (1970) have studied the major nutritional elements, Agnihotri (1967) has studied trace elements.

Pathogenicity

Reports of findings of *A. nidulans* in pathogenic situations are not generally included in this article, but some studies on the influence of various mutants and of ploidy on the pathogenicity of this fungus when injected into mice may be of interest to geneticists (Dorn *et al.*, 1967; Purnell and Martin, 1971, 1973*a,b;* Purnell, 1973).

Metabolism

Studies in metabolism are two numerous to mention individually, but Table 4 lists the genetic loci concerned with particular fields and also gives references to other biochemical studies.

Genetic Regulation

The existence of diploids confers definite advantages on *A. nidulans* for regulation studies (reviewed by Cove, 1970). For instance, quantitative estimates of dominance of constitutive and loss mutants at the regulator (*nirA*) locus for nitrate and nitrite reductases are essential in determining the nature of this positive regulation system (Cove, 1966, 1967, 1969; Pateman and Cove, 1967; Cove and Pateman, 1969; Pateman *et al.*, 1964, 1967).

In addition to nitrate reduction, which is the most thoroughly investigated system in this organism, many other pathways have been studied. The regulation of mannitol metabolism and the pentose phosphate pathway both appear to be connected with nitrate reductase (Hankinson and Cove, 1972*a,b*). Besides nitrate and nitrite reductases, other substrate-induced systems include amidases (Hynes, 1970, 1972; Hynes and Pateman, 1970*a,b,c;* Dunsmuir and Hynes, 1973), arginine catabolism (Weglenski, 1967; Cybis and Weglenski, 1973; Cybis *et al.*, 1970, 1972*a,b;* Bartnik *et al.*, 1973*a,b*), galactose utilization (Roberts, 1963*b*), isocitrate lyase and malate synthetase (Armitt *et al.*, 1970, 1971), lactose utilization (Paszewski *et al.*, 1970; Fantes and Roberts, 1973) and purine catabolism (Scazzocchio and Darlington, 1967, 1968; Scazzocchio, 1973; Scazzocchio *et al.*, 1973).

Nearly all of the systems studied in *A. nidulans* also involve repression. Aryl sulfatase (Harada and Spencer, 1962, Siddiqi 1966; Apte and Siddiqi, 1971) and choline sulfatase (Scott and Spencer, 1968) are repressed by sulfur sources. Glutamine (Pateman, 1969) and methionine biosynthesis (Pieniazek *et al.*, 1973*b*) are end-product repressed. Sugar

TABLE 4. Metabolism of Aspergillus nidulans

Metabolic system	Genetic loci	References[a]
Amino acid biosynthesis	ab, arg, aro, fpa, glu, his, ile, ind, lu, lys, mec, meth, orn, ota, phen, ppa, pro, su-pro, tryp, tyr	
Antibiotics	npe, pen	Dulaney; 1974; Lafont et al., 1970
Carbon metabolism	ani, cre, gdhC, suFpro	
acetate	alc, ald, fac	
arabinose	galB	
galactose	gal, gam, bga	Chojnacki et al., 1969
glyoxalate cycle	mas, icl	
lactose	lac	
mannitol-1-phosphate dehydrogenase		Hankinson and Cove, 1972b
pentose phosphate pathway	ppp	
pyruvate	pyc, pdh	
sugars	fr, gal, gam, lac, mal, sb, sor	
Esters		Cerny et al., 1972
Lipids	cho, ino	Naguib and Saddik, 1960, 1961, Singh and Walker, 1955, 1956; Singh et al., 1955
Metals		Agnihotri, 1967
copper	yg	
chromium	sB	
molybdenum and tungsten	cnx, galC, gam, mol, pac, pal, palc, sB, su-pal	Arst, 1968; Arst and Cove, 1970; Arst et al., 1970; Downey, 1973a
Nitrogen metabolism		
amides	amd, fmd	
amines	mau, mea	
amino acid catabolism	ani, ota, prn, su-pro	Coll and Leal, 1972
ammonium	amr, are, gdhA, mea, amd T, xprD	
creatinine		Selvam and Shanmugasundar-am, 1972
glutamine synthetase		Pateman, 1969
nitrate	cnx, nia, nii, nir, rin	

TABLE 4. Continued

Metabolic system	Genetic loci	References[a]
protease	xpr	Brotskaya, 1958, 1960; Verbina, 1958
purines	aa, ad, al, alp, apl, cnx, hx, hscn, oxp, pur, ua, uap	
pyrimidines	pyr	
urea	u, uru	
Phenol oxidases	ivo, mel, y, yg	Loginova, 1961; Bull and Carter, 1973
Phosphatases	pac, pal, palc, su-pal	Dorn, 1967b, 1968; Dorn and Rivera, 1966; Harsanyi and Dorn, 1972
Selenium		Arst, 1968; Pieniazek et al., 1973a
Sulfur	csu, s, meth, mec	Hockenhull, 1949; Nakamura, 1962; Nakamura and Sato, 1960, 1962; Shepherd, 1956; Yoshimoto et al., 1961, 1967
aryl sulfatase		Apte and Siddiqi, 1971; Saddiqi et al., 1966b
choline sulfatase		Arst, 1971; Scott and Spencer, 1968
Uptake	aau, alp, azg, fpa, fr, pur, sB, sorA, uru, uap	
Vitamins	an, aro, bi, nic, paba, panto, pu, pyro, ribo, rin, thi	
biotin deficiency		Rao and Modi, 1968, 1972; Roper, 1966a; Strigini and Morpurgo, 1961
riboflavin degradation		R. Shanmugasundaram and Shanmugasundaram, 1965
bioassays		Princivalle, 1958; Princivalle and Caradonna, 1962; Mastropietro and Princivalle, 1963

[a] These references are supplementary to those for particular loci which are given in Table 1.

uptake and metabolism (Romano and Kornberg, 1968, 1969) and NAD-glutamate dehydrogenase (Kinghorn and Pateman, 1973*a,c*) are carbon repressible, while nitrate reduction, purine catabolism, ammonium and methylammonium uptake (Pateman *et al.* 1973), and urea uptake (Dunn and Pateman, 1972) are ammonium repressible. In addition to these systems, amidases, arginine catabolism, NADP-glutamate dehydrogenase (Pateman, 1969; Kinghorn and Pateman, 1973*a*, 1974), and acidic amino acid uptake (Hynes, 1973*a;* Pateman and Kinghorn, 1974; Robinson *et al.*, 1973*a,b*) require both ammonium and glucose for repression; proteases (Cohen, 1972, 1973*b,c*) are only repressed if nitrogen, carbon, and sulfur sources are present.

Ammonium repression is clearly complex, and mutants at a number of loci have overlapping specificities for derepression of different systems. These loci are designated *meaA, meaB, gdhA* and *areA* (the last is probably the same as *amdT* and *xprD*) (Arst and Cove, 1969, 1973; Arst and Page, 1973; Cohen, 1972, 1973*b*; Kinghorn and Pateman, 1973*a*; Hynes, 1973*b*; Pateman *et al.*, 1973).

Mutants causing derepression of carbon-repressed systems have also been described—*cre, gdhC* and *suFpro* (Arst and Cove, 1973; Kinghorn and Pateman, 1973*c*; Bartnik *et al.* 1973*b*). Carbon starvation may also affect systems which are not repressed by glucose by accelerating enzyme decay (Hynes, 1973*c*).

Nucleic Acids

Relatively little work has been done in this field with *A. nidulans.* DNA labeling studies with tritiated thymidine are prevented by the absence of thymidine kinase (Grivell and Jackson, 1968), but this problem has been circumvented by the use of labeled adenine in conjunction with ribonuclease in studies of the cell cycle and of the nonrandom segregation of DNA strands during mitosis (Rosenberger and Kessel, 1967, 1968; Kessel and Rosenberger, 1968). Holloman (1970) and Bainbridge (1971) have also looked at the timing of synthesis of nucleic acids and proteins during spore germination, and the stringency of control of nucleic acid synthesis during various adverse conditions has been examined by Arst and Scazzocchio (1971, 1972). *Aspergillus* DNAases have been investigated by Winder and Campbell (1973). Cybis and Weglenski (1973) have estimated the half life of arginase messenger RNA to be 2.7 minutes.

A rather different topic, the unique species of RNA found in fungal ribosomes, has been studied by Edelman *et al.* (1970, 1971) and Verma *et al.* (1970, 1971).

Nuclei

Vegetative nuclear division as described by Robinow and Caten (1969) occurs without breakdown of the nuclear membrane and involves a primitive form of spindle. A more complex view of the process is presented by Weijer and Weisberg (1966) and Weisberg and Weijer (1968). Meiosis appears to be more conventional (Elliott, 1960a), and eight chromosomes corresponding to the eight linkage groups can be counted.

The distribution of nuclei in haploids, diploids, and heterokaryons has been studied by Clutterbuck and Roper (1966) and Clutterbuck (1969b); in diploid hyphae, the spacing of nuclei is doubled although the hyphal dimensions are unchanged. There is no evidence for long-distance nuclear migration in this species, but elongated nuclei are frequently seen and there is evidence of short-distance nuclear movements both from labeling studies (Rosenberger and Kessel, 1968) and from migration-deficient *aps* mutants (Clutterbuck, 1969b and unpublished).

Nuclear division (Rosenberger and Kessel, 1967) and DNA synthesis (Kessel and Rosenberger, 1968) are normally synchronized when the fungus is growing on rich medium and nuclear division is followed by the formation of a series of septa (Clutterbuck, 1970a). The timing of nuclear division during conidial germination has been followed by Bainbridge (1971). Nuclear division can be inhibited by griseofulvin (Crackower, 1972) and methyl benzimidazole carbamate (Davidse, 1973), and is also sensitive to the various agents causing haploidization (see above).

Hyphal Structure and Development

In addition to studies made in connection with growth rate and nuclear behavior (see above), more detailed studies of hyphae have been made. Hyphal-wall components have been analyzed by Bull (1970) and Zonneveld (1971), and the incorporation of galactose and hexosamine into cell walls has been followed in appropriate mutants by Cohen et al. (1969) and Katz and Rosenberger (1970a,b, 1971a,b; Sternlicht et al., 1973). Threlfall (1968, 1972) has found that pentachloronitrobenzene causes an increase in hexosamine content of cell walls, and that a mutant resistant to this agent is deficient in hexosamine. A particular wall component, α-1,3-glucan, is probably a storage product required for cleistothecium development (Zonneveld, 1972a,b, 1973). Another polysaccharide, composed basically of galactose, is mainly extracellular (Gorin and Eveleigh, 1970). Melanin, which may have a protective role in cell walls (Bull, 1970a,b) is also excreted into the medium (Rowley and Pirt,

1972). Mutants affecting hyphal melanization have been studied by Bull (1970a) and Martinelli and Bainbridge (1974), while the associated enzyme tyrosinase has been purified by Bull and Carter (1973). The melanization of conidiophores is apparently an independent process (Clutterbuck, 1973).

Hyphal branching has been analyzed by Trinci (1970a) and Katz *et al.* (1972). Branching can be induced by various shock treatments which also induce a redistribution of wall material (Katz and Rosenberger, 1971a). Cytochalasin B has also been shown to affect hyphal morphogenesis, possibly by inhibition of transport of wall materials to their deposition sites (Oliver, 1973a).

Protoplasts have been used for investigations of cell permeability (Elorza *et al.*, 1969) and also in studies of wall synthesis (Peberdy and Gibson, 1971; Gibson and Peberdy, 1972; Peberdy and Buckley 1973).

A common type of spontaneous or induced mutant in *A. nidulans* has an excess of aerial mycelium, often to the extent of complete suppression of the formation of conidia and cleistothecia. Some such "fluffy" mutants have been treated as ordinary recessive nuclear mutants (Kwiatowski and Bohdanowicz, 1962; Ball and Azevedo, 1964), while Roper (1958) concluded that his "mycelial" mutants were the result of interaction of nuclear and cytoplasmic factors. The fluffy mutants studied by Dorn *et al.* (1967) and Dorn (1970) were invasive of other colonies, in some cases dominant, and under complex genetic control suggestive of cytoplasmic genes or virus infection in addition to nuclear factors.

Morphological variants of other types often behave as simple nuclear mutants, but may have complex phenotypes in terms of growth, respiration, and biochemistry (Brotskaia, 1958; Katsatkina, 1961; Kuzyurina, 1959b, Loginova, 1960, 1961; Verbina, 1958, 1959, 1960; Bainbridge and Trinci, 1969).

Conidia

Conidial development has been studied by means of conidiation-deficient mutants by Clutterbuck (1969c), Martinelli and Clutterbuck (1971), and Axelrod (1972), Axelrod *et al.* (1973), and Barbata *et al.* (1973). Clutterbuck (1972) has studied the production of a spore-specific *p*-diphenol oxidase concerned with conidial pigmentation in some of these mutants; a second enzyme specifically concerned with conidiophore melanization is also being studied (Clutterbuck, 1973b). The ultrastructure of the conidial apparatus is described by Oliver (1972). In this and other studies (Weisberg and Turian, 1971) lomasomes and plasmalemmasomes were prominent. Attention has also been drawn to the distribution of sulfhydryl groups during development (Oliver, 1974).

The occurrence of conidiation in submerged liquid culture has been described by Saxena and Sinha (1973). A number of authors have described the ultrastructure of dormant and germinating conidia: Hess and Stocks (1969), Border and Trinci (1970), and Florance *et al.* (1972). The germination process has been followed from a biochemical point of view by Shepherd (1957), and Holloman (1970), and Bainbridge (1971). At high densities some auto-inhibition of spore germination has been found (Trinci and Whittaker, 1968; Scott *et al.*, 1972).

Cleistothecia

Few details of cleistothecium formation are known, but some physiological factors involved have been studied. Cleistothecium-promoting media have been developed by Acha and Villanueva (1961) and Leal and Villanueva (1962). Houghton (1970, 1971) has found some respiratory-deficient noncleistothecial (*sgp*) mutants, and two of the types of conidiation mutants described by Clutterbuck (1969c) were also noncleistothecial. The frequency of cleistothecium formation is also affected by naturally occurring cytoplasmic variation (see below), as well as quantitative nuclear factors (Baracho and Azevedo, 1972). Zonneveld (1972a,b, 1973) found that in various mutants cleistothecium formation was correlated with the production of α-1,3-glucan and the corresponding glucanase.

The genetics of mutants affecting ascospore and cleistothecial pigments has been investigated by Apirion (1963).

Toxicity and Resistance

A. nidulans itself produces a number of antibiotics (Lafont *et al.*, 1970; Argoudelis *et al.*, 1965), including penicillin (Dulaney, 1947; Holt and Macdonald, 1968a,b).

The effects of various toxic agents on growth and development have been described by Bull and Faulkner (1964), Trinci and Gull (1970), Ditchburn and Macdonald (1971), Zonneveld (1973), and Verma and Sinha (1973). In a few special cases, mutants may be phenotypically "cured" by protein-synthesis inhibitors (Ball, 1967; Dorn, 1970).

Resistance to toxic agents has been used as a means of obtaining a variety of mutants with specific properties related to uptake or regulation, and references to these will be found in the appropriate sections. In some cases, resistant mutants have been found to have lost the ability to metabolize the toxic compound and also the normal metabolite of which it is an analog, e.g., fluoroacetate-resistant mutants (Apirion, 1962, 1965), purine analogs (Darlington and Scazzocchio, 1967). Alternatively, resistance to an analog may be due to overproduction of the normal compound, e.g., *p*-fluorophenylalanine resistance (Sinha, 1967); in other cases, although specific metabolic changes are found in the mutants, their relation to the phenomenon of resistance is less clear, e.g., sorbose resistance (Elorza and Arst, 1971), molybdate resistance (Arst *et al.*, 1970; Arst and Cove, 1970), and pentachloronitrobenzene resistance (Threlfall, 1968, 1972). In many other instances, however, little is in fact known about the mutants apart from their genetic properties and cross-resistance patterns, e.g., resistance to acridines (Roper and Käfer, 1957; Ball and Roper, 1966), iodoacetate, actidione, malachite green, and teoquil (Warr and Roper, 1965), metal ions (Elorza, 1969), and benzimidazole (Hastie and Georgopoulos, 1971).

Oligomycin resistance is a special case in that it appears to depend on both nuclear and extranuclear genes for its inheritance (Rowlands and Turner, 1973).

Resistant mutants have a special function in genetic analysis since it is possible to select for a homozygous diploid or haploid mutant from a heterozygote; this is an invaluable tool in diploid analysis (Pontecorvo and Käfer, 1956, 1958).

Radiation Sensitivity

Sensitivity to radiation depends on many factors, including spore pigments (Wright and Pateman, 1970), inhibitors (Georgopoulos and Georgadis, 1969), temperature (Ashwood-Smith and Horne, 1972), spore hydration (Wilson and Powers, 1970) and germination (Jansen, 1970*a*; Scott *et al.*, 1972), radiation dosage (Kuzyurina, 1959*a*), photoreactivation (Fortuin, 1971*c*), and dark repair systems. Repair systems can be lost by mutation at a number of *uvs* loci (Lanier *et al.*, 1968; Shanfield and Käfer, 1969; Wright and Pateman, 1970; Kameneva and Romanova, 1969; Jansen, 1970*a*; Fortuin, 1971*a*; Kameneva and Evseeva, 1972). Radiation sensitivity may also be determined by cytoplasmic factors (Arlett, 1966*a,b*). Many of the *uvs* mutants also have defects in recombination which may be accompanied by sexual sterility, but the phenomenon can still be investigated by means of mitotic recombination (see Table 2). Mutants affecting radiation sensitivity are also generally affected in mutability (Arlett, 1966*b*; Wohlrab and Tuveson, 1969; Jansen, 1972).

Metabolite Uptake

The commonest method for obtaining uptake-defective mutants employs toxic analogs. Mutants resistant to them are often unable to take up the natural metabolite as well as the analog. This approach has been used for purines (Darlington and Scazzocchio, 1967), amino acids (Sinha, 1969; Pateman and Kinghorn, 1973), ammonia (Arst and Cove, 1969; Arst, 1972; Pateman *et al.*, 1973*a,b*), and urea (Dunn and Pateman, 1972). The amino acids lysine and arginine are natural analogs of each other and complete for uptake (Pontecorvo, 1950; Pontecorvo *et al.*, 1953; Cybis and Weglenski, 1969). There is also competition for uptake between other amino acids (Sinha, 1969; Dhillon and Garber, 1970). Sorbose can be regarded as a toxic analog of glucose, however, *sorA* mutants are specifically defective in uptake of sorbose and deoxyglucose, but not of other sugars (Elorza and Arst, 1971). Sugar uptake has also been studied largely from a biochemical angle by Romano and Kornberg (1969), Brown and Romano (1969), and Mark and Romano (1971). Similarly, amino acid uptake systems have been investigated by Benko *et al.* (1967), Robinson *et al.* (1973*a,b*), Pateman and Kinghorn (1974), Hynes (1973*a*).

Other uptake mutants are detected by their inability to grow on particular sources of one element, e.g., *sB* mutants are unable to take up sulfate or thiosulfate as sulfur sources, (Arst, 1968; Bradfield *et al.*, 1970; Gravel *et al.* 1970; Spencer and Moore, 1973). Mutants altered in specificity for uptake of the sulfate analogs chromate, molybdate, tungstate, and selenate can also be obtained (Arst *et al.*, 1970; Jansen, 1972; Lukaszkiewicz and Pieniazek, 1972).

Elorza *et al.* (1969) used protoplasts to show that the cell wall plays no important part in the uptake systems they studied.

An unusual uptake system is suggested by the data of Scazzocchio and Darlington (1967) and Scazzocchio (1970) in which differences between diploids and heterokaryons indicate that the control of xanthine dehydrogenase in *A. nidulans* requires the functions of a nuclear permease.

Genetic and Phenotypic Instabilities

Unstable mutants giving rise to variegated or sectored colonies are often conspicuous in filamentous fungi; the causes of such instability, however, are not always easy to discover. Somatic segregation of cytoplasmic variants is one possible cause (see below), while unstable aneuploidy is another (Käfer, 1961; Upshall, 1966; Käfer and Upshall, 1973). Virus infection (Dorn *et al.,* 1967) is a third, as yet hypothetical, cause.

Phenotypic, as opposed to genetic, instability may be due to variegated position effects (Ball, 1967; Clutterbuck, 1970*b*).

One cause of instability which has been the subject of some study originates from duplicated chromosomal material which is derived from a translocation (Bainbridge and Roper, 1966; Millington-Ward, 1967; Ball, 1967; Bainbridge, 1970). This duplicated material is unstable, and the term "mitotic nonconformity" has been applied to the process by which it is lost (Nga and Roper, 1968, 1969; Roper and Nga, 1969). This loss is accelerated by trypan blue (Cooke *et al.,* 1970), caffeine (Roper *et al.,* 1972), and the presence in the strain of a *uvs* mutant (Burr *et al.,* 1971). In addition to the initial instability, this system gives rise to further successive and genetically transposable variants (Azevedo and Roper, 1970).

Cytoplasmic Inheritance

In *Aspergillus,* cytoplasmic inheritance can readily be distinguished from other phenomena by the heterokaryon test (Jinks, 1963). Although a female role is probably distinguishable in cleistothecium formation (Apirion, 1963), it may be taken by either parent, and maternal inheritance of cytoplasmic characters has not been demonstrated. Vegetative instability is one criterion of cytoplasmic inheritance, but it may also have other causes (see above).

Cytoplasmic variation in the balance of propensities for formation of cleistothecia or conidia occurs in the wild and has been investigated by various workers: Jinks (1954, 1958), Mahoney and Wilkie (1958, 1962), Arlett (1960), and Croft (1966). Quantitative variation in growth rate is also cytoplasmically inherited (Jinks, 1956).

Induced cytoplasmic mutants have been studied by Arlett (1957), Arlett *et al.* (1962), Faulkner and Arlett (1964), and Grindle (1964). The "mycelial" and "fluffy" mutants of Roper (1958) and Dorn *et al.* (1967) which behave as if due to nucleo–cytoplasmic interactions have been described above (see Hyphal Structure and Development). Most of these cytoplasmic mutants are distinguished by morphological characters alone, however, one described by Arlett (1966*a,b*) showed increased resistance to ultraviolet light and reduced mutability, while cytoplasmically inherited oligomycin resistance has been reported by Rowlands and Turner (1973), and a cytoplasmic cold-sensitive mutant has been obtained by Waldron and Roberts (1973).

Population Genetics

Studies of wild populations of *A. nidulans* have shown the existence of a number of genetically determined heterokaryon groups (Grindle, 1963*a,b;* Jinks and Grindle, 1963). This incompatibility is not a complete barrier to crossing, however (Butcher, 1968), but studies of quantitative inheritance of growth rate and cleistothecium density do suggest that the groups are partially isolated (Jinks *et al.,* 1966; Butcher, 1969; Butcher *et al.,* 1972). A population study of amylases and other enzymes showed two patterns typical of *A. nidulans,* while other patterns were considered to indicate sibling species (Kurzeja and Garber, 1973).

Literature Cited

Acha, I. G. and J. R. Villanueva, 1961 A selective medium for the formation of ascospores by *Aspergillus nidulans. Nature (Lond.)* **189:**328.

Agnihotri, V. P., 1961 Utilization of sugars in mixtures by some ascosporic members of the *Aspergillus nidulans* group. *Flora Allg. Bot. Zeit.* **151:**159–161.

Agnihotri, V. P., 1962*a* Studies on Aspergilli. VII. Utilization of oligo- and polysaccharides by some ascosporic members of the *Aspergillus nidulans* group. *Lloydia* **25:**94–99.

Agnihotri, V. P., 1962*b* Studies on Aspergilli. V. Utilization of monosaccharides by some ascosporic members of the *Aspergillus nidulans* group. *Flora Allg. Bot. Zeit.* **152:**81–90.

Agnihotri, V. P., 1963*a* Studies on Aspergilli. XIII. Carbon requirements of some ascosporic members of the *Aspergillus nidulans* group. *Acta Biol. Acad. Sci. Hung.* **14:**45–50.

Agnihotri, V. P., 1963*b* Studies on Aspergilli. VIII. Sulphur requirements of some ascosporic members of the *Aspergillus nidulans* group. *Pathol. Microbiol.* **26:**810–816.

Agnihotri, V. P., 1964 Studies on Aspergilli. XVI. Effect of *p*H, temperature and carbon and nitrogen interaction. *Mycopath. Myc. Appl.* **24:**305–314.

Agnihotri, V. P., 1967 Role of trace elements in the growth and morphology of five ascosporic Aspergillus species. *Can. J. Bot.* **45:**73–79.

Agnihotri, V. P. and B. S. Mehrotra, 1961 The amino acid composition of some ascosporic members of the *Aspergillus nidulans* group. *Lloydia* **24:**41–44.

Alderson, T., 1969 Spontaneous and induced reversion of ICR-170-induced xanthine dehydrogenase mutants of *Aspergillus nidulans. Mutat. Res.* **8:**521–529.

Alderson, T. and A. M. Clark, 1966 Interlocus specificity for chemical mutagens in *Aspergillus nidulans. Nature (Lond.)* **210:**593–595.

Alderson, T. and M. J. Hartley, 1969 Specificity for spontaneous and induced forward mutation at several gene loci in *Aspergillus nidulans. Mutat. Res.* **8:**255–264.

Alderson, T. and C. Scazzocchio, 1967 A system for the study of interlocus specificity for both forward and reverse mutation at least eight gene loci in *Aspergillus nidulans. Mutat. Res.* **4:**567–577.

Alderson, T. and B. R. Scott, 1970   The photosensitizing effect of 8-methoxypsoralen on the inactivation and mutation of Aspergillus conidia by near ultraviolet light. *Mutat. Res.* **9:**569–578.

Alderson, T. and B. R. Scott, 1971   Induction of mutation by gamma-irradiation in the presence of oxygen or nitrogen. *Nat. New Biol.* **230:**45–48.

Apirion, D., 1962   A general system for the automatic selection of auxotrophs from prototrophs and vice versa in micro-organisms. *Nature (Lond.)* **195:**959–961.

Apirion, D., 1963   Formal and physiological genetics of ascospore colour in *Aspergillus nidulans. Genet. Res.* **4:**276–283.

Apirion, D., 1965   The two-way selection of mutants and revertants in respect of acetate utilization and resistance to fluoroacetate in *Aspergillus nidulans. Genet. Res.* **6:**317–329.

Apirion, D., 1966   Recessive mutants at unlinked loci which complement in diploids but not in heterokaryons of *Aspergillus nidulans. Genetics* **53:**935–941.

Apirion, D., G. L. Dorn and E. Forbes, 1963   The VIII linkage group. *Aspergillus Newsl.* **4:**15–16.

Apte, B. N. and O. Siddiqi. 1971   Purification and properties of arylsulphatase of *Aspergillus nidulans. Biochim. Biophys. Acta* **242:**129–140.

Argoudelis, A. D., J. H. Coats and R. R. Herr, 1966   Isolation and characterization of a new antibiotic. *Antimicrob. Agents Chemotherap. 1965* pp. 801–803.

Arlett, C. F., 1957   Induction of cytoplasmic mutants in *Aspergillus nidulans. Nature (Lond.)* **179:**1250–1251.

Arlett, C. F., 1960   A system of cytoplasmic variation in *Aspergillus nidulans. Heredity* **15:**377–388.

Arlett, C. F., 1966a   The radiation sensitivity of a cytoplasmic mutant of *Aspergillus nidulans. Int. J. Radiat. Biol.* **10:**539–550.

Arlett, C. F., 1966b   The influence of the cytoplasm on mutation in *Aspergillus nidulans. Mutat. Res.* **3:**410–419.

Arlett, C. F., 1966c   The interaction between ultraviolet and gamma irradiation in *Aspergillus nidulans. Int. J. Radiat. Biol.* **11:**313–320.

Arlett, C. F., M. Grindle and J. L. Jinks, 1962   The "red" cytoplasmic variant of *Aspergillus nidulans. Heredity* **17:**197–209.

Armitt, S., C. F. Roberts and H. L. Kornberg, 1970   The role of isocitrate lyase in *Aspergillus nidulans. FEBS (Fed. Eur. Biochem. Soc.) Lett.* **7:**231–233.

Armitt, S., C. F. Roberts and H. L. Kornberg, 1971   Mutants of *Aspergillus nidulans* lacking malate synthase. *FEBS (Fed. Eur. Biochem. Soc.) Lett.* **12:**276–278.

Arst, H. N., 1968   Genetic analysis of the first steps of sulphate metabolism of *Aspergillus nidulans. Nature (Lond.)* **219:**268–270.

Arst, H. N., 1971   Mutants of *Aspergillus nidulans* unable to use choline-O-sulphate. *Genet. Res.* **17:**273–277.

Arst, H. N. and D. J. Cove, 1969   Methylammonium resistance in *Aspergillus. J. Bacteriol.* **98:**1284–1293.

Arst, H. N. and D. J. Cove 1970   Molybdate metabolism in *Aspergillus nidulans.* II. Mutations affecting phosphatase activity and galactose utilization. *Mol. Gen. Genet.* **108:**146–153.

Arst, H. N. and D. J. Cove, 1973   Nitrogen metabolite repression in *Aspergillus nidulans. Mol. Gen. Genet.* **126:**111–141.

Arst, H. N. and D. W. MacDonald, 1973   A mutant of *Aspergillus nidulans* lacking NADP-linked glutamate dehydrogenase. *Mol. Gen. Genet.* **122:**261–265.

Arst, H. N. and B. M. Page, 1973   Mutants of *Aspergillus nidulans* altered in the transport of methylammonium and ammonium. *Mol. Gen. Genet.* **121**:239–245.

Arst, H. N. and C. Scazzocchio, 1971   RNA synthesis in *Aspergillus nidulans. Heredity* **26**:346.

Arst, H. N. and C. Scazzocchio, 1972   Control of nucleic acid synthesis in *Aspergillus nidulans. Heredity* **29**:131.

Arst, H. N., D. W. MacDonald and D. J. Cove, 1970   Molybdate metabolism in *Aspergillus nidulans*. I. Mutations affecting nitrate reductase and/or xanthine dehydrogenase. *Mol. Gen. Genet.* **108**:129–145.

Ashwood-Smith, M. J. and B. Horne, 1972   Response of *Aspergillus* and *Penicillium* spores to ultraviolet radiation at low temperatures. *Photochem. Photobiol.* **15**:89–92.

Aspen, A. J. and A. Meister, 1962   Conversion of α-aminoadipic acid to L-pipecolic acid by *Aspergillus nidulans. Biochemistry* **1**:606–612.

Axelrod, D. E., 1972   Kinetics of differentiation of conidiophores and conidia by colonies of *Aspergillus nidulans. J. Gen. Microbiol.* **73**:181–184.

Axelrod, D. E., M. Gealt and M. Pastushok, 1973   Gene control of developmental competence in *Aspergillus nidulans. Develop. Biol.* **34**:9–15.

Ayling, P. D., 1969   Methionine suppressors in *Aspergillus nidulans:* their genetics and behaviour in heterokaryons and diploids. *Genet. Res.* **14**:275–289.

Azevedo, J. L., 1965   The centromere of chromosome VII of *Aspergillus nidulans. Aspergillus Newsl.* **6**:7.

Azevedo, J. L., 1970   Recessive lethals induced by nitrous acid in *Aspergillus nidulans. Mutat. Res.* **10**:111–117.

Azevedo, J. L. and J. A. Roper, 1967   Lethal mutations and balanced lethal systems in *Aspergillus nidulans. J. Gen. Microbiol.* **49**:149–155.

Azevedo, J. L. and J. A. Roper, 1970   Mitotic nonconformity in *Aspergillus nidulans:* successive and transposable genetic changes. Genet. Res. **16**:79–93.

Bainbridge, B. W., 1966   Table of located or partially located mutants and revised map of linkage group III. *Aspergillus Newsl.* **7**:19–21.

Bainbridge, B. W., 1970   Genetic analysis of an unequal chromosomal translocation in *Aspergillus nidulans. Genet. Res.* **15**:317–326.

Bainbridge, B. W., 1971   Macromolecular composition and nuclear division during spore germination in *Aspergillus nidulans. J. Gen. Microbiol.* **66**:319–325.

Bainbridge, B. W. and J. A. Roper, 1966   Observations on the effects of a chromosome duplication in *Aspergillus nidulans. J. Gen. Microbiol.* **42**:417–424.

Bainbridge, B. W. and A. P. J. Trinci, 1969   Colony and specific growth rates of normal and mutant strains of *Aspergillus nidulans. Trans. Br. Mycol. Soc.* **53**:473–475.

Bainbridge, B. W., H. Dalton and J. H. Walpole, 1966   Identification of the arginosuccinase gene. *Aspergillus Newsl.* **7**:18.

Bainbridge, B. W., A. T. Bull, S. J. Pirt, B. I. Rowley and A. P. J. Trinci, 1971   Biochemical and structural changes in non-growing maintained and autolysing cultures of *Aspergillus nidulans. Trans. Br. Mycol. Soc.* **56**:371–385.

Ball, C., 1967   Chromosome instability related to gene suppression in *Aspergillus nidulans. Genet. Res.* **10**:173–183.

Ball, C. and J. L. Azevedo, 1964   A "fluffy" mutant in *Aspergillus nidulans. Aspergillus Newsl.* **5**:9.

Ball, C. and J. A. Roper, 1966   Studies on the inhibition and mutation of *Aspergillus nidulans* by acridines. *Genet. Res.* **7**:207–221.

Bandiera, M., G. Morpurgo and L. Volterra, 1970 "Barriers" to intragenic mitotic crossing-over, *Mutat. Res.* **9**:213-217.

Bandiera, M., D. Armaleo and G. Morpurgo, 1973 Mitotic intragenic recombination as a consequence of heteroduplex formation in *Aspergillus nidulans. Mol. Gen. Genet.* **122**:137-148.

Baracho, I. R. and J. L. Azevedo, 1972 A quantitative analysis of cleistothecia production in *Aspergillus nidulans. Experientia (Basel)* **28**:855-856.

Baracho, I. R., R. Vancovsky and J. L. Azevedo, 1970 Correlations between size and hybrid or selfed state of cleistothecia in *Aspergillus nidulans. Trans. Br. Mycol. Soc.* **54**:109-116.

Barbata, G., L. Valdes and G. Sermonti, 1973 Complementation among developmental mutants in *Aspergillus nidulans. Mol. Gen. Genet.* **126**:227-232.

Barratt, R. W., G. B. Johnson and W. N. Ogata, 1965 Wild type and mutant stocks of *Aspergillus nidulans. Genetics* **52**:233-246.

Barron, G. L. and B. H. MacNeill, 1962 A simplified procedure for demonstrating the parasexual cycle in *Aspergillus. Can. J. Bot.* **40**:1321-1327.

Bartnik, E., P. Weglenski and M. Piotrowska, 1973*a* Ammonium and glucose repression of the arginine catabolic enzymes in *Aspergillus nidulans. Mol. Gen. Genet.* **126**:75-84.

Bartnik, E., J. Guzewska and P. Weglenski, 1973*b* Mutations simultaneously affecting ammonium and glucose repression of the arginine catabolic enzymes in *Aspergillus nidulans. Mol. Gen. Genet.* **126**:85-92.

Beccari, E., P. Modigliani and G. Morpurgo, 1967 Induction of inter- and intragenic mitotic recombination by 5-fluorodeoxyuridine and 5-fluorouracil in *Aspergillus nidulans. Genetics* **56**:7-12.

Benko, P. V., T. C. Wood and I. H. Segal, 1967 Specificity and regulation of methionine transport in filamentous fungi. *Arch. Biochem. Biophys.* **122**:783-804.

Berlyn, M., 1967 Gene–enzyme relationships in histidine biosynthesis in *Aspergillus nidulans. Genetics* **57**:561-570.

Border, D. J. and A. P. J. Trinci, 1970 Fine structure of the germination of *Aspergillus nidulans* conidia. *Trans. Br. Mycol. Soc.* **54**:143-152.

Bradfield, G., D. Somerfield, T. Meyn, M. Holby, D. Babcock, D. Bradley, and I. H. Segal, 1970 Regulation of sulphate transport in filamentous fungi. *Plant Physiol.* **46**:720-727.

Brotskaia, S. Z., 1958 The morphology of variants of *Aspergillus nidulans* produced by ultraviolet irradiation. *Mikrobiologiia* (Eng. transl.) **27**:45-51.

Brotskaia, S. Z., 1960 Effect of ultraviolet irradiation in varying dosage on production of *Aspergillus nidulans* variants with active proteases. *Mikrobiologiia* (Eng. transl.) **29**:264-266.

Brown, C. E. and A. H. Romano, 1969 Evidence against necessary phosphorylation during hexose transport in *Aspergillus nidulans. J. Bacteriol* **100**:1198-1203.

Bull, A. T., 1970*a* Chemical composition of wild-type and mutant *Aspergillus nidulans* cell walls. The nature of polysaccharide and melanin constituents. *J. Gen. Microbiol.* **63**:75-94.

Bull, A. T., 1970*b* Inhibition of polysaccharases by melanin: enzyme inhibition in relation to mycolysis. *Arch. Biochem. Biophys.* **137**:345-356.

Bull, A. T. and B. L. A. Carter, 1973 The isolation of tyrosinase from *Aspergillus nidulans*, its kinetic and molecular properties and some consideration of its activity *in vivo. J. Gen. Microbiol.* **75**:61-73.

Bull, A. T. and B. M. Faulkner, 1964   Physiological and genetic effects of 8-azaguanine. *Nature (Lond.)* **203**:506–507.

Burr, K. W., H. M. Palmer and J. A. Roper, 1971   Mitotic non-conformity in *Aspergillus nidulans*: The effect of reduced DNA repair. *Heredity* **27**:487.

Butcher, A. C., 1968   The relationship between sexual outcrossing and heterokaryon incompatibility in *Aspergillus nidulans*. *Heredity* **23**:443–452.

Butcher, A. C., 1969   Non-allelic interactions and genetic isolation in wild populations of *Aspergillus nidulans*. *Heredity* **24**:621–631.

Butcher, A. C., J. Croft and M. Grindle, 1972   Use of genetic–environmental interaction analysis in the study of natural populations of *Aspergillus nidulans*. *Heredity* **29**:263–283.

Calef, E., 1957   Effect on linkage maps of selection of crossovers between closely linked markers. *Heredity* **11**:265–279.

Calvori, C. and G. Morpurgo, 1966   Analysis of induced mutations in *Aspergillus nidulans*. I. UV- and HNO_2-induced mutations. *Mutat. Res.* **3**:145–151.

Carter, B. L. A. and A. T. Bull, 1969   Studies of fungal growth and intermediary carbon metabolism under steady and non-steady state conditions. *Biotechnol. Bioeng.* **11**:785–804.

Carter, B. L. A. and A. T. Bull, 1971   The effect of oxygen tension in the medium on the morphology and growth kinetics of *Aspergillus nidulans*. *J. Gen. Microbiol.* **65**:265–273.

Carter, B. L. A., A. T. Bull, S. J. Pirt and B. I. Rowley, 1971   Relationship between energy, substrate utilization and specific growth rate in *Aspergillus nidulans*. *J. Bacteriol.* **108**:309–313.

Cerny, A., A. Capek and M. Semonsky, 1972   Antineoplastisch wirksame stoffe: 48 Mikrobielle katabolite der Aethylester von N-[8-(6-Purinylthio)-valeryl]-glycine, -diglycin und -triglycin sowie der entsprechenden saere. *Pharmazie* **27**:298–299.

Chang, L. T., J. E. Lennox and R. W. Tuveson, 1968   Induced mutation in UV-sensitive mutants of *Aspergillus nidulans* and *Neurospora crassa*. *Mutat. Res.* **5**:217–224.

Chojnacki, T., A. Paszewski and T. Sawicka, 1969   The formation of UDP glucose and UDP galactose in wild-type and mutants of *Aspergillus nidulans*. *Acta Biochim. Pol.* **16**:185–191.

Clutterbuck, A. J., 1965   A fawn conidia mutant in *Aspergillus nidulans*. *Aspergillus Newsl.* **6**:12.

Clutterbuck, A. J., 1968a   New conidial colour mutants in *Aspergillus nidulans*. *Aspergillus Newsl.* **9**:14.

Clutterbuck, A. J., 1968b   Gene symbols and nomenclature: Proposals and notes on them. *Aspergillus Newsl.* **9**:26–29.

Clutterbuck, A. J., 1969a   Stock list of *Aspergillus nidulans* strains held at the Department of Genetics, University of Glasgow. *Aspergillus Newsl.* **10**:30–37.

Clutterbuck, A. J., 1969b   Cell volume per nucleus in haploid and diploid strains of *Aspergillus nidulans*. *J. Gen. Microbiol.* **55**:291–299.

Clutterbuck, A. J., 1969c   A mutational analysis of conidial development in *Aspergillus nidulans*. *Genetics* **63**:317–327.

Clutterbuck, A. J., 1969d   Further comments on gene symbols. *Aspergillus Newsl.* **10**:26–28.

Clutterbuck, A. J., 1970*a* Synchronous nuclear division and septation in *Aspergillus nidulans. J. Gen. Microbiol.* **60**:133–135.

Clutterbuck, A. J., 1970*b* A variegated position effect in *Aspergillus nidulans. Genet. Res.* **16**:303–316.

Clutterbuck, A. J., 1972 Absence of laccase from yellow-spored mutants of *Aspergillus nidulans. J. Gen. Microbiol.* **70**:423–435.

Clutterbuck, A. J., 1973*a* Gene symbols in *Aspergillus nidulans. Genet. Res.* **21**:291–296.

Clutterbuck, A. J., 1973*b* Interrelations between development and pigmentation during conidiation of *Aspergillus nidulans. Genetics* **74**:s50.

Clutterbuck, A. J. and D. J. Cove, 1974 The genetic loci of *Aspergillus nidulans.* In *Handbook of Microbiology,* edited by H. Lechevalier, Chemical Rubber Co., Cleveland, Ohio.

Clutterbuck, A. J., and J. A. Roper, 1966 A direct determination of nuclear distribution in heterokaryons of *Aspergillus nidulans. Genet. Res.* **7**:185–194.

Cohen, B. L., 1972 Ammonium repression of extracellular protease in *Aspergillus nidulans. J. Gen. Microbiol.* **71**:293–299.

Cohen, B. L., 1973*a* Growth of *Aspergillus nidulans* in a thin liquid layer. *J. Gen. Microbiol.* **76**:277–283.

Cohen, B. L., 1973*b* The neutral and alkaline proteases of *Aspergillus nidulans. J. Gen. Microbiol.* **77**:521–528.

Cohen, B. L., 1973*c* Regulation of intracellular and extracellular neutral and alkaline proteases in *Aspergillus nidulans. J. Gen. Microbiol.* **79**:311–320.

Cohen, J., D. Katz and R. F. Rosenberger, 1969 Temperature sensitive mutant of *Aspergillus nidulans* lacking amino sugars in its cell wall. *Nature (Lond.)* **244**:713–715.

Coll, J. and J. A. Leal, 1972 The utilization of L-tryptophan as nitrogen source by *Fusarium culmorum, Aspergillus nidulans* and *Penicillium italicum. Can. J. Microbiol.* **18**:1353–1356.

Cooke, P., J. A. Roper and W. Watmough, 1970 Trypan blue-induced deletion strains of *Aspergillus nidulans. Nature (Lond.)* **226**:276–277.

Cove, D. J., 1966 The induction and repression of nitrate reductase in the fungus *Aspergillus nidulans. Biochim. Biophys. Acta* **113**:51–56.

Cove, D. J., 1967 Kinetic studies of the induction of nitrate reductase and cytochrome *c* reductase in the fungus *Aspergillus nidulans. Biochem. J.* **104**:1033–1039.

Cove, D. J., 1969 Evidence for a near limiting intracellular concentration of a regulator substance. *Nature (Lond.)* **224**:272–273.

Cove, D. J., 1970 Control of gene action in *Aspergillus nidulans. Proc. R. Soc. Lond. Ser. B Biol. Sci.* **176**:267–275.

Cove, D. J. and A. Coddington, 1965 Purification of nitrate reductase and cytochrome *c* reductase from *Aspergillus nidulans. Biochim. Biophys. Acta* **110**:312–318.

Cove, D. J. and J. A. Pateman, 1963 Independently segregating genetic loci concerned with nitrate reductase activity in *Aspergillus nidulans. Nature (Lond.)* **198**:262–263.

Cove, D. J. and J. A. Pateman, 1969 Autoregulation of the synthesis of nitrate reductase in *Aspergillus nidulans. J. Bacteriol.* **97**:1374–1378.

Crackower, S. H. B., 1972 The effect of griseofulvin on mitosis in *Aspergillus nidulans. Can. J. Microbiol.* **18**:683–687.

Croft, J. H., 1966 A reciprocal phenotypic instability affecting development in *Aspergillus nidulans. Heredity* **21**:565–579.

Cybis, J. and P. Weglenski, 1969   Effects of lysine on arginine uptake and metabolism in *Aspergillus nidulans. Mol. Gen. Genet.* **104**:282–287.

Cybis, J. and P. Weglenski, 1973   Arginase induction in *Aspergillus nidulans.* The appearance and decay of the coding capacity of messenger. *Eur. J. Biochem.* **30**:262–268.

Cybis, J., M. Piotrowska and P. Weglenski, 1970   Control of ornithine transcarbamylase formation in *Aspergillus nidulans. Bull. Acad. Pol. Sci. Ser. Sci. Biol.* **18**:669–672.

Cybis, J., M. Piotrowska and P. Weglenski, 1972a   Genetic control of the arginine pathways in *Aspergillus nidulans.* Common regulation of anabolism and catabolism. *Mol. Gen. Genet.* **118**:273–277.

Cybis, J., M. Piotrowska and P. Weglenski, 1972b   The genetic control of the arginine pathway in *Aspergillus nidulans.* Mutants blocked in arginine biosynthesis. *Acta Microbiol. Pol. Ser. A Microbiol. Gen.* **4**:163–169.

Da Cunha, P. R. 1970   A study of aspects of heterokaryosis in *Aspergillus nidulans. Mem. Inst. Oswaldo Cruz Rio de J.* **68**:119–167.

Darlington, A. J. and C. Scazzocchio, 1967   The use of analogues and the substrate-sensitivity of mutants in analysis of purine uptake and breakdown in *Aspergillus nidulans. J. Bacteriol.* **93**:937–940.

Darlington, A. J., and C. Scazzocchio, 1968   Evidence for an alternative pathway of xanthine oxidation in *Aspergillus nidulans. Biochim. Biophys. Acta* **166**:569–571.

Darlington, A. J., C. Scazzocchio and J. A. Pateman, 1965   Biochemical and genetical studies of purine breakdown in *Aspergillus. Nature (Lond)* **206**:599–600.

Davidse, L. C., 1973   Antimitotic activity of methyl benzimidazol-2-YL carbamate (MBC) in *Aspergillus nidulans. Pest. Biochem.* **3**:317.

Devi, C. S. S. and E. R. B. Shanmugasundaram, 1969   Genetics and biochemistry of a riboflavin requiring mutant. *Curr. Sci. (Bangalore)* **38**:193–195.

Dhillon, T. S. and E. D. Garber, 1970   Methionine-sensitive leucine-requiring mutants of *Aspergillus nidulans. Z. Biol.* **116**:349–353.

Ditchburn, P. and K. D. Macdonald, 1971   The differential effects of nystatin on growth of auxotrophic and prototrophic strains of *Aspergillus nidulans. J. Gen. Microbiol.* **67**:299–306.

Dorn, G. L. 1965a   Genetic analysis of the phosphatases in *Aspergillus nidulans. Genet. Res.* **6**:13–26.

Dorn, G. L., 1965b   Phosphatase mutants in *Aspergillus nidulans. Science (Wash., D.C.)* **150**:1183–1184.

Dorn, G. L. 1967a   A revised linkage map of the eight linkage groups of *Aspergillus nidulans. Genetics* **56**:619–631.

Dorn, G. L., 1967b   Purification of two alkaline phosphatases from *Aspergillus nidulans. Biochim. Biophys. Acta* **132**:190–193.

Dorn, G. L., 1968   Purification and characterization of phosphatase I from *Aspergillus nidulans. J. Biol. Chem.* **243**:3500–3506.

Dorn, G. L. 1970   Genetic and morphological properties of undifferentiated and invasive variants of *Aspergillus nidulans. Genetics* **66**:267–279.

Dorn, G. L., 1972   Computerized meiotic mapping of *Aspergillus nidulans. Genetics* **72**:595–605.

Dorn, G. L. and W. Rivera, 1965   Supplementary list of located or partially located mutants in *Aspergillus nidulans. Aspergillus Newsl.* **6**:13–15.

Dorn, G. L. and W. Rivera, 1966   Kinetics of fungal growth and phosphatase formation in *Aspergillus nidulans. J. Bacteriol.* **92**:1618–1622.

Dorn, G. L., G. M. Martin and D. M. Purnell, 1967   Genetic and cytoplasmic control of undifferentiated growth in *Aspergillus nidulans. Life Sci.* **6**:629–633.

Downey, R. J., 1971   Characterization of the reduced nicotinamide adenine dinucleotide phosphate-nitrate reductase of *Aspergillus nidulans. J. Bacteriol.* **105**:759–768.

Downey, R. J., 1973*a*   The role of molybdenum in formation of the NADPH-nitrate reductase by *Aspergillus nidulans. Biochem. Biophys. Res. Commun.* **50**:920–925.

Downey, R. J., 1973*b*   The multimeric nature of NADPH-nitrate reductase from *Aspergillus nidulans. Microbios* **7**:53–60.

Downey, R. J. and D. J. Cove, 1971   Attempts to detect an alternative vital role for the reduced nicotinamide adenine dinucleotide phosphate-nitrate reductase structural gene in *Aspergillus nidulans. J. Bacteriol.* **106**:1047–1049.

Duarte, F. A. M., 1972   Efeitos mutagenicos de alguns esteres de acidos inorganicos em *Aspergillus nidulans* (Eidam) Winter. *Cienc. Cult. (Sao Paulo)* **24**:42–52.

Dulaney, E. L., 1947   Some aspects of penicillin production by *Aspergillus nidulans. Mycologia* **39**:570–581.

Dunn, E. and J. A. Pateman, 1972   Urea and thiourea uptake in *Aspergillus nidulans. Heredity* **29**:129.

Dunsmuir, P. and M. J. Hynes, 1973   Temperature-sensitive mutants affecting the activity and regulation of the acetamidase of *Aspergillus nidulans. Mol. Gen. Genet.* **123**:333–346.

Edelman, M., I. M. Verma and U. Z. Littauer, 1970   Mitochondrial ribosomal RNA from *Aspergillus nidulans:* Characterization of a novel molecular species. *J. Mol. Biol.* **49**:67–83.

Edelman, M., I. M. Verma, R. Herzog, E. Galun and U. Littauer, 1971   Physiochemical properties of mitochondrial ribosomal RNA from fungi. *Eur. J. Biochem.* **19**:372–378.

Elliott, C. G., 1956   Triploid *Aspergillus nidulans. Microb. Genet. Bull.* **13**:7.

Elliott, C. G., 1960*a*   The cytology of *Aspergillus nidulans. Genet. Res.* **1**:462–476.

Elliott, C. G., 1960*b*   Non-localized negative interference in *Aspergillus nidulans. Heredity* **15**:247–262.

Elorza, M. V., 1969   Toxicidad de los iones metalicos para *Aspergillus nidulans. Microbiol. Espan.* **22**:131–138.

Elorza, M. V. and H. N. Arst, 1971   Sorbose-resistant mutants of *Aspergillus nidulans. Mol. Gen. Genet.* **111**:185–193.

Elorza, M. V., H. N. Arst, D. J. Cove and C. Scazzocchio, 1969   Permeability properties of *Aspergillus nidulans* protoplasts. *J. Bacteriol.* **99**:113–115.

Fantes, P. A. and C. F. Roberts, 1973   β-galactosidase activity and lactose utilization in *Aspergillus nidulans. J. Gen. Microbiol.* **77**:471–486.

Faulkner, B. M. and C. F. Arlett, 1964   The "minute" cytoplasmic variant of *Aspergillus nidulans. Heredity* **19**:63–73.

Florance, E. R., W. C. Denison and T. C. Allen, 1972   Ultrastructure of dormant and germinating conidia of *Aspergillus nidulans. Mycologia* **64**:115–123.

Foley, J. M., N. H. Giles and C. F. Roberts, 1965   Complementation at the adenylosuccinase locus in *Aspergillus nidulans. Genetics* **52**:1247–1263.

Forbes, E., 1959*a*   Use of mitotic segregation for assigning genes to linkage groups in *Aspergillus nidulans. Heredity* **13**:67–80.

Forbes, E., 1959*b*   Recombination in the *pro* region in *Aspergillus nidulans. Microb. Genet. Bull.* **13**:9–11.

Forbes, E. and U. Sinha, 1966   Location of some temperature-sensitive mutants. *Aspergillus Newsl.* **7**:17.

Fortuin, J. J. H., 1971a Another two genes controlling mitotic intragenic recombination and recovery from UV damage in *Aspergillus nidulans*. I. UV sensitivity, complementation and location of six mutants. *Mutat. Res.* **11**:149–162.

Fortuin, J. J. H., 1971b Another two genes controlling mitotic intragenic recombination and recovery from UV damage in *Aspergillus nidulans* II. Recombination behaviour and X-ray sensitivity of *uvsD* and *uvsE* mutants. *Mutat. Res.* **11**: 265–277.

Fortuin, J. J. H., 1971c Another two genes controlling mitotic intragenic recombination and recovery from UV damage in *Aspergillus nidulans*. III. Photoreactivation of UV damage in *uvsD* and *uvsE* mutants. *Mutat. Res.* **13**:131–136.

Fortuin, J. J. H., 1971d Another two genes controlling mitotic intragenic recombination and recovery from UV damage in *Aspergillus nidulans*. IV. Genetic analysis of mitotic intragenic recombinants from *uvs+/uvs+*, *uvsD/uvsD* and *uvsE/uvsE* diploids. *Mutat. Res.* **13**:137–148.

Fratello, B., G. Morpurgo and G. Sermonti, 1960 Induced somatic segregation in *Aspergillus nidulans*. *Genetics* **45**:785–800.

Gajewski, W. and J. Litwinska, 1968 Methionine loci and their suppressors in *Aspergillus nidulans*. *Mol. Gen. Genet.* **102**:210–220.

Gajewski, W., J. Litwinska, A. Paszewski and T. Chojnacki, 1972 Isolation and characterization of lactose non-utilizing mutants of *Aspergillus nidulans*. *Mol. Gen. Genet.* **116**:99–106.

Garber, E. D., G. W. Bryan, B. Capon, L. B. Liddle and N. W. Miller, 1961 Evidence for parthenogenesis in *Aspergillus nidulans*. *Am. Nat.* **95**:309–313.

Georgopoulos, S. G. and E. Georgadis, 1969 Iodoacetate resistance and radiosensitization of conidia of *Aspergillus nidulans*. *Radiat. Bot.* **9**:69–73.

Gibson, R. K. and J. F. Peberdy, 1972 Fine structure of protoplasts of *Aspergillus nidulans*. *J. Gen. Microbiol.* **72**:529–538.

Gorin, P. A. J. and D. E. Eveleigh, 1970 Extracellular 2-acetamido-2-deoxy-D-galacto-D-galactan from *Aspergillus nidulans*. *Biochemistry* **9**:5023–5027.

Gravel, R. A., E. Käfer, A. Niklewicz-Borkenhagen and P. Zambryski, 1970 Genetic and accumulation studies in sulphite-requiring mutants of *Aspergillus nidulans*. *Can. J. Genet. Cytol.* **12**:831–840.

Grindle, M., 1963a Heterokaryon incompatibility of unrelated strains in the *Aspergillus nidulans* group. *Heredity* **18**:191–204.

Grindle, M., 1963b Heterokaryon incompatibility of closely related wild isolates of *Aspergillus nidulans*. *Heredity* **18**:397–405.

Grindle, M. 1964 Nucleo-cytoplasmic interaction in the "red" cytoplasmic variant of *Aspergillus nidulans*. *Heredity* **19**:75–95.

Grivell, A. R. and J. F. Jackson, 1968 Thymidine kinase: evidence for its absence from *Neurospora crassa* and some other microorganisms and the relevance of this to the specific labeling of deoxyribonucleic acid. *J. Gen. Microbiol.* **54**:307–317.

Hankinson, O. and D. J. Cove, 1972a Genetic regulation of the pentose phosphate pathway of *Aspergillus nidulans*. *Heredity* **28**:276.

Hankinson, O. and D. J. Cove, 1972b The effect of nitrate on the activity of the D-mannitol-1-phosphate dehydrogenase of *Aspergillus nidulans*. *Heredity* **29**:121.

Harada, T. and B. Spencer, 1962 The effect of sulphate assimilation on the induction of arylsulphatase synthesis in fungi. *Biochem. J.* **82**:148–156.

Harsanyi, Z. and G. L. Dorn, 1972 Purification and characterization of acid phosphatase V from *Aspergillus nidulans*. *J. Bacteriol.* **110**:246–255.

Hartley, M. J., 1969   Reversion of non-nitrate utilizing (*niaD*) mutants of *Aspergillus nidulans. Mutat. Res.* **7**:163–170.

Hartley, M. J., 1970a   Contrasting complementation patterns in *Aspergillus nidulans. Genet. Res.* **16**:123–125.

Hartley, M. J., 1970b   The frequency of reverse mutation at the XDH loci of *Aspergillus nidulans. Mutat. Res.* **10**:175–183.

Hastie, A. C., 1970   Benlate-induced instability of *Aspergillus* diploids. *Nature (Lond.)* **226**:771.

Hastie, A. C., and S. G. Georgopoulos, 1971   Mutational resistance to fungitoxic benzimidazole derivatives in *Aspergillus nidulans. J. Gen. Microbiol.* **67**:371–373.

Heagy, F. C. and J. A. Roper, 1952   Deoxyribonucleic acid content of haploid and diploid *Aspergillus* conidia. *Nature (Lond.)* **170**:713–714.

Hess, W. M. and D. L. Stocks, 1969   Surface characteristics of *Aspergillus* conidia. *Mycologia* **61**:560–571.

Hockenhull, D. J. D., 1948   Mustard gas mutation in *Aspergillus nidulans. Nature (Lond.)* **161**:100.

Hockenhull, D. J. D., 1949   The sulfur metabolism of mold fungi: The use of "biochemical mutant" strains of *Aspergillus nidulans* in elucidating the biosynthesis of cystine. *Biochim. Biophys. Acta* **3**:326–335.

Hockenhull, D. J. D., 1950   Studies in the metabolism of mold fungi. Preliminary study of the metabolism of carbon, nitrogen and sulphur by *Aspergillus nidulans. J. Exp. Bot.* **1**:194–200.

Holl, F. B., 1971   Immunochemical analysis of nitrate reductase in *Aspergillus nidulans. Heredity* **27**:311.

Holl, F. B. and C. Scazzocchio, 1970   Immunological differences between inducible and constitutive xanthine dehydrogenases in *Aspergillus nidulans. FEBS (Fed. Eur. Biochem. Soc.) Lett.* **12**:51–53.

Holloman, D. W., 1970   Ribonucleic acid synthesis during fungal spore germination. *J. Gen. Microbiol.* **62**:75–87.

Holt, G. and K. D. Macdonald, 1968a   Isolation of strains with increased penicillin yield after hybridisation in *Aspergillus nidulans. Nature (Lond.)* **219**:636–637.

Holt, G. and K. D. Macdonald, 1968b   Penicillin production and its mode of inheritance in *Aspergillus nidulans. Antonie Van Leeuwenhoek J. Microbiol. Serol.* **34**:409–416.

Houghton, J. A., 1970   A new class of slow growing non-perithecial mutants of *Aspergillus nidulans. Genet. Res.* **16**:285–292.

Houghton, J. A., 1971   Biochemical investigations of the slow growing non-perithecial (*sgp*) mutants of *Aspergillus nidulans. Genet. Res.* **17**:237–244.

Hussey, C., B. A. Orsi, J. Scott and B. Spencer, 1965   Mechanism of choline sulphate utilization in fungi. *Nature (Lond.)* **207**:632–634.

Hutter, R. and J. A. DeMoss, 1967   Enzyme analysis of the tryptophan pathway of *Aspergillus nidulans. Genetics* **55**:241–247.

Hynes, M. J., 1970   Induction and repression of amidase enzymes in *Aspergillus nidulans. J. Bacteriol.* **103**:482–487.

Hynes, M. J., 1972   Mutants with altered glucose repression of amidase enzymes in *Aspergillus nidulans. J. Bacteriol.* **111**:717–722.

Hynes, M. J., 1973a   Alterations in the control of glutamate uptake in mutants of *Aspergillus nidulans. Biochem. Biophys. Res. Commun.* **54**:685–689.

Hynes, M. J., 1973b   Pleiotropic mutants affecting the control of nitrogen metabolism in *Aspergillus nidulans. Mol. Gen. Genet.* **125**:99–107.

Hynes, M. J., 1973c   The effect of lack of a carbon source on nitrate-reductase activity in *Aspergillus nidulans*. *J. Gen. Microbiol.* **79**:155–157.

Hynes, M. J. and J. A. Pateman, 1970a   The genetic analysis of regulation of amidase synthesis in *Aspergillus nidulans*. I. Mutants able to utilize acrylamide. *Mol. Gen. Genet.* **108**:97–106.

Hynes, M. J. and J. A. Pateman, 1970b   The genetic analysis of regulation of amidase synthesis in *Aspergillus nidulans*. II. Mutants resistant to fluoroacetamide. *Mol. Gen. Genet.* **108**:107–116.

Hynes, M. J. and J. A. Pateman, 1970c   The use of amides as nitrogen sources by *Aspergillus nidulans*. *J. Gen. Microbiol.* **63**:317–324.

Jansen, G. J. O., 1964   UV-induced mitotic recombination in the *paba1* region of *Aspergillus nidulans*. *Genetica (The Hague)* **35**:127–131.

Jansen, G. J. O., 1967   Some properties of the *uvs1* mutant of *Aspergillus nidulans*. *Aspergillus Newsl.* **8**:20–21.

Jansen, G. J. O., 1970a   Survival of *uvsB* and *uvsC* mutants of *Aspergillus nidulans* after UV-irradiation. *Mutat. Res.* **10**:21–32.

Jansen, G. J. O., 1970b   Abnormal frequencies of spontaneous mitotic recombination in *uvsB* and *uvsC* mutants of *Aspergillus nidulans*. *Mutat. Res.* **10**:33–41.

Jansen, G. J. O., 1972   Mutator activity in *uvs* mutants of *Aspergillus nidulans*. *Mol. Gen. Genet.* **116**:47–50.

Jinks, J. L., 1954   Somatic selection in fungi. *Nature (Lond.)* **174**:409–410.

Jinks, J. L., 1956   Naturally occurring cytoplasmic changes in fungi. *C. R. Trav. Lab. Carlsberg* **26**:183–203.

Jinks, J. L., 1958   Cytoplasmic differentiation in fungi. *Proc. R. Soc. Lond. Ser. B Biol. Sci.* **148**:314–321.

Jinks, J. L., 1963   Cytoplasmic inheritance in fungi. In *Methodology in Basic Genetics,* edited by W. J. Burdette, pp. 325–354, Holden-Day, San Francisco, Calif.

Jinks, J. L. and M. Grindle, 1963   The genetical basis of heterokaryon incompatibility in *Aspergillus nidulans*. *Heredity* **18**:407–411.

Jinks, J. L., C. E. Caten, G. Simchen and J. H. Croft, 1966   Heterokaryon incompatibility in *Aspergillus nidulans*. *Heredity* **21**:227–239.

Käfer, E., 1958   An eight-chromosome map of *Aspergillus nidulans*. *Adv. Genet.* **9**:105–145.

Käfer, E., 1960   High frequency of spontaneous and induced somatic segregation in *Aspergillus nidulans*. *Nature (Lond.)* **186**:619–620.

Käfer, E., 1961   The processes of spontaneous recombination in vegetative nuclei of *Aspergillus nidulans*. *Genetics* **46**:1581–1609.

Käfer, E., 1962   Translocations in stock strains of *Aspergillus nidulans*. *Genetica (The Hague)* **33**:59–68.

Käfer, E., 1963   Radiation effects and mitotic recombination in diploids of *Aspergillus nidulans*. *Genetics* **48**:27–45.

Käfer, E., 1965   The origins of translocations in *Aspergillus nidulans*. *Genetics* **52**:217–232.

Käfer, E., 1969   Effects of ultraviolet irradiation on heterozygous diploids of *Aspergillus nidulans*. II. Recovery from UV-induced mutation in mitotic recombinant sectors. *Genetics* **63**:821–841.

Käfer, E. and T. L. Chen, 1964   Translocations and recessive lethals induced in *Aspergillus nidulans* by ultraviolet light and gamma rays. *Can. J. Genet. Cytol.* **6**:249–254.

Käfer, E. and J. A. DeMoss, 1973   Formation of hybrid anthranilate synthetase *in vitro* from components of *Aspergillus* and *Neurospora. Biochem. Genet.* **9**:203–211.

Käfer, E. and A. Upshall, 1973   The phenotypes of the eight disomics and trisomics of *Aspergillus nidulans. J. Hered.* **64**:35–38.

Kameneva, S. V. and G. V. Evseeva, 1972   Genetic control of the sensitivity to mutagenic factors in *Aspergillus nidulans.* II. Sensitivity of *uvs* mutants to different mutagens. *Genetika* **8**(3):72–78.

Kameneva, S. V. and Y. M. Romanova, 1969   Genetic control of sensitivity to mutagenic factors in *Aspergillus nidulans.* I. Obtaining of mutants sensitive to UV light. *Genetika* **5**(11):196–198.

Katsatkina, I. D., 1959   Biochemical mutants of *Aspergillus nidulans* produced by irradiation with ultraviolet rays. *Mikrobiologiia* (Eng. transl.) **28**:751–757.

Katsatkina, I. D., 1961   The morphology of aminoacid deficient variants of *Aspergillus nidulans* as a function of the composition of the medium. *Mikrobiologiia* (Eng. transl.) **29**:367–370.

Katz, D. and R. F. Rosenberger, 1970*a*   The utilization of galactose by an *Aspergillus nidulans* mutant lacking galactose phosphate-UDP glucose transferase and its relation to cell wall synthesis. *Arch. Mikrobiol.* **74**:41–51.

Katz, D. and R. F. Rosenberger, 1970*b*   A mutation in *Aspergillus nidulans* producing hyphal walls which lack chitin. *Biochim. Biophys. Acta* **208**:452–460.

Katz, D. and R. F. Rosenberger, 1970*c*   The effect of CO_2 on the purine requirement of *Aspergillus nidulans ad3* mutants. *Biochim. Biophys. Acta* **224**:279–281.

Katz, D. and R. F. Rosenberger, 1971*a*   Hyphal wall synthesis in *Aspergillus nidulans:* Effect of protein synthesis inhibition and osmotic shock on chitin insertion and morphogenesis. *J. Bacteriol.* **108**:184–190.

Katz, D. and R. F. Rosenberger, 1971*b*   Lysis of an *Aspergillus nidulans* mutant blocked in chitin synthesis and its relation to wall assembly and wall metabolism. *Arch. Mikrobiol.* **80**:284–292.

Katz, D., D. Goldstein and R. F. Rosenberger, 1972   Model for branch initiation in *Aspergillus nidulans* based on measurements of growth parameters. *J. Bacteriol.* **109**:1097–1100.

Kessel, M. and R. F. Rosenberger, 1968   Regulation and timing of deoxyribonucleic acid synthesis in hyphae of *Aspergillus nidulans. J. Bacteriol.* **95**:2275–2281.

Kilbey, B. J., 1960   'Sage': A colour modifier in *Aspergillus nidulans. Nature (Lond.)* **186**:906–907.

Kinghorn, J. R. and J. A. Pateman, 1973*a*   NAD- and NADP-glutamate dehydrogenase activity and ammonium regulation in *Aspergillus nidulans. J. Gen. Microbiol.* **78**:39–46.

Kinghorn, J. R. and J. A. Pateman, 1973*b*   Nicotinamide-adenine dinucleotide phosphate-linked glutamate dehydrogenase activity and ammonium regulation in *Aspergillus nidulans. Biochem. Soc. Trans.* **1**:672–674.

Kinghorn, J. R. and J. A. Pateman, 1973*c*   The regulation of nicotinamide–adenine dinucleotide-linked glutamate dehydrogenase in *Aspergillus nidulans. Biochem. Soc. Trans.* **1**:675–676.

Kinghorn, J. R. and J. A. Pateman, 1974   The effect of carbon source on ammonium regulation in *Aspergillus nidulans. Mol. Gen. Genet.* **128**:95–98.

Klimczuk, J., 1970   Spontaneous and induced reversions of *meth1* mutant of *Aspergillus nidulans. Genet. Pol.* **11**:313–319.

Kovalenko, S. P., 1964   Determination of the mutagenic activity of certain alkylating reagents by the method of back mutations with *Aspergillus nidulans. Dokl. Akad. Nauk. SSSR (Engl. Trans.)* **58**:684–685.

Kovalenko, S. P., 1972   High mutagenic effect of phenethyl nitrogen mustard and ethyleneiminopyrimidines in *Aspergillus nidulans. Mutat. Res.* **14**:115–118.

Kovalenko, S. P. and E. M. Tkachenko, 1973   A comparison of activities of nitrogen mustards in the induction of mitotic crossing-over in a diploid strain of *Aspergillus nidulans. Genetika* **9**:97–101.

Kovalenko, S. P., V. K. Panchenko and L. B. Rapp, 1969a   Comparison of the mutagenic action of chemically similar bifunctional nitrogen mustards on *Aspergillus nidulans. Doklady Biol. Sci. (Eng. Trans. Dokl. Akad. Nauk. SSSR Ser. Biokhim.)* **187**:548–550.

Kovalenko, S. P., P. E. Vavrish and V. K. Panchenko, 1969b   Mutagenic activity of some nitrogen mustards on *Aspergillus nidulans. Tsitol. Genet.* **3**:252–254.

Kovalenko, S. P., V. K. Panchenko and L. B. Rapp, 1970a   The dependence of mutagenic activity of N-benzyl-N,N-di2chloroethylamine homologues on their chemical structure. *Tsitol. Genet.* **4**:283.

Kovalenko, S. P., G. V. Shishkin, V. K. Panchenko and L. B. Rapp, 1970b   The influence of aromatic cycles and their substituents on the mutagenic activity of nitrogen mustards. *Genetika* **6**:103–109.

Kovalenko, S. P., V. K. Panchenko and L. B. Rapp, 1971   The mutagenic properties of chlorethyl derivatives of phenethylamine. *Genetika* **7**:160–162.

Kurzeja, K. C. and E. D. Garber, 1973   A genetic study of electrophoretically variant extracellular amylolytic enzymes of wild-type strains of *Aspergillus nidulans. Can. J. Genet. Cytol.* **15**:275–287.

Kuzyurina, L. A., 1959a   The resistance of *Aspergillus nidulans* and *Aspergillus niger* conidia to ultraviolet rays. *Mikrobiologiia* (Eng. transl.) **28**:33–39.

Kuzyurina, L. A., 1959b   Production of mutants by ultraviolet light. II. Morphological characteristics of *Aspergillus nidulans* variants obtained through irradiation with different doses of ultraviolet rays. *Mikrobiologiia* (Eng. transl.) **28**:625–631.

Kwiatowski, Z. A. 1962   Radiation action on the mitotic crossing-over in *Aspergillus nidulans. Acta Microbiol. Pol.* **11**:3–11.

Kwiatowski, Z. A. 1965   Studies on the mechanism of gene recombination in *Aspergillus*. I. Analysis of the stimulating effect of the removal of some metallic ions on mitotic recombination. *Acta Microbiol. Pol.* **14**:3–13.

Kwiatowski, Z. A. and K. Bohdanowicz, 1962   New mycelial mutants in *Aspergillus nidulans. Acta Microbiol. Pol* **11**:17–20.

Kwiatowski, Z. and K. Grad, 1965   A comparison of the ultraviolet effect on the mitotic recombination in two cistrons of *Aspergillus nidulans. Acta Microbiol. Pol.* **14**:15–18.

Lafont, P., J. Lafont and L. Frayssinet, 1970   La nidulotoxine: toxine d'*Aspergillus nidulans* Wint. *Experientia (Basel)* **26**:61–62.

Lanier, W. B., 1967   Apparently aberrant segregation of nutritional markers in *Aspergillus nidulans. Bot. Gaz.* **128**:16–31.

Lanier, W. B., R. W. Tuveson and J. E. Lennox, 1968   A radiation-sensitive mutant of *Aspergillus nidulans. Mutat. Res.* **5**:23–31.

Leal, J. A. and J. R. Villanueva, 1962   An improved selective medium for the formation of ascospores by *Aspergillus nidulans. Nature (Lond.)* **193**:1106.

Lhoas, P. 1961   Mitotic haploidisation by treatment of *Aspergillus niger* diploids with *p*-fluorophenylalanine. *Nature (Lond.)* **190**:744.

Lhoas, P. 1968 Growth rate and haploidisation of *Aspergillus niger* on medium containing *p*-fluorophenylalanine. *Genet. Res.* **12**:305–315.

Lilly, L. J., 1965 An investigation of the suitability of the suppressors of *meth1* in *Aspergillus nidulans* for the study of induced and spontaneous mutation. *Mutat. Res.* **2**:192–195;

Loginova, L. G., 1960 On the activity of hydrolytic enzymes in the *Aspergillus nidulans* variant produced by irradiation with ultraviolet rays. *Mikrobiologiia* (Eng. transl.) **29**:493–498.

Loginova, L. G., 1961 The activity of some oxidative enzymes in an *Aspergillus nidulans* variant obtained by means of ultraviolet irradiation. *Mikrobiologiya* (Eng. transl.) **29**:607–609.

Luig, N. H., 1962 Recessive suppressors in *Aspergillus nidulans* closely linked to an auxotrophic mutant which they suppress. *Genet. Res.* **3**:331–332.

Lukaszkiewicz, Z. and N. J. Pieniazek, 1972 Mutations increasing the specificity of the sulphate permease of *Aspergillus nidulans*. *Bull. Acad. Pol. Sci.* **20**:833–836.

McCully, K. S. and E. Forbes, 1965 The use of *p*-fluorophenylalanine with "master strains" of *Aspergillus nidulans* for assigning genes to linkage groups. *Genet. Res.* **6**:352–359.

Mackintosh, M. E. and R. H. Pritchard, 1963 The production and replica plating of micro-colonies of *Aspergillus nidulans*. *Genet. Res.* **4**:320–322.

Mahoney, M. and D. Wilkie, 1958 An instance of cytoplasmic inheritance in *Aspergillus nidulans*. *Proc. R. Soc. Lond. Ser. B Biol. Sci.* **148**:359–361.

Mahoney, M. and D. Wilkie, 1962 Nucleo–cytoplasmic control of perithecial formation in *Aspergillus nidulans*. *Proc. R. Soc. Lond. Ser. B Biol. Sci.* **156**:524–532.

Mark, C. G. and A. H. Romano, 1971 Properties of the hexose transport systems of *Aspergillus nidulans*. *Biochim. Biophys. Acta* **249**:216–226.

Martinelli, S. D. and B. W. Bainbridge, 1974 Phenol oxidases in wild type and mutant strains of *Aspergillus nidulans*. in manuscript.

Martinelli, S. D. and A. J. Clutterbuck, 1971 A quantitative survey of conidiation mutants in *Aspergillus nidulans*. *J. Gen. Microbiol.* **69**:261–268.

Mastropietro, M. and M. Princivalle, 1963 Dossaggio microbiologico di alcune vitamine del gruppo B. VII. Un nuovo metodo per la titolazione della riboflavina. *Rend. Ist. Super Sanita* **26**:845–852.

Mehrotra, B. S. and V. P. Agnihotri, 1961 Utilization and synthesis of oligosaccharides by some ascosporic members of the *Aspergillus nidulans* group. *Phyton (Argentina)* **16**:195–205.

Mehrotra, B. S. and V. P. Agnihotri, 1962 Nitrogen requirements of some ascoporic members of the *Aspergillus nidulans* group. *Sydowia Ann. Mycol.* **16**:106–114.

Millington-Ward, A. M. 1967 A vegetative instability in *Aspergillus nidulans*. *Genetics* **38**:191–207.

Millington-Ward, A. M. 1970 Recombination and transcription in the *hisB* and *pabaA-1* loci of *Aspergillus nidulans*. *Genetica (The Hague)* **41**:557–574.

Millington-Ward, A. M., F. B. J. Koops and C. Van der Mark-Iken, 1971 Further data on the polarity of the *paba1* locus of *Aspergillus nidulans*. *Genetica (The Hague)* **42**:13–24.

Morpurgo, G., 1962 A new method for estimating forward mutations in fungi: resistance to 8-azaguanine and *p*-fluorophenylalanine. *Sci. Rep. Super. Sanita* **2**:9–12.

Morpurgo, G., 1963a Induction of mitotic crossing over in *Aspergillus nidulans* by bifunctional alkylating agents. *Genetics* **48**:1259–1263.

Morpurgo, G., 1963*b*   Somatic segregation induced by *p*-fluorophenylalanine. *Aspergillus Newsl.* **4**:8.

Morpurgo, G. and L. Volterra, 1966   Fine analysis of mitotic intracistron crossing-over in *Aspergillus nidulans. Ann. Ist. Super. Sanita* **2**:426–428.

Morpurgo, G. and L. Volterra, 1968   The nature of mitotic intragenic recombination in *Aspergillus nidulans. Genetics* **58**:529–541.

Naguib, K., 1959   The growth and metabolism of *Aspergillus nidulans* Eidam in surface culture. *Can. J. Bot.* **37**:353–364.

Naguib, K. and K. Saddik, 1960   Growth and metabolism of *Aspergillus nidulans* Eidam on different nitrogen sources in synthetic media conducive to fat formation. *Can. J. Bot.* **38**:613–622.

Naguib, K. and K. Saddik, 1961   The use of ammonium source of nitrogen in the metabolism of *Aspergillus nidulans* Eidam. *Can. J. Bot.* **39**:955–964.

Nakamura, T., 1962   Biochemical genetical studies on the pathway of sulphate assimilation in *Aspergillus nidulans. J. Gen. Microbiol.* **27**:221–230.

Nakamura, T. and R. Sato, 1960   Cysteine-s-sulphonate as an intermediate in microbial synthesis of cysteine. *Nature (Lond.)* **185**:163–164.

Nakamura, T. and R. Sato, 1962   Accumulation of s-sulphocysteine by a mutant strain of *Aspergillus nidulans. Nature (Lond.)* **193**:481–482.

Nakamura, T. and R. Sato, 1963   Synthesis from sulfate and accumulation of S-sulfocysteine by a mutant strain of *Aspergillus nidulans. Biochem. J.* **86**:328–335.

Nga, B. H. and J. A. Roper, 1968   Quantitative intrachromosomal changes arising at mitosis in *Aspergillus nidulans. Genetics* **58**:193–209.

Nga, B. H. and J. A. Roper, 1969   A system generating spontaneous intrachromosomal changes at mitosis in *Aspergillus nidulans. Genet. Res.* **14**:63–70.

Noronha, L. 1970   Genetic investigation of tryptophan-requiring mutants of *Aspergillus nidulans. Indian J. Exp. Biol.* **8**:298–301.

Ogata, W. N. 1962   Preservation of *Neurospora* stock cultures with anhydrous silica gel. *Neurospora Newsl.* **1**:13.

Oliver, P. T. P., 1972   Conidiophore and spore development in *Aspergillus nidulans. J. Gen. Microbiol.* **73**:45–54.

Oliver, P. T. P., 1973   Influence of cytochalasin B on hyphal morphogenesis of *Aspergillus nidulans. Protoplasma* **76**:279–281.

Oliver, P. T. P., 1974   Ultrastructural localization of free sulphydryl groups in developing conidiophores of *Aspergillus nidulans.* in manuscript.

Page, M. M. and D. J. Cove, 1972   Alcohol and amine catabolism in the fungus *Aspergillus nidulans. Biochem. J.* **127**:17P.

Panicker, R. H. and E. R. B. Shanmugasundaram, 1962   Temperature-independent riboflavineless mutants of *Aspergillus nidulans. Am. J. Bot.* **49**:555–559.

Paszewski, A. and J. Grabski, 1973 β-cystathionase and O-acetylhomoserine sulphhydrylase as the enzymes of alternative methionine biosynthetic pathways in *Aspergillus nidulans. Acta Biochim. Pol.* **20**:159–168.

Paszewski, A., T. Chojnacki, J. Litwinska and W. Gajewski, 1970   Regulation of lactose utilization in *Aspergillus nidulans. Acta Biochim. Pol.* **17**:385–391.

Patel, K. S., 1973   Occurrence of barrage phenomenon in *Aspergillus nidulans. Curr. Sci.* **42**:144.

Pateman, J. A., 1969   Regulation of synthesis of glutamate dehydrogenase and glutamine synthetase in micro-organisms. *Biochem. J.* **115**:769–775.

Pateman, J. A. and D. J. Cove, 1967   Regulation of nitrate reduction in *Aspergillus nidulans*. *Nature (Lond.)* **215**:1234–1237.

Pateman, J. A. and J. R. Kinghorn, 1974   Glutamic and aspartic acid uptake in *Aspergillus nidulans*. *J. Bacteriol*. in press.

Pateman, J. A., D. J. Cove, B. M. Rever and D. B. Roberts, 1964   A common cofactor for nitrate reductase and xanthine dehydrogenase which also regulates the synthesis of nitrate reductase. *Nature (Lond.)* **201**:58–60.

Pateman, J. A., B. M. Rever and D. J. Cove, 1967   Genetic and biochemical studies of nitrate reduction in *Aspergillus nidulans*. *Biochem. J.* **104**:103–111.

Pateman, J. A., J. R. Kinghorn, E. Dunn and E. Forbes, 1973   Ammonium regulation in *Aspergillus nidulans*. *J. Bacteriol.* **114**:943–950.

Peberdy, J. F. and C. E. Buckley, 1973   Adsorption of fluorescent brighteners by regenerating protoplasts of *Aspergillus nidulans*. *J. Gen. Microbiol.* **74**:281–288.

Peberdy, J. F. and R. K. Gibson, 1971   Regeneration of *Aspergillus nidulans* protoplasts. *J. Gen. Microbiol.* **69**:325–330.

Pees, E., 1965   Polarized negative interference in the *lys-51* region of *Aspergillus nidulans*. *Experientia (Basel)* **21**:514–515.

Pees, E., 1966   Lysine, histidine and isoleucine mutants. *Aspergillus Newsl.* **7**:11–12.

Pees, E. 1967   Genetic fine structure and polarized negative interference of the *lys-51* (FL) locus of *Aspergillus nidulans*. *Genetica (The Hague)* **38**:275–304.

Pieniazek, N. J., P. P. Stepien and A. Paszewski, 1973*a*   An *Aspergillus nidulans* mutant lacking cystathionine β-synthase: Identification of L-serine sulfhydrylase with cystathionine β-synthase and its distinctness from O-acetyl-L-serine sulfhydrylase. *Biochim. Biophys. Acta* **297**:37–47.

Pieniazek, N. J., I. M. Kowalska and P. P. Stepien, 1973*b*   Deficiency in methionine adenosyl transferase resulting in limited repressibility of methionine biosynthetic enzymes in *Aspergillus nidulans*. *Mol. Gen. Genet.* **126**:367–374.

Piotrowska, M., M. Sawacki and P. Weglenski, 1969   Mutants of the arginine–proline pathway in *Aspergillus nidulans*. *J. Gen. Microbiol.* **55**:301–305.

Pirt, S. J., 1973   Estimation of substrate affinities (K_s values) of filamentous fungi from colony growth rates. *J. Gen. Microbiol.* **75**:245–247.

Pollard, R., E. Käfer and M. Johnston, 1968   Influence of translocations on meiotic and mitotic nondisjunction in *Aspergillus nidulans*. *Genetics* **60**:743–757.

Pontecorvo, G. 1947   Genetic systems based on heterokaryosis. *Cold Spring Harbor Symp. Quant. Biol.* **11**:193–201.

Pontecorvo, G., 1949   Auxanographic techniques in biochemical genetics. *J. Gen. Microbiol.* **3**:122–126.

Pontecorvo, G., 1950   New fields in the biochemical genetics of micro-organisms. *Biochem. Soc. Symp.* **4**:40–50.

Pontecorvo, G. 1952*a*   Genetic formulation of gene structure and function. *Adv. Enzymol.* **13**:121–149.

Pontecorvo, G., 1952*b*   Genetic analysis of cell organization. *Symp. Soc. Exp. Biol.* **6**:218–229.

Pontecorvo, G., 1954   Mitotic recombination in the genetic system of filamentous fungi. *Caryologia Suppl.* **6**:192–200.

Pontecorvo, G., 1955   Gene structure and action in relation to heterosis. *Proc. R. Soc. Lond. Ser. B Biol. Sci.* **144**:171–177.

Pontecorvo, G., 1956   The parasexual cycle. *Annu. Rev. Microbiol.* **10**:393–400.

Pontecorvo, G., 1958   Self reproduction and all that. *Symp. Soc. Exp. Biol.* **12**:1–5.

Pontecorvo, G., 1959   *Trends in Genetic Analysis,* Oxford University Press, London.

Pontecorvo, G., 1962   Methods of microbial genetics in an approach to human genetics. *Br. Med. Bull.* **18**:81–84.

Pontecorvo, G. 1963   Microbial genetics: retrospect and prospect. *Proc. R. Soc. Lond. Ser. B Biol. Sci.* **158**:1–23.

Pontecorvo, G. and E. Käfer, 1956   Mapping the chromosome by means of mitotic recombination. *Proc. R. Phys. Soc. Edinb.* **25**:16–20.

Pontecorvo, G. and E. Käfer, 1958   Genetic analysis based on mitotic recombination. *Adv. Genet.* **9**:71–104.

Pontecorvo, G. and J. A. Roper, 1956   Resolving power of genetic analysis. *Nature (Lond.)* **178**:83–84.

Pontecorvo, G., J. A. Roper, D. W. Hemmons, K. D. Macdonald and A. W. Bufton, 1953   The genetics of *Aspergillus nidulans. Adv. Genet.* **5**:141–238.

Pontecorvo, G., E. Tarr-Gloor and E. Forbes, 1954   Analysis of mitotic recombination in *Aspergillus nidulans. J. Genet.* **52**:226–237.

Prasad, I., 1970   Mutagenic effects of the herbicide 3′–4′ dichloropropionanilide and its degradation products. *Can. J. Microbiol.* **16**:369–372.

Princivalle, M., 1958   Microbiologic assay of some vitamins of the B group. IV. Titration of p-aminobenzoic acid (PABA). *Rend. Ist. Super Sanita* **21**:928–933.

Princivalle, M. and C. Caradonna, 1962   Dossaggio microbiologico di alcune vitamine del gruppo B. VI. Un nuovo metodo per la titolazione della vitamina PP. *Ann. Chim.* **52**:1248–1253.

Pritchard, R. H., 1954   Ascospores with diploid nuclei in *Aspergillus nidulans. Caryologia (Florence)* 6, Suppl. **1**:1117.

Pritchard, R. H., 1955   The linear arrangement of a series of alleles of *Aspergillus nidulans. Heredity* **9**:343–371.

Pritchard, R. H., 1960*a*   Localized negative interference and its bearing on models of gene recombination. *Genet. Res.* **1**:1–24.

Pritchard, R. H., 1960*b*   The bearing of recombination analysis at high resolution on genetic fine structure in *Aspergillus nidulans* and the mechanism of recombination in higher organisms. *Symp. Soc. Gen. Microbiol.* **10**:155–180.

Pritchard, R. H., 1968   Experiments with *Aspergillus nidulans.* In *Experiments in Microbial Genetics,* edited by R. C. Clowes and W. Hayes, Blackwell, Oxford.

Purnell, D. M., 1973   The effects of specific auxotrophic mutations on the virulence of *Aspergillus nidulans* for mice. *Mycopath. Mycol. Appl.* **50**:195–203.

Purnell, D. M. and G. M. Martin, 1971   *Aspergillus nidulans:* Association of certain alkaline phosphatase mutants with decreased virulence in mice. *J. Infect. Dis.* **123**:305–306.

Purnell, D. M. and G. M. Martin, 1973*a*   Heterozygous diploid strains of *Aspergillus nidulans:* enhanced virulence for mice in comparison to a prototrophic haploid strain. *Mycopath. Mycol. Appl.* **49**:307–319.

Purnell, D. M. and G. M. Martin, 1973*b*   A morphologic mutation in *Aspergillus nidulans* associated with increased virulence in mice. *Mycopath. Mycol. Appl.* **51**:75–79.

Putrament, A., 1964   Mitotic recombination in the *paba1* cistron of *Aspergillus nidulans. Genet. Res.* **5**:316–327.

Putrament, A., 1966   Diepoxybutane-induced mitotic recombination in *Aspergillus nidulans. Proceedings of the Symposium on Mutational Process,* Prague, pp. 107–114, Academia, Prague.

Putrament, A., 1967*a* On the mechanism of mitotic recombination in *Aspergillus nidulans*. I. Intragenic recombination and DNA replication. *Mol. Gen. Genet.* **100**:307–320.

Putrament, A., 1967*b* On the mechanism of mitotic recombination in *Aspergillus nidulans*. II. Simultaneous recombination within two very closely linked cistrons. *Mol. Gen. Genet.* **100**:321–336.

Putrament, A., J. Guzewska and D. Pieniazek, 1970 Further characteristics of methionine mutants and their suppressors in *Aspergillus nidulans*. *Mol. Gen. Genet.* **109**:209–218.

Putrament, A., T. Rozbicka and K. Wojciecowska, 1971 The highly polarized recombination pattern within the *methA* gene of *Aspergillus nidulans*. *Genet. Res.* **17**:125–131.

Radha, K. and E. R. B. Shanmugasundaram, 1962 Genetics and biochemistry of riboflavin auxotrophs of *Aspergillus nidulans*. *Nature (Lond.)* **193**:165–166.

Rao, K. K. and V. V. Modi, 1968 Metabolic changes in biotin-deficient *Aspergillus nidulans*. *Can. J. Microbiol.* **14**:813–815.

Rao, K. K. and V. V. Modi, 1970 Effect of ammonium ions on the growth of *Aspergillus nidulans*. *Experientia (Basel).* **26**:590–591.

Rao, K. K. and V. V. Modi, 1972 Biochemical changes in biotin deficient *Aspergillus nidulans*. *Ind. J. Exp. Biol.* **10**:385–388.

Raper, K. B. and D. I. Fennell, 1965 *The Genus Aspergillus,* Williams and Wilkins, Baltimore.

Roberts, C. F., 1959 A replica plating technique for the isolation of nutritionally exacting mutants of a filamentous fungus (*Aspergillus nidulans*). *J. Gen. Microbiol.* **20**:540–548.

Roberts, C. F., 1963*a* The genetic analysis of carbohydrate utilization in *Aspergillus nidulans*. *J. Gen. Microbiol.* **31**:45–48.

Roberts, C. F., 1963*b* The adaptive metabolism of D-galactose in *Aspergillus nidulans*. *J. Gen. Microbiol.* **31**:285–295.

Roberts, C. F., 1964 Complementation in balanced heterokaryons and heterozygous diploids of *Aspergillus nidulans*. *Genet. Res.* **5**:211–229.

Roberts, C. F., 1967 Complementation analysis of the tryptophan pathway of *Aspergillus nidulans*. *Genetics* **55**:233–239.

Roberts, C. F., 1968 Further analysis of the group E mutants in *Aspergillus nidulans*. *Heredity* **23**:467.

Roberts, C. F., 1969 Isolation of multiple aromatic mutants in *Aspergillus nidulans*. *Aspergillus Newsl.* **10**:19–20.

Roberts, C. F., 1970 Enzyme lesions in galactose non-utilizing mutants of *Aspergillus nidulans*. *Biochim. Biophys. Acta* **201**:267–283.

Robinow, C. F. and C. E. Caten, 1969 Mitosis in *Aspergillus nidulans*. *J. Cell Sci.* **5**:403–431.

Robinson, J. H., C. Anthony and W. T. Drabble, 1973*a* The acidic amino-acid permease of *Aspergillus nidulans*. *J. Gen. Microbiol.* **79**:53–63.

Robinson, J. H., C. Anthony and W. T. Drabble, 1973*b* Regulation of the acidic amino-acid permease of *Aspergillus nidulans*. *J. Gen. Microbiol.* **79**:65–80.

Romano, A. H. and H. L. Kornberg, 1968 Regulation of sugar utilization by *Aspergillus nidulans*. *Biochim. Biophys. Acta.* **158**:491–493.

Romano, A. H. and H. L. Kornberg, 1969 Regulation of sugar uptake by *Aspergillus nidulans*. *Proc. R. Soc. Lond. Ser. B Biol. Sci.* **173**:475–490.

Roper, J. A., 1950   Search for linkage between genes determining a vitamin requirement. *Nature (Lond.)* **166**:956.

Roper, J. A. 1952   Production of heterozygous diploids in filamentous fungi. *Experientia (Basel)* **8**:14–15.

Roper, J. A., 1958   Nucleo-cytoplasmic interactions in *Aspergillus nidulans. Cold Spring Harbor Symp. Quant. Biol.* **23**:141–154.

Roper, J. A., 1961   The steps in the parasexual cycle. In *Recent Advances in Botany,* pp. 375–379, University of Toronto Press, Toronto.

Roper, J. A. 1966a   Culture temperature and biotin requirement in *Aspergillus. Aspergillus Newsl.* **7**:22.

Roper, J. A., 1966b   Mechanisms of inheritance: The parasexual cycle. In *The Fungi,* Vol. 2, pp. 589–617, edited by G. C. Ainsworth and A. S. Sussman, Academic Press, New York.

Roper, J. A., 1971   Aspergillus. In *Chemical Mutagens,* edited by A. Hollaender, Vol. 2, Ch. 12, pp. 343–363. Plenum Press, New York.

Roper, J. A. and J. L. De Azevedo, editors, 1970   Questionaire on gene symbols. *Aspergillus Newsl.* **11**:18–19.

Roper, J. A. and E. Käfer, 1957   Acriflavine-resistant mutants of *Aspergillus nidulans. J. Gen. Microbiol.* **16**:660–667.

Roper, J. A. and B. H. Nga, 1969   Mitotic non-conformity in *Aspergillus nidulans:* The production of hypodiploid and hypohaploid nuclei. *Genet. Res.* **14**:127–163.

Roper, J. A. and R. H. Pritchard, 1955   The recovery of the complementary products of mitotic crossing over. *Nature (Lond.)* **175**:639.

Roper, J. A., H. M. Palmer and W. A. Watmough, 1972   Mitotic non-conformity in *Aspergillus nidulans:* The effects of caffeine. *Mol. Gen. Genet.* **118**:125–133.

Rosenberger, R. F. and M. Kessel, 1967   Synchrony of nuclear replication in individual hyphae of *Aspergillus nidulans. J. Bacteriol.* **94**:1464–1469.

Rosenberger, R. F. and M. Kessel, 1968   Non-random sister chromatid segregation and nuclear migration in hyphae of *Aspergillus nidulans. J. Bacteriol.* **96**:1208–1213.

Rowlands, R. T. and G. Turner, 1973   Nuclear and extranuclear inheritance of oligomycin resistance in *Aspergillus nidulans. Mol. Gen. Genet.* **126**:201–216.

Rowley, B. I. and A. T. Bull, 1973   Chemostat for the cultivation of moulds. *Lab. Pract.* **22**:286–289.

Rowley, B. I. and S. J. Pirt 1972   Melanin production by *Aspergillus nidulans* in batch and chemostat cultures. *J. Gen. Microbiol.* **72**:553–563.

Sadasivam, S., R. Shanmugasundaram and E. R. B. Shanmugasundaram, 1969   The pinkish-red pigment produced by an adenineless mutant of *Aspergillus nidulans. Indian J. Biochem.* **6**:237.

Sadique, J., R. Shanmugasundaram and E. R. B. Shanmugasundaram, 1966a   Formation of 4,5-diaminouracil in a riboflavineless mutant of *Aspergillus nidulans. Naturwissenschaften* **53**:282.

Sadique, J., R. Shanmugasundaram and E. R. B. Shanmugasundaram, 1966b   Isolation of 6,7-dimethyl-8-ribityl lumazine from a riboflavineless mutant of *Aspergillus nidulans. Experientia (Basel)* **22**:32.

Sadique, J., R. Shanmugasundaram and E. R. B. Shanmugasundaram, 1966c   Isolation of 5-amino-4-ribitylaminouracil from a riboflavineless mutant of *Aspergillus nidulans. Biochem. J.* **101**:2C–3C.

Sadique, J., R. Shanmugasundaram and E. R. B. Shanmugasundaram, 1966d   A pair of pteridine derivatives in a heterokaryon of two mutants of *Aspergillus nidulans. Naturwissenschaften* **53**:179.

Saxena, R. K. and U. Sinha, 1973   Conidiation of *Aspergillus nidulans* in submerged liquid culture. *J. Gen. Appl. Microbiol.* **19**:141–146.

Scazzocchio, C., 1970   Nuclear compartmentalisation in the control of gene action in *Aspergillus nidulans. Heredity* **25**:683.

Scazzocchio, C., 1973   The genetic control of molybdoflavoproteins in *Aspergillus nidulans*. II. Use of the NADH dehydrogenase activity associated with xanthine dehydrogenase to investigate substrate and product induction. *Mol. Gen. Genet.* **125**:147–155.

Scazzocchio, C. and A. J. Darlington, 1967   The genetic control of xanthine dehydrogenase and urate oxidase synthesis in *Aspergillus nidulans. Bull. Soc. Chim. Biol.* **49**:1503–1508.

Scazzocchio, C. and A. J. Darlington, 1968   The induction and repression of the enzymes of purine breakdown in *Aspergillus nidulans. Biochim. Biophys. Acta* **166**:557–568.

Scazzocchio, C., F. B. Holl and A. I. Foguelman, 1973   The genetic control of molybdoflavoproteins in *Aspergillus nidulans*. Allopurinol-resistant mutants constitutive for xanthine-dehydrogenase. *Eur. J. Biochem.* **36**:428–445.

Scott, B. R. and T. Alderson, 1971   The random (non-specific) forward mutational response of gene loci in *Aspergillus* conidia after photosensitisation to near ultraviolet light (365 nm) by 8-methoxypsoralen. *Mutat. Res.* **12**:29–34.

Scott, B. R., T. Alderson and D. G. Papworth, 1972   The effect of radiation on the *Aspergillus* conidium. I. Radiation sensitivity and a "germination inhibitor" *Radiat. Bot.* **12**:45–50.

Scott, B. R., T. Alderson and D. G. Papworth, 1973   The effect of plating densities on the retrieval of methionine suppressor mutations after ultraviolet or gamma irradiation of *Aspergillus. J. Gen. Microbiol.* **75**:235–239.

Scott, J. M. and B. Spencer, 1968   Regulation of choline sulphatase synthesis and activity in *Aspergillus nidulans. Biochem. J.* **106**:471–477.

Selvam, R. and K. R. Shanmugasundaram, 1972   Absence of creatinine metabolism in the fungus *Aspergillus nidulans. Curr. Sci. (Bangalore)* **41**:144.

Sermonti, G., 1968   List of proposed symbols. *Aspergillus Newsl.* **9**:24–26.

Shanfield, B. and E. Käfer, 1969   UV-sensitive mutants increasing mitotic crossing over in *Aspergillus nidulans. Mutat. Res.* **7**:485–487.

Shanfield, B. and E. Käfer, 1971   Chemical induction of mitotic recombination in *Aspergillus nidulans. Genetics* **67**:209–219.

Shanmugasundaram, R. and E. R. B. Shanmugasundaram, 1965   Studies on the heterokaryotic vigour in the decomposition of riboflavin. *Curr. Sci. (Bangalore)* **33**:747–748.

Sharma, R. P., 1970   Combined effect of physical and chemical mutagens on mutation frequency in *Aspergillus nidulans. Indian J. Genet. Plant Breed.* **30**:199–211.

Shepherd, C. J., 1956   Pathways of cysteine synthesis in *Aspergillus nidulans. J. Gen. Microbiol.* **15**:29–38.

Shepherd, C. J. 1957   Changes occurring in the composition of *Aspergillus nidulans* conidia during germination. *J. Gen. Microbiol.* **16**:i.

Siddiqi, O. H., 1962a   Mutagenic action of nitrous acid on *Aspergillus nidulans. Genet. Res.* **3**:303–314.

Siddiqi, O. H., 1962b   The fine genetic structure of the *paba1* region of *Aspergillus nidulans. Genet. Res.* **3**:69–89.

Siddiqi, O. H. and A. Putrament, 1963   Polarized negative interference in the *paba1* region of *Aspergillus nidulans. Genet. Res.* **4**:12–20.

Siddiqi, O. H., B. N. Apte and M. P. Pitale, 1966   Genetic regulation of aryl sulphatases in *Aspergillus nidulans. Cold Spring Harbor Symp. Quant. Biol.* **31**:381–382.

Singh, J. and T. K. Walker, 1955   Influence of *p*H of the medium on the characteristics and composition of *Aspergillus nidulans* fat. *J. Sci. Ind. Res. Sect. C.* **15**:222–224.

Singh, J. and T. K. Walker, 1956   Changes in the composition of the fat of *Aspergillus nidulans* with age of the culture. *Biochem. J.* **62**:286–289.

Singh, J., T. K. Walker and M. L. Meara, 1955   The component fatty acids of the fat of *Aspergillus nidulans. Biochem. J.* **61**:85–88.

Sinha, U., 1967   Aromatic amino acid biosynthesis and para-fluorophenylalanine resistance in *Aspergillus nidulans. Genet. Res.* **10**:261–272.

Sinha, U., 1969   Genetic control of the uptake of amino acids in *Aspergillus nidulans. Genetics* **62**:495–505.

Sinha, U., 1970   Competition between leucine and phenylalanine and its relation to *p*-fluorophenylalanine-resistant mutations in *Aspergillus nidulans. Arch. Mikrobiol.* **72**:308–317.

Sinha, U., 1972   Studies with *p*-fluorophenylalanine-resistant mutants of *Aspergillus nidulans. Beitr. Biol. Pflanz.* **48**:171–180.

Skinner, V. M. and S. Armitt, 1972   Mutants of *Aspergillus nidulans* lacking pyruvate carboxylase. *FEBS (Fed. Eur. Biochem. Soc.) Lett.* **20**:16–18.

Sneath, P. H. A., 1955   Putrescine as an essential growth factor for a mutant of *Aspergillus nidulans. Nature (Lond.)* **175**:818.

Sorger, G. J., 1963   TPNH-cytochrome *c* reductase and nitrate reductase in mutant and wild-type *Neurospora* and *Aspergillus. Biochem. Biophys. Res. Comm.* **12**:395–401.

Spencer, B. and B. G. Moore, 1973   Specific sulphate binding in *Aspergillus nidulans* during sulphate transport. *Biochem. Soc. Trans.* **1**:304–306.

Spencer, B., E. C. Hussey, B. A. Orsi and J. M. Scott, 1968   Mechanism of choline O-sulphate utilization in fungi. *Biochem J.* **106**:461–469.

Sternlight, E., D. Katz and R. F. Rosenberger, 1973   Subapical wall synthesis and wall thickening induced by cycloheximide in hyphae of *Aspergillus nidulans. J. Bacteriol.* **114**:819–823.

Stevens, L. and A. Heaton, 1973   Induction, partial purification and properties of ornithine transaminase from *Aspergillus nidulans. Biochem. Soc. Trans.* **1**:749–751.

Strickland, W. N., 1958*a*   Abnormal tetrads in *Aspergillus nidulans. Proc. R. Soc. Lond. Ser. B Biol. Sci.* **148**:533–542.

Strickland, W. N., 1958*b*   An analysis of interference in *Aspergillus nidulans. Proc. R. Soc. Lond. Ser. B Biol. Sci.* **149**:82–101.

Strigini, P. and G. Morpurgo, 1961   Biotin requirement and carbon and sulphur sources in *Aspergillus* and *Neurospora. Nature (Lond.)* **190**:557.

Strigini, P., C. Rossi and G. Sermonti, 1963   Effects of disintegration of incorporated ^{32}P in *Aspergillus nidulans. J. Mol. Biol.* **7**:683–699.

Tector, M. A. and E. Käfer, 1962   Radiation-induced chromosomal aberrations and lethals in *Aspergillus nidulans. Science (Wash., D.C.)* **136**:1056–1057.

Threlfall, R. J., 1968   The genetics and biochemistry of mutants of *Aspergillus nidulans* resistant to chlorinated nitrobenzenes. *J. Gen. Microbiol.* **52**:35–44.

Threlfall, R. J., 1972   Effect of pentachloronitrobenzene (PCNB) and other chemicals on sensitive and PCNB-resistant strains of *Aspergillus nidulans. J. Gen. Microbiol.* **71**:173–180.

Trinci, A. P. J., 1969   A kinetic study of the growth of *Aspergillus nidulans* and other fungi. *J. Gen. Microbiol.* **57**:11–24.

Trinci, A. P. J., 1970*a*   Kinetics of apical and lateral branching in *Aspergillus nidulans* and *Geotrichum lactis. Trans. Br. Mycol. Soc.* **55**:17–28.

Trinci, A. P. J., 1970*b* Kinetics of the growth of mycelial pellets of *Aspergillus nidulans*. *Arch. Mikrobiol.* **73**:353–367.

Trinci, A. P. J., 1971 Influence of the width of the peripheral growth zone on the radial growth rate of fungal colonies on solid media. *J. Gen. Microbiol.* **67**:325–344.

Trinci, A. P. J. and K. Gull, 1970 Effect of actidione, griseofulvin and triphenyltin acetate on the kinetics of fungal growth. *J. Gen. Microbiol.* **60**:287–292.

Trinci, A. P. J. and C. Whittaker, 1968 Self-inhibition of spore germination in *Aspergillus nidulans*. *Trans. Brit. Mycol. Soc.* **51**:594–596.

Upshall, A., 1966 Somatically unstable mutants of *Aspergillus nidulans*. *Nature (Lond.)* **209**:1113–1115.

Upshall, A., 1971 Phenotypic specificity of aneuploid states in *Aspergillus nidulans*. *Genet. Res.* **18**:167–171.

Van Arkel, G. A., 1962 A new colour mutant "pale." *Aspergillus Newsl.* **3**:4.

Van Arkel, G. A., 1963 Sodium arsenate as an inducer of somatic reduction. *Aspergillus Newsl.* **4**:9.

Verbina, N. M., 1958 On some peculiarities of development of *Aspergillus nidulans* variants produced by ultraviolet irradiation. *Mikrobiologiia* (Eng. transl.) **27**:164–171.

Verbina, N. M., 1959 Biomass accumulation in greatly altered variants of *Aspergillus nidulans* under various conditions of cultivation. *Mikrobiologiia* (Eng. transl.) **28**:355–361.

Verbina, N. M., 1960 Respiration of greatly modified *Aspergillus nidulans* variants obtained by ultraviolet irradiation. *Mikrobiologiia* Eng. transl.) **29**:144–146.

Verma, S. and U. Sinha, 1973 Inhibition of growth by amino acid analogues in *Aspergillus nidulans*. *Beitr. Biol. Pflanzen* **49**:47–58.

Verma, I. M., M. Edelman, M. Herzberg and U. Z. Littauer, 1970 Size determination of mitochondrial ribosomal RNA from *Aspergillus nidulans* by electron microscopy. *J. Mol. Biol.* **52**:138–140.

Verma, I. M., M. Edelman and U. Z. Littauer, 1971 A comparison of nucleotide sequences from mitochondrial and cytoplasmic RNA of *Aspergillus nidulans*. *Eur. J. Biochem.* **19**:124–129.

Waldron, C. and C. F. Roberts, 1973 Cytoplasmic inheritance of a cold-sensitive mutant in *Aspergillus nidulans*. *J. Gen. Microbiol.* **78**:379–381.

Warr, J. R. and J. A. Roper, 1965 Resistance to various inhibitors in *Aspergillus nidulans*. *J. Gen. Microbiol.* **40**:273–281.

Weglenski, P., 1966 Genetical analysis of proline mutants and their suppressors in *Aspergillus nidulans*. *Genet. Res.* **8**:311–321.

Weglenski, P., 1967 The mechanism of action of some proline suppressors in *Aspergillus nidulans*. *J. Gen. Microbiol.* **47**:77–85.

Weijer, J. and S. H. Weisberg, 1966 Karyokinesis of the somatic nuclei of *Aspergillus nidulans*. I. The juvenile chromosome cycle (Feulgen staining). *Can. J. Genet. Cytol.* **8**:361–374.

Weisberg, S. H. and G. Turian, 1971 Ultrastructure of *Aspergillus nidulans* conidia and conidial lomasomes. *Protoplasma* **72**:55–67.

Weisberg, S. H. and J. Weijer, 1968 Karyokinesis of the somatic nucleus of *Aspergillus nidulans*. II. Nuclear events during hyphal differentiation. *Can. J. Genet. Cytol.* **10**:699–722.

Wilson, J. D. and E. L. Powers, 1970 X-ray sensitivity and modifying effects of water in conidia of *Aspergillus nidulans*. *Radiat. Res.* **43**:698–710.

Winder, F. G. and G. R. Campbell, 1973 The deoxyribonucleases of *Aspergillus nidulans*. *Heredity* **31**:423.

Wohlrab, G. and R. W. Tuveson, 1969   Effects of liquid holding on the induction of mutations in an ultraviolet-sensitive strain of *Aspergillus nidulans*. *Mutat. Res.* **8**:265–275.

Wood, S. and E. Käfer, 1967   Twin-spots as evidence for mitotic crossing-over in *Aspergillus nidulans*. *Nature (Lond.)* **216**:63–64.

Wood, S. and E. Käfer, 1969   Effects of ultraviolet irradiation on heterozygous diploids of *Aspergillus nidulans*. I. UV-induced mitotic crossing over. *Genetics* **62**:507–518.

Wright, P. J. and J. A. Pateman, 1970   Ultraviolet-light sensitive mutants of *Aspergillus nidulans*. *Mutat. Res.* **9**:579–587.

Yoshimoto, A., T. Nakamura and R. Sato, 1961   A sulphite reductase from *Aspergillus nidulans*. *J. Biochem.* **50**:553–554.

Yoshimoto, A., T. Nakamura and R. Sato, 1967   Isolation from *Aspergillus nidulans* of a protein catalyzing the reduction of sulphite by reduced violagen dyes. *J. Biochem.* **62**:756–766.

Zaudy, G., 1969   The location of some multiple aromatic mutants in *Aspergillus nidulans*. *Aspergillus Newsl.* **10**:22.

Zonneveld, B. J. M., 1971   Biochemical analysis of the cell wall of *Aspergillus nidulans*. *Biochim. Biophys. Acta* **249**:506–514.

Zonneveld, B. J. M., 1972a   A new type of enzyme, an exo-splitting α-1,3-glucanase from non-induced cultures of *Aspergillus nidulans*. *Biochim. Biophys. Acta* **258**:541–547.

Zonneveld, B. J. M., 1972b   The significance of α-1:3-glucan of the cell wall and α-1:3-glucanase for cleistothecium development. *Biochim. Biophys. Acta* **273**:174–184.

Zonneveld, B. J. M., 1973   Inhibitory effect of 2-deoxyglucose on cell wall α-1, 3-glucan synthesis and cleistothecium development in *Aspergillus nidulans*. *Develop. Biol.* **34**:1–8.

27

Neurospora crassa

Raymond W. Barratt

Introduction

Neurospora is a genus of fungi belonging to the Ascomycetes, subclass Pyrenomycetes. In nature, Neurospora exists as a soil saprophyte—the four recently described species were all isolated from soil samples—though the original isolations were obtained from sugar cane bagasse, carbonized vegetation, or from bakeries. Prior to the routine use of mold inhibitors in commercial bread, Neurospora was an important pest on bread and flour in bakeries, leading to its characterization as the red bread mold (Shear and Dodge, 1927). Although ten species have been described (Shear and Dodge, 1927; Dodge, 1930, 1932, 1935, 1942; Tai, 1935, 1936; Gochenaur and Backus, 1962; Nelson *et al.*, 1964; Nelson and Backus, 1968; Frederick *et al.*, 1969; Mahoney *et al.*, 1969) genetic investigation has been confined nearly exclusively to three species—*Neurospora crassa, Neurospora sitophila,* and *Neurospora tetrasperma*. Primarily, *N. crassa,* the most widely studied species, will be described in this review.

Vegetative Structures, Nutrition, and Sexual Reproduction

A brief illustrated description of the life cycle of the eucaryotic organism *N. crassa* appears in most standard genetics textbooks (see especially *General Genetics* by Srb, Owen and Edgar (1965) p. 99 and following pages and *Genetics* by Herskowitz (1965) p. 27 and pp. 124–125.

Raymond W. Barratt—Dean, School of Science and Director, Fungal Genetics Stock Center, California State University, Humbolt, Arcata, California.

Vegetative Stages

A growing vegetative culture (clone) is normally haploid and consists of branched hyphal filaments some 5 μ or less in diameter. Growth of a hypha occurs at the tip, with side branches occurring some distance back. The rate of hyphal growth under optimal conditions may reach 5 mm/hr (Ryan *et al.*, 1943). Collectively, a mass of hyphae are called a mycelium (pl. mycelia). As in all filamentous ascomycetous fungi, each hypha is sub-divided by crosswalls called septa. The resulting compartments are often referred to as cells, though this is incorrect since the septa contain pores of sufficient size to allow nuclear and cytoplasmic passage. Strictly speaking *Neurospora* is acellular, correctly referred to as a coenocyte (Davis, 1966). Nuclear division is not associated in a one-to-one relation with septum formation; this, in part, accounts for the multinucleate conditions found in each compartment. Nuclear division apparently occurs by mitosis (Bakers-pigel, 1959, 1969; Somers *et al.*, 1960; Ward and Ciurysek, 1962; Bianchi and Turian, 1967; Van Winkel *et al.*, 1971) though, because of the small size of nuclei, some differences in interpretation exist (Weijer *et al.*, 1965; Keeping, 1969).

Hyphae growing on the surface of a substratum send up aerial branches that develop into conidiophores which in turn give rise to asexual spores, called macroconidia. Macroconidia (6–8 μ in diameter) are produced abundantly in chains and are the chief means of vegetative re-production in nature. They are rich in carotenoids and in mass result in most of the bright orange color observed in a culture. Macroconidia contain one to ten or more haploid nuclei, though considerable variation occurs depending upon the strain and environment (Huebschman, 1952; Pittenger, 1967; Koski and Hedman, 1971). Specialized media for enhanced macroconidial production have been described (Wainwright, 1959; Barratt, 1963; Turian, 1964; Tuveson *et al.*, 1967). Macroconidia germinate readily in a few hours when transferred to a suitable culture medium (Ryan, 1948; Bradford and Gibgot, 1963).

In addition to the abundant production of macroconidia, mycelia produce microconidia, which are probably vestigal male elements (Dodge, 1930, 1932), on microconidiophores by a budding-like process from the side of the hyphae (Lowry *et al.*, 1967). Microconidia (2–3 μ) are mar-kedly smaller than macroconidia (Gillie, 1967), are over 99 percent uninucleate (Barratt and Garnjobst, 1949; Baylis and DeBusk, 1967), are more short-lived under ordinary laboratory conditions (Barratt, 1964), and germinate more slowly than macroconidia. Certain genetic strains, which are exclusively microconidial (Barratt and Garnjobst, 1949; Turian *et al.*, 1967) are especially useful for some types of genetic investigations. In mass, microconidia have a gray-green to brown color. Substitution of glycerol for sucrose as a carbon source in the culture medium and lowering the incubation temperature to 25 °C markedly enhances microconidiation (Barratt and Garnjobst, 1949). Reproduction by either macro- or microconidia is a vegetative or clonal form of reproduction.

In common with many fungi, Neurospora also possesses a growth phase in which genetically different haploid nuclei can exist within the same cytoplasm in a hyphae. This phenomenon, known as heterokaryosis, is equivalent in many ways to the diploid but differs since more than one nucleus takes part in the determination of the phenotype (Davis, 1966). Heterokaryons can originate by direct fusion between two hyphae as they touch (a process called anastomosis) and the subsequent flowing of nuclei from one hyphae to the other, or by mutation of a nucleus in a homokaryotic culture (Srb and Infanger, 1964). Demonstration of heterokaryosis requires the isolation of two or more nuclear types from a macroconidium or hyphal tip (Dodge, 1942; Beadle and Coonradt, 1944). While heterokaryon stability is controlled by a number of factors, including mating type (Sansome, 1946; Whitehouse, 1949; Gross, 1952) and several incompatibility genes (Garnjobst and Wilson, 1956; Wilson and Garnjobst, 1966), stable heterokaryons are readily obtainable in Neurospora, and provide a powerful tool for certain types of genetic analysis. Nuclear ratios in heterokaryons and their effect on the phenotype have been investigated fairly extensively (Beadle and Coonradt, 1944; Barratt and Garnjobst, 1949; Pittenger and Atwood, 1956; Pittenger, 1964; Davis, 1966).

Nutrition and Culture Media

The nutrition of Neurospora is simple. Wild-type strains may be grown on a suitable mixture of mineral salts in water plus a carbon source (sucrose) and the vitamin biotin (Beadle and Tatum, 1941, 1945; Anonymous, 1966). Still or aerated liquid cultures or agar-slant cultures are routinely used. Agar-slant cultures are usually used for culture maintenance and are incubated at 25–34 °C (Ryan *et al.*, 1943). Subcultures are fully grown including conidiation within 3–5 days. Both growth on race tubes (Ryan *et al.*, 1943) and dry weights of mycelia (Beadle and Tatum, 1941) are useful for accurate measurement of growth, the former measuring rate and the latter terminal growth mass. In liquid culture, yields of 2.6 g (dry weight) of mycelium/liter of medium per day can be obtained (Garrick, 1967). Various specialized media have been developed [see the section *Methods, Growth* in Bachmann and Strickland (1965) and *Growth Methods* in Bachmann (1970)].

The rapid linear growth rate of Neurospora mycelia is a serious drawback when individual colonies need to be isolated. This problem was circumvented by the finding that when certain chemicals, notably the ketohexose sorbose, are added to the culture media a restricted colonial growth habit results (Tatum *et al.*, 1949). The importance of the ratio between other carbon sources and sorbose has been investigated (Tatum *et al.*, 1949; Brockman and de Serres, 1964; Pittenger, 1964). The colonizing effect of sorbose when plating microconidia, macroconidia, or ascospores (Newmeyer, 1954) has permitted the handling of Neurospora with routine microbiological techniques. A temperature-sensitive colonial strain or suitable gene combinations resulting in conidiating colonial strains have also proved useful for plating (Barratt and Garnjobst, 1949; Ressig, 1956; Maling, 1960; Perkins, 1971, 1972*b*; Littlewood and Munkres, 1972).

Sexual Reproduction

While the fungi as a group exhibit a great variety of life cycles and sexual arrangements, *N. crassa* exists in two mating types, designated A and a. A pure strain (either A or a) is unable to undergo sexual reproduction. Sexual fruiting bodies, called perithecia, are produced only when the two mating types are brought together. There is no morphological difference between A and a strains, and either can form reproductive structures abundantly, called perithecial fundaments or protoperithecia, when grown on a suitable agar crossing medium (Shear and Dodge, 1927; Westergaard and Mitchell, 1947; Anonymous, 1966). A protoperithecium consists of a coiled multicellular hypha, enclosed in a knotlike aggregation of hyphae. Protoperithecia are translucent and can be seen in four- or five-day old agar-slant cultures at 25°C either with an experienced naked eye or the aid of the lower power of a dissection microscope. From the protoperithecium project several very slender receptive hyphae, called trichogynes. Fertilization and subsequent stages of perithecial development occur when a nucleus of the opposite mating type enters the trichogyne and migrates down the trichogyne into the ascogonium. In making crosses in the laboratory, the protoperithecial parent is fertilized after incubation of a culture at 25°C for 6 days in the dark (Westergaard and Mitchell, 1947; Strickland, 1960; Thomas and Frost, 1966). Contact of the trichogyne by a macroconidium, microconidium, or a hyphal filament and the resultant fusion by anastomosis provide the nucleus (Dodge, 1935). The resulting pair of opposite-mating-type nuclei, called a dikaryon (one from the ascogenous hyphae and the one passing down the trichogyne), become associated and begin to divide synchronously. Synchronous division continues until a mass of ascogenous hyphae is developed. The process of division is complex, involving crozier formation as in the Basidiomycetes, and results in a mass of binucleate cells (Mc-Clintock, 1945; Singleton, 1953). After proliferation of the ascogenous hyphae is complete (up to 300 pairs of nuclei may be produced in one perithecium) the pairs of nuclei fuse, resulting in a diploid nucleus in each cell. Each of these fusion cells, now called ascus initial, without further proliferation, immediately undergoes meiosis. There is only one diploid nucleus in the life cycle of Neurospora. The two daughter nuclei from the first meiotic division move to the opposite ends of a developing ascus. Each meiosis is confined to one ascus. The second meiotic division follows. The spindles of the two second meiotic divisions occurring within the elongating ascus do not overlap [except in rare cases (Howe, 1956)] and result in four linearly arranged nuclei, each adjacent pair being derived from the second meiotic division. Each nucleus is now haploid and contains a set of seven chromosomes. The ascus continues to elongate and a mitotic division ensues, again without overlapping spindles, resulting in 8 nuclei (4 pairs). Around each nucleus an ascospore is organized, developing a thick ribbed wall as it matures (thus Neurospora = nerved spores). The walls become impregnated with melanin and the ascospores turn olivaceous, at first, and ultimately dark brown to black at maturity. As melanization proceeds, one additional mitotic division follows within each ascospore, resulting in asci each containing 8 binucleate haploid ascospores ($27–30 \times 14–15 \mu$). Fertilization stimulates further growth, the differentiation of perithecial wall tissue to enclose the mass, and the development of a black heavy-walled perithecium $400–600 \mu$ in diameter. The entire sexual cycle occupies about 10–14 days, depending upon environmental conditions (Westergaard and Mitchell, 1947). As the perithecium matures, a small beak (ostiole) develops in the end distal to the substratum, through which, in nature, the ascospores are shed.

Ordered tetrad analysis of individual meiosis is accomplished by squeezing a perithecium with a pair of watchmaker's forceps, spreading the contents onto the surface of 4-percent-water agar, teasing out individual asci, dissecting the ascospores from the ascus in order with the aid of a fine needle, and transferring them to individual culture tubes (Ryan,

1950). The entire operation is carried out with the aid of a dissection microscope. With experience 10–15 asci can be dissected per hour. A modification of the procedure for isolating the individual asci, which employs certain genetic strains, was developed by Maling (1960). The rapid analysis of an unordered tetrad is possible as a result of the observation that under suitable conditions ascospores from individual asci are forcibly ejected from the ostiole of the perithecium in groups of eight (Strickland, 1960; Strickland and Thorpe, 1963; Perkins, 1966*b*). Random spore analysis is usually performed on mature crosses in which the ascospores have been ejected from the perithecia onto the glass of the cross culture tube. Samples of ascospores are removed from the tube onto 4-percent-water agar and individually transferred to fresh culture media. Germination of ascospores requires heat (Shear and Dodge, 1927; Lindegren, 1932) or chemical (Emerson, 1948, 1954) activation (for exception, see Faull, 1930). Routinely, ascospores are heat-activated at 50–60°C for 30 minutes either on the surface of agar or in a water suspension. The treatment has the added advantage of killing any conidia which may be present as contaminants.

Although begun by Dodge (Dodge, 1928), genetic investigation with *N. tetrasperma* has lagged behind investigations with *N. crassa*. *N. tetrasperma* differs from *N. crassa* and *N. sitophila* by having four-spored asci and being secondarily homothallic (also called mixochimeric, facultatively heterothallic, or pseudo-heterothallic (Dodge, 1957)). During meiosis in this species, the orientation of the first meitotic spindle is longitudinal, and the two daughter nuclei come to rest one somewhat above the other. Orientation of the second-meiotic-division spindle occurs in two ways, but both produce the same results. In one, the two spindles are oriented longitudinally, with a spindle near each end of the ascus. In the other, the two spindles are oriented obliquely and lie nearly parallel to each other near the center of the ascus. Both however, result in a pair of nonsister nuclei becoming situated near each end of the developing ascus (see illustrations in Howe, 1963). The four spindles in the third (mitotic) division are obliquely or nearly transversely oriented. The resulting eight nuclei assume nonsister pairwise arrangements, after which a pair of nonsister nuclei is included within each developing ascospore. Nuclear passage normally occurs at both the second and third division (Howe, 1963), resulting predominantly in ascospores heterokaryotic for mating type.

Within the past decade four other species have been described which are truly homothallic (Gochenaur and Backus, 1962; Nelson *et al.*, 1964; Mahoney *et al.*, 1969).

Sources of Information, Genetic Strains, and Stock Cultures

Sources of Information

The beginning student will probably wish to consult some of the reviews on the biology, genetics, and biochemistry of Neurospora. An earlier treatment of Neurospora genetics is included in *The Genetics of Microorganisms* (Catcheside, 1951). For more recent reviews which include a linkage map of most of the important genetic loci in Neurospora, and a survey of the life cycles of the genetically important fungi, mapping methods, cytological studies, and many of the important contributions of Neurospora to genetic and biochemical theory, as well as information on nonnuclear inheritance, the reader is directed to *Fungal Genetics* (Fincham and Day, 1963). Genetic and microbiological research techniques for Neurospora have been recently reviewed by Davis and de Serres (1970). The reviews in *The Fungi* on morphogenesis in Ascomycetes (Turian, 1966), mechanisms of inheritance (Emerson, 1966), and heterokaryosis (Davis, 1966) will provide additional background. Some interesting historical perspective and personal insights into Neurospora genetics can be gleaned from a recent series of short papers (Beadle, 1973; Catcheside, 1973; Horowitz, 1973; Lindegren, 1973; Robbins, 1973; Srb, 1973) and from the Nobel prize lectures of G. B. Beadle and E. L. Tatum (Herskowitz, 1965).

Neurospora is probably genetically and biochemically the most thoroughly studied of any eukaryotic organism. Some 4300 research papers have been published on the genus Neurospora. References to the first 2310 papers published through 1963 are included in the *Neurospora Bibliography and Index* (Bachmann and Strickland, 1965). This publication includes an author index and an extremely valuable subject index. Some 1300 papers published between 1964 and 1969 are referenced in *Neurospora Bibliography and Index,* number two (Bachmann, 1970), which also includes an author index and a detailed subject index. A bibliography of papers on Neurospora published during 1970 is included in *Neurospora Newsletter* **18** (Bachmann, 1971), those published during 1971 in *Neurospora Newsletter* **19** (Bachmann, 1972), and those published in 1972 in *Neurospora Newsletter* **20** (Bachmann, 1973).

Neurospora Newsletter is an informal publication which includes research and technical notes, linkage data and maps, summaries of the bian-

nual Neurospora conferences, stock lists, as well as the previously mentioned annual bibliography of research papers. Twenty issues have been
published between its inception in 1962. The newsletter is edited by Dr.
B. J. Bachmann, Department of Microbiology, Yale University School of
Medicine, New Haven, Connecticut and published by the Fungal
Genetics Stock Center, Humboldt State University, Arcata, California
95521.

Wild-Type Strains

Unfortunately wild-type strains of *N. crassa* of different origins were
used in the early genetic investigation of Neurospora (Beadle and Tatum,
1945; Tatum *et al.*, 1950). This problem of strain heterogeneity led to incompatibility, poor fertility in crosses, meiotic abnormalities, differing
recombination distances, etc. (Barratt, 1954; Holloway, 1954; Frost,
1961). More recently several standard wild-type strains of *N. crassa* of
known genetic origin have been developed [for pedigrees see Barratt
(1962) and Case *et al.* (1965)]. Most laboratories now use as standard
wild-type strains 74-OR8-1a and 74-OR23-1A (Case *et al.*, 1965). Over
370 wild-type strains of Neurospora were collected in 1968–1969 from
nature by Dr. D. D. Perkins (1970).

Reference strains, authentic material or derivatives of type material,
are available for several species of Neurospora—*N. crassa* (Perkins,
1972*b*), *N. tetrasperma* (see Barratt and Ogata, 1972), *N. sitophila*
(Perkins, 1972*c*), *N. intermedia* (Perkins and Turner, 1973).

Mutant Strains and Cultures

Over 350 loci controlling morphology, nutrition, drug resistance, etc.
have been described and mapped (Barratt and Radford, 1969, 1970*a,b*;
Radford, 1972). For references to specific mutants, the reader is again
directed to the Mutant Index (part 2) of Bachmann and Strickland (1965)
and to Section 3, Mutants and Mutations, of Bachmann (1970). Many
loci have been subjected to fine-structure analysis. Cultures of genetic
strains of Neurospora are available for research from the Fungal Genetics
Stock Center, Humboldt State University, Arcata, California, one of
several genetic stock centers supported, in part or totally, by grants from
the National Science Foundation. Currently the stock center maintains
about 2200 strains for distribution, including single-mutant, multiple-
mutant, extrachromosomal-mutant, wild-type, and aberration stocks. The

stock center maintains its stocks on anhydrous silica gel (Ogata, 1962; Perkins, 1962) and in lyophil in order to avoid the accumulation of secondary mutations. The stock center publishes a stock list biannually of genetic and wild-type strains available (Barratt and Ogata, 1972). In addition to characterizing the genotype (locus and allele) of each strain, the stock list includes information regarding linkage group and arm, genetic background, mutagen employed, and selected references pertinent to the locus.

Cytology, Chromosomes, Linkage Groups, and Maps

The basis for cytogenetic research with *N. crassa* began with the classic work of McClintock, (1945, 1947). Despite the extreme small size and difficulty of the material, she developed methods for and investigated chromosome behavior in the devloping ascus. She demonstrated that seven chromosomes make up the haploid set, established the relative length of each, studied the first translocations, and investigated the cytology of meiosis. A study of somatic mitosis also demonstrated seven morphologically distinct chromosomes at metaphase (Ward and Ciurysek, 1962). The cytological events at pachytene were confirmed and extended to *N. sitophila* and *N. tetrasperma* by Singleton (1953). Cytological techniques have been summarized (Barry, 1966). The seven chromosomes of *N. crassa* are identified by Arabic numerals 1–7 in the order of decreasing length (McClintock, 1945), and the seven linkage groups (see below) are identified by Roman numerals I–VII in the chronological order in which they were discovered (Barratt *et al.*, 1954). Correlation between the linkage groups and chromosomes is incomplete, and has been studied via cytogenetic studies on aberration strains with known linkage anomalies. Linkage group I is located in chromosome 6, with the mating-type locus in the short arm; linkage group II is located in chromosome 1; linkage group VII is located in chromosome 7; and linkage group V in chromosome 2 (Barratt *et al.*, 1954; Barry, 1967; Phillips, 1967; Barry and Perkins, 1969). Over 70 chromosomal aberrations, including duplications (Perkins *et al.*, 1969; Newmeyer, 1970; Perkins, 1972a), inversions (Newmeyer and Taylor, 1967; Turner *et al.*, 1969), and translocations have been reported (Perkins, 1966a), and their special characteristics and uses investigated.

Linkage Groups and Maps

Linkage was first demonstrated in *Neurospora* by Lindegren (1933). Subsequently, he published linkage maps of linkage group I (Lindegren,

1936), and II (Lindegren and Lindegren, 1939). For references to establishment of the remaining linkage groups, the first set of linkage maps, and a descriptive index of mutants see the review on map construction (Barratt *et al.*, 1954). Nearly all the early linkage data were obtained from ordered tetrad analysis (Barratt *et al.*, 1954). The merits of tetrad and random-strand analysis for linkage detection were compared and the conclusion reached that single-strand analysis is theoretically and practically more efficient than tetrad analysis (Perkins, 1953). Except for the analysis of special problems, random-strand analysis has been used almost exclusively since 1954 for linkage detection and mapping. Recent linkage maps and a characterization of mutants are published in *Neurospora Newsletter* (Barratt and Radford, 1969, 1970*a*; Radford, 1972) (see also Barratt and Radford, 1970*b*, 1972).

Detection and Isolation of Mutant Strains

The original Neurospora mutants were isolated by treating conidia with physical mutagens (ultraviolet light, x rays, or neutrons), crossing the treated conidia onto the opposite mating type, establishing cultures from random ascospore isolates, and testing the genotype (Beadle and Tatum, 1945). From over 80,000 individual ascospore isolates, a wide variety of auxotrophic and morphological strains were obtained. Subsequently, techniques have been developed which are equally applicable to physical and chemical mutagens and which eliminate the necessity for crossing. One employs direct plating of microconidia (Barratt and Tatum, 1958), another enriches for mutant types (or a specific mutant type) by means of differential germination between wild-type and mutant macroconidia (Woodward *et al.*, 1954), and a third utilizes the reduced death rate of germinating macroconidia of an inositol-requiring strain when a secondary mutation is present (Lester and Gross, 1959). Simplified replica-plating methods permitting the ready characterization of colonial strains have also been described (Schroeder, 1970; Perkins, 1971; Littlewood and Munkres, 1972). Other special detection and scoring techniques are given by Ryan (1950) and in the previously described bibliographies and indexes [see Methods in Bachmann and Strickland (1965) and Bachmann (1970)].

Research Topics

As a consequence of the unique advantages of Neurospora for research—simple nutritional requirements, wide diversity of mutant genotypes produced and available, rapid growth of mycelia, ease of cloning from conidia, rapidity and ease of handling of the sexual cycle, and the ability to work with the organism using ordinary microbiological techniques—it has been intensively studied by geneticists, biochemists, and microbiologists. Neurospora has contributed to the understanding of diverse biological phenomena.

The following list of selected significant research topics is designed to acquaint the reader with most of the major areas which have been investigated, and to provide sufficient names of workers to enable interested persons to consult the author or subject indices in the two bibliographies referred to previously in order to find specific references of interest. Specific references are given for review articles.

Investigations Primarily of Biochemical Phenomena

Amino acid biosynthesis: D. Bonner; J. R. S. Fincham; R. B. Flavell; R. H. Davis; H. K. Mitchell; E. L. Tatum (see also the review by Roberts *et al.*, 1955).

Mitochondrial characterization growth and division: R. J. Eakin; D. J. L. Luck; D. Woodward (see also the review by Linnane *et al.*, 1972).

Metabolic channeling: R. H. Davis and co-workers, J. A. De Moss and co-workers.

Enzyme localization: R. H. Davis; D. Woodward.

Exo- and intramural enzyme formation, localization, and activity: W. K. Bates; B. M. Eberhardt; A. G. DeBusk and co-workers; H. E. Lester and co-workers; R. L. Metzenberg; E. Roboz.

Permeability and uptake studies; genetic control of permeases, amino acid—carbohydrate and inorganic ion uptake and transport: A. G. DeBusk and co-workers; W. Klingmüller; C. W. Slayman and E. L. Tatum; W. M. Thwaites; W. R. Wiley and co-workers.

Enzyme regulation: (see Regulation, individual enzyme in Bachmann, 1970).

RNA studies: W. E. Barnett and co-workers; N. H. Horowitz; S. R. Gross; D. J. L. Luck and co-workers; R. Storck and co-workers; S. D. Wainwright.

DNA studies (chromosomal and mitochondrial): S. K. Dutta and co-workers; D. J. L. Luck; E. Reich; R. Storck.

Biosynthesis of B-vitamins: D. Bonner and co-workers; E. L. Tatum and co-workers (see individual vitamins in Bachmann and Strickland, 1965 and Bachmann, 1970).

Carotenoid biosynthesis: R. W. Harding; P. C. Huang and H. K. Mitchell; F. Haxo; A. Karunakaran; R. E. Subden; M. Zalokar.

Carbohydrate metabolism: glycolysis, hexose monophosphate shunt, tricarboxylic acid cycle: S. Brody; R. B. Flavell; B. S. Strauss; E. L. Tatum.

Enzyme regulation (induction, derepression): R. W. Barratt; M. E. Case and A. Giles; S. R. Gross; N. H. Horowitz; R. L. Metzenberg and co-workers; W. N. Strickland (see also the review by Horowitz and Metzenberg, 1966).

Investigations Primarily of Genetic Phenomena

Nonreciprocal recombination (gene conversion): M. E. Case and N. H. Giles; C. C. Lindegren; H. K. Mitchell; M. B. Mitchell; D. R. Stadler and A. M. Towe (also see the review by Emerson, 1966).

Polarized intragenic recombination: N. E. Murray; D. R. Stadler.

Heterokaryon stability, incompatability mechanisms, and genetic loci: L. Garnjobst; D. Newmeyer; D. D. Perkins; J. E. Wilson and co-workers (see also the reviews by Davis, 1966 and Horowitz and Metzenberg, 1966).

Origin and nature of pseudowild types: H. Bertrand; T. H. Pittenger; M. B. Mitchell.

Extranuclear inheritance (mitochondria): H. Bertrand and T. H. Pittenger; E. G. Diacumakos, L. Garnjobst and E. L. Tatum; M. B. Mitchell; E. Reich and D. J. L. Luck; D. O. Woodward and K. D. Munkres (see also Diacumakos *et al.,* 1965).

Nature and inheritance of chromosome aberrations: translocations, inversions, duplications: E. G. Barry; B. McClintock; D. Newmeyer; D. D. Perkins; R. L. Phillips; P. St. Lawrence.

Genes effecting recombination: D. G. Catcheside.

Mapping functions for tetrads and linkage mapping methods: D. D. Perkins.

Suppressors: mechanism of action and supersuppressors: M. E. Case; N. H. Giles; K. J. McDougall; H. K. Mitchell and M. B. Mitchell; T. W. Seale, S. K. Suskind.

Gene clusters and enzyme aggregates: D. G. Catcheside; J. A. De Moss and co-workers; N. H. Giles and co-workers; S. R. Gross; R. P. Wagner.

Allelic complementation and complementation mapping: H. Bernstein; M. E. Case; D. G. Catcheside and co-workers; A. Coddington; F. J. de Serres; J. R. S. Fincham; N. H. Giles; S. R. Gross; R. M. Lacy; N. E. Murray; J. A. Pateman; D. Woodward; V. W. Woodward.

Gene action: (see review by D. G. Catcheside, 1965).

Abnormal genetic segregation: M. B. Mitchell.

Chromatid and chromosome interference: M. R. Emerson; C. C. Lindegren; D. D. Perkins and co-workers; D. R. Stadler.

Crossing over at 4-strand stage: C. C. Lindegren; H. L. K. Whitehouse.

Linkage detection methods: R. W. Barratt; D. D. Perkins.

Gene action and morphology: S. Brody and co-workers; A. M. Srb and co-workers; E. L. Tatum and co-workers; G. Turian.

Mutagenesis: action of physical and chemical mutagens, environmental mutagens: C. Auerbach; G. W. Beadle; F. J. de Serres and co-workers; N. H. Horowitz; A. Radford (see also Mutagenesis in Bachmann and Strickland, 1965 and Bachmann, 1970).

Conditional mutants: S. Brody; D. D. Perkins.

Investigations Primarily of General Biological Phenomena

Bioassay: F. J. Ryan (see also review by Snell, 1948) (also see Methods, bioassay in Bachmann and Strickland, 1965).

Radiation sensitivity: A. L. A. Schroeder; R. W. Tuveson; I. Uno.

Protoplasts: B. J. Bachmann; M. S. Manocha.

Circadian rhythms: M. O. Berliner and R. W. Neurath; M. L. Sargent; A. S. Sussman and co-workers; S. Brody.

Conidial (macro- and micro-) formation, genetic control, germination, morphogenesis, and ultrastructure: R. W. Barratt; R. J. Lowry; A. N. Namboodiri; A. S. Sussman; G. Turian; J. C. Urey.

Hyphal and cell-wall structure and ultrastructure: E. M. Crook; L. Glaser; A. S. Sussman; E. L. Tatum and co-workers.

Ascospore dormancy, activation, and ultrastructure: M. Emerson; D. R. Goddard; R. J. Lowry; A. S. Sussman.

Toxic agents, their action, and resistant mutants: K. K. Jha; K. S. Hsu; S. E. A. McCallan and co-workers; N. C. Mishra and E. L. Tatum; D. U. Richmond; D. R. Stadler.

Virus-like particles in an extranuclear mutant: E. G. Diacumakos, L. Garnjobst and E. L. Tatum; H. Küntzel, Z. Barath, I. Ali, J. Kind, and H. H. Althaus (see also Diacumakos *et al.*, 1965 and Küntzel *et al.*, 1973).

Literature Cited

Anonymous, 1966   Neurospora media. *Neurospora Newsl.* **10**:34–35.

Bachmann, B. J., 1970   Neurospora Bibliography and Index, number two. *Neurospora Newsl.* **17**:3–80.

Bachmann, B. J., 1971   Neurospora Bibliography 1970. *Neurospora Newsl.* **18**:32–38.

Bachmann, B. J., 1972 Neurospora Bibliography ·1971. *Neurospora Newsl.* **19**:106–112.

Bachmann, B. J., 1973   Neurospora Bibliography 1972. *Neurospora Newsl.* **20**:57–69.

Bachmann, B. J. and W. N. Strickland, 1965   *Neurospora Bibliography and Index,* Yale University Press, New Haven, Conn.

Bakerspigel, A., 1959   The structure and manner of division of the nuclei in the vegetative mycelium of *Neurospora crassa. Am. J. Bot.* **46**:180.

Bakerspigel, A., 1969   Migrating and dividing nuclei in somatic cells of *Neurospora. Neurospora Newsl.* **14**:5.

Barratt, R. W., 1954   A word of caution: genetic and cytological effects of Abbott stocks in *Neurospora crassa. Microb. Genet. Bull.* **11**:5–8.

Barratt, R. W., 1962   Origin of important wild-type stocks of *N. crassa. Neurospora Newsl.* **2**:24–25.

Barratt, R. W., 1963   Effect of environmental conditions on the NADP-specific glutamic acid dehydrogenase in *Neurospora crassa. J. Gen. Microbiol.* **33**:33–42.

Barratt, R. W., 1964   Viability of microconidia. *Neurospora Newsl.* **6**:6–7.

Barratt, R. W. and L. Garnjobst, 1949   Genetics of a colonial microconidiating mutant strain of *Neurospora crassa. Genetics* **34**:351–369.

Barratt, R. W. and W. N. Ogata, 1972   Neurospora stock list, sixth revision. *Neurospora Newsl.* **19**:34–105.

Barratt, R. W. and A. Radford, 1969   Genetic markers, linkage groups and enzymes in *Neurospora crassa. Neurospora Newsl.* **14**:27–38.

Barratt, R. W. and A. Radford, 1970a   Genetic markers, linkage groups, and enzymes in *Neurospora crassa*—Supplement 1. *Neurospora Newsl.* **16**:19–22.

Barratt, R. W. and A Radford, 1970b   Genetic markers, linkage groups and enzymes in *Neurospora crassa.* In *Handbook of Biochemistry,* second edition, edited by H. A. Sober, pp. I-69–78. Chemical Rubber Co., Cleveland, Ohio.

Barratt, R. W. and A. Radford, 1972   Linkage groups: plants. Part II. *Neurospora crassa.* In *Biology Data Book,* Vol. 1, second edition, edited by P. L. Altman and D. S. Dittmer, pp. 61–81, Federation of American Society for Experimental Biology, Bethesda, Maryland.

Barratt, R. W. and E. L. Tatum, 1958   Carcinogenic mutations. *Ann. N.Y. Acad. Sci.* **71**:1072–1084.

Barratt, R. W., D. Newmeyer, D. D. Perkins and L. Garnjobst, 1954   Map construction in *Neurospora crassa. Adv. Genet.* **6**:1–93.

Barry, E. G., 1966   Cytological techniques for meiotic chromosomes in *Neurospora. Neurospora Newsl.* **10**:12–13.

Barry, E. G., 1967   Chromosome aberrations in *Neurospora,* and the correlations of chromosomes and linkage groups. *Genetics* **55**:21–32.

Barry, E. G. and D. D. Perkins, 1969   Position of linkage group V markers in chromosome 2 of *Neurospora crassa. J. Hered.* **60**:120–125.

Baylis, J. R. and A. G. DeBusk, 1967   Estimation of the frequency of multinucleate conidia in microconidiating strains. *Neurospora Newsl.* **11**:9.

Beadle, G. W., 1973   Comments on Bernard O. Dodge. *Neurospora Newsl.* **20**:13.

Beadle, G. W. and V. L. Coonradt, 1944   Heterocaryosis in *Neurospora crassa. Genetics* **29**:291–308.

Beadle, G. W. and E. L. Tatum, 1941   Genetic control of biochemical reactions in Neurospora. *Proc. Natl. Acad. Sci. USA* **27**:499–506.

Beadle, G. W. and E. L. Tatum, 1945 *Neurospora.* II. Methods of producing and detecting mutations concerned with nutritional requirements. *Am. J. Bot.* **32:**678–686.

Bianchi, D. E. and G. Turian, 1967 Nuclear division in *Neurospora crassa* during conidiation and germination. *Experientia (Basel)* **23:**192–197.

Bradford, S. W. and B. I. Gibgot, 1963 Rapid production of *Neurospora* hyphae. *Neurospora Newsl.* **4:**17–19.

Brockman, H. E. and F. J. de Serres, 1964 "Sorbose toxicity" in *Neurospora. Am. J. Bot.* **50:**709–714.

Case, M. E., H. E. Brockman and F. J. de Serres, 1965 Further information on the origin of the Yale and Oak Ridge wild-type strains of *Neurospora crassa. Neurospora Newsl.* **8:**25–26.

Catcheside, D. G., 1951 *The Genetics of Micro-organisms,* Pitman, London.

Catcheside, D. G., 1965 Gene action. In *The Fungi: An Advanced Treatise,* Vol. 1, edited by G. C. Ainsworth and A. S. Sussman, pp. 659–693, Academic Press, New York.

Catcheside, D. G., 1973 *Neurospora crassa* and genetics. *Neurospora Newsl.* **20:**6–8.

Davis, R. H., 1966 Mechanisms of inheritance. 2. Heterokaryosis. In *The Fungi: An Advanced Treatise,* Vol. 2, edited by G. C. Ainsworth and A. S. Sussman, pp. 567–588, Academic Press, New York.

Davis, R. H. and F. J. de Serres, 1970 Genetic and microbiological research techniques for *Neurospora crassa. Methods Enzymol.* **27:**79–143.

Diacumakos, E. G., L. Garnjobst and E. L. Tatum, 1965 A cytoplasmic character in *Neurospora crassa. J. Cell. Biol.* **26:**427–443.

Dodge, B. O., 1928 Spore formation in asci with fewer than eight spores. *Mycologia* **20:**18–21.

Dodge, B. O., 1930 Breeding albinistic strains of the Monilia bread mold. *Mycologia* **22:**9–38.

Dodge, B. O., 1932 The non-sexual and sexual functions of microconidia in *Neurospora. Bull. Torrey Bot. Club* **59:**347–360.

Dodge, B. O., 1935 The mechanics of sexual reproduction in *Neurospora. Mycologia* **27:**418–438.

Dodge, B. O., 1942 Heterocaryotic vigor in *Neurospora. Bull Torrey Bot. Club* **69:**75–91.

Dodge, B. O., 1957 Rib formation in ascospores of *Neurospora* and questions of terminology. *Bull Torrey Bot. Club* **84:**182–188.

Emerson, M. R., 1948 Chemical activation of ascospore germination in *Neurospora crassa. J. Bacteriol.* **55:**327–330.

Emerson, M. R., 1954 Some physiological characteristics of ascospore activation in *Neurospora crassa. Plant Physiol.* **29:**418–428.

Emerson, S., 1966 Mechanisms of inheritance. 1. Mendelian. In *The Fungi: An Advanced Treatise,* Vol. 2, edited by G. C. Ainsworth and A. S. Sussman, pp. 513–566, Academic Press, New York.

Esser, K. and R. Kuenen, 1967 Genetics of Fungi, Springer Verlag, New York.

Faull, A. F., 1930 On the resistance of *Neurospora crassa. Mycologia* **22:**288–303.

Fincham, J. R. S. and P. R. Day, 1963 *Fungal Genetics,* F. A. Davis Co., Philadelphia.

Frederick, L., F. A. Uecker and C. R. Benjamin, 1969 A new species of *Neurospora* from soil of West Pakistan. *Mycologia* **61:**1077–1084.

Frost, L. C., 1961 Heterogeneity in recombination frequencies in *Neurospora crassa. Genet. Res.* **2:**43–62.

Garnjobst, L. and J. F. Wilson, 1956   Heterocaryosis and protoplasmic incompatibility in *Neurospora crassa. Proc. Natl. Acad. Sci. USA* **42**:613–618.

Garrick, M. D., 1967   A simple expedient for obtaining large quantities of *Neurospora Neurospora Newsl.* **11**:5.

Gillie, O. J., 1967   Use of Coulter counter to measure the number and size distribution of macroconidia and microconidia of *Neurospora crassa. Neurospora Newsl.* **11**:16.

Gochenaur, S. E. and M. P. Backus, 1962   A new species of *Neurospora* from Wisconsin lowland soil. *Mycologia* **54**:555–562.

Gross, S. R., 1952   Heterokaryosis between opposite mating types in *Neurospora crassa. Am. J. Bot.* **39**:574–577.

Herskowitz, I. J., 1965   *Genetics,* second edition., Little, Brown and Co., Boston, Mass.

Holloway, B. W., 1954   Segregation of the mating type locus in *Neurospora crassa. Micro. Genet. Bull.* **10**:15–16.

Horowitz, N. H., 1973   *Neurospora* and the beginnings of molecular genetics. *Neurospora Newsl.* **20**:4–6.

Horowitz, N. H. and R. L. Metzenberg, 1966   Biochemical aspects of genetics. *Annu. Rev. Biochem.* **34**:527–564.

Howe, H. B., 1956   Crossing over and nuclear passage in *Neurospora crassa. Genetics* **41**:610–622.

Howe, H. B., 1963   Markers and centromere distances in *Neurospora tetrasperma. Genetics* **48**:121–131.

Huebschman, C., 1952   A method for varying the average number of nuclei in the conidia of *Neurospora crassa. Mycologia* **44**:599–604.

Keeping, E. S., 1969   Letter to the editor. *Neurospora Newsl.* **8**:27.

Koski, E. M. and S. C. Hedman, 1971   Optimal length of exposure to light for conidiation. *Neurospora Newsl.* **18**:11.

Küntzel, H., Z. Barath, I. Ali, J. Kind and H. Althaus, 1973   Virus-like particles in an extranuclear mutant of *Neurospora crassa. Proc. Natl. Acad. Sci. USA* **70**:1574–1578.

Lester, H. E. and S. R. Gross, 1959   Efficient method for selection of auxotrophic mutants of *Neurospora. Science (Wash., D.C.)* **129**:572.

Lindegren, C. C., 1932   The genetics of *Neurospora.* I. The inheritance of response to heat-treatment. *Bull. Torrey Bot. Club* **59**:85–102.

Lindegren, C. C., 1933   The genetics of *Neurospora.* III. Pure-bred stocks and crossing over in *Neurospora crassa. Bull. Torrey Bot. Club* **60**:133–154.

Lindegren, C. C., 1936   A six-point map of the sex chromosome of *Neurospora crassa. J. Genet.* **32**:243–256.

Lindegren, C. C., 1973   Reminiscences of B. O. Dodge and the beginnings of *Neurospora* genetics. *Neurospora Newsl.* **20**:13–14.

Lindegren, C. C. and G. Lindegren, 1939   Non-random crossing over in the second chromosome of *Neurospora crassa. Genetics* **24**:1–7.

Linnane, A. W., J. M. Haslam, H. B. Lukins and P. Nagley, 1972   The biogenesis of mitochondria in microorganisms. *Annu. Rev. Microbiol.* **26**:163–198.

Littlewood, R. K. and K. D. Munkres, 1972   Simple and reliable method for replica plating *Neurospora crassa. J. Bacteriol.* **110**:1017–1021.

Lowry, R. J., T. L. Durkee and A. S. Sussman, 1967   Ultrastructural studies of microconidium formation. *Neurospora Newsl.* **11**:9.

McClintock, B., 1945   *Neurospora.* I. Preliminary observations of the chromosomes of *Neurospora crassa. Am. J. Bot.* **32**:671–678.

McClintock, B., 1947 Cytogenic studies of maize and *Neurospora. Carnegie Inst. Wash. Year Book* **46**:146–152.

Mahoney, D. P., L. H. Huang and M. P. Backus, 1969 New homothallic *Neurospora* from tropical soils. *Mycologia* **61**:264–272.

Maling, B. O., 1960 Replica plating and rapid ascus collection of *Neurospora. J. Gen. Microbiol.* **23**:257–260.

Nelson, A. C. and M. P. Backus, 1968 Ascocarp development in two homothallic *Neurospora. Mycologia* **60**:16–28.

Nelson, A. C., R. O. Novak and M. P. Backus, 1964 A new species of *Neurospora* from soil. *Mycologia* **56**:384–392.

Newmeyer, D., 1954 A plating method for genetic analysis in *Neurospora. Genetics* **39**:604–618.

Newmeyer, D., 1970 A suppressor of the heterokaryon-incompatibility associated with mating type in *Neurospora crassa. Can. J. Genet. Cytol.* **12**:914–926.

Newmeyer, D. and C. W. Taylor, 1967 A pericentric inversion in *Neurospora,* with unstable duplication progeny. *Genetics* **56**:771–791.

Ogata, W. N., 1962 Preservation of *Neurospora* stock cultures with anhydrous silica gel. *Neurospora Newsl.* **1**:13.

Perkins, D. D., 1953 The detection of linkage in tetrad analysis. *Genetics* **38**:187–197.

Perkins, D. D., 1962 Preservation of *Neurospora* stock cultures with anhydrous silica gel. *Can. J. Microbiol.* **8**:592–594.

Perkins, D. D., 1966*a* Preliminary characterization of chromosomal rearrangements using shot asci. *Neurospora Newsl.* **9**:10–11.

Perkins, D. D., 1966*b* Details for collection of asci as unordered groups of eight projected ascospores. *Neurospora Newsl.* **9**:11.

Perkins, D. D., 1970 Genetics of *Neurospora* populations collected from nature. *Year Book Am. Phil. Soc.* **1970**:333–334.

Perkins, D. D., 1971 Conidiating colonial strains that are homozygous fertile and suitable for replication. *Neurospora Newsl.* **18**:12.

Perkins, D. D., 1972*a* Presumptive new alleles of *het-c* detected by the use of partial diploids. *Neurospora Newsl.* **19**:27–28.

Perkins, D. D., 1972*b* Special purpose *Neurospora* stocks. *Neurospora Newsl.* **19**:30–32.

Perkins, D. D., 1972*c* Reference strains of *Neurospora sitophila. Neurospora Newsl.* **19**:28–29.

Perkins, D. D. and B. C. Turner, 1973 Reference strains of *Neurospora intermedia. Neurospora Newsl.* **20**:41–42.

Perkins, D. D., D. Newmeyer, C. W. Taylor and D. E. Bennett, 1969 New markers and map sequences in *Neurospora crassa,* with a description of mapping by duplication coverage, and of multiple translocation stocks for linkage testing. *Genetica (The Hague)* **40**:247–278.

Phillips, R. L., 1967 The association of linkage group V with chromosome 2 in *Neurospora crassa. J. Hered.* **58**:263–265.

Pittenger, T. H., 1964 Conidial plating techniques and the determination of nuclear ratios in heterokaryotic cultures. *Neurospora Newsl.* **6**:23–26.

Pittenger, T. H., 1967 Distribution of nuclei in conidia. *Neurospora Newsl.* **11**:10–12.

Pittenger, T. H. and K. C. Atwood, 1956 Stability of nuclear proportions during growth of *Neurospora* heterokaryons. *Genetics* **41**:227–241.

Radford, A., 1972   Revised linkage maps of *Neurospora crassa. Neurospora Newsl.* **19**:25–26.

Reissig, J. L., 1956   Replica plating with *Neurospora crassa. Microb. Genet. Bull.* **14**:31–32.

Robbins, W. J., 1973   Bernard Ogilvie Dodge 1872–1960. *Neurospora Newsl.* **20**:10–12.

Roberts, R. B., P. H. Abelson, D. B. Cowie, E. T. Bolton and R. J. Britten, 1955   Studies of biosynthesis in *Escherichia coli*: Synthesis of amino acids by *Neurospora crassa* and *Torulopsis utilis. Carnegie Inst. Washington Publ.* **607**:207–217.

Ryan, F. J., 1948   The germination of conidia from biochemical mutants of *Neurospora. Am. J. Bot.* **35**:497–503.

Ryan, F. J., 1950   Selected methods of *Neurospora* genetics. *Methods Med. Res.* **3**:51–75.

Ryan, F. J., G. W. Beadle and E. L. Tatum, 1943   The tube method of measuring the growth rate of *Neurospora. Am. J. Bot.* **30**:784–799.

Sansome, E. R., 1946   Heterokaryosis, mating-type factors, and sexual reproduction in *Neurospora. Bull. Torrey Bot. Club* **73**:397–409.

Schroeder, A. L., 1970   Ultraviolet-sensitive mutants of *Neurospora*. I. Genetic basis and effect on recombination. *Mol. Gen. Genet.* **107**:291–304.

Shear, C. L., and B. O. Dodge, 1927   Life histories and heterothallism of the red bread-mold fungi of the *Monilia sitophila* group. *J. Agric. Res.* **34**:1019–1042.

Singleton, J. R., 1953   Chromosome morphology and the chromosome cycle in the ascus of *Neurospora crassa. Am. J. Bot.* **40**:124–144.

Snell, E. E., 1948   Use of microorganisms for assay of vitamins. *Physiol. Rev.* **28**:255–282.

Somers, C. E., R. P. Wagner and T. C. Hsu, 1960   Mitosis in vegetative nuclei of *Neurospora crassa. Genetics* **45**:801–810.

Srb, A. M., 1973   Beadle and *Neurospora*, some recollections. *Neurospora Newsl.* **20**:8–9.

Srb, A. M. and A. M. Infanger, 1964   Formation of heterokaryons by fusion of isolated hyphal tips on solid medium in petri plates. *Neurospora Newsl.* **6**:26.

Srb, A. M., R. D. Owen and R. S. Edgar, 1965   *General Genetics*, second edition, W. H. Freeman and Co., San Francisco, Calif.

Strickland, W. N., 1960   A rapid method for obtaining unordered *Neurospora* tetrads. *J. Gen. Microbiol.* **22**:583–588.

Stickland, W. N. and D. Thorpe, 1963   Sequential ascus collection in *Neurospora crassa. J. Gen. Microbiol.* **33**:409–412.

Tai, F. L., 1935   Two new species of *Neurospora. Mycologia* **27**:328–330.

Tai, F. L., 1936   Sex-reaction linkage in *Neurospora. Mycologia* **28**:24–30.

Tatum, E. L., R. W. Barratt and V. M. Cutter, 1949   Chemical induction of colonial paramorphs in *Neurospora* and *Syncephalastrum. Science (Wash., D.C.)* **109**:509–511.

Tatum, E. L., R. W. Barratt, N. Fries and D. Bonner, 1950   Biochemical mutant strains of *Neurospora* produced by physical and chemical treatment. *Am J. Bot.* **37**:38–46.

Thomas, D. Y. and L. C. Frost, 1966   Temperature and fertility cf wild type in *Neurospora. Neurospora Newsl.* **10**:3.

Turian, G., 1964   Synthetic conidiogenous media for *Neurospora crassa. Nature (Lond.)* **202**:1240.

Turian, G., 1966   Morphogenesis in Ascomycetes. In *The Fungi: An Advanced Treatise,* Vol. 2, edited by G. C. Ainsworth and A. S. Sussman, pp. 339–385, Academic Press, New York.

Turian, G., N. Oulevey and F. Tissot, 1967 Preliminary studies on pigmentation and ultrastructure of microconidia of *Neurospora crassa. Neurospora Newsl.* **11**:17.

Turner, B. C., C. W. Taylor, D. D. Perkins and D. Newmeyer, 1969 Non-duplication-generating inversions in *Neurospora. Can. J. Genet. Cytol.* **11**:622–638.

Tuveson, R. W., D. J. West and R. W. Barratt, 1967 Glutamic acid dehydrogenases in quiescent and germinating conidia of *Neurospora crassa. J. Gen. Microbiol.* **48**:235–248.

Van Winkle, W. B., J. J. Biesele and R. P. Wagner, 1971 The mitotic spindle apparatus of *Neurospora crassa. Can. J. Genet. Cytol.* **13**:873–887.

Wainwright, S. D., 1959 On the development of increased tryptophan synthetase activity by cell free extracts of *Neurospora crassa. Can. J. Biochem. Physiol.* **37**:1417–1430.

Ward, E. W. B. and K. W. Ciurysek, 1962 Somatic mitosis in *Neurospora crassa. Am. J. Bot.* **49**:393–399.

Weijer, J., A. Koopmans and D. L. Weijer, 1965 Karyokinesis of somatic nuclei of *Neurospora crassa*: III. The juvenile and maturation cycles (Feulgen and crystal violet staining). *Can. J. Genet. Cytol.* **7**:140–163.

Westergaard, M. and H. K. Mitchell, 1947 *Neurospora*. V. A synthetic medium favouring sexual reproduction. *Am. J. Bot.* **34**:573–577.

Whitehouse, H. L. K., 1949 Heterothallism and sex in the fungi. *Biol. Rev. (Camb.)* **24**:411–447.

Wilson, J. F. and L. Garnjobst, 1966 A new incompatibility locus in *Neurospora crassa. Genetics* **53**:621–631.

Woodward, V. W., J. R. de Zeeuw and A. M. Srb, 1954 The separation and isolation of particular biochemical mutants of *Neurospora* by differential germination of conidia, followed by filtration and plating. *Proc. Natl. Acad. Sci. USA* **40**:192–200.

28

Podospora anserina

Karl Esser

Introduction

The main purpose of this contribution is to acquaint more biologists with an organism that within the last twenty years has been used increasingly for a great variety of genetic and biochemical studies. The research with *Podospora anserina* was initiated with the fundamental studies of Rizet (1939, 1940, 1941*a–g*, 1942, 1943) and Rizet and Engelmann (1949). This fungus was subsequently used as an experimental subject by various laboratories, especially in France and Germany. The genus *Podospora*, which is also known under a variety of synonymous names (i.e., *Bombardia, Philocopra, Pleurage, Schizotheca, Schizothecium, Sordaria,* and *Sphaeria*), contains many species. It belongs to the family of Sordariaceae and the order of Sphaeriales. *P. anserina* can be distinguished from most of these since it possesses four linearly arranged ascospores. It is a close relative of *Neurospora crassa*, the favorite subject of fungal genetics.

I have previously introduced the fungus in *P. anserina* in a brief article and in a literature compilation, both of which have been published in the *Neurospora Newsletter* (Esser, 1969*b*, 1971*c*, respectively). This material is integrated and completed in this article. In the "Literature Cited" section I will present the complete literature list on *Podospora*, despite the fact that not all papers have been quoted in the text. This unusual procedure however, might help any student or scholar to learn more about *P. anserina*.

Karl Esser—Lehrstuhl für allgemeine Botanik, Ruhr-Universität Bochum, Bochum, Germany.

Life Cycle and Genetic Behavior

The natural habitat of *P. anserina* is the dung of herbivores. Numerous wild strains have been isolated from horse, cow, rabbit, and goat dung collected in France and Germany. Its life cycle resembles that of *N. crassa* with a few differences. The mycelia grow at a rate of 7 mm/day at 27°C and have a growth habit similar to the colonial strains of *Neurospora*. Macroconidia are not formed, and the microconida which are produced generally do not germinate. Since microconidia act as male gametes and are able to fertilize ascogonia via trichogynes, they are called spermatia. The fact that spermatia do not germinate under normal conditions was a serious handicap for years. However, a technique has been developed recently (Esser and Prillinger, 1972) that allows germination of spermatia at a rate of one in 100–500.

The asci develop only four spores. At the beginning of its development, each spore contains the two nonsister nuclei of the

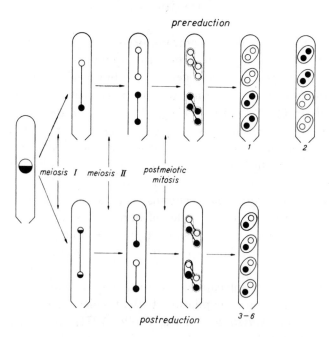

Figure 1. A diagram illustrating ascospore formation in P. anserina. The original nucleus is derived from a unifactoral cross involving a mutant and a wild-type allele controlling spore color. Postreductional disjunction occurs 98 percent of the time. Since the spindles of the postmeiotic mitoses overlap, the binucleate spores formed are heterokaryotic in cases where postreduction has occurred (from Esser and Kuenen, 1967).

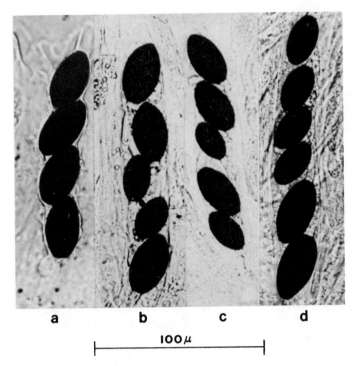

a   b   c   d

|——— 100 μ ———|

*Figure 2. Photomicrographs of asci of P. anserina. (a) A normal
ascus with four binucleate spores; (b,c) abnormal asci with a pair of
uninucleate spores in place of one of the binucleate spores; (d) an
abnormal ascus with two pairs of uninucleate spores (from Esser
and Kuenen, 1967).*

postmeiotic mitosis. Therefore, each spore contains the genetic information of half a tetrad (see Figure 1). Due to this phenomenon, spores
are heterokaryotic for all factors that exhibit second-division segregation,
and therefore it is not necessary to isolate the spores in order for tetrad
analysis. The size of these spores is 19 × 37 μ. Therefore they are
considerably larger than the ascospores of *N. crassa* (14 × 28 μ). The
mating behavior of *P. anserina* is controlled, as in the case of *Neurospora*,
by the bipolar mechanism of homogenic incompatibility, which generates
two hermaphroditic mating types (+ and −). The normal binucleate
spores are heterokaryotic for the mating type alleles, because the +/−
alleles undergo postreduction disjunction with a frequency of 0.98. The
mycelia originating from normal spores are mostly self-fertile because the
+ ascogonia may be fertilized by the − spermatia (and *vice versa*) which
are both present in the same mycelium. This phenomenon is called
pseudocompatibility, and it also occurs in *Neurospora tetrasperma*

and *Gelasinospora tetrasperma*. However, in 1–2 percent of the asci of most strains (especially in young perithecia) a pair of spores is formed, each originating from a single nucleus (see Figure 2). The mycelia growing from them are self-incompatible since they possess either the + or the − mating-type allele; they act like *N. crassa* with respect to their mating behavior. Tetrad analyses can easily be performed upon abnormal asci containing 5 or 6 spores. Further details of the life cycle and genetics of *Podospora* are given in Ames (1932, 1934), Dodge (1936), Dowding (1931), Rizet and Engelmann (1949), and Esser (1956c, 1959a).

Media and Methods for Podospora Culture and Manipulation

Media

Minimal Medium 1. (Modified Westergaard's Medium). Each liter of solution contains 20 ml of mineral concentrate, 20 g fructose, and 200 μg of thiamine. The solution is brought to pH 6.4–6.6 with about 2 ml of a 10-percent KOH solution. If a solid medium is desired, 20 g of Bacto-agar is added. The mineral concentrate contains per 1000 ml of solution: 25 g $MgSO_4 \cdot 7H_2O$, 5 g NaCl, 5 g $CaCl_2 \cdot 6H_2O$, 50 g KH_2PO_4, 50 g KNO_3, and 5 ml of trace-element concentrate. The trace-element concentrate contains per 100 ml of solution: 5 g ascorbic acid $\cdot 1H_2O$, 5 g $ZnSO_4 \cdot 7H_2O$, 1 g $Fe(NH_4)_2(SO_4)_2 \cdot 6H_2O$, 0.25 g $CuSO_4 \cdot 5H_2O$, 0.05 g $MnSO_4 \cdot 1H_2O$, 0.05 g H_3BO_4, and 0.05 g $Na_2MoO_4 \cdot 2H_2O$. The mineral and trace-element concentrate may be stored at room temperature if a few milliliters of chloroform are added to prevent contamination.

Minimal Medium 2. Each liter of solution contains 0.25 g KH_2PO_4, 0.3 g K_2HPO_4, 0.25 g $MgSO_4 \cdot 7H_2O$, 0.84 g $NaNO_3$, 2.5 μg biotin, 50 μg thiamine, 6 g fructose, and 0.1 ml trace-element concentrate.

Complete Medium (Modified Rizet's Medium). To a liter of cornmeal extract add 1.5 g malt extract and 20 g agar. To obtain the cornmeal extract, 250 g cornmeal is extracted in 10 liters of water and kept at 60°C overnight. The supernatant is poured off and the remaining cornmeal is discarded. Difco cornmeal agar may be used in place of the cornmeal extract.

Defined Complete Medium. This has the same composition as minimal medium 2 except that $NaNO_3$ is replaced by urea (0.5 g/liter) and fructose is replaced by yellow dextrin (10 g/liter). Growth is best on this medium at 70 percent relative humidity.

Moist Horse Dung. This is used for crossing some very weak mutant strains.

Spore-Germination Medium. This consists of 0.44 percent ammonium acetate in complete medium, or a one-percent solution of Bacto-peptone, or a one-percent fructose solution (for detecting nutritional mutants).

Medium for the Mass Production of Microconidia. This contains 2 g sorbose, 2 g yeast extract, and 1 g glucose per liter.

Medium for Germination of Microconidia. This contains 3 g sorbose, 2 g yeast extract, and 20 g Bacto-agar per liter.

Methods

For biochemical studies, large quantities of the fungus can be grown in aerated liquid medium. The mycelia are then pressed dry between sheets of filter paper and weighed. In a 10 liter carboy, after 4–5 days of incubation in complete medium, one obtains about 50 g of mycelia, and this can be used for inoculating fermentation vessels of still larger volume.

Spore germination can be observed a few hours after inoculation since *Podospora* spores do not have a dormancy phase. Spore germination also occurs on the other media mentioned above and even in water droplets. However, a spore-germination rate of nearly 100 percent can only be obtained on the spore germination medium. The fructification of *Podospora* can be markedly enhanced by light. Open shelves in a culture room or incubators with glass windows which allow entrance of rays from an ordinary fluorescent light are sufficient. The optimal temperature is 27°C, and the life cycle is completed in 8–15 days, depending on the strain and the culture medium.

Crosses are performed either by confrontation of the monokaryotic mycelia or by spermatization. The latter involves pouring a suspension of microconidia that have been filtered through fritted glass of 10–20-micron pore size over a monokaryotic mycelium that is used as the female parent. This technique allows the genetic analysis of dikaryotic, self-fertile strains and the comparison of reciprocal crosses.

Spores are isolated under a dissecting microscope. Ripe perithecia are cracked open with a watchmaker's forceps, and the contents are transferred to a Petri dish containing aqueous 5-percent agar gel. Under a dissecting microscope asci in various developmental stages are observed; with some practice, individual asci may be dissected using small steel pins or needles, and the spores then isolated.

Mass isolation of spores may be performed as follows: The lids are removed from cultures containing ripe perithecia and are replaced by Petri dishes half filled with aqueous 2–5-percent-agar gel; the cultures are then placed under a light source. This stimulates the perithecia to shoot their asci against the cover plates. In most cases the four spores of a tetrad stick together. The spore-shooting phenomenon may be enhanced by gently tapping the plates on the table once they have been illuminated for an hour.

It is advisable to use uninucleate microconidia when trying to recover mutations rather than polynucleate ascospores, since the latter may contain as many as 32 nuclei when mature. The screening techniques developed for *Neurospora* have been successfully adapted to *Podospora*.

Protoplasts can be easily obtained by the method of Bachmann and Bonner (1959), and filtration through fritted glass of 20–40-micron pore size removes all fragments of living mycelia.

Stock cultures may be grown on any of the solid media (except germination medium) and stored at 4°C.

Mutants and Linkage Groups

All the mutant strains of *Podospora* have been obtained from (or isogenized with) the same geographic race that was first isolated and described by Rizet (1952). Fungi from such strains bear the gene S or its allele s. According to Franke (1962), the haploid chromosome number for *P. anserina* is 7. There are seven well-defined linkage groups, plus a few genes which are probably independent (see Table 1). I have also noted in this table some of the more interesting genes that have not been mapped. With the exception of some mutants affecting spore size, color, or shape, which are named with numbers, the other morphological mutants have been "baptized" with Latin words that describe the main characteristic of the mutants. The biochemical mutants are named in the conventional way. The wild-type genes carry + as a superscript, i.e., z^+. Incompatibility genes are distributed among nine loci, each designated by a small letter. Bernet's symbols for the incompatibility genes are given in parenthesis. Many of the strains mentioned in the table are kept in my collection in Bochum. The spore mutants and some morphological mutants are in the collection at Professor D. Marcou's laboratory at the Laboratoire de Génétique, Faculté des Sciences, 91405 Orsay/Essonne, France. About 20 geographic races that are not mentioned in the table are kept by Professor J. Bernet at the Laboratoire de Génétique, Faculté des Sciences, 351 Cours de la Liberation, 33-Talence, France.

TABLE 1. The Mutants of Podospora anserina

Linkage group[a]	Locus	Name	Number of alleles including wild type	Frequency of postmeiotic reduction	Characteristics[b]
I	485		4	85	Spores green
	rib-1	riboflavin-1	2	70	Riboflavin requiring, spores yellow, pale and sterile mycelium
	k		2	12	Incompatibility factor
	6		2	6	Spores brown
	rib-2	riboflavin-2	2	0	Riboflavin requiring, spores dark green, mycelium sterile
	mei-2	meiosis-2	4	0	Meiosis stopped
	my-2	mycelial-2	2	0	Mycelium not dense, growth a little slow, poorly fertile
	122		28	2	Spores colorless
	a-su-1		2	25	Antisuppressor of $su1$
	f	flexuosa	5	81	Mycelium flat, no aerial hyphae
	pa	pallida	2	97	Mycelium pale and sterile
	+/−	mating-type	2	98	
II	52		2	10	Spores brown, mycelium pale
	14	albospora = as	200	0.5	Spores colorless to dark green mycelium colorless to wild type
	ci	circulosa	2	10	Clock mutant
	p	pumila	2	19	Spores smaller, some spermatia sterile
	z	zonata	2	83	Mycelium brown, clock mutant, no ascogonia
	385		4	83	Spores green
III	7		2	90	Spores brown and irregular, mycelium brown, irregular growth
	187		2	78	Spores green
	let-1	lethal-1	2	75	No spores when homokaryotic
	mod-2		2	11	Modifier of incompatibility in nonallelic systems

TABLE 1. *Continued*

Linkage group[a]	Locus	Name	Number of alleles including wild type	Frequency of postmeiotic reduction	Characteristics[b]
III	s		4	11	Heterogenic incompatibility, allelic mechanism from strain S
	su-m		2	4	Suppressor for m (linkage group IV)
	b	(Bernet: C)	16	4	Heterogenic incompatibility, nonallelic mechanism
	t	(Bernet: B)	2	13	Heterogenic incompatibility, allelic mechanism
	2		2	33	Spores yellow
	my-3	mycelial-3	2	50	Mycelium colorless, no ascogonia, many spermatia
	I	incoloris	36	80	Mycelium colorless, sterile, dominant
	96		4	86	Spores round, semidominant
IV	su-4		2	70	Non-cistron-specific suppressor
	mei-3	meiosis-3	3	65	Meiosis stopped
	a		16	18	Heterogenic incompatibility, nonallelic mechanism
	m	minor	2	17	Smaller perithecia
	49		2	15	Spores small, no germination
	g	glaber	2	0.2	Mycelium smooth
	su-1		50	30	Non-cistron-specific suppressor
	82		4	35	Spores green
	oct-1	octospora-1	2	68	Many uninucleated spores
	64		3	73	Spores having persistant appendix
	un	undulata	2	75	Clock mutant
	mei-4	meiosis-4	2	82	Meiosis stopped

TABLE 1. Continued

Linkage group[a]	Locus	Name	Number of alleles including wild type	Frequency of postmeiotic reduction	Characteristics[b]
V	*v*	(Bernet: R)	2	40	Heterogenic incompatibility, allelic and nonallelic mechanism
	mei-1	meiosis-1	5	<1	Meiosis stopped
	su-3		2	0	Non-cistron-specific suppressor
	154		3	0	Spores having persistent appendix
	sp	splendida	2	0.2	Mycelium glossy, sterile
	la	lanosa	3	77	Mycelium velvety
	lb	lano-alba	2	87	Mycelium velvety, white
VI	*5*		5	0.1	Spores brown, mycelium pale, few perithecia, few asci
	ta	tarda	2	1.7	Slow growth, clock mutant
	111		3	10	Spores having persistent appendix
	68		2	11	Spores smaller, no germination
	110		3	15	Spores yellow, perithecia nearly sterile
	my-1	mycelial-1	2	15	Slow growth, mycelium glossy and sterile
	631		2	15	Auxotrophic, perhaps *ad*⁻
	491		2	36	Spores green
	121		3	40	Spores yellow
	63		2	42	Spores having persistent appendix, no germination
	lg	lanuginosa	3	45	Mycelium velvety, sterile, slow growth
	l	lenta	3	47	Mycelium velvety, sterile, very slow growth
VII	*su5*		2	75	Non-cistron-specific suppressor
	su-2		15	70	Non-cistron-specific suppressor
	ao	albo-lana	2	62	Mycelium velvety, white, sterile

TABLE 1. *Continued*

Linkage group[a]	Locus	Name	Number of alleles including wild type	Frequency of postmeiotic reduction	Characteristics[b]
VII	*try*	tryptophan-less	2	36	Tryptophan-requiring, slow and abnormal mycelial growth on complete medium
	30		13	0–0.5	Spores dark green, of variable size
	u	(Bernet: Q)	2	5	Heterogenic incompatibility, allelic mechanism
	100		2	33	Spores yellow
	401		2	66	Spores green, white mycelium, nearly sterile
	lp	lano-pallida	3	86	Mycelium velvety, white

Genes independent from these seven groups

?	*4*		2	56	Spores green
	sp-1		2	78	Spores having persistent appendix
	311		3	40	Spores round, semidominant

Genes not localized[c]

	al	alba	2	84	Mycelium bright
	fu	fulva	2	83	Spores bright, mycelium brown
	fl	fluctuosa	2	45	Clock mutant
	c	(Bernet: P)	2	73	Heterogenic incompatibility, nonallelic mechanism
	d	(Bernet: D)		80	Heterogenic incompatibility, allelic and nonallelic mechanisms
	va	vacua	2	60	Both mating types form sterile perithecia

[a] Next to the number of each linkage group a straight line is drawn to symbolize the chromosome. The centromere is represented by an open circle. In the case of chromosomes I and V lines are drawn to genes used to mark the centromeres.

[b] Only the main properties distinguishing the mutant from the wild phenotype (black spores, dark green-black mycelium with aerial hyphae, and male and female sex organs) are given.

[c] There are more than 100 others (biochemical or morphological).

Main Lines of Genetic Investigation with Podospora

Basic Genetics

The results of studies on the life cycle and of the peculiar distribution of nuclei during the formation of ascospores have already been mentioned. Investigations of chromosome and chromatid interference have shown that positive chromosome interference is almost complete. This phenomenon is responsible for the occurrence of postreduction-disjunction frequencies up to 98 percent. Chromatid interference has not been detected. Corrected maps of the seven chromosomes have been published by Kuenen (1962a,b), who developed special mapping functions; cytological studies have confirmed these genetic data (Franke, 1959, 1962). Details of the genetic control and the light- and electron-microscopic cytology of meiosis with special reference to the division spindle have been published by Beckett and Wilson (1968), Zickler (1970), and Simonet and Zickler (1972).

Genetic Control of Sexual and Vegetative Reproduction

Genetic analyses of two geographic races led to the discovery of heterogenic incompatibility. This phenomenon refers to the inability of different incompatibility genes to coexist in a common protoplast. Heterogenic incompatibility results during the sexual phase in the inhibition of ascospore formation, and during the vegetative phase in the prevention of heterokaryon formation (Rizet and Esser, 1953; Esser, 1954a,b, 1956c, 1959a,b). Subsequent studies of other geographic races have confirmed and extended these concepts (Bernet, 1965, 1967;Belcour and Bernet, 1969; Bernet and Begueret, 1968; Bernet and Belcour, 1967).

Heterogenic incompatibility seems to result from the rupture of lysosomes after the fusion of incompatible cells. Catabolic enzymes such as aminopeptidases are liberated which destroy the hybrid protoplasts (Blaich and Esser, 1970, 1971). Similar experiments by Begueret (1972) have revealed alterations of proteolytic activities accompanying the incompatibility reaction.

Heterogenic incompatibility determines whether or not genetic recombination can take place, and thus it plays a decisive role in the speciation and further evolution of the genus. This phenomenon may also serve as a model for the mechanism determining the histoincompatibility reactions that occur following tissue transplantations in mammals (Esser, 1971a,b).

Intragenic Recombination

Spore-color mutants are very well suited to studies of genetic fine structure. Among others, locus 14 (white spores, linkage group II) has been analyzed for both intragenic recombination and intragenic complementation. These experiments have revealed that this locus is composed of several cistrons, the lengths of which could be defined by the frequencies of recombinational events (Marcou and Picard, 1967*a,b*; Marcou, 1969; Touré and Marcou, 1970; Picard, 1971, 1973; Touré, 1972*a,b*).

A variety of suppressor genes for locus 14 have been studied. One class suppresses nonsense mutations, and another class suppresses both nonsense and missense mutations. The working hypothesis advanced is that the members of the first class are mutants of genes controlling the tRNA's and those of the second class are mutants that affect genes controlling the production of ribosomal proteins. Finally, an antisuppressor (a gene suppressing the action of a suppressor gene) has been detected (Bennoun-Picard, 1973).

The Genetics and Physiology of Rhythmic Mycelial Growth

Noncircadian rhythmical growth occurs in *P. anserina* and is influenced by both genetic and environmental factors (Esser, 1956*b*; 1969*a*; Lysek and Esser 1970). The morphology of rhythmically growing strains has been studied by Nguyen Van (1968) and by Esser and Minuth (1972). Biochemical and physiological experiments have revealed that rhythmic growth is correlated with an increased carbohydrate degradation that shifts the energy sources into catabolic rather than anabolic pathways. Normal growth can be obtained by decreasing the respiratory cycle of rhythmically growing strains (Esser, 1972; Lysek and Esser, 1971; Lysek, 1971*a,b*, 1972).

Extrachromosomal Inheritance

A single case of heterogenic incompatibility concerning only the vegetative phase is known to be caused by an interaction of genetic and cytoplasmic factors (Rizet, 1952). This interaction between the allelic genes *S* and *s* and some unknown genetic elements in the cytoplasm is not yet understood in detail. However, two alternative explanations have been offered: an episomal model and a hypothesis involving cytoplasmic gene inactivation (Beisson-Schecroun, 1962).

Senescence occurring spontaneously after uninterrupted vegetative growth leads to a reduction of mycelial growth and finally to cellular death. According to Rizet and Marcou (1954) and Marcou (1961), this syndrome is caused by infectious particles of hitherto unknown nature.

Biochemical Genetics of Phenol Oxidases

Under suitable conditions, *P. anserina* forms a tyrosinase (EC 1.10.3.1) with a molecular weight of about 40,000 daltons. It also synthesizes at least three different forms of laccase (EC 1.10.3.2), each having a different molecular weight (laccase I, 390,000; laccase II, 70, 000; laccase III, 80,000). Biochemical and ultrastructural studies have shown that the high-molecular-weight laccase consists of 4 subunits. Each of these subunits is about 6 × 7 nm and is similar to the low-molecular-weight laccases. The ratio of the three laccases varies according to environmental and genetic conditions. The wild strain of *P. anserina* produces mostly laccase I during its stationary growth phase. The other enzymes are more common in the active growth phase, in the autolytic phase, and in some mutants. An increase in enzymatic activity occurs as the subunits aggregate into the tetramer. Laccase I also shows different substrate specificities from laccase II and III (Esser, 1963*a,b,c,* 1966; Esser and Minuth, 1970, 1971; Esser *et al.,* 1964; Herzfeld and Esser, 1969; Molitoris and Esser, 1970, 1971; Molitoris *et al.,* 1972; Schánĕl and Esser, 1971).

Meiotic Mutants

Cytogenetic studies have been made of several mutants affecting meiosis. Three loci are known that generate blocks during meiotic prophase. In the case of *mei-2,* it is possible to obtain a few asci from the mutant × mutant cross. In the progeny the frequencies of recombination in some regions of different chromosomes is modified. The mutant influences crossing over rather than conversion and is also characterized by an elevated rate of spontaneous mutation.

The mutant *mei-4* affects recombination in only chromosome IV. However, this effect is only seen under heterozygous conditions. When homozygous, *mei-4* is characterized by abnormal spindles during the meiotic divisions (Simonet, 1973).

Advantages of Podospora as an Object of Genetic Studies

Compared to the "standard object" of fungal genetics, *N. crassa, Podospora* has many advantages which compensate for its inconveniences, as may be seen from the following compilation:

1. *Podospora* has a short life cycle (8–12 days). There is no dormancy of spores, they germinate immediately after isolation.

2. Because of their larger size, ascospores are easier to isolate than *Neurospora* spores.

3. There is no problem in obtaining quantities of uninucleated microconidia for mutation studies, since there are no contaminating macroconidia.

4. Since microconidia germinate only under special conditions, there is no danger of contamination of cultures by conidia.

5. Heterokaryons occur naturally. Each postreductional disjunction event of a heterozygous gene pair leads to a heterokaryotic spore. The resulting mycelia start growing with equal proportions of the two types of nuclei. Thus nature performs the complementation test.

6. It is easy to obtain mutations affecting spore color, and these facilitate studies of intragenic recombination where one needs only to check the segregation of spore-color genes in the spores of individual asci.

7. The presence of uninucleate microconidia containing a very small amount of cytoplasm makes it possible to investigate nucleo–cytoplasmic relations in detecting differences in reciprocal crossings. Furthermore, anastomoses between mycelial fragments occur without nuclear migration.

8. *Podospora* is particularly suitable for the study of heterogenic incompatibility since both vegetative and sexual incompatibility occur without formation of lethal sexual progeny.

9. Because of the pseudocompatibility, every wild race is isogenic. The existence of uninucleate spores of either + or − mating types makes, in fact, this organism heterothallic.

Acknowledgments

I am grateful to Professor D. Marcou for providing me with data from unpublished studies of various French workers for incorporation into this contribution. This chapter is dedicated on the occasion of his 60th birthday to Professor Georges Rizet, who introduced me to *Podospora*.

Literature Cited[*]

Ames, L. M., 1932   A hermaphrodite self-sterile condition in *Pleurage anserina. Bull. Torrey Bot. Club* **59**:341–345.
Ames, L. M., 1934   Hermaphroditism involving self-sterility and cross-fertility in the ascomycete *Pleurage anserina. Mycologia* **26**:392–414.

[*] These works have not been cited in the text. However, they are presented here to provide a complete listing of the literature of *Podospora*.

*Auviti, M., 1968   Dominance apicale et sénescence chez le *Podospora anserina*. *C. R. Hebd. Seances Acad. Sci. Ser. D Sci. Nat.* **267**:1705–1708.

Bachmann, B. J. and D. M. Bonner, 1959   Protoplasts from *Neurospora crassa*. *J. Bacteriol.* **78**:550–556.

Beckett, A. and I. M. Wilson, 1968   Ascus cytology of *Podospora anserina*. *J. Gen. Microbiol.* **53**:81–87.

*Begueret, J., 1967   Sur la réparation en groupes de liaison de gènes concernant la morphologie des ascospores chez le *Podospora anserina*. *C. R. Hebd. Seances Acad. Sci. Ser. D Sci. Nat:* **264**:462–465.

*Begueret, J., 1969   Sur le mécanisme des réactions d'incompatibilité chez le *Podospora anserina*. *C. R. Hebd. Seances Acad. Sci. Ser. D Sci. Nat.* **269**:458–461.

Begueret, J., 1972   Protoplasmic incompatibility: possible involvement of proteolytic enzymes. *Nat. New Biol.* **235**:56–58.

Beisson-Schecroun, J., 1962   Incompatibilité cellulaire et interactions nucléocytoplasmiques dans les phénomènes de "barrage" chez le *Podospora anserina*. Thèse Fac. Sci. Univ. Paris, Sér A. Orsay No. 2 et *Ann. Genet.* **4**:3–50.

*Belcour, L., 1971   Mode d'action des gènes d'incompatibilité chez le *Podospora anserina*. I. Analyse de deux mutations à effect pléiotrope. *Mol. Gen. Genet.* **112**:263–274.

Belcour, L. and J. Bernet, 1969   Sur la mise en évidence d'un gène dont la mutation supprime spécifiquement certaines manifestations d'incompatibilité chez le *Podospora anserina*. *C. R. Hebd. Seances Acad. Sci. Ser. D Sci. Nat.* **269**:712–714.

Bennoun-Picard, M., 1973   Mise en évidence d'une unité de transcription polygénique et de suppresseurs informationnels chez l'Ascomycète *Podospora anserina*: analyse génétique. Thèse Univ. Paris-Sud, Centre d'Orsay, France.

*Bernet, J., 1963a   Action de la température sur les modifications de l'incompatibilité cytoplasmique et les modalités de la compatibilité sexuelle entre certaines souches de *Podospora anserina*. *Ann. Sci. Nat. Bot. Biol. Veg. Ser. 12* **4**:205–223.

*Bernet, J. 1963b   Sur les modalités d'expression de gènes pouvent conduire à une incompatibilité cytoplasmique chez le champignon *Podospora anserina*. *C. R. Hebd. Seances Acad. Sci. Ser. D Sci. Nat.* **256**:771–773.

Bernet, J. 1965   Mode d'action des gènes de "barrage" et relation entre l'incompatibilité cellulaire et l'incompatibilité sexuelle chez *Podospora anserina*. *Ann. Sci. Nat. Bot. Biol. Veg.* **6**:611–768.

*Bernet, J., 1967   Les systèmes d'incompatibilité chez le *Podospora anserina*. *C. R. Hebd. Seances Acad. Sci. Ser. D Sci. Nat.* **265**:1330–1333.

*Bernet, J., 1971   Sur un cas de suppression de l'incompatibilité cellulaire chez le champignon *Podospora anserina*. *C. R. Hebd. Seances Acad. Sci. Ser. D Sci. Nat.* **273**:1120.

Bernet, J. and J. Begueret, 1968   Sur les propriété et la structure des facteurs d'incompatibilité chez *Podospora anserina*. *C. R. Hebd. Seances Acad. Sci. Ser. D Sci. Nat.* **266**:716–719.

Bernet, J. and L. Belcour, 1967   Sur la possibilité de sélectionner des mutants de génes d'incompatibilité chez le *Podospora anserina* et sur les propriétés des premiere alleles obtenus. *C. R. Hebd. Seances Acad. Sci. Ser. D Sci. Nat.* **265**:1536–1539.

*Bernet, J. and J. Labarère, 1969   Effet de la dihydrostreptomicine sur l'incompatibilité chez le champignon *Podospora anserina*. *C. R. Hebd. Seances Acad. Sci. Ser. D Sci. Nat.* **269**:59–62.

*Bernet, J., K. Esser, D. Marcou and J. Schecroun, 1960   Sur la structure génétique de l'espéce *Podospora anserina* et sur l'intérêt de cette structure pour certaines recherches de génétique. *C. R. Hebd. Seances Acad. Sci. Ser. D Sci. Nat.* **250**:2053–2055.

Bernet, J., J. Begueret and J. Labarère, 1973   Incompatibility in the fungus *Podospora anserina*. Are the mutations abolishing, the incompatibility reaction ribosomal mutations? *Mol. Gen. Genet.* **124**:35–50.

*Blaich, R., 1970   Somatische Rekombination in einem Heterokaryon des Ascomyceten *Podospora anserina*. *Naturwissenschaften* **57**:47.

*Blaich, R. and K. Esser, 1969   Der cis-trans-Positionseffekt in Heterokaryen morphologischer Mutanten von *Podospora anserina*. *Arch. Mikrobiol.* **68**:201–209.

Blaich, R. and K. Esser, 1970   The incompatibility relationships between geographical races of *Podospora anserina*. IV. Biochemical aspects of the heterogenic incompatibility. *Mol. Gen. Genet.* **109**:186–192.

Blaich, R. and K. Esser, 1971   The incompatibility relationships between geographical races of *Podospora anserina*. V. Biochemical characterization of heterogenic incompatibility on cellular level. *Mol. Gen. Genet.* **111**:265–272.

*Cain, R. F., 1962   Studies on coprophilous ascomycetes. VIII. New species of *Podospora*. *Can. J. Bot.* **40**:447–490.

*Chevaugeon, J. and H. Nguyen Van, 1969   Internal determinism of hyphal growth rhythms. *Trans. Br. Mycol. Soc.* **53**:1–14.

Dodge, B. O., 1936   Spermatia and nuclear migration in *Pleurage anserina*. *Mycologia* **28**:284–291.

Dowding, E. S., 1931   The sexuality of the normal, giant and dwarf spores of *Pleurage anserina* (Ces.) Kuntze. *Ann. Bot. (Lond.)* **45**:1–14.

Esser, K., 1954a   Sur le déterminisme génétique d'un nouveau type d'incompatibilité chez *Podospora*. *C. R. Hebd. Seances Acad. Sci. Serb Sci. Nat.* **238**:1731–1733.

Esser, K., 1954b   Genetische Analyse eines neuen Incompatibilitätstypes bei dem Ascomyceten *Podospora anserina*. *C. R. Eighth Int. Bot. Congr. Paris Sect* **10**:72–77.

*Esser, K., 1955   Genetische Untersuchungen an *Podospora anserina*. *Ber. Dtsch. Bot. Ges.* **68**:143–144.

*Esser, K., 1956a   Wachstum, Fruchtkörper- und Pigmentbildung von *Podospora anserina* in synthetischen Nährmedien. *C. R. Trav. Lab. Carlsberg Ser. Physiol.* **26**:103–116.

Esser, K., 1956b   Gen-Mutanten von *Podospora anserine* (CES.) REHM mit männlichem Verhalten. *Naturwissenschaften* **43**:284.

Esser, K., 1956c   Die Incompatibilitätsbeziehungen zwischen geographischen Rassen von *Podospora anserina* (CES.) REHM. I. Genetische Analyse der Semi-Incompatibilität. *Z. Indukt. Abstammungs-Vererbungsl.* **87**:595–624.

*Esser, K., 1958   The significance of semi-incompatibility in the evolution of geographic races in *Podospora anserina*. *Proc. Int. Congr. Genet.* **2**:76–77.

Esser, K., 1959a   Die Incompatibilitätsbeziehungen zwischen geographischen Rassen von *Podospora anserina* (CES.) REHM. II. Die Wirkungsweise der Semi-Incompatibilitäts-Gene. *Z. Vererbungsl.* **90**:29–52.

Esser, K., 1959b   Die incompatibilitätsbeziehungen zwischen geographischen Rassen von *Podospora anserina* (CES.) REHM. III. Untersuchungen zur Genphysiologie der Barragebildung und der Semi-Incompatibilität. *Z. Vererbungsl.* **90**:445–456.

Esser, K., 1963a   Die Wirkungsweise von Antiseren auf die Phenoloxydasen von *Podospora anserina*. *Naturwissenschaften* **50**:576–577.

Esser, K., 1963b   Die Phenoloxydasen des Ascomyceten *Podospora anserina*. I. Identifizierung von Laccase und Tyrosinase beim Wildstamm. *Arch. Mikrobiol.* **46**:217–226.

Esser, K., 1963c   Quantitatively and qualitatively altered phenoloxydases in *Podospora anserina*, due to mutation at non-linked loci. *Proc. XI Int. Congr. Genet.* **1**:51–52.

Esser, K., 1966   Die Phenoloxydasen des Ascomyceten *Podospora anserina*. III. Quantitative und qualitative Enzymunterschiede nach Mutation an nicht gekoppelten Loci. *Z. Vererbungsl.* **97**:327–344.

*Esser, K., 1968a   Phenoloxidases and morphogenesis in *Podospora anserina*. *Genetics* **60**:281–288.

*Esser, K., 1968b   Genetic control of laccase formation in the ascomycete *Podospora anserina*. *Proc. XII. Int. Congr. Genet.* **1**:21.

Esser, K., 1969a   The influence of pH on rhythmic mycelial growth in *Podospora anserina*. *Mycologia* **61**:1008–1011.

*Esser, K., 1969b   An introduction to *Podospora anserina*. *Neurospora Newsl.* **15**:27–31.

Esser, K., 1971a   Breeding systems in fungi and their significance for genetic recombination. *Mol. Gen. Genet.* **110**:86–100.

Esser, K., 1971b   Radiation and radioisotopes for industrial microorganisms: Application and importance of fungal genetics for industrial research. *Int. At. Energy Agency Proc. Ser.* **1971**:83–91.

Esser, K., 1971c   Neurospora Newsl. **18**:19–21.

Esser, K., 1972   Genetic and biochemical analysis of rhythmically grown mycelia in the Ascomycete *Podospora anserina*. *J. Interdiscipl. Cycle Res.* **3**:123–128.

Esser, K., and R. Kuenen, 1967   *Genetics of Fungi*, translated by Erich Steiner, Springer-Verlag, New York.

Esser, K. and W. Minuth, 1970   The phenoloxidases of the ascomycete *Podospora anserina*. VI. Genetic regulation of the formation of laccase. *Genetics* **64**:441–458.

Esser, K. and W. Minuth, 1971   The phenoloxidases of the Ascomycete *Podospora anserina* IX. Microheterogeneity of laccase II. *Eur. J. Biochem.* **23**:484–488.

Esser, K. and W. Minuth, 1972   Abnormal cell wall structure in rhythmically growing mycelia of *Podospora anserina*. *Cytobiologie* **5**:319–322.

Esser, K. and H. Prillinger, 1972   A new technique to use spermatia for the production of mutants in *Podospora*. *Mutat. Res.* **16**:417–419.

Esser, K., S. Dick and W. Gielen, 1964   Die Phenoloxydasen des Ascomyceten *Podospora anserina*.II. Reinigung und Eigenschaften der Laccase. *Arch. Mikrobiol.* **48**:306–318.

Franke, G., 1959   Die Cytologie der Ascusentwicklung von *Podospora anserina*. *Z. Indukt. Abstammungs. Vererbungsl.* **88**:159–160.

Franke, G., 1962   Versuche zur Genomverdoppelung des Ascomyceten *Podospora anserina*. *Z. Vererbungsl.* **93**:109–117.

Herzfeld, F. and K. Esser, 1969   Die Phenoloxydasen des Ascomyceten *Podospora anserina*. IV. Reinigung und Eigenschaften der Tyrosinase. *Arch. Mikrobiol.* **65**:146–162.

*Hodgkiss, I. J. and R. Harvey, 1971   Factors affecting fruiting of *Pleurage anserina* in culture. *Trans. Br. Mycol. Soc.* **57**:533–536.

*Khan, R. S. and R. F. Cain, 1972   Five new species of *Podospora* from East Africa. *Can. J. Bot.* **50**:1649–1661.

*Kuenen, R., 1962a   Crossover- und Chromatiden-Interferenz bei *Podospora anserina* (CES.) REHM. *Z. Vererbungsl.* **93**:66–108.

*Kuenen, R., 1962b   Ein Modell zur Analyse der Crossover-Interferenz. *Z. Vererbungsl.* **93**:35–65.

Lysek, G., 1971*a* Rhythmic mycelial growth in *Podospora anserina*. III. Effect of metabolic inhibitors. *Arch. Mikrobiol.* **78**:330–340.

Lysek, G., 1971*b* Rhythmische Wuchsformen bei höheren Pilzen. *Präparator* **17**:105–111.

Lysek, G., 1972 Rhythmisches Mycelwachstum bei *Podospora anserina*. IV. Rhythmischer Verlauf der Trockengewichtsproduktion. *Arch. Mikrobiol.* **81**:221–233.

Lysek, G. and K. Esser, 1970 Rhythmic mycelial growth in *Podospora anserina*. I. The pleiotropic phenotype of a mutant is caused by a point mutation. *Arch. Mikrobiol.* **73**:224–230.

Lysek, G. and K. Esser, 1971 Rhythmic mycelial growth in *Podospora anserina*. II. Evidence for a correlation with carbohydrate metabolism. *Arch. Mikrobiol.* **75**:360–373.

*Mainwaring, H. R. and I. M. Wilson, 1968 The life cycle and cytology of an apomictic *Podospora*. *Trans. Br. Mycol. Soc.* **51**:663–677.

*Marcou, D., 1954*a* Sur la longévité des souches de *Podospora anserina* cultivées à diverses températures. *C. R. Hebd. Seances Acad. Sci. Ser. D Sci. Nat.* **239**:895–897.

*Marcou, D., 1954*b* Sur la rajeunissement par le froid des souches de *Podospora anserina*. *C. R. Hebd. Seances Acad. Sci. Ser. D Sci. Nat.* **239**:1153–1155.

*Marcou, D., 1957 Rajeunissement et arrêt de croissance chez *Podospora anserina*. *C. R. Hebd. Seances Acad. Sci. Ser. D Sci. Nat.* **244**:661–663.

*Marcou, D., 1958 Sur le déterminisme de la sénescence observée chez l'Ascomycète *Podospora anserina*. *Proc. X. Int. Congr. Genet.* **2**:179.

Marcou, D., 1961 Notion de longévité et nature cytoplasmique du déterminant de la sénescence chez quelques champignon. *Ann. Sci. Nat. Bot. Biol. Veg. Sér.* 12, **2**:653–673.

*Marcou, D., 1963 Sur l'influence du mode d'association des gènes sur les propriétés de certains hétérocaryotes du *Podospora anserina*. *C. R. Hebd. Seances Acad. Sci. Ser. D Sci. Nat.* **256**:768–770.

Marcou, D., 1969 Sur la nature des recombinaisons intracistroniques et sur leurs répercussions sur la ségrégation de marqueurs extérieurs chez le *Podospora anserina*. *C. R. Hebd. Seances Acad. Sci. Ser. D Sci. Nat.* **269**:2362–2365.

Marcou, D. and M. Picard, 1967*a* Sur l'interêt du *Podospora anserina* pour l'étude des problèmes de la recombinaison. *C. R. Hebd. Seances Acad. Sci. Ser. D Sci. Nat.* **264**:759–762.

Marcou, D. and M. Picard, 1967*b* Sur la structure d'un locus complexe chez le *Podospora anserina:* les relations entre la carte génétique et la carte de complémentation. *C. R. Hebd. Seances Acad. Sci. Ser. D Sci. Nat.* **265**:1962–1965.

*Mirza, J. H. and S. I. Ahmed, 1970 A new species of *Podospora* from Pakistan. *Mycologia* **62**:1003–1007.

*Mirza, J. H. and R. F. Cain, 1969 Revision of the genus *Podospora*. *Can. J. Bot.* **47**:1999–2048.

*Molitoris, H. P., 1972 Zur Struktur der Laccasen des Ascomyceten *Podospora anserina*. *Hoppe-Seyler's Z. Physiol. Chem.* **353**:736.

Molitoris, H. P. and K. Esser, 1970 Die Phenoloxydasen des Ascomyceten *Podospora anserina*. V. Eigenschaften der Laccase I nach weiterer Reinigung. *Arch. Mikrobiol.* **72**:267–296.

Molitoris, H. P. and K. Esser, 1971 The phenoloxidases of the Ascomycete *Podospora anserina*. VII. Quantitative changes in the spectrum of phenoloxidases during growth in submerged culture. *Arch. Mikrobiol.* **77**:99–110.

Molitoris, H. P., J. F. L. Van Breemen, E. F. J. Van Bruggen and K. Esser, 1972   The phenoloxidases of the Ascomycete *Podospora anserina*. X. Electron microscopic studies on the structure of laccases I, II, and III. *Biochim. Biophys. Acta* **281**:286–291.

*Monnot, F., 1953a   Sur la localisation du gène S et sur quelques particularités du crossing-over chez *Podospora anserina*. *C. R. Hebd. Seances Acad. Sci. Ser. D Sci. Nat.* **236**:2330–2332.

*Monnot, F., 1953b   Sur la possibilité de réaliser des croisements réciproque chez l'Ascomycète *Podospora anserina*. *C. R. Hebd. Seances Acad. Sci. Ser. D Sci. Nat.* **236**:2263–2264.

*Moreau, F. and M. Moreau, 1951   Observations cytobiologiques sur les Ascomycètes du genre *Pleurage*. *Rev. Mycol. (Paris)* **16**:198–207.

*Nguyen Van, H., 1962   Rôle des facteurs internes et externes dans la manifestation de rythmes de croissance chez l'ascomycète *Podospora anserina*. *C. R. Hebd. Seances Acad. Sci. Ser. D Sci. Nat.* **254**:2646–2648.

*Nguyen Van, H., 1967   Les relations intercellulaires au cours de la croissance en "vogue" chez le *Podospora anserina*. *C. R. Hebd. Seances Acad. Sci. Ser. D Sci. Nat.* **264**:280–283.

Nguyen Van, H., 1968   Etude de rythmes internes de croissance chez le *Podospora anserina*. *Ann. Sci. Nat. Bot. Biol. Veg.* **12**:257.

*Padieu, E. and J. Bernet, 1967   Môde d'action des gènes responsables de l'avortement de certains produits de la méiose chez l'Ascomycète *Podospora anserina*. *C. R. Hebd. Seances Acad. Sci. Ser. D Sci. Nat.* **264**:2300–2303.

*Perham, J. E., 1961   The development of a new biochemical genetic tool: *Podospora anserina*. Ph.D. Thesis, Florida State University, Tallahassee, Florida.

*Picard, M., 1970   Sur les données génétiques suggérant l'existence d'une unité de transcription polycistronique au sein du locus 14 chez le *Podospora anserina*. *C. R. Hebd. Seances Acad. Sci. Ser. D Sci. Nat.* **270**:498–501.

Picard, M., 1971   Genetic evidences for a polycistronic unit of transcription in the complex locus "14" in *Podospora anserina*. I. Genetic and complementation maps. *Mol. Gen. Genet.* **111**:35–50.

Picard, M., 1973   Genetic evidence for a polycistronic unit of transcription in the complex locus "14" in *Podospora anserina*. II. Genetic analysis of informational suppressors. *Genet. Res.* **21**:1–15.

Rizet, G., 1939   De l'hérédité du caractère absence de pigment dans le mycelium d'un ascomycète du genre *Podospora*. *C. R. Hebd. Seances Acad. Sci. Ser. D Sci. Nat.* **209**:771–774.

Rizet. G., 1940   Sur les diverses formes de la distribution des "sexes" chez le *Podospora minuta*. *C. R. Seances Soc. Biol. Fil.* **133**:31–33.

Rizet, G., 1941a   Sur l'analyse génétique des asques du *Podospora anserina*. *C. R. Hebd. Seances Acad. Sci. Ser. D Sci. Nat.* **212**:59–61.

Rizet, G., 1941b   La ségrégation des sexes et de quelque caractère somatiques chez le *Podospora anserina*. *C. R. Hebd. Seances Acad. Sci. Ser. D Sci. Nat.* **213**:42–45.

Rizet, G., 1941c   La valeur génétique de périthèce nés sur des souches polycaryotiques chez le *Podospora anserina*. *Bull. Soc. Bot. Fr.* **88**:517–520.

Rizet, G., 1941d   Quelques caractères remarquables des mutations observées chez le *Podospora anserina*. *C. R. Seances Soc. Biol. Fil.* **135**:1080–1082.

Rizet, G., 1941e   La formation d'asques hybrides dans les confrontations de souches "self-steriles" et de souches "self-fertiles" chez le *Podospora anserina*. *Rev. Mycol. (Paris)* **6**:128–133.

Rizet, G., 1941*f*   Sur l'hérédité d'un caractère de croissance et du pouvoir germinatif des spores dans une lignée de l'Ascomycète *Podospora anserina*. *Bull. Soc. Linn. Normandie (Ninth Sér.)* **2**:131–137.

Rizet, G., 1941*g*   Les résultats d'ordre génétique et le problème de la sexualité chez le *Podospora anserina*. *Rev. Mycol. (Paris)* **7**:97–100.

Rizet, G., 1942   Sur l'obtention d'irrégularités dans les asques et les spores de Pyrénomycètes du genre *Podospora* par l'action de quelques anastèsiques. *Bull. Soc. Linn. Normandie* **2**:123–130.

Rizet, G., 1943   L'existence de races génétiques chez le *Podospora anserina*. *Bull. Soc. Linn. Normandie* **3**:14–15.

Rizet, G., 1952   Les phénomènes de barrage chez *Podospora anserina*. I. Analyse génétique des barrages entre souches S and s. *Rev. Cytol. Biol. Veg.* **13**:51–92.

*Rizet, G., 1953*a*   Sur la multiplicité des mécanismes génétiques conduisant à des barrages chez *Podospora anserina*. *C. R. Hebd. Seances Acad. Sci. Ser. D Sci. Nat.* **237**:666–668.

*Rizet, G., 1953*b*   Sur l'impossibilité d'obtenir la multiplication végétative ininterrompue et illimitée de l'Ascomycète *Podospora anserina*. *C. R. Hebd. Seances Acad. Sci. Ser. D Sci. Nat.* **237**:838–840.

*Rizet, G., 1953*c*   Sur la longévité des souches de *Podospora anserina*. *C. R. Hebd. Seances Acad. Sci. Ser. D Sci. Nat.* **237**:1106–1109.

*Rizet, G., 1957   Les modifications qui conduisent à la sénescence chez *Podospora anserina* sont-elles de nature cytoplasmique. *C. R. Hebd. Seances Acad. Sci. Ser. D Sci. Nat.* **244**:663–665.

*Rizet, G., and G. Delannoy, 1950   Sur la production par des hétérozygotes monofactoriels de *Podospora anserina* de gamétophytes phénotypiquement differents des gamétophytes parentaux. *C. R. Hebd. Seances Acad. Sci. Ser. D Sci. Nat.* **231**:588–590.

Rizet, G. and C. Engelmann, 1949   Contribution à l'étude génétique d'un Ascomycète tétraspore: *Podospora anserina* (CES.) REHM. *Rev. Cytol. Biol. Veg.* **11**:201–304.

Rizet, G. and K. Esser, 1953   Sur des phénomènes d'incompatibilité entre souches d'origines différentes chez *Podospora anserina*. *C. R. Hebd. Seances Acad. Sci. Ser. D Sci. Nat.* **237**:760–761.

Rizet, G. and D. Marcou, 1954   Longévité et senescence chez l'Ascomycète *Podospora anserina*. *C. R. 8th Int. Congr. Bot. Paris Sect.* **10**:121–128.

*Rizet, G., and J. Schecroun, 1959   Sur les facteurs cytoplasmiques associés au couple de gènes S-s chez le *Podospora anserina*. *C. R. Hebd. Seances Acad. Sci. Ser. D Sci. Nat.* **249**:2392–2394.

*Rizet, G. and G. Sichler, 1950*a*   Sur le camouflage durable du génotype chez *Podospora anserina*. *C. R. Hebd. Seances Acad. Sci. Ser. D Sci. Nat.* **231**:630–631.

*Rizet, G. and G. Sichler, 1950*b*   Sur le démasquage du génotype des souches s^s de *Podospora anserina*. *C. R. Hebd. Seances Acad. Sci. Ser. D Sci. Nat.* **231**:719–721.

*Rizet, G., D. Marcou and J. Schecroun, 1958   Deux phénomènes d'hérédité cytoplasmique chez l'Ascomycète *Podospora anserina*. *Bull. Soc. Fr. Physiol. Veg.* **4**:136–149.

Schánĕl, L. and K. Esser, 1971   The phenoloxidases of the Ascomycete *Podospora anserina*. VIII. Substrate specificity of laccase with different molecular structures. *Arch. Mikrobiol.* **77**:111–117.

*Schecroun, J., 1958*a*   Sur la réversion progoqués des souches s^s en s chez *Podospora anserina*. *C. R. Hebd. Seances Acad. Sci. Ser. D Sci. Nat.* **246**:1268–1270.

*Schecroun, J., 1958*b*   Sur la réversion provoquée d'une modification cytoplasmique chez *Podospora anserina*. *Proc. X. Int. Congr. Genet.,* **2**:252.

*Schecroun, J., 1959   Sur la nature de la différence cytoplasmique entre souches s and s[s] de *Podospora anserina*. *C. R. Hebd. Seances Acad. Sci Ser. D Sci. Nat.* **248**:1394–1397.

Simonet, J. M. and D. Zickler, 1972   Mutations affecting meiosis in *Podospora anserina*. I. Cytological studies. *Chromosoma* **37**:327–351.

Simonet, J. R., 1973   Mutations affecting meiosis in *Podospora anserina*. II. Effect of *mei2* mutants on recombination. *Mol. Gen. Genet.* **123**:263–281.

*Smith, J. R., 1970   A genetic study of development of "senescence" in *Podospora anserina*. Ph.D. Thesis, Yale University, New Haven, Conn.

*Tavlitzki, J., 1954   Sur la croissance de *Podospora anserina* en milieu synthétique. *C. R. Hebd. Seances Acad. Sci. Ser. D Sci. Nat.* **238**:2341–2343.

Touré, B., 1972*a*   Double reversal of gene conversion polarity and multiple conversion events in the locus "14" in *Podospora anserina*. *Mol. Gen. Genet.* **117**:267–280.

Touré, B., 1972*b*   Consequences of double cross-over detections on the functional interpretation of segment "29" in *P. anserina*. *Genet. Res.* **19**:313–319.

Touré, B. and D. Marcou, 1970   Nature, dimension et limites des événements de recombinaison génétiques à l'intérieur d'une de transcription polycistronique chez le *Podospora anserina*. *C. R. Hebd. Seances Acad. Sci. Ser. D Sci. Nat.* **270**:619:621.

*Touré, B. and M. Picard, 1972   Consequences of double cross-over detections on the functional interpretation of segment "29" in *Podospora anserina*. *Genet. Res.* **19**:313–319.

*Weber, P. and K. Esser, 1964   Die Aminosäurezusammensetzung von Laccase aus *Podospora anserina*. *Naturwissenschaften* **51**:491–492.

Zickler, D., 1970   Division spindle and centrosomal plaques during mitosis and meiosis in some ascomycetes. *Chromosoma* **30**:287–304.

29

Sordaria

Lindsay S. Olive

Introduction

The earlier studies on the genetics of *Sordaria fimicola* (Olive, 1956) were begun with the object of establishing a procedure for ordered tetrad analysis in a homothallic species. This was accomplished by first obtaining spore-color mutants by ultraviolet irradiation. Any of these mutants when paired with wild type would yield some perithecia containing hybrid asci, visibly recognizable by their four dark wild-type and four lighter mutant spores (Figure 1). As in the 8-spored ascus of *Neurospora,* the products of meiosis are found in an orderly series representing an exact account of meiotic segregation. Since *S. fimicola* is a homothallic species in which any haploid nucleus in the mycelium is compatible with any other nucleus, selfed perithecia are produced abundantly along with the hybrid ones. Indeed, some clusters of asci from a single perithecium may contain both hybrid and homozygous asci.

In the life cycle of *S. fimicola* (Figure 1) there does not appear to be a male sexual organ. Heterokaryosis results when the hyphae of two different isolates come into contact and anastomose, thus permitting the migration of nuclei from one mycelium into the other. The process has been found to be a unidirectional one rather than one of mutual exchange (Carr and Olive, 1959) and is apparently controlled by a number of unidentified genetic factors. Within a few days after cultures are started, ascogonia (female sexual organs) appear on the hyphae and become the centers of perithecial development. Those that contain only like nuclei will give rise to

Lindsay S. Olive—Department of Botany, University of North Carolina, Chapel Hill, North Carolina.

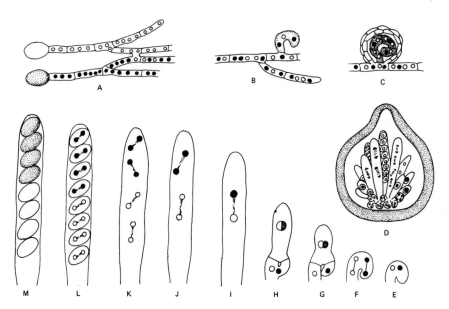

Figure 1. Life cycle of S. fimicola. (A) Heterokaryon formation, (B) development of heterokaryotic ascognium, (C,D) stages in perithecial development, (E–M), stages in crozier and ascus development, with segregation for an ascospore-color factor (Olive, 1963).

homozygous asci, while heterokaryotic ascogonia may give rise to hybrid asci, or sometimes a mixture of both hybrid and homozygous asci, by means of the usual ascogenous system of proliferating croziers and conjugate nuclear divisions (Figure 1). As in *Neurospora*, the haploid chromosome complement is comprised of seven morphologically distinct chromosomes (Carr and Olive, 1958). The entire life cycle is completed within 8 days.

Conidia are not produced by *S. fimicola*. While this removes a source of cross-contamination, it offers a disadvantage to mutagenic procedures. Most mutants used in the experiments described here have been obtained by ultraviolet treatment of young mycelia (Olive, 1956) and irradiation of ascospores with x rays (El-Ani *et al.*, 1961). The mutants studied have been mostly morphological ones readily identifiable in the ascus or by the growth habit of the mycelium, including mutants rendered self-sterile but remaining cross-fertile when paired with wild-type or other self-sterile mutants. Recently, nitrosoguanidine has been found to be an effective mutagenic agent for *S. fimicola* (K. K. Jha and Olive, unpublished).

Materials and Methods

The minimal medium for *Neurospora* also supports the growth and fruiting of *S. fimicola*. However, hybridization is best accomplished on an enriched cornmeal agar described by Kitani and Olive (1967).* The two cultures to be paired are inoculated about 4 cm apart on the plate; they will then grow out and meet at the center of the plate. Most of the hybrid perithecia will be found in the peripheral areas of this line of contact, along with the two selfed types.

In making direct analysis of asci heterozygous for a spore-color factor, several perithecia from the line of contact are transferred to a drop of water on a slide and gently squashed under a cover slip. The cover slip is then removed, and clusters of hybrid asci are pulled across into a nearby drop of water, thus ridding them of perithecial debris. When a cover slip is gently lowered onto these clusters, the latter flatten out into rosettes in which segregation patterns can be easily counted. In a relatively short time, one may analyze several hundred hybrid asci, recording first- and second-division segregations and then determining the gene-to-centromere distance. The method is especially advantageous for class demonstration in which it is desirable to illustrate Mendelian segregation without the necessity of ascus dissection.

In crosses requiring ascus dissection and further analysis of the progeny, it is necessary to add 0.7 percent sodium acetate to the medium; otherwise very few spores will germinate. Incubation at 30°C will hasten germination somewhat. After the clusters of asci have remained on the agar surface for about 15 minutes the ascus walls soften, and the spores can be readily pulled out in correct order with a micromanipulator needle. They should be sufficiently separated from each other to permit their being cut out after germination begins, each on a small block of agar, and transferred to nonacetate medium.

The mutants† described here are *gray-spored* or g_1 (Olive, 1956), *tan-spored* or t_1 (El-Ani and Olive, unpublished), and *restricted* or r_1 (El-Ani *et al.*, 1961).

Determining Gene-to-Centromere Distances by Direct Observation of Asci

Cross $t_1 \times t_1 +$ (Tan-Spored × Wild Type) The hybrid asci of this cross contain four dark-brown wild-type ascospores and four light-tan ones. Direct analysis of these asci reveals that approximately 56 percent of them show second-division (M II) segregation (2:2:2:2 and 2:4:2); the remaining 44 percent show first-division (M I) segregation (4:4). The

---

* Crossing Medium: 10 g sucrose, 7 g glucose, 1 g yeast extract, 0.1 g KH_2PO_4, 17 g Difco cornmeal agar, 1 liter water.
† Wild-type and mutant cultures are available from the American Type Culture Collection, 12301 Parklawn Drive, Rockville, Maryland.

same principles apply here as in *Neurospora* for determining the gene-to-centromere distance. Since ascus counts represent numbers of tetrads in which each M II ascus represents one pair of homologs with recombination between mutant locus and centromere and the other pair which is nonrecombinant, it is necessary, in order to obtain gene-to-centromere values comparable to those resulting from random analysis, to divide the M II percentage by 2. In this cross, the resultant gene-to-centromere distance is about 28 map units.

Cross $g_1 \times g_1 +$ (Gray-Spored $\times$ Wild Type)   Even a cursory examination of a cluster of hybrid asci from this cross will show that first-division-segregation asci are considerably outnumbered by second-division-segregation asci. When the asci have been counted and classified, it will be found that the M II asci constitute two-thirds or approximately 66.7 percent of the total, the M I group making up the remaining 33.3 percent. The high percentage of M II asci means, of course, that a great deal of crossing over between homologs occurs in the interval between the *g* locus and the centromere (Figures 2, 3), and, consequently, that the locus is a considerable distance from the centromere. In fact, this cross may be used to illustrate the point that, unless there is significant crossover interference, crosses involving loci beyond a certain distance (33.3 crossover units) on a chromosome will not furnish exact information on the position of the locus unless intervening markers are used. With the aid of such markers, it has been demonstrated that the *g* locus is approximately 56 map units from the centromere (Kitani and Olive, 1967). The apparent discrepancy is explained by the fact that the greater the

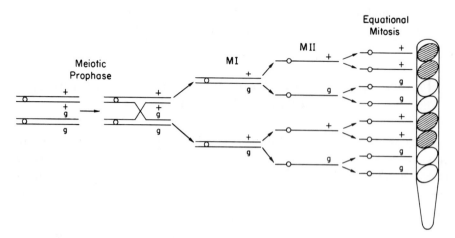

Figure 2. Diagram of meiotic prophase, with a crossover and second-division segregation of an ascospore-color factor (gray).

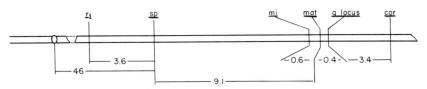

Figure 3. The gray spore linkage group in S. fimicola. Most of the markers are morphological ones affecting mycelial growth (Kitani and Olive, 1967).

distance between a locus and its centromere, the greater the probability of multiple crossovers in the same interval. Thus, asci resulting from 2-strand and 4-strand double crossovers will appear as M I asci (as though no crossing over had occurred), while those resulting from 3-strand doubles will produce the M II arrangement. Therefore, a point on the chromosome is reached beyond which there is no further increase in the M II class. This point, in the absence of interference, is obtained at about 33.3 map units, at or beyond which the proportion of M II asci will remain at 66.7 percent.

In all crosses, occasional spores appear misplaced in the ascus due to spindle overlap in the third (equational) division, but most of these can be properly classified. Thus, an ascus with the spore order $3+$, $1g_1$, $1+$, $3g_1$ has generally resulted from overlap of the two inner spindles in the third division and should be classified M I.

Linkage Relationships

In crosses heterozygous for two different mutant loci, three types of hybrid asci are produced and their proportions indicate whether the two loci are linked on the same chromosome or segregate independently. In the dihybrid cross, four genotypes are found among the progeny and three major types of asci may be distinguished. For example, in the cross of a *gray-spored* mutant X a *tan-spored* mutant, one finds the following types: *parental ditype* (PD) with 4 gray and 4 tan spores, *recombinant ditype* (RD) with 4 wild-type and 4 double-mutant (g_1t_1) spores, and *tetratype* (T) with 2 gray, 2 tan, 2 wild-type, and 2 double-mutant ascospores. The arrangement of the spore pairs in the ascus is not of importance in this experiment.

If the *gray* and *tan* loci are on nonhomologous chromosomes and are therefore inherited independently of each other, the number of PD and RD asci should be approximately equal. If the two are linked on the same chromosome, then it is clear that a special type of crossing over—4-strand double—would be required to produce an RD ascus, while a single crossover would produce the tetratype (T). Clearly, then, unless linkage is very distant, RD asci will be less in number than PD, the difference being more striking the closer together the loci are. At the same time, the percentage of T asci will indicate the distance (within the 66.7-percent limit previously described) between the two loci, just as the percentage of M II asci in the foregoing experiments indicates the gene-to-centromere distance.

Absence of Linkage: $g_1t_1+ \times g_1+ t_1$ (Gray-Spored $\times$ Tan-Spored). An ascus count here quickly shows that PD and RD asci occur in equal proportions, indicating absence of linkage. The T asci are very abundant but of no interest if linkage is not involved. *Gray* and *tan* are on different chromosomes.

Presence of Linkage: $g_1r_1 \times g_1+ r_1+$ (Gray-Spored Restricted Growth Mutant $\times$ Wild Type). The effect of r_1 is to produce a slower-than-normal mycelial growth, detectable even in early stages of development, but it has no effect on spore color. Therefore, it is necessary to dissect the asci, recording first the spore color and later the type of mycelial development of the progeny. Since both members of each spore pair are usually genetically alike, it is sufficient to remove the spores in successive pairs (or select one spore of each pair). After separating the progeny an adequate distance apart on a firm cornmeal–acetate medium and allowing about 6–8 hours for germination to begin, the 4 spore pairs may then be transferred to the nonacetate medium for observation of mycelial growth. When the data on spore color and growth rate are combined, it will be apparent that there are very few RD asci in proportion to PD asci. In addition, a tetratype percentage of around 26 percent shows that the two loci are about 13 map units apart.

Random spore analysis may also be used to demonstrate linkage here, if a few precautions are observed. Ascus clusters containing only hybrid asci and no homozygous ones may be transferred to cornmeal–acetate medium and left undisturbed for a couple of hours, allowing the acetate to weaken the walls. Now, if a small drop of water is added to the mass of asci and the latter gently macerated with the blunt end of a sterile stirring rod, many of the spores will be freed. A few milliliters of water may then be added to the surface of the medium, and the plate then may be swirled to distribute the spores over the agar surface. After the spores have had time to settle, the excess water may be poured off. The plate is then left undisturbed until germination occurs, after which germinating spores are cut out and transferred to fresh cornmeal agar. When data on spore color and growth rate are combined, it will be found that spores of the two parental genotypes (g_1r_1 and g_1+r_1+) are approximately equal in number and much more common than the recombinant types (g_1r_1+ and g_1+r_1), thus demonstrating linkage. The percentage of recombinant spores will be equivalent to recombination values obtained by random analysis in other fungi and higher organisms. Therefore, the percentage of recombinant spores (about 13 percent in this case) is a direct measure of the distance (about 13 map units) between g_1 and r_1 (see Figure 3).

Aberrant Segregation

In the cross *gray* × wild type, an occasional ascus (ca. 1 out of 430) will be found which does not fit into any Mendelian pattern (Olive, 1959; Kitani and Olive, 1967). Such asci may have 6 spores of one color and 2 of the other or 5 of one and 3 of the other, or a 4:4 ascus may have a heterogeneous spore pair in each end of the ascus. Since mutation of gray to wild type occurs extremely rarely if at all, the aberrant asci are best explained as products of *gene conversion*. If an extra marker such as r_1 is included in the cross, it can be shown by ascus dissection that both members of an odd spore pair in 5:3 and aberrant 4:4 asci are alike with regard to the r_1 allele. Since segregation of the spore-color marker occurred in these pairs during the third (equational) nuclear division, Whitehouse has referred to it as *postmeiotic segregation*.

While the exact nature of gene conversion is still unknown, most modern explanations of it are based on models proposed by Holliday (1964) and Whitehouse (1964). Both models involve the formation of short segments of hybrid DNA in the region of the pertinent locus in paired homologs during meiotic prophase. For example, one way in which hybrid DNA segments might develop would be for the two strands of DNA in each of the two involved chromatids to become disassociated in the region of the locus studied and to exchange positions in the two homologous chromatids before becoming re-annealed (Holliday, 1964) (Figure 4). The other pair of chromatids in the tetrad remains uninvolved. There is then a tendency for correction to take place in the aberrant base pairing that has occurred, but with correction occurring either to wild type or mutant. If correction in the same direction occurred in both hybrid strands, a 6:2 ascus would result in which the majority type could be either wild type or mutant. It is thought that 5:3 asci have resulted from correction in base pairing in one chromatid but not in the other, thus leading to the presence of one odd spore pair. The

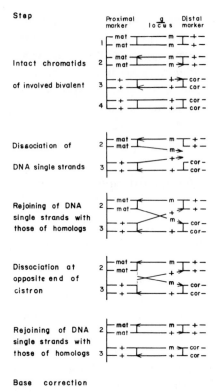

Figure 4. Hypothetical model (based on Holliday, 1964) showing one method by which gene conversion might occur (Kitani and Olive, 1969).

aberrant 4:4 asci with two odd spore pairs would result from failure of correction in both chromatids.

The *gray* × wild type cross in *S. fimicola* is particularly useful for demonstrating gene conversion, since correction at the *g* locus fails to occur frequently enough that 5:3 asci, along with a few aberrant 4:4's, may be found among the 6:2 types. The 5:3 and aberrant 4:4 types have only rarely been reported in other ascomycetes. (See Kitani and Olive, 1967, 1969.)

Other Species of Sordaria

A considerable amount of genetic information has been made available for other homothallic species of *Sordaria*, especially *Sordaria macrospora* in which ascospore-color mutants have also been obtained, by Heslot (1958) and Esser and Straub (1958). As in *S. fimicola* (Carr and Olive, 1959; El-Ani, 1964), these authors obtained self-sterile mutants that were cross-fertile, though only in *S. fimicola* has a stable condition resembling true heterothallism been obtained.

In the heterothallic species *Sordaria brevicollis*, seven linkage groups have been identified with the aid of numerous spore-color mutants (Chen, 1965; Chen and Olive, 1965). The species is characterized by biased segregation due to spindle overlap during the second meiotic division. Since this is easily corrected, it does not interfere with ascus analysis. This species has the advantage of possessing microconidia that are useful in mutagenic treatment, but one disadvantage is found in the tendency of the mycelium during vegetative growth to mutate spontaneously at a locus affecting its fertility in crosses.

Much of the genetic research on *Sordaria* has been reviewed in books by Fincham and Day (1965) and Esser and Kuehnen (1967).

Literature Cited

Carr, A. J. H. and L. S. Olive, 1958   Genetics of *Sordaria fimicola*. II. Cytology. *Am. J. Bot.* **45**:142–150.

Carr, A. J. H. and L. S. Olive, 1959   Genetics of *Sordaria fimicola*. III. Crosscompatibility among self-fertile and self-sterile cultures. *Am. J. Bot.* **46**:81–91.

Chen, K. C., 1965   The genetics of *Sordaria brevicollis*. I. Determination of seven linkage groups. *Genetics* **51**:509–517.

Chen, K. C. and L. S. Olive, 1965   The genetics of *Sordaria brevicollis*. II. Biased segregation due to spindle overlap. *Genetics* **51**:761–766.

El-Ani, A. S., 1964   Self-sterile auxotrophs and their relation to heterothallism in *Sordaria fimicola*. *Science (Wash., D.C.)* **145**:1067–1068.

El-Ani, A. S., L. S. Olive and Y. Kitani, 1961   Genetics of *Sordaria fimicola*. IV. Linkage group I. *Am. J. Bot.* **48**:716–723.

Esser, K. and R. Kuehnen, 1967   *Genetics of Fungi*, Springer-Verlag, New York.

Esser, K. and J. Straub, 1958   Genetische Untersuchungen an *Sordaria macrospora* Auersw., Kompensation und Induktion bei genbedingten Entwicklungsdefekten. *Z. Indukt. Abstammungs. Vererbungsl.* **89**:729–746.

Fincham, J. R. S. and P. R. Day, 1965   *Fungal Genetics*. F. A. Davis Co., Philadelphia, Pa.

Heslot, H., 1958   Contribution a l'étude cytogénétique et génétique des Sordariacées. *Rev. Cytol. Biol. Vég.* **19**(Suppl. 2):1–235.

Holliday, R., 1964   A mechanism for gene conversion in fungi. *Genet. Res.* **5**:282–304.

Kitani, Y. and L. S. Olive, 1967   Genetics of *Sordaria fimicola*. VI. Gene conversion at the *g* locus in mutant × wild-type crosses. *Genetics* **57**:767–782.

Kitani, Y. and L. S. Olive, 1969   Genetics of *Sordaria fimicola*. VII. Gene conversion at the *g* locus in interallelic crosses. *Genetics* **62**:23–66.

Olive, L. S., 1956   Genetics of *Sordaria fimicola*. I. Ascospore color mutants. *Am. J. Bot.* **43**:97–107.

Olive, L. S., 1959   Aberrant tetrads in *Sordaria fimicola*. *Proc. Natl. Acad. Sci. USA* **45**:727–732.

Olive, L. S., 1963   Genetics of homothallic fungi. *Mycologia* **55**:93–103.

Whitehouse, H. L. K., 1964   A theory of crossing-over and gene conversion involving hybrid DNA. *Proc. 11th Int. Congr. Genet.* **2**:87–88.

30

Ascobolus

Bernard Decaris, Jacqueline Girard, and

Gérard Leblon

Introduction

The genus *Ascobolus* belongs to the Ascomycetes subclass Pezizomycetidae. It is entirely coprophilous, composed mainly of octosporus bipolar heterothallic species such as *Ascobolus stercorarius* (Bull.) Schroet. (Bulliard, 1791; Schroeter, 1893; Dowding, 1931; Green, 1931; Bistis, and Olive, 1954), *Ascobolus immersus* (Rizet, 1939) and *Ascobolus magnificus* (Dodge, 1912, 1920; Gwynne-Vaughan and Williamson, 1932; Dodge and Seaver, 1946; Yu Sun, 1954). *A. stercorarius* produces tetrads which are often well ordered; whereas the spores of *A. immersus* are placed in a haphazard manner. The latter is utilized much more frequently for genetic studies due to the fact that its large ascospores (55–65 μ) can easily be manipulated under a binocular dissecting microscope.

A. stercorarius and *A. immersus* can be easily isolated in nature from various herbivorous excrements, particularly cow dung. If portions of dung are placed on sterilized horse dung at 20–25 °C, numerous fructifications of *A. immersus* appear in 8 or 10 days. Their ascospores are shot out in groups of 8 on an agar film. It is then sufficient to collect from each culture one ascus, which leads to 8 stock strains, the totality of which constitutes what we call a "breed group."

Bernard Decaris, Jacqueline Girard, and Gérard Leblon—Laboratoire de Génétique, Université de Paris-Sud, Centre d'Orsay, Orsay, France.

As for *A. immersus*, the germination medium now utilized is convenient for practically all of the stocks collected in this way (Yu Sun, 1954; Lissouba, 1961; Makarewicz, cited by Lissouba *et al.*, 1962). The technique consists of subjecting the spores to thermic shock at 39–40°C for at least 2 hours on a medium containing 1 liter of water, 12.5 g Difco Bacto-agar, and 12.5 g Difco Bacto-peptone with added sodium hydroxide (1.5 g). Germination is near 100 percent. For *A. stercorarius*, the medium described by Yu Sun (1954) does not produce such a high percentage of germination.

On the other hand, the nutritional requirements vary enormously. While certain strains can only fructify on horse dung (Lissouba, 1961), others fructify as well or better still on artificial gelose media (Bistis, 1956*a*; Yu Sun, 1964*b*). Therefore, the synthetic media described as convenient for fructification in one strain are not suitable for all.

The various isolated breeds show other forms of polymorphism. There are a number of totally or partially intersterile breed groups; when the crosses between these breeds are possible, one may frequently note anomalies in the morphology, the pigmentation, and the arrangement of the spores.

The spontaneous mutability varies greatly according to the breeds, so much so that before 1967 the study of mutagenesis was practically impossible on the breed utilized in our lab. It was very often observed that for spore color mutations, in the crosses between 2 wild stocks of same breed group, a frequency of tetrads bearing 4 white spores: 4 wild spores was greater than 1×10^{-3} (Lissouba, 1961). At the present time we possess stocks with a low spontaneous mutability (2.5×10^{-4}) (Leblon, 1970) and a stock in which the majority of the ascospore coloration mutants appear in asci bearing 1 mutant spore: 7 wild spores.

The cycle of *Ascobolus* is very similar to that of *Neurospora crassa*, which is well known to geneticists (see Chapter 27 by Barratt).

The cytological study of mitosis and meiosis has recently been the subject of work by Zickler (1967, 1969, 1970, 1971, 1973*a, b, c*). In *A. stercorarius* the process of fecundation has been made the subject of a precise and elegant study (Bistis, 1956*b*, 1957; Bistis and Raper, 1963). Each + or − stock can bear female and male elements (oïdia). The oïdia act chemotactically on a stock of compatible mating type to induce the formation of the female organs while, serving the same time as fecunding agents. One may also effect crosses by confrontation and spermatization, and this offers the possibility of obtaining reciprocal crosses.

In *A. immersus*, the + and − filaments are incompatible as in *N. crassa*, but this incompatibility does not appear at the level of the sexual organs. In *A. immersus*, the male elements are mycelian filaments, and there may be several nuclei of male or female origin implicated in a single fructification (Jupin, 1966). It is also possible to realize spermatization by utilizing as male elements, fragmented vegetative filaments which have been grown in a liquid medium (Lewis and Decaris, 1974). Fructification may be induced by exposure to an intense light stimulus for a minimum of 5 seconds, and supplementary light is essential for maturation. Under these conditions a partial synchronization of meiosis is obtained (Lewis, unpublished).

Stocks are maintained on gelosed media (Lissouba, 1961) at various temperatures. It would seem that for *A. immersus* the dessication of cultures at ambiant temperature favors the revival of growth after rehydration (Lewis, private communication), but the preservation of stocks for long periods remains difficult.

Induction of Mutants

Several thousands of spontaneous mutants can and have been obtained for *Ascobolus*. The easiest to find are the morphological mutants of ascospores with autonomous expression in spores (Rizet *et al.,* 1960a). Mutant ascospores may be larger or smaller than the wild type ones (Paquette, 1972) or they may have lessened or aberrant pigmentation (pink, rough, or granulous spores), or the number of ascospores in the ascus may be abnormal (plurinucleate spores) (Rossignol, 1967; Jupin, 1966). Nonpigmented spores are a common mutant.

Numerous growth mutants have also been found, and among these are rhythmic growth mutants (Chevaugeon, 1959a,b,c) and slow growth mutants (Yu Sun, 1964a). Some of these are thermosensitive (Yu Sun, 1969). A mutant bearing spores with no dormancy is also known (Decaris and Lefort, cited by Zickler, 1973c).

Genes that suppress spore-coloration mutations are known in breed group 28. The study of their action spectrum has shown that they are allele specific, but not gene specific (Leblon, 1972b). Nonautonomous ascospore mutants (Makarewicz, 1966), the "selfer" mutants (Paszewski and Surzycki, 1964), and several very unstable "genes" (Decaris *et al.,* 1972) are also being investigated.

Using stocks with low spontaneous mutability, great numbers of mutants have been induced by uv, x rays, n-methyl-N´-nitro-N-nitroso-guanidine, ethyl methanesulfonate, and 2-methoxy-6-chloro-9-[(3-ethyl-2-chloroethyl)-aminopropylamino] acridine dihydrochloride (ICR-170) (Leblon, 1970, 1972a,b; Stadler *et al.,* 1970; Stadler and Towe, 1971). Treatment may be applied to the mycelium, but one may also utilize the uninucleate ascospores prior to their germination (Zickler, 1973*a*). Following such treatments, a few biochemical (Yu Sun, 1964*a*) and drug-resistant mutants (Stadler *et al.,* 1970) have also been obtained.

Many genetic studies on this organism utilize the white spore mutants, and a few other mutants have served as markers. The mutant which produces plurinucleate spores has made complementation tests possible (Mousseau, 1963; Rossignol, 1967; Mousseau, 1967).

Gene Mapping

The preliminary studies of white spore mutants resulted in the formation of gene maps. Mutants which, when intercrossed, produce a large majority of parent ditype asci form "series groups." There are about 30 of these series or loci. White spore mutants lack pigment for a variety of reasons. Pigment synthesis may be blocked at different stages, e.g., the pigment precursor may not attach to the membrane, structural anomalies of the membrane may prevent attachment, etc. In many cases, at same locus, different phenotypic mutants are known (white or pink spores). Most granulous mutants belong to the same locus.

The locus of mating-type factors is practically always prereduced at the meiosis. The same is true for many ascospore coloration mutants, and this allows—at least in principle—the localization of other genes with regard to their centromere (Lefort, cited by Makarewicz, 1964; Yu Sun, 1964*a*). However, the number of cases of linkage known is very small. Judging from the great number of chiasmata (Zickler, 1967), this is perfectly understandable. On the other hand, the chromosome maps which have been established are incomplete (Yu Sun, 1964*a*, 1966; Paszewski *et al.,* 1966; Paquette, 1972). Moreover, it is often impossible to compare them, since they may have been established in different stocks which are intersterile.

The existence of ascospore-coloration mutants has made *A. immersus* a choice material for the study of the intragenic recombination (Rizet *et al.,* 1960*a*). Among the descendants of a cross between two mutant strains altered in the same gene, the recombination events leading to wild spores can readily be detected. From a cross of mutant (m) by wild type $(+)$, conversions affecting a whole chromatid (WCC) (6m:2+ and 2m:6+) and those affecting a half chromatid (HCC) (5m:3+ and 3m:5+) are easily selected. The ease of such determinations explains the very extensive results concerning intragenic meiotic recombination. However, only the most important results obtained from a study of ten loci will be discussed here. Many loci of breed group S have been investigated. These include:

(1) locus *46* (Lissouba and Rizet 1960; Rizet and Rossignol, 1963; Rossignol, 1964; Gajewski, *et al.*, 1968), (2) locus *19* (Mousseau, 1967), (3) locus *75* (Rossignol, 1967, 1969), (4) locus *726* (Makarewicz, 1964), (5) locus *Y* (Kruszewska and Gajewski, 1967), (6) locus *84W* (Paszewski, 1967; Paszewski and Prazmo, 1969; Paszewksi *et al.*, 1971), and (7) locus *164* (Baranowska, 1970). Locus *W17* of the Pasadena breed group has also been studied (Stadler *et al.*, 1970; Stadler and Towe, 1971). In breed group 28, the loci b_1 and b_2 have been investigated (Leblon, 1972*a,b*; Leblon and Rossignol, 1973), and it has been shown that locus b_2 of breed group 28 is homologous to locus *75* of breed group S. The results obtained by these investigators are summarized below.

Nonreciprocal Intragenic Recombination

The analysis of 6m:2+ asci taken from the descendants of a mutant × mutant cross has shown that such asci result either from crossing over or conversion. Conversion is often the more frequent phenomenon (Lissouba *et al.*, 1962; Makarewicz, 1964; Kruszewska and Gajewski, 1967). However, crossing over is more important in loci *19, 75,* and *726* than in locus *46*. In a given locus, the relative frequencies of crossing over and conversion also vary with the mutant alleles which are studied (Rizet and Rossignol, 1966).

Variable Conversion Frequencies

In a given stock, conversion frequencies may vary considerably according to the locus considered. For example, in stocks taken from breed group 28 the induced mutants at the b_1 locus give conversion frequencies from 0.005 to 0.064 in mutant × wild crosses. Those at the b_2 locus give frequencies between 0.105 and 0.284 (Leblon, 1972*a*). Even at a given locus, the frequencies are also very variable according to the mutant allele studied. A correlation between the position of the mutant site and the conversion frequency has been established for many genes.

Conversion Polarity

The clearest example in *Ascobolus* of polarized behavior is at locus *46* (Lissouba, 1961; Lissouba and Rizet, 1960). Within this locus, the mutants may be arranged in a linear manner on the basis of the frequency of 6m:2+ asci obtained from the descendants of the mutant × mutant crosses. The analysis of these asci shows that they are practically always the result of conversion. This conversion affects the site localized to the right of the map established by Lissouba (1961). Similarly, in the monofactorial crosses, the closer the site is localized to the right of the map, the higher the conversion frequency (Lissouba, 1961; Rossignol, 1964). The term "polaron" has been proposed by Lissouba and Rizet (1960) to refer to a chromosomal segment within which polarized genetic recombination takes place by gene conversion.

In the other loci, the polarity may not be noticeable (locus Y; Kruszewska and Gajewski, 1967). In general, the conversion maxima are situated either at one end or at both ends of the locus (locus *19*; Mousseau, 1967). But this phenomenon can be masked, as in locus *75*, when the mutants show important dissymmetries in the relative conversion frequencies toward mutant and wild (Rossignol, 1967).

Conversion Dissymmetry

With different mutants one may note important variations in relative frequencies of WCC toward mutant (asci 6m:2+) and toward wild type (asci 2m:6+). The relation between these two frequencies has been called the dissymmetry coefficient. For gene *75*, each mutant may be placed in one of five classes according to the value of the dissymmetry coefficient (Rossignol, 1967). In breed group 28 this dissymetry has been shown to be related to the nature of the mutation Frame-shift mutations convert either preferentially towards mutant or preferentially towards wild according to the mechanism of their genetic alteration (addition or deletion of base pairs) (Leblon, 1972*b*).

Half-Chromatid Conversions (HCC)

In the mutant × wild crosses one may detect HCC by the appearance of asci of unitary segregation (5m:3+ and 3m:5+). Whereas in stock S, HCC are the characteristic of only a few of the spontaneously obtained mutants (Gajewski *et al.*, 1968), all the spontaneous mutants studied by Emerson and Yu Sun (1967) produce them abundantly. In stock 28 it has been shown that only the substitution-type mutants produce HCC (Leblon, 1972b). Generally, the same dissymmetry exists among WCC and among HCC. However the *W62* (Emerson, 1966) and $EMSb_1$ *9E* mutants (Leblon, 1972*a*) do not show this characteristic.

Simultaneous Conversions

The existence of conversions simultaneously involving many sites and appearing with a frequency higher than that expected if they occurred independently, has been already demonstrated in loci *46, 19, 75,* and *Y* of *Ascobolus*. Their frequency depends on the conversion frequency at the level of each of the heterozygote sites (Kruszewska, and Gajewski, 1967; Rossignol, 1964) and shows an inverse relationship with the distance which separates the sites (Rossignol, 1967).

Map Expansion

The phenomenon of map expansion is observed when the recombination frequency in a given interval is greater than the sum of the frequencies of its component intervals. The most important expansion has been observed in locus *19* (Mousseau, 1967).

The Association between Conversion and Crossing Over

The analysis of three-factor crosses shows that intragenic crossing over can be associated with conversions (Rossignol, 1964; Mousseau, 1967; Rizet and Rossignol, 1966). This correlation only exists when the crossing over and conversion are restricted to the same two chromatids. The same association is noted with intergenic crossing over (Baranowska, 1970; Stadler *et al.*, 1970).

Correction Phenomena

Among the models that have been proposed for intragenic recombination are those that involve a correction mechanism (Whitehouse, 1963; Holliday, 1964). It has been possible to test the existence of such mechanisms on locus b_2 by utilizing the characteristics of intragenic suppression. In this way, a specific mutant correction has been demonstrated, as has also the fact that this correction may overlap other mutant sites (Leblon and Rossignol, 1973).

Hereditary Factors Affecting Recombination Frequencies

Several authors have described 3- to 5-fold variations in conversion frequencies at a mutant site (Emerson and Yu Sun, 1967; Stadler *et al.*,

1970), and a 100-fold increase was observed in breed group 28 (Rizet *et al.*, 1969). The existence of a pair of factors, *cv2A* and *cv2B* has been demonstrated, specifically affecting recombination in locus b_2. When *cv2* is present at the heterozygous state, the frequencies of conversion and crossing over are lowered considerably. This pair of factors is linked to the region affected by it, and the present results can be explained if it is assumed that they correspond to a region implicated in the specific control of the recombination of this locus. The same situation seems to be true in loci b_1, b_4, and b_6 (Girard, 1973; Girard and Rossignol, 1974).

Summary

The study on *Ascobolus* has been focused on intragenic recombination. The phenomena of conversion and crossing over have been thoroughly described owing to the utilization of numerous spontaneous ascospore-coloration mutants of *A. immersus*. The great diversity of the *A. immersus* strains collected from nature has also led to low mutability strains, and because of the specific action of ICR-170 and nitrosoguanidine on *A. immersus*, it is now possible to utilize mutants with known genetic alterations. The study of natural populations of *A.immersus* has also revealed an interesting source for the study of other genetic problems.

Acknowledgments

We wish to thank Dr. G. Rizet for his useful discussions and Dr. L. A. Lewis for his help in translating the manuscript.

Literature Cited

Baranowska, H., 1970 Intragenic recombination within *164* locus of *Ascobolus immersus* in the presence of outside markers. *Genet. Res.* **16**:185–206.

Bistis, G. N., 1956a Studies on the genetics of *Ascobolus stercorarius* (Bull.) Schröt. *Bull. Torrey Bot. Club.* **83**:35–61.

Bistis, G. N., 1956b Sexuality in *Ascobolus stercorarius*. I. Morphology of the ascogonium; plasmogamy; evidence for a sexual hormonal mechanism. *Am. J. Bot.* **43**:389–394.

Bistis, G. N., 1957 Sexuality in *Ascobolus stercorarius*. II. Preliminary experiments on various aspects of the sexual process. *Am. J. Bot.* **44**:436–443.

Bistis, G. N. and L. S. Olive, 1954 Ascomycete spore mutants and their use in genetic studies. *Science Wash., D.C.* **120**:3107.

Bistis, G. N. and J. R. Raper, 1963 Heterothallism and sexuality in *Ascobolus stercorarius*. *Am. J. Bot.* **50**:880–891.

Bulliard, P., 1791   Histoire des Champignons de la France. **I:**1–368.

Chevaugeon, J., 1959*a*   La zonation du thalle, phénomène périodique autonome chez l'*Ascobolus immersus*. *C.R. Hebd. Seances Acad. Sci. Ser. D Sci. Nat.* **248:**1381–1384.

Chevaugeon, J., 1959*b*   Sur le déterminisme interne du rythme de croissance chez un mutant vague de l'*Ascobolus immersus*. *C.R. Hebd. Seances Acad. Sci. Ser. D Sci. Nat.* **248:**1841–1844.

Chevaugeon, J., 1959*c*   Influence de quelques substances sur la manifestation du rythme de croissance chez *Ascobolus immersus*. *C.R. Hebd. Seances Acad. Sci. Ser. D Sci. Nat.* **249:**1703–1705.

Decaris, B., C. Lefort and G. Rizet, 1972   Etude génétique d'un mutant instable chez le champignon *Ascobolus immersus*. *Commun. Symp. Soc. Génét. Fr. (Strasbourg)* pp. 10.

Dodge, B. O., 1912   Artificial culture of *Ascobolus* and *Aleuria*. *Mycologia* **4:**218–222.

Dodge, B. O., 1920   The life history of *Ascobolus magnificus*. *Mycologia* **12:**115–134.

Dodge, B. O. and F. J. Seaver, 1946   Species of *Ascobolus* for genetic study. *Mycologia* **38:**639–651.

Dowding, E. S., 1931   The sexuality of *Ascobolus stercorarius* and the transportation of the oïdia by mites and flies. *Ann. Bot. (Lond.)* **45:**621–638.

Emerson, S., 1966   Quantitative implications of the DNA repair model of gene conversion. *Genetics* **53:**475–485.

Emerson, S. and C. C. C. Yu Sun, 1967   Gene conversion in the Pasadena strain of *Ascobolus immersus*. *Genetics* **55:**39–47.

Gajewski, W., A. Paszewski, A. Dawidowicz and B. Dudzinka, 1968   Postmeiotic segregation in locus "46" of *Ascobolus immersus*. *Genet. Res.* **11:**311–317.

Girard, J., 1973   Etude d'un cas de suppression localisée de la conversion chez *Ascobolus immersus*: mise en évidence d'un nouveau type de facteur affectant la recombinaison. Thèse de Doctorat 3ème cycle. Université de Paris-Sud, Orsay, France.

Girard, J. and J-L. Rossignol, 1974   The suppression of gene conversion and intragenic crossing-over in *Ascobolus immersus*: evidence for modifiers acting at the heterozygous state. *Genetics* **76:**221–243.

Green, E., 1931   Observations on certain Ascobolaceae. *Trans. Br. Mycol. Soc.* **15:**321–332.

Gwynne-Vaughan, H. C. I. and H. S. Williamson, 1932   The cytology and development of *Ascobolus magnificus*. *Ann. Bot. (Lond.)* **46:**653–670.

Holliday, R., 1964   A mechanism for gene conversion in fungi. *Genet. Res.* **5:**282–304.

Jupin, H., 1966   Etude préliminaire des phénomènes de sexualité chez l'Ascomycète *Ascobolus immersus*. Diplôme d'études supérieures de sciences naturelles, Université de Paris.

Kruszewska, A. and W. Gajewski, 1967   Recombination within the *Y* locus in *Ascobolus immersus*. *Genet. Res.* **9:**159–177.

Leblon, G., 1970   Sur l'existence d'une corrélation entre l'agent mutagène et le spectre de conversions des mutants induits chez *Ascobolus immersus*. *C.R. Hebd. Seances Acad. Sci. Ser. D Sci. Nat.* **271:**196–199.

Leblon, G., 1972*a*   Mechanism of gene conversion in *Ascobolus immersus*. I. Existence of a correlation between the origin of mutants induced by differents mutagens and their conversion spectrum. *Mol. Gen. Genet.* **115:**36–48.

Leblon, G., 1972*b*   Mechanism of gene conversion in *Ascobolus immersus*. II. The relationships between the genetic alterations in b_1 or b_2 mutants and their conversion spectrum. *Mol. Gen. Genet.* **116:**322–335.

Leblon, G. and J-L. Rossignol, 1973   Mechanism of gene conversion in *Ascobolus immersus*. III. The interaction of heteroalleles in the conversion process. *Mol. Gen. Genet.* **122**:165–182.

Lewis, L. A., and B. Decaris, 1974   The induction of apothecial formation in *Ascobolus immersus* by a spermatization technique. in manuscript

Lissouba, P., 1961   Mise en évidence d'une unité génétique polarisée et essai d'analyse d'un cas d'interférence négative. Thèse d´Etat, Université de Paris, et *Ann. Sci. Nat. Bot. Biol. Vég.* **44**:641–720.

Lissouba, P. and G. Rizet, 1960   Sur l'existence d'une unité génétique polarisée ne subissant que des échanges non réciproques. *C. R. Hebd. Seances Acad. Sci. Ser. D Sci. Nat.* **250**:3408–3410.

Lissouba, P., J. Mousseau, G. Rizet and J-L. Rossignol, 1962   Fine structures of genes in the ascomycete *Ascobolus immersus. Adv. Genet.* **11**:343–380.

Makarewicz, A., 1964   First results of genetic analysis in series 726 of *Ascobolus immersus. Acta Soc. Bot. Pol.* **33**:1–8.

Makarewicz, A., 1966   Colourless mutants in *Ascobolus immersus* with alternative phenotypes. *Acta Soc. Bot. Pol.* **35**:175–179.

Mousseau, J., 1963   On the possibility of carrying out the functionnal test for allelism in the Ascomycete *Ascobolus immersus. Proc. XI Int. Congr. Genet.* **1**:36.

Mousseau, J., 1967   Analyse de la structure fine d'un gène chez *Ascobolus immersus*. Contribution à l'étude de la recombinaison méiotique. Thèse de Doctorat d'Etat, Université de Paris.

Paquette, N., 1972   Cartographie préliminaire des mutants morphologiques de la souche S_2 d'*Ascobolus immersus* et mise en évidence d'un nouveau système génétique pour l'étude de la recombinaison intragénique avec marqueurs externes chez ce champignon. *Mém. Univ. Montréal.*

Paszewksi, A., 1967   A study on simultaneous conversion in linked genes in *Ascobolus immersus. Genet. Res.* **10**:121–126.

Paszewski, A. and W. Prazmo, 1969   The bearing of mutant and cross specificity on the pattern of intragenic recombination. *Genet. Res.* **14**:33–43.

Paszewski, A. and S. Surzycki, 1964   "Selfers" and high mutation rate during meiosis in *Ascobolus immersus. Nature (Lond.)* **204**:809.

Paszewski, A., S. Surzycki and M. Mankowska, 1966   Chromosome maps in *Ascobolus immersus* (Rizet's strain). *Acta Soc. Bot. Pol.* **35**:181–188.

Paszewski, A., W. Prazmo and E. Jaszczuk, 1971   Multiple recombinational events within the *84W* locus of *Ascobolus immersus. Genet. Res.* **18**:199–214.

Rizet, G., 1939   Sur les spores dimorphes et l'hérédité de leurs caractères chez un nouvel *Ascobolus* hétérothallique. *C.R. Hebd. Seances Acad. Sci. Ser. D Sci. Nat.* **208**:1669.

Rizet, G. and J-L. Rossignol, 1963   Sur la dissymétrie de certaines conversions et sur la dimension de l'erreur de copie chez *Ascobolus immersus. Rev. Biol. Lisb.* **3**:261–268.

Rizet, G. and J-L. Rossignol, 1966   Sur la dimension probable des échanges réciproques au sein d'un locus complexe d'*Ascobolus immersus. C.R. Hebd. Seances Acad. Sci. Ser. D Sci. Nat.* **262**:1250–1253.

Rizet, G., N. Engelman, C. Lefort, P. Lissouba and J. Mousseau, 1960a   Sur un Ascomycète intéressant pour l'étude de certains aspects du problème de la structure du gène. *C.R. Hebd. Seances Acad. Sci. Ser. D Sci. Nat.* **250**:2050–2052.

Rizet, G., P. Lissouba and J. Mousseau, 1960*b*. Les mutations d'ascospores chez l'ascomycète *Ascobolus immersus* et l'analyse fine de la structure des gènes. *Bull. Soc. Fr. Physiol. Veg.* **6**:175–193.

Rizet, G., J-L. Rossignol and C. Lefort, 1969 Sur la variété et la spécificité des spectres d'anomalies de ségrégation chez *Ascobolus immersus*. *C.R. Hebd. Seances Acad. Sci. Ser. D. Sci. Nat.* **269**:1427–1430.

Rossignol, J-L., 1964 Phénomènes de recombinaison intragénique et unité fonctionnelle d'un locus chez l'*Ascobolus immersus*. Thèse de Doctorat 3è cycle, Université de Paris.

Rossignol, J-L., 1967 Contribution à l'étude des phénomènes de recombinaison intragénique. Thèse de Doctorat d'Etat, Université de Paris.

Rossignol, J-L., 1969 Existence of homogeneous categories of mutants exhibiting various conversion patterns in gene 75 of *Ascobolus immersus*. *Genetics* **63**:795–805.

Schroeter, J., 1893 Die Pilze Schlesiens II. *Jahresber. Schles. Gesellsch. Vaterl. Cultur.* 1–256.

Stadler, D. R. and A. Towe, 1971 Evidence for meiotic recombination involving only one member of a tetrad. *Genetics* **68**:401–413.

Stadler, D. R., A. Towe and J-L. Rossignol, 1970 Intragenic recombination of ascospore colour and its relationship to the segregation of outside markers. *Genetics* **66**:429–447.

Whitehouse, H. L. K., 1963 A theory of crossing-over by means of hybrid deoxyribonucleic acid. *Nature (Lond.)* **199**:1034–1040.

Yu Sun, C. C. C., 1954 The culture and spore germination of *Ascobolus* with emphasis on *Ascobolus magnificus*. *Am. J. Bot.* **41**:21.

Yu Sun, C. C. C., 1964*a* Biochemical and morphological mutants of *Ascobolus immersus*. *Genetics* **50**:987–998.

Yu Sun, C. C. C., 1964*b* Nutritional studies of *Ascobolus immersus*. *Am. J. Bot.* **51**:231–237.

Yu Sun, C. C. C., 1966 Linkage groups in *Ascobolus immersus*. *Genetics* **37**:569–580.

Yu Sun, C. C. C., 1969 Temperature-sensitive mutants of *Ascobolus immersus*. *Am. J. Bot.* **56**:341–343.

Zickler, D., 1967 Analyse de la méiose du champignon Discomycète *Ascobolus immersus*. *C. R. Hebd. Seances Acad. Sci. Ser. D Sci. Nat.* **265**:198–201.

Zickler, D., 1969 Sur l'appareil cinétique de quelques Ascomycètes. *C. R. Hebd. Seances Acad. Sci. Ser. D Sci. Nat.* **268**:3040–3042.

Zickler, D., 1970 Division spindle and centrosomal plaques during mitosis and meiosis in some Ascomycetes. *Chromosoma* **30**:287–304.

Zickler, D., 1971 Déroulement des mitoses dans les filaments en croissance de quelques Ascomycètes. *C.R. Hebd. Seances Acad. Sci. Ser. D Sci. Nat.* **273**:1687–1689.

Zickler, D., 1973*a* Fine structure of chromosome pairing in ten Ascomycetes: Meiotic and Premeiotic (mitotic) synaptonemal complexes. *Chromosoma* **40**:401–416.

Zickler, D., 1973*b* Evidence for the presence of DNA in the centrosomal plaques of *Ascobolus*. *Histochemie* **34**:227–238.

Zickler D., 1973*c* La méiose et les mitoses au cours du cycle de quelques ascomycètes. Thèse de Doctorat d'Etat, Université de Paris-Sud, Centre d'Orsay, Orsay, France.

31

Ustilago maydis

Robin Holliday

Introduction

Fungi which have yeastlike growth have considerable advantages for genetic studies over filamentous fungi. The replica plating of colonies with velvet greatly simplifies the isolation of many types of mutants and the classification of the phenotypes of the progeny of crosses. In addition, the growth of single, uninucleate cells in liquid media facilitates many physiological and biochemical experimental procedures. A particular reason for choosing a yeastlike smut fungus of the genus *Ustilago* was the number of reports in the literature that the haploid chromosome number was two; this would have been a considerable advantage for genetic studies. However, subsequent genetic analysis in *Ustilago maydis* and *Ustilago violacea* has shown that these cytological reports are untrue. In the latter species at least ten linkage groups have been identified (Day and Jones, 1969). Although the vegetative cells grow vigorously on synthetic media, the sexual stage of smut fungi is parasitic. *U. maydis* (De Candolle) Corda was chosen for detailed studies because a few days after infection it produces diploid teliospores (brandspores) in vegetative parts of the host *Zea mays*, and the life cycle is completed in less than two weeks. Many other species produce teliospores only in the inflorescence of the host, many weeks or months after inoculation.

Early studies on the genetics of *U. maydis* were hindered by the absence of suitable markers, but this problem was solved by the isolation of biochemical mutants (Perkins, 1949). Subsequently, techniques for

Robin Holliday—National Institute for Medical Research, Mill Hill, London, England.

analyzing the progeny of crosses were developed (Holliday, 1961a); the discovery of mitotic crossing over in vegetative diploids (Holliday, 1961b) followed soon after. A major advance was made when it was discovered that dikaryons and diploids could be readily obtained on agar medium, as this greatly facilitated complementation tests (Puhalla, 1968; Day and Anagnostakis, 1971a). *U. maydis* is particularly suitable for genetic studies of pathogenicity as the genetics of the host is so well known.

Linkages between biochemical or other markers have not been easy to find, and the mapping of the genome has been largely neglected. Instead, a handful of markers have been routinely used in experimental studies of genetic recombination and repair. A number of mutants have been isolated which are deficient in these processes or which are defective in DNA synthesis, and progress has been made in the study of enzymes such as DNA polymerase or deoxyribonucleases, which are known or suspected to be absent or altered in certain of these mutants. It is hoped that a thorough investigation of the genetic properties of the various mutants, together with a detailed biochemical study of the enzymes they lack, will throw light on the mechanisms of recombination, repair, mutation, and replication in this simple eukaryotic organism.

General Methods

Cell Growth

Haploid cells (sporidia) are uninucleate, 12–18 μ long, and 4–5 μ wide. They divide by producing buds at either end of the cell. The nucleus moves into the bud and divides there; one nucleus passes back into the parent cell. Nuclear division is immediately followed by the S phase during which the cells separate. Therefore, for most of the division cycle the cells are in the G2 phase. Stationary-phase cells are also in G2. In rich liquid media at 32°C cells divide every 80 minutes; in synthetic media they divide every 120 minutes. Diploid cells are larger but have less than twice the volume of haploid ones. They grow and divide in the same way. Cells are conveniently counted with a Coulter particle counter, and standard microbiological techniques are used for handling cells and colonies on plates. Colonies can be counted after 2–3 days and replica plated with velvet after 3–4 days, when the colonies are 2–3 mm in diameter. The organism is particularly suitable for replica plating, as only a few hundred cells are transferred.

Media

U. maydis will grow on a wide variety of rich or synthetic media. Yeast extract peptone (2 percent glucose, 2 percent bactopeptone, 1 percent yeast extract) is a convenient complete medium for many experiments, but the cells tend to autolyse if held in stationary phase in liquid or on plates. Czapek Dox appears to be a suitable simple minimal medium, but it has not been thoroughly tested under a variety of experimental situations. The following media have been routinely used in this laboratory (quantities are per liter of medium).

1. *Minimal medium.* 10 g glucose, 3 g KNO_3, and 62.4 ml salt solution.
2. *Complete medium.* 10 g glucose, 2.5 g hydrolyzed casein, 5 ml hydrolyzed nucleic acid (see Pontecorvo, 1953), 10 ml vitamin solution, 1 g yeast extract, 1.5 g NH_4NO_3, and 62.5 ml salt solution.
3. *Salt solution.* 16 g KH_2PO_4, 4 g Na_2SO_4, 8 g KCl, 2 g $MgSO_4$, 1 g $CaCl_2$, and 8 ml trace elements.
4. *Trace element solution.* 60 mg H_3BO_3, 140 mg $MnCl_2 \cdot 4H_2O$, 400 mg $ZnCl_2$, 40 mg $Na_2MoO_4 \cdot 2H_2O$, 100 mg $FeCl_3 \cdot 6H_2O$, and 400 mg $CuSO_4 \cdot 5H_2O$.
5. *Vitamin solution.* 100 mg thiamin, 50 mg riboflavin; 50 mg pyridoxin, 200 mg calcium pantothenate, 50 mg p-aminobenzoic acid, 200 mg nicotinic acid, 200 mg choline chloride, and 1000 mg inositol.
6. *Mating medium.* Day and Anagnostakis (1971a) use double-strength complete medium with the addition of 1 percent activated charcoal. In this laboratory liquid complete medium is supplemented with 1.7 percent Difco cornmeal agar, 1 percent agar, and 1 percent activated charcoal.

The *p*H of complete and minimal media is adjusted to 7.0 before autoclaving. For solid media, 2 percent agar is added. Individual growth factors are added to minimal medium as required: 100 mg amino acids, 10 mg purines and pyrimidines, and vitamins at the concentrations given per liter. In order to simplify the complete medium, the vitamin solution and the hydrolyzed nucleic acid can be omitted and the yeast extract increased from 0.1 to 1 percent.

Preservation of Stocks

Freshly grown streaks are harvested with a loop and transferred to tubes of anhydrous sterile silica gel. To recover stocks, the lumps of dried cells are broken up and a few granules scattered on a plate or slant. Stocks have so far been preserved for up to 10 years by this method.

Crosses

The "tetrapolar" mating system is controlled by different alleles at two loci. Two alleles occur at the *a* locus, and multiple alleles have been identified at the unlinked *b* locus. Fifteen *b* alleles were identified by Rowell (1955); later, 18 were identified by Puhalla (1970), who estimated that at least 25 are likely to exist in wild populations. He designated these *b*A, *b*B, *b*C, . . . *b*R. In this laboratory only 2 *b* alleles have been routinely used, b_1 and b_2. On mating medium haploid cells with different *a* alleles fuse to form dikaryotic cells, and from these diploids can be selected. Dikaryotic cells with different *b* alleles form an aerial mycelium of "infection hyphae." Crosses are made by mixing suspensions of haploid cells with different alleles at both loci and injecting 4–5-day-old maize seedlings. About 5–6 days after inoculation, neoplastic growths, or galls, appear on the leaves or stem base of the seedlings, and a few days after this diploid teliospores appear within the gall tissue. These can be used immediately or dried and stored in silica gel until required. Gall tissue is ground in a small mortar and the debris removed by filtering through glass wool. Brandspores are suspended in 1.5 percent $CuSO_4$ solution overnight to remove contaminants and vegetative cells. Meiosis occurs when the spores are incubated on complete media. Diploid strains heterozygous for *a* and *b*,or only for *b*, are solopathogenic, i.e., when inoculated alone they infect the host and produce teliospores which germinate with meiosis. Diploids tend to produce smaller galls and fewer spores than two haploids of opposite mating type.

Isolation of Mutants

The compact colony growth and the use of replica plating with velvet makes it very easy to obtain mutant strains. Nitrosoguanidine has been found to be a very effective mutagen. In experiments in which temperature-sensitive and auxotrophic mutants are scored, up to 10 percent of the cells surviving treatment are mutant (Unrau and Holliday, 1970, and unpublished). The mutant yield after uv irradiation is about 10 times lower.

A selective method exists which makes it possible to isolate spontaneous mutations. When inositol-requiring auxotrophs are incubated in minimal medium without inositol, they die rapidly after about 24 hours unless the cells happen to have an additional biochemical requirement. Therefore, populations of cells starved of inositol for 2–4 days and then plated on complete medium yield a low frequency of survivors, a high proportion of which are auxotrophs, mainly amino acid requirers (Holliday, 1962*b*). This method could also be used to obtain temperature-sensitive mutants.

Auxotrophic Mutants

The following types of mutant have been isolated (Holliday, 1961*a*, and unpublished; Perkins, 1949, private communication): amino acid requirements—arginine, methionine, lysine, leucine, histidine, glycine (or serine), isoleucine, and phenylalanine; vitamin requirements—nicotinic acid, anthranilic acid, inositol, pyridoxine, p-aminobenzoic acid, thiamine, choline, pantothenic acid, and biotin; purine or pyrimidine requirements—adenine and cytidine (or uridine); others—thiosulfate and fatty acids.

Nitrate Reductase and Nitrite Reductase Mutants

About a quarter of auxotrophic mutants which are unable to grow on minimal medium containing nitrate as the sole source of nitrogen are blocked in the nitrate assimilatory pathway. One class, designated *nir*, are unable to reduce nitrite. In the presence of nitrate, a large amount of nitrite accumulates in the medium (Resnick and Holliday, 1971; Holliday, 1971). It is presumed that these mutants lack nitrite reductase, but this has never been proved by enzyme studies. A second class is designated *nar* mutants, and these all lack functional nitrate reductase. These mutants are particularly easy to obtain as they are resistant to chlorate, an analog of nitrate. The mutants fall into 6 complementation groups, and it is clear that one of these, *nar*1 (*nar*A), is the structural gene for the enzyme (Holliday, 1966; Lewis and Fincham, 1970*b*). This locus is closely linked to the *nir*1 gene.

Nitrate reductase is an inducible enzyme, and a number of constitutive mutants have been isolated. Single mutants are partially derepressed, but when these are crossed, fully derepressed double-mutant strains can be isolated (Lewis and Fincham, 1970*b*). One of the constitutive mutants studied is very closely linked to one of the *nar* loci (*nar*E) which might, therefore, be the structural gene for the repressor (Lewis and Fincham, 1970*b*).

Temperature-Sensitive Mutants

Many hundreds of mutants able to grow at 22°C, but not at 32°C, have been isolated. These mutants lack an indispensible cellular function at the restrictive temperature and are invaluable for studies of macromolecular synthesis or other biochemical processses. Several have been found which are unable to synthesize DNA at the restrictive temperature, and these are designated *tsd* mutants (Unrau and Holliday, 1970). One of them (*ts*92A) has been shown to contain a temperature-sensitive DNA polymerase, and it has therefore been renamed *pol*1-1 (Jeggo *et al.*, 1973).

Radiation-Sensitive and Recombination-Defective Mutants

The first mutants of this type to be isolated in a eukaryotic organism were obtained in *U. maydis* by a simple replica-plating method. Replicas from colonies derived from mutagenized cells are treated with a low dose of uv or ionizing radiation which is not sufficient to kill many wild-type cells. Sensitive colonies, therefore, give gaps in the replica. These mutants are almost as frequent as auxotrophs or temperature-sensitive mutants. In searches for particular mutants or their alleles, over 200 radiation-sensitive strains have been isolated. Some auxotrophs are radiation sensitive (Holliday, 1965*b*), and a particularly interesting group of these are pyrimidine requirers. Certain radiation-sensitive mutants have been shown to be defective in genetic recombination (see below).

DNase-Deficient Mutants

Colonies digest DNA in the medium on which they are growing; this can be shown by precipitating DNA with acid. This makes it possible to detect DNase-deficient mutants by the absence of a halo or digested material around the colony (Holliday and Halliwell, 1968; Badman, 1972). Such mutants lack extracellular enzyme activity. Secondary mutants lacking intracellular enzyme may then be obtained by treating colonies on DNA medium with toluene for several hours before the DNA is precipitated. Toluene treatment allows intracellular enzymes to diffuse from the cells, and, therefore, deficient mutants can again be detected. These mutants appear to be defective in genetic recombination (Badman, 1972) (see below).

Drug-Resistant Mutants

Apart from one resistant to parafluorophenylalanine, none of these drug-resistant mutants have been used so far in genetic experiments. Mutants resistant to sulfanilamide, nystatin, or cycloheximide are easily obtainable.

Mutants of the *b* Mating-Type Locus

By using diploids homozygous for the *b* locus, which form yeastlike colonies on mating medium, mutants in the *b* locus could be isolated because they produce white infection hyphae and are solopathogenic (Day *et al.*, 1971). The properties of these mutants will be described in a later section.

Senescent Mutants

In connection with studies on the mechanism of clonal aging, temperature-sensitive mutants have been isolated which undergo a limited number of divisions at the restrictive temperature before the death of all the cells occurs. These were isolated as weak suppressors of a missense *nar2* mutation. This procedure was used because evidence was available that the senescent mutant natural death (*nd*) in *Neurospora* synthesized defective proteins and also acted as a weak suppressor of certain adenine auxotrophs (Holliday, 1969; Lewis and Holliday, 1971). One of the *Ustilago* senescent mutants is resistant to cycloheximide, which suggests that it may have altered ribosomes.

Genetic Analysis

Complementation Tests

These are carried out, as in yeasts, by synthesizing appropriate diploid strains. The two mutants to be tested for complementation must be in strains of opposite *a* mating type and carrying forcing auxotrophic or temperature-sensitive markers which allow selection of a diploid. (The *b* alleles, which appear to be without influence on the fusion of haploids, can be the same or different.) Suppose the complementation of two radiation-sensitive mutants is to be examined. For instance, one could be in a *pan⁻* a_1 strain and the other in a temperature-sensitive a_2 strain. These haploids are streaked on mating medium at the permissive temperature (22°C). After 2 days, the diploids are selected from the dikaryotic hyphae by restreaking or by replica plating to minimal medium and incubating at the restrictive temperature (32°C). Only diploid isolates grow vigorously, and the complementation of the two radiation-sensitive mutants is tested by measuring the radiation sensitivity of one of these diploids. Complementation in dikaryons can in some cases also be scored by removing the binucleate hyphae with a sharp needle under a dissecting microscope and testing their phenotype directly (Holliday and Resnick, 1969).

Mitotic Recombination

Two types of diploid are routinely used for genetic analyses. Heterozygous diploids carry two or more recessive markers. Heteroallelic diploids carry noncomplementing or weakly complementing pairs of alleles as well as at least two recessive markers, which we use for selecting the diploid strains. The segregation of markers from heterozygous diploids is usually detected by replica plating. Colonies are grown on complete medium and replicated to whatever growth conditions will allow the detection of various homozygous derivatives (e.g., minimal medium, restrictive temperature, or exposure to low radiation doses). Individual segregant colonies are picked for fuller testing. Spontaneous segregation occurs at the frequency of 10^{-3}–10^{-4}, depending on the distance of the marker from its centromere. This frequency is increased 10- to 100-fold by prior treatment of the diploid with uv, ionizing radiation, nitrosoguanidine, or other recombinagens such as mitomycin C or 5-fluorodeoxyuridine (Holliday, 1961b, 1964; Esposito and Holliday, 1964; Holliday, 1965a). The segregation which is observed is largely due to mitotic crossing over, which in half the cases produces homozygosis from the position of the exchange to the end of the arm. The best evidence for this comes from the detection of mosaic colonies which contain in roughly equal proportions the reciprocal products of an exchange (Holliday, 1961b).

Apart from single mitotic crossovers, segregants can arise in other ways. Double exchanges can make a proximal marker homozygous while leaving a distal one heterozygous. The same phenotype could be produced by mitotic gene conversion (nonreciprocal recombination). Another possible source of segregation is the loss of pieces of chromosome arms leading to the production of strains hemizygous for particular recessive markers. All three possibilities may, in principle, be detected by further analyses. Unfortunately, haploidization does not occur in *U. maydis* either spontaneously or after treatment with p-fluorophenylalanine. Moreover meiotic analysis of the segregant is often difficult, if not impossible, owing to infertility of many strains which express an auxotrophic marker. For these reasons, the exact contributions of crossovers, gene conversions, and chromosome breaks has not been accurately determined.

Allelic recombination can be studied with heteroallelic diploids. In this case the recombinants contain the dominant wild-type allele and can therefore be selected on the appropriate medium. Spontaneous allelic recombination is about 1000-fold rarer than mitotic crossing over between widely spaced markers, but it is very strongly stimulated by low doses of uv or ionizing radiation, as well as by other mutagens. For uv, a dose which kills only a small proportion of cells may stimulate allelic recombination at least 100-fold, whereas a comparable dose would only increase mitotic crossing over about 10-fold. A comparison of homoallelic with heteroallelic diploids makes it certain that almost all selected colonies are recombinants rather than reverse mutations of either of the mutant alleles. As in yeast, allelic recombination appears to be largely the result of gene conversion. This was shown by analyzing diploid cells (half-tetrads) containing a recombinant chromosome by the examination of meiotic products from such cells. In one study, 34 recombinants were analyzed, and no double-mutant chromosome (reciprocal recombination product) was detected (Holliday, 1966). Usually, about 12 percent of allelic recombinants are homozygous for a distal marker; this indicates that about one quarter of mitotic gene-conversion events are associated with crossing over.

Meiotic Analysis

The methods used in *U. maydis* are very rapid, but unconventional in that individual products of meiosis are not themselves isolated; instead, their asexual progeny are characterized. When teliospores are plated or are spread on complete medium, they germinate with meiosis after about 20 hours. The 4 haploid products from a promycelial cell are budded off as vegetative cells which immediately start growing and dividing. The members of the tetrad are not necessarily formed synchronously, so one of them may have budded off 2 or 3 cells before the last member of the tetrad has itself appeared as a bud. If left undisturbed, each germinating teliospore gives rise to a colony consisting of up to 4 haploid clones; each clone generally consists of a different number of cells.

To analyze the random products of meiosis, large numbers of teliospores are spread on plates of complete medium. When the colonies are just visible to the naked eye, 36–48 hours later, all the cells are washed off the plate with sterile water. The resulting suspension is diluted and 50–100 cells are spread on plates of complete medium. After incubation, the colonies are replica plated on to a single omission series of media or any other necessary test media. Scoring is carried out by numbering all isolated colonies and checking their growth responses. In general, 1:1 segregation ratios are almost always observed for mutant and wild-type alleles. If an excess of a wild-type allele is recorded, this is usually due to the presence of a few percent unreduced diploid cells among the haploid progeny. Certain markers, such as *leu1*, should be avoided, as they appear

TABLE 1. *Mutants Used in Genetic Studies*

Locus and allele	Previous description	Phenotype
ad 1–1	*ad* 1	Growth requirement for adenine
me 1–1	*me* 1	Growth requirement for methionine
me 1–2	*me* 15	
leu 1–1	*leu* 1	Growth requirement for leucine
nic 1–1	*nic* 3	Growth requirement for nicotinic acid
nic 1–2	*nic* 17	
nic 2–1	*nic* 10	
inos 1–1	*inos* 1	Growth requirement for inositol
inos 1–2	*inos* 2	
inos 1–3	*inos* 3	
inos 1–4	*inos* 4	
inos 1–5	*inos* 5	
pan 1–1	*pan* 1	Growth requirement for pantothenic acid
pdx 1–1	*pdx* 2	Growth requirement for pyridoxine
cho 1–1	*cho* 6	Growth requirement for choline
pyr 1–1	*pyr* 7	Growth requirement for cytidine or uridine; radiation sensitive
pyr 1–2	—	
pyr 1–3	—	
pyr 2–1	*pyr* 2	
pyr 2–2	—	
pyr 2–3	—	
pyr 2–4	—	
pyr 3–1	—	
pyr 4–1	—	
nir 1–1	*nir* 1	Lacks nitrite reductase; growth requirement for NH_4^+
nar 1–1[a]	*nar* 6, *nar* A6	Mutants in structural gene for nitrate reductase; growth requirement for NH_4^+ or NO_2^-
nar 1–2	*nar* 9	
nar 1–3	*nar* 10, *nar* A10	
nar 1–4	*nar* 11	
nar 1–5	*nar* 12	
nar 1–6	*nar* 13	
nar 1–12	—	
nar 1–13	—	
nar 2–1[a]	*nar* 1, *na* 1	Lacks nitrate reductase; growth requirement for NH_4^+ or NO_2^-
pfp 1–1	*pfp* 1	Resistant to p-fluorophenylalanine
tsd 1–1	*tsd* 1	DNA synthesis inhibited at 32°C
pol 1–1	*ts* 92A	DNA synthesis inhibited at 32°C; mutant in structural gene for DNA polymerase
rec 1–1	*uvs* 1	Radiation sensitive; recombination deficient
rec 1–2	—	
rec 2–1	*uvs* 2	Radiation sensitive; recombination deficient

TABLE 1. Continued

Locus and allele	Previous description	Phenotype
uvs3–1	uvs3	uv sensitive
nuc1–1	—	
nuc1–2	—	Deficient in extracellular DNase
nuc1–3	—	
nuc2–1	—	
nuc2–2	—	Deficient in intracellular DNase; recombination defective
nuc2–3	—	
nuc2–4	—	
bGmut1	—	
bGmut2	—	Defective *b* allele function
bDmut	—	
bImut	—	

a In addition, 28 *nar* mutants were assigned to 6 complementation groups, A-F, and 12 mutants constitutive for nitrate reductase (*c* mutants) were assigned to 4 complementation groups, I-IV. Complementation groups E and I are the same, and *nar*1 corresponds to complementation group A (Lewis and Fincham, 1970*b*).

to upset meiotic division and lead to disturbed segregation ratios for all markers in the cross. [For this reason, it was falsely reported (Holliday, 1961*a*) that analysis of the random products of meiosis was an unreliable method.]

The method routinely used for analysis of tetrads depends on the fact that the individual products of meiosis or their descendents form colonies which are morphologically distinguishable. Teliospores are spread on complete medium at several dilutions. When the spores have germinated and produced colonies of about 50–100 cells, plates are selected which contain a few hundred colonies. Using a dissecting microscope and a sharp needle, individual isolated colonies are removed by cutting out small pieces of agar, and are then transferred to fresh plates of complete medium. A drop of sterile water is added and the cells spread out over the plate with a spatula. (Originally four teliospore colonies were spread per plate, but it is more satisfactory to spread only one or two per plate.) After several days of incubation, preferably in the light, it may be seen that each spread contains up to 4 clearly distinguishable morphological types (see Holliday, 1961*a*). Then, 1 or 2 each of these is transferred to a fresh plate of complete medium (6–8 tetrads per plate), and the phenotypes are determined by replica plating to the appropriate test media. It is not clear why the segregation of biochemical or other markers should so strongly affect colony morphology. Results from early experi-

TABLE 2. *Summary of Linkage Data*[a]

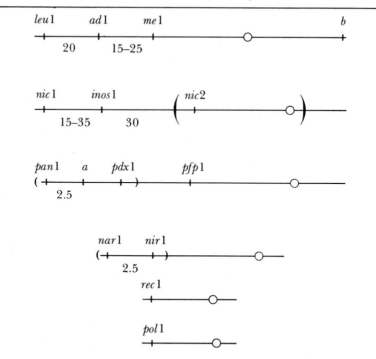

[a] Brackets indicate that the order of markers relative to the centromere is unknown.

ments on morphological mutation in *U. maydis* were interpreted to mean that the species was very variable (Stakman *et al.*, 1943). It now seems more likely that the surface texture, color, and shape of colonies is affected by a large number of diverse mutations, none of which would in the usual sense be described as morphological mutants.

This method of unordered tetrad analysis is quick and quite satisfactory for most purposes. Its main disadvantage is that the 4 products of meiosis do not always appear among the spread teliospore colonies. In many crosses about 20 percent of the germinated teliospores yield complete tetrads and about 40 percent yield triads in which the genotype of the missing member can be deduced. In other words, about 65 percent of the meiotic products survive. Day and Anagnostakis (1971*b*) reported a survival of 87–95 percent of the meiotic products in crosses between wild-type strains. Another method of tetrad analysis, devised by Perkins (1949) and extensively used in early studies (Holliday, 1961*a*), has been found to be unreliable and has been discontinued.

A list of mutants which have been used in genetic studies is given in Table 1, and the known linkages are given in Table 2. As in some other fungi, the frequency of recombination between linked markers varies in different crosses, presumably because of the existence of segregating factors which influence these frequencies. The positions of centromeres relative to certain linked markers is known from mitotic crossing over in vegetative diploids.

Summary of Experimental Studies

Genetic Recombination

Investigations into the mechanism of recombination have been mainly carried out with heterozygous and/or heteroallelic diploid strains. Using the *nar*1 locus, mitotic allelic recombination has been shown to be due to gene conversion, which can be polarized in that one site is preferentially converted. The frequency of conversion is enormously increased by low doses of radiation, particularly when cells are irradiated during the S phase in the cell cycle (Holliday, 1966). This observation suggests that if lesions such as pyrimidine dimers are replicated, they then have to be repaired by recombination; at other periods in the cell cycle most of the lesions would be removed by an excision repair process. This interpretation is supported by studies with certain radiation-sensitive mutants (Holliday, 1967, and unpublished; Unrau, 1974). Diploids homozygous for the mutant *uvs*3-1, which is defective in the excision of pyrimidine dimers, have an enhanced frequency of uv-induced gene conversion, whereas diploids homozygous for either of 2 radiation-sensitive mutants, *rec*1-1 or *rec*2-1, which have normal excision of dimers, show very little or no increase in recombination after irradiation. These 2 mutants also have other effects on recombination. Heteroallelic diploids homozygous for both have a *rec*⁻ phenotype, in that no allelic recombination, spontaneous or induced, is detectable. The *rec*2-1 mutant blocks meiosis at the promycelial stage, and no haploid products are produced. (This effect should not be confused with mere sterility. For instance, many auxotrophs when selfed are sterile in that no teliospores are produced.) Crosses between *rec*1 mutants are also defective in meiosis. In this case cells are budded off the promycelium, but 90 percent of them are inviable, and among the rest are many aneuploids and diploids. The meiotic recombination frequency appears to be within the normal range, but it is presumed that chromosome breakage and nondisjunction commonly occurs. The effect of *rec*1 mutants on mitotic allelic recombination has been examined in some detail (Holliday *et al.*, 1974). For instance, in one study a diploid heteroallelic for *nar*1, *nic*1, *inos*1, and *rec*1 was synthesized. This mutant has a higher than normal frequency of spontaneous recombination at each locus and a much lower than normal frequency of radiation-induced recombination. If a radiation-resistant ($+/rec$1) recombinant is selected, this has the normal frequency of spontaneous and induced recombination for each locus. In this strain, 85–90 percent of the *inos*$^+$ recombinants remain heteroallelic for the distal *nic*1 locus. However, in the radiation-sensitive diploid, 50 percent of *inos*$^+$ recombinants are homozygous and 25 percent are hemizygous for the distal *nic* locus. This demonstrates that normal gene conversion, without recombination of outside markers, cannot be occurring. Rather, the results indicate that allelic recombination in *rec*1 diploids is due to defective crossing over in which one recombinant chromatid is often formed with a broken one.

Other recombination-defective mutants were isolated initially as DNase-deficient strains (*nuc*1 and *nuc*2) (Badman, 1972). Allelic recombination at 3 tested loci does not occur in diploids homozygous for *nuc* mutants, whereas crossing over between widely spaced markers occurs at the normal frequency. One of the DNases present in only a small amount in certain *nuc* strains has been purified and characterized. It is specific for denatured DNA and cuts this into oligonucleotides as if it were an endonuclease, but it in fact requires free ends for activity (Holloman, 1973).

A biochemical method for measuring the time and frequency of allelic recombination has been developed (Holliday, 1971). Diploids heteroallelic for *nar*1 and homoallelic for *nir*1-1 can recombine to give cells which convert nitrate to nitrite. In nitrate minimal medium the appearance of *nar*$^+$ recombinants which can synthesize active nitrate reductase is signaled by the detection of small quantities of nitrite in the medium. This method is sensitive enough to detect a recombination frequency of 10^{-5}. After irradiation, recombinants appear after about 4 hours, and, from the kinetics of recombinant formation with increasing dose, it can be deduced that recombination occurs only in viable cells. This in turn suggests that cells which are unable to recombine after irradiation are unlikely to survive, presumably because the recombination repair mechanism has not come into operation. It is believed that the reason for this is that recombination may normally be repressed in mitotic cells but induced after irradiation (see Holliday, 1971). With increasing doses, the induction of the recombination mechanism is progressively blocked, as is known to be the case for the inducible enzyme nitrate reductase (Resnick and Holliday, 1971).

Repair Mechanisms

Wild-type cells are very resistant to radiation damage (Holliday, 1971), and the isolation of radiation-sensitive mutants indicates that efficient genetic repair mechanisms exist (Holliday, 1965b). Of many mutants available, so far only a few have been examined in any detail. The *rec*1, *rec*2, and *uvs*3 mutants are likely to have defects in different pathways, as the 3 double-mutant strains are much more sensitive than any of the singles, and the triple-mutant is more sensitive than the doubles. This triple-mutant strain is about 1000 times more sensitive to ionizing radiation than wild-type cells. It is slow growing, gives rise to variable colonies, and produces a high proportion (70–80 percent) of nonviable cells during growth in liquid medium. It is very likely that this strain fails to repair spontaneous genetic lesions.

Procedures have now been developed for measuring uv-induced pyrimidine dimers in DNA containing labeled cytidine and thymidine (Unrau *et al.*, 1972). After irradiation of wild-type cells, most of these dimers are rapidly removed from the DNA, presumably by an excision repair mechanism. The recombination-defective mutants *rec*1 and *rec*2 are capable of this excision of dimers, whereas the *uvs*3 strain, which is specifically sensitive to uv, retains dimers in DNA for a period of about 2 hours (Unrau, 1973). Their subsequent removal may well be the result of another repair mechanism, possibly involving recombination.

Pyrimidine auxotrophs are a class of radiation-sensitive mutants which appear to be unique to *U. maydis*. Their sensitivity is largely independent of the pyrimidine concentration in the growth medium. Some of the mutants have very unusual survival curves: with low doses, there is exponential killing, but with increasing dose there is actual recovery of viability, in that the survival curve follows that of wild-type cells (P.D. Moore, private communication, 1972). This effect is perhaps analogous to uv reactivation of bacteriophage lambda. It was suspected that the sensitivity of *pyr* mutants might be due to abnormal levels of DNA precursors, and this has in part been confirmed. The thymidine triphosphate pool is very much lower than in wild-type cells, whereas other nucleoside triphosphate pools are normal (P.D. Moore, private communication, 1972). It may well be that the enzyme responsible for repair synthesis functions inefficiently when the concentration of DNA precursors is disturbed.

Additional evidence that at least one repair mechanism is very rapid has also been obtained by examining the effect of uv light on the induction of nitrate reductase. Again, use was made of the *nir* mutant, which makes it possible to measure accurately enzyme activity by the rate of accumulation of nitrite. Low doses of uv light result in a brief delay in enzyme synthesis after induction, but this delay is much longer in radiation-sensitive strains, e.g., *uvs*3. This delay effect is due to damage to the DNA, as it is fully reversed by treatment with visible light (photoreactivation). The results strongly indicate that repair of the structural gene occurs before transcription can take place. In *rec*1 *uvs*3 strains it is probable that defective repair occurs, and that this leads to errors in transcription, since after induction an altered enzyme is synthesized. This is shown by the presence of enzyme cross-reacting material and by the heat lability of the enzyme (Resnick and Holliday, 1971).

Genetic Replication

A number of temperature-sensitive mutants defective in DNA synthesis have been examined for DNA polymerase activity. One, designated *pol*1-1, was found to contain 10–25 percent of the wild-type activity after the cells had been held at the restrictive temperature for several hours. Moreover, the partially purified enzyme isolated from this mutant grown at the permissive temperature is heat labile. It is therefore probable that *pol*1 is the structural gene for the enzyme (Jeggo *et al.*, 1973). The mutant is not particularly radiation sensitive, and it is not recombination deficient. The DNA polymerase is not required for mitochondrial DNA replication, as cells made permeable incorporate TTP (thymidine triphosphate) into mitochondrial DNA at the restrictive temperature as efficiently as do wild-type cells (Banks, 1973). It may well be required for chromosome replication. Another temperature-sensitive mutant blocked in DNA synthesis, *tsd*1-1, becomes radiation sensitive and undergoes mitotic recombination in diploid strains after some hours at the restrictive temperature. This mutant has been used to show that mitotic recombination occurs at the 4-strand stage (G2), presumably as a result of breakage and reunion of chromatids (Unrau and Holliday, 1972).

Earlier experiments with synchronized populations demonstrated that when diploids were irradiated with uv early in the S period, mitotic crossing over was induced in distally marked intervals; but when they were irradiated late in the S period, crossing over mainly occurred between the proximal marker and the centromere (Holliday, 1965*a*). If replicated lesions stimulate or are repaired by recombination, as was suggested in the previous section, then the observation may mean that replication proceeds from the ends of the chromosome arms to the centromere.

Mutation

After uv treatment of heterozygous diploids, the reciprocal products of mitotic crossing over can often be recovered as mosaic colonies. This phenomenon has been exploited in mutation experiments with diploids homozygous for *ad*1-1, *inos*1-2, or *inos*1-3. Reverse mutations were selected after uv treatment under conditions in which both surviving daughter cells could be identified by crossing over. It could be shown that a high proportion of the induced mutations were transmitted to both daughters, and this result was interpreted to mean that the mutation had occurred in both strands of the DNA. It was suggested that this might occur as a result of the misrepair of uv lesions, where, for instance, incorrect bases were inserted opposite a pyrimidine dimer (Holliday, 1962*a*).

Evidence that recombination repair may be involved in induced mutation comes from studies on reversion of *ad*1-1 in haploid *rec*1-1 strains. Although the spontaneous mutation frequency is about 5 times higher than in *rec*$^+$ diploids, there is no increase in mutation after uv treatment. On the other hand, the excision-defective *uvs*3-1 strain has a high frequency of induced mutation. These results are in line with much more detailed studies in *Escherichia coli*.

Genetic Regulation

The synthesis of the enzyme nitrate reductase is induced by its substrate. Ammonium represses this synthesis and the addition of ammonium after induction leads to the breakdown of the enzyme (Lewis and Fincham, 1970*a*). Constitutive mutants have been isolated which synthesize the enzyme in the presence of ammonium. These are not fully derepressed, as they form up to 50 percent of the fully induced enzyme level in wild-type cells and they themselves can be induced by nitrate to synthesize the normal amount of enzyme. There are 12 mutants, and they fall into 4 complementation groups; all are recessive. One of the mutants, *c*-8, was used to isolate a fully derepressed strain. This strain was shown to contain 2 mutations, each on its own being partially constitutive (Lewis and Fincham, 1970*b*). Wild-type cells will not grow on nitrate minimal medium which contains cesium ions, as they constitute an analog of ammonium and presumably repress nitrate reductase. Mutants resistant to cesium can easily be obtained, and these are probably derepressed for nitrate reductase (J.E. Pugh, private communication 1973).

The lifetime of the RNA messenger was determined by a novel method. It has been mentioned that low doses of uv light block transcription. If cells which are synthesizing nitrate reductase are irradiated, synthesis stops almost immediately and resumes again after about 20 minutes; repair of the damaged gene is believed to occur during this time. Since the uv lesions are primarily in the DNA, this result means that the lifetime of the messenger is very short, probably not more than 2.5 minutes (Resnick and Holliday, 1971).

Cytoplasmic Inheritance

Certain wild-type strains inhibit the growth of others on plates (Puhalla, 1968). Isolates which produce the inhibitor were designated P1; those that did not might either be sensitive to the inhibitor, P2*s* strains, or resistant, P2*r* strains. It was shown that the difference between *s* and *r* was determined by a single pair of alleles. Later, another strain, P3, was found with a phenotype which was like P2*r*, but which could be shown to carry the *s* allele. The absence of Mendelian segregation in crosses involving P1, P2, and P3 (apart from *s* and *r*) clearly shows that these effects are due to different cytoplasmic states. Since P1 and P3 are both unstable and can give rise to P2 strains, whereas P2 itself is completely stable, it has been concluded that P2 lacks some cytoplasmic factor which is present in the other strains and that this factor prevents the expression of the *s* allele (Puhalla, 1968). The situation shows strong resemblance to the much more fully analyzed "killer" characteristic in yeast, which is also controlled by nucleo–cytoplasmic interactions (Bevan and Somers, 1969; Somers and Bevan, 1969).

Mating Types and Pathogenicity

The *a* locus appears to control the fusion of haploid cells, and the *b* locus controls the pathogenicity of the resulting dikaryon. Diploids have been synthesized which are homozygous *a* and heterozygous *b* (*a=b≠*). These are solopathogenic and produce teliospores which will germinate with meiosis. They cannot be crossed with haploids of opposite *a* mating type (Holliday, 1961*b*). Diploids which are heterozygous *a* and homozygous *b* (*a≠b=*) do not produce infection hyphae on mating medium and are nonpathogenic (Puhalla, 1970). This feature has been used to detect mutants in such diploids at the *b* locus which do produce such hyphae (Day *et al.*, 1971). These uv-induced mutants do not carry *b* specificities; rather, the *b* locus has impaired function. Diploids containing the mutants are pathogenic, although they form smaller galls and fewer spores than normal *a≠b≠* diploids. Moreover, when tetrads from germinated spores are analyzed, almost all of them have only one or two products. This suggests that two different wild-type *b* alleles are necessary for normal meiotic division. One of the mutants, *bGmut*1, was recovered as a prototrophic haploid which was itself pathogenic, producing teliospores which germinated to produce haploids of the same genotype. Like *a=b≠* diploids, this mutant could not be crossed with haploids of opposite mating type *a*. Two other mutants, *bDmut* and *bImut*, when crossed with normal strains which had *b* alleles different from those from which the mutants were derived, did not affect meiosis. However, when these two mutants were crossed, although normal tetrads were formed, many teliospore colonies contained more than four phenotypes or had aberrant segregation for particular markers. These were probably derived from spores containing two diploid nuclei, each of which underwent meiotic division. No normal *b*-factor recombinants were recovered among over 5000 progeny from crosses between these 2 *b* mutants. In other basidiomycetes the *b* locus has been shown to consist of two subunits, and recombination between these subunits generates new mating-type specificities. In crosses between normal *b*-allele strains in *U. maydis*, however, no new *b*-allele specificities have been recovered in over 6000 progeny (Puhalla, 1970).

Literature Cited

Badman, R., 1972   Deoxyribonuclease-deficient mutants of *Ustilago maydis* with altered recombination frequencies. *Genet. Res.* **20**:213–229.

Banks, G. R., 1973   Mitochondrial DNA synthesis in permeable cells. *Nat. New Biol.* **245**:196–198.

Bevan, E. A. and J. M. Sommers, 1969   Somatic segregation of the killer (k) and neutral (n) cytoplasmic determinants in yeast. *Genet. Res.* **14**:71–77.

Day, A. W. and J. K. Jones, 1969   Sexual and parasexual analysis of *Ustilago violacea*. *Genet. Res.* **14**:195–221.

Day, P. R. and S. L. Anagnostakis, 1971*a*   Corn smut dikaryon in culture. *Nat. New Biol.* **231**:19–20.

Day, P. R. and S. L. Anagnostakis, 1971*b*   Meiotic products from natural infections of *Ustilago maydis*. *Phytopathology* **61**:1020–1021.

Day, P. R. S. L. Anagnostakis and J. E. Puhalla, 1971   Pathogenicity resulting from mutation at the *b* locus of *Ustilago maydis*. *Proc. Natl. Acad. Sci. USA* **68**:533–535.

Esposito, R. E. and R. Holliday, 1964   The effect of 5-fluorodeoxyuridine on genetic replication and mitotic crossing over in synchronized cultures of *Ustilago maydis*. *Genetics* **50**:1009–1017.

Holliday, R., 1961*a*   The genetics of *Ustilago maydis*. *Genet. Res.* **2**:204–230.

Holliday, R., 1961*b*   Induced mitotic crossing-over in *Ustilago maydis*. *Genet. Res.* **2**:231–248.

Holliday, R., 1962*a*   Mutatation and replication in *Ustilago maydis*. *Genet. Res.* **3**:472–486.

Holliday, R., 1962*b*   Selection of auxotrophs by inositol starvation in *Ustilago maydis*. *Microb. Genet. Bull.* **18**:28–30.

Holliday, R., 1964   The induction of mitotic recombination by mitomycin C in *Ustilago* and *Saccharomyces*. *Genetics* **50**:323–335.

Holliday, R., 1965*a*   Induced mitotic crossing-over in relation to genetic replication in synchronously dividing cells of *Ustilago maydis*. *Genet. Res.* **6**:104–120.

Holliday, R., 1965*b*   Radiation sensitive mutants of *Ustilago maydis*. *Mutat. Res.* **2**:557–559.

Holliday, R., 1966   Studies on mitotic gene conversion in *Ustilago*. *Genet. Res.* **8**:323–337.

Holliday, R., 1967   Altered recombination frequencies in radiation-sensitive strains of *Ustilago*. *Mutat. Res.* **4**:275–288.

Holliday, R., 1969   Errors in protein synthesis and clonal senescence in fungi. *Nature* (*Lond.*) **221**:1224–1228.

Holliday, R., 1971   Biochemical measure of the time and frequency of radiation-induced allelic recombination in *Ustilago*. *Nature* (*Lond.*) **232**:233–236.

Holliday, R. and R. E. Halliwell, 1968   An endonuclease-deficient strain of *Ustilago maydis*. *Genet. Res.* **12**:95–98.

Holliday, R. and M. A. Resnick, 1969   Components of the genetic repair mechanism are not confined to the nucleus. *Nature* (*Lond.*) **222**:480–481.

Holliday, R., R. E. Halliwell, V. Rowell and M. W. Evans, 1974   Abnormal recombination in *rec*-1 strains of *U. maydis*. in preparation.

Holloman, W. K., 1973   Studies on a nuclease from *Ustilago maydis*. II. Substrate specificity and mode of action of the enzyme. *J. Biol. Chem.* **248**:8114–8119.

Holloman, W. K. and R. Holliday, 1973   Studies on a nuclease from *Ustilago maydis*. I. Purification, properties and implication in recombination of the enzyme. *J. Biol. Chem.* **248**:8107–8113.

Jeggo, P. A., P. Unrau, G. R. Banks and R. Holliday, 1973   A temperature sensitive DNA polymerase mutant of *Ustilago maydis*. *Nat. New Biol.* **242**:14–15.

Lewis, C. M. and J. R. S. Fincham, 1970*a*   Regulation of nitrate reductase in the basidiomycete *Ustilago maydis*. *J. Bacteriol.* **103**:55–61.

Lewis, C. M. and J. R. S. Fincham, 1970*b*   Genetics of nitrate reductase in *Ustilago maydis*. *Genet. Res.* **16**:151–163.

Lewis, C. M. and R. Holliday, 1971   Mistranslation and ageing in *Neurospora*. *Nature* (*Lond.*) **228**:877–880.

Perkins, D. D., 1949   Biochemical mutants in the smut fungus *Ustilago maydis*. *Genetics* **34**:607–626.

Pontecorvo, G., 1953   The genetics of *Aspergillus nidulans*. *Adv. Genet.* **5**:141–238.

Puhalla, J. E., 1968   Compatibility reactions on solid medium and interstrain inhibition in *Ustilago maydis*. *Genetics* **60**:461–474.

Puhalla, J. E. 1970   Genetic studies of the *b* incompatibility locus of *Ustilago maydis*. *Genet. Res.* **16**:229–232.

Resnick, M. A. and R. Holliday, 1971   Genetic repair and the synthesis of nitrate reductase in *Ustilago maydis* after UV irradiation. *Mol. Gen. Genet.* **111**:171–184.

Rowell, J. B., 1955   Functional role of compatibility factors and an *in vitro* test for sexual compatibility with haploid lines of *Ustilago zeae*. *Phytopathology* **45**:370–374.

Somers, J. M. and E. A. Bevan, 1969   The inheritance of the killer character in yeast. *Genet. Res.* **13**:71–83.

Stakman, E. C., N. F. Kernkamp, H. K. Thomas and W. J. Martin, 1943 Genetic factors for mutability and mutant characters in *Ustilago zeae. Am. J. Bot.* **30:**37–48.

Unrau, P., 1974 The excision of pyrimidine dimers in wild-type and mutant strains of *U. maydis.* in preparation.

Unrau, P. and R. Holliday, 1970 A search for temperature-sensitive mutants of *Ustilago maydis* blocked in DNA synthesis. *Genet. Res.* **15:**157–169.

Unrau, P. and R. Holliday, 1972 Recombination during blocked chromosome replication in temperature-sensitive strains of *Ustilago maydis. Genet. Res.* **19:**145–155.

Unrau, P., R. Wheatcroft and B. S. Cox, 1972 Methods for the assay of ultraviolet light-induced pyrimidine dimers in *Saccharomyces cerevisiae. Biochim. Biophys. Acta* **269:**311–321.

32

Schizophyllum commune

JOHN R. RAPER AND ROBERT M. HOFFMAN

Historical Perspective

Early studies of *Schizophyllum*, as well as of other members of the Basidiomycetes, were concerned first with the general features of the life cycle and later with the details of sexuality. About a century ago, it was found that two distinct vegetative mycelial stages occurred: first, the homokaryon stage, typically consisting of uninucleate cells with simple septa, and second, the dikaryon stage, constituted of binucleate cells with buckle-shaped appendages at the septa. It was also known that sporulating fruiting bodies were typically formed only on the dikaryon, but the origin of the dikaryotic mycelium remained a mystery until 1917. In that year, Matilde Bensaude of Nemours, France, and Hans Kniep of Würtzburg, Germany, independently discovered that the dikaryon was the heterokaryotic product of the sexual interaction of two compatible homokaryotic mycelia in *Coprinus lagopus* and *Schizophyllum commune*, respectively (Bensaude, 1918; Kniep, 1920).

The discovery of sexuality in these two species stimulated a flurry of investigations of related forms during the 1920's. A number of species was found to be homothallic or self-fertile, with the homokaryotic mycelium spontaneously converting to the dikaryotic phase, but a majority of species was found to be heterothallic, with each requiring the interaction of two compatible homokaryons to complete the sexual cycle. Two distinct types of heterothallism, however, were soon recognized: "bipolar sexuality" in

JOHN R. RAPER—The Biological Laboratories, Harvard University, Cambridge, Massachusetts. ROBERT M. HOFFMAN—Genetics Unit, Massachusetts General Hospital, Boston, Massachusetts.

which each fruiting body produced progeny of two self-sterile, cross-fertile classes, and "tetrapolar sexuality" in which each of four classes of self-sterile progeny was cross-fertile with only one of the other three. The distribution of forms displaying the three patterns of sexuality had no evident phylogenetic significance, as they appeared to be randomly scattered throughout the various taxa of the class to the level of the species, each of which exhibited only one particular pattern.

The tetrapolar pattern was the first to be described in detail, and the author (Kniep, 1920) correctly interpreted it as the result of the independent assortment and segregation of "alleles" of two unlinked characters. Bipolarity was later described (Brunswick, 1924) as resulting from the segregation of alleles of a single character. These "sexual" characters were subsequently termed *incompatibility factors* (Bauch, 1930).

The first unusual feature of the incompatibility factors was found by Kniep (1922) in the tetrapolar-form *Schizophyllum*: although each homokaryotic isolate would interact only with one quarter of its siblings, it was usually compatible with all progeny of fruiting bodies of different origins. Multiple "alleles" (i.e., factors) of a single series in the bipolar forms and of two series in the tetrapolar forms are a universal feature of these systems. The complete interfertility of progeny of fruits of different origins was originally attributed to evolutionary divergence due to isolation, i.e., "geographical races." It was soon learned, however, with the demonstration of full intercompatibility of progeny from mushrooms on opposite sides of a piece of horse dung, that geography was of small consequence.

The origin of the numerous alternate factors in any single species has been a mystery that remains yet to be solved. In an early study, Kniep maintained a culture of *Schizophyllum*, inoculated originally with a single dikaryon, for a period of years and periodically determined the mating behavior of its progeny. In addition to the original factors, he occasionally found isolates with two new factors of each of the two series. He interpreted the new factors as mutations from the originals (Kniep, 1930), and the matter of the origin of the extensive series of factors was considered solved for some two decades until Papazian (1951) showed Kniep's "mutations" actually to be recombinants between two distinct loci of complex factors. Meanwhile, Quintanilha (1935) had discovered at least one valid mutation of an incompatibility factor, but its effect was not to specify a new factor but to alter the process of sexual interaction regulated by the factor.

Although *Schizophyllum* and related fungi are acceptable objects for most of the same types of genetic study for which filamentous fungi have been used, the incompatibility system with its profound effect upon sexual morphogenesis is the unique feature that has received primary emphasis in genetic work with these forms. The following summary reflects this bias. The incompatibility system has been demonstrated in a large number of species, but few of these have been extensively studied; in these, the system is remarkably constant and differs only in detail. The account to follow will deal primarily with *Schizophyllum* with few references to other forms, almost exclusively *Coprinus*.

Life Cycle and Developmental History

The life cycle of *Schizophyllum* and related forms is a simple progression consisting of the following essential phases: spore → haploid homokaryotic mycelium → dikaryotic mycelium → fruiting body → basidia (nuclear fusion and meiosis) → spore.

Haploid Phase. The haploid spore germinates to produce a mold-like homokaryotic mycelium that is typically made up of uninucleate cells and is capable of indefinite vegetative growth. Each homokaryon possesses specific incompatibility factors of two kinds, A and B, that determine its mating type, which may be symbolized as $AxBx$. Each factor, A and B, exists in an extensive multiple series, and two homokaryons are compatible and interact to establish the fertile dikaryon only when they carry different A and different B factors, i.e., $AxBx$ and $AyBy$. When any two homokaryons grow together, cells of the two fuse to establish heterokaryotic cells, but interactions between homokaryons sharing A or B factors yield only infertile heterokaryons (see below).

Nuclear Migration. In the compatible or sexually fertile interaction, nuclei of either mate move out of the fusion cells and migrate throughout the preestablished mycelium of the other mate (Buller, 1931; Snider, 1963*b*). The process of nuclear migration involves two unusual features: (1) Nuclei move in an oriented manner at rates far in excess of the rate of radial mycelial growth, and the rate of migration is strongly dependent on temperature, ca. 1 mm/hr at 23°C and 6 mm/hr at 33°C (Snider and Raper, 1958). The forces and mechanisms that move the nuclei are unknown. It is known only that microtubules and microfibrils are associated with nuclei in cells in which nuclei are characteristically in motion (Raudaskoski, 1970*b*, 1972*a*, *b*, 1973; Girbardt, 1968). (2) Nuclear migration requires the rapid dissolution or disruption of a succession of septa (Giesy and Day, 1965), each of which is a complex structure, e.g., the dolipore septum (Moore and McAlear, 1962; Girbardt, 1965; Jersild *et al.*, 1967). The cells are approximately 100 μ long; nuclei in rapid migration must thus traverse a cell and a disrupted septum at the rate of one each minute.

Dikaryotic Phase. The reciprocal exchange and migration of nuclei rapidly bring migrant nuclei to the growing hyphal tips, where in each apical cell a migrant nucleus pairs with the resident nucleus. The paired nuclei henceforth divide synchronously (conjugate division) to provide each subsequently formed cell with two nuclei, one from each parental strain. In the process of conjugate division, a buckle-shaped appendage, the clamp connection, is formed at each new septum. The dikaryotic product of this process is in most essential respects the genetic and physiologic equivalent of the diploid phase of other organisms, although the dikaryon is capable of a number of significant interactions normally lacking in diploids (J. R. Raper and Flexer, 1970). The dikaryon, like the homokaryon, is capable of prolonged or indefinite vegetative growth but usually initiates the process of fruiting shortly after vigorous growth is established.

Fruiting. The origin of the fruiting bodies is uncertain. The primordium of the fruiting body for *Coprinus* was described almost a century ago (Brefeld, 1877) as a weft of filaments originating from the rapid proliferation of a single cell of the dikaryon. A recent report for *Coprinus* (Matthews and Niederpruem, 1972), however, gives substantial evidence of local but multicellular origins of the primordium. Whatever the precise origins of the primordia, their initiation requires a competent genome (J. R. Raper and Krongelb, 1958; Perkins and Raper, 1970), a somewhat higher concentration of thiamine than for vegetative growth, a brief period of illumination with blue light (Perkins, 1969; Perkins and Gordon, 1969), and a less than 4-percent concentration of CO_2 (Niederpruem, 1963).

After induction, the development of fruiting bodies proceeds quickly through five stages (Leonard and Dick, 1968): (1) the initial loose, microscopic tuft; (2) a compact mass of undifferentiated cells that elongates into (3) a cylindrical stalk with a small pit in its free end; (4) the lateral expansion of the pit with the formation of a layer of spore-producing cells; and (5) the fully expanded, mushroomlike, fruiting body with its radiating, spore-bearing gills. The gills of *Schizophyllum* are unique in that each is made up of two lamellae, which split and curl apart upon dessication, a character that accounts for the generic name.

A palisade layer of spore-producing cells, the basidia, covers the gills and the entire lower surface of the fruiting body. Each club-shaped basidium initially contains a pair of compatible nuclei, and these fuse to produce a diploid nucleus, which almost immediately undergoes the two meiotic divisions. Four spores are then formed exogenously on the top of the basidium, and into each of these migrates one of the four haploid meiotic nuclei. Once in the spore, the nucleus divides mitotically, and shortly thereafter the spores are discharged and in nature disseminated by the wind.

Techniques and Methodologies

Culture Methods. The culture and maintenance of stocks of *Schizophyllum* are simple and differ only in detail from routine practices with other filamentous fungi. A suitable complete medium consists of 20 g glucose, 2 g yeast extract, 2 g peptone, 0.46 g KH_2PO_4, 1 g K_2HPO_4, and 0.5 g $MgSO_4$ per liter of distilled water; minimal medium substitutes 1.5 g L-asparagine or 1.5 g $(NH_4)_2HPO_4$ and 120 μg thiamine for the yeast extract and peptone. Growth of prototrophic strains is 0.5–0.7 cm per day on either medium at 23°C. The rate of growth is approximately doubled at 32°C, and halved at 15°C. Stock cultures remain viable for several years on complete medium at 1–4°C, but with 5–10-percent recovery upon transfer of a fast-growing morphological mutant, *thin*. Preliminary tests have shown lyophilization and storage on silica gel to be promising, but resources have not been available for long-term testing of these techniques.

Mating Test. An important diagnostic tool in the work on *Schizophyllum* is the mating-type test to determine the *A* and *B* constitution of homokaryotic strains. This consists simply of inoculating the unknown strain in close proximity (1–2 mm) to inocula of each of a series of tester strains of known constitutions and determining at the appropriate time the pattern of interactions of the unknown homokaryon with the tester strains. A number of interactions may occur (cf. Incompatibility Factors and Sexual Morphogenesis, below), and these are discernible at 2–3 days at 33°C and 5–7 days at 22°C.

Fruiting and Sampling. Fruiting of most dikaryons occurs in 7–14 days at 22°C. Production of spores is heavy and continuous for several days, and the spores germinate in high percentage within 18–24 hours when streaked on complete medium. Isolation of single spores in the germling stage is facilitated by a mechanical device attached to the microscope (J. R. Raper, 1963). Homokaryons derived from spores are large enough to be used in tests, matings, etc., 3–5 days after isolation. An entire sexual generation can thus be achieved in as little as 10 days.

Macerates. The only spores normally produced by *Schizophylum* are the immediate products of meiosis in biparental crosses, and the spores cannot be used to establish populations of genetically identical individuals. There are two situations, however, in which homokaryons are induced to fruit and produce homogenous progeny: by a substance produced by a number of fungi (Leonard and Dick, 1968) and by certain mutations (J. R. Raper *et al.*, 1965; C. A. Raper and Raper, 1966). It has accordingly been necessary in most cases to develop procedures to use mycelial macerates. Maceration of a liquid culture in a blender for 1–2 minutes yields a heterogeneous mixture of fragments that can be made somewhat less heterogeneous by decanting the suspension after brief settling. Macerates can be used for quantitative work of various types, i.e., induction of mutations, estimation of nuclear ratios, determination of nuclear migration, etc., when allowance is made for the lack of homogeneity (Snider, 1963*a*).

Mutagenesis. Procedures for mutagenesis reflect only minimal adaptations of common procedures. Concentrated suspensions of mycelial fragments or spores in rare cases are treated with mutagenic agents, diluted in water, and surface-spread upon or pour plated (at 50°C) in complete, minimal, or appropriately supplemented media. Useful mutagenic agents include (with dosage in the optimal range): unfiltered 100-kV x rays, ca. 200,000 r; ultraviolet light, 2500 ergs/cm²; N-methyl-N′-nitro-N-nitrosoguanidine, 0.1 percent for 50 minutes; and ethyl methanesulfonate, 1 percent for 12 hours (J. R. Raper *et al.*, 1965).

Selective Systems. In connection with treatment for induction of mutations, various procedures and selective systems have been used that are appropriate to the types of mutations sought. Unfortunately, no effective selective system for auxotrophic mutations has been developed to the point of practical usefulness, and these have been identified in total samples inoculated to complete and minimal media.

Selection for mutated *A* and *B* factors depend upon the appearance of dikaryotic sectors and fruiting bodies in treated common-factor heterokaryons. For example, in a common-*B* heterokaryon ($A{\neq}B{=}$), the mutation of a *B* factor to a new state compatible with the original *B* factor results in a dikaryotic sector. Such mutations of the *A* and *B* factors in all cases have resulted in self-compatibility and distinctive morphological phenotypes. The distinctive morphologies of the *A*-mutant and *B*-mutant homokaryons as compared to that of the wild homokaryon provide a visible basis for selection of revertants, suppressors, or mutations modifying the mutant-*A* or mutant-*B* phenotypes. Auxotrophic mutations have been extensively used as markers to trace migrant nuclei (Snider and Raper, 1958; Snider, 1963*a*, 1968; Ellingboe, 1964), as nutritional forcing agents in the establishment of heterokaryons (C. A. Raper and

Raper, 1964, 1966), and for selection of recombinants in various systems. A number of morphological mutations have similarly been used as markers (*puff, fir, streak, blue*), as agents to restrict growth (*A-dom, B-dom*, compact mutants linked to the *A-* and *B*-factors, respectively), and to enhance fruiting (*bug*). Other mutations serve to restrict nuclear migration (modifier *MIV* and certain mutations of the *B* factor). These and the compact mutations are particularly useful for selective systems in which contact between colonies and ensuing heterokaryosis are undesirable.

Membrane Cultures. Growth of mycelia on permeable membranes overlying nutrient media has proved useful for various studies. Peripheral regions of such mycelia mounted in 14 percent gelatin are particularly favorable for observation with phase-contrast microscopy, for fixation and staining for semipermanent microscopic preparations (J. R. Raper, 1966), and for preparations for electron microscopy (Koltin and Flexer, 1969). Membrane-grown mycelia have also provided ideal material for studies on the dynamics of growth (C. S. Deppe, unpublished) and respirometry (Hoffman and Raper, 1971).

Isogenization. Much of the biochemical work to date has been devoted to a search for correlations between proteins, isozymes, and the activity of the incompatibility factors, e.g., in homokaryons vs. dikaryons. To minimize differences other than those related to developments regulated by the *A* and *B* factors insofar as possible, only strains made coisogenic by 10 or more generations of backcrosses have been used. These procedures have proved very effective in damping variability (J. R. Raper and Esser, 1961; Wang and Raper, 1969, 1970).

Large-Scale Culture. Mycelia can be grown in a synthetic medium with glucose and ammonium phosphate. In order to prevent the production of viscous polysaccharide during growth, the mycelia are successively transferred to liquid media of increasing amounts (Wang and Raper, 1969; Hoffman and Raper, 1971, 1972). With ethanol as the sole carbon source, the polysaccharide production is also reduced (Leonard and L. Phillips, unpublished). The mycelia are macerated in a Waring blender at each transfer, and the cultures are incubated on rotary or reciprocating shakers. The optimum temperature for growth is in the region of 30°. The mycelia can be harvested either by centrifugation or filteration, and the amount of growth can be monitored by dry-weight measurements.

Disruption of Mycelia. For gentle disruption, e.g., to obtain intact mitochondria, a French pressure cell may be used. The mycelia are suspended in a suitable buffer with w/v ratio of 1:2. Pressures as low as 2000 psi give significant breakage (Hoffman and Raper, 1972). For more drastic breakage, e.g., to obtain soluble protein, the Raper–Hyatt pressure cell may be used (J. R. Raper and Hyatt, 1963). The cell and mycelia are frozen in dry ice and the mycelia broken at a pressure of 18,000 psi.

Protoplasts. A procedure for the release of viable, osmotically sensitive protoplasts of *Schizophyllum* utilizes a lytic enzyme preparation from *Trichoderma viride* (de Vries and Wessels, 1972)

Isolation of Phosphorylating Mitochondria. Mitochondria demonstrating P/O ratios of two or more with citrate (Hoffman and Raper, 1974) and able to increase their respiration rate twofold upon addition of ADP have been obtained using the above growth methods and the French pressure cell. After breakage, mitochondria are isolated by differential centrifugation (Hoffman and Raper, 1972). These methods have also been used to assay the mitochondrial ATPase which is stimulated by magnesium and 2,4-dinitrophenol (Hoffman and Raper, 1974).

Isolation of Soluble Proteins. The above growth and breakage steps are used in the initial phases. The homogenates are then extracted with appropriate buffers and centrifuged at 37,000 $\times$ *g*. The supernatant yields the crude soluble extract which may be further purified (Wang and Raper, 1969).

Isolation of RNA Polymerase. The mycelia are broken gently in the French pressure cell and the homogenate is filtered through miracloth. The filtrate is centrifuged at 6000 $\times$ *g* for 20 minutes. The pellet contains the polymerase activity, and the enzymes can be solubilized for further purification by using a buffer which contains 0.4M KCl (Hoffman and J. P. Rosenbusch, unpublished).

Respirometry of Intact Mycelia. Mycelia incubated in liquid medium on a shaker grow as very tiny, uniform pellets, if the inoculum is heavy. These pellets can be quantitatively pipetted into Warburg manometer flasks (Niederpruem and Hackett, 1961; Hoffman and Raper, 1971). Respiration rates, sensitivity to inhibitors, and molar ratios of O_2 taken up or CO_2 evolved to glucose utilized are easily measured. To do these studies on relatively unstable heterokaryons, mycelia are grown on cellophane membranes overlying agar media. The colonies are easily washed off membranes into the respirometer flaskes (Hoffman and Raper, 1971).

Specialized techniques have also been developed for respiratory studies (Bonitati, *et al.*, 1970).

Preparation of Ribosomes and Nucleic Acids. Lyophilized mycelia have been used to prepare functional ribosomes and polysomes from *Schizophyllum* (Leary, *et al.*, 1969). DNA and ribosomal RNA have been prepared by disrupting the cells with mortar and pestle after freezing the mycelium with liquid nitrogen (Lovett and Haselby, 1971; Storck and Alexopoulos, 1970).

Incompatibility Factors and Sexual Morphogenesis

Mycelial Interactions. Early work on *Schizophyllum* recognized only two reactions as the results of matings of homokaryotic strains: dikaryosis or no interaction. It was later shown that four different reactions yielding four distinct heterokaryotic products correspond with four basic combinations of *A* and *B* factors carried by the two mates: (1) $A \neq B \neq$, *compatible,* with unlike *A* and unlike *B* factors; (2) $A = B \neq$, *hemicompatible-a,* with shared *A* factor and unlike *B* factors; (3) $A \neq B =$, *hemicompatible-b,* with unlike *A* factors and shared *B* factor, and (4) $A = B =$, *incompatible,* with shared *A* and *B* factors. Each type of mating yields a heterokaryon that is distinctive and characteristic, although the latter two interactions, $A \neq B =$ and $A = B =$, require nutritional forcing for the recovery and maintenance of the heterokaryons.

Sexual Morphogenesis. The initial step in interstrain reactions is the fusion of cells of the two mates, and this occurs regardless of the factor combinations, although the frequency and initial result may differ significantly in the various combinations (Ahmad and Miles, 1970; Sicari and Ellingboe, 1967). The interaction in $A \neq B \neq$ matings consists of a series of steps that establish the dikaryon (J. R. Raper and Raper, 1968). These are traced by the central vertical line of Figure 1:

1. Reciprocal nuclear migration
2. Pairing of migrant with resident nuclei in apical cells
3. Formation of lateral branch proximal to the pair of nuclei
4. Conjugate division of nuclei, a daughter of one nucleus moving into the lateral branch, the hook cell
5. Septation of cell at base of the hook and septation of the hook
6. Fusion of hook cell with subterminal cell and migration of nucleus from hook cell to restore the pair of nuclei in each cell

Stages 3–6 occur at each subsequent cellular division to maintain in the dikaryon a ratio of 1:1 of nuclei of the two parental types.

This sequence of events requires unlike *A* and unlike *B* factors. In matings between strains with shared *A* and *B* factors, none of the sequence occurs, and a nutritionally forced $A = B =$ heterokaryon is morphologically indistinguishable from the homokaryon. A single shared factor, either *A* or *B*, however, results in either case in the expression of only a part of the sequence. When only the *B* factors are different, nuclear

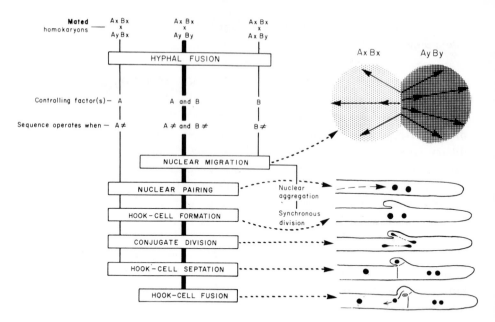

Figure 1. Control by the A and B incompatibility factors of sexual morphogenesis in Schizophyllum. The progression comprises two distinct and complementary series of events, the A sequence and the B sequence, traced by vertical lines at left and right and regulated by A and B factors, respectively. Operation of either sequence requires interaction of compatible factors, i.e., A≠ (Ax × Ay) or B≠. Morphogenesis is completed only when both sequences are operative (central heavy vertical line).

migration is essentially normal, but no subsequent developmental event occurs (right line in Figure 1). In fact, the progression appears to be stalled in continual septal disruption and synthesis and nuclear movement between cells to bring about nuclear aggregation; cell walls are also weakened, and extrusion of blobs of protoplasm and intercellular fusions are common. Conversely, when only the *A* factors are different, there is no nuclear migration, but all subsequent events occur save the final fusion of the hook cell (left line of Figure 1). This final step of the process requires different *A* and *B* factors. The process of dikaryosis is thus revealed as a developmental sequence made up of two partial and complementary sequences, one under the control of the *A* factor, the other under the control of the *B* factor. The entire process comprising the *A* sequence and *B* sequence has been termed *sexual morphogenesis* (J. R. Raper and Raper, 1968).

   Heterokaryosis.   The partial control exerted by either factor imposes highly specific characters on the heterokaryotic products of $A \neq B =$ and $A = B \neq$ interactions. Because the fusion of the hook cell requires *B*-

factor interaction, the $A \neq B =$ heterokaryon consists of binucleate apical cells, uninucleate unfused hook cells at the septa, and uninucleate non-apical cells. The heterokaryon is quite unstable and rapidly sectors into its component strains when not maintained by nutritional forcing. The characteristics of the $A \neq B =$ heterokaryon thus directly reflects the specific controls exerted by the A factor in the morphogenetic sequence. By analogy, the $A = B \neq$ heterokaryon should be a mycelium lacking any visible features that would distinguish it from the homokaryon. This is essentially the case in *Coprinus* (Swiezynski and Day, 1961), but in *Schizophyllum*, the operation of the B sequence alone results in a partial uncoupling of energy metabolism that renders the $A = B \neq$ heterokaryon markedly subnormal. It is slow-growing with little aerial growth, and its filaments are gnarled and irregularly branched. Cell walls are apparently imperfect, as extrusions of protoplasm from the cells are common. These features combine to impart a very distinctive phenotype that has been termed "flat."

Factor Identification. The distinctiveness of the $A \neq B \neq$ and $A = B \neq$ interactions from each other and from $A \neq B =$ and $A = B =$ interactions (which in the absence of nutritional forcing cannot be easily distinguished) provide the basis for the identification of A and B factors in mating-type tests (see Table 1).

The Buller Phenomenon. An unusual feature of the higher fungi is the competence of the dikaryon to provide fertilizing nuclei to homokaryotic strains. This procedure, termed the Buller Phenomenon (Quintanilha, 1937) in honor of its discoverer (Buller, 1931), occurs in three configurations *vis-a-vis* the incompatibility factors of the opposed dikaryon and homokaryon: (1) *legitimate, compatible*: both nuclei of dikaryon compatible with that of homokaryon, e.g., $(AxBx + AyBy) \times$

TABLE 1. *Identification of A and B Factors in Mating-Type Tests*

Unknown strains	Tester strains[a]				Assigned mating type
	$AxBx$	$AxBy$	$AyBx$	$AyBy$	
1	−	F	−	+	$AxBx$
2	F	−	+	−	$AxBy$
3	−	+	−	F	$AyBx$
4	+	−	F	−	$AyBy$
5	+	−	+	−	$AzBy$
6	+	+	+	+	$AzBz$

[a] +, dikaryon; F, $A=B\neq$ heterokaryon; −, no discernible reaction, i.e., either $A\neq B=$ or $A=B=$.

AzBz; (2) *legitimate, hemicompatible*: one nucleus of dikaryon compatible with homokaryon, e.g., $(AxBx + AyBy) \times Ax\ Bx$; and (3) *illegitimate, incompatible*: neither nucleus of dikaryon compatible with homokaryon, e.g., $(AxBx + AyBy) \times Ax\ By$.

Dikaryotization in the first two cases necessarily involves only the passage of compatible nuclei into the homokaryon via fusion cells. In fully compatible interactions, however, the choice of the dikaryotic partner to migrate is not random chance: in each mating, the frequency with which the two nuclear types dikaryotize the homokaryon may vary over the entire range of 1:0 to 0:1. The basis for the discrimination is uncertain (Ellingboe and Raper, 1962*a*), but it is possibly correlated with the degree of heterozygosity of the incompatibility factors in the dikaryon to those of the homokaryon (Crowe, 1963). Dikaryotization in incompatible matings requires either the replacement of the homokaryotic nuclei by the two genomes of the dikaryon or the formation of nuclei recombinant for mating type. Actually, both of these processes occur, not only in illegitimate matings but in legitimate ones as well (Papazian, 1950; Crowe, 1960). The origin of recombinant nuclei has been the subject of much investigation and controversy and may involve either parasexual processes or specific factor exchange, or both (see J. R. Raper, 1966, for review).

Genetic Structure of the Incompatibility Factors

Three features define the basic genetics of the independently assorting *A* and *B* incompatibility factors: (1) Each factor is constituted of two linked loci α and β, separable in most cases by recombination. (2) Each of the four incompatibility genes has a series of multiple alleles. (3) Two factors of either series, *A* or *B*, are compatible if they have different alleles at either α or β, or at both α and β. The factor phenotype with respect to mating competence is thus specified by a unique combination of alleles at α and β, and the number of distinct factors is thus the product of the alleles at the two loci.

Number and Distribution of Factors. In the paper that told of the discovery of sexuality in *Schizophyllum*, Kniep (1920) reported complete interfertility of 14 progeny of a fruiting body from Würzburg with 11 of one from Holland, and two years later he (Kniep, 1922) found complete interfertility between the progeny of 9 fruiting bodies collected at different locations. A rough estimate of the extent of the series of "multiple alleles" at the mating "loci" was given by Whitehouse (1949) as of the order of 100; a decade later a study to determine the number of factors in the two series gave projected estimates of 339 (5-percent limits, 217 and

562) *A* factors and 64 (5-percent limits, 53 and 79) *B* factors (J. R. Raper *et al.*, 1958*b*). An important corollary of this study was the finding that the many factors of both series were randomly distributed wherever woody plants occur with apparent disregard of climate and geographical location.

Structure of Factors. The demonstration that the *A* factor was a complex of two or more distinct genes (Papazian, 1951) was followed by an intensive study of the structure of the *A* and *B* factors. Each factor was revealed to be constituted of two loci, α and β, each with multiple alleles (J. R. Raper *et al.*, 1958*a*). The allelic constitution of 56 native *A* factors was determined, and the sample contained 9 *A*α's and 26 *A*β's (J. R. Raper *et al.*, 1960). Projections based on the frequency of repeats of alleles gave estimates of 9 *A*α's and 50 *A*β's. A more recent evaluation of the same data that accounts for the factor-repeats in the sample revises the estimate of *A*β's to 32 (Stamberg and Koltin, 1972). No single irregularity in the structure of the *A* factor has ever been reported.

The nature of the *B* factor has proved to be quite dissimilar from that of the *A* factor in a number of ways. Although the *B* factor in all cases is made up of α and β components, there are three distinct classes (Koltin *et al.*, 1967; Koltin and Raper, 1967*a,b*): Class-I factors combine seven *B*α's with seven *B*β's, but recombination does not occur between two *B*α's and two *B*β's in any combination (Stamberg and Koltin, 1971). Class-II factors combine the seven *B*α alleles of Class-I factors with two *B*β alleles *(B*β'*)* that are different from the *B*β alleles of Class I, and no intra-*B*-factor recombination has been achieved with Class-II factors (Koltin, 1969). Class-III factors combine two distinctive *B*α alleles (*B*α') with *B*β alleles of Class I. Recombination of Class-III factors with Class-I factors has been achieved, but no intraclass recombination has been reported (Koltin and Raper, 1967*b*). The allelic constitution of a number of *B* factors has not yet been determined, and it is uncertain whether there also exist Class-IV factors combining *B*α' and *B*β' alleles (Parag and Koltin, 1971).

The number of *A*- and *B*-factor phenotypes calculated as the product of α alleles $\times$ β alleles for each factor is in reasonable agreement with the earlier estimates for intact factors.

The full significance of the dual-locus structure of the *A* and *B* factors is not entirely clear (J. R. Raper, 1966; Simchen, 1967*a*; Koltin *et al.*, 1972). One feature of the factors, however, has a very clear bearing on the control they exert. Because interaction between two different α's or two different β's can fully activate the relevant *A* or *B* sequence, there is total duplication of control at the initial level, but whether the duplication extends beyond this level into subsequent events of the two sequences is unknown. Apparently in the species a large number of factor phenotypes has critical utility, and the extensive series of factors can be maintained in "bottleneck populations" far more effectively with two genes than with one (Simchen, 1967a).

Mutations Affecting Sexual Morphogenesis

Three categories of mutations have been reported that have profound effects upon the course of sexual interactions and the characteristics of the resulting heterokaryons. Mutations of each type may arise either spontaneously or by induction. Mutations of two of the types occur within the incompatibility loci: (1) initial or *primary mutations* of wild-type alleles that "turn on" or make constitutive the sequence normally controlled by the affected factor, and (2) subsequent or *secondary mutations* in primary mutant alleles that "turn off" the constitutive sequence of the primary mutation and further affect the course of sexual development in various ways. (3) Mutations of the third type, *modifier mutations*, lie outside the incompatible loci and are expressed as disruptions or alterations of the normal development controlled by the A or B factor.

Primary Mutations in Incompatibility Loci. Mutations in the incompatibility factors were independently reported in 1960 for the $B\beta$ locus of *Schizophyllum* (Parag and Raper, 1960; Parag, 1962) and the $A\beta$ locus of *Coprinus* (Day 1960, 1963). Recovery of the mutations in each case was from fruiting bodies that occurred spontaneously in cultures of common-factor heterokaryons. The common-factor heterokaryons, with or without mutagenic treatment, were conceived as natural selective systems for the recovery of anticipated mutations from one wild allele to new, compatible wild alleles. Such a mutation, either spontaneous or induced, has never been reported. Instead, every primary mutation described to date has lost the discrimination of self-incompatibility for the mutated factor, i.e., it is self-fertile and universally compatible with all wild factors of the same series, the progenitor included, and "turns on" or makes constitutive the morphogenetic sequence normally controlled by the factor and therefore mimics the effect of the interaction of two compatible wild factors of the same series. These characteristics as expressed in the mutant-B homokaryon result in a "flat" phenotype that mimics the $A = B \neq$ heterokaryon in microscopic morphology (Parag, 1962), in nuclear distribution (J. R. Raper, 1966; C. A. Raper and Raper, 1966), in ultrastructure (Jersild *et al.*, 1967; Koltin and Flexer, 1969), and in physiology (Wessels, 1969a; Hoffman and Raper, 1971, 1972). A dozen or more primary mutations have been induced in the B factor, and all that were specifically located occurred in the $B\beta$ locus save one in the $B\alpha'$ locus of a Class III factor (Koltin and Raper, 1966; Koltin, 1968).

Six mutations in the $A\beta$ locus were later recovered (J. R. Raper *et al.*, 1965), and each conformed to the generalized characteristics of the primary *B*-factor mutations with the result that they closely mimicked the $A \neq B$ = heterokaryon. Because the mutant-*B* factor gives the same effect in the homokaryon as the normal interaction of two compatible *B* factors in the heterokaryon and the mutant-*A* factor similarly behaves as two compatible *A* factors, the association of the mutant-*A* and -*B* factors in the homokaryon very closely mimics the $A \neq B \neq$ heterokaryon, the dikaryon. The mimicry is extremely close, with the mutant-*A*–mutant-*B* homokaryon completing the life cycle with the production of fruiting bodies and spores. This is one of the rare cases in which abundant samples of identical spores can be obtained.

When a macerate of a mutant-$B\beta$ strain is treated with mutagenic agents and plated, a lawn of thin, sickly mycelium is established. Against this background, clumps of vigorously growing mycelium with normal homokaryotic morphology are quite conspicous (J. R. Raper *et al.*, 1965). These clumps, upon isolation, prove to be mutations of two types: secondary mutations in the $B\beta$ locus and modifier mutations in other loci that suppress the flat phenotype of the *B* mutation.

Secondary Mutations in Incompatibility Loci. All secondary $B\beta$ mutations share a number of characteristics in addition to wild-type morphology: all are self-incompatible, all are compatible with the primary *B* mutant, and they are completely incompatible among themselves, except in two cases: one is a reversion to the original allele (C. A. Raper and Raper, 1974), the other involves secondaries derived from primaries of two different wild alleles (J. R. Raper and Raudaskoski, 1968). Otherwise the secondary $B\beta$ mutations can be differentiated into a dozen or so classes that vary widely in frequency of occurrence and in the details of their effects on the sexual interaction (Koltin and Raper, 1966; Raudaskoski, 1970, 1972*b*, 1973; C. A. Raper and Raper, 1973).

The characteristics of secondary $B\beta$ mutations are revealed in test matings that compare each $B\beta$ mutation with the entire series of wild $B\beta$ alleles. Two classes of secondary mutations are compatible with all wild $B\beta$ alleles, the original progenitor allele included, but they differ in their control of nuclear migration and hook-cell fusion in interactions with wild-type mates. A number of other classes reveal various degrees of loss of function, e.g., incompatibility with only certain wild-type $B\beta$ alleles, to the end-point of no discernible $B\beta$ function. A number of the latter have also lost all $B\alpha$ activity. Such strains are consequently sterile with all wild

B factors, but they will cross with primary *B*-mutant strains to yield sterile and "flat" progeny (C. A. Raper and Raper, 1973).

Rationalization of the many variations in mating interactions imposed by the primary and secondary mutations of *Bβ* appears to involve at least three functions for the *Bβ* gene: (1) determination of specificity, i.e., the recognition of self *vs.* nonself, (2) control of nuclear migration, and (3) control of hook-cell fusion. These functions reside in separately mutable sites of a complex gene or gene-complex, and one of the functions, the control of nuclear acceptance in migration, is separable by recombination (C. A. Raper and Raper, 1973).

Modifier Mutations. Modifier mutations have only a single common characteristic: they have no expression in the otherwise wild-type homokaryon and are distinguishable only by their disruptive effects in sexual morphogenesis. The catalog of the effects of the 13 classes of modifiers now known is extremely varied and includes, among several others, the suppression of the "flat" phenotype of *B*-mutant strains and $A = B \neq$ heterokaryons, the total inhibition of nuclear migration, the blocking of hook-cell formation, the blocking of hook-cell fusion (C. A. Raper and Raper, 1966), and the prompt establishment of uniformly diploid mycelia in $A \neq B \neq$ matings (Koltin and Raper, 1968; Gladstone, 1972). Certain of the modifiers are dominant to their wild alleles, others are recessive, and the interactions between different mutations are complex, with epistasis of one mutation to another a common feature. In one common type of modifier (a suppressor of "flat" phenotype), crosses between some pairs of noncomplementing mutations yield no wild-type progeny, whereas others regularly give rise to wild-type segregants (Koltin, 1970; Dubovoy, 1973). Several of these mutations are closely linked; others are independent.

The relative frequency of mutations of the three classes described here presents an interesting comparison. Primary mutations of the incompatibility loci are rare, and even after effective mutagenic treatment have been recovered at a frequency of only about 10^{-7}. Furthermore, all primary mutations of the incompatibility loci known to date are of a single general type. It should be noted, however, that the only selective system employed for primary mutations would not have revealed mutations such as the secondaries. Additional types of primary mutations, e.g., to sterility, might be recovered in appropriate selective systems. By contrast, secondary mutations occur in treated material in a total frequency of about 10^{-5}, and they represent no less than 12 distinguishable classes. Modifier mutations occur even more frequently by a factor of about 10 following mutagenic induction, i.e., at a frequency of about 10^{-4}, and they belong to a minimum of 13 known classes.

Biochemical Studies

Protein Specificity. A biochemical basis of the control of development by the incompatibility factors was first sought in a serological comparison of soluble proteins from isogenic mycelia in the different stages of sexual morphogenesis (J. R. Raper and Esser, 1961). The proteins were later separated on polyacrylanide gel by disk electrophoresis, and the banding patterns compared. The banding patterns of homokaryons isogenic except for the *A* and *B* factor were almost identical, but the pattern was markedly different in the dikaryon formed by the mating of the homokaryons. The dikaryon is closely mimicked in banding pattern as well as in morphology by the *A*-mutant–*B*-mutant strain. Thus the coordinated release of control by both *A* and *B* factors either by the interaction of different factors or by mutation results in the appearance of many new bands and the disappearance of others. Correspondingly, when only one or the other factor is active in mutant-*A* or mutant-*B* strains, patterns intermediate between those of the homokaryon and dikaryon are obtained. A type-IV modifier, which blocks the operation of the *B* sequence, however, induces a normal homokaryotic banding pattern in a *B*-mutant strain (Wang and Raper, 1969).

Isozyme Pattern. In a similar study, isozyme patterns were examined from mycelia in the various stages of sexual morphogenesis controlled by the *A* and *B* factors. The banding patterns again were of soluble proteins separated by disk gel electrophoresis. Differences in isozyme patterns in 14 enzymes, i.e., NADH dehydrogenases, NADPH dehydrogenases, a number of NAD- and NADP-dependent dehydrogenases, acid phosphatase, leucine aminopeptidases, and esterases, were correlated with the operation or inactivity of the *A* and *B* sequences of sexual morphogenesis. In only a single instance, phenolases, could no marked differences be correlated with sexual morphogenesis. None of the enzymes examined are thought to be primary effectors of morphogenesis. These results and the large changes in the total soluble-protein spectra outlined above thus seem to indicate that the release of control by the *A* and *B* factors may induce a shift in the classes of genes to be expressed (Wang and Raper, 1970).

Uncoupling of Energy Metabolism in the "Flat" Phenotype. The *B* sequence of sexual morphogenesis has been the object of additional biochemical studies that have revealed two aspects of *B*-factor control: metabolic malfunction and septal disruption. To what extent the two are related is unknown.

Operation of the sequence of morphogenesis controlled by the *B* factor, either through interacting *B* factors or the presence of a primary *B* mutation with the *A* sequence inactive, results in a sick, poorly growing mycelium (J. R. Raper and San Antonio, 1954; Hoffman and Raper, 1971). The basis of this poor growth is a partial uncoupling of energy-yielding and -conserving metabolism, with substrate energy being largely dissipated instead of utilized for growth. Mitochondria isolated from the *B* mutant demonstrate a respiratory system that responds to ADP only to half the extent of mitochondria from a normal homokaryon. This alteration of energy metabolism seems unique from any other known (Hoffman and Raper, 1971, 1972).

In compatible matings, the *B* sequence seems to be activated transiently before the *A* sequence, and energetic uncoupling may play a critical role in sexual morphogenesis.

Later work has shown that the B-factor mutant utilizes glucose in the production of cellular material with an efficiency of about 9 percent of that of wild-type mycelium. The production of ATP by mitochondria isolated from the mutant is equal to that of mitochondria of normal mycelium, however, and mitochondrial ATPase activity appears the same from mycelia of the two types. Other comparative studies show the mutant to achieve approximately the same yield on ethanol and glucose, whereas wild type grows considerably less on ethanol. The mutant is much more sensitive than wild type to temperature and to inhibition by Krebs-cycle intermediaries. Both mutant and wild type are sensitive to the phosphorylation-inhibiting agent DCCD (Hoffman and Raper, 1974).

Degradation of Septa and Nuclear Migration. Nuclear migration occurs during formation of the dikaryon and in the establishment of the $A = B \neq$ heterokaryon. The septa seem to disintegrate to accomodate the passage of migrant nuclei. Partial septa are common in the $A = B \neq$ heterokaryon and *B*-mutant homokaryon and almost certainly constitute a transient phenomenon in dikaryosis (Jersild *et al.*, 1967; Koltin and Flexer, 1969; Niederpruem and Wessels, 1969). It accordingly seems that *B*-factor activity induces septal disruption and that this is a basic controlling mechanism for nuclear migration. High activities of a glucanase for a glucan fraction of the cell wall occur in the $A = B \neq$ heterokaryon and *B*-mutant homokaryon. The activity of this enzyme decreases this glucan fraction in "flat" mycelia, and it is thought that the disrupted septa found in the "flat" phenotype are due in part to this glucanase. Cell-wall preparations with intact septa can be prepared devoid of cytoplasm, and in such preparations from wild homokaryon, the septa are completely broken down upon incubation with this glucanase and chitinase (Janszen and Wessels, 1970). The *B* factor thus appears to control septal breakdown by the induction of glucanase (Niederpruem and

Wessels, 1969; Wessels and Niederpruem, 1967; Wessels, 1969*a,b*, 1971).

Nucleic Acids and Ribosomes.   *Schizophyllum* has been the subject of few studies at the molecular level. Nuclear DNA has been determined to have a G + C content of 61 percent by buoyant density (Villa and Storck, 1968) and 58 percent by chemical analysis (Vanyushin *et al.*, 1960). The former authors found mitochondrial DNA to be 28 percent G + C. Ribosomes have sedimentation constants of 80S, with 60S, and 40S subunits, and the ribosomes have also been found in polysomes (Leary *et al.*, 1969). Ribosomal RNA's having sedimentation constants of 25S and 18S with molecular weights of 1.3×10^6 and 0.73×10^6 daltons have also been measured (Lovett and Haselby, 1971).

Other Aspects

Although most of the genetic work with *Schizophyllum* has dealt with the incompatibility system, other lines of inquiry have yielded significant results.

Control of Recombination.   In the work that delineated the two-locus structure of the *A* and *B* factors, great differences were found in the frequency of $A\alpha-A\beta$ recombination (J. R. Raper *et al.*, 1958*a*). Subsequent work has shown the variation in frequency of recombination in the $A\alpha-A\beta$ and $B\alpha-B\beta$ regions, as well as other intergenic regions both linked to and independent of the *A* and *B* factors, to be under genetic control. The imposed control affects specific restricted regions and is of the type designated "fine control" by Simchen and Stamberg (1969*a*). The major features of these controls have been established.

1. Controlled regions and controlling elements are distinct and usually assort independently (Stamberg, 1968, 1969*a*).
2. Different regions, even adjacent segments in the same linkage group, have different controlling elements (Simchen and Stamberg, 1969*b*).
3. The effects of temperature on recombination are highly region-specific and genome-specific, e.g., recombination at 32°C for $A\alpha-A\beta$ in different crosses may be higher, lower, or the same as at 22° (Stamberg, 1968; Stamberg and Simchen, 1970).
4. The controlling elements are typically polygenic, but the degree of heterogeneity and the pattern of segregation are cross-specific (Simchen, 1967*a*; Stamberg, 1968).

5. A controlling element may affect two distinct regions either similarly or differently, e.g., control of recombination in $A\alpha$–$B\beta$ and $B\alpha$–$B\beta$ is independent in crosses of strains from different locations but negatively correlated in a single population (Simchen and Stamberg, 1969b; Schaap and Simchen, 1971).

6. Low frequency of recombination between the component loci of A and B factors is dominant to high frequency. Low frequency appears to have a selective advantage, as a majority of wild-type strains recombine at low frequency (Stamberg, 1969b; Simchen, 1967a; Simchen and Connolly, 1968).

Controlling and controlled elements [cf. (1) above] for recombination between $B\alpha$ and $B\beta$ of the B factor have recently been reported. Two alleles of a gene, *B-rec-1*, located 9 centimorgans from $B\beta$, determine high and low levels of recombination in the segment (Koltin and Stamberg, 1973). Control of the exact level of recombination frequency has been found to reside in an element located within the affected region. This appears to consist of a number of sites with additive effects and is responsive to the outlying controlling gene, *B-rec-1* (Stamberg and Koltin, 1973).

Somatic Recombination and Diploidy. Since the first demonstration of somatic recombination in the higher fungi (Quintanilha, 1939), much effort and controversy have been devoted to the underlying mechanism. Until recently, the objects of study were dikaryon–homokaryon, or *di–mon* matings (Papazian, 1950), in which the homokaryons served as selective agents for recombinant nuclei (see J. R. Raper, 1966, for review). No single mechanism could be identified as responsible in all cases, but certain findings were of interest: (1) recombination between linked loci approached the frequency of meiotic recombination (Crowe, 1960), (2) markers from all three original genomes were often associated in the derived recombinants (Ellingboe and Raper, 1962b), and (3) the recombinant nuclei often differed from the original homokaryon only in the A and B factors, one of which was derived from either member of the dikaryon ("specific transfer") (Ellingboe, 1963). The results of none of the studies with *Schizophyllum* appeared to fit the requirements of "classic" parasexuality (Pontecorvo, 1956).

More recently, somatic recombination has been studied in $A=B=$ and $A\neq B\neq$ diploids, and the mechanisms in the two cases appear to be quite different. In the $A=B=$ diploid, haploidization via aneuploidy of rare diploid nuclei (Middleton, 1964) and mitotic crossing over (Mills and Ellingboe, 1971) represent aspects of a routine parasexual cycle. In the $A\neq B\neq$ diploid, no evidence of breakdown via aneuploidy to haploidy has been found, but mitotic recombination results in the occurrence of partial homozygotes (Gladstone, 1972). Of more interest and possible significance, however, is the pattern of breakdown of the diploid in diploid–homokaryon matings. $A{\neq}B{\neq}$ diploids are compatible with all homokaryons, and dikaryons are readily formed; these can usually be brought to fruiting, but they yield only haploid progeny. The diploid nucleus is unstable and breaks down in growth and fruitification—sometimes prior to and sometimes during the development of the fruiting body—in a process that cannot be distinguished by its products from meiosis (Gladstone, 1972).

Diploidy. Diploidy itself is of some interest in a form such as *Schizophyllum* that has evolved a stable dikaryon as a diploid substitute (Raper and Flexer, 1970). Diploidy resulting from rare fusions of haploid nuclei as known in other fungi are known to occur sporadically in $A{\neq}B{=}$ and $A{=}B{=}$ heterokaryons (Parag and Nachman, 1966; Mills and Ellingboe, 1969).

Genetically determined diploidy involving most or all nuclei in $A{\neq}B{\neq}$ matings was first reported to be due to a double dose of a recessive mutation, dik^- (Koltin and Raper, 1968), but later work has shown dik^- to be dominant (Gladstone, 1972). Diploids caused by dik^- differ from previously known sporadic diploids in one important respect: the morphology is that of the homokaryon rather than that of the corresponding heterokaryon.

Fruiting. Normal fruiting in *Schizophyllum* depends upon the prior process of dikaryosis, but the genetics of the two developmental phases are quite distinct. Once the dikaryon is formed, it may or may not fruit depending upon genetic factors other than A and B factors. Two independent studies (J. R. Raper and Krongelb, 1958; Jürgens, 1958) concluded that the competence to fruit as well as the time and abundance of fruiting were determined as polygenic characters. Otherwise, a number of single-gene characters (mostly mutations) are known that alter the form of the fruiting body (Zattler, 1924; J. R. Raper and Krongelb, 1958), prevent the initiation of fruiting (Perkins and Raper, 1970), and allow normal fruiting in total darkness (Jürgens, 1958).

Fruiting of homokaryons, however, is no rare occurrence, but it seems to be of questionable biological significance, as homokaryotic fruiting bodies are usually few, tardily produced, and sporulate sparsely, and only a few of the spores germinate. Only certain homokaryons are competent to fruit, and the genetic basis for this appears to be distinct from that of dikaryotic fruiting (J. R. Raper and Krongelb, 1958).

Homokaryotic fruiting may be induced in certain strains by an extrinsic factor, FIS (Fruiting-inducing substance), originally from an im-

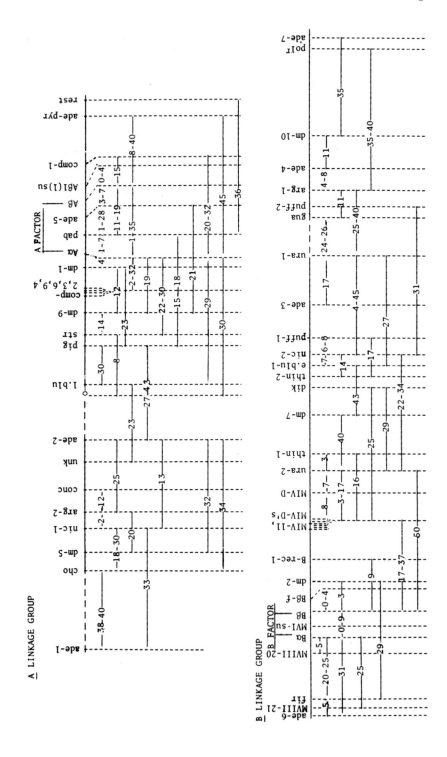

Figure 2. The A- and B-linkage groups of Schizophyllum. *Ranges in frequency of recombination in different crosses are as indicated wherever two values are given.*

Loci marked with auxotrophic mutations are conventionally designated arg-1, ade-5, ura-2, unknown (unk), etc.; morphological mutations are early blue (e.blu), late blue (l.blu) (indigo), chocolate (cho), compact (comp), chocolate (cho) (indigo), concentric (conc), dome (dm)-1, -10, fir, pigmy (pig), puff-1,-2, restricted (rest), streak (str), thin-1,-2. Other mutations are dikaryon (dik) specifying dikaryon by wild allele–vegetative diploid by dominant mutant allele, polymyxin resistance (polr); A β1 (1)-su, suppressor of mutant allele Aβ 1(1); Bβ-f, blocker of Bβ-regulated B sequence; M IV-11, -D's, and M VIII-20,-21, modifier mutations affecting B-factor controlled morphogenesis; and M VI-su, suppressor of M VI modifier mutation; B-rec-I is a gene controlling recombination in the B factor. Linkages indicated by dashes require confirmation.

Linkage maps prepared by Carlene A. Raper.

perfect fungus, *Hormodendrum* (Leonard and Dick, 1968), but later obtained from fruiting bodies of *Schizophyllum* and the commercial mushroom, *Agaricus bisporus*. Inducibility vs. noninducibility was found to depend upon alleles of a single gene (Leonard and Raper, 1969).

Phenoloxidase activity has been found to be correlated with fruiting competence of dikaryons and homokaryons (Leonard, 1972). For example, *B*-mutant homokaryons have no phenoloxidases and do not fruit; a second mutation, a type-V modifier (C. A. Raper and Raper, 1966), results in enzyme activity and copious fruiting; and a third mutation, *cohesiveless* (Perkins and Raper, 1970; Leonard, 1972), "turns off" both enzymatic activity and fruiting.

Morphological Mutations. The aging, sickly $A=B\neq$ heterokaryon proves to be an effective selective system for numerous morphological mutations in addition to the modifier mutations described above (J. R. Raper *et al.*, 1958*c*; Dick and Raper, 1961). A majority of these mutations affect the mating response, many are slow-growing, and a number have been shown to effect a quantitative alteration of cell-wall composition (Wang *et al.*, 1968). Strains carrying one such mutation, *blue*, deposit indigotin; this is interpreted as a process of detoxification to remove indole that otherwise would accumulate to lethal levels (Miles *et al.*, 1956; Swack and Miles, 1960).

Variation and Population Genetics. A limited number of investigations relating to variability and other aspects of population genetics has shown *Schizophyllum* to be a suitable object for biometrical analyses (Simchen and Jinks, 1964; Simchen, 1966*a,b*; Brasier, 1970). A number of characters, such as growth rate of homokaryons and dikaryons, time and abundance of fruiting, and morphology of fruiting bodies, have been studied in detail and correlations sought among the several traits. These characters are typically polygenic and display continuous variation. Positive correlations exist between certain characters, whereas other pairs are independent. An analysis of growth rates of dikaryons constituted of sibling vs. nonsibling homokaryons indicated that the alleles of the underlying polygenic systems had probably evolved independently within distinct gene pools (Simchen, 1967*b*).

Mapping. No intensive effort has been devoted to establishing linkage maps *per se* for *Schizophyllum*. This together with the central emphasis on the *A* and *B* incompatibility factors has restricted mapping to two extensive linkage groups that contain the two independently assorting incompatibility factors and to a few fragmentary groups of two to a few genes. These latter plus a large number of unmapped loci approximately equal those in the *A* and *B* linkage groups.

Provisional maps for the *A* and *B* linkage groups as given here (Figure 2) represent the most probable ordering of loci that can be made with available data. The linkage distances, however, reflect the extreme variability of frequency of recombination in this material as reported by different workers, recombination frequency being highly cross specific and temperature sensitive (see above). Where multiple values have been reported, the range is given.

Selected strains that carry genetic characters of interest have been maintained, and these are generally available upon request from the various workers in the field.

Literature Cited

Ahmad, S. S. and P. G. Miles, 1970   Hyphal fusions in the wood-rotting fungus *Schizophyllum commune.* I. The effects of incompatibility factors. *Genet. Res.* **15**:19–28.

Bauch, R., 1930   Die genetischen Grundlagen der multipolaren Sexualität der Pilze. *Züchter.* **2**:174–182.

Bensaude, M. 1918   Recherches sur le cycle évolutif et la sexualité chex les Basidiomycêtes. Thesis. Nemours.

Bonitati, J., P. G. Miles and W. B. Elliot, 1970   New techniques for sample preparation for respirometry of filamentous fungi. *Plant Physiol.* **45**:708–711.

Brasier, C. M., 1970   Variation in a natural population of *Schizophyllum commune.* *Am. Nat.* **104**:191–205.

Brefeld, O., 1877   *Untersuchungen über Schimmelpilze,* Vol. 3, Arthur Felix, Leipzig.

Brunswik, H., 1924   Untersuchungen über die Geschlechts- und Kernverhältnisse bei der Hymenomyzetengattung, *Coprinus, Bot. Abh. K. Goebel* **5**:1–152.

Buller, A. H. R., 1931   *Researches on Fungi, IV,* Longman's Green & Co., London.

Crowe, L. K., 1960   The exchange of genes between nuclei of a dikaryon. *Heredity* **15**:397–405.

Crowe, L. K., 1963   Competition between compatible nuclei in the establishment of a dikaryon in *Schizophyllum commune. Heredity* **18**:525–533.

Day, P. R., 1960   Mutations affecting the *A* mating-type locus of *Coprinus lagopus Heredity* **15**:457.

Day, P. R., 1963   Mutations affecting the *A* mating-type locus in *Coprinus lagopus.* *Genet. Res.* **4**:55–65.

de Vries, O. M. H. and J. G. H. Wessels, 1972   Release of protoplasts of *Schizophyllum commune* by a lytic enzyme preparation from *Trichoderma viride. J. Gen. Microbiol.* **73**:13–22.

Dick, S. and J. R. Raper, 1961   Origin of expressed mutations in *Schizophyllum commune. Nature (Lond.)* **189**:81–82.

Dubovoy, C., 1973   A class of genes controlling *B*-factor regulated development in *Schizophyllum.* Dissertation, Harvard University, Cambridge, Mass.

Ellingboe, A. H., 1963   Illegitimacy and specific factor transfer in *Schizophyllum commune. Proc. Natl. Acad. Sci. USA* **49**:286–292.

Ellingboe, A. H. 1964 Nuclear migration in dikaryotic-homokaryotic matings in *Schizophyllum commune. Am. J. Bot.* **51**:133–139.

Ellingboe, A. H., and J. R. Raper, 1962a The Buller Phenomenon in *Schizophyllum commune*: nuclear selection in fully compatible dikaroytic-homokaryotic matings. *Am. J. Bot.* **49**:454–459.

Ellingboe, A. H. and J. R. Raper, 1962b Somatic recombination in *Schizophyllum commune. Genetics* **47**:85–98.

Giesy, R. M. and P. R. Day, 1965 The septal pores of *Coprinus lagopus* (Fr.) *sensu* Buller in relation to nuclear migration. *Am. J. Bot.* **52**:287–293.

Girbardt, M., 1965 Perspektiven elektronischer Zellforschung. *Naturwiss. Rundsch.* **18**:345–349.

Girbardt, M., 1968 Ultrastructure and dynamics of the moving nucleus. Aspects of cell motility. *XXII Symposium of the Society for Experimental Biology, Oxford,* pp. 249–259, Cambridge University Press, London.

Gladstone, P. R., 1972 Genetic studies on heritable diploidy in *Schizophyllum.* Dissertation, Harvard University, Cambridge, Mass.

Hoffman, R. M. and J. R. Raper, 1971 Genetic restriction of energy conservation in *Schizophyllum. Science (Wash., D.C.)* **171**:418–419.

Hoffman, R. M. and J. R. Raper, 1972 Lowered respiratory response to ADP of mitochondria isolated from a mutant-*B* strain of *Schizophyllum commune. J. Bacteriol.* **110**:780–781.

Hoffman, R. M. and J. R. Raper, 1974 Genetic impairment of energy conservation in *Schizophyllum*: Efficient mitochondria in energy-starved cells. *J. Gen. Microbiol.,* unpublished.

Janszen, F. H. A. and J. G. H. Wessels, 1970 Enzymatic dissolution of hyphal septa in a basidiomycete. *Antonie Van Leeuwenhoek J. Microbiol. Serol.* **36**:255–257.

Jersild, R., S. Mishkin and D. J. Niederpruem, 1967 Origin and ultrastructure of complex septa in *Schizophyllum commune* development. *Arch. Mikrobiol.* **57**:20–32.

Jürgens, C., 1958 Physiologische und genetische Untersuchungen über die Fruchtkorperbildung bei *Schizophyllum commune. Arch. Mikrobiol.* **30**:409–432.

Kniep, H., 1920 Über morphologische und physiologische Geschlechtsdifferenzierung. (Untersuchungen an Basidiomyzeten.) *Verh. Phys. Med. Ges. Würzburg,* **46**:1–18.

Kniep, H., 1922 Über. Geschlechtsbestimmung und Reduktionsteilung. *Verh. Phys. Med. Ges. Würzburg* **47**:1–28.

Kniep, H., 1930 Uber Selektionswirkungen in fortlaufenden Massenaussaaten von *Schizophyllum. Z. Bot.* **23**:510–536.

Koltin, Y., 1968 The genetic structure of the incompatibility factors of *Schizophyllum commune*: comparative studies of primary mutations in the *B* factor. *Mol. Gen. Genet.* **102**:196–203.

Koltin, Y., 1969 The structure of the incompatibility factors of *Schizophyllum commune*: class II factors. *Mol. Gen. Genet.* **103**:380–384.

Koltin, Y., 1970 Studies on mutations disruptive to nuclear migration in *Schizophyllum commune. Mol. Gen. Genet.* **106**:155–161.

Koltin, Y. and A. S. Flexer, 1969 Alterations in nuclear distribution in *B*-mutant strains of *Schizophyllum commune. J. Cell. Sci.* **4**:739–749.

Koltin, Y. and J. R. Raper, 1966 *Schizophyllum commune*: new mutations in the *B* incompatibility factor. *Science (Wash., D.C.)* **154**:510–511.

Koltin, Y. and J. R. Raper, 1967a The genetic structure of the incompatibility factor of *Schizophyllum commune*: three functionally distinct classes of *B* factors. *Proc. Natl. Acad. Sci. USA* **58**:1220–1226.

Koltin, Y. and J. R. Raper, 1967*b*   The genetic structure of the incompatibility factors of *Schizophyllum commune:* resolution of class III *B* factors. *Mol. Gen. Genet* **100**:275–282.

Koltin, Y. and J. R. Raper, 1968   Dikaryosis: Genetic determination in *Schizophyllum. Science (Wash., D.C.)* **160**:85–86.

Koltin, Y. and J. Stamberg, 1973   Genetic control of recombination in *Schizophyllum commune:* location of a gene controlling *B*-factor recombination. *Genetics* **74**:55–62.

Koltin, Y., J. R. Raper and G. Simchen, 1967   Genetic structure of the incompatibility factors of *Schizophyllum commune:* the *B* factor. *Proc. Natl. Acad. Sci. USA* **57**:55–63.

Koltin, Y., J. Stamberg, and P. A. Lemke, 1972   Genetic structure and evolution of the incompatibility factors in higher fungi. *Bacteriol. Rev.* **36**:156–171.

Leary, J. V., A. J. Morris and A. H. Ellingboe, 1969   Isolation and functional ribosomes and polysomes from lyophilized fungi. *Biochim. Biophys. Acta* **182**:113–120.

Leonard, T. J., 1971   Phenoloxidase activity and fruiting body formation in *Schizophyllum commune. J. Bacteriol.* **106**:162–167.

Leonard, T. J., 1972   Phenoloxidase activity in mycelia carrying modifier mutations that affect sporocarp development in *Schizophyllum commune. J. Bacteriol.* **111**:292–293.

Leonard, T. J. and S. Dick, 1968   Chemical induction of haploid fruiting in *Schizophyllum commune. Proc. Natl. Acad. Sci. USA* **59**:745–751.

Leonard, T. J. and J. R. Raper, 1969   *Schizophyllum commune:* gene controlling induced haploid fruiting. *Science (Wash., D.C.)* **165**:190.

Lovett, J. S. and J. A. Haselby, 1971   Molecular weights of the ribosomal ribonucleic acid of fungi. *Arch. Mikrobiol.* **80**:191–204.

Matthews, T. R. and D. J. Niederpruem, 1972   Differentiation in *Coprinus lagopus.* I. Control of fruiting and cytology of initial events. *Arch. Mikrobiol.* **87**:257–268.

Middleton, R. B., 1964   Sexual and somatic recombination in common-*AB* heterokaryons of *Schizophyllum commune. Genetics* **50**:701–710.

Miles, P. G., H. Lund and J. R. Raper, 1956   The identification of indigo as a pigment produced by a mutant strain of *Schizophyllum commune. Arch. Biochem. Biophys.* **62**:1–5.

Mills, D. I. and A. H. Ellingboe, 1969   A common-*AB* diploid of *Schizophyllum commune. Genetics* **62**:271–279.

Mills, D. I. and A. H. Ellingboe, 1971   Somatic recombination in the common-*AB* diploid of *Schizophyllum commune. Mol. Gen. Genet.* **110**:67–76.

Moore, R. T. and J. H. McAlear, 1962   Fine structure of Mycota. 7. Observations on septa of Ascomycetes and Basidiomycetes. *Am. J. Bot.* **49**:86–94.

Niederpruem, D. J., 1963   Role of carbon dioxide in the control of fruiting of *Schizophyllum commune. J. Bacteriol.* **85**:1300–1308.

Niederpruem, D. J. and D. P. Hackett, 1961   Cytochrome system in *Schizophyllum commune. Plant Physiol.* **36**:79–84.

Niederpruem, D. J. and J. G. H. Wessels, 1969   Cytodifferentiation and morphogenesis in *Schizophyllum commune. Bacteriol. Rev.* **33**:505–535.

Papazian, H. P., 1950   Physiology of the incompatibility factors in *Schizophyllum commune. Bot. Gaz.* **112**:143–163.

Papazian, H. P., 1951   The incompatibility factors and a related gene in *Schizophyllum commune. Genetics* **36**:441–459.

Parag, Y., 1962   Mutations in the *B* incompatibility factor in *Schizophyllum commune. Proc. Natl. Acad. Sci. USA* **48**:743–750.

Parag, Y. and Y. Koltin, 1971   The structure of the incompatibility factors of *Schizophyllum commune:* constitution of the class III factors. *Mol. Gen. Genet.* **112:**43–48.

Parag, Y. and B. Naçhman, 1966   Diploidy in the tetrapolar heterothallic Basidiomycete *Schizophyllum commune. Heredity* **21:**151–154.

Parag, Y. and J. R. Raper, 1960   Genetic recombination in a common-B cross of *Schizophyllum commune. Nature (Lond.)* **188:**765–766.

Perkins, J. H., 1969   Morphogenesis in *Schizophyllum commune.* I. Effects of white light. *Plant. Physiol.* **44:**1706–1711.

Perkins, J. H. and S. A. Gordon, 1969   Morphogenesis in *Schizophyllum commune.* II. Effects of monochromatic light. *Plant. Physiol.* **44:**1712–1716.

Perkins, J. H. and J. R. Raper, 1970   Morphogenesis in *Schizophyllum commune.* III. A mutation that blocks initiation of fruiting. *Mol. Gen. Genet.* **106:**151–154.

Pontecorvo, G., 1956   The parasexual cycle in fungi. *Annu. Rev. Microbiol.* **10:**393–400.

Quintanilha, A., 1935   Cytologie et génétique de la sexualité chez les Hymenomycètes. *Bol. Soc. Broteriana* **10:**289–332.

Quintanilha, A., 1937   Contribution a l'étude génétique du phénomèe de Buller. *C. R. Hebd. Seances Acad. Sci. Ser. D Sci. Nat.* **205:**745–747.

Quintanilha, A., 1939   Étude génétique du phenomene de Buller. *Bol. Soc. Broteriana* **13:**425–486.

Raper, C. A. and J. R. Raper, 1964   Mutations affecting heterokaryosis in *Schizophyllum commune. Am. J. Bot.* **51:**503–513.

Raper, C. A. and J. R. Raper, 1966   Mutations modifying sexual morphogenesis in *Schizophyllum. Genetics* **54:**1151–1168.

Raper, C. A. and J. R. Raper, 1973   Mutational analysis of a regulatory gene for morphogenesis in *Schizophyllum. Proc. Natl. Acad. Sci. USA* **70:**1427–1431.

Raper, J. R., 1963   Device for the isolation of spores. *J. Bacteriol.* **86:**342–344.

Raper, J. R., 1966   *Genetics of Sexuality in Higher Fungi,* Ronald Press, New York.

Raper, J. R. and K. Esser, 1961   Antigenic differences due to the incompatibility factors in *Schizophyllum commune. Z. Vererbungsl* **92:**439–444.

Raper, J. R. and A. S. Flexer, 1970   The road to diploidy with emphasis on a detour. In *Organization and Control in Prokaryotic and Eukaryotic Cells,* edited by H. P. Charles and B. C. J. G. Knight, pp. 401–432, Cambridge University Press, London.

Raper, J. R. and E. A. Hyatt, 1963   Modified press for disruption of microorganisms. *J. Bacteriol.* **85:**712–713.

Raper, J. R. and G. S. Krongelb, 1958   Genetic and environmental aspects of fruiting in *Schizophyllum commune* Fr. *Mycologia* **50:**707–740.

Raper, J. R. and C. A. Raper, 1968   Genetic regulation of sexual morphogenesis in *Schizophyllum commune. J. Elisha Mitchell Sci. Soc.* **84:**267–273.

Raper, J. R. and M. Raudaskoski, 1968   Secondary mutations at the $B\beta$ locus of *Schizophyllum. Heredity* **23:**109–117.

Raper, J. R. and J. P. San Antonio, 1954   Heterokaryotic mutagenesis in Hymenomycetes. I. Heterokaryosis in *Schizophyllum commune. Am. J. Bot.* **41:**69–86.

Raper, J. R., M. G. Baxter and R. B. Middleton, 1958a   The genetic structure of the incompatibility factors in *Schizophyllum commune. Proc. Natl. Acad. Sci. USA* **44:**889–900.

Raper, J. R., G. S. Krongelb and M. G. Baxter, 1958b   The number and distribution of incompatibility factors in *Schizophyllum commune. Am. Nat.* **92:**221–232.

Raper, J. R., J. P. San Antonio and P. G. Miles, 1958c   The expression of mutations in common-*A* heterokaryons of *Schizophyllum commune*. *Z. Vererbungsl* **89**:540–558.

Raper, J. R., M. G. Baxter and A. H. Ellingboe, 1960   The genetic structure of the incompatibility factors of *Schizophyllum commune*: the *A* factor. *Proc. Natl. Acad. Sci. USA* **46**:833–842.

Raper, J. R., D. H. Boyd and C. A. Raper, 1965   Primary and secondary mutation at the incompatibility loci in *Schizophyllum*. *Proc. Natl. Acad. Sci. USA* **53**:1324–1332.

Raudaskoski, M., 1970   A new secondary *Bβ* mutation in *Schizophyllum* revealing functional differences in wild *Bβ* alleles. *Hereditas* **64**:259–266.

Raudaskoski, M., 1970b   Occurrence of microtubules and microfilaments, and origin of septa in dikaryotic hyphae of *Schizophyllum commune*. *Protoplasma* **70**:415–422.

Raudaskoski, M., 1972a   Occurrence of microtubules in the hyphae of *Schizophyllum commune* during intercellular nuclear migration. *Arch. Mikrobiol.* **86**:91–100.

Raudaskoski, M., 1972b   Secondary mutations at the *Bβ* incompatibility locus and nuclear migration in the basidiomycete *Schizophyllum commune*. *Hereditas* **72**:175–182.

Raudaskoski, M., 1973   Light- and electron-microscope study of unilateral mating between a secondary mutant and a wild-strain of *Schizophyllum commune*. *Protoplasma* **76**:35–48.

Schaap, T. and G. Simchen, 1971   Inbreeding and the genetic control of recombination in a natural population of *Schizophyllum commune*. *Genetics* **68**:67–75.

Sicari, L. M. and A. H. Ellingboe, 1967   Microscopical observations of initial interactions in various matings of *Schizophyllum commune* and *Coprinus lagopus*. *Am. J. Bot.* **54**:437–439.

Simchen, G., 1966a   Monokaryotic variation and haploid selection in *Schizophyllum commune*. *Heredity* **21**:241–263.

Simchen, G., 1966b   Fruiting and growth rate among single wild isolates of *Schizophyllum commune*. *Genetics* **53**:1151–1165.

Simchen, G., 1967a   Genetic control of recombination and the incompatibility system in *Schizophyllum commune*. *Genet. Res.* **9**:195–210.

Simchen, G., 1967b   Independent evolution of a polygenic system in isolated populations of the fungus *Schizophyllum commune*. *Evolution* **21**:310–315.

Simchen, G. and V. Connolly, 1968   Changes in recombination frequency following inbreeding in *Schizophyllum*. *Genetics* **58**:319–326.

Simchen, G. and J. L. Jinks, 1964   The determination of dikaryotic growth-rate in the Basidiomycete, *Schizophyllum commune*: a biometrical analysis. *Heredity* **19**:629–649.

Simchen, G. and J. Stamberg, 1969a   Fine and course control of genetic recombination. *Nature (Lond.)* **222**:329–332.

Simchen, G. and J. Stamberg, 1969b   Genetic control of recombination in *Schizophyllum commune*: specific and independent regulation of adjacent and non-adjacent chromosomal regions. *Heredity* **24**:369–381.

Snider, P. J., 1963a   Estimation of nuclear ratios directly from heterokaryotic mycelia in *Schizophyllum*. *Am. J. Bot.* **50**:255–262.

Snider, P. J., 1963b   Genetic evidence for nuclear migration in Basidiomycetes. *Genetics* **48**:47–55.

Snider, P. J., 1968   Nuclear movements in *Schizophyllum*. In *Aspects of Cell Motility*. *XXII Symposium of the Society of Experimental Biology*, pp. 261–283, Cambridge University Press, London.

Snider, P. J. and J. R. Raper, 1958 Nuclear migration in the basidiomycete *Schizophyllum commune. Am. J. Bot.* **45**:538–546.

Stamberg, J., 1968 Two independent gene systems controlling recombination in *Schizophyllum commune. Mol. Gen. Genet.* **102**:221–228.

Stamberg, J., 1969a Genetic control of recombination in *Schizophyllum commune:* separation of the controlled and controlling loci. *Heredity* **24**:306–309.

Stamberg, J. 1969b Genetic control of recombination in *Schizophyllum commune:* the occurrence and significance of natural variation. *Heredity* **24**:361–368.

Stamberg, J. and Y. Koltin, 1971 Selectively recombining *B* incompatibility factors of *Schizophyllum commune. Mol. Gen. Genet.* **113**:157–165.

Stamberg, J. and Y. Koltin, 1972 The organization of the incompatibility factors in higher fungi: The effects of structure and symmetry on breeding. *Heredity* **30**:15–26.

Stamberg, J. and Y. Koltin, 1973 Genetic control of recombination in *Schizophyllum commune:* evidence for a new type of regulatory site. *Genet. Res.* **22**:101–111.

Stamberg, J. and G. Simchen, 1970 Specific effects of temperature on recombination in *Schizophyllum commune. Heredity* **25**:41–52.

Storck, R. and C. J. Alexopoulos, 1970 Deoxyribonucleic acid of fungi. *Bacteriol. Rev.* **34**:126–154.

Swack, N. S. and P. G. Miles, 1960 Conditions affecting growth and indogotin production by strain 130 of *Schizophyllum commune. Mycologia* **52**:574–583.

Swiezynski, K. M. and P. R. Day, 1961 Heterokaryon formation in *Coprinus lagopus. Genet. Res.* **1**:114–128.

Vanyushin, B. F., A. N. Belozersky and S. L. Bogdanova, 1960 A comparative study of the nucleotide composition of ribonucleic acid and deoxyribonucleic acid in some fungi and myxomycetes. *Dokl. Akad. Nauk. SSSR Ser Biol. Sci. Sect. (Engl. Transl.)* **134**:1222–1225.

Villa, V. D. and R. Storck, 1968 Nucleotide composition of nuclear and mitochondrial deoxyribonucleic acid of fungi. *J. Bacteriol.* **96**:184–190.

Wang, C.-S. and J. R. Raper, 1969 Protein specificity and sexual morphogenesis in *Schizophyllum commune. J. Bacteriol.* **99**:291–297.

Wang, C.-S. and J. R. Raper, 1970 Isozyme-patterns and sexual morphogenesis in *Schizophyllum. Proc. Natl. Acad. Sci. USA* **66**:882–889.

Wang, C.-S., M. N. Schwalb and P. G. Miles, 1968 A relationship between cell-wall composition and mutant morphology in the basidiomycete *Schizophyllum commune. Can. J. Microbiol.* **14**:809–811.

Wessels, J. G. H., 1969a Biochemistry of sexual morphogenesis in *Schizophyllum commune:* effects of mutations affecting the incompatibility system on cell-wall metabolism. *J. Bacteriol.* **98**:697–704.

Wessels, J. G. H., 1969b A β-1,6-glucan glucanohydrolase involved in hydrolysis of cell-wall glucan in *Schizophyllum commune. Biochim. Biophys. Acta* **178**:191–193.

Wessels, J. G. H., 1971 Cell wall metabolism and morphogenesis in *Schizophyllum. Int. Congr. Microbiol. (Mexico)* **10**:141–146.

Wessels, J. G. H. and D. J. Niederpruem, 1967 Role of a cell-wall glucan-degrading enzyme in mating of *Schizophyllum commune. J. Bacteriol.* **94**:1594–1602.

Whitehouse, H. L. K., 1949 Multiple-allelomorph heterothallism in the fungi. *New Phytol.* **48**:212–244.

Zattler, F., 1924 Verebungstudien an Hutpilzen (Basidiomyzeten). *Z. Bot.* **16**:433–499.

33

Coprinus

Jean Louis Guerdoux

The existence of a dikaryotic phase during which two haploid nuclear types coexist in the same cytoplasm is one of the characteristics of most Basidiomycetes. This phase is very stable but it can be interrupted by various experimental procedures, thus leading to cells having haploid nuclei of only one type in a mixed cytoplasm. Such a system can be very useful for the study of nuclear–cytoplasmic interactions since it offers an opportunity to compare the behavior of genetic material located in different nuclei, during the dikaryotic phase, to that of the same material enclosed in a single nucleus in diploid cells.

The study of *Coprinus radiatus* was initiated in 1958. This species was chosen among other Hymenomycetes mainly because of its ability to produce fruiting bodies rapidly and reliably.

The Life Cycle of *Coprinus radiatus*

A spore of *C. radiatus* will germinate and give rise to a mycelium containing a single type of haploid nuclei (called a haploid monokaryotic mycelium). The number of nuclei in a cell varies from one to ten. However this mycelium is also capable of producing, during vegetative growth, uninucleated cells called oïdia. When two "compatible" monokaryotic mycelia meet and fuse, a dikaryotic mycelium results, with one nucleus of each haploid type per cell.

Jean Louis Guerdoux—Centre de Génétique Moléculaire, Centre National de la Recherche Scientifique, Gif sur Yvette, France.

Compatibility is controlled by two independent allelic series, *A* and *B*. Two haploid monokaryotic mycelia give rise to a dikaryon only when they are heteroallelic for both loci. Wild strains covering a wide variety of alleles at both the *A* and the *B* loci have been collected, thus making *C. radiatus* a good material for the study of incompatibility phenomena. Homothallic strains have also been found.

Dikaryotization of a haploid monokaryotic culture can be obtained by starting with a single nucleus. This is possible because the oïdia, although unable to germinate, can function in fertilization. The "invading" nucleus multiplies rapidly until a numerical equilibrium between the two types of nuclei is reached. The equilibrium is then maintained through synchronous division. The dikaryotic thalli can be differentiated from the monokaryotic ones macroscopically by their radiating growth habit and microscopically by the presence of clamp connections. The formation of carpophores takes place after 11 days of growth under optimal conditions.

The diploid phase is reduced to the young basidium in which meiosis occurs. The cytological study of meiosis in *C. radiatus* is rather difficult because of the small size of the chromosomes. Nevertheless, some interesting conclusions have been reached from preliminary studies. After 11 days, the carpophores contain 4-spore tetrads which can be collected spore-by-spore with the help of a micromanipulator. One can easily collect up to 200 tetrads a day. Random spore sampling can be done simply by laying the carpophores on a glass slide; after approximately one hour the spores which have fallen from the hymenal gills can be collected. One can readily obtain up to 10^7 spores in this manner, a number suitable for genetic fine-structure analysis.

Spores germinate easily under suitable conditions, and after about three hours a "germinative vesicle" can be observed. Its nucleus is visible after four hours. Twelve hours later, a septate mycelium having about twenty nuclei can be observed; this has the typical morphology of haploid monokaryotic mycelium.

Methodology

Most of the techniques employed are described in G. Prévost's thesis (Prevost, 1962).

Growth Conditions. *C. radiatus* can be grown on synthetic or natural media. Both liquid and solid media yield large amounts mycelia. The inoculum generally used consists of small pieces of mycelium, about 0.2-mm thick, cut out from the thallus and picked up with a sterile platinum needle.

Growth Measurements. A rapid determination of amount of growth can be made by measuring the diameter of the thallus growing on solid medium. In liquid medium the dry weight of the mycelium will have to be determined; this requires that the mycelial mass be washed first with distilled water then with 95-percent alcohol before drying. One can also estimate the amount of extractable protein.

Extraction of Cellular Components. Cellular extracts for growth measurements or enzymatic studies are routinely obtained by suspending the mycelium in a minimal volume of phosphate buffer and cooling the mixture to a temperature of −60°C. The preparation is then thawed and vortexed at high speed; this ensures the rupture of the hyphal filaments. The homogenate is then centrifuged. The supernatant contains 10–25 mg of protein per ml, depending on the strain and growth conditions employed. The DNA concentration in the supernatant is about 30 μg/ml.

Mating Procedures. Dikaryotization occurs when two compatible mycelia come in contact; it is also possible to obtain dikaryons by using a vegetative haploid mycelium and oïdia from another compatible mycelium. Oïdia are harvested by washing a monokaryotic culture grown on a synthetic glucose medium. Mycelial fragments are disposed of by centrifugation or by filtration. Fertilization is obtained by placing a microdrop of the suspension on the border of a thallus of compatible mating type.

Dikaryotic mycelium can be unequivocally recognized by the presence of clamp connections. If required, the two haploid constitutants of a vegetative dikaryotic mycelium can be separated and recovered; this is performed by disrupting the mycelium in a vortex blender.

Fruiting bodies are obtained from dikaryotic mycelium grown on horse dung at 22°C.

Mutants. Most of the mutants have been obtained by irradiation of spores with x rays (Prevost, 1962) or uv light (Gans and Masson, 1969), or by exposure to nitrosoguanidine or nitrous acid (Babillot, 1963). After the mutagenic treatment, the spores are either spread directly on a natural germination medium or induced to germinate in liquid if they are to be plated on a synthetic medium. Screening for biochemical mutants is tedious since it requires individual testing of spores on different media. However, by filtering a spore suspension that has been induced to germinate, it is possible to enrich the filtrate in auxotrophic spores which, growing more slowly, stand a better chance of escaping retention by the filter (Prevost, 1962). The number of biochemical mutant strains of *C. radiatus* is relatively large. Mutants are known for most of the biochemical steps of the metabolic pathways of adenine, arginine, uracil, and tryptophan. Several mutants resistant to inhibitors, such as sorbose and fluorouracil, have been obtained, morphological mutants have also been described.

Selection of mutants that cannot be compensated for by supplementing the medium is also possible. This can be done by taking advantage of the properties of dikaryons. After mutagenesis a dikaryon is cloned; within each clone a sample of mycelium is dissociated into its haploid constituents, and the absence of one of the two parental types indicates the presence of a recessive lethal gene.

Several chromosomal aberrations are known. Two of them are interesting for the study of incompatibility; the first is a duplication of the *B* locus (compatible with *B1* and *B2*), the second is a translocation that separates the two subunits of factor *B* (Brygoo, 1971).

Complementation tests are performed by confronting the two haploid types under study. A dikaryon is obtained, samples of which are tested on different media. Complementation is said to occur if the phenotype of the dikaryon is different from that of the haploids.

Genetic Mapping.   In his thesis Prevost (1962) presented the first genetic maps for *C. radiatus*. Although ordered tetrads are not available, the distance between genes and their centromeres can be estimated by measuring the frequency of tetratype tetrads. A graphical method devised by Prevost makes calculations unnecessary. Eight linkage groups have been found and about 40 genes have been located on them. However, cytological studies made by D. Zickler (private communication) suggest that there are more than eight chromosomes.

Research Results

Although some studies have been devoted to the analysis of nuclear migration (Gans *et al.*, 1958) and to the biosynthesis of tryptophan (Henke *et al.*, 1973), most investigations with *Coprinus* have been focused on three areas: somatic recombination, genetic instability, and the biosynthesis of uracil and arginine.

Somatic Recombination.   When three mycelial types, unable to produce dikaryons in pairwise combinations, are confronted, the occasional occurrence of dikaryotic growth sectors is nevertheless observed. By using genetic markers it can be shown that these sectors result from the association of one of the parental haploids with a newly produced recombinant nuclear type. The latter may be produced in either of two ways. The first type of event begins with the formation of diploid nuclei, followed by the random loss of various chromosomes. In support of this pathway, diploid strains have indeed been isolated from such confrontations. They appear to be rather stable, but can become progressively haploid with a frequency depending on the strains. The existence of a second type of event leading to the occurrence of the recombinant nuclei is indicated by an excess of parental associations in the recombinants of markers located on separate chromosomes. This type of event has been shown to involve only one pair of homologous chromosomes, the recombinants resulting from the haploidization of a disomic. The chromosomal transfer has been demonstrated to occur at an early stage in triheterokaryons. The frequency of this second type of event is variable, ranging from very rare to a frequency comparable to that of the classical parasexual cycle described as the first event.

In addition to demonstrating the existence of these two parasexual processes, this (Prud'homme, 1970) work uncovered an efficient method for selecting diploid strains starting with a normal dikaryon. Thus the differences in the functional interaction of genes, when located in two different nuclei or in the same diploid nucleus, could be explored.

An interesting feature of diploid strains is their ability to exhibit alternatively a haploid or a dikaryotic morphology (Prud'homme, unpublished). This differentiation may well be worthy of further study.

Genetic Instability. In the course of the studies on tryptophan metabolism and on the biosynthesis of nicotinic acid, it was found that the *nicotinic 2* mutant (*nic-2*) had several peculiar genetic properties (Guerdoux, 1972). Several states of the mutant have been identified, all of them derived from the same mutant strain. The different states are functionally allelic, and they all map in the same region of chromosome II, five recombination units away from the *adenine 9* (*ad-9*) gene. All of them entail the requirement for nicotinic acid during vegetative growth. They revert with a high frequency at meiosis. The meiotic reversion frequency allows the characterization of two states, one, *nic-2.F*, which reverts with a frequency of 23 percent, the other *nic-2.fn*, which reverts with a frequency of 2 percent. The prototrophic revertants do not seem to differ from the standard wild-type strain.

With the help of the marker gene *ad-9* it can be shown that the reversion of *nic-2.F* and *nic-2.fn* is independent; a dikaryon (*ad-9 nic-2.F/ ad$^+$ nic-2.fn*) yields 23 percent of *nic$^+$* among *ad$^-$* spores and 2 percent of *nic$^+$* among *ad$^+$* spores.

A third state of the *nic-2* mutant, *nic-2.fa*, reverts with a comparatively low frequency of 1 percent, but in contrast to the other states, it imposes its own meiotic reversion frequency on the allele associated with it in the dikaryon. In addition to this trans effect, *nic-2.fa* has the peculiar property of being unstable during meiosis, when it switches to the *fn* state.

It has been shown that all the states of the *nic-2* mutation mentioned above are stable during vegetative growth, including the diploid state. Reversion of *nic-2* is strictly meiotic.

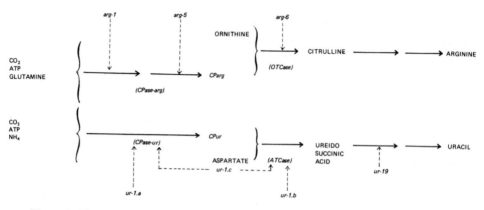

Figure 1. Metabolic pathways of arginine and uracil in Coprinus radiatus. Abbreviations: arg-1, arg-5, arg-6, arginine auxotrophs; ur-1.a, ur-1.b, ur-1.c, ur-19, uracil auxotrophs; →, enzymatic activities; ---→, gene control of enzymatic activities; CPase-arg, carbamyl phosphate synthetase of arginine pathway; CPase-ur, carbamyl phosphate synthetase of uracil pathway; ATCase, aspartic transcarbamylase; OTCase, ornithine transcarbamylase; CParg, CPur, carbamyl phosphate for arginine or uracil pathway; |, substrates of enzymatic reactions.

The reversion cannot be attributed to the segregation of a complex polygenic system since it occurs under conditions of complete homozygosity in homothallic strains. Neither is it correlated with reciprocal recombination, a fact which precludes any strict homology with the mechanism proposed for the reversion of the *Bar* mutation in *Drosophila* (involving unequal reciprocal exchanges).

As matters stand, the preferred hypothesis is that the *nic-2* mutation results from the insertion in the *nic-2* gene of a nucleotide sequence whose nature, either repetitive or symmetrical, allows instrastrand pairing. This in turn is responsible for its removal by some mechanism operative during meiosis; its loss causes the observed high reversion rates.

The Biosynthetic Pathways of Arginine and Uracil. The pathways of arginine and uracil have a common precursor, carbamyl phosphate (CP). In *C. radiatus* the CP produced by the arginine metabolic pathway (CP arg) is not normally available for uracil biosynthesis, and reciprocally, the CP produced for uracil biosynthesis (CP ura) is not available for the pathway of arginine (See Figure 1). The CP arg mutants, *arg-1* and *arg-5*, are thus auxotrophic for arginine, and the CP ur mutant (*ur-1.a*) is auxotrophic for uracil. This channeling of CP can disappear as a consequence of certain easily selected mutations. First, when CP arg is liberated, it becomes available for uracil biosynthesis, and the need for this metabolite by a *ur-1.a* strain is suppressed. The release of CP arg occurs when the synthesis of ornithine is reduced or when the enzyme ornithine transcarbamylase (OTCase) is modified. Selection for suppressors of *ur-1a* has led to the identification of five genes affecting the synthesis of ornithine and to the isolation of several alleles at the *arg-6* locus, the gene for *OTCase* (e.g., *su-16*).

On the other hand, the release of CP ura abolishes the arginine need of *arg-5* mutants. This release occurs, for example, when the activity of aspartic transcarbamylase (ATCase) is low or absent. Therefore, among the suppressors of *arg-5*, one can obtain *ura-1.b* mutants. The *ura-1.b* gene codes for ATCase. The *ur-1.b* and *ur-1.c* mutants can also be obtained by taking advantage of the fact that *ur-19 su-16* strains are poisoned on an uracil-supplemented medium because of the accumulation of ureidosuccinic acid (USA). These mutations, by blocking the production of USA, act as suppressors of the lethal effect of uracil. They account for 40 percent of all suppressors obtained by this means of selection (Motta, 1967). In contrast, the *ur-1.a* mutants cannot suppress the lethality of uracil for *ur-19 su-16* strains. It is thus possible to obtain readily a large number of *ur-1.a ur-1.c* or *ur-1.a ur-1.b* double mutants by starting with an *ur-1.a* strain, and thus one can carry out fine-structure mapping of the *ur-1* locus by three-point crosses within the gene (Gans and Masson, 1969). It has been shown that *ur-1.a* mutants deficient for uracil carbamyl phosphate synthetase (*CPase-ura*) are located at one extremity of the locus, while *ur-1.b* mutants deficient for ATCase are located at the other. The two groups of mutants complement each other perfectly. The *ur-1.c* mutants, deficient for both enzymes, complement neither the *ur-1.a* nor the *ur-1.b* mutants, and they behave in some ways as polar mutants (Masson, 1971). These findings suggest that the *ur-1* locus is composed of two genes, *ur-1.a* and *ur-1.b*, that are transcribed together.

Molecular Association and Localization of Enzymes. The CPase and ATCase enzymes specified by the *ur-1* locus are localized in the cytosol and are probably part of a molecular complex (Hirsch, 1968; Jullien, 1969). In contrast, the genes governing the synthesis (*arg-1, arg-5*) and the utilization (arg-6) of CParg are not genetically linked. It is possible that the enzymes they specify also make up a molecular complex. However, most of them are located in the mitochondria. This difference in localization of the two enzyme complexes possibly constitutes the basis of the observed compartmentation of CP (Callen, 1971).

Regulation. The double-mutant strain *ur-1a arg-6* will grow on media containing low concentrations of arginine. High concentrations, however, inhibit growth of this strain. This observation can be explained by a regulatory effect of arginine on the synthesis of CParg; however, it should be kept in mind that in such a strain the CP produced by the arginine pathway is required for the uracil pathway. Mutants that are resistant to arginine have been obtained (Cabet *et al.*, 1967). Three such mutants show a twofold increase in CPase level over that of the reference strain, and this level is not reduced by the addition of arginine. These mutants appear to be derepressed for CPase-arg (Cabet-Busson, unpublished).

As mentioned above, by selecting for suppressors of *arg-5*, one mainly obtains *ur-1.b* mutants. In contrast to what is observed in the case of CParg production, the synthesis of CPase ur is not sensitive to repression by uracil, and this class of *arg-5* suppressors can function at high concentrations of uracil. Nevertheless, uracil-sensitive, arginine-5 suppressors located at the *ur-1* locus have been obtained. The studies performed on one of these, *ur-67*, revealed a low level of ATCase activity. This was enhanced by uracil, and therefore the sensitivity to uracil of the suppressor effect could not result from an increase of CP utilization by ATCase at high uracil concentrations. The *ur-67* mutant maps in the *ur-1.a* gene. A possible explanation for the uracil effect could be hypersensitivity of the *ur-1.a*-specified CPase to feedback control by uracil (Masson, 1971).

Biosynthesis of Uridin Monophosphate from Orotic Acid. Orotic acid is not a suitable source of pyrimidines in *Coprinus* except for strains mutated in gene *r*, whose function is unknown.

Strains resistant to 5-fluorouracil (5Fu) and to 5-fluorouridine have been selected. Some excrete uracil or uridine into the medium. The enzymes responsible for the step in the transformation of uracil to uridine and UMP have been characterized. A general scheme of the interconvertibility in *Coprinus* of bases to nucleosides and to nucleotides has been made available through studies of enzyme levels and their modifications in 5Fu-resistant mutants.

The uptake and metabolism of various pyrimidines and the antagonistic effects of pyrimidines and fluoropyrimidines in wild-type or in mutant strains have been studied. Of special interest is the fact that uridine is a more suitable source of pyrimidines than uracil. This can be explained by the more efficient transformation of uridine to UMP by the uridine-inducible uridine kinase (Le Hegarat, 1973).

Conclusion

The main biological interest of *C. radiatus* is the existence of dikaryotic and diploid strains. The dikaryotic strain can be used either to study the nucleocytoplasmic relationships or to select noncompensable mutants (see the section **Mutants**). The comparison of dikaryotic and diploid strains was one of the purposes of the study of *C. radiatus*. This comparison is now possible: on one hand, diploids strains have been obtained (see the section **Somatic Recombination**); and on the other, numerous mutant types have been selected, in particular those affecting enzyme regulation or metabolite channeling (see the section **The Biosynthetic Pathways of Arginine and Uracil**). The finding that a diploid strain can alternatively exhibit haploid or dikaryotic morphologies (see the section **Somatic Recombination**) makes *C. radiatus* a choice material for workers interested in cellular differentiation.

Literature Cited

Babillot, C., 1963   Induction par l'acide nitreux et étude génétique de mutants tryptophane exigeant. Diplôme d'études supérieures de Sci. Nat. Faculté des Sciences de Paris, Paris.

Brygoo, Y., 1971   Contribution à l'étude de l'incompatibilité chez *Coprinus radiatus*: structure du locus *B*. Thèse de Doctorat 3ème Cycle Génétique, Université Paris XI, Paris.

Cabet, D., R. Motta and G. Prévost, 1967   Interaction entre les chaînes de biosynthèse de l'arginine et de l'uracile et son exploitation en vue de la sélection de gènes mutés chez le coprin. *Bull. Soc. Chim. Biol.* **49**:1537–1543.

Callen, J. C., 1971   Sur la régulation de métabolisme de l'arginine chez le *Coprinus radiatus*. Thèse Doctorat 3ème Cycle Génétique Université Paris XI, Paris.

Gans, M. and M. Masson, 1969   Structure fine du locus *ur-1* chez *Coprinus radiatus Mol. Gen. Genet.* **106**:164–181.

Gans, M., N. Prud'homme and G. Prévost, 1958   Problèmes posés par l'étude génétique des basidiomycètes. *Bull. Soc. Fr. Physiol. Vég.* **4**:149–157.

Guerdoux, J. L., 1972   Transformations spontanées du mutant nicotinique deux du *Coprinus radiatus*: Etude de souches *nic-2* homothalliques ou diploides. *Mol. Gen. Genet.* **119**:119–129.

Henke, A., J. L. Guerdoux and R. Hütter, 1973   Relation entre les gènes et les enzymes de la biosynthèse du tryptophane chez le *Coprinus radiatus. C. R. Hebd. Seances Acad. Sci. Ser. D Sci. Nat.* **276**:2001–2004.

Hirsch, M. L., 1968   Séparation des enzymes jouant un rôle dans la compartimentation entre les chaînes de biosynthèse de l'uracile et de l'arginine chez le *Coprinus radiatus. C. R. Hebd. Seances Acad. Sci. Ser. D Sci. Nat.* **267**:1473–1476.

Jullien, M., 1969   Etude des mécanismes Moléculaires d'une compartimentation cellulaire chez le basidiomycète *Coprinus radiatus*. Thèse Doctorat 3ème Cycle Faculté des Sciences de Paris, Paris.

Le Hegarat, F., 1973   Etude génétique et physiologique de la biosynthèse des pyrimidines chez Coprinus radiatus. Thèse Doctorat ès-Sciences Naturelles Université Paris XI, Paris.

Masson, M., 1971   Etude génétique et physiologique du locus *ur-1* chez *Coprinus radiatus*. Thèse Doctorat ès-Sciences Naturelles Université Paris VI, Paris.

Motta, R., 1967 Méthode de sélection de mutants uracile exigeants au locus *ur-1* de *Coprinus radiatus. C. R. Hebd. Seances Acad. Sci. Ser. D Sci. Nat.* **264**:654–657.

Prévost, G., 1962 Etude génétique d'un basidiomycète: *Coprinus radiatus.* Thèse Sciences Naturelles Faculté des Sciences Paris, Paris.

Prud'homme, N., 1970 Recombinaisons mitotiques chez un basidiomycètes: *Coprinus radiatus. Mol. Gen. Genet.* **107**:256–271.

Editor's Note

An excellent illustrated account of the biology of another species of *Coprinus* is also available. See G. E. Anderson's "The Life History and Genetics of *Coprinus lagopus*," published in 1971 by Philip Harris International Ltd., Lynn Lane, Shenstone, Staffordshire, England.

Author Index*

Abbondandolo, A., 437, 438, 440, 442
Abd-el-al, A., 216
Abe, M., 56
Abelson, J., 266, 267, 307, 440
Abelson, J. N., 331
Abelson, P. H., 528
Abrass, I. B., 201
Acha, I. G., 488
Achtman, M., 161
Adams, A., 105
Adams, J. M., 289
Adams, M. H., 130
Adelberg, E., 41, 107, 131, 132, 217, 218, 355, 388
Adhya, S., 167, 307, 321
Adler, K., 167
Admiraal, W., 109
Adondi, G., 437
Agardh, C. A., 354
Agnihotri, V. P., 488, 501
Aharonowitz, Y., 105
Ahmad, S. S., 621
Ahmad-Zadeh, C., 201
Ahmed, S. I., 548
Ainsworth, G. C., 506, 525, 528
Ajl, S. J., 220
Akaboshi, E., 338
Alderson, T., 255, 488, 489, 507
Aldrich, C., 331

Alexander, R. R., 217
Alexopoulos, C. J., 39, 357, 626
Algranati, I. D., 293
Ali, A. M. M., 437
Ali, I., 526
Alikhanian, S. I., 253, 255
Allen, T. C., 495
Allison, D. P., 131
Althaus, H., 526
Althaus, M., 437
Altman, P. L., 39, 524
Altman, S., 437
Amagase, S., 338
Amaldi, F., 195
Amati, P., 217
Ames, B. N., 217, 218, 220, 221, 235
Ames, G. F., 217
Ames, L. M., 544
Ammann, J., 289
Anagnostakis, S. L., 593
Anagnostopoulos, C., 105, 106, 109, 110, 113
Anderegg, J. W., 290
Andersen, L., 180
Anderson, B., 220
Anderson, D. L., 266, 268
Anderson, P., 194, 196
Anderson, T. F., 268, 269
Anderson, W., 167
Andoh, T., 178
Angehrn, P., 437, 440
Anthony, C., 505

Anton, D. N., 217
Aono, H., 178
Apirion, D., 194, 195, 196, 197, 199, 200, 489
Applebury, M. L., 178, 179
Apte, B. N., 489, 508
Aragón, C. M. G., 355
Archer, L. J., 106, 112, 113, 114
Arditti, R., 167
Argetsinger-Steitz, J., 267, 289, 291
Argoudelis, A. D., 489
Arlett, C. F., 489, 495
Armaleo, D., 491
Armentrout, R. W., 106
Armitt, S., 489, 508
Armstrong, E. C., 57
Armstrong, F. B., 217
Armstrong, R. L., 106
Arnold, C. A., 39
Arst, H. N., 489, 495
Ascensio, C., 268
Ashwood-Smith, M. J., 490
Ashworth, J. M., 254
Aspen, A. J., 490
Atkins, C. G., 217
Attardi, G., 195
Atwood, K. C., 199, 202, 527
Audit, C., 106
Auerbach, C., 437, 444
Auerbach-Rubin, F., 54
August, J. T., 289, 292

* This lists authors *only* where they appear in the bibliographies following the chapters.

637

Subject Index

Contents of Other Volumes

Volume 2: Plants, Plant Viruses, and Protists

The Plants

Insects of Genetic Interest

The German Cockroach, Blattella germanica
  Mary H. Ross and Donald G. Cochran (Virginia Polytechnic
  Institute)

The Domesticated Silkmoth, Bombyx mori
  Yataro Tazima (National Institute of Genetics), Hiroshi Doira
  (Kyushu University) and Hiromu Akai (National Sericultural
  Experiment Station)

The Mediterranian Meal Moth, Ephestia kühniella
  Ernst W. Caspari (University of Rochester) and Frederick J. Gottlieb
  (University of Pittsburgh)

The Flour Beetles, Tribolium castaneum and T. confusum
  Alexander Sokoloff (California State College)

The Honey Bee, Apis mellifera
  Walter C. Rothenbuhler (Ohio State University)

The Wasps, Habrobracon and Mormoniella
  Joseph D. Cassidy (Northwestern University)

Lower Diptera with Giant Chromosomes

Rhynchosciara
  C. Pavan (University of Texas), A. Brito da Cunha and
  P. F. Sanders (Universidade de São Paulo)

Sciara
  Natalia Gabrusewycz-Garcia (State University of New York)

Chironomus
  Klaus Hägele (Ruhr-Universität Bochum)

Glyptotendipes
  Ludwig Walter (Ruhr-Universität Bochum)

Mosquitoes and Flies of Genetic Interest

Anopheline Mosquitoes
  Mario Coluzzi (University of Rome) and James B. Kitzmiller
  (University of Illinois)

Aedes
  Karamjit S. Rai (University of Notre Dame) and W. Keith Hartberg
  (Georgia Southern College)

Volume 4: Vertebrates of Genetic Interest

Amphibia